Advances in Natural and Technological Hazards Research

Volume 41

More information about this series at http://www.springer.com/series/6362

Geoffroy Lamarche • Joshu Mountjoy
Suzanne Bull • Tom Hubble • Sebastian Krastel
Emily Lane • Aaron Micallef
Lorena Moscardelli • Christof Mueller
Ingo Pecher • Susanne Woelz
Editors

Submarine Mass Movements and their Consequences

7th International Symposium

Springer

Editors
Geoffroy Lamarche
National Institute of Water
and Atmospheric Research (NIWA)
Wellington, New Zealand

Joshu Mountjoy
National Institute of Water
and Atmospheric Research (NIWA)
Wellington, New Zealand

Suzanne Bull
GNS Science
Lower Hutt, New Zealand

Tom Hubble
School of Geosciences
The University of Sydney
Sydney, NSW, Australia

Sebastian Krastel
Institute of Geosciences
Christian-Albrechts-University at Kiel
Kiel, Germany

Emily Lane
National Institute of Water
and Atmospheric Research (NIWA)
Christchurch, New Zealand

Aaron Micallef
University of Malta
Msida, Malta

Lorena Moscardelli
Bureau of Economic Geology
Austin, TX, USA now at Statoil RDI,
Austin, TX, USA

Christof Mueller
GNS Science
Lower Hutt, New Zealand

Ingo Pecher
The University of Auckland
Auckland, New Zealand

Susanne Woelz
National Institute of Water
and Atmospheric Research (NIWA)
Wellington, New Zealand

ISSN 1878-9897 ISSN 2213-6959 (electronic)
Advances in Natural and Technological Hazards Research
ISBN 978-3-319-20978-4 ISBN 978-3-319-20979-1 (eBook)
DOI 10.1007/978-3-319-20979-1

Library of Congress Control Number: 2015952210

Springer Cham Heidelberg New York Dordrecht London

Printed on acid-free paper

Springer International Publishing AG Switzerland is part of Springer Science+Business Media (www.springer.com)

Preface

This sixth volume of *Submarine Mass Movements and Their Consequences* presents an impressive collection of papers presented at the 7th International Symposium held in Wellington, New Zealand, in November 2015. The meeting is the cornerstone for the submarine landslide research community. Both this community and society at large owe much to the support of the International Union of Geological Science (IUGS) and the United Nations Educational, Scientific and Cultural Organization (UNESCO) who have now sponsored three International Geoscience Programmes in this research area: IGCP-511 (SMMTC, Submarine Mass Movements and Their Consequences); IGCP-585 (E-Marshal, Earth's Continental MARgins: aSsessing the geoHazard from submarine Landslides); and, as of 2015, IGCP-640 (S^4SLIDE, Significance of Modern and Ancient Submarine Slope LandSLIDEs). It is through this support that the research community continues to flourish in terms of collaborative research, student support, and integration of developing nations. It is our pleasure to bring you this volume of fine research papers that capture the exciting state of research in our field. We believe this volume provides both a snapshot of the state of research in submarine mass movements and confirmation of the strong directions in which the field is moving and growing.

Wellington, New Zealand	Geoffroy Lamarche
Wellington, New Zealand	Joshu Mountjoy
Lower Hutt, New Zealand	Suzanne Bull
Lower Hutt, New Zealand	Christof Mueller
Sydney, NSW, Australia	Tom Hubble
Kiel, Germany	Sebastian Krastel
Christchurch, New Zealand	Emily Lane
Msida, Malta	Aaron Micallef
Austin, TX, USA	Lorena Moscardelli
Auckland, New Zealand	Ingo Pecher
Wellington, New Zealand	Susanne Woelz

Contents

Chapter 1
Submarine Mass Movements and Their Consequences: Progress and Challenges

Geoffroy Lamarche, Joshu Mountjoy, Suzanne Bull, Tom Hubble, Sebastian Krastel, Emily Lane, Aaron Micallef, Lorena Moscardelli, Christof Mueller, Ingo Pecher, and Susanne Woelz

Abstract This sixth edition of the *Submarine Mass Movements and Their Consequences* volume, coincident with the seventh eponymous conference includes 61 papers that span a variety of topics and are organized into nine parts as follows: (1) Submarine mass movement in margin construction and economic significance; (2) Failure dynamics from landslide geomorphology; (3) Geotechnical aspects of mass movement; (4) Multidisciplinary case studies; (5) Tectonics and mass movement processes; (6) Fluid flow and gas hydrates, (7) Mass transport deposits in modern and outcrop sedimentology; (8) Numerical and statistical analysis; and, (9) Tsunami generation from slope failure. The breath and quality of this body of work underpins a positive outlook and our enthusiasm for the future direction of research in this area of science as it moves towards ever more detailed analysis and monitoring. We also emphasize in this volume the need to look at mountain-scale

G. Lamarche (✉) • J. Mountjoy • S. Woelz
National Institute of Water and Atmospheric Research (NIWA), Wellington, New Zealand
e-mail: geoffroy.lamarche@niwa.co.nz

S. Bull • C. Mueller
GNS Science, Lower Hutt, New Zealand

T. Hubble
School of Geosciences, The University of Sydney, Sydney, NSW, Australia

S. Krastel
Institute of Geosciences, Christian-Albrechts-University at Kiel, Kiel, Germany

E. Lane
National Institute of Water and Atmospheric Research (NIWA), Christchurch, New Zealand

A. Micallef
University of Malta, Msida, Malta

L. Moscardelli
Bureau of Economic Geology, Austin, TX, USA now at Statoil RDI, Austin, TX, USA

I. Pecher
The University of Auckland, Auckland, New Zealand

G. Lamarche et al. (eds.), *Submarine Mass Movements and their Consequences*, Advances in Natural and Technological Hazards Research 41,
DOI 10.1007/978-3-319-20979-1_1

outcrops to better understand our seismic imaging, to carry out statistical studies that draw on global data sets to better constrain broad behavioural characteristics, and to undertake numerical modelling to understand the sensitivity of a range of natural slopes.

1.1 Introduction

Research on submarine mass movements remains highly topical and vibrant, whether it focuses on the consequences that such events may have on human and natural environments or on process-oriented science. The consequences of submarine landslides on infrastructure and coastal communities is reasonably well-known, thanks to a few notorious events that made global headlines, such as: the 1929 Grand Banks submarine landslide, Newfoundland (Piper and Aksu 1987) that ruptured a series of submarine cables and triggered a tsunami; the collapse of reclaimed land at Nice airport in 1979 (Kopf et al. 2011), which also generated a local tsunami; the dramatic landslide-generated Papua New Guinea tsunami in 1998, which claimed more than 2,000 lives (Tappin et al. 2001); or the hyperpycnal flow that ruptured communication cables south of Taiwan following the 2009 typhoon (Carter et al. 2012).

In order to improve our understanding of the potential impacts of submarine landslides, the research has focused on a variety of processes, approaches and methodologies. Field surveys undertaken immediately after catastrophic events have provided valuable information of magnitude-frequency relationship that underpin the development of hazard studies (Tappin et al. 2001). But arguably, investigations of mass-transport deposits (MTD) and source areas has represented the richest research effort pertinent to submarine landslides. MTD vary tremendously in geometry and size, ranging over at least five orders of magnitude in volume. There is a wide diversity of physical and mechanical processes that govern initiation and emplacement of submarine landslides, which has called for a plethora of research techniques. These include geophysical surveying, (e.g., seismic reflection, multibeam bathymetry), geological sampling (e.g., coring) and geotechnical investigations, analogue and numerical modelling. More recently, state-of-the-art technologies such as the use of remotely-operated vehicle (ROV) and autonomous underwater vehicles (AUV) have unveiled unprecedented details about the morphology of MTD (Huvenne et al. 2016).

The present volume is the sixth of its kind, and is associated to the seventh conference on Submarine Mass Movements and their Consequences. It contains 61 papers, spanning a variety of topics that are organized in nine parts.

The geographic focus of the research papers produced in this volume is predominantly the European and North American continental margins (Fig. 1.1), with 27 and 10 studies herein, respectively. Oceania, with the margins of New Zealand, eastern Australia and Indonesia, features strongly, possibly in part

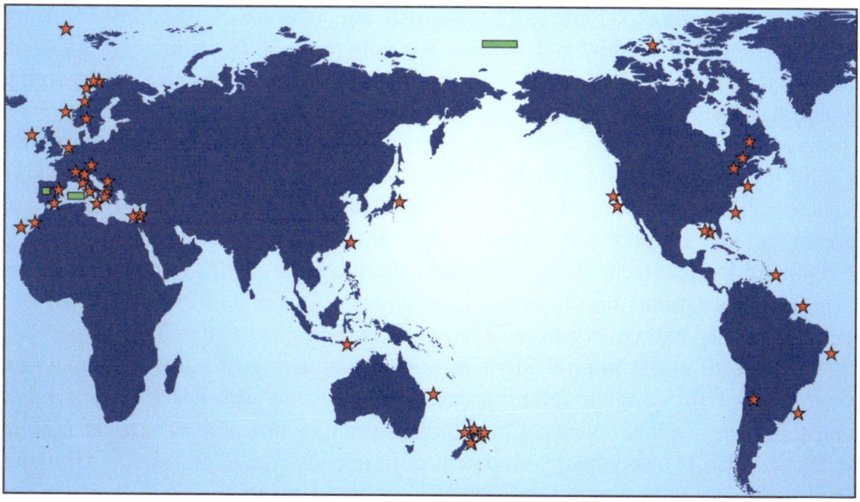

Fig. 1.1 Geographic distribution of paper focus for the submarine mass movement and their consequences 2015 volume. Local (*red stars*) and regional (*green rectangles*) studies are indicated

because the conference will be held for the first time in the southern Hemisphere, but more likely and most importantly because the southeastern Pacific region is highly tectonically active, which likely facilitate the occurrence of submarine landslides, at least those seismically triggered. Paradoxically, western North America is represented by only two papers, and western South America, southwest Africa and the Indian Ocean have no papers in this volume. The paucity of baseline data from some of these areas is an issue that the submarine landslide-research community should attempt to address. Another conspicuous absence is mid-ocean ridges, which is somewhat disappointing as recent studies have demonstrated that these regions present some interesting examples of submarine landslides (Ferrini et al. 2013; Gao 2006; Tucholke 2016).

Contributions from high latitude regions solely concern the northern hemisphere, specifically on the northern Atlantic, including two papers on the Canadian Arctic (Normandeau et al. 2016; Priest and Grozic 2016).

There is an increase from previous years in the number of papers on landslides affecting lakes. These studies provide valuable observations and conclusions on the tsunami hazard associated with such events as well as paleoseismicity indicators (Locat et al. 2016). Perhaps "submarine" is too exclusive, and should be replaced by "subaqueous" if only to encourage such lacustrine studies to develop. Likewise studies of submarine MTD outcrops provide a means to gather new evidence on emplacement processes.

Submarine landslides occur on a wide range of scales. However, this year we note a lesser focus on large-scale "giant" features in comparison to previous issues. This may correlate with the increase in social, economic and hazard relevance of the research, which tend to concentrate less on rare giant submarine events.

If comparing the 2003 and 2015 volume is any measure of a trend, we surmise that the research is healthy and vibrant. We note since 2003, an increase focus on lakes, as mentioned above, but also an intensification in the use of emerging technology such as AUV and ROV, numerical modelling, and progress on event observation. Although the collection of in situ observations, let alone measurements, of mass movements remains a very challenging task, it is now a conceivable objective that is attracting strong interest from of some research groups, as e.g., to assess the role of pore pressure in slope stability. These ambitions are being invigorated by new technology deployed from the ocean surface and mechanisms to make observatories on the ocean floor directly. Better multibeam bathymetry, including micro bathymetry from underwater vehicles and viable 3D-seismic data sets, has led to major improvement in our understanding of landslide dynamics, internal geometries and the failure process, and it is now well-accepted that major landslides are multiple events. Concentrated seafloor mapping efforts, at margin scales, has also allowed the development of frequency-size relationships. Of major importance was the scientific drilling of landslide deposits (e.g. IODP 308 in the Gulf of Mexico, and IODP 340) which allowed a quantitative assessment of specific pre-conditioning factors for individual landslides and an understanding of recurrence rates at targeted sites.

1.2 The 2015 Volume

1.2.1 Part I: Submarine Mass Movement in Margin Construction and Economic Significance

The Society for Underwater Technology estimates that the cost of damage to pipelines caused by submarine mass movements is about $400 million annually (Mosher et al. 2010). This direct economic risk to exploration and production platforms is most relevant to small events with short reoccurrence intervals, and warrants ongoing research investment to enhance our understanding of processes and risk prediction. The study of submarine landslides also has an important geopolitical connotation as, their geomorphology has been used to define the distal toe of the continental slope and Extended Continental Shelf (ECS) by the UN Convention for the Law of the Sea (UNCLOS) (Mosher et al. 2016).

While it is generally accepted that submarine mass movements have a role in the development of petroleum systems (in particular in generating traps and seals), the understanding of the processes involved remains limited. It is estimated that ~90 % of submarine landslides are mud-prone but recent exploration has found some sand-rich MTD that could be considered petroleum reservoirs (Beaubouef and Abreu 2010; Meckel III 2011). Submarine landslides have proven to play a role defining hydrocarbon reservoir location and geometry, facies distributions, development of stratigraphic traps and integrity of seal intervals. In this volume, a specific case study in the Gulf of Mexico is analysed by studying fabric development and pore-throat reductions of a local MTD that is clearly linked to a reservoir interval (Cardona et al. 2016). Such approaches, where seismic scale observations are combined with rock properties, provides a glimpse into the future study of MTD in the context of hydrocarbon systems.

It is noticeable that with the global increase in deep-water hydrocarbon exploration and production, the need to study and understand, let alone forecast, the occurrence of submarine landslides will continue to be the focus of the community (Omeru et al. 2016). Issues such as security of offshore infrastructure, and economic relevance of hazard and risk are the focus of contributions in following parts of this volume.

1.2.2 Part II: Failure Dynamics from Landslide Geomorphology

The complexity and variety of physical mechanisms involved in submarine failures is still up for debate. Whilst the consequence on human life is of paramount importance, the potential impact that any submarine landslides may have on seafloor or coastal infrastructures is increasingly concerning, especially with the increasing development of high seas infrastructures associated with deep-sea petroleum exploitation, communication cable and deep sea mining. There is also a rising awareness about the potential impact that submarine landslides can have on the natural environment. For the first time to our knowledge, these consequences could be ecologically positive, as suggested in the case of the Ionian Sea where hard strata and irregular small-scale topography resulting from landslides are likely to have facilitated cold water coral (re-)colonization (Savini et al. 2016). This shows linkages developing between submarine landslides research and economic, social and environmental issues.

1.2.3 Part III: Geotechnical Aspects of Mass Movement

Advances in geotechnical technology has been driven by the need for quality data required to quantify slope sensitivity response to deformation. Such progress in the recent past is being driven by engineering concerns around seafloor stability with respect to infrastructure hazard (Forsberg et al. 2016). Unprecedented insights into mass movement are now being offered by recent technological advances, in the laboratory (Torbahn et al. 2016), through the comparison between laboratory to in-situ observations (Kluger et al. 2016), and by correlating geophysical and in-situ geotechnical experiments (Stegmann et al. 2016). The confines offered by fjords and estuaries make perfect natural laboratories to study detailed geotechnical aspects of mass movements that can then provide insight into less accessible submarine processes (Forsberg et al. 2016; Kluger et al. 2016; Mastbergen et al. 2016; Turmel et al. 2016).

1.2.4 Part IV: Multidisciplinary Case Studies

High resolution bathymetric imaging combined with advanced geotechnical methods are truly providing significant advances in the field. Shipboard and AUV multibeam bathymetry remains unarguably the most used tool in submarine landslide investigations. Great advances in this technique mean that very high resolution studies are increasingly becoming routine. However, the significance of these data is further amplified when combined with multidisciplinary field investigations. We are very pleased to see outstanding contributions from investigations in the very challenging environment of the northern Arctic, revealing insights of previously unsurveyed areas (Forwick et al. 2016; Laberg et al. 2016; Normandeau et al. 2016). As well as a case study examining micro scale controls on mass movement complexes (Kuhlmann et al. 2016), another study reveals the long-term failure history of a confined subaqueous environment (Lindhorst and Krastel 2016).

1.2.5 Part V: Tectonics and Mass Movement Processes

The hazard and risk associated with submarine mass movement remains a fundamentally important focus of the scientific community. Intense efforts have been made to improve our understanding of the relation between tectonics and mass movement occurrence (Katz et al. 2016; Locat et al. 2016; Moore and Strasser 2016). The effect of tectonic activity on seismic strengthening is a hypothesis that is being supported by a comparison of shear strength in cores from active and passive margin settings (DeVore and Sawyer 2016). The use of submarine landslides to develop paleoseismic records at local or regional scales is building momentum and is providing some beneficial outcomes for constraining earthquake recurrence intervals and improving seismic hazard assessment (Chapron et al. 2016; Gracia et al. 2013). Other studies use a generic modelling approach to tackle the overarching question on the role of earthquakes in slope destabilization (Trandafir and Popescu 2016).

1.2.6 Part VI: Fluid Flow and Gas Hydrates

Understanding the role that fluids plays in pre-conditioning slope instability remains a significant challenge. New contributions made herein on the role of gas in slope failures (Gwiazda et al. 2016; Micallef et al. 2016) shed light on the possibility that fluids have a controlling influence on landslide basal failure planes. These observations are largely the result of recent technological developments. Similarly, the role of gas hydrates in submarine landslides remains a controversial one. Crutchley et al. (2016) present a series of different slope failure scenarios on the New Zealand continental margin that highlight a range of processes related to gas hydrates that could weaken sediments.

Priest and Grozic (2016) picks up on the widespread observations that climate change is destabilizing gas hydrate deposits in the Arctic regions. Using a modelling approach, they provide a thought provoking thesis on the role that climate change could play in inducing large slope failures.

1.2.7 Part VII: Mass Transport Deposits in Modern and Outcrop Sedimentology

Marine geophysical (seismic, multibeam echosounder) and geological (sediment cores, seafloor operations) still lack the fine-scale information and cross-sectional insights on mass failure processes that can be preserved and yielded from the study of fossilized landslides in outcrops onland. Outcrop investigations provide detailed insight into the kinematics (Ogata et al. 2016) and internal mixing and structure (Sobiesiak et al. 2016) of MTD. Sympathetic to these approaches are fine-scale sedimentological studies that provide a modern context of present day seafloor processes (Postma et al. 2016; Sawyer and Hodelka 2016). The link between onland outcrop studies of MTD and detailed marine studies is one that deserves to be strengthened given the benefits of cross pollination between these disciplines.

1.2.8 Part VIII: Numerical and Statistical Analysis

Statistical models allow us to develop rules and quantify relationships that encompass our analysis. Better understanding of magnitude-frequency relationships for submarine landslides is imperative if we are to develop robust hazard models (Clare et al. 2016; Mueller et al. 2016). GIS models developed to predict where slope failures can occur often become the underpinning driver for hazard assessments (Borrell et al. 2016; Dabson et al. 2016; Haneberg 2016), while advanced numerical modelling can be the key to understanding the sensitivity of individual slopes to external perturbations (Huhn et al. 2016). Numerical modelling and statistical analysis theme continues to run strongly in this volume, and reflect the state-of-the-art in terms of our general understanding of spatial causes of submarine landslides.

1.2.9 Part IX: Tsunami Generation from Slope Failure

Tsunamis are one the biggest hazards to coastal environments caused by submarine landslides. In this volume we have papers covering a broad range of landslide-related tsunami generation, including volcanic generation (De Lange and Moon 2016), landslides into an ice-covered lake (Leblanc et al. 2016), and advanced models that couple landslide deformation processes to wave generation (Wang et al. 2016). With increasing coastal populations and on-going climate change, the need to continually refine and test robust tsunami hazard assessments is more pressing than ever. These papers capture some of the diverse range of processes that need to be considered in these assessments.

1.3 Looking to the Future

Of excellent augur to the future is the announcement in 2015 of the IGCP-640 S^4SLIDE (Significance of modern and ancient submarine slope landslides – https://sites.google.com/a/utexas.edu/s4slide/). This project builds on the success of IGCP-511 and IGCP-585, and is the third IGCP project that focuses on submarine landslide-related science. This will no doubt continue to promote the research and foster relationships across the globe.

The S^4SLIDE project intends to create an international and multidisciplinary platform to allow geoscientists from academia and industry to sustain a dialogue conducive to the integration of findings from different fields into a more cohesive understanding of submarine landslides. The initial project submission, listed the following series of key scientific questions as those needing to be addressed in the near future:

1. What is the nature of the interaction between local structural controls, seafloor morphology, sediment supply, relative sea-level fluctuations, current-controlled sedimentation and submarine landslides in lacustrine and marine environments?
2. What role do transient, turbulent-laminar flows play in the formation of submarine landslides?
3. What is the impact of climatic variations on the occurrence of submarine landslides?
4. What is the role of submarine landslides in the realm of hydrocarbon exploration and production?
5. Do we understand the hazard that submarine landslides poses in mid-oceanic ridges with regard to the developing deep-sea mining activities?
6. What are the indicators of incipient submarine slope instability in different geological-tectonic settings?
7. How do submarine landslides evolve after initiation and how do their dynamics affect their impact on coastal areas, including the generation of tsunamis?
8. Is it possible to develop a submarine landslide early-warning system?

The initial strategy was included in the S^4SLIDE proposal to address the above key questions:

1. Continue of increase investment in sub-seafloor characterization and sampling to reduce data uncertainty in the third dimension.
2. Generate statistically significant submarine landslide data sets to allow robust probabilistic stability and hazard assessment.
3. Repeat multibeam surveys to address dynamic slope behaviour during failure and transient phenomena in slope failure preconditioning and triggering.

4. Advance the capability to model submarine landslide triggered tsunamis, particularly by better constraining the disintegration of material during failure.
5. Promote multiscale quantitative geomorphological investigations by integrating field measurements, remote sensing, and numerical models.

We are very enthusiastic about the future direction of research in Submarine Mass Movements and Their Consequences. As we move toward ever more detailed analysis with surgical seafloor operations, we also need to step back and take in the broad context offered by mountain scale outcrops of our seismic imaging, carry out unifying statistical studies that draw on global data sets, and explore numerical modelling to understand the sensitivity of natural slopes.

New IODP drilling proposals may also offer unprecedented opportunities. A couple of completed IODP projects on landslides have already contributed to great gains in our understanding of failure processes. However with at least two proposals in the system to date, and the linkages developing through S^4SLIDE, we expect to see some major paradigm shifts in our understanding of landslide architecture and controlling influences. Installing monitoring equipment in IODP holes may offer that opportunity to monitor landslide controls and we encourage the whole community to consider how an IODP proposal could advance this science over the next decade.

The rich scientific literature on the topic of submarine mass movement is testimony to the strength and agility of the research community, exemplified by the biannual Submarine Mass Movements and Their Consequences symposium and the high standard of material included in this book. We greatly look forward to contributing to the sixth volume and seeing just how much our understanding of these fascinating and societally important processes have advanced.

Acknowledgments We would like to thanks all those who accepted to review the manuscript for this volume. Indeed the list is too long to itemise here. Thank you to Alan Orpin (NIWA) and Gareth Crutchley (GNS Science) for reviewing this manuscript, and to Janet George (Absolutely Organised) for organising the SMMTC2015 portal.

References

Beaubouef RT, Abreu V (2010) MTCs of the Brazos-Trinity slope system; thoughts on the sequence stratigraphy of MTCs and their possible roles in shaping hydrocarbon traps. In: Mosher DC, Shipp RC, Moscardelli L, Chaytor JD, Baxter CDP, Lee HJ, Urgeles R (eds) Submarine mass movements and their consequences. Springer, Dordrecht

Borrell N, Somoza L, León R, Medialdea T, Gonzalez F, Gimenez-Moreno C (2016) GIS catalogue of submarine landslides in the Spanish continental shelf: potential and difficulties for susceptibility assessment. In: Lamarche G, Mountjoy J (eds) Submarine mass movements and their consequences. Springer, Dordrecht, pp 499–508

Cardona S, Wood L, Day-Stirrat R, Moscardelli L (2016) Fabric development and pore-throat reduction in a mass-transport deposit in the Jubilee gas field, eastern gulf of Mexico: cnsequences for the Sealing Capacity of MTDs. In: Lamarche G, Mountjoy J (eds) Submarine mass movements and their consequences. Springer, Dordrecht, pp 27–38

Carter L, Milliman JD, Talling PJ, Gavey R, Wynn RB (2012) Near-synchronous and delayed initiation of long run-out submarine sediment flows from a record-breaking river flood, offshore Taiwan. Geophys Res Lett 39:L12603

Chapron nE, Simonneau A, Ledoux G, Arnaud F, Lajeunesse P, Albéric P (2016) French alpine foreland Holocene paleoseismicity revealed by coeval mass wasting deposits in glacial lakes. In: Lamarche G, Mountjoy J (eds) Submarine mass movements and their consequences. Springer, Dordrecht, pp 341–350

Clare M, Talling P, Challenor P, Hunt J (2016) Tempo and triggering of large submarine landslides – statistical analysis for hazard assessment. In: Lamarche G, Mountjoy J (eds) Submarine mass movements and their consequences. Springer, Dordrecht, pp 509–518

Crutchley G, Mountjoy J, Pecher I, Gorman A, Henrys S (2016) Submarine slope instabilities coincident with shallow gas hydrate systems: insights from New Zealand examples. In: Lamarche G, Mountjoy J (eds) Submarine mass movements and their consequences. Springer, Dordrecht, pp 401–410

Dabson O, Barlow J, Moore R (2016) Morphological controls on submarine slab failures. In: Lamarche G, Mountjoy J (eds) Submarine mass movements and their consequences. Springer, Dordrecht, pp 519–528

De Lange W, Moon V (2016) Volcanic generation of tsunamis: two New Zealand palaeo-events. In: Lamarche G, Mountjoy J (eds) Submarine mass movements and their consequences. Springer, Dordrecht, pp 559–568

DeVore J, Sawyer D (2016) Shear strength of siliciclastic sediments from passive and active margins (0–100 meters below seafloor): insights into seismic strengthening. In: Lamarche G, Mountjoy J (eds) Submarine mass movements and their consequences. Springer, Dordrecht, pp 173–180

Ferrini VL, Shillington DJ, Gillis K, MacLeod CJ, Teagle DAH, Morris A, Cazenave PW, Hurst S, Tominaga M (2013) Evidence of mass failure in the Hess Deep Rift from multi-resolutional bathymetry data. Mar Geol 339:13–21

Forsberg CF, Heyerdahl H, Solheim A (2016) Underwater mass movements in lake Mjøsa, Norway. In: Lamarche G, Mountjoy J (eds) Submarine mass movements and their consequences. Springer, Dordrecht, pp 191–200

Forwick M, Laberg J,S, Husum K, Gales J (2016) Submarine mass wasting on Hovgaard Ridge, Fram Strait, European Arctic. In: Lamarche G, Mountjoy J (eds) Submarine mass movements and their consequences. Springer, Dordrecht, pp 253–264

Gao D (2006) Gravitational sliding on the Mid-Atlantic Ridge at the Kane transform: implications for submarine basin-slope degradation and deformation. AAPG Bull 90:159–176

Gracia E, Lamarche G, Nelson H, Pantosti D (2013) Preface: marine and lake paleoseismology. Natl Hazard Earth Sci Process 13:3469–3478

Gwiazda R, Paull C, Caress D, Lorenson T, Brewer P, Peltzer E, Walz P, Anderson K, Lundsten E (2016) Eel canyon slump scar and associated fluid venting. In: Lamarche G, Mountjoy J (eds) Submarine mass movements and their consequences. Springer, Dordrecht, pp 411–418

Haneberg W (2016) Incorporating correlated variables into GIS-based probabilistic submarine slope stability assessments. In: Lamarche G, Mountjoy J (eds) Submarine mass movements and their consequences. Springer, Dordrecht, pp 529–536

Huhn K, Strozyk F, Kock I (2016) Quantifying the key role of slope material peak strength – using discrete element simulations. In: Lamarche G, Mountjoy J (eds) Submarine mass movements and their consequences. Springer, Dordrecht, pp 537–546

Huvenne V, Georgiopoulou A, Chaumillon L, Lo Iacono C, Wynn R (2016) Novel method to map the morphology of submarine landslide headwall scarps using Remotely Operated Vehicles. In: Lamarche G, Mountjoy J (eds) Submarine mass movements and their consequences. Springer, Dordrecht, pp 135–144

Katz O, Reuven E, Elfassi Y, Paldor A, Gvirtzman Z, Aharonov E (2016) Spatial and temporal relation of submarine landslides and faults along the Israeli continental slope, eastern Mediterranean. In: Lamarche G, Mountjoy J (eds) Submarine mass movements and their consequences. Springer, Dordrecht, pp 351–360

Kluger MO, Kreiter S, L'Heureux JS, Stegmann S, Moon V, Mörz T (2016) In situ cyclic softening of marine silts by vibratory CPTU at Orkdalsfjord test site, mid Norway. In: Lamarche G, Mountjoy J (eds) Submarine mass movements and their consequences. Springer, Dordrecht, pp 201–210

Kopf A, Kasten S, Blees J (2011) Geochemical evidence for groundwater-charging of slope sediments: the Nice airport 1979 landslide and tsunami revisited. In: Mosher D, Shipp RC, Moscardelli L, Chaytor JD, Baxter CDP, Lee HJ, Urgeles R (eds) Submarine mass movements and their consequences. Springer, Dordrecht

Kuhlmann J, Huhn K, Ikari MJ, Krastel S (2016) Do embedded volcanoclastic layers serve as potential glide planes? – an integrated analysis from the Gela Basin offshore southern Sicily. In: Lamarche G, Mountjoy J (eds) Submarine mass movements and their consequences. Springer, Dordrecht, pp 273–280

Laberg JS, Baetenb N, Vannestec M, Forsberg CF, Forwick M, Haflidason H (2016) Sediment failure affecting muddy contourites on the continental slope offshore northern Norway – lessons learned and some outstanding issues. In: Lamarche G, Mountjoy J (eds) Submarine mass movements and their consequences. Springer, Dordrecht, pp 281–290

Leblanc J, Turmel D, Therrien J, Locat J (2016) Observations of coastal landslide-generated tsunami under an ice cover: the case of Lac-des-Seize-Îles, Québec, Canada. In: Lamarche G, Mountjoy J (eds) Submarine mass movements and their consequences. Springer, Dordrecht, pp 607–614

Lindhorst K, Krastel S (2016) Mass wasting history within lake Ohrid Basin (Albania/Macedonia) over the last 600ka. In: Lamarche G, Mountjoy J (eds) Submarine mass movements and their consequences. Springer, Dordrecht, pp 291–300

Locat J, Turmel D, Habersetzer M, Trottier AP, Lajeunesse P, St-Onge G (2016) Earthquake induced landslides in Lake Éternité, Québec, Canada. In: Lamarche G, Mountjoy J (eds) Submarine mass movements and their consequences. Springer, Dordrecht, pp 361–370

Mastbergen D, van den Ham G, Cartigny M, Koelewijn A, de Kleine M, Clare M, Hizzett J, Azpiroz M, Vellinga A (2016) Multiple flow slide experiment in the Westerschelde Estuary, The Netherlands. In: Lamarche G, Mountjoy J (eds) Submarine mass movements and their consequences. Springer, Dordrecht, pp 241–249

Meckel III LD (2011) Reservoir characteristics and classification of sand-prone submarine mass-transport deposits, SEPM Special Publication 96, 432–452

Micallef A, Mountjoy J, Krastel S, Crutchley G, Koch S (2016) Shallow gas and the development of a weak layer in submarine spreading, Hikurangi margin (New Zealand). In: Lamarche G, Mountjoy J (eds) Submarine mass movements and their consequences. Springer, Dordrecht, pp 419–426

Moore G, Strasser M (2016) Large mass transport deposits in Kumano Basin, Nankai Trough, Japan. In: Lamarche G, Mountjoy J (eds) Submarine mass movements and their consequences. Springer, Dordrecht, pp 371–380

Mosher DC, Shipp RC, Moscardelli L, Chaytor JD, Baxter CD P, Lee HJ, Urgeles R (2010) Submarine mass movements and their consequences – 4th International Symposium. In: Advances in natural and technological hazards research. Springer, Dordrecht

Mosher D, Laberg JS, Murphy A (2016) The role of submarine landslides in the Law of the Sea. In: Lamarche G, Mountjoy J (eds) Submarine mass movements and their consequences. Springer, Dordrecht, pp 15–26

Mueller C, Mountjoy J, Power W, Lane E, Wang X (2016) Towards a spatial probabilistic submarine landslide hazard model for submarine canyons. In: LamarcheG, Mountjoy J (eds) Submarine mass movements and their consequences. Springer, Dordrecht, pp 589–598

Normandeau A, Joyal G, Lajeunesse P, Francus P, Lamoureux S, Lapointe F (2016) Late-Holocene Mass Movements in High Arctic East Lake, Melville Island (Western Canadian Arctic Archipelago). In: Lamarche G, Mountjoy J (eds) Submarine mass movements and their consequences. Springer, Dordrecht, pp 311–320

Ogata K, Pini GA, Festa A, Poga?nik Ž, Lucente C (2016) Meso-scale kinematic indicators in exhumed mass transport deposits: definitions and implications. In: Lamarche G, Mountjoy J (eds) Submarine mass movements and their consequences. Springer, Dordrecht, pp 461–468

Omeru T, Cartwright J, Bull S (2016) Kinematics of submarine slope failures in the deepwater Taranaki Basin, New Zealand. In: Lamarche G, Mountjoy J (eds) Submarine mass movements and their consequences. Springer, Dordrecht, pp 61–70

Piper DJW, Aksu AE (1987) The source and origin of the 1929 grand banks turbidity current inferred from sediment budgets. GeoMar Lett 7:177–182

Postma G, Hoyal D, Abreu V, Cartigny M, Demko T, Fedele J, Kleverlaan K, Pederson KH (2016) Morphodynamics of supercritical turbidity currents in the channel-lobe transition zone. In: Lamarche G, Mountjoy J (eds) Submarine mass movements and their consequences. Springer, Dordrecht, pp 469–478

Priest J, Grozic J (2016) Stability of fine-grained sediments subject to gas hydrate dissociation in the Arctic continental margin. In: Lamarche G, Mountjoy J (eds) Submarine mass movements and their consequences. Springer, Dordrecht, pp 427–436

Savini A, Marchese F, Verdicchio G, Vertino A (2016) Submarine slide topography and the distribution of vulnerable marine ecosystems: a case study in the Ionian Sea (eastern Mediterranean). In: Lamarche G, Mountjoy J (eds) Submarine mass movements and their consequences. Springer, Dordrecht, pp 163–170

Sawyer D, Hodelka B (2016) Tiny fossils, big impact: the role of foraminifera-enriched condensed section in arresting the movement of a large retrogressive submarine landslide in the Gulf of Mexico. In: Lamarche G, Mountjoy J (eds) Submarine mass movements and their consequences. Springer, Dordrecht, pp 479–486

Sobiesiak M, Kneller B, Alsop GI, Milana JP (2016) Inclusion of substrate blocks within a mass transport deposit: a case study from Cerro Bola, Argentina. In: Lamarche G, Mountjoy J (eds) Submarine mass movements and their consequences. Springer, Dordrecht, pp 487–496

Stegmann S, Kreiter S, L'Heureux JS, Vanneste M, Völker D, Baeten NJ, Knudsen S, Rise L, Longva O, Brendryen J, Haflidason H, Chand S, Mörz T (2016) First results of the geotechnical in situ investigation for soil characterisation along the upper slope off Vesterålen – Northern Norway. In: Lamarche G, Mountjoy J (eds) Submarine mass movements and their consequences. Springer, Dordrecht, pp 211–220

Tappin DR, Watts P, McMurtry GM (2001) The Sissano, Papua New Guinea tsunami of July 1998: offshore evidence on the source mechanism. Mar Geol 175:1–23

Torbahn L, Strege S, Kwade A, Trandafir A (2016) A novel micro-shear tester for failure analysis of fine and cohesive granular matter. In: Lamarche G, Mountjoy J (eds) Submarine mass movements and their consequences. Springer, Dordrecht, pp 221–228

Trandafir A, Popescu M (2016) Insights into effectiveness of simplified seismic displacement procedures to evaluate earthquake behavior of a deepwater slope. In: Lamarche G, Mountjoy J (eds) Submarine mass movements and their consequences. Springer, Dordrecht, pp 381–390

Tucholke BE (2015) Massive submarine rockslide in the rift-valley wall of the Mid-Atlantic Ridge. Geology 20:129–132

Turmel D, Locat J, Parker G, Konrad JM (2016) Knickpoint migration induced by landslide: evidence from laboratory to field observations in Wabush Lake. In: Lamarche G, Mountjoy J (eds) Submarine mass movements and their consequences. Springer, Dordrecht, pp 229–239

Wang X, Mountjoy J, Power W, Lane E, Mueller C (2016) Coupled modelling of the failure and tsunami of a submarine debris avalanche offshore central New Zealand. In: Lamarche G, Mountjoy J (eds) Submarine mass movements and their consequences. Springer, Dordrecht, pp 599–606

Part I
Submarine Mass Movement in Margin Construction and Economic Significance

Chapter 2
The Role of Submarine Landslides in the Law of the Sea

David C. Mosher, Jan Sverre Laberg, and Alain Murphy

Abstract Article 76 of the United Nations Convention on the Law of the Sea prescribes two approaches that a nation may employ to determine the extent of its' legal continental shelf: (1) 60 nautical miles (M) seaward of the foot of the continental slope (FoS), or, (2) to a point seaward of the FoS where the sediment thickness is 1 % of the distance from the FoS. In both of these formulae, the "foot of the continental slope" is a critical metric. Article 76 defines the "foot of the continental slope" as: "*In the absence of evidence to the contrary, the foot of the continental slope shall be determined as the point of maximum change in the gradient at its base*". Geomorphologic complexity or low gradients (<1°) of continental slopes rarely permit a ready determination of the maximum change in gradient; particularly at a position that a geologist might qualitatively recognize as the base-of-slope zone. Recognizing that submarine mass movement is a slope process that also influences the shape of the continental margin, several nations have successfully argued that the downslope termination of mass transport deposits assist in distinguishing the continental slope from the rise and abyssal plain. The Commission on the Limits of the Continental Shelf have now made recommendations for a number of coastal States with rift margins, transform margins and subduction margins where the extents of surficial mass transport deposits were used to help delineate the base of slope zone within which the foot of the continental slope is chosen.

D.C. Mosher (✉)
Geological Survey of Canada, Natural Resources Canada Bedford Institute of Oceanography, 1 Challenger Dr., Dartmouth, NS B2Y 4A2, Canada

Center for Coastal and Ocean Mapping, University of New Hampshire, 24 Colovos Rd., Durham, NH 03824, USA
e-mail: dmosher@nrcan.gc.ca; david.mosher@unh.edu

J.S. Laberg
Department of Geology, University of Tromsø – The Arctic University of Norway, Dramsveien 201, 9037 Tromsø, Norway
e-mail: jan.laberg@uit.no

A. Murphy
GeoLimits Consulting (formerly Department Communications, Energy and Natural Resources, Ireland), 41 Condor Rd., Bedford, Nova Scotia, Canada

G. Lamarche et al. (eds.), *Submarine Mass Movements and their Consequences*, Advances in Natural and Technological Hazards Research 41,
DOI 10.1007/978-3-319-20979-1_2

2.1 Introduction

The Commission on the Limits of the Continental Shelf (CLCS); the body that reviews extended continental shelf submissions under the UN Convention on the Law of the Sea, in its guidelines and decisions to date, has encouraged coastal States to use scientific arguments in defining the elements of a continental margin and thereby delimit its outer edges. One of the critical metrics in this process is establishment of the "foot of the continental slope" (FoS) and submarine landslides can assist in identifying this metric.

The outer edge of the legal continental shelf is defined in article 76 of the United Nations Convention on the Law of the Sea as the submerged prolongation of the land mass of the coastal State, consisting of the seabed and subsoil of the shelf, the slope and the rise. It does not include the deep ocean floor. The outer edge of the legal continental shelf is determined by either 60 nautical miles (M) from the foot of the continental slope (known as the Hedberg formula (Hedberg 1981)), or not more than the distance in which the thickness of sedimentary rocks is 1 % of the shortest distance from the foot of the continental slope (the Gardner formula) (Fig. 2.1). One can appreciate from these formulae that the foot of the continental slope is an essential metric in definition of the continental margin and therefore entitlement to an extended continental shelf, yet the language is simple and ambiguous. The foot of the continental slope is defined within article 76 as 'the point of maximum change in the gradient at its base'.

Geologic complexities of continental margins rarely allow for the identification of a single point of maximum change in gradient, and if so, these points would rarely coincide with the area that a geologist would identify as the base of slope. As a result of this and other ambiguities, the CLCS adopted a set of scientific and technical guidelines to assist State's in the preparation of their submission (CLCS 1999). These guidelines provide that identification of the foot of the continental slope first requires identification of the region defined as the base of the continental slope; and then the determination of the location of the point of maximum change in gradient within the base of the continental slope (see Fig. 2.1). In order to identify the base of the continental slope, one must first distinguish the various elements of the margin.

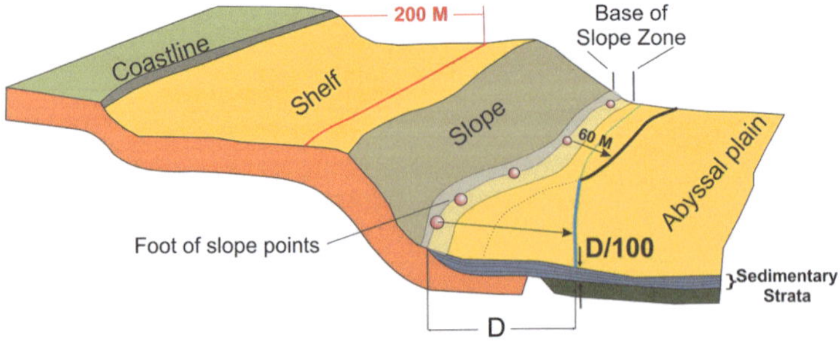

Fig. 2.1 Elements prescribed in article 76 for defining the outer edge of the legal continental shelf: (1) the foot of the continental slope plus 60 nautical miles (M) (Hedberg formula), or (2) the distance (D) from the foot of the continental slope where it is equal to 1 % of the thickness of sedimentary rocks (D/100) (the Gardner formula)

The simple concept of a margin comprising shelf, slope, rise and deep ocean floor distinguished by slope angle or bathymetry alone, as described by Heezen et al. (1959) is rarely observed in nature. As a result, the CLCS has allowed the use of arguments based on geological and geophysical observations of natural processes to distinguish these components of a margin. Submarine landslides, for example, are a slope phenomenon and thereby characterize the continental slope and distinguish it from the shelf, rise and abyssal plain. To date, the CLCS has accepted arguments from a number of State submissions regarding establishment of the foot of the continental slope with submarine landslide processes, including examples from passive rift (Norway, Ireland), transform (French Guiana), and active (Sumatra) margins. This paper will demonstrate how submarine landslides are used in application of article 76 to establish the outer edge of the continental margin by way of presenting these case studies.

2.2 Case Studies

2.2.1 Rift Margins

2.2.1.1 Ireland, Porcupine Bank

Porcupine Bank offshore of Ireland extends the Irish Shelf more than 300 km westwards from the coast. West of Porcupine Bank is the Porcupine Abyssal Plain lying in water depths greater than 4000 m (Fig. 2.2). The western edge of Porcupine Bank is a rounded shoulder, providing a gradual and indistinct shelf break from 150 to >500 m water depth. This shoulder evolves into an upper slope segment that is more than 10° down to a depth of about 3000–3500 m. At this depth, there is a distinct gradient change of the slope to between 2 and 4°. This segment is referred to as the lower slope. The angles of this lower slope then decrease as water depths increase to the deep ocean floor. The maximum change in gradient of the slope occurs at the initial gradient change in 3000–3500 m water depth, but the seafloor clearly continues to dip toward the deep ocean. There is no subsequent distinctive change in gradient to separate the 'slope' from a 'rise' from the 'deep ocean floor' (Porcupine abyssal plain).

In its interaction with the CLCS, Ireland argued for a two-segment continental slope and that the base of slope region did not include the local maximum between the upper and lower slope segments, but rather laid seaward (Ireland 2005). In particular, Ireland presented multichannel seismic and vintage scanned seismic data showing mass slumps and slides along the Porcupine Bank margin (van Weering et al. 2003), which combined with the multibeam data demonstrated the nature of the two segment slope. The base of the continental slope was shown to lie at the outer edge of the complex lower slope formed by mass-transport depositional processes associated with slope failure and erosion of the Porcupine Bank margin and not at the more landward maximum change in regional gradient (i.e. the base of the upper slope). Foot of the continental slope (FoS) points were chosen at the maximum change in gradient within this seaward zone, therefore (Fig. 2.2).

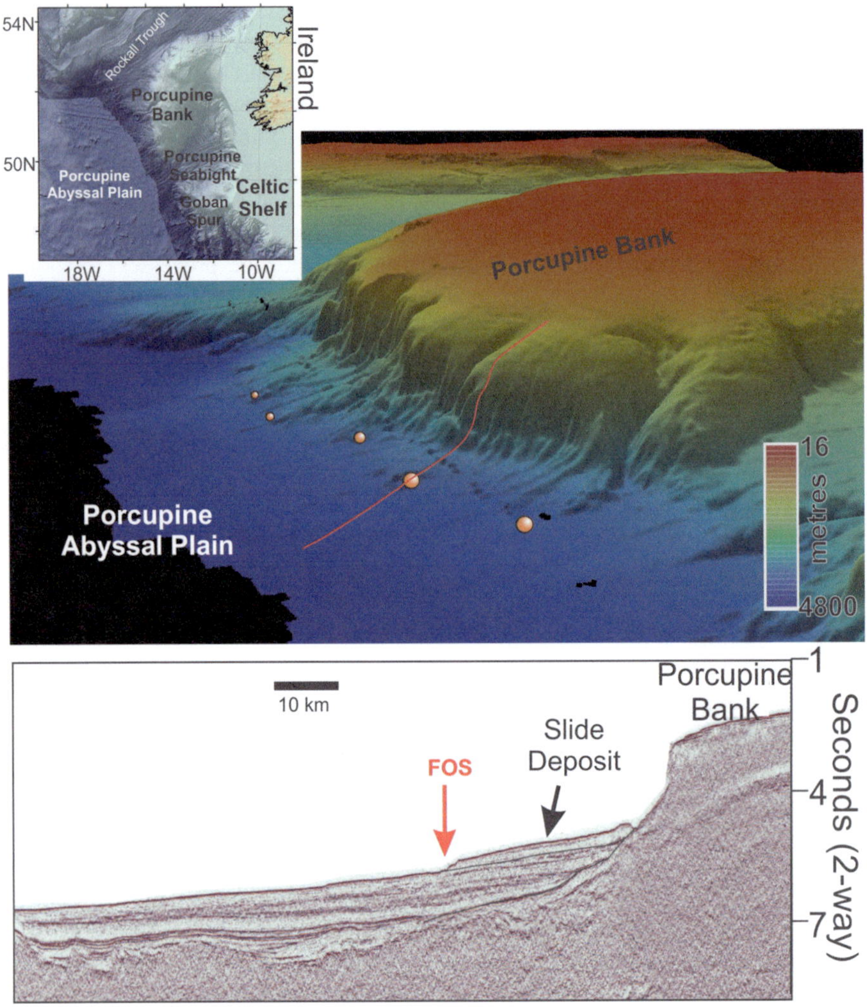

Fig. 2.2 The upper panel is a perspective view of Porcupine Bank offshore Ireland. A few of the foot of the continental slope (FoS) points off of Porcupine Bank are indicated showing that they are located at the base of a two-segmented continental slope; not at the base of the steepest section but rather at the base of the lower slope formed by mass transport deposits. The lower panel is a seismic reflection profile showing a surficial submarine landslide and the FoS is located at the seaward toe of this feature. Multibeam data for the seafloor render were provided by the Geological Survey of Ireland, INSS/INFOMAR projects (Seismic data courtesy of Dept. of Communications, Energy and Natural Resources, Ireland)

In its recommendations to Ireland, adopted on the 5th April 2007 (CLCS 2007), the CLCS identified the importance of slope failure/erosional features and the complexities that can exist with the identification of the continental slope. The CLCS accepted Ireland's geomorphological analysis and concluded that "slope failure/erosion of the margin of Porcupine Bank has produced a lower slope composed of coalescing slump, slide and debris flow deposits separated by channels and gulleys", and most significantly that "these depositional features are characteristic of slopes and not rises". As noted by the CLCS, "this distinction is consistent with paragraphs 5.4.5 and 5.4.6, as well as paragraphs 5.4.12 and 6.3.12 of the Guidelines" (CLCS 1999).

2.2.1.2 Norway, Bear Island (Bjørnøya) and Franz-Victoria Trough Mouth Fans (TMF)

The Barents Shelf is a broad shallow platform that extends from the northern tip of Norway across to Novaya Zemlya of Russia and includes the islands of Svalbard (Norway) and Franz Josef Land (Russia). The northern part of the shelf borders Nansen Basin of the Arctic Ocean and the southern segment from Svalbard to mainland Norway borders the Lofoten Basin of the Norwegian-Greenland Sea. Fram Strait connects the Arctic Ocean to the Greenland Sea and lies between Svalbard and Greenland.

The Barents Shelf and its margins were glaciated numerous times during the Pleistocene with dramatic effect on sediment delivery and construction of the modern margin morphology (Vorren et al. 2003; Jakobsson et al. 2012; Laberg et al. 2012).

The broad Barents Shelf (part of the Eurasia plate) has a number of deep (<500 m) troughs crossing the platform, formed by glacial ice streams. Two of the most significant are the Franz-Victoria Trough and the Bear Island Trough. Seaward of these troughs, beyond the edge of the shelf, are glacial-marine trough-mouth fans, including: (1) Franz-Victoria TMF, lying between Svalbard and Franz-Josef Land and extending into Nansen Basin, and (2) Bear Island (Bjørnøya) TMF extending into Lofoten Basin between mainland Norway and Svalbard (Fig. 2.3). As a result of the significant sediment supply to these fans, the continental slope has an overall concave morphology with relatively low gradients (<2°). There is a near constant change in gradient from the upper slope to its base. Consequently, in these areas, the location of the base of the continental slope is not readily identifiable solely on the basis of morphology.

Large mass failures are known to have occurred throughout the Pleistocene within the Bear Island (Bjørnøya) fan complex (Laberg and Vorren 1996; Dowdeswell et al. 2003). Although these processes are important in shaping the modern seafloor, the CLCS puts no importance on the nature of the bulk of the sedimentary section. It is, however, concerned with the shape of the modern seafloor and its composition. It was shown that glacigenic debris flow (GDF) deposits form the bulk of the seafloor (and shallow subbottom) of the modern Bear Island TMF (Fig. 2.4). Norway was able to establish the base of the continental slope zone by following a regional change in seafloor gradient that is formed by the seaward edge of lobe-shaped deposits of the outermost edge of these glacigenic debris flow deposits. The CLCS "...agreed with the new approach adopted by Norway to define the base of the continental slope associated with the Bear Island TMF. In particular, based on the morphological and bathymetric evidence, supplemented by other geological and geophysical data provided by Norway, the

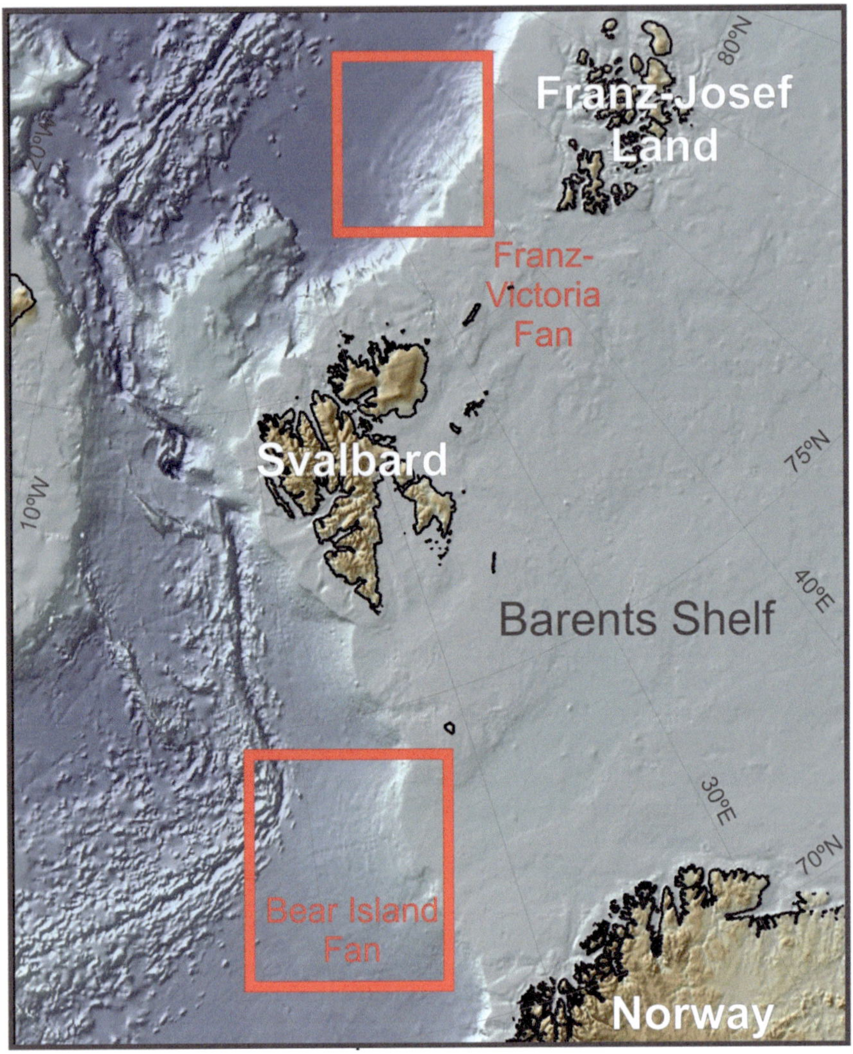

Fig. 2.3 Location of Bear Island Trough Mouth Fan (TMF) and Franz-Victoria TMF

CLCS agreed with the way the locations of new critical foot of continental slope points . . . were established" (CLCS 2009a).

For the region of the Franz-Victoria TMF, Norway based the initial foot of the continental slope positions on multichannel reflection data which lacked sufficient resolution to image surficial GDF deposits. Initially, the CLCS expressed the view that there were insufficient geological and geophysical data to support these points and advised Norway to explore more landward possibilities associated with regionally significant inflection points in the gradient of the seafloor. In response, Norway used a line of high-resolution sub-bottom profiler data (Parasound) that showed lobes of GDF deposits down the slope of the fan. They were able to establish a foot

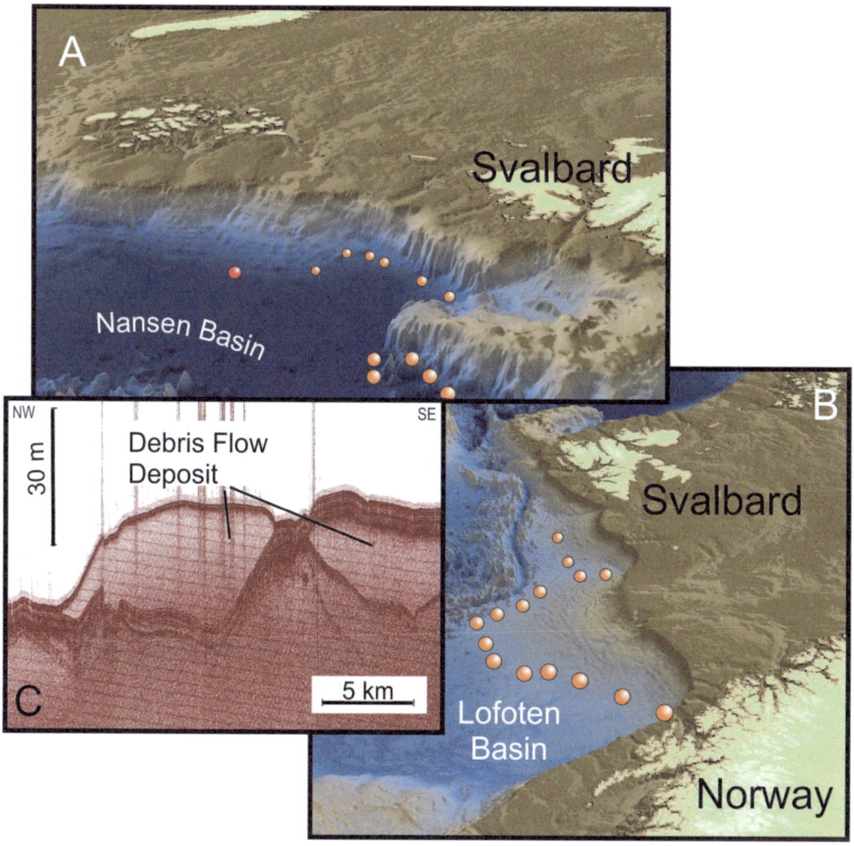

Fig. 2.4 (**a**), (**b**) Are perspective views of the Franz-Victoria and Bear Island Trough Mouth Fans, respectively. The *dots* represent approximate positions of the foot of the continental slope. The *red dot* in (**a**) shows the FoS point at the base of the Franz-Victoria Fan. In (**b**), the *dots* reflect the base of the Bear Island Fan. (**c**) is a subbottom reflection profile showing debris flow deposits as they appear on these glacio-marine trough mouth fans (data were provided by the University of Bergen. Seafloor renders in (**a**) and (**b**) are from the IBCAO bathymetric grid (Jakobbson et al. 2012))

of the continental slope point that was in fact seaward of its original position (Fig. 2.4a). The CLCS agreed with the general approach adopted by Norway to define the base of the continental slope associated with the Franz-Victoria Fan, and the location of the revised FOS point. Critical elements to this agreement were "…the newly presented high resolution, Parasound, sub-bottom profiler data; consistency [in argument] with the base of slope location on the Bear Island TMF; and its location at a regional change in gradient at the base of the debris flow apron of the fan that is seen from the Parasound data to underlie all of the continental slope from water depths of more than 4000 m back to the shelf break" (CLCS 2009a).

2.2.2 *Transform Margins*

2.2.2.1 France: French Guiana

The margin of French Guiana represents a transform margin setting. The northern segment of its margin includes the Demerara Plateau. The southern segment is referred to as the eastern Guiana margin. Demerara Plateau is conjugate to the Guinea Plateau off West Africa, separated by the Guinea Fracture Zone. As a transform margin, the flank of the Demerara Plateau is steep (>10°) and the base of the continental slope is straight-forward to recognize. The eastern Guiana margin, however, is heavily sedimented and has lower slope gradients from 2 to 4°. It is under the distal influence of the Amazon fan.

France, on behalf of French Guiana, needed to establish only one critical foot of the continental slope point along its eastern segment. This point was determined employing morphological and geological information supported by geophysical evidence (CLCS 2009b). The CLCS was not initially convinced with the supporting evidence concerning this point and a series of exchanges between France and the CLCS took place, including submission of additional information and of refereed papers (Loncke et al. 2009; Gaullier et al. 2011).

The authors of these papers demonstrated that the Eastern Guiana slope is affected by repeated destabilizations, with slope failures generating large debris flow lobes that are apparent in multibeam seafloor maps and subbottom profiles. For example, Fig. 2.5 shows a N-S elongated debris flow cut on both sides by active submarine valleys. Seismic imaging confirms the presence of several large stacked recent mass-movement deposits. France located the foot of the continental slope at a change in gradient at the base of these features.

The CLCS subsequently agreed with the approach adopted to define the base of the continental slope associated with the French Guiana region (CLCS 2009b). This fact underscores the value not only of submarine mass failures in helping determine the base and foot of the continental slope, but also of supporting evidence in peer-reviewed journals.

2.2.3 *Active Margins*

2.2.3.1 Indonesia: Northwest Sumatra

Subduction of the Indian Plate has led to construction of a complex accretionary prism on the over-riding plate adjacent to Sumatra that has a steep (>8°) slope at the frontal thrust region (Henstock et al. 2007). Regionally, the subduction trench forms a clear and abrupt change in gradient of the margin. Indonesia argued, however, that hemipelagic sediments, volcanic ash or volcaniclastic sediments and detritus sediments from the uplifted accretionary prism are accumulating along the trench's lower slope as apron, small fan and wedge- shaped sedimentary bodies due to slumping, debris flow or turbidite sedimentary processes

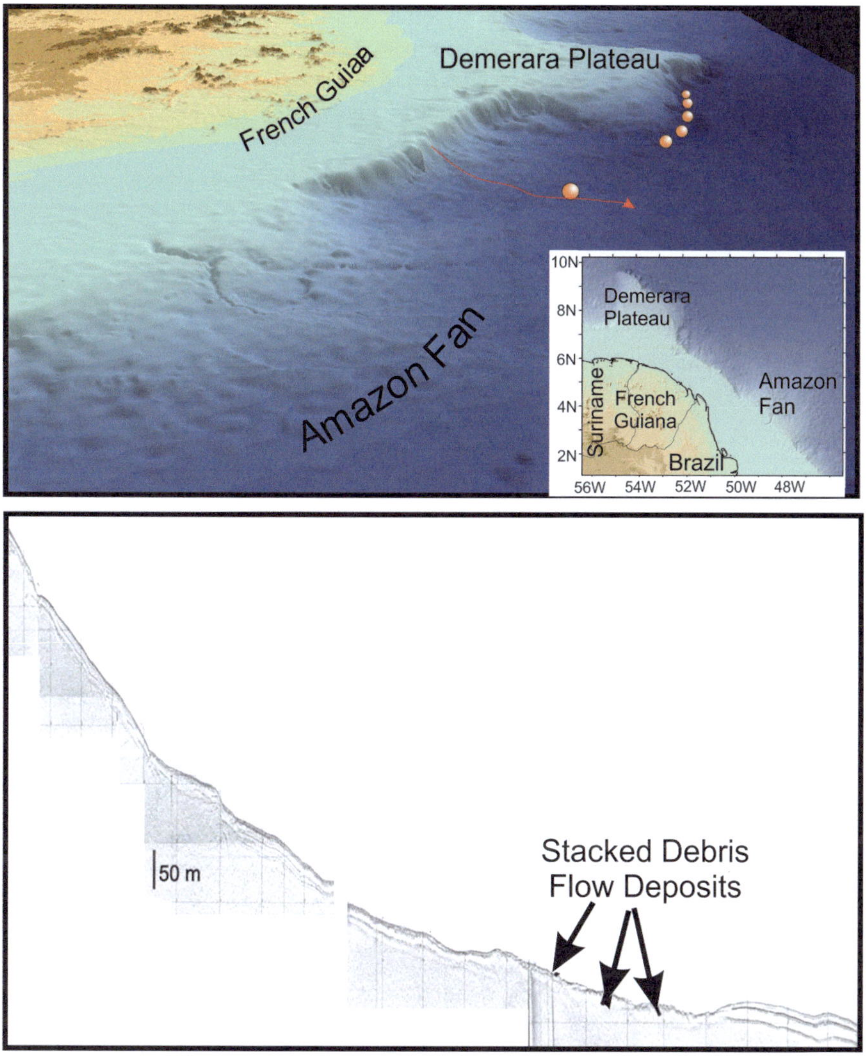

Fig. 2.5 *Top* is a perspective view (from south to north) of the French Guiana margin. The *orange dots* show the approximate positions of the foot of continental slope points. One point was critical along the eastern Guiana margin. It was select on the basis of the seaward limit of debris flow deposits, as exemplified in the lower profile (from Loncke et al. 2009). Seafloor render is from the GEBCO bathymetric data grid

(Tappin et al. 2007; Mosher et al. 2008). Indonesia chose one of their critical foot of the continental slope points on the outward edge of a submarine landslide in order to maximize the seaward extent of their continental margin.

In recommendations to Indonesia for Northwest Sumatra, the CLCS accepted that one critical foot of the continental slope (FOS #38) is located beyond the base of the accretionary prism (CLCS 2011). Indonesia supplied morphological and bathymetric evidence, supplemented by other geological and geophysical data to support the location of the foot of the continental slope. The foot of the continental slope position was acceptable to the CLCS as a sedimentary apron or wedge shaped body due to small slump-slide or debris flow (Fig. 2.6).

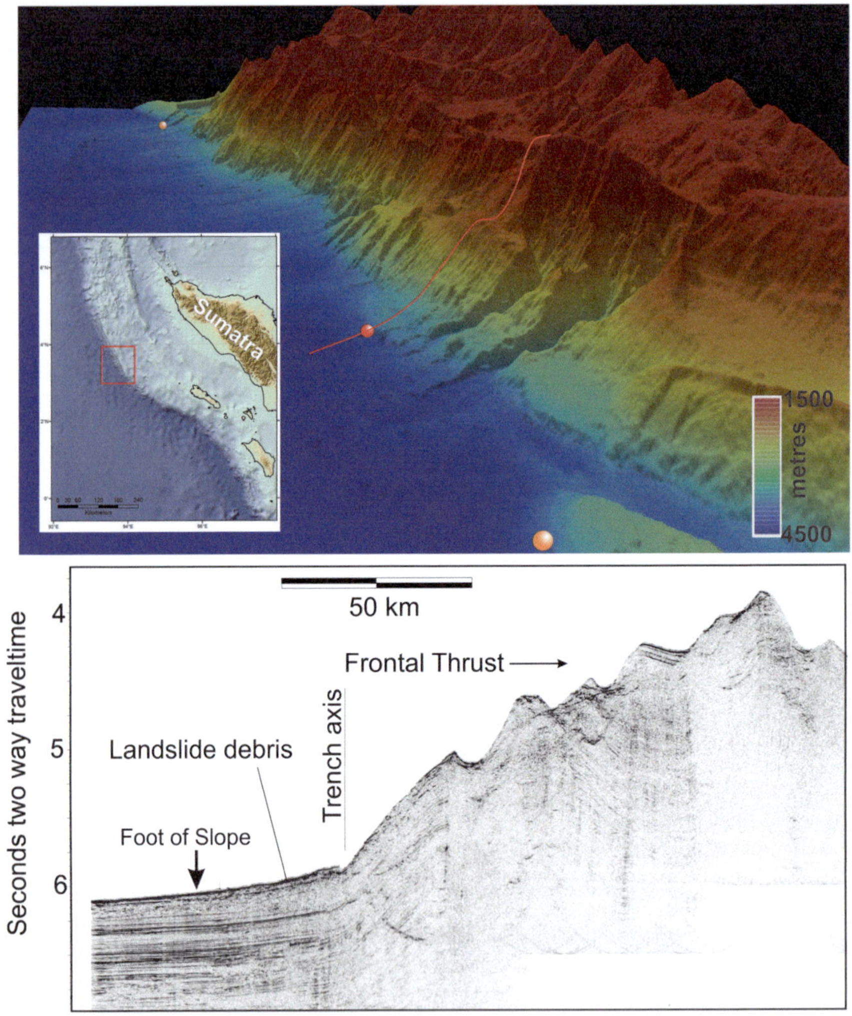

Fig. 2.6 Sumatra accretionary prism. *Top* is a perspective view looking north showing the base of the frontal thrust region. The *red dot* shows the approximate position of the critical foot of the slope point that is at the base of landslide debris. *Below* is a seismic profile transecting the frontal thrust and out to the basin floor of the subducting plate. It shows the foot of the slope is seaward of the trench axis (Multibeam data for the seafloor render were provided by Henstock et al. (2007). Seismic data were provided by Mosher et al. (2008))

2.3 Conclusions

In its guidelines and recommendations to date, the Commission on the Limits of the Continental Shelf has rendered decisions that recognize submarine landslides as characteristic of slopes and not rises or deep ocean floor. These decisions allow coastal States to make scientific arguments for foot of continental slope positions, as opposed to determinations made with morphological measurements alone. They have stressed that these geologic processes must be recent (surficial) and not part of the geologic past. They have furthermore emphasized the need for corroborating data types that support geologic interpretations (e.g. subbottom profiler and backscatter). As a result, identification of submarine mass movements has become an important criterion for extended continental shelf determinations for many coastal States.

Acknowledgements The authors would like to express their appreciation to C. Berndt, L. Moscardelli and S. Krastel for their reviews and valuable editorial comments. We also express thanks to our many colleagues who have entertained dialogue about this subject and most specifically P. Symonds and H. Brekke who, in their role on the CLCS, understood the importance of geological processes and implemented their consideration in the CLCS process.

References

CLCS (1999) Scientific and technical guidelines of the Commission on the Limits of the Continental Shelf. http://www.un.org/depts/los/clcs_new/commission_documents.htm. Accessed 9 Mar 2015

CLCS (2007) Recommendations of the Commission on the Limits of the Continental Shelf in regard to the partial submission made by Ireland on 25 May 2005 on the proposed outer limit of its continental shelf beyond 200 nautical miles in the area abutting the Porcupine Abyssal Plain. http://www.un.org/depts/los/clcs_new/submissions_files/irl05/irl_rec.pdf. Accessed 9 Mar 2015

CLCS (2009a) Summary of recommendations of the Commission on the Limits of the Continental Shelf in regard to the submission made by Norway in respect of the areas in the Arctic Ocean, the Barents Sea and the Norwegian Sea on 27 November 2006. http://www.un.org/depts/los/clcs_new/submissions_files/nor06/nor_rec_summ.pdf. Accessed 9 Mar 2015

CLCS (2009b) Summary of recommendations of the Commission on the Limits of the Continental Shelf in regard to the submission made by France in respect of French Guiana and New Caledonia regions on 22 May 2007. http://www.un.org/depts/los/clcs_new/submissions_files/fra07/COM_REC_FRA_02_09_2009_summary.pdf. Accessed 9 Mar 2015

CLCS (2011) Summary of recommendations of the Commission on the Limits of the Continental Shelf in regard to the submission made by Indonesia in respect of the area north west of Sumatra on 16 June 2008. http://www.un.org/depts/los/clcs_new/submissions_files/idn08/Summary%20Recommendations%20for%20Indonesia.pdf. Accessed on 9 Mar 2015

Dowdeswell J, Kenyon NH, Taylor J (2003) Slides and debris flows: GLORIA long-range side scan sonar mosaic of the north Norwegian margin (67–75°N). In: Meinert J, Weaver P (eds) European margin sediment dynamics. Springer, Berlin, pp 31–43

Gaullier V, Loncke L, Droz L, Basile C, Maillard A, Patriat M, Roest WR, Loubrieu B, Folens L, Carol F (2011) Slope instability on the French Guiana transform margin from swath-bathymetry and 3.5 kHz echograms. In: Mosher DC et al (eds) Submarine mass movements and their consequences IV; Adv Nat Tech Hazards Res 28: 569–579

Hedberg HD (1981) Geomorphic basis for national-international boundaries on the ocean floor. Proceedings of the 2nd American Association of Petroleum Geologists Circum Pacific Energy and Mineral Resources Conference, Honolulu, 1 Aug 1978, pp 441–464

Heezen BC, Tharp M, Ewing M (1959) The floors of the oceans: I: the north Atlantic. Geological Society of America, Boulder, CO

Henstock TJ, McNeill LC, Tappin DR (2007) Seafloor morphology of the Sumatran subduction zone: surface rupture during megathrust earthquakes? Geology 34:485–488

Ireland (2005) Executive Summary Ireland Submission to the Commission on the Limits of the Continental Shelf pursuant to Article 76, paragraph 8 of the United Nations Convention on the Law of the Sea 1982 in respect of the area abutting the Porcupine Abyssal Plain. http://www.un.org/depts/los/clcs_new/submissions_files/submission_irl.htm. Accessed 9 Mar 2015

Jakobsson, M and many others (2012) The International Bathymetric Chart of the Arctic Ocean (IBCAO) version 3.0. Geophys Res Lett 39: 6 pp

Laberg JS, Vorren TO (1996) The middle and late Pleistocene evolution of the bear island trough mouth fan. Global Planet Change 12:309–330

Laberg JS, Andreassen K, Vorren TO (2012) The late Cenozoic erosion of the high-latitude southwestern Barents Sea shelf revisited. Geol Soc Am Bull 124:77–88. doi:10.1130/B30340.1

Loncke L, Droz L, Gaullier V, Basile C, Patriat M, Roest W (2009) Slope instabilities from echo-character mapping along the French Guiana transform margin and Demerara abyssal plain. Mar Pet Geol 26:711–723

Mosher DC, Austin JA Jr, Fisher D, Gulick SPS (2008) Deformation of the northern Sumatra accretionary prism from high-resolution seismic reflection profiles and ROV observations. Mar Geol 252:89–99

Tappin DR, McNeil LC, Henstock T, Mosher DC (2007) Mass wasting processes – offshore Sumatra. In: Lykousis V, Dimitris S, Locat J (eds) Submarine mass movements and their consequences III. Springer, The Netherlands, pp 327–336

Van Weering TCE et al (2003) Images of sliding and slumping along the Porcupine and SW Rockall Trough margins. In: Mienert J, Weaver P (eds) European margin sediment dynamics. Springer, Berlin, pp 173–178

Vorren TO, Sejrup HP, Dowdeswell JA (2003) The Norwegian margin. In: Meinert J, Weaver P (eds) European margin sediment dynamics. Springer, Berlin, pp 19–30

Chapter 3
Fabric Development and Pore-Throat Reduction in a Mass-Transport Deposit in the Jubilee Gas Field, Eastern Gulf of Mexico: Consequences for the Sealing Capacity of MTDs

Sebastian Cardona, Lesli J. Wood, Ruarri J. Day-Stirrat, and Lorena Moscardelli

Abstract Mass-transport deposits (MTDs) are important stratigraphic elements in many deepwater basins. MTDs have traditionally been identified as seals but can also act as migration pathways. Studying the character of deposits within a MTD from proximal to distal, in a framework of seismically identifiable morphologies provides a template for using seismic character to predict the petrophysical properties of such deposits. During failure and subsequent transport, MTDs are exposed to shear deformation and remolding that can enhance clay alignment and destroy large pore-throats thus creating potential seal quality facies. Deformation in the various MTD morpho-domains can be quantified by measuring the degree of clay-fabric alignment. In this study we investigate a MTD acting as the top-seal in the Jubilee gas field, Gulf of Mexico, by integrating 3D-seismic, core, and well-log data to characterize clay fabrics. X-ray-texture goniometry analysis was performed using core material from the top-seal MTD to determine the degree of clay fabric alignment. Final results indicate that samples have an anomalously high clay-fabric orientation not correlated with burial depth or diagenesis. We conclude that

S. Cardona (✉)
Bureau of Economic Geology, Jackson School of Geosciences, The University of Texas, Austin, TX, USA now at Colorado School of Mines
e-mail: sebastian.cardona@utexas.edu

L.J. Wood
Colorado School of Mines, Golden, CO, USA

R.J. Day-Stirrat
Bureau of Economic Geology, Austin, TX, USA now at Shell International E and P, Houston, TX, USA

L. Moscardelli
Bureau of Economic Geology, Austin, TX, USA now at Statoil RDI, Austin, TX, USA

G. Lamarche et al. (eds.), *Submarine Mass Movements and their Consequences*, Advances in Natural and Technological Hazards Research 41,
DOI 10.1007/978-3-319-20979-1_3

these zones with high clay-fabric alignment in the MTD are the result of shear deformation as the gravity flow moved downslope. Recognition of zones with enhanced microfabrics has important implications for shallow geohazards as well as sealing potential evaluation. This technique—although in its infancy—could be used to identify sealing MTD facies in core samples and outcrop studies.

3.1 Introduction

MTDs are volumetrically important in modern and ancient deepwater basins (Posamentier and Kolla 2003). In terms of their significance to oil and gas exploration, MTDs can act as reservoirs, seals, or significantly modify the nature of pre- and post-MTD deposited strata (Moscardelli et al. 2006; Meckel 2011; Armitage et al. 2009; Alves et al. 2014). The Jubilee gas field, Gulf of Mexico (GOM), U.S.A. (Fig. 3.1) is an area where the hydrocarbon-trapping role of MTDs has been proven—as of 2014 the field has produced an estimated of 198 billion cubic feet of natural gas.

It is often difficult to identify why MTDs seal and still more difficult to predict when such sealing character will occur (Aplin and Macquaker 2011). It is possible that zones with enhanced clay alignment and reduced pore-throat structure within MTDs as a result of shear deformation and remolding are important in creating the high seal quality in MTDs (Day-Stirrat et al. 2013). Moreover, different types of MTDs (attached vs. detached sensu stricto Moscardelli and Wood 2008) as well as different morpho-domains within the same MTD (Bull et al. 2009) may exhibit distinctive shear deformation that can be more likely to develop quality seal facies. Attempts in determining the degree of deformation/alignment of clay particles in MTDs using microscopic methods have been pioneered by Yamamoto and Sawyer (2012) and Day-Stirrat et al. (2013). This paper, however, is the first to examine the fabric of a deeply buried MTD acting as the top-seal of a hydrocarbon field. We will focus on the petrophysical characteristics of the top-seal MTD in the Jubilee field at a depth of 2,523 m below sea floor (mbsf). A dataset of three dimensional (3D) seismic, core, and well data was utilized to establish practical observations about the attributes of an effective seal-MTD.

3.2 Geologic Setting

The Jubilee field is located in the Atwater Valley area of the GOM approximately 190 km off the coast at a water depth of ~2,600 m (8,830 ft) (Fig. 3.1). The discovery well was drilled to the target depth of 5,580 m (18,310 ft) with respect to sea level. The field area is relatively structureless with deep deformed autochthonous salt in a compactional drape trap style (Weimer and Bouroullec 2013). The

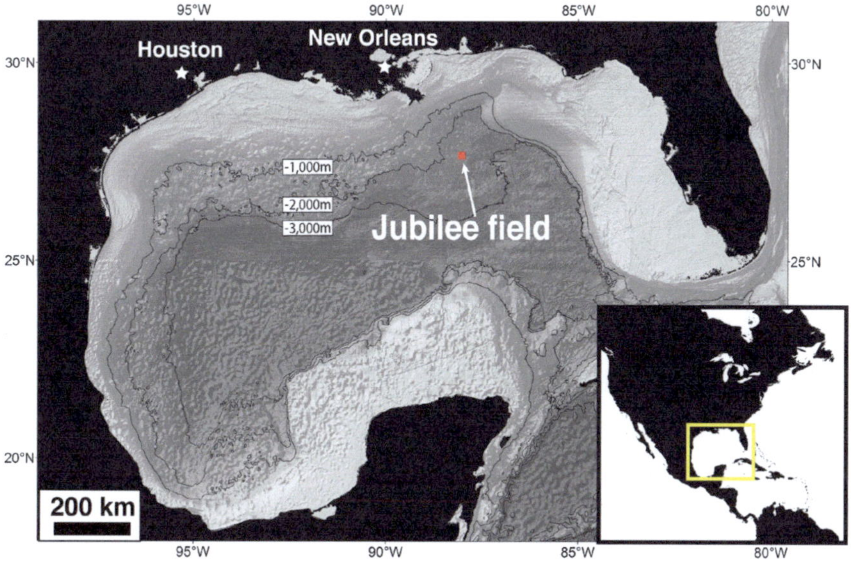

Fig. 3.1 Present-day bathymetry of the Gulf of Mexico, U.S.A. with the study area

reservoir is composed by high-density turbidites and frontal-splay deposits of Upper Miocene age (5.3–7.2 Ma), compensationally stacked in an abyssal plain environment. MTDs overlying the reservoir are interpreted as being triggered by gravitational failures on the flank of a salt mass. The deposits of the Jubilee field exhibit a typical succession laid down in a deepwater setting (Posamentier and Kolla 2003) whereby a MTD at the base is overlain by turbidite deposits of a frontal splay, a channel complex, and subsequently capped by a later MTD (Fig. 3.2a).

3.3 Available Data and Methodology

The available data for this study are a 3D industry-quality reflection seismic survey of 141 km^2 over the field, 56 m (185 ft) of core from 2,509 to 2,565 mbsf (17,040–17,125 ft subsea) through the top-seal MTD and underlying reservoir interval **of** the AT 349-2 well, and several conventional wireline well logs from wells located closely adjacent to the cored well (Fig. 3.2a). In total nine samples from AT 349-2 well core, four of which are directly from the top-seal MTD (Fig. 3.3), were analyzed using X-Ray Texture Goniometry (XRTG) methods to determine the degree of alignment of clay minerals. In addition to XRTG, other microscopic methodologies used to characterize the petrophysical properties of the top-seal MTD were X-Ray Diffraction (XRD) and Mercury Injection Capillary Pressure

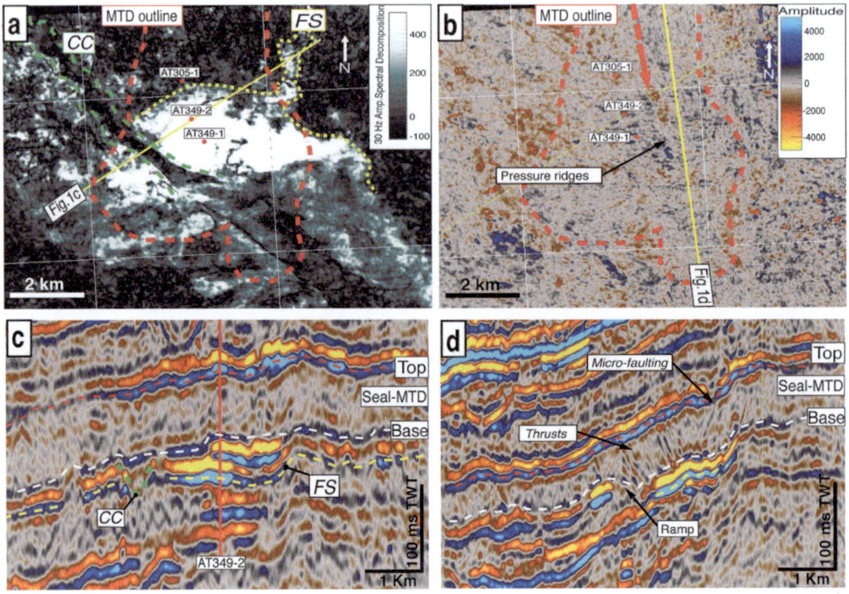

Fig. 3.2 (**a**) Map of 30 Hz spectral decomposition amplitude extraction of frontal splay (FS) horizon. *Red circles* are location of wells in the Jubilee field. The main depositional elements are outlined; frontal splay (FS), mud-filled channel complex (CC), and top-seal MTD. Location of core at AT349-2 well, (**b**) Seismic amplitude map of seal MTD base horizon. Arcuate to linear features interpreted as pressure ridges. *Red arrow* indicates interpreted transport direction of MTD, (**c**) Oblique-seismic section showing chaotic seismic facies of seal-MTD and bright amplitude of gas-charged FS deposits, (**d**) Dip-seismic section showing extensional micro-faulting and compressional thrusts downdip in the seal-MTD

(MICP). We have also conducted a detailed seismic geomorphologic interpretation of the 3D seismic volume with emphasis on the Jubilee reservoir interval and the near-reservoir MTDs (Fig. 3.2).

3.4 Subsurface Expression of the Top-Seal MTD

The seismic expression of MTDs is well documented by a variety of authors, including but not limited to Posamentier and Kolla (2003), Moscardelli et al. (2006), Bull et al. (2009), Gamboa et al. (2010), and Alves et al. (2014). The top-seal MTD associated with the Jubilee trap shows many of the basic seismic characteristics of MTDs, including chaotic seismic facies expression, weak seismic amplitude, micro-faulting, and pressure ridge development (Fig. 3.2). Unlike other MTDs, the Jubilee top-seal MTD is virtually devoid of large transported blocks

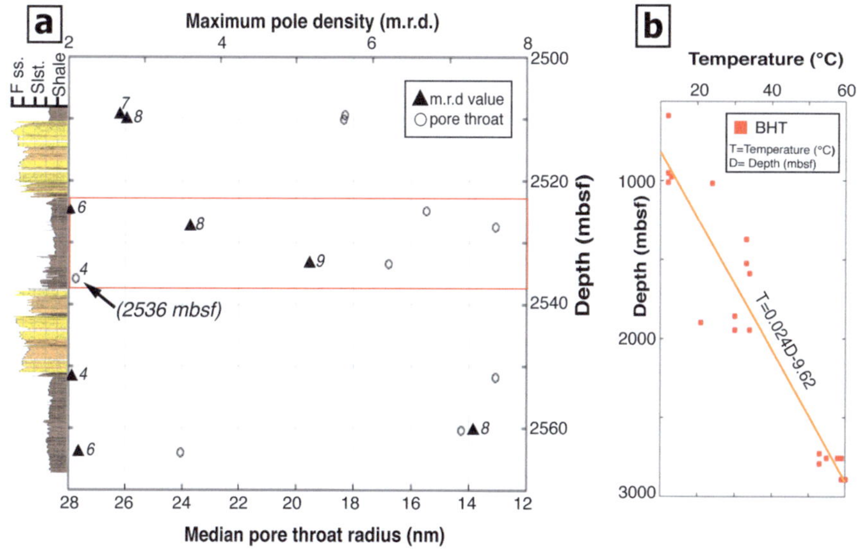

Fig. 3.3 (**a**) Values of m.r.d. and median pore-throat radius plotted versus depth for all the samples analyzed in the core. Core diagram on left with general lithology; shale, siltstone (Slst.), fine sandstone (F ss.). *Red square* corresponds to the top-seal MTD overlying the turbidite reservoir. *Numbers* indicate % of illite in mixed layer illite-smectite. Notice deeper buried samples with weak fabrics (small m.r.d.). No m.r.d. value for sample at 2,536 mbsf could be measured due to lack of fabric because of the relatively high silt content, (**b**) *Bottom hole* temperatures (BHT) from wells in the Jubilee field plotted versus depth

(Fig. 3.2b). Presence of rafted blocks in MTDs can create points of fluid bypass when buried in low permeability strata as documented by Gamboa and Alves (2015).

The top-seal MTD over the field equates ~60–80 ms TWT and has an interpreted area of about 35 km^2. The part of the MTD acting as the effective top seal in the field is identified as occurring in the transitional area between the translational and compressional toe morpho-domains (Bull et al. 2009). A single 2D seismic line was tied to the 3D seismic survey, showing the top-seal MTD being sourced from a nearby salt diapir—a typical slope failure driver in the area. The MTD has a run-out distance of about 12 km to the location of the AT349-2 well. Kinematic indicators (Bull et al. 2009) interpreted from the top-seal MTD suggest a transport direction from the north to the south-southeast (Fig. 3.2b). Based on these characteristics, the top-seal MTD is interpreted to be of detached style according to Moscardelli and Wood (2008), and its low-amplitude and chaotic seismic character is suggestive of a mud-prone composition (Posamentier and Kolla 2003).

3.5 Clay Fabric of the Top-Seal MTD: XRD, MICP, and XRTG Results

XRD results from the AT349-2 core samples suggest that the composition of the top-seal MTD is the typical composition of a GOM mudstone (Milliken, personal communication). The bulk samples are dominated by quartz (26–47 %) and illite + illite-smectite (30–49 %) with smaller amounts of plagioclase, K-feldspar (2.5–4.8 %), calcite, dolomite, siderite, halite, muscovite, kaolinite and chlorite. The <2 μ clay fraction is dominated by mixed-layer illite-smectite (84–92 %) with smaller amounts of illite (4–7 %) (Fig. 3.3a), kaolinite and chlorite. The average temperature for the top-seal MTD interval is ~52 °C calculated using maximum bottom hole temperatures (BHT) (Fig. 3.3b).

MICP methods measure pore volume by forcing mercury into the pore space. The maximum size of hydrocarbon column a lithology can seal is limited by the largest interconnected pore-throats (Krushin 1997). Therefore, high gas columns in the top-seal MTD are associated with small pore-throat sizes (Fig. 3.4a). Pore size distribution from MICP data shows the preferential loss of the large pores in the top-seal MTD except in the silt-rich sample at 17,129 ft subsea (2,536 mbsf) (Fig. 3.4c). Silt particles shield the larger pore throats and inhibit compaction (Day-Stirrat et al. 2010). No pore throat radius in the mud-rich section of the top-seal MTD is greater than ~100 nm and median pore throat radii are between 13.1 and 27.7 nm (Fig. 3.4). Consequently, MICP-derived permeabilities are in the microdarcy range as shown in Fig. 3.4b. Another quantitative value to represent the

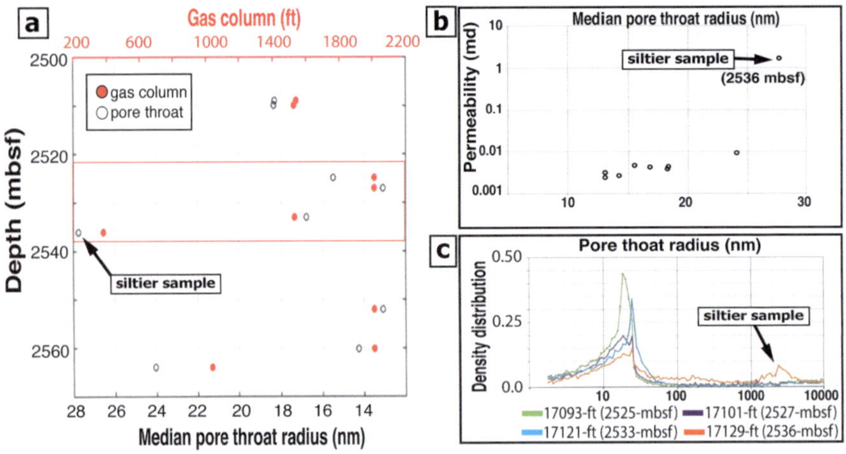

Fig. 3.4 (**a**) Values of gas column height at 35 % saturation and median pore throat radius versus depth. Note correlation between smaller pores and higher gas columns. *Red square* corresponds to top-seal MTD, (**b**) Median pore throat vs. Permeability. Silt-content significantly increases permeability of sample at 17,129 ft (2,536 mbsf), (**c**) Pore throat radius vs. Pore density distribution plot showing dominantly monomodal pore-throat distribution except in silt-rich sample

pore geometry from MICP data is the pore-throat sorting (PTS) number (Jennings 1987). The number ranges from 1.0 (perfect sorting) to 8.0 (essentially no sorting). The PTS numbers of samples in the mud-rich section of the top-seal MTD at 17,093, 17,101, and 17,121 ft subsea are 1.6, 1.7, and 1.8, respectively. In contrast, the silt-rich sample at 17,129 ft subsea has a PTS value of 5.1 suggesting poorly pore-throat sorting.

XRTG method is described by van der Pluijm et al. (1994). Deformation in clay minerals can be quantified by using the degree of phyllosilicate mineral preferred orientation expressed as maximum pole densities in multiples of random distribution, m.r.d., in which higher values reflect higher degrees of preferred orientation or fabric (Wenk 1985). Previous work on mudstones has generated maximum pole densities between 1 and 7 m.r.d., where very weak fabrics have values of <2.5 m.r.d. and strong fabrics of >4 m.r.d. Typical m.r.d. values of GOM mudstones exposed to intense mechanical compaction and/or diagenesis range from 1.74 to 4.05 m.r.d., and are mainly <3 m.r.d. (Aplin et al. 2006). An anomalously high degree of clay orientation has been recognized in several MTDs studied in the GOM (Sawyer et al. 2009; Yamamoto and Sawyer 2012; Day-Stirrat et al. 2013). Samples from the top-seal MTD have an average value of 3.6 m.r.d., yet one sample has a value as high as 5.18 m.r.d. indicative of high fabric (Fig. 3.5).

3.6 Discussion and Conclusion

The top-seal MTD in the Jubilee field—identified as a detached-MTD—has the composition of a typical GOM mudstone, pore-throat radii smaller than 100 nm with a monomodal distribution, permeability in the microdarcy range that increases with silt content, and an anomalously high clay fabric. We notice no systematic burial depth correlation with the values of m.r.d. widely varying in a relatively short depth-interval (Fig. 3.3a). However, an incipient relationship between the degree of clay fabric and pore-throat size exists where higher fabric is associated with small pores. We acknowledge the difficulty in characterizing MTDs deeply buried as recognized by Ogiesoba and Hammes (2012). At depths greater than 7,000 ft (2,133 m), some of the diagnostic features are obliterated because of sediments are compacted and end up paralleling the bedding plane. Nonetheless, the unusual high degree of clay-fabric in the top-seal MTD cannot be explained only by normal compaction and/or diagenesis. Some samples with similar clay composition located deeper in the core exhibit weak fabrics (smaller m.r.d. values). Additionally, the relatively low temperature (~52 °C), high content of K-feldspar, and low content of illite in the samples indicate minimal diagenesis impact (Hower et al. 1976).

The shearing processes and remolding associated with slope failure for significant distances (>~10 km) can enhance the alignment of clay particles (Bennett

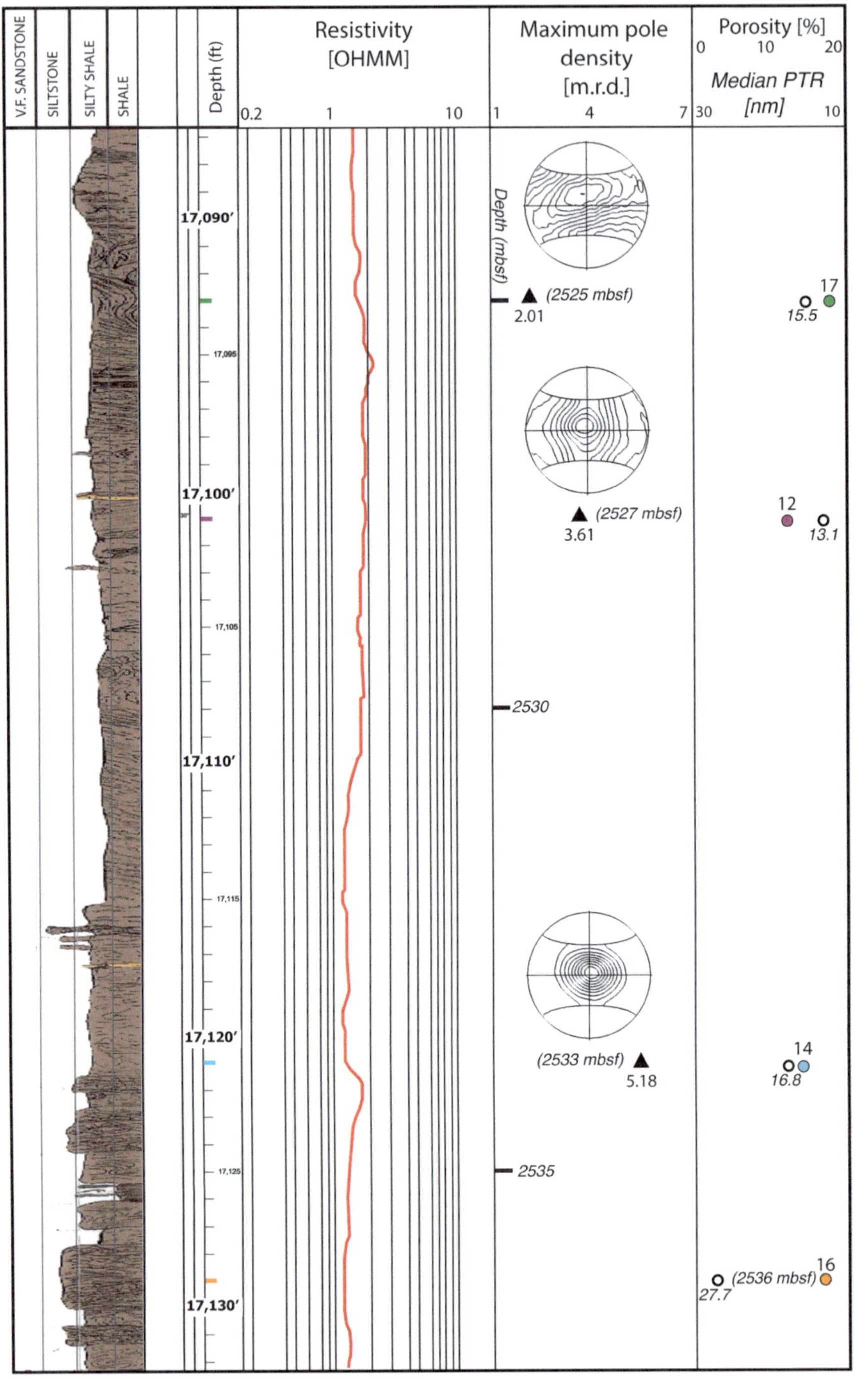

Fig. 3.5 Core of AT349-2 well through top-seal MTD. Depth scale in subsea feet and meters below sea floor (mbsf). Resistivity curve from AT349-1 well. Values of m.r.d from core material with their respective pole figure. No m.r.d. value for sample at 17,129 ft subsea (2,536 mbsf) could

et al. 1991; Day-Stirrat et al. 2013), and preferentially destroy large pore throats and pore-interconnections (Hedberg 1926) resulting in an overall decrease in permeability and porosity (Ikari and Saffer 2012; Ogiesoba and Hammes 2012; Alves et al. 2014). This reduction in permeability and porosity is even more dramatic when the MTD lithology is mud-rich. These mud-rich MTDs are highly plastic, and plasticity is noted to play an important role in seal integrity (Downey 1984). We conclude that the initially sheared material in the top-seal MTD promoted the development of a higher degree of fabric with increasing burial. Haines et al. (2009) carried out physical experiments comparing clay samples subjected solely to compression and samples exposed to shearing before compression. They noted that shear strain is important for the development of stronger fabrics. Samples only subjected to compression exhibit uniformly weak fabrics (1.6–1.8 m.r.d.) even when compressed at high normal stresses of ~150 MPa. Although a larger dataset is desired, we propose that clay mineral alignment can indicate degree of shearing during mass transport. As such, the extent of clay alignment reflects the potential seal quality. Several studies in the GOM mudstones (Freed and Peacor 1989; Ho et al. 1999; Aplin et al. 2003) show the concomitant relation between sealing effect and enhanced-clay fabric. Therefore, where diagenesis is not dominant, zones of enhanced microfabrics in mud-prone MTDs can play an important role in creating an effective sealing surface for fluids in the subsurface.

Acknowledgments We thank Anadarko for providing the Quantitative Clastic Laboratory at the Bureau of Economic Geology with the core and log data and specifically William Winters for suggesting the study. We thank Western Geco for providing us access to their 3D seismic survey over the Jubilee field. We thank ION Geophysical for providing access to 2D lines over the area via their Gulfspan survey. Landmark is thanked for providing software via the Landmark University Grants program. Kitty Milliken and Dallas Dunlap are thanked for their collaboration with microscopic and seismic data, respectively. Constructive and insightful reviews by Dr. Greg Browne, Dr. Tiago Alves, and Dr. Suzanne Bull significantly improved the final manuscript and are gratefully acknowledged.

References

Alves TM, Kurtev K, Moore GF, Strasser M (2014) Assessing the internal character, reservoir potential, and seal competence of mass-transport deposits using seismic texture: a geophysical and petrophysical approach. Am Assoc Pet Geol Bull 98:793–824. doi:10.1306/09121313117

Aplin AC, Macquaker JHS (2011) Mudstone diversity: origin and implications for source, seal, and reservoir properties in petroleum systems. Am Assoc Pet Geol Bull 95:2031–2059. doi:10.1306/03281110162

Fig. 3.5 (continued) be measured due to lack of fabric because of the relatively high silt content. Porosity and median pore throat radius (PTR) are given for each sample. Porosity is relatively high for these depths, however no trend with pore-throat size is noted

Aplin AC, Matenaar IF, van der Pluijm BA (2003) Influence of mechanical compaction and chemical diagenesis on the microfabric and fluid flow properties of Gulf of Mexico mudstones. J Geochem Explor 79:653–657. doi:10.1016/S0375-6742)

Aplin AC, Matenaar IF, McCarty DK, van der Pluijm BA (2006) Influence of mechanical compaction and clay mineral diagenesis on the microfabric and pore-scale properties of deep-water Gulf of Mexico mudstones. Clays Clay Minerals 54:500–514. doi:10.1346/CCMN.2006.0540411

Armitage DA, Romans BW, Covault JA, Graham SA (2009) The influence of mass-transport-deposit surface topography on the evolution of turbidite architecture: the Sierra Contreras, Tres Pasos Formation (Cretaceous), southern Chile. J Sediment Res 79:287–301. doi:10.2110/jsr.2009.035

Bennett RH, Bryant WR, Hulbert MH (1991) Microstructure of fine-grained sediments. Front Sediment Geol. doi: 10.1007/978-1-4612-4428-8

Bull S, Cartwright J, Huuse M (2009) A review of kinematic indicators from mass-transport complexes using 3D seismic data. Mar Pet Geol 26:1132–1151. doi:10.1016/j.marpetgeo.2008.09.011

Day-Stirrat RJ, Dutton SP, Milliken KL et al (2010) Fabric anisotropy induced by primary depositional variations in the silt: clay ratio in two fine-grained slope fan complexes: Texas Gulf Coast and northern North Sea. Sedimentol Geol 226:42–53. doi:10.1016/j.sedgeo.2010.02.007

Day-Stirrat RJ, Flemings PB, You Y, van der Pluijm BA (2013) Modification of mudstone fabric and pore structure as a result of slope failure: Ursa Basin, Gulf of Mexico. Mar Geol 341:58–67. doi:10.1016/j.margeo.2013.05.003

Downey MW (1984) Evaluating seals for hydrocarbon accumulations. Am Assoc Pet Geol Bull 68:1752–1763

Freed RL, Peacor DR (1989) Geopressured shale and sealing effect of smectite to illite transition. Am Assoc Pet Geol Bull 73:1223–1232

Gamboa D, Alves TM (2015) Three-dimensional fault meshes and multi-layer shear in mass-transport blocks: implications for fluid flow on continental margins. Tectonophysics. doi:10.1016/j.tecto.2015.02.007

Gamboa D, Alves T, Cartwright J, Terrinha P (2010) MTD distribution on a "passive" continental margin: the Espírito Santo Basin (SE Brazil) during the Palaeogene. Mar Pet Geol 27:1311–1324. doi:10.1016/j.marpetgeo.2010.05.008

Haines SH, Van Der Pluijm BA, Ikari MJ et al (2009) Clay fabric intensity in natural and artificial fault gouges: implications for brittle fault zone processes and sedimentary basin clay fabric evolution. J Geophys Res Solid Earth. doi:10.1029/2008JB005866

Hedberg HD (1926) The effect of gravitational compaction on the structure of sedimentary rocks. Am Assoc Pet Geol Bull 10:1035–1072

Ho N, Peacor D, Van der Pluijm B (1999) Preferred orientation of phyllosilicates in gulf coast mudstones and relation to the smectite-illite transition. Clays Clay Miner 47:495–504

Hower J, Eslinger EV, Hower ME, Perry EA (1976) Mechanism of burial metamorphism of argillaceous sediment: 1. Mineralogical and chemical evidence. Bull Geol Soc Am 87:725–737. doi:10.1130/0016-7606(1976)87<725:MOBMOA>2.0.CO;2

Ikari MJ, Saffer DM (2012) Permeability contrasts between sheared and normally consolidated sediments in the Nankai accretionary prism. Mar Geol 295–298:1–13. doi:10.1016/j.margeo.2011.11.006

Jennings JB (1987) Capillary pressure techniques: application to exploration and development geology. Am Assoc Pet Geol Bull 71:1196–1209

Krushin JT (1997) Seal capacity of nonsmectite shale. Seals, traps. In: Surdam R (ed) Seals, traps, and the petroleum system. American Association of Petroleum Geologists, Tulsa

Meckel LD (2011) Reservoir characteristics and classification of sand-prone submarine mass-transport deposits. In: Shipp RC, Weimer P, Posamentier HW (eds) Mass-transport deposits in deepwater settings. SEPM Special Publication, Tulsa

Moscardelli L, Wood L (2008) New classification system for mass transport complexes in offshore Trinidad. Basin Res 20:73–98. doi:10.1111/j.1365-2117.2007.00340.x

Moscardelli L, Wood L, Mann P (2006) Mass-transport complexes and associated processes in the offshore area of Trinidad and Venezuela. Am Assoc Pet Geol Bull 90:1059–1088. doi:10.1306/02210605052

Ogiesoba O, Hammes U (2012) Seismic interpretation of mass-transport deposits within the upper Oligocene Frio formation, south Texas Gulf Coast. Am Assoc Pet Geol Bull 96:845–868. doi:10.1306/09191110205

Posamentier HW, Kolla V (2003) Seismic geomorphology and stratigraphy of depositional elements in deep-water settings. J Sediment Res 73:367–388. doi:10.1306/111302730367

Sawyer DE, Flemings PB, Dugan B, Germaine JT (2009) Retrogressive failures recorded in mass transport deposits in the Ursa Basin, Northern Gulf of Mexico. J Geophys Res 114:B10102. doi:10.1029/2008JB006159

Van der Pluijm BA, Ho N-C, Peacor DR (1994) High-resolution X-ray texture goniometry. J Struct Geol 16:1029–1032. doi:10.1016/0191-8141(94)90084-1

Weimer P, Bouroullec R (2013) Petroleum geology of the Mississippi Canyon, Atwater Valley, Western Desoto Canyon, and Western Lloyd Areas, Northern Deep Gulf of Mexico: traps, reservoirs, and their timing. New Underst. Pet. Syst. Cont. Margins World 32nd Annu. Society of Economic Paleontologists and Mineralogists, pp 110–132

Wenk H-R (1985) Measurement of pole figures. In: Wenk R-H (ed) Preferred orientation in deformed metals and rocks: an introduction to modern texture analysis. Academic Press, Orlando

Yamamoto Y, Sawyer DE (2012) Systematic spatial variations in the fabric and physical properties of mass-transport deposits in the Ursa Region, Northern Gulf of Mexico. Submar Mass Mov Conseq. doi:10.1007/978-94-007-2162-3_58

Chapter 4
Seismic Geomorphology of the Israel Slump Complex in the Levant Basin (SE Mediterranean)

Ovie Emmanuel Eruteya, Murad Safadi, Nicolas Waldmann, Yizhaq Makovsky, and Zvi Ben-Avraham

Abstract Mass transport complexes (MTCs) are significant constituents of the post-evaporitic overburden in the Levant Basin, offshore Israel. Analysis of a new 3D seismic dataset offshore central Israel reveals that the Israel Slump Complex (ISC) consists of three stacked mass transport deposits (MTDs). The MTDs vary in lateral extent from between ~351 and 752 km^2 with thicknesses between ~190 and 325 m, accounting for a remobilised sediment volume of ~35–94 km^3. Interestingly, each MTD is unique, exhibiting different geometries, internal architectures, and halokinetic-related imprints. We document a novel palm-like erosional morphology, blocky facies, arcuate facies, mounded structure, syn-depositional thrust systems, and a channelised geometry. These configurations indicate different transport distance, mechanics, and kinematic history for each MTD within the complex, and may suggest different trigger events. The results of this study shed light on the interplay of multiple MTDs within a greater MTC. It also provides new insights into the nature and formation of the ISC in the offshore area of central Israel, which arose from at least three short-timed mass wasting events during the Late Pliocene. Likewise it may serve as an analogue to understand the configurations of MTDs in basins with well-developed evaporite layers.

O.E. Eruteya (✉) • M. Safadi • N. Waldmann • Y. Makovsky
Dr. Moses Strauss Department of Marine Geosciences, Leon H. Charney School of Marine Sciences (CSMS), University of Haifa, Haifa 3498838, Israel
e-mail: oeruteya@campus.haifa.ac.il

Z. Ben-Avraham
Dr. Moses Strauss Department of Marine Geosciences, Leon H. Charney School of Marine Sciences (CSMS), University of Haifa, Haifa 3498838, Israel

Department of Geophysics and Planetary Sciences, Tel-Aviv University, Ramat-Aviv, Tel-Aviv 69978, Israel

G. Lamarche et al. (eds.), *Submarine Mass Movements and their Consequences*, Advances in Natural and Technological Hazards Research 41,
DOI 10.1007/978-3-319-20979-1_4

4.1 Introduction

The post-evaporitic succession in the Levant Basin is characterised by the emplacement of several mass transport complexes (MTCs) (Frey-Martinez et al. 2005). The largest of these MTCs is the enigmatic buried Israel Slump Complex (ISC), first outlined a decade ago by Frey-Martinez et al. (2005) (Fig. 4.1b). Unfortunately, incomplete 3D seismic coverage over the entire ISC has hampered an inclusive understanding of its architecture and related processes in the offshore of central and north Israel. We investigated a northern portion of the ISC, mapped here in a new 3D seismic data as three stacked MTDs approximately 50 km offshore central Israel. This chapter describes the morphology, internal architecture and depositional processes relating to these MTDs. Our findings bring a new perspective to the current understanding of the anatomy of the ISC, and better document the kinematic history of the ISC offshore central Israel.

4.2 Geological Setting

The Levant Basin located in the south-easternmost portion of the Mediterranean Sea was formed during a Mesozoic episode of rifting (Fig. 4.1a) (Gardosh et al. 2010). During the latter part of the Miocene (Messinian), approximately 1.5 km of evaporites were deposited during the Messinian Salinity Crisis (~5.96-5.33 Ma) (see Roveri et al. 2014). Subsequently, substantial volumes of siliciclastics were delivered into the basin from its periphery (Roveri et al. 2014). The Plio–Pleistocene succession is characterised by successive MTCs emplaced during repeated episodes of slope failures and also by gravitational tectonics initiated in the Late Pliocene (Fig. 4.2) (Frey Martinez et al. 2005). The ISC spans ~4800 km^2 accounting for ~1000 km^3 of failed Late Pliocene deep-water claystone. Head scar and toe domain of the ISC are situated between 2–20 km and 100–140 km offshore Israel, respectively (Fig. 4.1b). Probable triggering mechanisms include a combination of fluid migration, slope over-steepening and regional seismicity (Frey-Martinez et al. 2005). Detailed 3D seismic analysis of the ISC were limited to the southern margin (Fig. 4.1b). A basin-wide mapping of the ISC was permitted only by coarse resolution 2D seismic lines, albeit with associated aliasing of its geomorphology in the offshore area of central and northern Israel (Fig. 4.1b).

4.3 Dataset and Methodology

We analysed an exploration 3D seismic reflection dataset spanning 1350 km^2 in water depths ranging 1100–1500 m off central Israel (Fig. 4.1b). The dataset is depth-migrated, with spatio-temporal resolution of ~6 m. We then applied a seismic

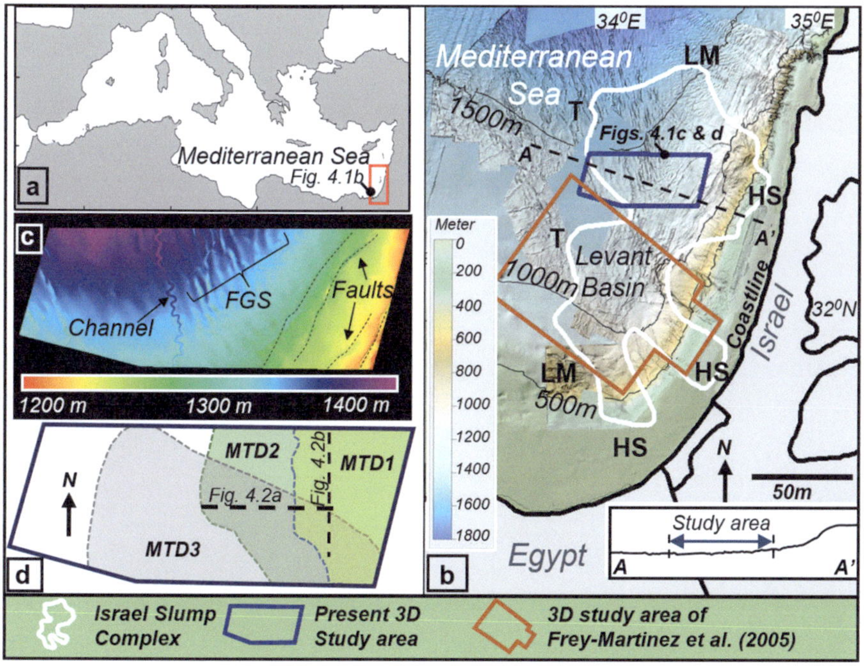

Fig. 4.1 (**a**) Map showing location of the study area in the south-easternmost portion of the Mediterranean Sea. (**b**) Bathymetric map of the Levant Basin (modified from Gvirtzman et al. [2015]) showing location of the Israel Slump Complex (ISC) outlined by Frey-Martinez et al. (2005) (Location in Fig. 4.1a). Insert below is the margin profile for *line* A – A′. The lateral margin (LM), toe domain (T) and head scarp of the ISC are labelled after Frey-Martinez et al. (2005). (**c**) Seafloor structural map over the study area from the 3D seismic dataset showing the Levant turbidite channel (see Gvirtzman et al. 2015) and salt tectonics structures (folds and normal faults). *FGS* denotes fold gully system. (**d**) Spatial distribution of the three MTDs (MTD1, MTD2 and MTD3) mapped in this study as parts of the ISC

geomorphologic approach to resolve the character of the MTDs by honouring classical 3D mapping, in addition to geometric attribute analysis (coherency and root-mean-square attributes [RMS]) to illuminate the internal architecture of the MTDs (after Posamentier et al. 2007; Dunlap et al. 2010). Computed thickness within each MTD was used to infer depositional processes (after Dunlap et al. 2010; Jackson 2012). Existing stratigraphic frameworks from Frey-Martinez et al. (2005) and Eruteya et al. (2015) were used to infer the ISC. The high quality of the 3D dataset permitted a clear subdivision of the ISC into three stacked prograding MTDs, labelled from 1 to 3 (oldest – youngest), buried beneath ~120–500 m of sediments cover (Fig. 4.2). The MTDs are distinguished by prominent horizons H1, H2, H3, and H4 (Fig. 4.2). The MTDs commonly exhibit low amplitude, chaotic seismic reflections sandwiched between continuous undeformed reflections of the basin series (Fig. 4.2).

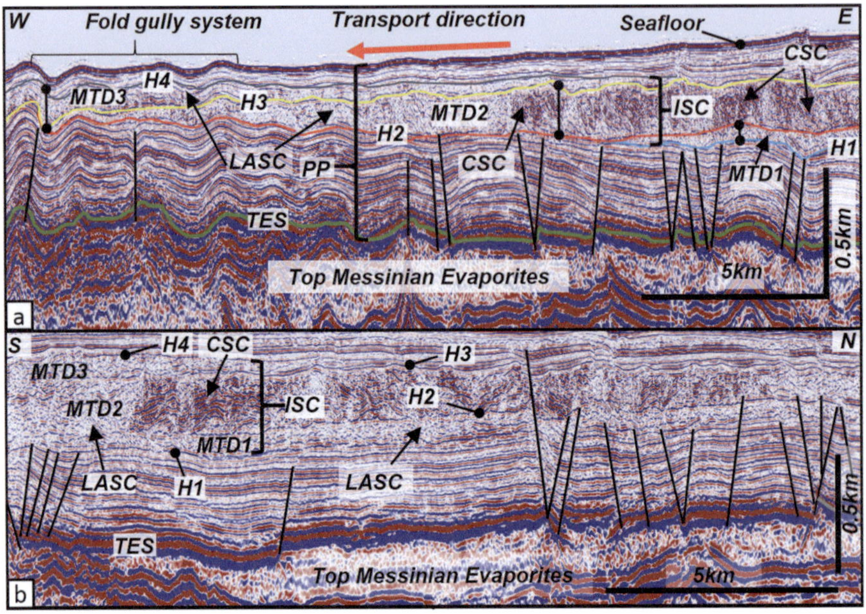

Fig. 4.2 (**a**) W-E interpreted and (**b**) S-N partly interpreted, seismic cross-section showing the three stacked MTDs (MTD1, MTD2 and MTD3) comprising the ISC (location in Fig. 4.1d). *PP* = Plio-Pleistocene unit, *LASC* = Low amplitude seismic character, *CSC* = Continuous seismic character and *TES* = Top erosional surface of the evaporitic basement. The *red arrow* in Figure 4.2a indicates transport direction of the MTDs

4.4 Geomorphology of the MTDs

MTD1 covers ~351 km^2, with a maximum thickness of up to 190 m and an estimated volume of ~34 km^3. Internally, this unit is characterised by bi-partite seismic facies (SF). SF1 is distinctive, appearing mound-like in section view and is characterised by the localisation of chaotic, moderate-high amplitude reflections. This contrasts with SF2, which is similarly chaotic, but with low-amplitude and semi-transparent reflections (Fig. 4.3c). Likewise some blocks, tapering upwards and up to ~100 m high and ~600 m wide, are documented within this unit (Fig. 4.3b).

MTD2 spans ~654 km^2 with thickness of up to 230 m, accounting for the largest volume of any of the three MTDs (~94 km^3) within the study area (Fig. 4.4a). MTD2 is characterised by continuous to semi-continuous high amplitude reflectors, as well as regions of low seismic amplitude (Fig. 4.2). The coherence slice in Fig. 4.4e shows this unit consists of regions with high coherency within a background of low coherency, as also seen on the RMS map (Fig. 4.4b). The coherent zones relate to blocky facies consisting of rafted blocks and other poorly deformed

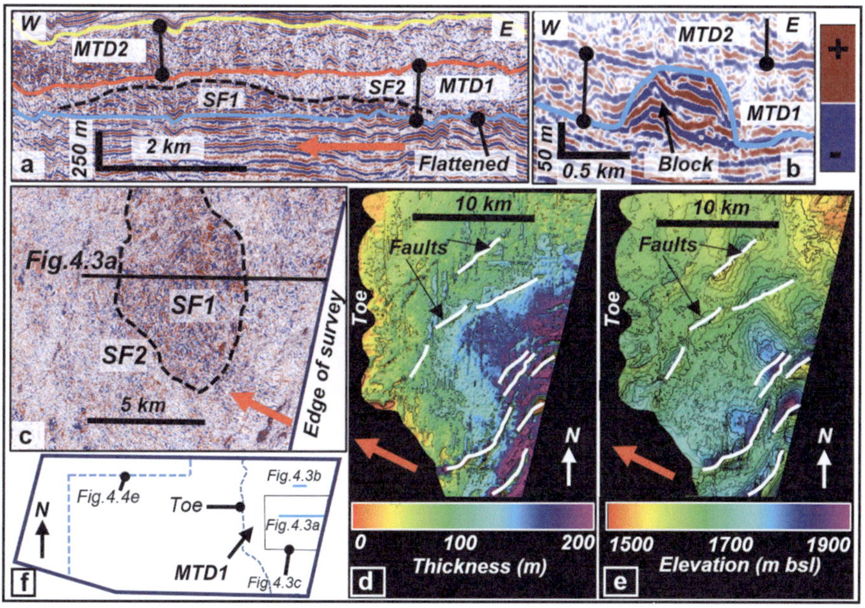

Fig. 4.3 Morphologic attributes of MTD1. (**a**) Flattened seismic section showing internal architecture of MTD1 to be composed of bipartite seismic facies (SF). SF1 appears mounded and chaotic, with moderate to high amplitude response in a background low-amplitude chaotic facies (SF2). (**b**) Block (Location in Fig. 4.3f). (**c**) Depth slice mid-way between MTD1 showing SF1 and SF2. (**d**) Thickness map (**e**) Structural map of the *top* surface. (**f**) Location map for Figs. 4.3a-e and 4.4e. The *red arrow* indicate transport direction of MTD1

materials within the deposit. Other regions with high coherency and RMS values arise from intervening hemipelagic sediments, while the background (low coherency) implies a higher degree of deformation within the failed mass. The deformed blocks are closely spaced and tilted above the basal shear surface (BSS) (H2). In some places, localised increases in thickness of MTD2 are observed close to the final location of some blocks (cf. Figs. 4.4a and e). The top of MTD2 is uneven, rugose and folded towards the western part of the deposit (Figs. 4.4c and d).

MTD3 is the most extensive MTD in the study area (~752 km^2), with thickness of up to 325 m and an estimated volume of 34 km^3. It is characterised by chaotic facies of low amplitude and in some cases moderate amplitude (Fig. 4.2). Internally it shows arcuate structures indicating a compressional regime during the flow (Fig. 4.4e [e2]). A unique anomalous thickness pattern of the failed mass is observed in what appears as a novel "palm-like" configuration (Fig. 4.5a). This region consists of four finger-like structures termed primary scours and labelled X, Y, Y′ and Z. These scours range in length from 3.2 to 5 km and have widths ranging from 1.5 to 2.2 km while extending between 65 and 100 m below the BSS (H3). Interestingly, the Z scour translates into a smaller scale scour (secondary scour). The secondary scour can be mapped for 9.5 km to the north-west, while ramping towards termination. Internally, the secondary scours display some evidence of

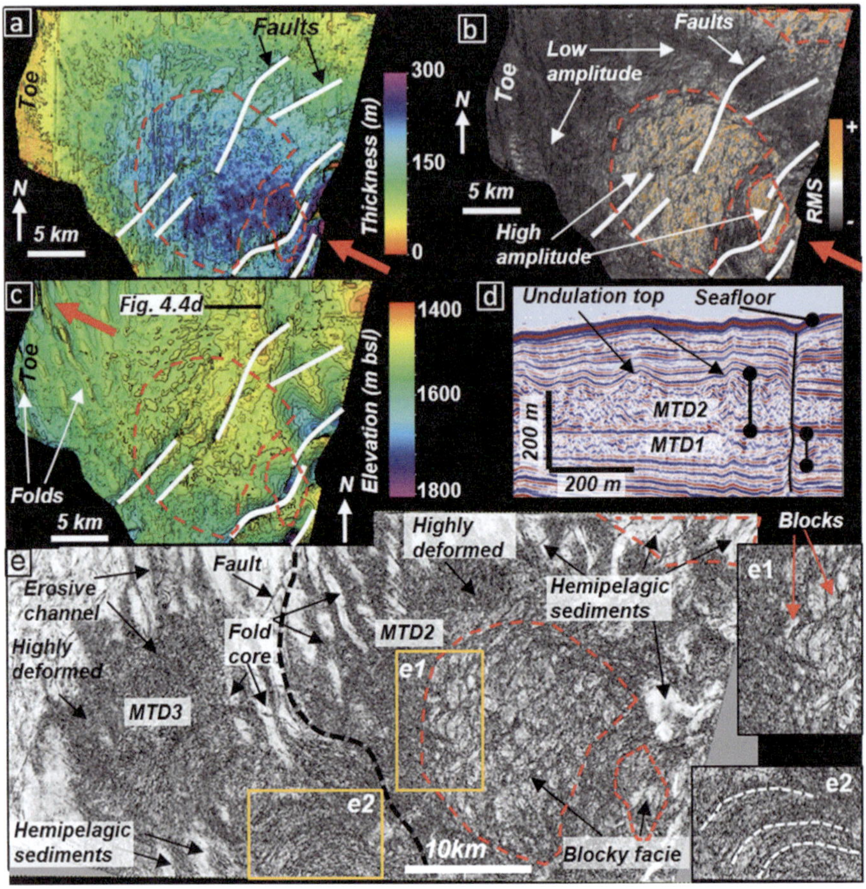

Fig. 4.4 Morphologic attributes of MTD2. (**a**) Thickness map (**b**) RMS map for MTD2. High amplitude corresponds to blocky seismic facies with the exception of areas with intervening hemipelagic sediments as labelled in Fig. 4.4e. (**c**) *Top* of MTD2 showing rugose and folded morphology. (**d**) Flattened seismic section showing the uneven and hummocky nature of the *top* surface (location in Fig. 4.4c). (**e**) Flattened coherence map mid-way between the combined *top* and *base* of MTD2 and MTD3 (location in Fig. 4.3f). *Light coloured* coherent patterns in MTD2 corresponds to blocks and other poorly deformed strata, reflecting the continuous undeformed nature of most of its composing strata. The *dark, lower-coherence* areas correspond to areas of higher deformation within MTD2. Note the arcuate seismic facies (e2) and a channelized morphology in MTD3. The *red arrow* indicates the transport direction of MTD2

syn-depositional thrusting towards the toe domain (Fig. 4.5c). The secondary scour incises up to 15 m of the underlying stratigraphy and has a width of ~800 m (Figs. 4.5a, c, and e). The top of MTD3 is uneven and likewise folded in the same region as the underlying MTD2 (Fig. 4.5b). Moreover, as seen on a RMS map, a remarkable ~10 km long northward trending high-amplitude sinusoidal paleo-channel morphology incised the top of MTD3 (Fig. 4.5d).

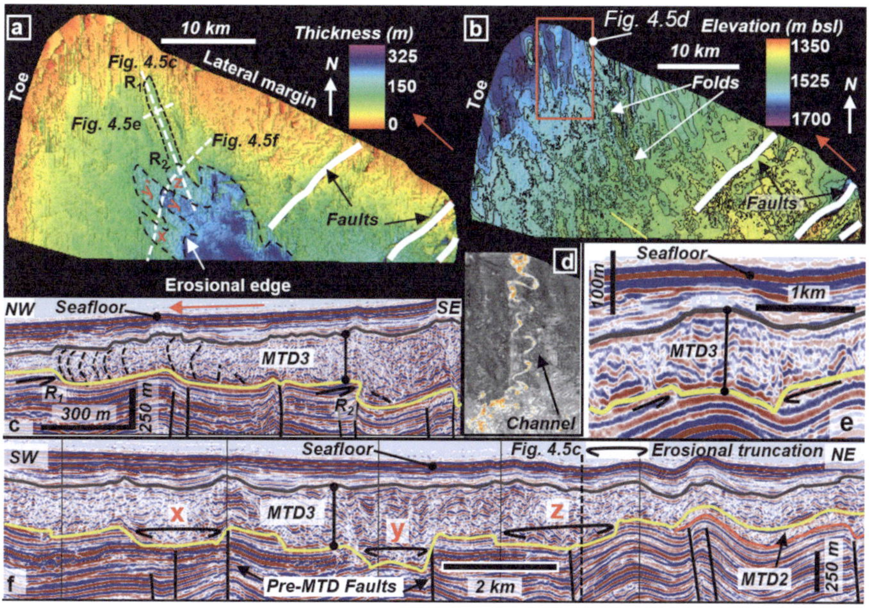

Fig. 4.5 Morphologic attributes of MTD3. (**a**) Thickness map revealing the location of the primary scours (x,y,y' and z), secondary scour and frontal ramps (R1 and R2) within MTD3. (**b**) Top surface of MTD3. (**c**) Seismic section parallel to scouring direction, also showing ramping, faint *black dashed lines* are syn-depositional thrust towards the toe. (**d**) RMS map computed between the *top* and *base* of MTD3 shows a high amplitude sinuous channel morphology trending north-ward incising the top of the MTD. (**e**) Seismic section across the secondary scour location in Fig. 4.5a. (**f**) Seismic profile across the primary scours (location in Fig. 4.5a). The *red arrow* indicates the transport direction of MTD3

4.5 Discussion and Conclusions

The stacking pattern of the three MTDs mapped in the study area as parts of the ISC raises obvious questions about their origin and mechanism of failure (e.g. Omeru and Cartwright 2015). We speculate the ISC emplaced offshore central Israel resulted from three closely-timed episodes of slope failure during the Late Pliocene. Although majority of the MTDs are not fully imaged by our dataset, the toe domain of all three MTDs are captured (see Fig. 4.1d). Based on the geometry of the MTDs mapped, the distal termination (head scarp) of MTD1 and MTD2 may lie towards the continental shelve, while that of MTD3 is most likely located around the continental slope (Fig. 4.1b). Yet, each MTD is characterised by distinct morphologies that imply varying kinematic regime during translation of the failed mass. Moreover, the configurations of the MTDs have been imprinted by juvenile salt tectonics manifested as normal faults and folds systems deforming the MTDs after emplacement (Figs. 4.3, 4.4, and 4.5).

The top surface of MTD1 (H2) is relatively even (Fig. 4.3e) when compared to the uneven and sometimes rugose/mounded morphologies arising from syn-depositional processes in the case of the topsurface of MTD2 and MTD3 (Figs. 4.4c and 4.5b). Internally, the mound-like structure with moderate to high amplitude chaotic facies documented in MTD1 can be discounted to arise from clastic injection or fluid sequestration as the BSS (H1) is relatively undeformed. One possible explanation for the formation of the mounded structure is a scenario where denser materials are segregated to the base of the flow during transport. The blocks documented within MTD1 retain some stratified continuity with the pre-failure stratigraphy suggesting draping by the flow (Fig. 4.2b) (e.g. Frey-Martinez et al. 2005; Moscardelli et al. 2006). In the case of MTD2, the background materials within the deposit characterised by low coherency typify highly deformed materials capable of suspending and aiding transportation of the blocks within the flow (Bull et al. 2009; Dunlap et al. 2010). The closeness of the rafted blocks point to shorter transport that was insufficient to sustain mechanical disaggregation of the materials (Bull et al. 2009). These blocks may be of potential risk to deep-water drilling within this region (sensu Jackson 2012). Erosional morphologies in MTD3 are comparable to the scours documented in the offshore area of Trinidad and Morocco (Fig. 4.5a) (Moscardelli et al. 2006; Dunlap et al. 2010). The scours arise from the cannibalisation of the underlying stratigraphy to a maximum depth where materials could be entrained in the flow (Bull et al. 2009). In our case, no failed blocks were documented distal to these scours to have tooled the BSS. In addition, no physiographic or bathymetric elements (e.g. mud diapirs) were documented along the paleo-slope template that could have provided confinement during the flow to promote gouging beneath the BSS. What is obvious are pre-MTD related faults that may have promoted weakness in proximity to the scours to enable the evacuation of sediments (Fig. 4.5f). The enigmatic development of frontal ramps in MTD3 along the secondary scour basin-wards may be a result of variation in the stress conditions of the failure and the mechanical property of the BSS, as well as those of the strata above the BSS (Bull et al. 2009). A more resistant material encountered by the failed mass along the BSS may have caused the failed mass to step-up stratigraphically and fail along a new weaker level. The syn-depositional thrust system documented close to the toe domain of MTD3 are very similar to those documented along the southern Levant margin (Frey-Martinez et al. 2005). These structures are consequent of a compressive regime within the flow in response to a barrier (R_1), resulting in the confinement of the failed mass and reduction in accommodation for the flow (Fig. 4.5c) (Frey-Martinez et al. 2005; Bull et al. 2009).

In conclusion, detailed geomorphological characterisation of a high resolution 3D seismic dataset for the first time, reveals the ISC emplaced offshore central Israel is a composite of three stacked MTDs. The uniqueness of each MTD in terms of their internal architecture imply the MTDs are derived from different mass wasting events during the Late Pliocene (at least three), and also different kinematic regime during their emplacement. It is crucial to assess the geohazard potential associated with paleo-mass-wasting in the Levant Basin in the light of hydrocarbon exploration.

Acknowledgments The authors are grateful to ILDC, Modi'in Energy and GGR for graciously granting permission to show the seismic dataset. We would like to thank Dallas Dunlap and Joe Cartwright for their extremely careful and constructive reviews. Special thanks also to guest editor Aaron Micallef for his useful comments. Ovie is grateful to the Graduate Studies Authority, University of Haifa and Marie Curie Career Integration Grants (CIG) FP7-PEOPLE-2011-CIG in the framework of Nicolas Waldmann's GASTIME project, for funding towards his Ph.D research. We are also grateful to Uri Schattner, Mathieu Rodriguez and Or Bialik for fruitful discussion. Schlumberger, Paradigm and IHS Kingdom are thanked for granting academic licenses of their respective software packages used as part of this study.

References

Bull S, Cartwright J, Huuse M (2009) A review of kinematic indicators from mass-transport complexes using 3D seismic data. Mar Pet Geol 26:1132–1151. doi:10.1016/j.marpetgeo.2008.09.011

Dunlap DB, Wood LJ, Weisenberger C, Jabour H (2010) Seismic geomorphology of offshore Morocco's east margin, Safi Haute Mer area. Bull Am Assoc Hist Nurs 94:615–642. doi:10.1306/10270909055

Eruteya OE, Waldmann N, Schalev D et al (2015) Intra- to Post-Messinian deep-water gas piping in the Levant Basin, SE Mediterranean. Mar Pet Geol. doi:10.1016/j.marpetgeo.2015.03.007

Frey-Martinez J, Cartwright J, Hall B (2005) 3D seismic interpretation of slump complexes: examples from the continental margin of Israel. Basin Res 17:83–108. doi:10.1111/j.1365-2117.2005.00255.x

Gardosh MA et al (2010) Tethyan rifting in the Levant region and its role in early Mesozoic crustal evolution. Geol Soc Lond Spec Publ 341(1):9–36

Gvirtzman Z et al (2015) Bathymetry of the Levant basin: interaction of salt-tectonics and surficial mass movements. Mar Geol. doi:10.1016/j.margeo.2014.12.00

Jackson CA-L (2012) The initiation of submarine slope failure and the emplacement of mass transport complexes in salt-related minibasins: a three-dimensional seismic-reflection case study from the Santos Basin, offshore Brazil. Bull Geol Soc Am 124:746–761

Moscardelli L, Wood L, Mann P (2006) Mass-transport complexes and associated processes in the offshore area of Trinidad and Venezuela. AAPG Bull 90:1059–1088

Omeru T, Cartwright JA (2015) Multistage, progressive slope failure in the Pleistocene pro-deltaic slope of the West Nile Delta (Eastern Mediterranean). Mar Geol 362:76–92. doi:10.1016/j.margeo.2015.01.012

Posamentier HW, Davies RJ, Cartwright JA, Wood LJ (2007) Seismic geomorphology – an overview. Geol Soc Spec Publ 277:1–14

Roveri M, Flecker R, Krijgsman W et al (2014) The Messinian salinity crisis: past and future of a great challenge for marine sciences. Mar Geol 352:25–58. doi:10.1016/j.margeo.2014.02.002

Chapter 5
Multiple Megaslide Complexes and Their Significance for the Miocene Stratigraphic Evolution of the Offshore Amazon Basin

Cleverson C. Silva, Antonio T. dos Reis, Rodrigo J. Perovano, Marcus A. Gorini, Marcos V.M. dos Santos, Izabel K. Jeck, Ana Angélica A. Tavares, and Christian Gorini

Abstract The interpretation of a new and extended 2D seismic database on the offshore Amazon Basin (Foz do Amazonas Basin) confirms the widespread presence of regional-scale mass-transport deposits (MTDs) that are important architectural elements of the Amazon Deep Sea Fan. These MTDs were deposited since the late Miocene and extend throughout an area of nearly 315,000 km^2. They are grouped into three megaslide complexes: the northwestern Amapá Complex, the Central Amazon Fan Complex and the southeastern Pará-Maranhão Complex. Each complex has multiple stacked MTDs with various internal seismic facies that are indicative of large downslope modification and disruption of the original stratigraphy. The majority of the MTDs show chaotic or transparent internal seismic facies that we interpret as indicative of debris flow deposits. Although we cannot determine the exact triggering mechanism(s) for the various sediment failures, these events appear to be related to the gravitational compression of fold-and-thrust belts created by gravity-tectonic processes on the upper Amazon Fan and to structurally-induced mobilization of large blocks on the upper continental slope in response to

C.C. Silva (✉)
Department of Geology and Geophysics, Fluminense Federal University, Av. Gen Milton Tavares de Souza s.n., Niterói, RJ 24.240-020, Brazil
e-mail: cguizan@id.uff.br

A.T. dos Reis
School of Oceanography, Rio de Janeiro State University, Rio de Janeiro, Brazil

R.J. Perovano • M.A. Gorini • M.V.M. dos Santos
MAG – Mar, Ambiente e Geologia Serviços Ltda, Rio de Janeiro, Brazil

I.K. Jeck • A.A.A. Tavares
Directorate of Hidrography and Navigation (DHN), Brazilian Navy, Nitéroi/RJ

C. Gorini
Laboratoire Evolution et Modélisation des Bassins Sédimentaires, Institut des Sciences de la Terre Paris (iSTeP) - UMR 7193 UPMC-CNRS, Sorbonne Universités, UPMC Univ Paris 06, Paris, France

G. Lamarche et al. (eds.), *Submarine Mass Movements and their Consequences*, Advances in Natural and Technological Hazards Research 41,
DOI 10.1007/978-3-319-20979-1_5

overpressure along impermeable surfaces. These processes were apparently more active during the Pleistocene in response to increased sedimentation rates on the fan.

5.1 Introduction

The sedimentary construction of passive continental margins in areas of high sediment input is commonly disrupted by gravitational mass-transport events (MTEs), which result in deformation and re-sedimentation of thick sedimentary successions. These processes are particularly prevalent on the Foz do Amazonas sedimentary basin of the Brazilian Equatorial Margin (Damuth and Embley 1981; Damuth et al. 1988; Manley and Flood 1988; Piper et al. 1997; Maslin and Mikkelsen 1997; Pirmez and Imran 2003; Maslin et al. 2005; Araújo et al. 2009; Silva et al. 2010; Reis et al. 2010). The extremely high sedimentation rates in the Amazon Fan, on average 100 cm ka^{-1} for the last 10 Ma (Cobbold et al. 2004), exceeded 5,000 cm ka^{-1} during the Quaternary sea-level lowstands (Mikkelsen et al. 1997; Piper et al. 1997). These high rates are considered important conditioning factors that increase overpressure and induce seabed instabilities (Piper et al. 1997; Maslin and Mikkelsen 1997; Maslin et al. 2005). On the northern and southern flanks of the Amazon Fan sedimentation rates are lower but despite these conditions two extensive megaslides as large as 90,000 km^2 are present (Araújo et al. 2009; Silva et al. 2010). The Amapá Megaslide Complex is located to the north of the Amazon Deep Sea Fan while the Pará-Maranhão Megaslide Complex is located to the south (Fig. 5.1). These mass transport complexes (MTCs) are mostly siliciclastic in nature and were generated by upslope remobilization of large sediment blocks that were detached over a regional Upper Miocene décollement level. This stratigraphic interval also acted as a regional décollement level for the gravity tectonics that is observed in the Amazon Fan (Araújo et al. 2009; Silva et al. 2010). Reis et al. (2010) also demonstrated a clear mechanism that coupled MTEs with the regional gravity tectonics that affected the Amazon Fan. According to this proposed mechanism, gravity tectonics generated fold-and-thrust belts that increased the steepness of slopes inducing seafloor instabilities. In the present study, we utilize additional 2D regional seismic data to estimate in detail the areal extension of these MTDs, and to determine the extent of their contribution to the construction of the continental margin and the Amazon Fan. The seismic character, areal distribution and stratigraphic character of the Amapá Megaslide Complex (AMC), the Central Amazon Fan Megaslide Complex (CAFMC) and the Pará-Maranhão Megaslide Complex (PMMC) are described in this work. In addition, we calculate the volumetrics of these MTDs. We define megaslide complex as a group of distinct stacked MTDs with different internal seismic facies.

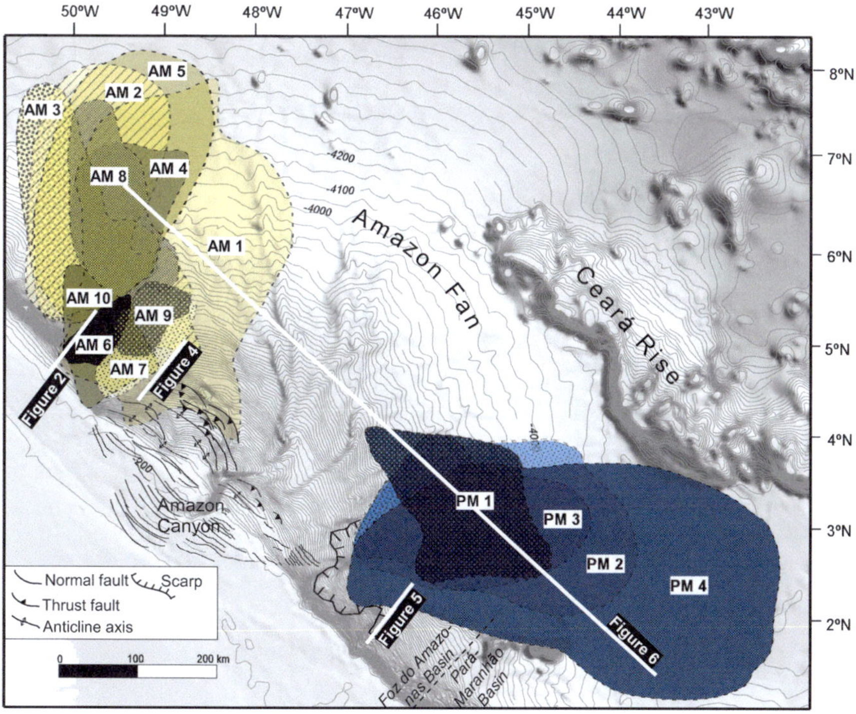

Fig. 5.1 Location of Amapá Megaslide Complex (AM 1 to AM 10, *yellow*, in the northwest) and Pará-Maranhão Megaslide Complex (PM 1 to PM 4, *blue*, in the southeast). Locations of seismic *lines* shown in Figs. 5.2, 5.4, 5.5 and 5.6

5.1.1 Database and Methods

The dataset for this study included a regional grid of 2D multi-channel seismic data from different origins and with different vertical resolution. The Brazilian Navy pro-vided the regional seismic data of the LEPLAC project (Brazilian Legal Continental Shelf Survey, Phases 1 and 2) (www.mar.mil.br/secirm/leplac.html) that was acquired in two different batches: a 100 km spaced grid, including lines recorded to 13 s (TWT) (vertical resolution of 20 m) and lines recorded to 7 s (TWT) (vertical resolution of 5 m). Additional seismic data provided by the Brazilian Petroleum Agency (ANP) and by different service companies (GAIA and FUGRO) were also available. These data were acquired with a vertical coverage of 10 s (TWT) (10 m vertical resolution). The stratigraphic positions of various MTDs were based on the recently published works by Gorini et al. (2014) and Reis et al. (in press), who revised the biostratigraphy of available exploratory wells and calibrated some of the seismic sections that are used in this study. An interval seismic velocity of 2,000 m/s was adopted for all estimates of sediment thickness.

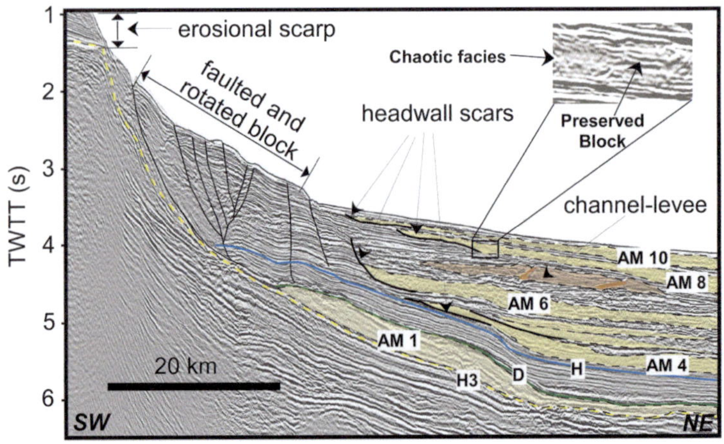

Fig. 5.2 Illustrative section of the upper slope in the Amapá Megaslide Complex. The *inset* shows the chaotic facies and preserved block inside AM8. Profile location shown in Fig. 5.1

5.2 Results

5.2.1 *The Amapá Megaslide Complex (AMC)*

The AMC is composed of ten stacked MTDs (AM1 to AM10 from base to top) and it is located in water depths that range from 600 to 4,500 m (Figs. 5.1 and 5.2; Table 5.1). The individual MTDs have variable dimensions (Table 5.1), with maximum thicknesses ranging from 214 m (AM7) to 1,000 m (AM1), areas varying from 3,900 km^2 (AM6) to 72,000 km^2 (AM1), and estimated volumes ranging from 591 km^3 (AM6) to 20,000 km^3 (AM1). AM7 corresponds to the "western debris flow/slump" complex identified by Damuth and Embley (1981). The Amapá Megaslide Complex comprises a total volume of 53,000 km^3 of remobilized sediments. Most deposits are characterized by internal transparent to chaotic seismic facies, which in this case represent debris-flow deposits. Some of these MTDs contain interbedded parallel units that are interpreted as undisturbed turbidites. The largest MTD is AM1 and is located directly above regional horizon H3 (Late Miocene ~10.5 Ma) and below horizon D (~3.4 Ma) (Figs. 5.1 and 5.2). AM1 extends basinward for more than 350 km to water depths in excess of 4,200 m and is clearly separated from the MTDs above by well preserved, plane-parallel sedimentary intervals.

The steep NW continental slope, with gradients of as much as 5.8° in water depths between 500 and 2,500 m, is the locus for the headwall scars for the upper MTDs (AM2 to AM10) (Fig. 5.2). Some of these MTDs (AM4, AM6, AM8 and AM10) have their headwall scars located downslope from a large (more than 30 km) faulted and rotated "block zone" that apparently slid from the upper continental slope. This "block zone" left behind a steep erosional headwall scarp (~750 m of vertical relief) where older sedimentary units from the upper continental

Table 5.1 Dimensions and volumes of MTD complexes

Amapá Megaslide Complex			
MTD	**Max. thickness (m)**	**Area (km²)**	**Volume (km³)**
AM10	398	12,039	1,831
AM9	336	10,111	1,695
AM8	405	28,862	4,401
AM7*	214	7,010	727
AM6	379	3,894	591
AM5	600	46,457	10,188
AM4	501	24,194	4,883
AM3	245	22,341	2,867
AM2	279	42,378	5,867
AMI	1,036	71,658	20,010
Central Amazon Fan Megaslide Complex			
MTD	**Max. thickness (m)**	**Area (km²)**	**Volume (km³)**
CF10	220	8,482	931
CF9**	464	9,258	1,325
CF8	237	7,797	792
CF7	209	2,480	217
CF6***	315	53,648	5,845
CF5	519	9,804	1,977
CF4	173	2,558	237
CF3	371	38,164	5,290
CF2	475	31,513	6,803
CF1	274	22,031	2,259
Pará-Maranhão Megaslide Complex			
MTD	**Max. thickness (m)**	**Area (km²)**	**Volume (km³)**
PM4	797	96,760	34,774
PM3	482	31,352	5,214
PM2	291	47,590	6,825
PM1	203	24,474	2,302

AM7* corresponds to the Western debris flow/slump complex of Damuth and Embley (1981)
CF9** corresponds to the Central debris flow/slump complex of Damuth and Embley (1981)
CF6*** corresponds to Unit R of Manley and Flood (1988) and to URMTD and BMTD of Piper et al. (1997).

slope were removed (Fig. 5.2). These upper MTDs (AM2 to AM10) are all located above a regional horizon "H" (~1.8 Ma), and are mostly characterized by chaotic to transparent seismic facies, including well-preserved blocks inside the disturbed mass (Fig. 5.2). Some chaotic units still have lens-shaped geometries which suggest the disruption and deformation of former channel-levee systems (Fig. 5.2). Regionally, in map view, the main flow direction of the MTDs in the AMC is from SW to NE (Fig. 5.1).

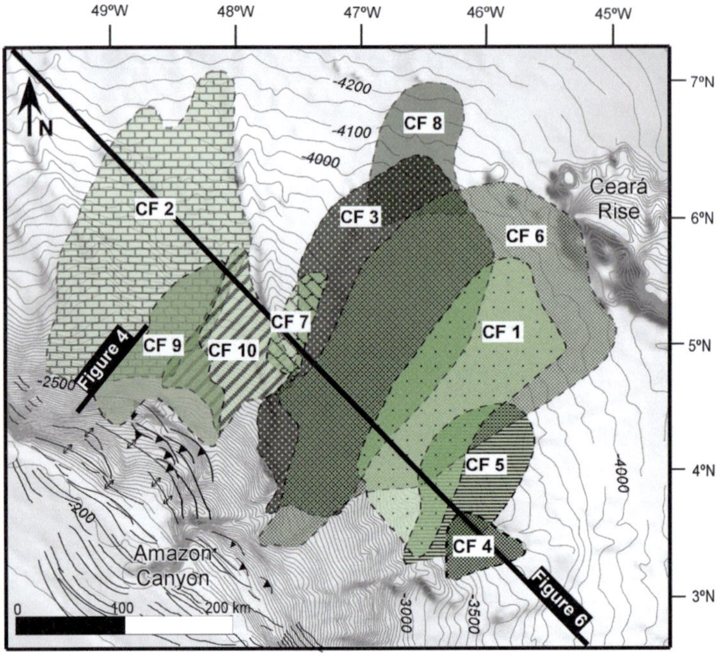

Fig 5.3 Central Amazon Fan Megaslide Complex. Profile locations of Figs. 5.4 and 5.6

5.2.2 *The Central Amazon Fan Megaslide Complex (CAFMC)*

The CAFMC is composed of ten stacked MTDs (CF1 to CF10 from base to top) that are located in water depths that range from 2,000 to 4,200 m (Fig. 5.3). These MTDs have variable dimensions with thicknesses that range between 209 (CF7) and 519 m (CF5), areal extents that vary from 2,480 (CF7) to 53,650 km^2 (CF6), and calculated volumes that range from 216 (CF7) to 6,800 km^3 (CF2) (Table 5.1). These MTDs are recognized by their transparent to chaotic seismic character. All are located above the "D" horizon (~3.4 Ma) and seaward of the gravity tectonics compressional domain (Figs. 5.3 and 5.4). AM1 is still preserved below the CAFMC, above horizon H3, however, due to the seismic resolution we could not map its extension inside the compressional gravity system (Fig. 5.4). The headwall scars associated with these MTDs are clearly traced; cutting well stratified sequences adjacent to the compressional front (Fig. 5.4). Altogether the displaced and deformed units of the CAFMC represent a total volume of 25,680 km^3. The CAFMC MTDs are widely separated in many cases by numerous well-developed channel-levee systems of the Amazon Fan that are located above horizon H (~1.8 Ma) (Fig. 5.4). There are clear indications that some of these channel-levee complexes were partially deformed by mass wasting processes (Fig. 5.4). These

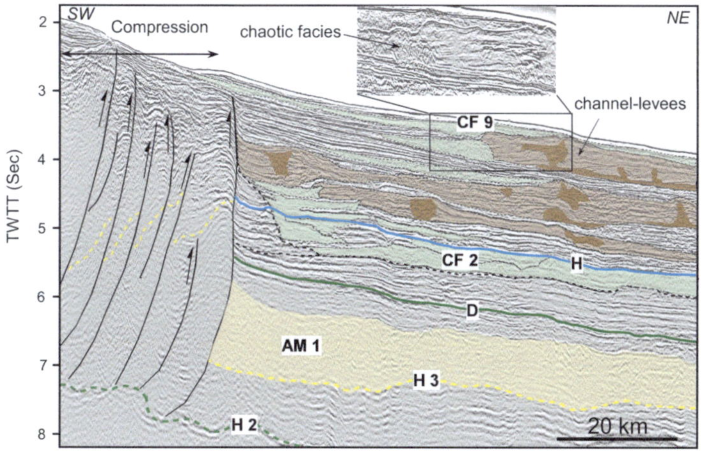

Fig. 5.4 Illustrative section of the compressional front and the MTDs of the Central Amazon Fan Megaslide Complex. The *inset* shows the chaotic facies of a partially deformed channel-levee complex. Profile location shown in Figs. 5.1 and 5.3

units include MTD CF9, that corresponds to the "western slump/debris flow complex" identified by Damuth and Embley (1981), and CF6 which is equivalent to "unit R" as described by Manley and Flood (1988) and later divided into "URMTD and BMTD" by Piper et al. (1997).

5.2.3 *The Pará-Maranhão Megaslide Complex (PMMC)*

The PMMC is composed of four stacked MTDs (from base to top PM1 to PM4) (Figs. 5.1 and 5.5). These MTDs originated on the steep continental slope, where modern gradients are up to 4° in water depths between 200 and 2,500 m, of the SW flank of the Amazon Fan. The MTDs are elongated with approximate W to E orientations, extending in modern water depths that range from 2,000 to 4,200 m. This geometry suggests flow in an easterly to northeasterly direction (Fig. 5.1) towards the flanks of the Ceará Rise and into the Pará-Maranhão sedimentary basin. These MTDs range from 200 (PM1) to 800 m (PM4) in thickness, with areas that vary from 24,500 (PM1) to 96,760 km^2 (PM4), and estimated sediment volumes between 2,300 (PM1) and 34,800 km^3 (PM4) (Table 5.1). Together, these MTDs have a total volume of 49,000 km^3. PM4 is the largest MTD within the PMMC containing 70 % of the total estimated volume.

The MTDs on the PMMC are all located basinward from a large slide which is up to 1,000 m thick and that encompasses an area of almost 12,500 km^2 on the upper continental slope (Fig. 5.5). This slide block lies over horizon H 3 (~10.5 Ma) and is detached from a prominent scarp (upper scarp ~1,000 m vertical relief) at the

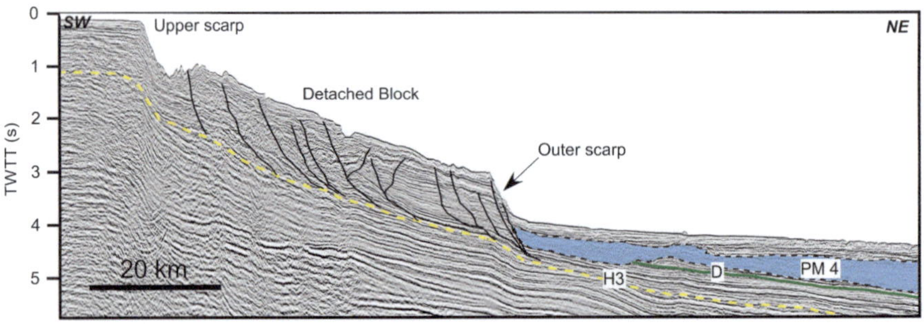

Fig. 5.5 Illustrative section of the detached block and MTD on the Pará-Maranhão Megaslide Complex. Profile location in shown in Fig. 5.1

shelf break (Fig. 5.5). The downslope limit of the detached block is also delineated by another prominent scarp (outer scarp, ~700 m vertical relief) that extends laterally for approximately 260 km and forms an amphitheater on the continental slope and southeastern flank of the Amazon Fan (Fig. 5.1). The head scar of PM4 is the most proximal to the outer scarp amphitheater (Fig. 5.5). The head scars of the other MTDs (PM1 to PM3) were eroded by PM4 and therefore only the distal portions of these earlier MTDs are preserved. MTDs PM1 and PM2 are stratigraphically located between horizons "D" (~3.4 Ma) and "H" (~1.8 Ma) (Fig. 5.6). PM3 and PM4 are located above horizon H, but PM4 eroded deeply into horizons H and D and it lies over horizon H3 in its proximal position near the outer scarp (Fig. 5.6).

5.3 Discussion

Based on our extended seismic database, we have mapped in more detail the pervasive distribution of MTDs in the Foz do Amazonas and Pará-Maranhão sedimentary basins. The magnitude and distribution of these units are evidence of the important role that MTD formation played in the stratigraphic configuration of this portion of the Brazilian continental margin. In this work, the original names of the Amapá Megaslide Complex (Araújo et al. 2009; Silva et al. 2010) and Pará-Maranhão Megaslide Complex are maintained; however, a new group of MTDs that are part of the Central Amazon Fan Megaslide Complex is introduced to describe the MTDs located downdip of the Amazon Fan compressional thrust front (Fig. 5.3). These new results provide us with updated estimates associated with the architecture and size of these units that were previously reported by Araújo et al. (2009) and Silva et al. (2010). In addition, availability of new age information (Gorini et al. 2014; Reis et al. in press) allowed us to better calibrate the stratigraphic position of these MTDs. The three megaslide complexes encompass a total sediment volume of ~128,000 km^3. The Foz do Amazonas Basin contains some of

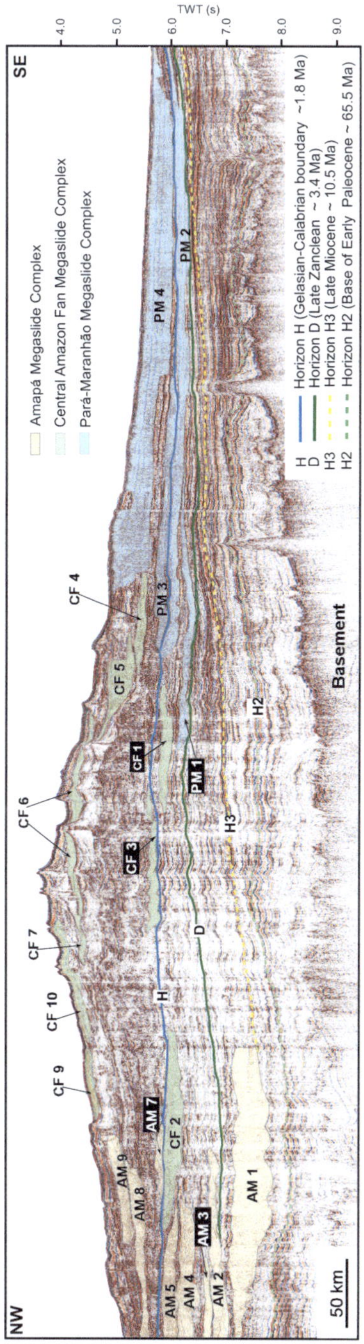

Fig. 5.6 Regional strike section illustrating the stratigraphy of the MTDs across the Amazon Fan. Location shown in Figs. 5.1 and 5.3

the most voluminous MTD deposits ever recognized in the world's oceans including PM4 (PMMC/~35,000 km^3) and AM1 (AMC/~20,000 km^3).

The AMC and PMMC, on the northwestern and southeastern flanks of the Amazon Fan originated downslope of steep continental slopes that are prone to sediment failure (Silva et al. 2010) (Figs. 5.2 and 5.5). The largest MTDs (more than 10,000 km^3) are located on the AMC (AM5 and AM1) and on the PMMC (PM4) and confirm the higher potential for large sediment failures along steep continental margins. However, we do not consider that seafloor over steepening alone was the main cause for sediment instabilities. We suggest that fluid overpressure at the base of horizon H3 created the necessary conditions for the regional seabed instabilities. Horizon H3 marks the base of the Amazon Fan, and is an important décollement surface for gravity-tectonic deformation in the Amazon Fan (Fig. 5.3) (Reis et al. 2010; Silva et al. 2009). Towards the flanks of the Amazon Fan, as suggested by Reis et al. (2010), this surface marks a detachment surface for the upper-slope displacement of large slide blocks and accommodates extensional normal faults which are observed on the proximal reaches of the AMC and PMMC (Figs. 5.2 and 5.5).

The reactivation of thrust faults along the Amazon Fan compressive front (Fig. 5.4) is suggested as the main mechanism that induced the mass movements of the CAFMC (Reis et al. 2010), generating large MTDs (~5,000 to ~7,000 km^3) (CF2, CF3, CF6). These results show that in addition to the previously reported upper Quaternary MTDs within the Amazon Fan (e.g., Damuth et al. 1988; Manley and Flood 1988; Flood and Piper 1997; Maslin and Mikkelsen 1997; Piper et al. 1997), deeper-rooted MTDs have occurred in this region since at least the late Miocene.

MTDs AM2 – AM6 from the AMC, CF1 and CF2 from the CAFMC, and PM1 and PM2 from the PMMC are located in the stratigraphic interval between horizons D and H (Fig. 5.6) and indicate that these events occurred in a time span from ~3.4 to 1.8 Ma. The total volume of these MTDs (~42,500 km^3) represent circa 33 % of all MTDs. The thickness of the sedimentary interval between horizons D and H increases towards the west and indicates that this portion of the margin was the preferred depocenter at that time interval (Silva et al. 2009) (Fig. 5.6). Possibly the higher sedimentation rates increased overpressures and induced sediment instabilities during this period on the AMC region.

Most MTDs (CF3–CF10) on the CAFMC occur above horizon H. We conclude that the displacements along the thrust faults, that triggered instabilities to form these MTDs, occurred during the last 1.8 Ma. Well preserved and partially deformed channel-levee systems, above Horizon H are interlayered with the MTDs and attest to the contemporaneous nature of turbidite and MTD processes during construction of the Amazon Deep-Sea Fan (Fig. 5.6).

The largest MTD on the entire Foz do Amazonas Basin, PM4 (~34,000 km^3), as well as MTDs PM3 and AM7 to AM10 are stratigraphically above Horizon H, and, thus indicate that all of these events occurred in the last ~1,8 Ma (Fig. 5.6). The total volume of MTDs above horizon H is ~65,256 km^3 and represents ~51 % of the total displaced sedimentary units in the Foz do Amazonas Basin. Given the extremely high sedimentation rates on the Amazon Fan, which exceed 5,000 cm ka^{-1} during the Quaternary sea-level lowstands (Mikkelsen et al. 1997; Piper et al. 1997), it is likely that this was one important conditioning factor for seabed instabilities during the last 1.8 Ma.

5.4 Conclusions

Huge volumes of sediments were displaced as part of mass wasting processes in the offshore Foz do Amazonas Basin. These MTDs were distributed as part of the Amapá Megaslide Complex (AMC), the Central Amazon Fan Megaslide Complex (CAFMC) and the Pará-Maranhão Megaslide Complex (PMMC). Our results confirm the importance of mass wasting processes during the margin construction of the Foz do Amazonas Basin. Some general conclusions:

(a) MTDs documented in this work were deposited during the Late Miocene (~10.5 Ma) onwards. This observation is evidence of a long history of seabed instabilities within this margin.
(b) Most documented MTDs occur above the Late Zanclean (~3.4 Ma) horizon. Remobilized sediment volumes increased during the last 1.8 Ma, representing nearly 50 % of the total estimated volume.
(c) An increase in sedimentation rates during the Pleistocene is regarded as an important conditioning factor that also favored slope instabilities.
(d) Reactivation of gravity tectonic deformation that was induced by an increase in sedimentation rates triggered sediment failures associated with the CAFMC.
(e) Fluid overpressure at the base of a well documented regional décollement surface (horizon H 3, ~10.5 Ma) favored the upper-slope displacement of large slide blocks that are associated with the genesis of regional extensive MTDs related to AMC and PMMC.
(f) The intercalation of MTDs and channel-levee systems demonstrates the contemporaneous nature of Newtonian flows and mass wasting processes during the construction of the Amazon Fan.

Acknowledgments Authors specially thank the Executive Committee of the LEPLAC PROGRAMME for permission to publish the acquired geophysical data. We also thank the Brazilian National Agency of Oil and Gas (ANP) for supplying extra seismic and well data. We equally gratefully acknowledge the Brazilian National Research Agency (CNPq) for providing research grants for the first and second authors. We also thank the reviewers Drs. John E. Damuth, Roger Flood and Lorena Moscardelli for their critical reviews that helped improve the overall quality of the manuscript.

References

Araújo ES, Silva CG, Reis AT et al (2009) Movimentos de massa multiescala na Bacia da Foz do Amazonas – Margem equatorial brasileira. Rev Bras Geofis 27(3):485–508

Cobbold PR, Mourgues K, Boyd K (2004) Mechanism of thin-skinned detachment in the Amazon Fan: assessing the importance of fluid overpressure and hydrocarbon generation. Mar Petrol Geol 21:1013–1025

Damuth JE, Embley RW (1981) Mass-transport processes on the Amazon Cone: western equatorial Atlantic. AAPG Bull 65:629–643

Damuth JE, Flood RD, Kowsmann RO et al (1988) Anatomy and growth pattern of Amazon deep-sea fan as revealed by long-range side-scan sonar (GLORIA) and high resolution seismic studies. AAPG Bull 72:885–911

Flood RD, Piper DJW (1997) Amazon Fan sedimentation: the relationship to equatorial climate change, continental denudation, and sea-level fluctuations. In: Flood RD, Piper DJW, Klaus A, Peterson LC (eds) Proceedings of the Ocean Drilling Program, Scientific Results, vol 155, Ocean Drilling Program, College Station, TX, p 653–675

Gorini C, Haq BU, dos Reis AT, Silva CG, Cruz AS, Grangeon D (2014) Late Neogene sequence stratigraphic evolution of the Foz do Amazonas Basin, Brazil. Terra Nova 26(3):179–185

Manley PL, Flood RD (1988) Cyclic sediment deposition within the Amazon deep-sea fan. Am Assoc Pet Geol Bull 72:912–925

Maslin M, Mikkelsen N (1997) Amazon Fan mass-transport deposits and underlying interglacial deposits: age estimates and fan dynamics. In: Flood RD, Piper DJW, Klaus A, Peterson LC (eds) Proceedings of the Ocean Drilling Program, Scientific Results, vol 155, Shipboard Scientific Party, pp 353–365

Maslin M, Vilela C, Mikkelsen N, Grootes P (2005) Causes of catastrophic sediment failures of the Amazon Fan. Q Sci Rev 24:2180–2193

Mikkelsen N, Maslin MA, Giraudeau J, Showers W (1997) Biostratigraphy and sedimentation rates of the Amazon Fan. In: Flood RD, Piper DJW, Klaus A, Peterson LC (eds) Proceedings of the Ocean Drilling Program, Scientific Results, vol 155, Ocean Drilling Program Texas A&M University, pp 577–594

Piper DJW, Pirmez C, Manley PL et al (1997) Mass-transport deposits of the Amazon Fan. In: Flood RD, Piper DJW, Klaus A, Peterson LC (eds) Proceedings of the Ocean Drilling Program, Scientific Results, vol 155, D. W. Friesen & Sons Ltd, pp 109–146

Pirmez C, Imran J (2003) Reconstruction of turbidity currents in Amazon Channel. Mar Petrol Geol 20:823–849

Reis AT, Perovano R, Silva CG et al (2010) Two-scale gravitational collapse in the Amazon Fan: a coupled system of gravity tectonics and mass-transport processes. J Geol Soc Lond 167:593–604

Reis AT, Araújo E, Silva CG et al (in press) Effects of a regional décollement level for gravity tectonics on late Neogene-Quaternary large-scale slope instabilities in the Foz do Amazonas Basin, Brazil. Mar Petrol Geol

Silva R, Reis AT, Silva CG et al (2009) O processo de colapso gravitacional e a estruturação da seção marinha da bacia da Foz do Amazonas – Margem equatorial brasileira. Rev Bras Geofis 27(3):459–484

Silva CG, Araújo EFS, Reis AT et al (2010) Megaslides in the Foz do Amazonas Basin, Brazilian equatorial margin. In: Mosher DC, Shipp RC, Moscardelli L, Chaytor JD, Baxter CDP, Lee HJ, Urgeles R (eds) Submarine mass movements and their consequences. Springer, New York, pp 581–591

Chapter 6
Kinematics of Submarine Slope Failures in the Deepwater Taranaki Basin, New Zealand

Tuviere Omeru, Joseph A. Cartwright, and Suzanne Bull

Abstract Mass transport deposits (MTDs) are formed by gravity driven processes whereby dominant transport direction is downslope. Here we use 2D seismic data from the deep water Taranaki Basin to describe two volumetrically extensive MTDs (MTD 1 and 2) that were emplaced within the Plio-Pleistocene succession. Key kinematic information, derived from the observed architectural relationships between the Aotea Seamount and MTD 2, suggest that this unit had a SW transport direction. This is in marked contrast to the NW transport direction derived for the underlying MTD 1. Given the geometry of these MTDs, we suggest that MTD 1 was triggered during the early stages of evolution of the Giant Foresets Formation when the system was prograding toward the north. On the other hand, the geometry of MTD 2 allowed us to infer that its headwall region was located toward the east (near the Northern Graben) in an area known to have been tectonically active at the time of deposition (c. 1.8 Ma to recent). These observations raise the likelihood of a tectonic trigger for MTD 2.

6.1 Introduction

Most examples of submarine slope failures in the literature are from unidirectional slopes that result in the formation of simple mass movements (e.g., Gulf of Corinth failures in Hasiotis et al. 2002). Few studies exist on the development of failures in areas with complex multi-directional slopes where failures are comprised of several, multi-directional flow components (e.g., Moscardelli et al. 2006; Ashabranner

T. Omeru (✉)
Department of Geosciences, University of Lagos, Lagos, Akoka, Nigeria
e-mail: tuviere.omeru@yahoo.com

J.A. Cartwright
Department of Earth Sciences, University of Oxford, South Parks Road, Oxford, UK

S. Bull
GNS Science, 1 Fairway Drive, Lower Hutt, New Zealand

G. Lamarche et al. (eds.), *Submarine Mass Movements and their Consequences*,
Advances in Natural and Technological Hazards Research 41,
DOI 10.1007/978-3-319-20979-1_6

et al. 2010). This study investigates two volumetrically extensive mass transport deposits (MTDs 1 and 2) from a basin margin where there is a major change in strike in the shelf-slope boundary to examine whether this physiography had any influence on slope failure mechanisms and kinematics. MTDs 1 and 2 were emplaced on a slope with a grossly northwesterly gradient within the Plio-Pleistocene Giant Foreset Formation in the deepwater Taranaki Basin (Fig. 6.1) and whose configuration clearly relates to the rapid progradation of the modern continental margin towards the northwest (Soenander 1992; Hansen and Kamp 2006). The latest of these MTDs (MTD 2) is of particular interest in that it exhibits an unusual propagation direction that contrasts with the underlying MTD 1 in exhibiting a SW transport direction, i.e. apparently orthogonal to the prevailing maximum slope gradient. Six large MTDs have been mapped within the Pleistocene succession in this area of the slope (Fig. 6.2a), but the youngest two of these are the focus of this study. The main aims of this study are (1) to describe the morphology, internal character, and depositional processes associated with these two MTDs, and (2) explore the relationship of the MTDs with the Aotea Seamount. The implications of the kinematics for possible triggering mechanisms are discussed briefly in the concluding section.

6.2 Data and Methods

This study is based on 34 Pre-Stack Time Migrated (PSTM) 2D seismic lines (total line length 6,440.4 km) covering an area of c. 16,380 km^2 in the SE of the deepwater Taranaki Basin, offshore New Zealand (Fig. 6.1). The data were acquired and processed by TGS NOPEC Geophysical Company in 2001. The dominant frequency of the data in the shallow interval of interest is 65 Hz. The vertical resolution of the data is in the order of c. 6–7 m using a near surface sediment velocity of 1,800 m/s.

6.3 Geological Framework

The Taranaki Basin is divided into two tectonic regions: the tectonically active Eastern Mobile Belt, which includes the northern Graben, and the tectonically quiescent and structurally simple Western Stable Platform (King and Thrasher 1996) (Fig. 6.1a). The structure and stratigraphy of the basin is complex, reflecting initial Late Cretaceous – Paleocene extension related to the Tasman Sea spreading and the Eastern Gondwana break up; Eocene passive margin post-rift thermal cooling and regional subsidence; and since then, persistence of passive margin conditions in the west and reactivation of large north-south striking normal faults in the east associated with the evolution of the Australia-Pacific convergent plate boundary (King and Thrasher 1996). The Plio-Pleistocene succession, on which this

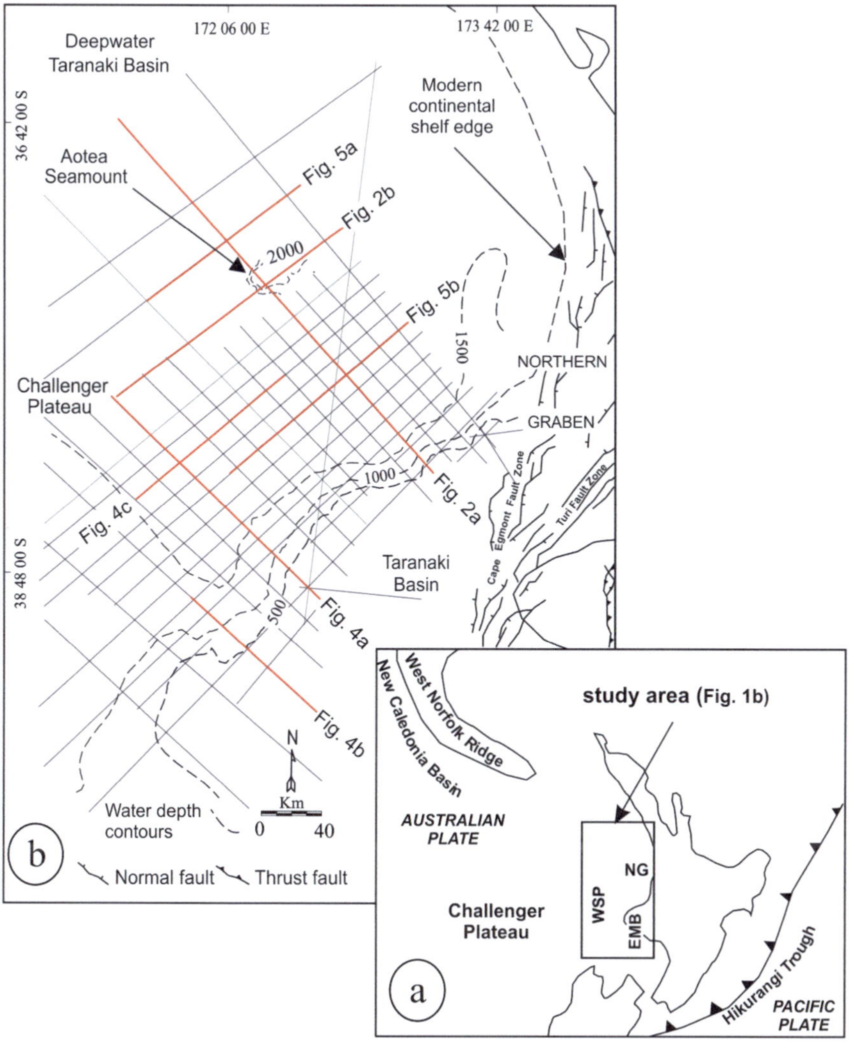

Fig. 6.1 (**a**) The Australian-Pacific plate boundary setting in the New Zealand region. The location of the study area is shown (*black rectangle*). *EMB* Eastern Mobile Belt, *NG* Northern Graben, *WSP* Western Stable Platform. (**b**) *Line* drawing of the study area showing the main structural elements, 2D seismic data coverage in the study area and seismic lines used in this study (*solid black lines* and *solid red lines* respectively). Seabed contour (*broken curved lines*) is drawn at 500 m interval

study focuses, is up to 2,000 m thick, reflecting very high sedimentation rates (c. 2 m/ka; King and Thrasher 1996) and rapid northward to northwestward progradation of the shelf margin (Hansen and Kamp 2002, 2006). The siliciclastic sediment making up the progradational wedge was derived from contemporary erosion of the Southern Alps in South Island (Tippett and Kamp 1995; Kamp

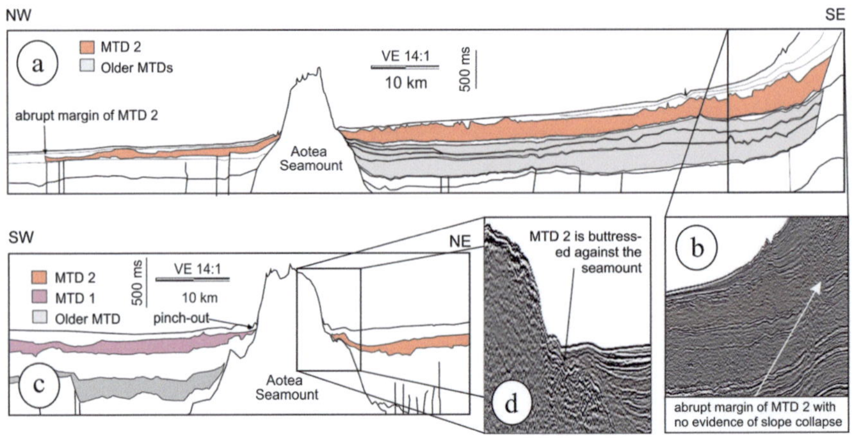

Fig. 6.2 (**a**) Interpreted representative dip-oriented seismic profile showing the MTDs in the study area (see Fig. 6.1 for location). (**b**) Enlarged seismic showing the abrupt margin of MTD 2. (**c**) Interpreted representative strike-oriented seismic profile in the study area (see Fig. 6.1 for location). (**d**) Enlarged seismic showing how MTD 2 is buttressed against the seamount

et al. 2004). The progradation of the Giant Foresets Formation was also contemporaneous with the active extension of the Northern Graben. The succession comprises several formations, including the muds, silts and sands of the Giant Foresets Formation, which has built up the modern continental shelf and slope as a series of outward building stacked clinoforms (Hansen and Kamp 2002, 2006).

6.4 Results and Interpretations

6.4.1 *MTD 1*

MTD 1 is located in the southwestern region of the study area and extends for c. 220 km from the shelf edge in a NNW direction (Fig. 6.3). It covers an area of c. 18,360 km^2 with an estimated volume of c. 2,600 km^3. The age of MTD 1 can be bracketed between 1.8 and 2.0 Ma (Omeru 2014). The deposit displays considerable lateral thickness variation, ranging from 350 to 650 m in the proximal southeastern headwall region (Figs. 6.3 and 6.4a). The thickness gradually decreases downslope to the NNW until it tapers to pinch-out in the distal margins of the deposit (Fig. 6.3). The buried headwall of MTD 1 is preserved beneath the present day shelf edge (Fig. 6.4a). The c. 40 km long NE-SW trending irregular headwall (Fig. 6.3a) has a relief of c. 350 m, with a westerly inclination ranging from 40 to 55° (Fig. 6.4a). A set of prograding clinoforms overlie the headwall and downlap onto the upper surface of MTD 1 (Fig. 6.4a, b). The basal surface of MTD 1 is defined by a generally continuous, low amplitude reflection that is smooth and

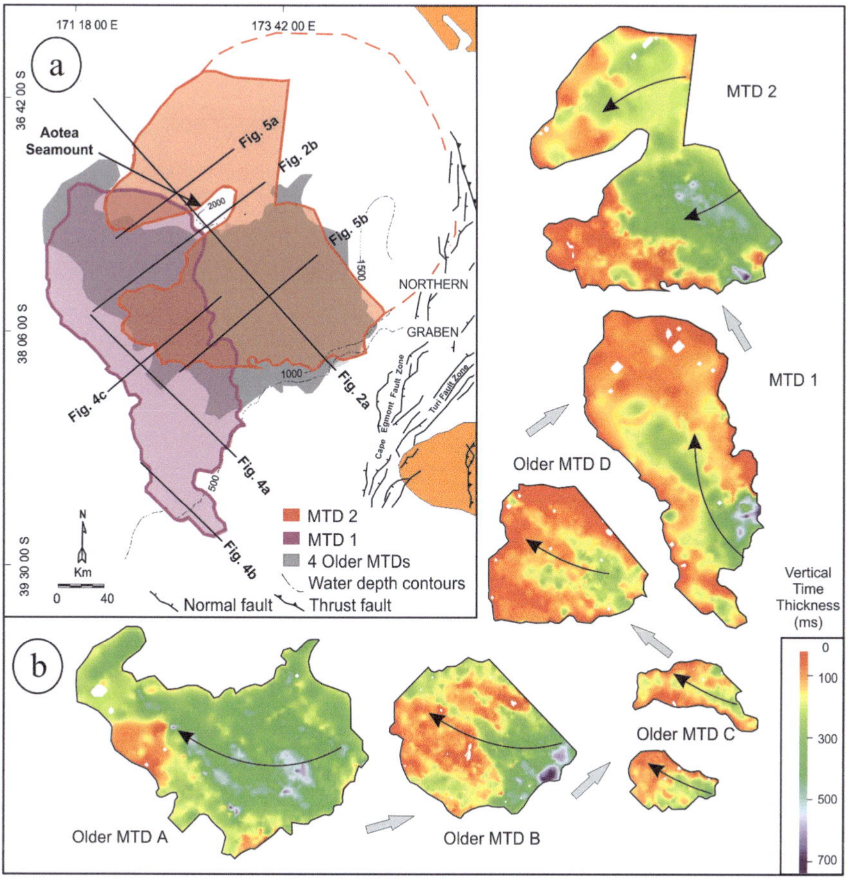

Fig. 6.3 (**a**) Outline of MTDs 1, 2 and the four older MTDs (*A–D*) in the study area. Notice the spatial relationship of the inferred updip limit of MTD 2 (indicated by *dashed red line* and the main structural elements in the Taranaki Basin (Modified from King and Thasher 1996). Line locations for Figs. 6.2, 6.4, and 6.5 are shown. (**b**) Isopach maps for all 6 MTDs showing interpreted transport directions (*black curved arrows*)

stratigraphically concordant for the most part (Figs. 6.2b and 6.4). However, locally it ramps up or down stratigraphy in a downslope direction, exhibiting a staircase geometry (Fig. 6.4b). The upper surface of MTD 1 is marked by a high amplitude, reflection that is generally continuous and sometimes parallel to the basal surface (Figs. 6.2b and 6.4). The upper surface has a smooth topography for the greater part of MTD 1, but is locally irregular in the headwall region (Fig. 6.4). The internal seismic character of MTD 1 is dominated by transparent and chaotic facies (Fig. 6.4), and coherent rotated, fault-bounded blocks in the headwall region (Fig. 6.4a).

The western and eastern lateral margins for much of the perimeter of MTD 1 are represented by an abrupt ramp that cuts up stratigraphy from the basal surface

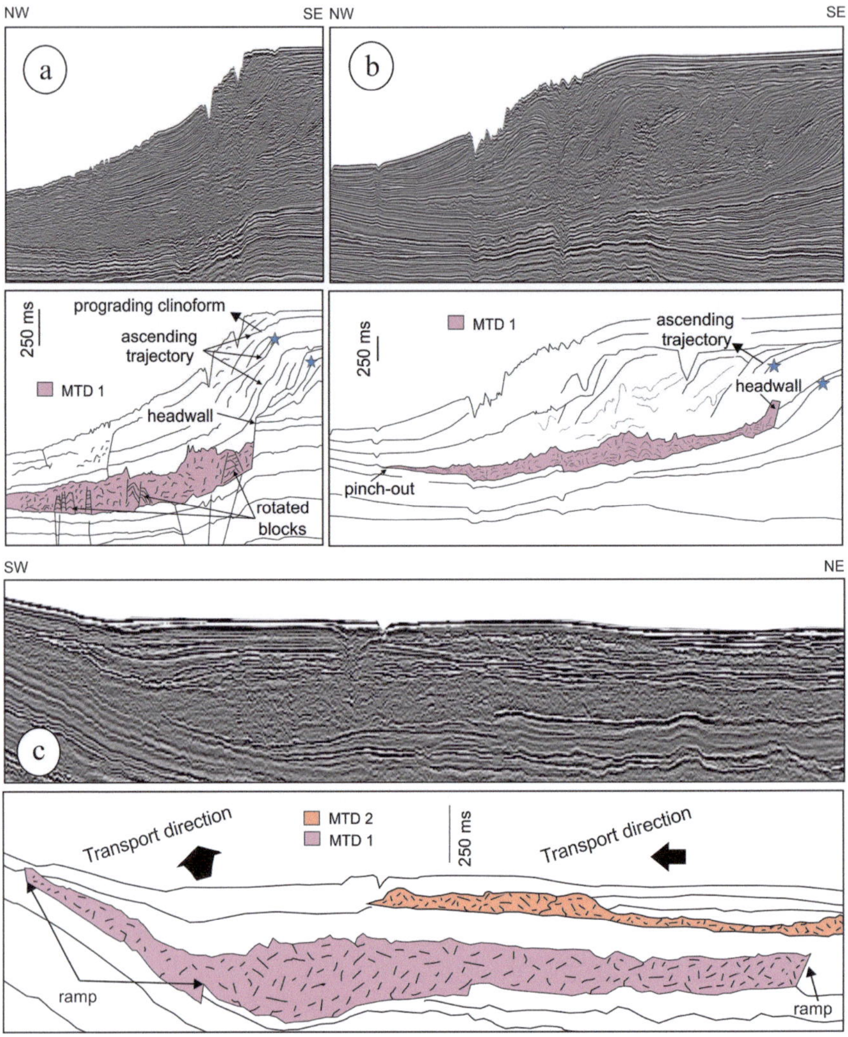

Fig. 6.4 (**a**) Uninterpreted and interpreted seismic profile showing MTD 1 headwall beneath prograding clinoform. The *blue* stars mark the position of the rollover points of the clinothemes encasing the MTD 1. (**b**) Uninterpreted and interpreted seismic profile showing MTD 1 headwall within prograding clinoform. (**c**) Uninterpreted and interpreted seismic profile showing the ramped lateral margins of MTD 1 (See Fig. 6.1 for seismic line location)

(Fig. 6.4b), and truncates adjacent undeformed strata. Both lateral margins have maximum heights of c. 150 m measured in the proximal southern region, but taper to a pinch-out distally. In map view, both lateral margins have segments with a NNW-SSE and a NW-SE orientation, suggesting a gross transport direction for MTD 1 in the range NW to NNW (cf. Bull et al. 2009).

6.4.2 MTD 2

MTD 2 is the youngest of the mass transport deposits with an estimated age younger than 1.8 Ma (Fig. 6.2a). It covers an area of c. 22,400 km^2 with an estimated volume of 3,700 km^3 (Fig. 6.3). MTD 2 forms a two lobed deposit separated by the Aotea Seamount. The southern lobe is the larger of the two covering an area of c. 12,400 km^2 while the northern lobe covers an area of c. 10,000 km^2 (Fig. 6.3). As with the MTD 1, MTD 2 exhibits considerable variation in thickness (Fig. 6.3).

The headwall of MTD 2 is not mapped because of limitations in data coverage toward the east. However, based on its observed geometry, it is inferred that the headwall of MTD 2 is located somewhere in the region of the present day shelf break (Fig. 6.3). The basal surface of MTD 2 is marked by a low amplitude reflection. In the northern lobe, it is continuous and truncates the underlying undeformed strata (Fig. 6.5a). In contrast, the southern lobe is mostly irregular and cuts along different stratigraphic levels creating staircase geometries (Fig. 6.5b). The upper surface of MTD 2 is represented by a high amplitude reflection with a smooth expression in the northern lobe where it is overlain by undeformed strata (Fig. 6.5a). In the southern lobe, the upper surface is highly irregular and characterized by positive features in the proximal region (Fig. 6.5b), that tend to coincide with underlying grooves. By analogy with Bull et al. (2009), these features are attributed to more coherent blocks within the deposit scouring the basal surface during their emplacement (Fig. 6.5b). The crudely E-W to NE-SW oriented southern lateral margin of MTD 2 is generally represented by an abrupt scarp that cuts up stratigraphy from the basal surface (Fig. 6.2a, b). The height of the scarp ranges from 30 to 300 m. The northern lateral margin is more modest in expression, tens of metres high, but nonetheless abrupt (Fig. 6.2a), striking NE-SW over a mapped distance of some 180 km. The lobate planform of MTD 2 is critical evidence bearing on transport direction. The bifurcation around the Aotea Seamount is only compatible with a SW direction of transport.

6.5 Discussion

The mapping of the two giant MTDs in the deepwater Taranaki Basin revealed that they are discrete and widespread with significant sediment volumes. More importantly the interpretation shows that not all these MTDs followed the progradation direction associated with the development of clinoforms linked to the Giant Foreset Formation. This critical observation raises obvious questions about the nature of the source area of these MTDs and their triggering mechanism. MTD1 has a well-defined headwall imaged on the seismic data (Fig. 6.4a) and the preservation of original relief allows the source area of the failure to be constrained to the south. MTD 1 was deposited during the Early Pliestocene (1.8–2.0 Ma) when clinoforms of the Giant Foresets Formation were prograding to the north-northwest. In

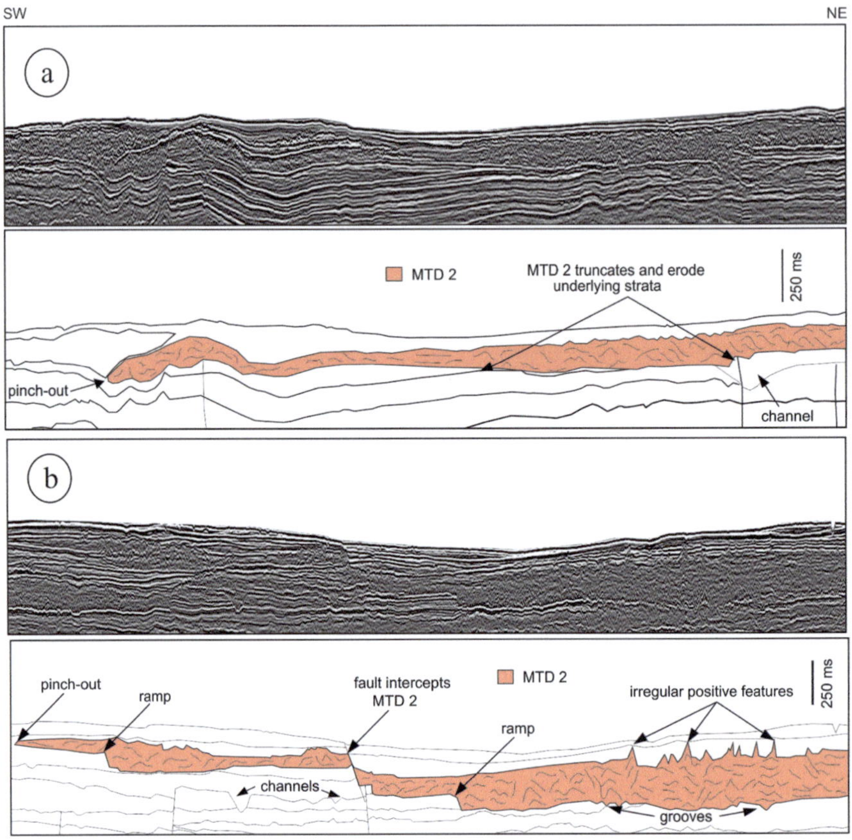

Fig. 6.5 (**a**) Uninterpreted and interpreted seismic profile through the distal region of MTD 2 (northern lobe, see Figs. 6.1 and 6.3 for location). Notice the erosional basal surface of the northern lobe. (**b**) Uninterpreted and interpreted seismic profile through the distal region of MTD 2 (southern lobe, see Figs. 6.1 and 6.3 for location). Notice the staircase geometry of the basal surface. Also notice the alignment of the grooves beneath the irregular positive features interpreted as resistant blocks

addition, Hayward (1984) reported high sedimentation rates (10 cm/100 year) during the Early Pleistocene that are consistent with Hansen and Kamp (2006) who observed slumping on their seismic record at this time. These observations suggest that high sedimentation rates and oversteepening of prograding clinoform morphologies (Hansen and Kamp 2002) could have triggered the formation of MTD 1 (Figs. 6.3 and 6.4). On the other hand, MTD 2 is observed on either sides of the Aotea Seamount and it pinches out against its flanks (Figs. 6.2a and 6.3). The nature of the relationship between the Aotea Seamount and MTD 2 strongly suggests that the transport direction was from the northeast to the southwest since the seamount is clearly generating a shadowing effect toward the southwest (Figs. 6.2b and 6.3). These relationships imply that the source area for MTD 2 is located toward the northeast near the Cape Egmont Fault Zone (Fig. 6.3). Since the age of MTD 2 is estimated to be younger than 1.8 Ma and the Northern Graben was an active structure during this time, it is possible to hypothesize that MTD 2 was triggered by seismicity associated with the graben formation.

6.6 Conclusions

A two-dimensional (2D) seismic reflection data from the deep water Taranaki Basin has allowed a detailed morphological analysis of two giant MTDs within the Plio-Pleistocene Giant Foreset Formation. The following conclusions can be made:

1. The volumes of MTDs 1 and 2 are c. 2,600 km^3 and 3,700 km^3 respectively, thus they are volumetrically significant.
2. MTD 1 has a northwesterly transport direction while MTD 2 has a south-westerly transport direction.
3. The failure of MTD 1 is believed to have been caused by rapid sedimentation on the shelf margin from Plio-Pleistocene to recent times coupled with concomitant rapid oversteppening of slope indicated by the development of headwall scarps along the bounding surfaces of progradational foresets. MTD 2 is believed to have been triggered by tectonic activity (e.g earthquake) given that it was emplaced in a translation direction different from MTD 1 in the study area and the inferred headwall lies in a tectonically active region.

Acknowledgments This work was conducted in the 3D seismic lab of Cardiff University, UK. The authors are grateful to the assigned editor Dr. Lorena Moscardelli and the reviewers Dr. Jose Frey-Martinez and Dr. Aggeliki Georgiopoulou for their constructive reviews. We also thank Uruski C. for providing the 2D seismic data.

References

Ashabranner L, Tripsanas E, Shipp R (2010) Multi-direction flow in a mass-transport deposit, Santos Basin, offshore Brazil. In: Mosher DC et al (eds) Submarine mass movements and their consequences. Springer, New York, pp 247–255

Bull S, Cartwright J, Huuse M (2009) A review of kinematic indicators from mass-transport complexes using 3D seismic data. J Mar Pet Geol 26(7):1132–1151

Hansen RJ, Kamp PJ (2002) Evolution of the Giant Foresets Formation, northern Taranaki Basin, New Zealand. 2002 New Zealand petroleum conference proceedings, pp 419–435

Hansen RJ, Kamp PJ (2006) Sequence stratigraphy and architectural elements of the Giant Foresets Formation, northern Taranaki Basin, New Zealand. New Zealand petroleum conference proceedings, pp 6–10

Hasiotis T, Papatheodorou G, Bouckovalas G et al (2002) Earthquake induced coastal sediment instabilities in the western Gulf of Corinth, Greece. J Mar Geol 186:319–335

Hayward BW (1984) Foraminiferal biostratigraphy of Wanui-1 offshore well, west Taranaki, New Zealand Geological Survey. Department of Scientific and Industrial Research

Kamp PJ, Vonk AJ, Bland KJ et al (2004) Neogene stratigraphic architecture and tectonic evolution of Wanganui, King Country, and eastern Taranaki Basins, New Zealand. N Z J Geol Geophys 47:625–644

King PR, Thrasher GP (1996) Cretaceous-Cenozoic geology and petroleum systems of the Taranaki Basin. Institute of Geological & Nuclear Sciences Ltd., Lower Hutt, New Zealand

Moscardelli L, Wood L, Mann P (2006) Mass-transport complexes and associated processes in the offshore area of Trinidad and Venezuela. Am Assoc Pet Geol Bull 90(7):1059–1088

Omeru T (2014) Mass transport deposits: implications for reservoir seals, unpublished PhD thesis, Cardiff University

Soenander H (1992) Seismic stratigraphy of the giant foreset formation, offshore north Taranaki Western Platform. In: Proceedings 1991 New Zealand petroleum conference proceedings ministry of commerce, pp 207–233

Tippett JM, Kamp PJ (1995) Geomorphic evolution of the southern Alps, New Zealand. Earth Surf Proc Land 20:177–192

Part II
Failure Dynamics from Landslide Geomorphology

Chapter 7
Postglacial Mass Failures in the Inner Hardangerfjorden System, Western Norway

Benjamin Bellwald, Berit Oline Hjelstuen, Hans Petter Sejrup, and Haflidi Haflidason

Abstract Based on high-resolution bathymetric records, sub-bottom profiles and sediment cores, we study postglacial mass transport deposits, slide scar complexes, cyclic steps and rockfall deposits as indicators of mass failures in the inner Hardangerfjorden system, western Norway. The stacked mass transport deposits show thicknesses of up to 4 m and witness that the inner Hardangerfjorden has been a site of repetitive mass failure events, potentially triggered by earthquakes related to glacioisostatic uplift. The cyclic steps, affecting an area of about 2 km^2 in the innermost fjord, have wave lengths of 40 m, heights of 5 m and are most likely related to fluvial sediment supply. Seven slide scar complexes, with stratigraphy-cutting scar heights of 6–34 m, are identified on the basin plain of the fjord. These are all associated with large depocenters along the fjord flanks, suggesting a link between locations of high sediment supply and mass failure. Although rapid deposition might, by itself, induce failures, an external trigger mechanism, such as an earthquake, should also be considered.

7.1 Introduction

Submarine mass failures are a common sedimentary process within fjord systems and represent a severe geohazard to infrastructures and populations. These mass failures are usually initiated along the fjord coastlines (e.g. L'Heureux et al. 2010). However, in spite of an increasing amount of studies of fjord systems, mass failure triggering mechanisms as well as sedimentary preconditioning factors are still under debate in many regions. In this study we utilise high resolution bathymetric records, TOPAS seismic profiles and gravity cores from the inner Hardangerfjorden, western Norway (Fig. 7.1) to (1) define seismic facies, (2) characterise seabed features and (3) describe sediment accumulation features in order to

B. Bellwald (✉) • B.O. Hjelstuen • H.P. Sejrup • H. Haflidason
Department of Earth Science, University of Bergen, NO-5007 Bergen, Norway
e-mail: Benjamin.Bellwald@uib.no

G. Lamarche et al. (eds.), *Submarine Mass Movements and their Consequences*, Advances in Natural and Technological Hazards Research 41,
DOI 10.1007/978-3-319-20979-1_7

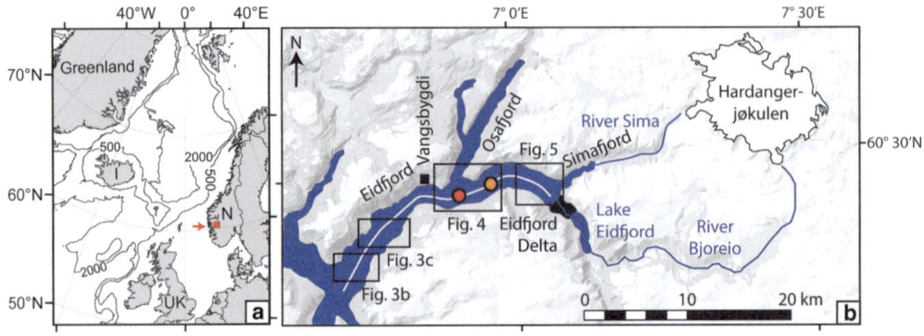

Fig. 7.1 (**a**) Regional overview map of the northern North Atlantic with location of study area (*red square*) indicated. (**b**) Innermost Hardangerfjorden system. Hardangerjøkulen Glacier, rivers fed by glacier meltwater (*blue lines*), TOPAS seismic profile (*white line*) and locations of gravity cores (*red dot* represents: GS14-187-05GC, yellow dot represents: GS14-187-04GC) are shown. Topographic data is from www.norgeibilder.no

discuss indicators of mass failures as well as pre-failure sediment conditions and possible trigger mechanisms within a fjord system.

7.2 Study Site and Geological Setting

The 20 km long Eidfjord (Fig. 7.1) reaches a maximum water depth of 710 m in its outermost parts, and is gradually shallowing to 250 m towards the northeast. The flat fjord bottom is 700–1200 m wide. The fjord walls, which are vulnerable to rockfall and snow avalanches (www.geonorge.no), have gradients of up to 45° and consist of Precambrian granite (Holtedahl and Dons 1960). Eidfjord and the surrounding areas are at present characterised by high precipitation and high (glacier) runoff (www.nve.no). In addition to River Sima and River Bjoreio, draining glacier meltwater from Hardangerjøkulen Glacier (Fig. 7.1b), 17 other major rivers are entering the fjord (Fig. 7.3). We note that Lake Eidfjord (Fig. 7.1b) is acting as a sediment sink for River Bjoreio. Compared to other regions in Scandinavia, seismic activity in Western Norway is high (Ahjos and Uski 1992). The recurrence interval for Norway and adjacent areas is estimated to 10 years for a M5 earthquake, and to 1100 years for a M7 earthquake (Bungum et al. 2005), noting that in the year of 2000, a M4.5 earthquake occurred in the outer part of the Hardangerfjorden system (Hicks and Ottemöller 2001). During the last glacial maximum Eidfjord was covered beneath the Fennoscandian Ice Sheet, which after a retreat in the Allerød time period readvanced during the Younger Dryas (Mangerud et al. 2011). At the end of the Younger Dryas time period, the ice sheet rapidly retreated with estimated rates of 240 ± 70 m/a (Mangerud et al. 2013). In Eidfjord, postglacial marine limits of 101–113 m asl are reflected in the deposition of a large glacifluvial delta (Fig. 7.1b, Anundsen and Simonsen 1967).

7.3 Data and Methods

This study is based on TOPAS high resolution seismic profiles, high resolution bathymetric records and gravity cores. 50 km of sub-bottom profiles (TOPAS PS 18, 0.5–6 kHz) were acquired and interpreted using PETREL v.2013. The gravity cores, GS14-187-04GC and GS14-187-05GC (Fig. 7.1b), were raised from water depths of 402 and 427 m, respectively, and have been analysed for sedimentological, geophysical and geotechnical properties all along their length of maximum 20 cm. The bathymetric data were collected by DOF Subsea AS, using a Kongsberg EM710 MBE system (70–100 kHz) and a ROV mounted Seabat 7125 (200 and 400 kHz) with spatial resolutions of 1×1 m and 0.2×0.2 m, respectively. Sediment volumes were calculated by using NaviModel (v.3.2.91) from EIVA.

7.4 Main Observations and Interpretations

Eidfjord is filled in with up to 160 ms (~120 m, using an assumed sediment velocity of 1500 m/s) of sediments, which onlap the acoustically non-structural acoustic basement (Fig. 7.2a). From the central part of Eidfjord, the seismic penetration decreases rapidly towards the head of the fjord. The ridge observed in the seismic profile nearby Vangsbygdi (Figs. 7.1b and 7.2) is separating steeper dipping seabed (~1.5°) from <1° dipping seabed of the outer and inner Eidfjord, respectively. In the very innermost part of Eidfjord, the fjord bottom gradient increases to 3–10° (Fig. 7.2).

The sediments can be sub-divided into two main seismic sequences. The lower sequence is dominated by acoustically laminated sediments (Figs. 7.2a and 7.4a) and is, in agreement with previous fjord studies (e.g. Hjelstuen et al. 2013), interpreted to represent postglacial glacimarine deposits. The upper seismic sequence, having a maximum thickness of 20 ms (~15 m), is characterized by acoustically transparent units with a thickness of about 5 ms (~4 m) and a length of up to 5 km. These units, suggested to represent stacked mass transport deposits (MTDs, compare St-Onge et al. 2012), are separated by continuous, high-amplitude reflectors, potentially acting as basal planes for the failed sediments (Fig. 7.4a).

The gently sloping seabed of the fjord is intersected by seven vertical steps, which are 6–34 m high, 20–35° steep and have a complex u-shaped pattern in the bathymetric data (numbers 1–7, Fig. 7.3). These stratigraphy-cutting steps in water depths of 335–645 m are interpreted as slide scars in other fjords (e.g. Aarseth et al. 1989). In front of the scars, disturbed sediment layers and isolated block-like features are identified (Fig. 7.4).

Twelve of the major rivers (Fig. 7.3) entering Eidfjord are associated with cone-shaped colluvial fans along the ~45° inclined submarine fjord flanks. These cones are locally producing a positive relief along the fjord bottom and are characterized by hummocky, acoustically homogenous sediment packages. They reach heights of

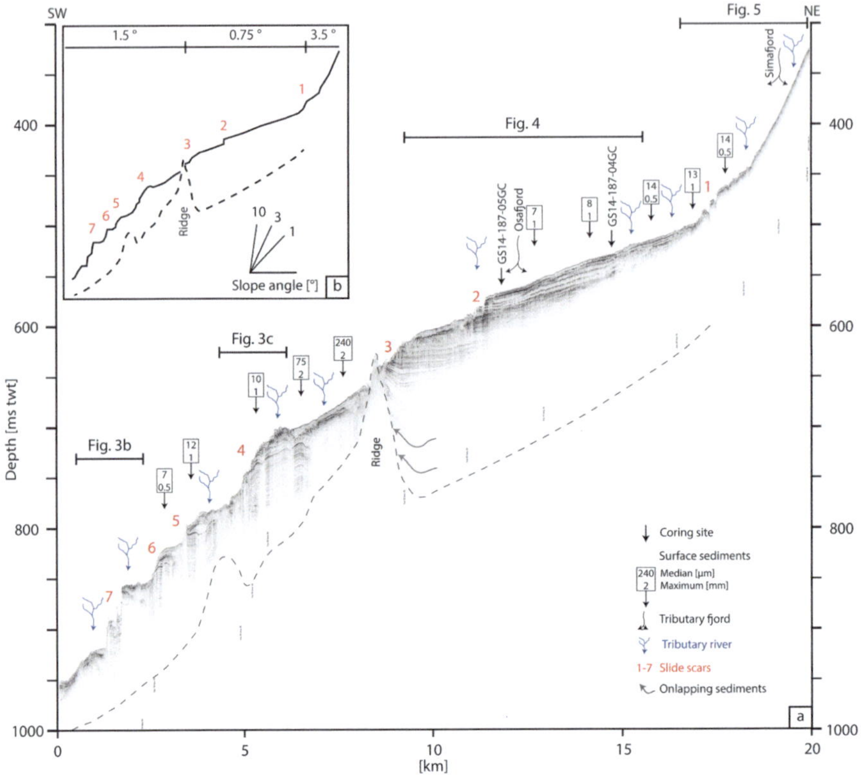

Fig. 7.2 (**a**) Seismic profile from Eidfjord (profile location in Fig. 7.1b) with tributary fjords, slide scars, rivers and surface sediment characteristics (from Holtedahl 1975) annotated. Acoustic basement is indicated by *stippled line*. (**b**) Seabed gradient for different parts of the studied fjord with observed slide scars. *Red* numbers refer to slide scars

Fig. 7.3 (**a**) Identified slide scar complexes (*yellow* numbers 1–7), cones and fans (*white* letters A-U), cyclic steps (*red lines*), tributary rivers (*white lines*) and locations of gravity cores (*red dot*: GS14-187-05GC, *yellow dot*: GS14-187-04GC). *Gray-shaded* bathymetric map of slide scar complex 7 and cone R (**b**) and slide scar complex 4 and cone M (**c**) are inserted (Bathymetric data from DOF Subsea Norway and Statnett. Topographic data from www.norgeibilder.no)

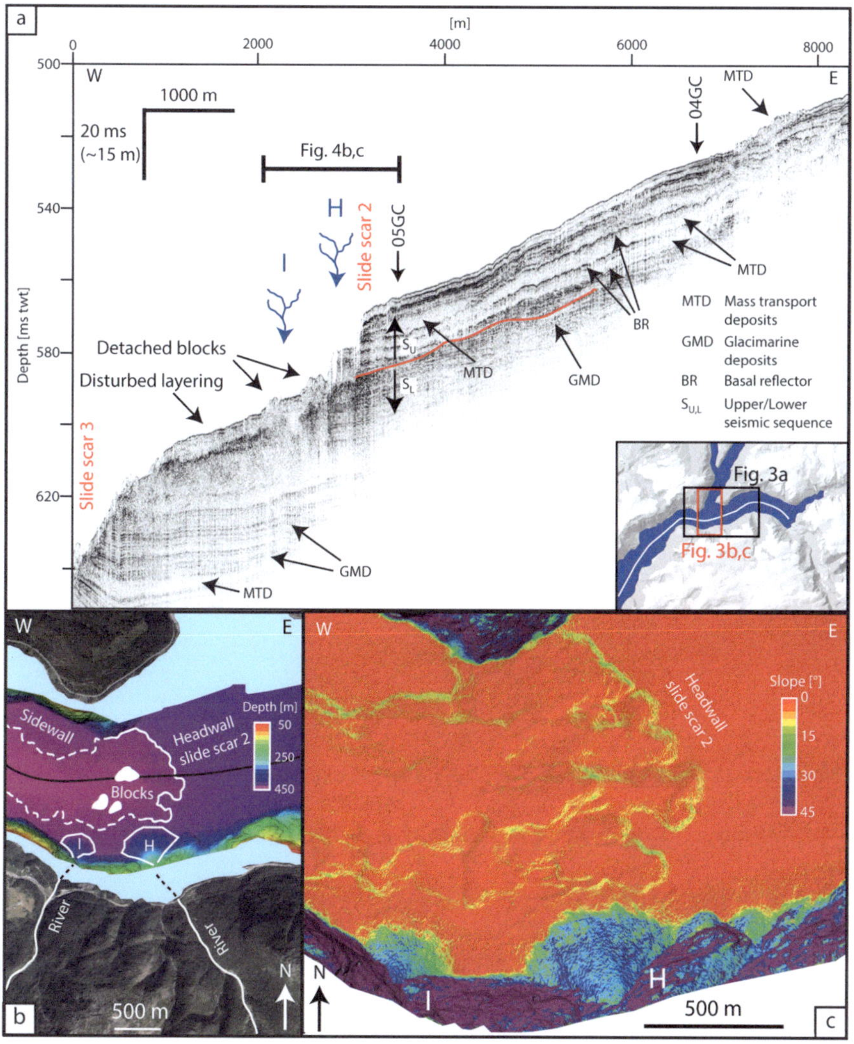

Fig. 7.4 (**a**) TOPAS seismic profile showing seismic facies characteristics across slide scar 2. Gravity core locations and tributary rivers (H and I) are indicated. Profile location shown in index map, lower right corner. (**b**) Interpreted bathymetric data including seismic profile (*black line*) and subaerial environment (from www.norgeibilder.no). Location shown in index map in Fig. 7.4a. (**c**) Gradient map of river cones correlating with slide scar complex 2 (Data source Fig. 7.4a, b: DOF Subsea Norway and Statnett)

90–440 m, have surface gradients of 20–35° and comprise sediment volumes up to 63×10^6 m^3. We note that the cone-shaped features are located along both fjord flanks, often nearby the identified scars (Fig. 7.3).

A fan-shaped deposit (T, Fig. 7.3) cannot be associated with any river. This deposit is 100 m long, 250 m wide and 30° steep. We also identify blocks nearby the innermost scar, which are standing more than 10 m above the surrounding fjord bottom (Fig. 7.5a–c). These blocks are up to 40 m long, 20 m wide, have an angular shape and are observed within a limited area. Because of their shape and location, these blocks might represent a rockfall deposit from the fjord flanks.

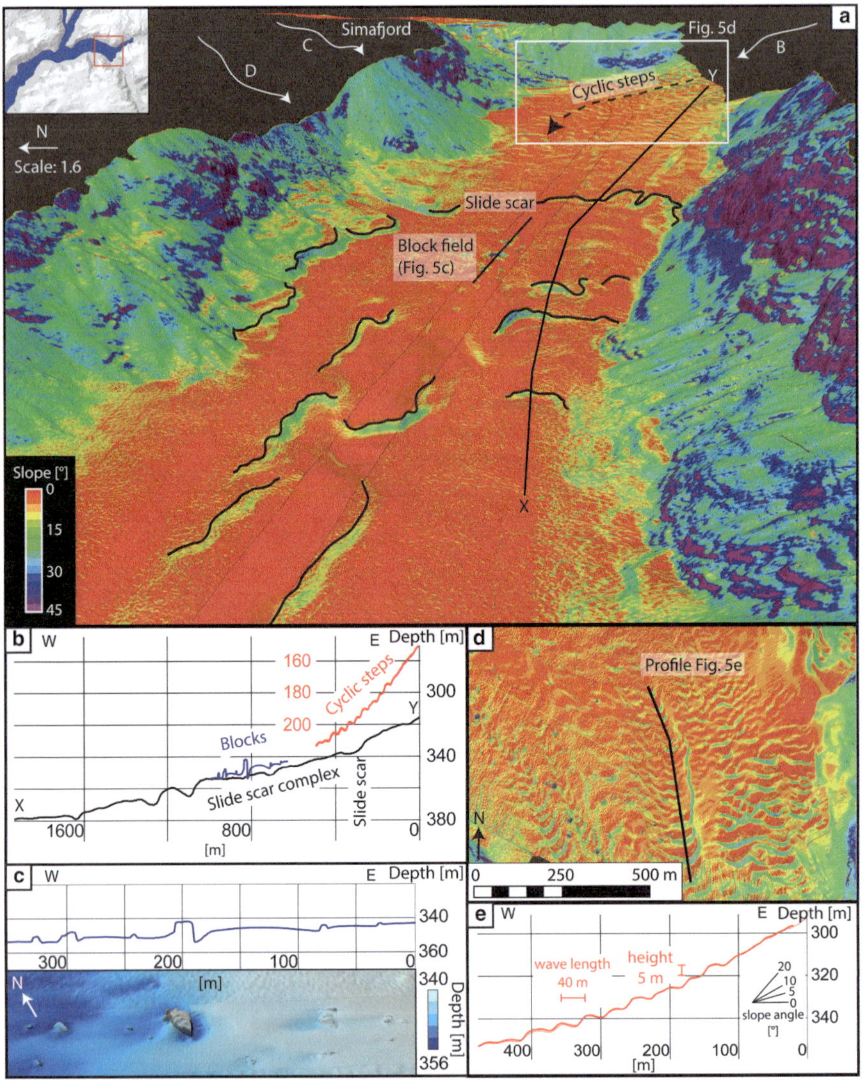

Fig. 7.5 Slide scar complex 1, innermost Eidfjord. (**a**) Gradient map, showing the retrogressive nature of the mass failure event in this area. Cyclic steps, block field and tributary rivers (B-D) are indicated. (**b**) Bathymetric profiles across slide scar complex (profile XY in Fig. 7.5a), cyclic steps (location in Fig. 7.5d) and block field (also shown in Fig. 7.4c). (**c**) *Upper* panel: Bathymetric profile across rockfall deposit (location in Fig. 7.5a). *Lower* panel: Bathymetric image of block field. (**d**) Gradient map of cyclic steps (location in Fig. 7.5a). (**e**) Bathymetric profile (location in Fig. 7.5d) across cyclic steps (Data source: DOF Subsea Norway and Statnett)

On the steeper terrain of the innermost Eidfjord, in water depths shallower than 350 m, we observe symmetrical ridges, having wave lengths of 40 m and heights of about 5 m, which are not changing with changing water depths (Figs. 7.3 and 7.5a, d, e). These structures, covering an area of about 2 km^2, are following channel-like depressions and are not observed west of slide scar 1. As the seismic penetration is limited, the expression of these features in the sub-bottom sediments is poorly resolved. These ridges could either be features related to cyclic steps (e.g. Hughes Clarke et al. 2014) or features related to sediment creep (e.g. Lee and Chough 2001). As the ridges are clearly following several feeder channels along the fjord head delta, they might be related to hyperpycnal currents and be the result of upslope migrating bedforms, what makes the cyclic step hypothesis more reasonable (compare Hughes Clarke et al. 2014).

The two sediment cores, 3.7 km apart, show large variation in the seabed grain-size distribution (GS14-187-04GC: 10 % gravel, 38 % sand, 52 % silt and clay. GS14-187-05GC: 0 % gravel, 22 % sand, 78 % silt and clay), noting that very coarse gravel and pebbles measuring up to 6 cm, were identified in GS14-187-04GC. The core analyses also show that the bulk density and shear strength of GS14-187-05GC are increasing within the uppermost 20 cm of the core, from 1.2 to 1.7 g/cm^3 and 7–16 kPa, respectively.

7.5 Discussion

We name the pattern of u-shaped steps in the bathymetric data a slide scar complex (Figs. 7.3, 7.4, and 7.5) and suggest that these complexes indicate retrogressive mass wasting, similarly to what is reported for the Storegga Slide Complex on the SE Nordic Sea margin (Kvalstad et al. 2005). The disturbed sediment layers (Fig. 7.4a), in front of the slide scars, are suggested to be related to sediment collapse taking place during the mass failure event. Furthermore, we interpret the identified block-like features as detached slide blocks, showing that the failed sediments did not fully disintegrate during failure (Fig. 7.4a, b). Similar features have commonly been reported from slides, both in fjords and on the Norwegian continental margin (Longva et al. 2003). We infer that the sediments involved in the slide failures consist both of older MTDs and glacimarine sediments, as the slide scars clearly cut off older MTDs (Fig. 7.4a).

The observation of submarine cone-shaped sediment features nearby six of the seven slide scars (number 1–2 and 4–7 in Fig. 7.3) strongly indicate a link between regions of high sediment accumulation and mass failure. However, as neither the slides nor the built-up of the sediment cones have been dated, a direct correlation is uncertain. Our sediment cores and previous studies (Holtedahl 1975, Fig. 7.2a) indicate coarser seabed sediments nearby major cones (Fig. 7.2). The fluvial deposition of these coarser sediments might also have influenced the sediment stability, acting as weak layers (Locat et al. 2014) with rupture along their surfaces.

The cyclic steps observed at the head of the fjord are probably related to enhanced sediment supply from River Bjoreio, and could negatively influence slope stability, acting as a pre-failure factor for mass wasting in the future. Block fields and fan-shaped sediment features are observed in regions vulnerable to subaerial mass failure (www.geonorge.no), and are thus probably triggered by rockfalls. As both the block field in the innermost (Fig. 7.5a) and the fan-shaped sediment feature in the central part of Eidfjord (T in Fig. 7.3) can be correlated with slide scar complexes, these processes could have been responsible for slope failure in the basin plain of the fjord.

The distribution of slide scars along the fjord bottom also indicates a relation between fjord geometry and the occurrence of mass failures. Five slide scars (numbers 3–7 in Fig. 7.3) are observed in the outer Eidfjord, where the fjord floor is steeper and the fjord basins are narrower compared to the wider inner Eidfjord, where the fjord bottom is only disrupted by two slide scars.

Even though sediment loading by cones and fans may lead to unfavourable conditions within the sediment package, causing sediment failures, another external trigger mechanism could be necessary. As western Norway is one of the most seismic active areas in Scandinavia (Bungum et al. 2005; Ahjos and Uski 1992), earthquakes could have acted as a final trigger for mass failures in Eidfjord, as seen for other fjords along the west- and mid-Norwegian coast (Bøe et al. 2004).

7.6 Conclusions

Based on TOPAS seismic profiles, bathymetric records, and sediment cores, we have studied indicators of mass failures and trigger mechanisms in Eidfjord, western Norway. The main conclusions are:

- Buried mass transport deposits, slide scar complexes, cyclic steps and rockfall deposits are identified as indicators of mass failures.
- Submarine landslides can originate in the basin plain of a fjord with very gentle inclination, and are not limited to the shore lines, as has commonly been observed in other fjord systems.
- Cones and fans along the fjord flanks are often associated with identified slide scar complexes. We speculate if there is a link between regions of high sediment accumulation and mass failure events.
- Cyclic steps, which may act as a pre-failure factor for mass failure in the future, are observed in the very innermost Eidfjord and are hypothesized to be related to sediment supply from River Bjoreio.
- There may be a link between the geometry of the fjord basin and the location of mass failures.
- Eidfjord is located in an area of high seismic activity. Thus, earthquakes may act as the final trigger mechanism of mass failure.

Acknowledgements We acknowledge Statnett and DOF Subsea Norway AS for access to high-resolution bathymetric data and thank Oddvar Longva and Jasper Moernaut for revision of the paper. The research leading to these results has received funding from the People Programme (Marie Curie Actions) of the European Union's Seventh Framework Programme FP7/2007–2013/ under REA grant agreement n° 317217. The research forms part of the GLANAM (GLAciated North Atlantic Margins) Initial Training Network.

References

Aarseth I, Løhne Ø, Giskeødegaard O (1989) Submarine slides in glaciomarine sediments in some western Norwegian fjords. Mar Geol 88:1–21

Ahjos T, Uski M (1992) Earthquakes in northern Europe in 1375–1989. Tectonophysics 207:1–23

Anundsen K, Simonsen A (1967) Et pre-borealt breframstøt på Hardangervidda og i området mellom Bergensbanen og Jotunheimen. Årbok for Universitetet i Bergen. Matematisk-naturvitenskapelige serie 7, 1–42

Bøe R, Longva O, Lepland A, Blikra LH, Sønstegaard E, Haflidason H, Bryn P, Lien R (2004) Postglacial mass movements and their causes in fjords and lakes in western Norway. Nor J Geol 84:35–55

Bungum H, Lindholm C, Faleide JI (2005) Postglacial seismicity offshore mid-Norway with emphasis on spatio-temporal-magnitudal variations. Mar Pet Geol 22:137–148

Hicks EC, Ottemöller LR (2001) The ML 4.5 Stord/Bømlo, southwestern Norway, earthquake of 12 August 2000. Nor J Geol 81:293–304

Hjelstuen BO, Kjennbakken H, Bleikli V, Ersland RA, Kvilhaug S, Euler C, Alvheim S (2013) Fjord stratigraphy and processes – evidence from the NE Atlantic Fensfjorden system. J Quat Sci 28:421–432

Holtedahl H (1975) The geology of the Hardangerfjord, West Norway. Norges Geol Unders 323:1–87

Holtedahl O, Dons JA (1960) Geological map of Norway. Norges Geol Unders 208

Hughes Clarke JE, Vidiera Marques CR, Pratomo D et al (2014) Imaging active mass-wasting and sediment flows on a Fjord delta, Squamish, British Columbia. In: Krastel S (ed) Submarine mass movements and their consequences, Advances in natural and technological hazards research. Springer, Dordrecht, pp 249–260

Kvalstad TJ, Andresen L, Forsberg CF, Berg K, Bryn P, Wangen M (2005) The Storegga slide: evaluation of triggering sources and slide mechanics. Mar Pet Geol 22:245–256

L'Heureux JS, Hansen L, Longva O, Emdal A, Grande LO (2010) A multidisciplinary study of submarine landslides at the Nidelva fjord delta, Central Norway – implications for geohazard assessment. Nor J Geol 90:1–20

Lee SH, Chough SK (2001) High-resolution (2–7 kHz) acoustic and geometric characters of submarine creep deposits in the South Korea Plateau, East Sea. Sedimentology 48:629–644

Locat J, Leroueil S, Locat A, Lee H (2014) Weak layers: their definition and classification from a geotechnical perspective. In: Krastel S et al (eds) Submarine mass movements and their consequences, Advances in natural and technological hazards research. Springer, Dordrecht, pp 3–12

Longva O, Janbu N, Blikra LH, Bøe R (2003) The 1996 Finneidfjord slide; seafloor failure and slide dynamics. In: Locat J, Mienert J (eds) Submarine mass movements and their consequences, Advances in natural and technological hazards research. Kluwer Academic, Dordrecht, pp 531–538. ISBN 19

Mangerud J, Gyllencreutz R, Lohne ØS, Svendsen JI (2011) Glacial history of Norway. In: Ehlers J et al (eds) Quaternary glaciations – extent and chronology. Elsevier, Amsterdam

Mangerud J, Goehring BM, Lohne ØS, Svendsen JI, Gyllencreutz R (2013) Collapse of marine-based outlet glaciers from the Scandinavian ice sheet. Quat Sci Rev 67:8–16

St-Onge G, Chapron E, Mulsow S, Salas M, Viel M, Debret M, Foucher A, Mulder T, Winiarski T, Desmet M, Costa PJM, Ghaleb B, Jaouen A, Locat J (2012) Comparison of earthquake-triggered turbidites from the Saguenay (Eastern Canada) and Reloncavi (Chilean margin) Fjords: implications for paleoseismicity and sedimentology. Sediment Geol 243–244:89–107

www.geonorge.no, Kartverket, The Norwegian mapping authority, Hønefoss, Norway, 25.11.2014

www.norgeibilder.no, Kartverket, The Norwegian mapping authority, Hønefoss, Norway, 25.11.2014

www.nve.no, Norges vassdrags- og energidirektorat (NVE), Norwegian water resources and energy directorate, Oslo, Norway, 10.12.2014

Chapter 8
Onshore and Offshore Geomorphological Features of the El Golfo Debris Avalanche (El Hierro, Canary Islands)

Ander Biain, Ricardo León, Roger Urgeles, Luis Somoza, Teresa Medialdea, Mercedes Ferrer, and Francisco Javier Gonzalez

Abstract This study shows an onshore-offshore morpho-structural characterization of the El Golfo flank collapse and debris avalanche on El Hierro (Canary Islands, Spain). Erosive and depositional features have been identified based on: LIDAR topography and geology from water galleries (onshore); and high-resolution 3.5 kHz and multichannel seismic reflection profiles, and multibeam data (offshore). The onshore headwall scarp shows a non-continuous profile formed by two semi-circular amphitheatres and extends offshore by a smooth chute. The chute ends at about 3000–3200 m water depth in the distal depositional area. Multichannel seismic profiles show two major subunits of chaotic reflectors in the debris avalanche deposits. Results suggest that the El Golfo debris avalanche event likely took place in multiple stages. Consequently, we suggest that the multistaged nature of El Golfo debris avalanche greatly reduces the tsunamigenic potential of these flank collapses.

8.1 Introduction

Landslides on the flanks of oceanic volcanic islands are considered among the largest gravity driven processes mobilising several hundred km^3 of rocks and sediment. These phenomena are inherent to the evolution of oceanic volcanic islands, and result from oversteepening of volcano slopes, seismicity, presence of weak layers, dyke injection, and/or volcanic expansion (Masson et al. 2002). The tsunamigenic potential of these huge landslides is a source of debate (Ward and Day 2001; Masson et al. 2006). Understanding how volcanic islands flanks collapse has

A. Biain • R. León • L. Somoza (✉) • T. Medialdea • M. Ferrer • F.J. Gonzalez
Geological Survey of Spain, IGME, Rios Rosas 23, 28003 Madrid, Spain
e-mail: l.somoza@igme.es

R. Urgeles
Institute of Marine Sciences, ICM-CSIC, Passeig Marítim de la Barceloneta, 37-49, 08003 Barcelona, Spain

G. Lamarche et al. (eds.), *Submarine Mass Movements and their Consequences*, Advances in Natural and Technological Hazards Research 41,
DOI 10.1007/978-3-319-20979-1_8

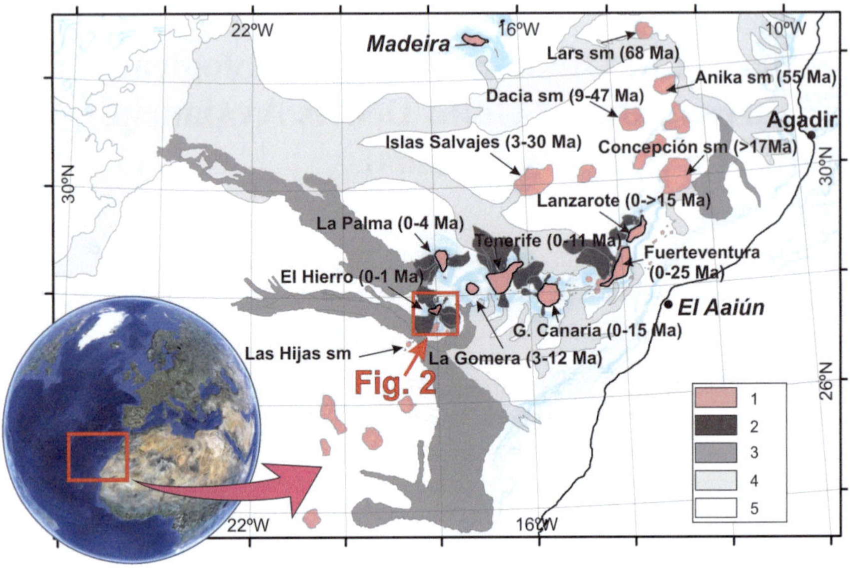

Fig. 8.1 Major sediment pathways in the study area. *1* volcanic edifices, *2* debris avalanches, *3* debris flow, *4* turbidity currents, *5* abyssal plain (Modified from data of Weaver et al. 2000)

critical implications for tsunamigenesis, particularly the style of failure and emplacement of related deposits on the sea floor (Watt et al. 2012), whether the failure is multistage or single block, retrogressive or progressive, submarine or sub-aerial, and disintegrative or not (Harbitz et al. 2006; Ward and Day 2001; Watt et al. 2012). The single- vs. multi-staged and the retrogressive character of island flank collapses (Garcia 1996; Masson et al. 2002; Hunt et al. 2013) controls the volume involved in a landslide and/or the duration and timing over which such volume is mobilized, which is a first order control on the initial tsunami wave amplitude (Harbitz et al. 2006). The aim of this study is to analyse both the onshore and offshore geomorphology and internal structure of the El Golfo landslide, on El Hierro to determine the nature of such flank collapses and help constrain their tsunamigenic potential.

8.1.1 Geological and Geomorphological Setting

The Canary Archipelago is the subaerial expression of the Canary ocean volcanic ridge (Carracedo 1994; Guillou et al. 1996). The ridge is a 1350 km long NE-SW oriented lineament (from Lars to Tropic) of volcanic islands and seamounts located 100 km off the northwest African coast (Fig. 8.1). El Hierro is the youngest and most south-westerly of the seven Canary Islands where the oldest subaerial rocks have been dated at 1.12 Ma (Guillou et al. 1996).

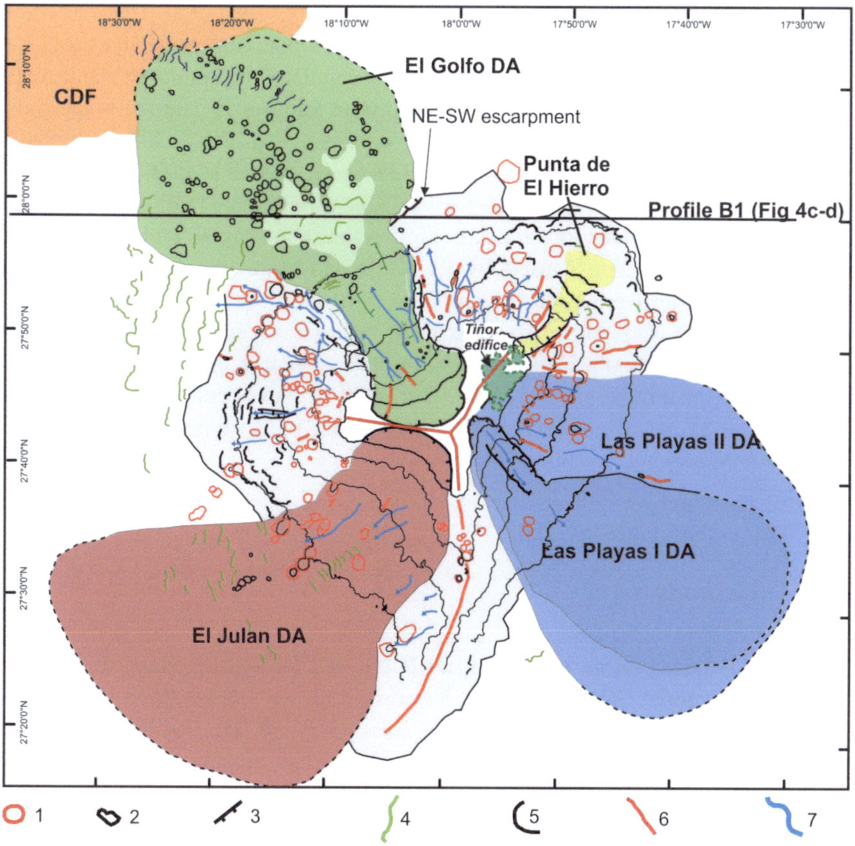

Fig. 8.2 Geomorphological schema of the El Hierro Island. *1* cone, *2* block, *3* collapse scars and escarpments, *4* bedforms, *5* terrace of lava, *6* Ridge, *7* channel or gully. Debris avalanche (DA) boundaries mapped from our multibeam data

The El Hierro Island rises about 5000 m from its base on the western margin and it is characterized by alternating eruptive episodes and major destructive events (Carracedo 1994). Three eruptive episodes have been recognized onshore El Hierro Island based on radiometric dating and paleomagnetism (Carracedo 1994; Guillou et al. 1996). In the last stage of construction (Rift Volcanism; 0.158–0.012 Ma), the island took its pseudo-triangular appearance and developed the El Golfo flank collapse. Lithostatic decompression caused by removal of volcanic material by the El Golfo landslide induced a new eruptive phase and lava flow emissions (240 m in thickness) towards the coast (Carracedo 1994).

At least five major landslides have been recognized in the El Hierro Island (Fig. 8.2) from both onshore geological evidence (Carracedo 1994) and offshore deposits (Urgeles et al. 1997; Masson et al. 2002). The oldest event (around 882 ka) involved the east sector of Tiñor Edifice (Fig. 8.2) (Carracedo 1994), although its related submarine deposit has not been yet identified (Masson et al. 2002). The El

Julan landslide (at least 160 ka old (Carracedo 1994; Guillou et al. 1996) affected the southwest flank of the island. Las Playas I (545–176 ka) and Las Playas II (176–145 ka) landslides (Masson et al. 2002), are located on the southeast flank of El Hierro. Finally, the El Golfo landslide (northwest flank of El Hierro) is the most recent event (15 ka) of the whole archipelago (Masson et al. 2002), although recent studies indicate a slightly older age (39 $\pm$ 13 ka ago; Longpré et al. 2011). Onshore, the landslide scar constitutes the most remarkable topographic feature of the island creating a 15 km wide and 1.4 km high horseshoe-shaped embayment. The lithostatic decompression caused by El Golfo landslide induced a rise of the magma and lava flow emissions towards the coast partially filling the embayment (Carracedo 1994). Offshore, two areas have been distinguished: a proximal erosional area with a smooth chute bounded by lateral scarps up to 600 m high that gradually diminish in height and fade out at 3200 m water depth; and a distal area characterised by an avalanche deposit with randomly distributed blocks (Urgeles et al. 1997; Masson et al. 2002).

8.1.2 Methods

This study is based on an onshore LIDAR derived DEM from Spanish Instituto Geografico Nacional (IGN) gridded at 5 m resolution (CNIG 2012) and the inventory of water well points in the El Golfo embayment (IGME 2012) and offshore multichannel seismic reflection and chirp profiles and multibeam echo-sounder data (bathymetry and backscatter). Multibeam echo-sounder data were acquired with a Simrad EM120 and an ATLAS Hydrosweep DS from the R/V Hesperides and R/V Sarmiento de Gamboa respectively. Ultra-high resolution chirp echosounder data were acquired using a Simrad TOPAS PS 18 and an ATLAS Parasound P35, again from the R/V Hespérides and R/V Sarmiento de Gamboa respectively. The TOPAS PS 18 system operated at 1.5–4.0 kHz and was digitally processed in real time. Data was acquired in high-penetration (chirp) mode. The shot interval was set to 5–7 s and the recording length was ~400 ms with a sampling frequency of 16 kHz. The ATLAS Parasound P35 operated on a secondary lower frequency of 3.5–4 kHz, was shot every 4–10 s and the recording length was 0.17–25 ms with a sampling frequency of 12.2 kHz.

Multichannel seismic reflection data was obtained in 1988 onboard R.V. TAU. It was acquired with a system of 34 airguns with a total capacity of 103 l, a shot spacing of 50 m and a pressure of 14 MPa. Data acquisition was performed using a 3000 m long GX-600 streamer with 120 channels.

8.2 Results

8.2.1 Morphology and Backscatter Mapping

The onshore morphology of El Hierro is characterized by a well-defined major amphitheatre-shaped structure that corresponds to the El Golfo flank collapse. The slip plane's upper portion of the collapse creates an arcuate scarp enclosing the El Golfo embayment. It is 15 km wide with three clearly defined sectors (Fig. 8.3a), producing the appearance that the central sector is indented within an already existing amphitheatre. The onshore amphitheatre scars surrounding El Golfo

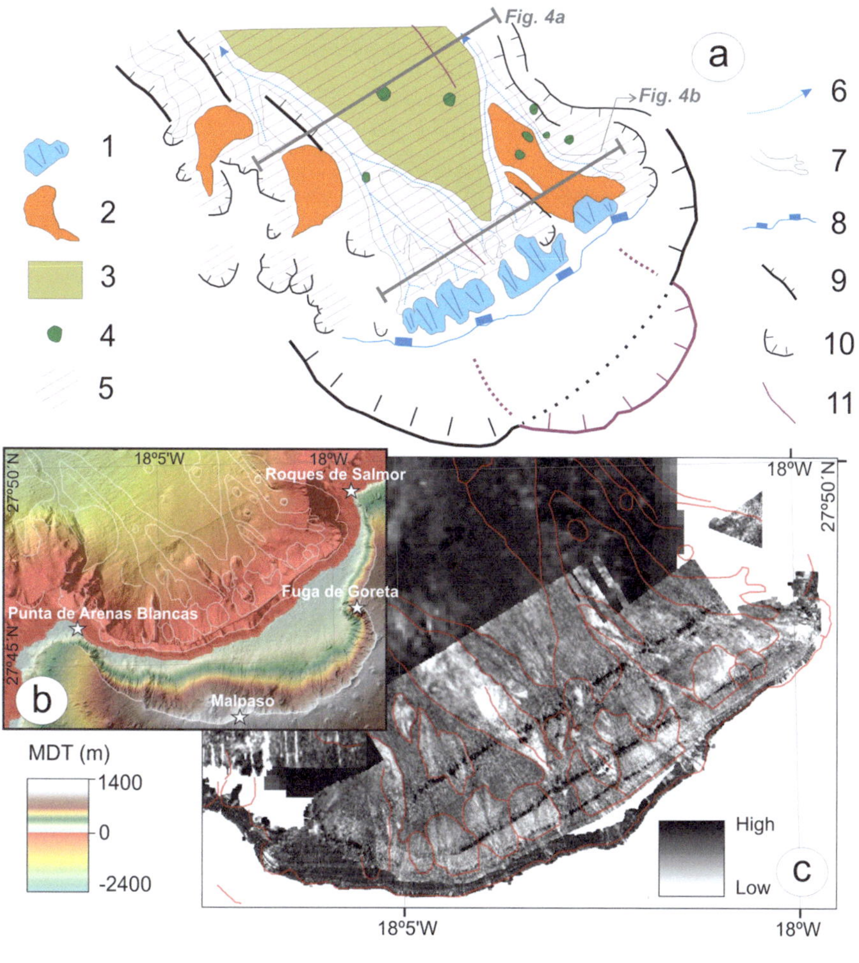

Fig. 8.3 Geomorphological sketch of the proximal offshore area of the El Golfo landslide. (**a**) Detail of the proximal area: Legend: *1* infra littoral sedimentary wedge, *2* recent minor gravitational instabilities, *3* Subunit "A", *4* block, *5* erosional surfaces, *6* direction of channel flows, *7* channel, *8* littoral scarp, *9* scarp, *10* headwall scar, *11* crest, 10. (**b**) Digital terrain model (bathymetry + topography). (**c**) Backscatter mosaic

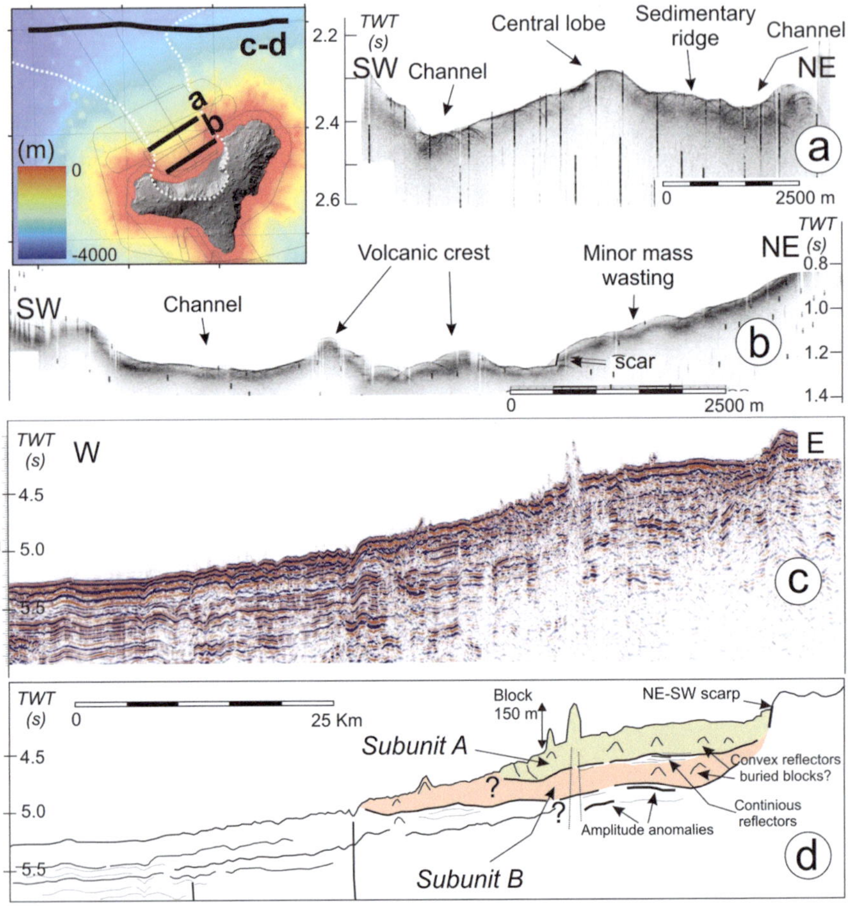

Fig. 8.4 Seismic dataset presented in this paper. (**a**). (**b**) Ultra-high seismic profiles, (**a**) *Line* H18_20120619205954 and (**b**) *Line* H16_20120619184010 of GAIRE cruise, respectively. (**c**) Multichannel seismic reflection profile (*Line* B of CANARIAS-88 cruise). (**d**) line drawing

embayment are laterally continuous with submarine scars and escarpments. These submarine scarps extend for about 6 km away from the coastline and trend mainly NW in direction. The maximum elevation of the scars (900 m) has been observed to the south-west of Roque de Salmor (Fig. 8.3b). The mean scar height gradually decreases from 300 m at 1500–2000 m water depth to complete disappearance at about 3000 m water depth. The SW scar has several V-shaped gullies (Figs. 8.3 and 8.4), with arcuate headwalls of ~1000 m in diameter.

The offshore escarpments are located on the sides of a chute evolving from 1000 to 2800 m water depth. Within this chute, the proximal area, from −700 to −2900 m water depth, has mainly an erosive character, but debris avalanche deposits are also present within the chute (Fig. 8.3). The surface of the chute shows relatively smooth relief, high backscatter values (−10 to −12 dB; Fig. 8.3c) and hyperbolic diffractions on TOPAS profiles. Within the chute there are five major feature types: steep sided crests and ridges, smooth sedimentary ridges, channels, minor debris lobes, and allochtonous blocks. The steep-sided crests and ridges are likely constituted of basement volcanic rocks exposed during the mass wasting processes. They are up to 120 m high and have flanks with slopes of up to 25°. Most of them show low backscatter values (−17 to −27 dB; Fig. 8.3) and are characterized by wide hyperbolic echoes in high-resolution TOPAS profiles (Fig. 8.4a, b).

The sedimentary ridges (Figs. 8.3 and 8.4a) are elongate positive relief features likely made of debris avalanche deposits. They show terraced profiles produced by the incision of channels on their flanks. The largest one, 7 km long, about 2 km wide and 100 m high, is located on the western side of the chute and is bounded by two channels with a gentle escarpment. Sedimentary ridges show transparent acoustic facies with hyperbolic diffractions (Fig. 8.4a). Within the chute, channels are present in the water depth range between 50 and 3000 m. They are mainly located at the margins of the chute and display high backscatter values (−8 to −10 dB) and semi-opaque acoustic facies with multiple low amplitude hyperbolae in high-resolution TOPAS profiles (Fig. 8.4). The narrower channels show incised V-shaped profiles, whereas the wider channels have flat-bottomed cross sections. A main central "U-shaped" channel, 4.5 km wide and 80 m deep, has been identified. It shows high backscatter values (−12 to −14 dB) and partially erodes the debris avalanche deposits from about 2000 to 3000 m water depth (Fig. 8.3).

Allochtonous blocks are interpreted to be large irregular fragments of failed volcanic rock involved in the failure of the volcanic edifice and transported inside the debris avalanche. Nearly a hundred of randomly distributed irregular blocks have been recognized over the El Golfo debris avalanche deposit. They are characterized by an irregular positive relief and relatively high backscatter values (−10 to −11 dB). Their size ranges widely from 200 m to 1.5 km in diameter, with an average diameter of 500 m. A majority of them are 50 m high and concentrate in the distal area of the deposit (from 3000 m water depth) despite some blocks are also present in relatively proximal parts of the chute. Between 500 and 1200 m depth, minor landslide fan-shaped deposits are also identified (Fig. 8.3). These debris lobes originate mostly from the gullies on the SW scar of *Roques de Salmor* and to the north of *Punta de las Arenas Blancas*. Most of the gullies where debris lobes initiate originate in 20–50 m water depth and are related to gravitational instabilities. However, between 100 and 600 m water depth partially eroded arcuate scars are also present. The debris lobes display transparent acoustic facies (Fig. 8.4a) and are 30–50 ms TWT thick. The lobes likely represent minor debris avalanche deposits.

8.2.2 Seismic Mapping

The debris avalanche deposits have been characterized based on multichannel seismic data. The main body unit is characterized by an upper irregular morphology bounded to the east by a 120 m high scarp that extends 6 km in a NE-SW direction.

This scarp has previously been interpreted as a possible relict scarp of the Canary Debris Flow (Urgeles et al. 1997; Masson et al. 2002) that originates in the area where the El Golfo Debris Avalanche was deposited.

Two subunits have been identified within the debris avalanche (Figs. 8.4c, d): Subunit "A" is the most superficial unit. It is characterized by high-amplitude and discontinuous reflectors. Hyperbolic diffractions are also observed, which likely indicate the presence of buried volcanic blocks. A group of relatively continuous reflectors at ~220 ms (TWTT) below seafloor represents the base of subunit A.

Subunit "B" is characterized by discontinuous and wavy reflectors, with regions of rather transparent acoustic response. As it occurs with subunit "A", its boundaries are more clearly marked in the eastern sector. The western boundary of the avalanche deposit is defined as the limit between presence/absence of surface volcanic blocks. Subunit B has a maximum thickness of 170 ms (~150 m using a p-wave velocity of 1750 ms^{-1}). The basal reflector is located at a mean depth of 440 ms TWTT (~400 m), is more continuous, and has larger amplitude than the reflectors separating both subunits. Amplitude anomalies are recognisable to 200 ms TWTT (~180 m) below the base of the subunit "B".

While it is not an easy task to assess the volume of these subunits with only one seismic profile, we have estimated a minimum volume of ~49 km^3 and ~210 km^3 for subunits A and B respectively. These volumes are based on the thickness of the units as measured from the seismic profile and the areal extent from geological mapping.

8.3 Discussion and Conclusions

The homogeneity of the debris avalanche deposits observed on previous seismic reflection profiles and the lack of evidence in swath bathymetric data to distinguish different episodes in the El Golfo Debris Avalanche has been considered to result from the fact that the debris avalanche occurred as a single major catastrophic event (Urgeles et al. 1997; Masson et al 2002). Wynn and Masson (2003) first proposed the idea that the El Golfo debris avalanche was a multistage landslide based on the character of the volcanoclastic turbidite "B" in the Agadir Basin. This turbidite "B", that directly relates to the genesis of the EL Golfo debris avalanche, displayed three subunits. Identification of subunits "A" and "B" within the debris avalanche in this work, provides the first direct evidence for the multistaged character of the El Golfo flank collapse. The similarity in surface morphology and internal acoustic response of subunits "A" and "B" suggests a similar sedimentary process originating from a similar source area and downslope pathway. Similar debris avalanche stacking patterns separated by thin packages of continuous reflectors have been described in other volcanic islands, such as in the island of Montserrat in the Lesser Antilles (Lebas et al. 2011; Watts et al. 2012). With evidence for two El Golfo debris avalanche deposits offshore, the morphology of the evacuation area can be more easily interpreted. The apparently single collapse scar is probably a composite scar with one amphitheatre superposed on an older one (near Fuga de Gorreta and Malpaso; Fig. 8.3b): an initial episode generated most of the scarp and a smaller subsequent episode created a second scarp that enlarged the El Golfo embayment. In this regard the sedimentary ridges that occur within the main offshore scars can be perhaps attributed to levees formed during the passage of this second event constrained by the offshore scars of the first event. Alternatively it could also be argued that the volcanism filling the El Golfo embayment buried the second collapse scar, which may explain the lack of piedmont materials in near-shore wells and galleries. This last hypothesis would imply that one of the landslides was almost entirely submarine.

We suggest that the El Golfo collapse is a two stage, or more likely, a multistage feature. We have estimated a minimum volume of ~49 km^3 and ~210 km^3 for the two well-identified events, respectively. The period of time separating them is difficult to assess due to the variability of the sedimentation rate over the submarine flanks of volcanic islands which range from 3 to 110 cm/ky (Le Friant et al. 2004). Distal turbidites in Madeira abyssal plain indicate at least three phases of failure for El Golfo landslide, closely spaced in time (hours to months) (Wynn and Masson 2003). One possible explanation for the offset in timescales between our observations on proximal avalanches and those from distal turbidites is that the continuous reflectors between the landslide deposits represent rapid sedimentation from sediment-gravity flows related to the landslide. Otherwise, a possible explanation for the offset in the number of landslide events is that some individual flows responsible for depositing the turbidite sub-units in the basin were generated by rapidly disintegrating landslides on the island flank that left no obvious landslide deposits in the upper slope region (Russell B Wynn, personal communication, Dec 10, 2014). In any case, it is apparent that the multistaged nature of El Golfo debris avalanche reduces the potential maximum size of a tsunami generated by the El Golfo avalanches, in turn reducing the regional risk due to this phenomenon.

Acknowledgments This work has been supported by the IGCP-640 S4SLIDE project and the Spanish projects SUBVENT (CGL2012-39524-C02), "Extension of the Spanish Continental Shelf West off Canary Islands" (CTM2010-09496-E) and MOVES (IGME.ref-2265). The authors would like express their appreciation to Thomas Hubble, Samantha Clarke, Thierry Mulder and Russell B. Wynn for their useful reviews and suggestions.

References

Carracedo JC (1994) The Canary Islands: an example of structural control on the growth of large oceanic-island volcanoes. J Volcanol Geotherm Res 60:225–241

CNIG (2012) Centro Nacional de Informacion Geográfica, Centro de Descargas. Instituto Geográfico Nacional. Modelo digital del terreno MDT05-LIDAR. Isla de El Hierro, Canarias. Hojas 1105/1108. Query: 20 Mars 2012. (http://centrodedescargas.cnig.es/CentroDescargas/buscadorCatalogo.do?codFamilia=MDT05)

Garcia MO (1996) Turbidites from slope failure on Hawaiian volcanoes. Geol Soc Spec Pub 110:281–294

Guillou H, Carracedo JC, Torrado FP, Badiola ER (1996) K-Ar ages and magnetic stratigraphy of a hotspot-induced, fast grown oceanic island: El Hierro, Canary Islands. J Volcanol Geotherm Res 73(1–2):141–155

Harbitz CB, Løvholt F, Pedersen G, Masson DG (2006) Mechanisms of tsunami generation by submarine landslides: a short review. Nor J Geol 86:255–264

Hunt JE, Wynn RB, Talling PJ, Masson DG (2013) Multistage collapse of eight western Canary Island landslides in the last 1.5 ma: sedimentological and geochemical evidence from subunits in submarine flow deposits. Geochem Geophys Geosyst 14:1–23

IGME (2012) Base de datos de aguas de Canarias. Query: 21 Mar 2012

Le Friant A, Harford CL, Deplus C, Boudon G, Sparks RSJ, Herd RA, Komorowski JC (2004) Geomorphological evolution of Montserrat (West Indies): importance of flank collapse and erosional processes. J Geol Soc 161(1):147–160

Lebas E, Le Friant A, Boudon G, Watt SFL, Talling PJ, Feuillet N, Deplus C et al (2011) Multiple widespread landslides during the long-term evolution of a volcanic island: insights from high-resolution seismic data, Montserrat, Lesser Antilles. Geochem Geophys Geosyst 12(5):Q05006

Longpré MA, Chadwick JP, Wijbrans J, Iping R (2011) Age of the El Golfo debris avalanche, El Hierro (Canary Islands): new constraints from laser and furnace $^{40}Ar/^{39}Ar$ dating. J Volcanol Geoth Res 203:76–80

Masson DG, Watts A, Gee MJ, Urgeles R, Mitchell N, Le Bas T, Canals M (2002) Slope failures on the flanks of the western Canary Islands. Earth-Sci Rev 57(1-2):1–35. doi:10.1016/S0012-8252(01)00069-1

Masson DG, Harbitz CB, Wynn RB, Pedersen G, Løvholt F (2006) Submarine landslides: processes, triggers and hazard prediction. Philos Trans R Soc Lond A: Math Phys Eng Sci 364:2009–2039. doi:10.1098/rsta.2006.1810

Urgeles R, Canals M, Masson D (1997) The most recent megalandslides of the Canary Islands: El Golfo debris avalanche and Canary debris flow. J Geophys Res 102:305–320

Ward SN, Day S (2001) Cumbre Vieja volcano – potential collapse and tsunami at La Palma, Canary Islands. Geophys Res Lett 28:3397–3400

Watt SFL, Talling PJ, Vardy ME, Masson DG, Henstock TJ, Hühnerbach V, Minshull T et al (2012) Widespread and progressive seafloor-sediment failure following volcanic debris avalanche emplacement: landslide dynamics and timing offshore Montserrat, Lesser Antilles. Mar Geol 323–325:69–94. doi:10.1016/j.margeo.2012.08.002

Weaver PPE, Wynn RB, Kenyon NH, Evans J (2000) Continental margins sedimentation, with special reference to the north-east Atlantic margin. Sedimentology 47:239–256

Wynn RB, Masson DG (2003) Canary island landslides and tsunami generation: can we use turbidite deposits to interpret landslide processes. In: Locat J, Mienert J (eds) Submarine mass movements and their consequences. Kluwer Academic, Dordrecht, pp 325–332

Chapter 9
New Insights on Failure and Post-failure Dynamics of Submarine Landslides on the Intra-slope Palmarola Ridge (Central Tyrrhenian Sea)

Daniele Casalbore, Alessandro Bosman, Francesco L. Chiocci, Michela Ingrassia, Leonardo Macelloni, Andrea Sposato, and Eleonora Martorelli

Abstract Newly collected multibeam and seismic data on the intra-slope Palmarola ridge show widespread pockmarks and landslide-related morphologies along its flanks. In detail, two main types of slope failures were identified: disintegrative-like and cohesive like landslides. The first type is characterized by a complex of small, nested scars affecting the steep and tectonically-controlled eastern flank of the ridge, suggesting a genesis related to retrogressive processes. The cohesive landslides affect the northern flank of the ridge and are characterized by larger scars, where material was not completely evacuated, and well-defined debris deposits at their base, with the development of pressure ridges. Tectonic activity and slope gradients represent the main controlling factors for the development of instabilities; moreover, we noted a relationship between pockmarks and landslide scars.

D. Casalbore (✉) • A. Bosman • A. Sposato • E. Martorelli
IGAG-CNR, Rome, Italy
e-mail: daniele.casalbore@igag.cnr.it

F.L. Chiocci
Sapienza, University of Rome, Rome, Italy

IGAG-CNR, Rome, Italy

M. Ingrassia
Sapienza, University of Rome, Rome, Italy

L. Macelloni
IGAG-CNR, Rome, Italy

NIUST, University of Mississippi, University, USA

G. Lamarche et al. (eds.), *Submarine Mass Movements and their Consequences*,
Advances in Natural and Technological Hazards Research 41,
DOI 10.1007/978-3-319-20979-1_9

9.1 Introduction

In the last years, the Italian marine geology community in collaboration with the Department of Civil Protection realized a big effort to extensively map most of the Italian continental margins in order to accomplish a first geohazard assessment of these areas (Chiocci and Ridente 2011). Particular attention has been devoted to the study of submarine flanks of active insular volcanoes (e.g. Chiocci and De Alteriis 2006; Casalbore et al. 2011; Romagnoli et al. 2012), where mass-movements are extremely common, ranging at different spatial and temporal scale. Recent investigations of Pontine Archipelago (Tyrrhenian Sea; Fig. 9.1 inset) have shown that mass-wasting processes are also widespread on the submarine flanks of ancient volcanoes, such as at Ventotene edifice (Casalbore et al. 2014) or nearby western Pontine Archipelago and its continental slope (Chiocci et al. 2003). The aim of this paper is to document for the first time extensive landslide scars and associated deposits along the flanks of the intra-slope Palmarola ridge located in the north-west of the Pontine Archipelago (Fig. 9.1, inset). We also discuss the variability of the identified landslide-related morphologies, providing some hints on their failure and post-failure behaviour.

9.2 Geological Setting

The Pontine Archipelago is located about 35 km offshore the Latium coastline (Fig. 9.1 inset) and is formed by two groups of volcanic islands: the western group Palmarola, Ponza and Zannone Islands and the eastern group Santo Stefano and Ventotene Islands. The two groups are characterized by different volcanological

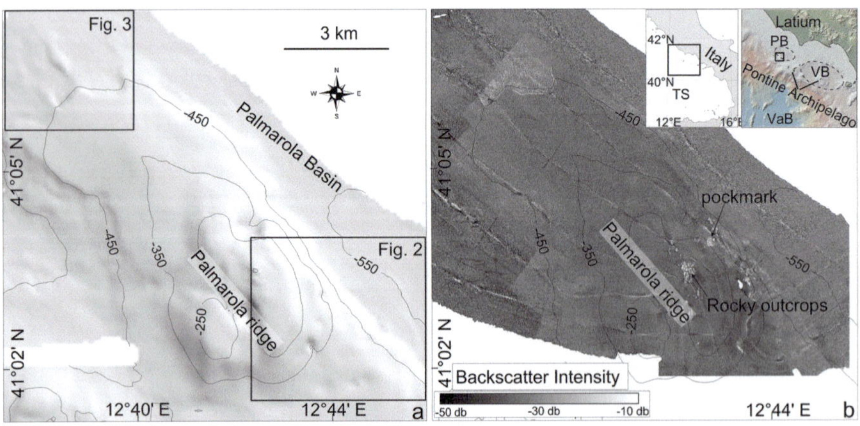

Fig. 9.1 Shaded relief (**a**) and backscatter (**b**) map of the Palmarola ridge, with location of Figs. 9.2 and 9.3 (*black box*); contours are in meters. The inset boxes show the location of the study area. *TS* Tyrrhenian Sea, *VB* Ventotene, *PB* Palmarola basin, *VaB* Vavilov basin

and structural evolution. In detail, the western group is located on a structural high developed during Plio-Pleistocene and separating the Palmarola and Ventotene basins (Fig. 9.1b inset, Zitellini et al. 1984). These islands were formed during two main volcanisms phases, dated 4.2–2.9 Ma and 1.6–0.9 Ma (Cadoux et al. 2005). Seismic activity is generally low with frequent but weak earthquakes (Favali et al. 2004).

North to Palmarola Island, lies the NW-SE oriented Palmarola ridge that limits, seaward, the Palmarola basin, where about 1000 m of Plio-Quaternary deposits are present (Zitellini et al. 1984). This intra-slope ridge is made up of folded Pleistocene sediments lying above a faulted basement (Marani and Zitellini 1986). Pleistocene normal faults affect ridge flanks (Zitellini et al. 1984).

9.3 Data and Methods

The data used for this work were collected during an oceanographic cruise carried out aboard the Research Vessel *Urania* (CNR) in June 2014. Multibeam data were acquired with a Kongsberg EM710 working at a frequency of 70–100 kHz. Details on data acquisition and processing are described in Casalbore et al. (2014). For this work, multibeam data were gridded to obtain a Digital Terrain Model with a cell-size of 7 m (Fig. 9.1a).

Multibeam backscatter signal was processed through SIPS SST (Side Scan Tools), applying radiometric and geometric corrections to the data. The multibeam backscatter mosaic was realized with a pixel size of 3 m (Fig. 9.1b). Details on data processing are described in Bosman et al. (2015).

High-resolution seismic reflection profiles were acquired with Sparker EG&G 1,5 KJ and Chirp III Datasonic profiler (frequency of 2–20 kHz); data were processed and interpreted through GeoSuite AllWorks software. A sound velocity of 1500 m/s was used for the water column and the first tens of meters of sediments for the time-depth conversion, whereas a sound velocity of 1800 m/s was used for deeper sediments.

9.4 Results

9.4.1 General Morphology of Palmarola Ridge

The Palmarola ridge (PR hereafter) is located between 160 and 510 m water depth (Fig. 9.1) and is elongated about 12 km in the NW-SE direction. The PR raises for about 230 m with respect to the surrounding seafloor and covers an area of 28 km^2. The general morphology of the ridge is quite smooth, except for the occurrence of pockmarks and landslide-related features (Figs. 9.2 and 9.3). Backscatter data also shows that medium-low backscatter values generally characterize the PR, except for a few small areas (Fig. 9.1b).

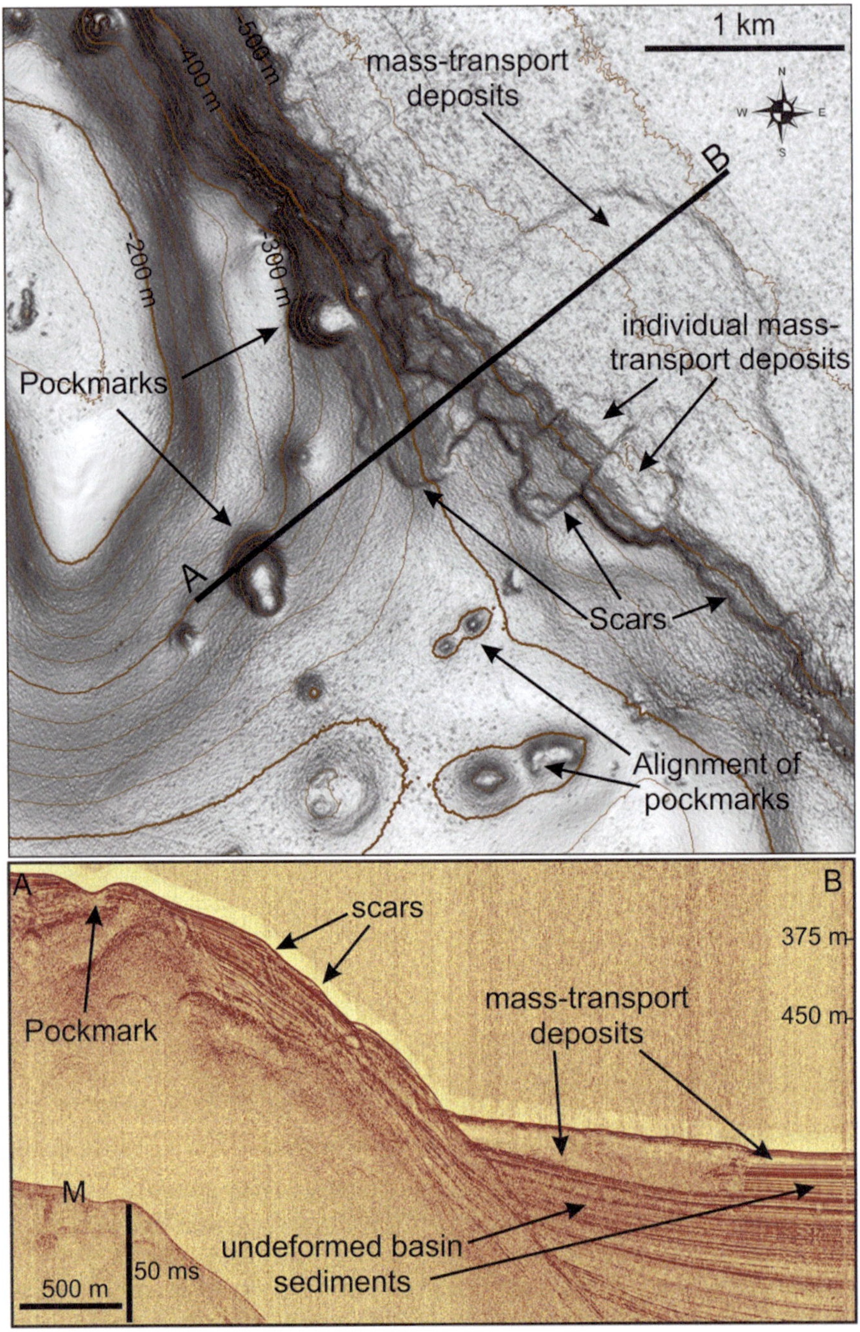

Fig. 9.2 Shaded relief map (azimuth of 90° and altitude of 90°) of the south-eastern flank of Palmarola ridge, largely affected by a complex of nested scars. A thick mass-transport deposit is present at the base of the flank, as recognizable on Sparker profile A-B and on multibeam data

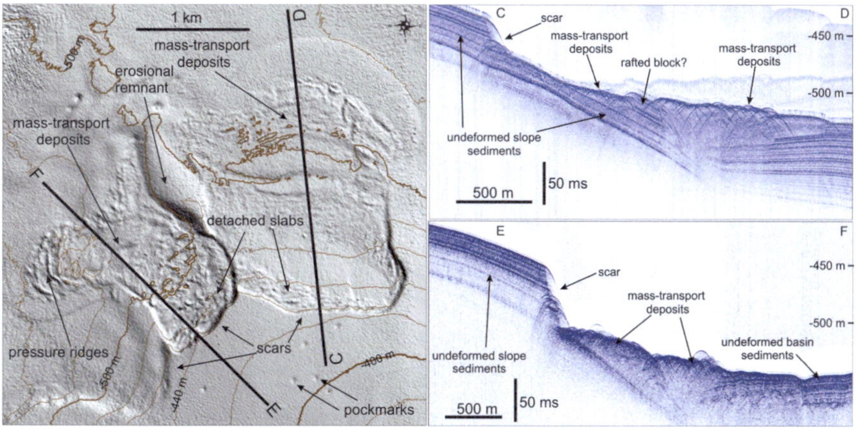

Fig. 9.3 *Left*: shaded relief map (azimuth of 90° and altitude of 90°) of the northern flank of Palmarola ridge, where two slide scars and related mass-transport deposits are present. *Right*: two CHIRP profiles show the erosive truncation of seismic reflectors at the scar headwall as well as the chaotic and semitransparent mass-transport units

Large pockmarks are present at the PR summit and in the southern part, at water depth between 200 and 400 m (Fig. 9.2). Pockmarks are variable in size, being 85–830 m wide, 90–1000 m long, and 9–147 m deep. Most of them are isolated, showing a circular plan-view shape, whereas some of them are aligned along WSW-ESE and SW-NE directions (Fig. 9.2). Small pockmarks, having an average diameter of 60 m, circular shape and negative relief of a few meters are present in the northern part of the PR (Fig. 9.3).

9.4.2 Landslide Scars and Deposits

Several landslide scars and related deposits were identified along the flanks and at the base of the PR between −350 and −520 m. The landslide morphologies can be divided into two main types:

(a) The first type (T1) is characterized by a complex of small (100-m wide) nested scars that mostly affect the steep eastern flank of PR, where slope gradients range between 12 and 20° (Fig. 9.2). Here, some scars show sidewalls roughly aligned to the chain of pockmarks. Except for a few cases, individual mass-transport deposits (Fig. 9.2) directly related to landslide scars are not recognizable on swath bathymetry. In contrast, a 2.5 km wide and smooth mass-transport deposits (i.e., lacking of any block features), is present at the base of the flank, covering an area of about 2.5 km^2 (Fig. 9.2)

Mass-transport deposits are seismically semi-transparent, with a significant seaward increasing thickness (from 20 to 50 m; Fig. 9.2). By comparing the scars and deposits surfaces, the mass-transport unit is not readily reconciled volumetrically with the material mobilized from each of the recognizable scars. This mass-transport unit has been interpreted as debris deposits related to an older scar, whose morphology is no longer recognizable on the present-day bathymetry.

(b) The second type (T2) is characterized by well-defined scars and associated deposits that affected the N flank of PR, where slope gradients range between 4 and 10°. The scars show marked headwalls, with heights of 20–25 m and slope gradients ranging between 28 and 44°. Both scars are characterized in their upper part by a series of step-forming inclined and detached slabs, up to 220 m long and 50 m wide, oriented perpendicular to the direction of slide motion (Fig. 9.3).

In detail, the eastern scar is roughly 1.5 km wide and 1 km long; related mass-transport deposits affect an area of 2.8 km^2 and show an uneven superficial morphology, characterized by a series of transverse ridges with different crest to crest distance (between 50 and 200 m), and resting above a smooth distal lobe. On CHIRP profiles a 10–25 m thick semi-transparent unit, with several hyperbolic echoes corresponds to the mass-transport deposits (profile C-D of Fig. 9.3). However, locally, deposits with a well-preserved original stratigraphy are also recognizable. The western scar is 800 m wide and 660 m long; landslide deposits affect an area of 1.3 km^2 and are characterized by an uneven morphology, made up by several blocks and elongated ridges with no recognizable orientation pattern. In the distal part, where landslide deposits are not confined by the headwall scar, they spread out in different directions, with a main tongue extending obliquely to the scar towards a WSW direction and following the seafloor dip. This tongue is characterized by some arcuate ridges, 400 m long and 5 m high, oriented perpendicular to the slide movement. Mass-transport deposits are acoustically characterized by a 10–20 m thick semi-transparent unit with hyperbolic echoes (profile E-F of Fig. 9.3).

An additional possibly older scar, similar in size to the previous ones, is present in the westernmost part of the PR (Fig. 9.3).

9.5 Discussions and Conclusions

The high-resolution morpho-bathymetric and seismic data depicted a quite large spectrum of instability processes that occur in a narrow area around a large ridge bordering the tectonically-controlled Palmarola basin. According to their morphology, internal architecture and occurrence of well-preserved slide deposits, two main types of instability phenomena can be distinguished: disintegrative-like and cohesive-like landslides (sensu McAdoo et al. 2000), corresponding to T1 and T2 respectively. The different behavior of the slides has been mainly related to differences in slope gradients, sediment type and trigger mechanisms (e.g. McAdoo et al. 2000; Yang et al. 2006; Minisini et al. 2007; Casas et al. 2011). T1 failures occur on the eastern flank of PR, and are characterized by a complex of small, nested scars, suggesting a genesis related to sequential instability events with retrogressive evolution (e.g. Mulder and Cochonat 1996; Prior and Suhayda 1979). These features are similar to those observed along the flanks of insular and oceanic islands (e.g. Casalbore et al. 2011; Quartau et al. 2014). The lacking of well-preserved mass-transport deposits directly related to landslide scars might be due to the fact that slided sediment either lost cohesion during failure or failed as a cohesive mass but disintegrated downslope due to the low mobilized volume and high energy (steep gradients). In both cases, landslides evolved into some kind of sedimentary gravity flows (Locat and Lee 2002). In fact, only for those scars located in the lower part of the steep PR flank, the related mass-transport deposits are still recognizable.

One of the main predisposing factors for their formation might be the steep gradients that would favor the development of small slides. These latter seem to occur more frequently on steep slopes than on gentler ones, where sediments can accumulate over large areas and for long time spans (e.g. McAdoo et al. 2000; Huhnerbach and Masson 2004). As the eastern flank of PR is tectonically-controlled, the possible role of tectonics for development of frequent slope failures can be also envisaged. In addition, the presence of pockmarks at or close to the landslide scars could suggest a possible relationship between the two processes.

Contrary to T1, T2 failures are characterized by larger scars and well-defined mas-transport deposits. These landslides are present in the northern part of PR (Fig. 9.3), where slope gradients are markedly lower than the eastern flank of PR. Similarly to the previous type, pockmarks are found close to the landslide scars and one of main predisposing factor might be represented by tectonic activity. In the upper part of the scar area, we observe the presence of a series of step-forming inclined and detached slabs, that can be interpreted as extensional ridges (Lastras et al. 2006). These features suggest that the mobilized sediment was not fully evacuated from the scar area and consists of material, which did not disintegrate completely. A possible rafted block identified in the eastern slide (profile C-D in Fig. 9.3), where the original slope stratigraphy was preserved, is another evidence of the limited deformation of deposits. Likewise, the acoustic character of the well-defined mass-transport units indicates that the moving sediment maintained a high degree of shear strength (e.g. Frey-Martínez et al. 2006; Minisini et al. 2007). Another evidence of the cohesive behaviour is the presence of the arcuate ridges on the distal part of the deposits (Fig. 9.3). These can be interpreted as pressure ridges produced by the loading and bulldozing of basin floor sediment (e.g. Lastras et al. 2006; Moscardelli et al. 2006).

The mobility of the T2 slided sediments has been inferred by measuring their apparent coefficient of friction (estimated by the ratio of maximum collapse height and runout distance, Legros 2002 and reference therein). It is 0.03 and 0.06 for the eastern and western landslides, respectively. These values are higher than most of known large-scale submarine mass-transport deposits and partially overlap the field of subaerial slides (Legros 2002). The possible explanation for the low-mobility of the studied landslides might be the relevant decrease of the slope gradients in this peculiar intra-slope setting as well as the remoulding of basin sediment with the formation of compressive features, implying the loss of momentum and decreased runout.

In summary, the present study shows that the Palmarola Ridge is largely affected by different spatial scale instability phenomena. These phenomena occur on a relatively narrow area but show a wide variability of morphologies, reflecting the different failure and post-failure behaviour. However, the acquisition of additional seismic profiles and cores is required to better understand the lithology and rheology of the different instabilities as well as for dating the sliding events. The available data suggest that most of landslides are relatively recent, because the scars are morphological "fresh" and the mass-transport deposits are covered by a thin or absent hemipelagic sedimentation, despite high-sedimentation rate in the Palmarola basin.

Acknowledgments The captain Vincenzo Lubrano and crew of R/V Urania are gratefully acknowledged as well as the people who participated in the sea survey. We gratefully acknowledge Sebastian Krastel, David Casas and Valeria Lo Presti for their suggestions that improved the quality of the paper.

This research was funded by National Research Council and carried out in the framework of "RITMARE" Project (SP4WP2 and WP1). L.M. has been supported through the Marie Curie Fellowship co-funded by the European Union under FP7-People – co-funding of Regional, National and International Programmes, GA n. 600407 and RITMARE Flagship Project.

References

Bosman A, Casalbore D, Anzidei M, Muccini F, Carmisciano C, Chiocci F (2015) The first ultra-high resolution marine digital Terrain model of the shallow-water sector around Lipari Island (Aeolian archipelago, Italy). Ann Geophys (in press)

Cadoux A, Pinti DL, Aznar C et al (2005) New chronological and geochemical constraints on the genesis and geological evolution of Ponza and Palmarola volcanic islands (Tyrrhenian Sea, Italy). Lithos 81:121–151

Casalbore D, Romagnoli C, Bosman A, Chiocci FL (2011) Potential tsunamigenic landslides at Stromboli Volcano (Italy): insight from marine DEM analysis. Geomorphology 126(1–2):42–50

Casalbore D, Bosman A, Martorelli E, Sposato A, Chiocci FL (2014) Mass wasting features on the submarine flanks of Ventotene volcanic edifice (Tyrrhenian Sea, Italy). In: Krastel S et al (eds) Submarine mass movements and their consequences, vol 37, Advances in natural and technological hazards research. Springer, Cham, pp 261–269. doi:10.1007/978-3-319-00972-8

Casas D, Ercilla G, Yenes M, Estrada F, Alonso B, García M, Somoza L (2011) The Baraza slide: model and dynamics. Mar Geophys Res 32(1–2):245–256

Chiocci FL, De Alteriis G (2006) The Ischia debris avalanche: first clear submarine evidence in the Mediterranean of a volcanic island prehistorical collapse. Terra Nova 18:202–209

Chiocci FL, Ridente D (2011) Regional-scale seafloor mapping and geohazard assessment. The experience from the Italian project MaGIC (Marine Geohazards along the Italian Coasts). Mar Geophys Res 32(1–2):13–23

Chiocci FL, Martorelli E, Bosman A (2003) Cannibalization of a continental margin by regional scale mass wasting: an example from the central Tyrrhenian Sea. In: Locat J, Miniert J (eds) Submarine mass movements and their consequences, vol 19, Advances in natural and technological hazards research. Springer, Dordrecht, pp 409–416

Favali P, Berenzoli L, Maramai A (2004) Review of the Tyrrhenian Sea seismicity: how much is still to be known? Mem Descr Carta Geol d'It XLIV:57–70

Frey-Martínez J, Cartwright J, James D (2006) Frontally confined versus frontally emergent submarine landslides: a 3D seismic characterisation. Mar Petrol Geol 23(5):585–604

Hühnerbach V, Masson DG (2004) Landslides in the North Atlantic and its adjacent seas: an analysis of their morphology, setting and behaviour. Mar Geol 213(1):343–362

Lastras G, Canals M, Amblas D, Ivanov M, Dennielou B, Droz L, Akhmetzhanov A, TTR-14 LEG 3 Shipboard Scientific Party (2006) Eivissa slides, western Mediterranean Sea: morphology and processes. Geo-Mar Lett 26(4):225–233

Legros F (2002) The mobility of long-runout landslides. Eng Geol 63(3):301–331

Locat, Lee (2002) Submarine landslides: advances and challenges. Can Geotech J 39:193–212

Marani M, Zitellini N (1986) Rift structures and wrenc tectonics along the continental slope between Civitavecchia and Circeo. Mem Soc Geol Ital 35:453–457

McAdoo BG, Pratson LF, Orange DL (2000) Submarine landslide geomorphology, US continental slope. Mar Geol 169:103–136

Minisini D, Trincardi F, Asioli A, Canu M, Foglini F (2007) Morphologic variability of exposed mass-transport deposits on the eastern slope of Gela Basin (Sicily channel). Basin Res 19(2):217–240

Moscardelli L, Wood L, Mann P (2006) Mass-transport complexes and associated processes in the offshore area of Trinidad and Venezuela. AAPG Bull 90(7):1059–1088

Mulder T, Cochonat P (1996) Classification of offshore mass movements. J Sediment Res 66(1):43–57

Prior DB, Suhayda JN (1979) Application of infinite slope analysis to subaqueous sediment instability, Mississippi Delta. Eng Geol 14(1):1–10

Quartau R, Hipolito A, Romagnoli C, Casalbore D, Madeira J, Tempera F, Roque C, Chiocci FL (2014) The morphology of insular shelves as a key for understanding the geological evolution of volcanic islands: insights from Terceira Island (Azores). Geochem Geophys Geosyst 15(5):1801–1826

Romagnoli C, Casalbore D, Chiocci FL (2012) La Fossa Caldera breaching and submarine erosion (Vulcano island, Italy). Mar Geol 303:87–98

Yang S, Solheim A, Kvalstad TJ, Forsberg CF, Schnellmann M (2006) Behaviour of the sediments in the Storegga Slide interpreted by the steady state concept. Norsk Geol Tidsskr 86(3):243–253

Zitellini N, Marani M, Borsetti M (1984) Post-orogene tectonic evolution of Palmarola and Ventotene basins (Pontine Archipelago). Mem Soc Geol Ital 27:121–131

Chapter 10
Assessment of Canyon Wall Failure Process from Multibeam Bathymetry and Remotely Operated Vehicle (ROV) Observations, U.S. Atlantic Continental Margin

Jason D. Chaytor, Amanda W.J. Demopoulos, Uri S. ten Brink, Christopher Baxter, Andrea M. Quattrini, and Daniel S. Brothers

Abstract Over the last few years, canyons along the northern U.S. Atlantic continental margin have been the focus of intensive research examining canyon evolution, submarine geohazards, benthic ecology and deep-sea coral habitat. New high-resolution multibeam bathymetry and Remotely Operated Vehicle (ROV) dives in the major shelf-breaching and minor slope canyons, provided the opportunity to investigate the size of, and processes responsible for, canyon wall failures. The canyons cut through thick Late Cretaceous to Recent mixed siliciclastic and carbonate-rich lithologies which impart a primary control on the style of failures observed. Broad-scale canyon morphology across much of the margin can be correlated to the exposed lithology. Near vertical walls, sedimented benches, talus slopes, and canyon floor debris aprons were present in most canyons. The extent of these features depends on canyon wall cohesion and level of internal fracturing, and resistance to biological and chemical erosion. Evidence of brittle failure over different spatial and temporal scales, physical abrasion by downslope moving flows, and bioerosion, in the form of burrows and surficial scrape marks provide insight into the modification processes active in these canyons. The presence of sessile fauna,

J.D. Chaytor (✉) • U.S. ten Brink
U.S. Geological Survey, Woods Hole Coastal and Marine Science Center, Woods Hole, MA, USA
e-mail: jchaytor@usgs.gov

A.W.J. Demopoulos
U.S. Geological Survey, Southeast Ecological Science Center, Gainesville, FL, USA

C. Baxter
Department of Ocean Engineering, University of Rhode Island, Narragansett, RI, USA

A.M. Quattrini
Cherokee Nation Technology Solutions, contracted to the US Geological Survey, Southeast Ecological Science Center, Gainesville, FL, USA

D.S. Brothers
U.S. Geological Survey, Pacific Coastal and Marine Science Center, Santa Cruz, CA, USA

G. Lamarche et al. (eds.), *Submarine Mass Movements and their Consequences*, Advances in Natural and Technological Hazards Research 41,
DOI 10.1007/978-3-319-20979-1_10

including long-lived, slow growing corals and sponges, on canyon walls, especially those affected by failure provide a critical, but as yet, poorly understood chronological record of geologic processes within these systems.

10.1 Introduction

Submarine canyon systems are pervasive morphologic features of the U.S. Atlantic continental margin (USAM) (Fig. 10.1). Major shelf-sourced (also referred to as shelf-breaching) canyons, such as Oceanographer, Hudson, and Norfolk canyons, are the primary conduits of sediments from the shelf and shallow slope to the deep-sea, especially during periods of lowered sea-level. The walls of these shelf-breaching canyon systems, and to some extent the smaller but more numerous slope-sourced canyons, not only expose a broad cross section of the continental margins' long geologic history, but also reveal the nature of the competing constructive and destructive processes that shape the margin over short- and long-time scales. Understanding the nature of destructive processes in submarine canyons is an important component of evaluating regional geohazard potential.

Strong currents enhancing the delivery of organic carbon, abundant exposed hard substrate, and protective complex morphology in submarine canyons facilitate the development of diverse faunal assemblages (Schlacher et al. 2007; McClain and Barry 2010; Vetter et al. 2010). Unfortunately, little baseline information exists

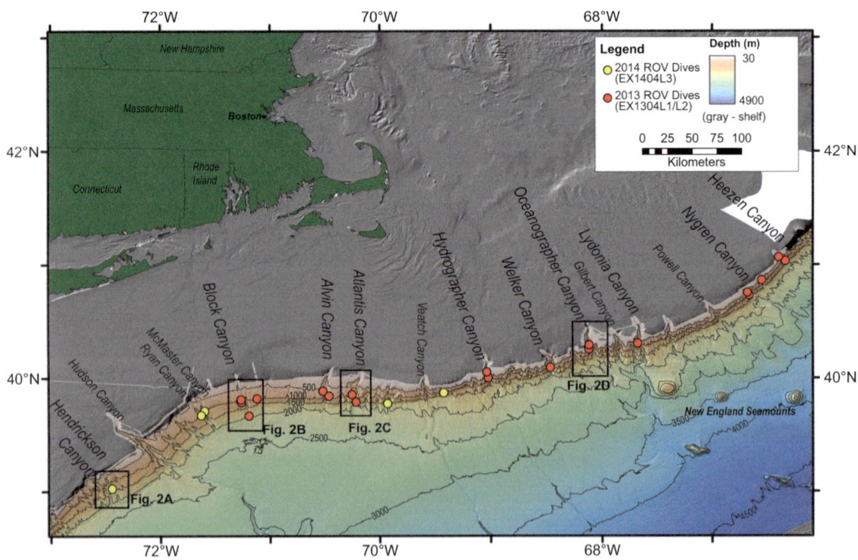

Fig. 10.1 Bathymetry map of the northern U.S. Atlantic margin (USAM) showing the locations of the major canyon systems and dives carried out in 2013 and 2014 by the NOAA Ship *Okeanos Explorer*/ROV Deep Discoverer. Contour interval is 500 m

from this region on which to robustly determine the drivers of faunal community changes and the relative roles of variables such as food availability, water mass conditions, and substrate stability. The presence, composition, size, density, and orientation of sessile fauna, particularly slow growing corals and sponges within canyons can provide clues to the long term stability of canyon walls and to the spatial extent and frequency of geologically-induced perturbations.

Here we provide a brief overview of the nature of failure and erosion processes within 14 submarine canyons along a ~600 km long segment of the USAM. We also discuss qualitatively the role of benthic communities in, and their response to, these processes. We couple high-resolution bathymetry with visual observations collected using a two-body ROV system in order to understand the geologic, physical and biologic controls, dimensions, and spatial pattern of these failure processes in the canyons.

10.2 Data

Visual investigations of the canyons detailed in this study were made during three tele-presence-enabled exploration legs of the NOAA Ship *Okeanos Explorer* in 2013 (EX1304L1 and EX1304L2) and 2014 (EX1404L3) using the Deep Discoverer (D2) ROV and Seirios camera platform/depressor system. High-resolution video and audio from D2, Seirios and the *Okeanos* collected during the 23 dives used in this study (Fig. 10.1) were transmitted to shore for analysis by participating scientists in near-real time. No samples were collected during any of the dives.

An extensive database of high-resolution multibeam data compiled from surveys conducted over the last decade, including additional data collected during the 2013 and 2014 *Okeanos* legs, exists for the US Atlantic continental margin, from the shelf edge (~200 m) to abyssal depths (~4,500–5,000 m) (Andrews et al. 2013). Bathymetry data for individual dive locations were gridded to a resolution of 25 m or higher.

10.3 Results

10.3.1 Canyon Morphology and Exposed Lithologies

As with other continental margins, canyon morphology along the upper slope of the USAM is a function of pre-existing geometry and stratigraphy of the margin and modification by recent sedimentary processes (Brothers et al. 2013). The canyons investigated in this study fall within three different general morphologic categories, dendritic-parallel (e.g., Hydrographer, Oceanographer, and Heezen canyons; Fig. 10.2a), trellis (e.g. Block, Ryan/McMaster canyons; Fig. 10.2b), and a mixed/transitional category between dendritic/parallel and trellis (e.g., Alvin,

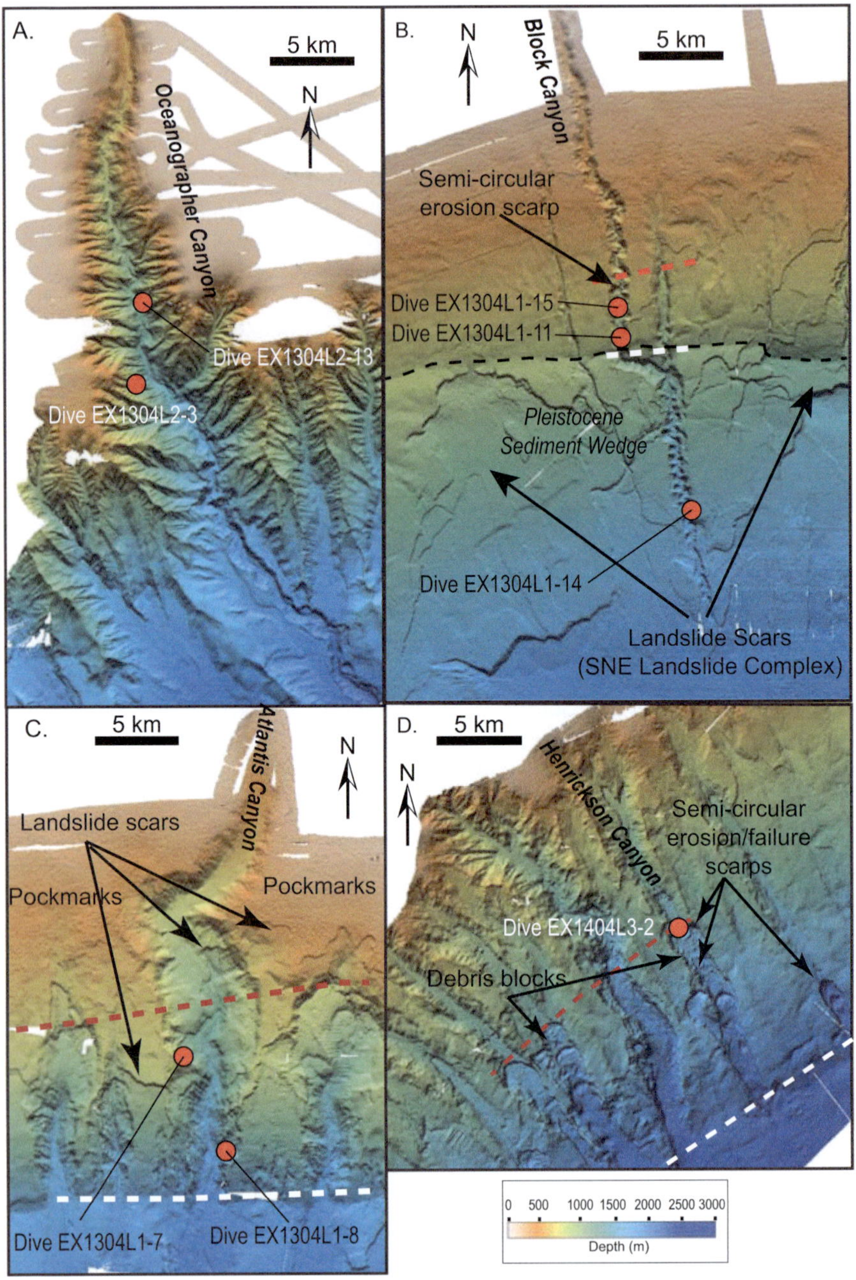

Fig. 10.2 Morphologic styles of canyons along the northern USAM (see Fig. 10.1 for locations), (**a**) Oceanographer Canyon, (**b**) Block Canyon, (**c**) Atlantis Canyon, (**d**) Hendrickson Canyon. *Red dashed line* indicates the location of the upper limit of exposure of Eocene chalk/carbonate rich lithologies, *white dashed line* marks the inferred lower limit

Atlantis, Hendrikson canyons; Fig. 10.2c, d). The canyons with strong trellis morphologies are limited to the southern New England margin segment where Pleistocene sediment was able to accumulate on the upper slope (Brothers et al. 2013) and intense mass wasting occurred in the late Pleistocene and Holocene (Twichell et al. 2009). Floors of the canyons are mostly U-shaped, partially filled by Pleistocene and Holocene hemipelagic sediments, displaced canyon wall material, and allochtonous mass transport deposits. Prominent circular scarps and increased canyon wall gradients appear to result from exposure and down-cutting through Eocene chalks and similar lithologies (Fig. 10.2b).

ROV dives explored canyon segments at depths between 580 and 2,100 m, focusing on steep (>30°) walls where rock outcrop and benthic communities would most likely be found. Based on existing samples and observations (Trumbull and McCamis 1967; Dillon and Zimmerman 1970; Weed et al. 1974; Valentine et al. 1980; McHugh et al. 1993), dives in the canyCanyon (Fig. 10.1) traversed exposures of interbedded siliciclastic and carbonate rocks (primarily chalks, some which appeared porcelanitic) of Eocene-Paleocene age. The deep dive in Block Canyon (~2,100 m) encountered Pleistocene-age gray mudstone/siltstones of the lower slope sediment wedge (O'Leary and Dobson 1992). Late Cretaceous (Maestrichtian) layered to massive mudstones/siltstones (Trumbull and McCamis 1967; Dillon and Zimmerman 1970; Weed et al. 1974; Ryan et al. 1978; Ryan and Miller 1981) were traversed in canyons between Hydrographer and Lydonia canyons. Mudstones and chalk(?), often heavily encrusted and Fe-Mn hydroxide coated, of Miocene to Cretaceous age (Weed et al. 1974; Ryan et al. 1978) were exposed along the walls in the canyons north of Lydonia. Hemipelagic drape thickness varied as a function of wall slope, location along the margin (distance from modern terrestrial sources), and depth.

10.3.2 Benthic Communities

Patches of unfailed sections of canyon walls and talus slopes-debris aprons with limited sediment drape were colonized by numerous species of sessile fauna, including corals, bivalves, anemones, and sponges. Scleractinians (*Desmophylum dianthus, Solenosmilia variabilis)*, octocorals (*Anthothela, Clavularia, Acanthogorgia*), sea pens, and bivalves (*Acesta* sp.) were often abundant under and around overhangs. Along some sections of the canyon walls, large coral colonies, including *Paragorgia arborea, Keratoisis sp., Paramuricea* spp., *and Primnoa resedaeformis* (Fig. 10.3a) were found. Crinoids (feather stars and sea lilies), crustaceans, octopods, sea pens, and ophiuroids (brittle stars) were found in association with the sessile fauna and on sedimented sections of the canyon walls. Chemosynthetic fauna (e.g., *Bathymodiolus* mussels, bacterial mats) were observed along the west wall of Nygren Canyon (~1,550 m water depth).

Fig. 10.3 (**a**) Dense population of pink and white *Paragorgia arborea* coral on Miocene(?) mudstone/siltstone, north wall of Heezen Canyon (~1,000 m water depth); (**b**) Talus slope on the east wall of Block Canyon composed of Pleistocene age semi-lithified mudstone (~2,000 m water depth); (**c**) Interbedded Eocene(?) chalk and porcellanite, east wall of Alvin Canyon (1,100 m water depth). The chalk is heavily bioeroded (burrows, borings, pits) while the porcellanite is relatively intact; (**d**) Collapse of angular slabs of porcellanite resulting from erosion of the underlying chalk. (**e**) Large (>5 m long) topple deposit of layered Cretaceous(?) age mudstone/siltstone from south wall of Heezen Canyon (~1,700 m water depth); (**f**) Spalling failure

10.3.3 Failure Processes and Erosion

The primary destructive processes occurring in USAM submarine canyons can be divided into two classes, (1) cohesive failure, and (2) erosion by non-cohesive sediment flows, dissolution, and biological activity. In the canyons, these processes act in continuum.

10.3.3.1 Cohesive Failure Processes

The cohesive nature of the consolidated and semi-consolidated canyon walls has led to rockfall being the primary failure process actively occurring in all canyons investigated. Although extensive talus slopes and base of slope debris aprons were traversed in several canyons, there was no clear indication that they were generated by a single catastrophic large-scale rock avalanche. Talus blocks varied in size and shape from pebble-sized "balls" of semi-consolidated mudstone to meter-scale tabular slabs and irregular-shaped boulders (Fig. 10.3b). Transport of the material from the steeper walls was largely unconfined, although confined flows of finer grained geologic and biologic material down rock exposures and sedimented slopes were observed. Sessile fauna attached to the canyon walls at the point of failure and in the downslope path of the failed material were dislodged and deposited with the debris blocks. Often, these fauna included dead scleractinian coral rubble and dislodged octocorals.

Erosion of less consolidated lithologies or those with chemistries susceptible to biological or chemical attack (e.g., chalks and other rocks with high $CaCO_3$ content) leads to undercutting and loss of support of overlying stronger units (Fig. 10.3c, d), driving a slow retrogressive process. Pre-existing joint sets and fractures and the development of surficial Fe-Mn hydroxide crusts may control the size and shape of the failed blocks and the directivity of failure progression along wall segments. Large blocks, some of which were composed of multiple layers and exceeding 5 m in length, were present in several canyons at the base of the walls (Fig. 10.3e). Detachment surfaces of these large blocks were not observed during 100–200 m transects up the canyon walls, suggesting that they possibly toppled from a significant height above.

Fig. 10.3 (continued) of surficial layers of Early Tertiary mudstone/siltstone, west wall of Oceanographer Canyon – the large antipatharian (*black coral*) and smaller octocorals, including *Paramuricea spp*., indicate areas of the wall that have remained stable over time; (**g**) Bioeroded spalling-failure detachment surface in Welker Canyon – the thickness of the failure appears to correspond to the depth of bioerosion of the unfailed sections of the wall; (**h**) A sea spider (Pycnogonida) with pits, burrows, scratches and other borings on the surface of an outcrop of Eocene chalk in Hendrickson Canyon; (**i**) Linear row of caves, west wall of a minor unnamed canyon east of Veatch Canyon; (**j**) Parallel, partially filled abrasion channels in Early Tertiary chalk on the west wall of Lydonia Canyon (~1,230 m water depth)

Spalling or exfoliation failure of surficial wall material (Fig. 10.3f, g) was observed in all canyons. The thickness of the removed material was commonly no more than a meter. In the $CaCO_3$-rich and chalky lithologies, spalling appeared to be limited to the depth of bioerosion of the outcrops. In extreme cases, large scale spalling of walls resulted in creation of isolated remnants of unfailed outcrop that were heavily encrusted with biological communities (Fig. 10.3d).

10.3.3.2 Erosion

In addition to detachment of large blocks of wall material, long-term movement of water through the canyons, biological activity, chemical dissolution, and periodic abrasion by turbidity currents and other sediment laden flows undoubtedly aid in the continuous destruction of the canyon walls. During the dives, strong bottom currents or suspension of sediment were rarely encountered, but the presence of sediment waves on canyon wall benches, smooth sculpting of outcrops, and the location and orientation of sessile fauna on exposed promontories, suggest episodically strong current activity.

Evidence of bioerosion (i.e. destruction of lithic substrate by direct action of organisms (Neumann 1966)), was pervasive in all investigated canyons. Bioerosion activities were observed primarily in the form of pits, scratches and scrapes, burrows and caves. Bioerosion affected all exposed lithologies, but was most intense in chalks and other rocks thought to have a significant $CaCO_3$ content. Pits, borings, and burrows were primarily circular with openings less than 10 cm in diameter and extended both into and laterally along outcrops (Fig. 10.3c, h). Excavation depths were difficult to determine visually, but were likely 10's of cm in most cases. Tool marks, such as scratches and scrapes were present at the openings of burrows and along "fresh" rock faces (Fig. 10.3h). Caves and cavities with wide openings and chambers that extended deep into outcrops (Fig. 10.3i) were found in a number of canyons, primarily associated with $CaCO_3$-rich lithologies. These caves were often home to crustaceans, fishes, and octopuses. Distinct ejecta piles were present at the mouths and downslope of many caves. It is unclear what controls the presence of caves, but bedding planes (Fig. 10.3i) and fractures perhaps provide zones of weakness that are more easily exploited.

Walls with extensive exposure of chalky lithologies also exhibited abrasion in the form of parallel, v-shaped channels caused by down-wall transport of sediment laden flows (Fig. 10.3j) similar to those described by McHugh et al. (1993). Across the series of dives, the phases of development of these rill-like morphologies were observed, from removal of drape sediments in either narrow linear or bifurcating patterns (creating a white line down the wall of the canyon), to deep linear incisions partially filled with sediment and larger debris (Fig. 10.3j). Where they intersected benches along the canyon walls, small depositional fans were present.

10.4 "Biomarkers" as Failure Timing and Magnitude Indicators

Submarine canyons along the USAM are complex bathymetric features constantly being shaped by geological, biological and oceanographic processes over different spatial and temporal scales. While geological and oceanographic processes are the dominant drivers of canyon evolution, the role of biological activity in the erosion of submarine rock outcrops has been identified as a significant contributor to the continued reshaping of submarine canyons (Warme et al. 1978; Hecker 1982). Bioerosion appears to have significant control over wall failure at all scales in USAM canyons through weakening of exposed rock faces. This weakening leads to localized spalling and removal of material from above and below layers less susceptible to differential erosion (e.g., porcelinite layers, Fig. 10.3c, d). Failure driven by this erosion leads to exposure of a "fresh" surface at the site of detachment as well as along the downslope travel and deposition path of the detached material.

The rate of occurrence, dimensions, and drivers of canyon wall failures are important, but poorly constrained, variables necessary for developing a better understanding of submarine canyon evolution. The intimate relationship between sessile fauna and the exposed substrate in submarine canyons provides a framework for determining the relative and, when physical samples are available, possibly the absolute age of these types of rockfall failures as well as provide evidence of past failure magnitude. The presence/absence, diversity, size and visual health of sessile communities on the canyon walls and displaced debris blocks provide indicators of relative timing of failure. Additionally, the extent of bioerosion (depth and percentage of outcrop affected by burrows, pits, etc.) provides another, but less easily constrained, age marker. As illustrated in Fig. 10.3, the presence of sessile fauna and differential bioerosion also provide robust constraints on dimensions of past failure events.

While visual analysis of sessile fauna and bioerosion is a very useful first-order relative aging technique, our understanding of the growth rates of deep-sea organisms, community structure and food-web dynamics, substrate preference, and larval dispersal mechanisms, and anthropogenic disturbance are limited in submarine canyon environments (McClain and Barry 2010; De Leo et al. 2010). Deep-sea corals are perhaps the most important organism to help estimate failure ages as they are firmly attached to the exposed substrate. Their orientation and location on substrates and morphology can provide insight into whether the underlying substrate has been stable overtime because certain species of corals exhibit specific growth forms and orientations, and settlement locations (Mortensen and Buhl-Mortensen 2005; Tong et al. 2013). Additionally, the dominant coral taxa found within these canyons are known to have slow growth rates, growing continuously for decades to thousands of years. Octocorals found in this region, including *Paragorgia arborea, Paramuricea* spp. and *Primnoa resedaeformis* have recorded life spans ranging from 80 to >600 years (Risk et al. 2002; Sherwood and Edinger

2009; Prouty et al. 2015). Maximum age estimates for the cup coral, *Desmophyllum dianthus*, are up to 200 years old (Risk et al. 2002). Characteristically slow linear extension/growth rates for these corals range from 0.01 to 2.6 cm/year; it may require at least a decade for some of these corals to be visible from an ROV. Thus, measurements and observations of coral size, shape, and location from visual surveys can provide a relative time estimate for failures.

10.5 Future Work

Interest in deep-sea habitat management is leading to an increasing availability of high-definition seafloor imagery for a number of regions along the USAM, Gulf of Mexico and Caribbean. These data are providing the opportunity to continue exploring the use of biological proxies for constraining failure characteristics of submarine escarpments (canyon walls, fault surfaces, landslide scarps) and in developing a better understanding of all the processes active in the evolution of submarine landscapes. Coupling of visual data with sampling of the exposed lithified and semi-consolidated substrate, adjacent sediments, and key benthic organisms in key locations will provide quantitative groundtruthing and allow further development of biological time markers.

Acknowledgments We would like to thank the captains and crews of the NOAA Ship Okeanos Explorer, ROV and telepresence engineers and scientists involved with Okeanos expeditions EX1304L1, EX1304L2, and EX1404L3. This manuscript benefited greatly from reviews provided by Laura Brother, Erika Lentz, Silvia Ceramicola, and Aaron Micallef. Any use of trade, product, or firm names is for descriptive purposes only and does not imply endorsement by the U.S. Government.

References

Andrews BD, Chaytor JD, ten Brink US, Brothers DS, Gardner JV (2013) Bathymetric terrain model of the Atlantic margin for marine geological Investigations. U.S. Geological Survey Open-File Report 2012–1266. U.S. Geological Survey, Washington D.C

Brothers DS, ten Brink US, Andrews BD, Chaytor JD (2013) Geomorphic characterization of the US Atlantic continental margin. Mar Geol 338:46–63

De Leo FC, Smith CR, Rowden AA, Bowden DA, Clark MR (2010) Submarine canyons: hotspots of benthic biomass and productivity in the deep sea. Proc R Soc Lond Ser B-Biol Sci 277:2783–2792

Dillon WP, Zimmerman HB (1970) Erosion by biological activity in two New England submarine canyons. J Sediment Petrol 40:542–547

Hecker B (1982) Possible benthic fauna and slope instability relationships. In: Saxov S, Nieuwenhuis JK (eds) Marine slides and other mass movements. Plenum Press, New York, pp 335–347

McClain CR, Barry JP (2010) Habitat heterogeneity, disturbance, and productivity work in concert to regulate biodiversity in deep submarine canyons. Ecology 91:964–976

McHugh CM, Ryan WB, Schreiber BC (1993) The role of diagenesis in exfoliation of submarine canyons. AAPG Bull 77:145–172
Mortensen P, Buhl-Mortensen L (2005) Morphology and growth of the deep-water gorgonians Primnoa resedaeformis and Paragorgia arborea. Mar Biol 147:775–788
Neumann AC (1966) Observations on coastal erosion in Bermuda and measurement of the boring rate of the sponge, Cliona lampa. Limnol Oceanogr 11:92–108
O'Leary DW, Dobson MR (1992) Southeastern New England continental rise: origin and history of slide complexes. In: Poag CW, de Grciansky PC (eds) Geologic evolution of Atlantic continental rises. van Nostrand Reinhold, New York, pp 214–265
Prouty NG, Fisher CR, Demopoulos AWJ, Druffel ER (2015) Growth rates and of deep-sea corals impacted by the Deepwater Horizon oil spill. Deep-Sea Res II Top Stud Oceanogr. doi:10.1016/j.dsr2.2014.10.021
Risk MJ, Heikoop JM, Snow MG, Beukens R (2002) Lifespans and growth patterns of two deep-sea corals: Primnoa resedaeformis and Desmophyllum cristagalli. Hydrobiologia 471:125–131
Ryan WBF, Miller EL (1981) Evidence of a carbonate platform beneath Georges Bank. Mar Geol 44:213–228
Ryan WBF, Cita MB, Miller EL, Hanselman D, Nesteroff WD, Hecker B, Nibbelink M (1978) Bedrock geology in New England submarine canyons. Oceanol Acta 1:233–254
Schlacher TA, Schlacher-Hoenlinger MA, Williams A, Althaus F, Hooper JNA, Kloser R (2007) Richness and distribution of sponge megabenthos in continental margin canyons off south-eastern Australia. Mar Ecol Prog Ser 340:73–88
Sherwood OA, Edinger EN (2009) Ages and growth rates of some deep-sea gorgonian and antipatharian corals of Newfoundland and Labrador Can. J Fish Aquat Sci 66:142–152
Tong RJ, Purser A, Guinan J, Unnithan V (2013) Modeling the habitat suitability for deep-water gorgonian corals based on terrain variables. Ecol Inform 13:123–132
Trumbull JVA, McCamis MJ (1967) Geological exploration in an east-coast submarine canyon from a research submersible. Science 158:370–372
Twichell DC, Chaytor JD, ten Brink US, Buczkowski B (2009) Morphology of La Quaternary submarine landslides along the U.S. Atlantic continental margin. Mar Geol 264:4–15
Valentine PC, Uzmann JR, Cooper RA (1980) Geology and biology of Oceanographer submarine canyon. Mar Geol 38:283–312
Vetter EW, Smith CR, De Leo FC (2010) Hawaiian hotspots: enhanced megafaunal abundance and diversity in submarine canyons on the oceanic islands of Hawaii. Mar Ecol 31:183–199
Warme JE, Slater RA, Cooper RA (1978) Bioerosion in submarine canyons. In: Stanley DJ, Kelling G (eds) Sedimentation in submarine canyons, fans, and trenches. Dowden, Hutchinson and Ross, Stroudsburg, pp 65–70
Weed EGA, Minard JP, Perry Jr WJ, Rhodehamel EC, Robbins EI (1974) Generalized pre-Pleistocene geologic map of the northern United States Atlantic continental margin. US Geological Survey Miscellaneous Investigations Series Map I-861. U.S. Geological Survey, Washington D.C

Chapter 11
The Chuí Megaslide Complex: Regional-Scale Submarine Landslides on the Southern Brazilian Margin

A. Tadeu dos Reis, Cleverson G. Silva, Marcus A. Gorini, Rafael Leão, Nara Pinto, Rodrigo Perovano, Marcos V.M. Santos, Josefa V. Guerra, Izabel K. Jeck, and Ana Angélica A. Tavares

Abstract The Brazilian Continental Shelf Survey Programme (LEPLAC) identified the occurrence of large-scale mass-transport deposits on the southernmost limit of the Brazilian margin, based mainly on analyses of acoustic imagery. The mass-transport deposits, named the Chuí Megaslide Complex, comprise a stack of large translational slides that spread from the shelf break ~650 km downslope to ~4,900 m water depth, cutting into Pliocene-Quaternary sedimentary successions and strongly affecting both the margin morphology and regional depositional processes. The main headwall scarp is U-shaped, 400–500 m high, and extends *c.* 80 km downslope as a large elongated evacuated scar, 50–85 km wide. Outside this main failure scar, external scarps evidence a large area of erosion and faulted blocks, indicating ongoing retrogressive sediment disruption. Slide masses occur as a combination of variably deformed failed masses and debris flows, covering an area of ~150,000 km^2. Main preconditioning parameters and the possible triggering mechanism for the Chui Megaslide Complex are likely a combination of a series of causative factors such as slope failure structurally-induced by gravity tectonics and high sediment influx into the shelf-edge and upper slope during the Early Miocene-Quaternary.

A.T. dos Reis (✉) • J.V. Guerra
School of Oceanography, Rio de Janeiro State University (UERJ), Rua São Francisco Xavier, 524, 4° Andar, Bl. E, Maracanã, Rio de Janeiro, RJ CEP 20.550-013, Brazil
e-mail: antonio.tadeu@pq.cnpq.br; tadeu.reis@gmail.com

C.G. Silva
Department of Geology and Geophysics, Fluminense Federal University (UFF), Av. Litorânea, s/n, Boa Viagem, Niterói, RJ CEP: 24.210-346, Brazil

M.A. Gorini • R. Leão • N. Pinto • R. Perovano • M.V.M. Santos
MAG – Mar, Ambiente e Geologia Serviços Ltda, Rio de Janeiro, Rua Visconde de Inhaúma, 37 - 21° andar, Centro - Rio de Janeiro, RJ CEP 20.091-007, Brazil

I.K. Jeck • A.A.A. Tavares
Directorate of Hidrography and Navigation (DHN), Brazilian Navy, Rua Barão de Jaceguaí - Ponta D'areia, Niterói, RJ CEP 24.048-900, Brazil

G. Lamarche et al. (eds.), *Submarine Mass Movements and their Consequences*, Advances in Natural and Technological Hazards Research 41,
DOI 10.1007/978-3-319-20979-1_11

11.1 Introduction and Backgrounds

The continental slope of the Southern Brazilian margin (Pelotas basin) is interrupted by pronounced morphological features, as for example the Santa Catarina Plateau, the Rio Grande Terrace and the prominent Miocene-Quaternary sedimentary feature known as the Rio Grande Cone – a prograding sedimentary wedge deformed by gravity tectonics (Zálan 2005; Pereira et al. 2010) (Figs. 11.1a, b). The continental slope is also affected by erosion of contour currents that transport Antarctic Intermediate Water, Upper Circumpolar Deep Water, North Atlantic Deep Water and Lower Circumpolar Deep Water (e.g., Preu et al. 2013). Besides that, the southward flowing Brazil Current, which becomes gradually deeper (thicker) and stronger as it moves southwestwards, roughly follows the orientation of the shelf break-upper slope as a contour current (Stramma and England 1999). The deep-water setting is under the influence of the deep Antartic Bottom Water circulation that comes freely as a bottom current from the south but is finally forced to form secondary currents and eddies in the water column across the Santa Catarina Plateau (Bassetto et al. 2000).

Throughout the Cenozoic, the persistent interplay between eustatic fluctuations and contour currents has significantly affected its architecture, as well the margin morphology (Pereira et al. 2010). As a result, drift deposits are ubiquitous across the entire margin, occurring either as large confined contourites, here named the Santa

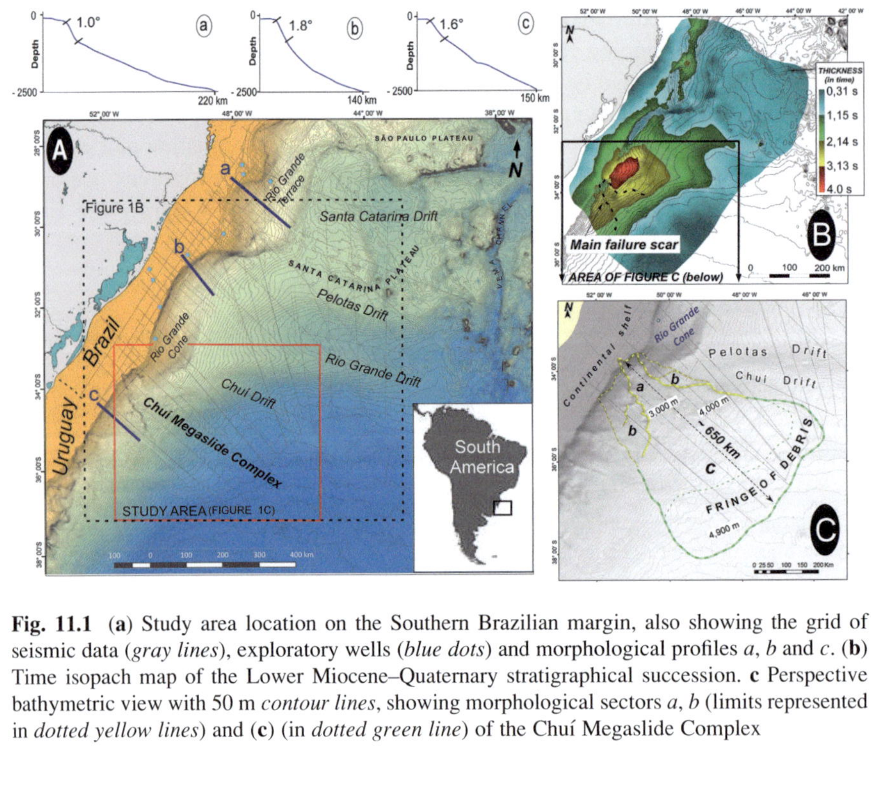

Fig. 11.1 (**a**) Study area location on the Southern Brazilian margin, also showing the grid of seismic data (*gray lines*), exploratory wells (*blue dots*) and morphological profiles *a*, *b* and *c*. (**b**) Time isopach map of the Lower Miocene–Quaternary stratigraphical succession. **c** Perspective bathymetric view with 50 m *contour lines*, showing morphological sectors *a*, *b* (limits represented in *dotted yellow lines*) and (**c**) (in *dotted green line*) of the Chuí Megaslide Complex

Catarina Drift, developed in the less subsided northern Pelotas Basin (Santa Catarina Plateau), or as gigantic elongated drifts across the southern Pelotas basin, which are deflected seawards as far as ~600 km perpendicular to the general lie of the margin, like the Rio Grande Drift, the Pelotas Drift (Pereira et al. 2010) and the Chuí Drift (Fig. 11.1a).

Only recently, during research projects within the scope of the Brazilian Continental Shelf Survey Programme (LEPLAC), features and deposits related to processes of slope disruption and mass-transport deposits were for the first time identified on the Southern Brazilian margin. These megaslides, grouped under the name of the Chuí Megaslide Complex, comprise a stack of large-scale submarine landslides (hereafter "landslides") spreading across the southernmost margin limits (Fig. 11.1a).

In the present paper our main objective is to introduce the Chuí Megaslide Complex, focusing our attention on the description of its geomorphological features and seismic architecture, as well as on the analysis of associated depositional features. We also present possible preconditioning factors and triggering mechanisms involved in the initiation and evolution of the megaslide complex.

11.2 Data and Methods

This study has been conducted using primarily three kinds of dataset: Along-track swath bathymetry, 3.5 kHz sub-bottom profiles and multichannel and single channel seismic data collected during LEPLAC (first and second phases). Extra data include 2D seismic lines made available by the Brazilian National Oil and Gas Agency (ANP), together with chronostratigraphic data from nine exploratory wells (Fig. 11.1a). Regional bathymetry was built by LEPLAC based on the compilation of information from various sources, including collected bathymetric data (swath and conventional bathymetry), the General Bathymetric Charts of the Oceans – GEBCO (www.gebco.net), PETROBRAS, and the Geophysical Data System – GEODAS (www.ngdc.noaa.gov/mgg/geodas).

11.3 Results and Discussions

11.3.1 Geomorphological Characterization

MTDs on the southern Brazilian margin are located on the southern flank of the Rio Grande Cone, spreading from the shelf break ~650 km downslope (Figs. 11.1b, c). Seabed features related to mass-transport processes can be divided into three main morphological sectors (**a**, **b** and **c**, in Fig. 11.1c), each one characterized by distinct morphologies and processes (Figs. 11.2 and 11.3).

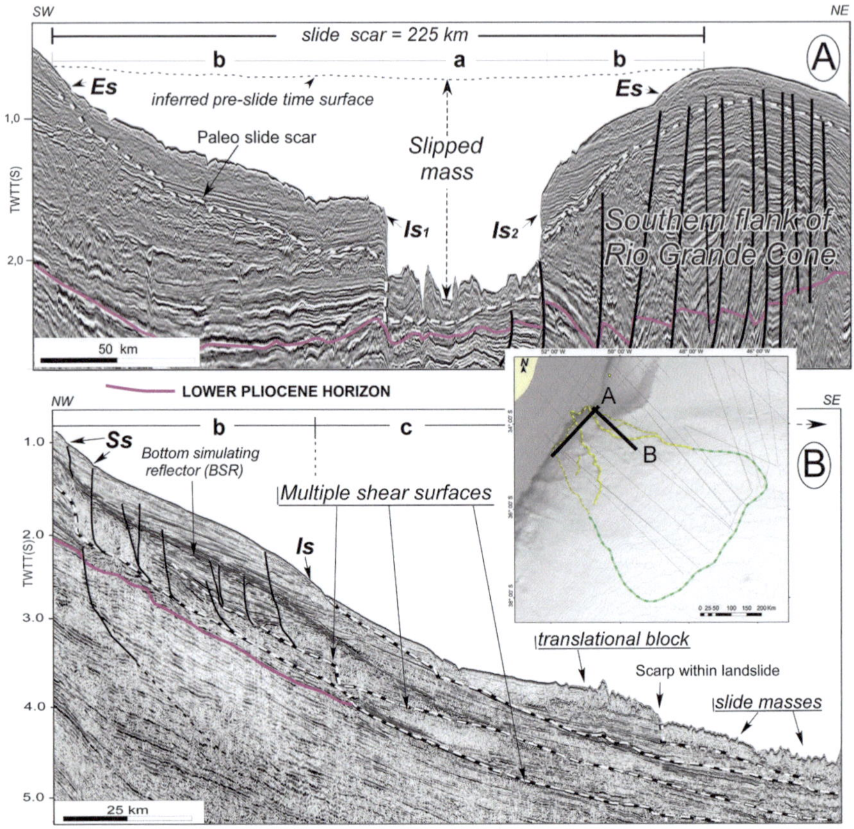

Fig. 11.2 Strike (**a**) and dip (**b**) seismic lines illustrating morphological features and seismic facies of the head area of the Chuí Megaslide Complex. *a*, *b* and *c* are morphological sectors across the submarine landslide complex; *Is* Internal scarps, *Ss* Secondary scarps

Morphological sectors **a** and **b** comprise the upslope erosive head area (Fig. 11.1c). Headwalls consist of multiple scarps which separate different levels of failure features between morphological sectors **a** and **b** (Fig. 11.2a). Morphological sector **a** is the main and deepest slide scar. It is a U-shaped failure scar with an along-slope orientation, extending ~180 km to ~3,000 m water depth, constituting a large morphological corridor some 50–85 km wide. Upslope it is bounded by internal scarps (**Is**), as high as 400–500 m, strongly affecting the margin morphology (Fig. 11.2a). Internal scarp **Is$_2$** lies in vertical continuity with listric faults that compose the gravity system of the Rio Grande Cone, thus constituting a lateral fault scarp (Fig. 11.2a). Downslope, relief of internal scarps is reduced to *c.* 100 m high (Fig. 11.2b). Morphological sector **a** is not totally devoid of displaced sediment and slide blocks, but its quite deep erosive depression (up to 1,100 m high) attests to a somewhat evacuated corridor that is obviously the main erosive area from which all MTDs are sourced from multiple-failure events (Fig. 11.2a). The morphological

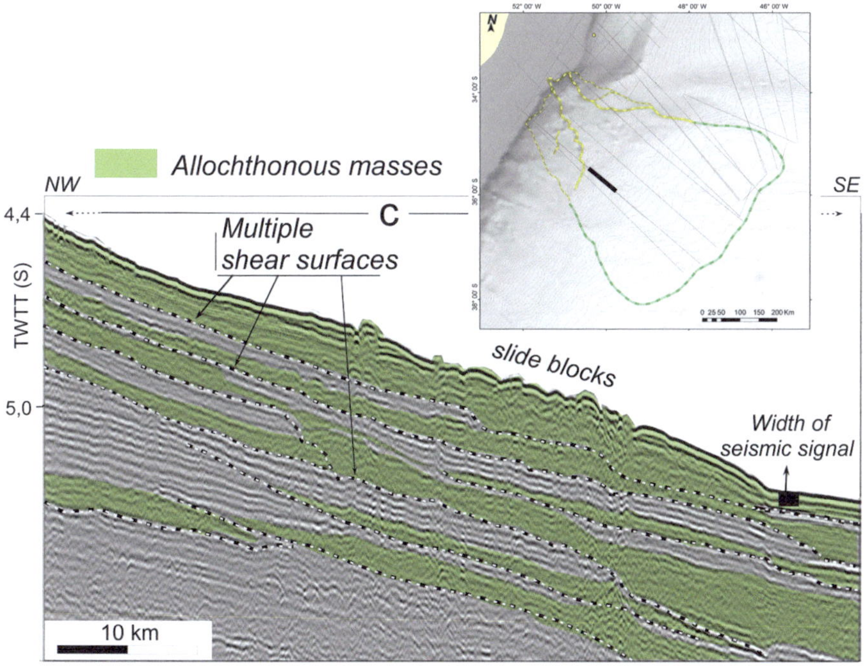

Fig. 11.3 Dip-oriented seismic line showing internal features and architectural elements of the Chuí Megaslide Complex

sector **b**, in its turn, comprises two separated areas: one along the northern flank of the main topographic low (sector **a**) and the other one along its southern flank (Figs. 11.2a, b). In both settings, sector **b** is bounded by external scarps (**Es**), located outside the main evacuated **a** slide scar, which exhibit reduced relief, as high as 100 m. A series of secondary inner scarps (**Ss**), up to *c.* 80 m high, give the sea-floor a general staircase appearance across sector **b** (Figs. 11.2b and 11.4a).

Depression **a** is open at its SE boundary and connects downslope with an unconfined slope area (sector **c**) dominated by an irregular rugged morphology up to tens of meters high, that tends to smooth downslopes (Fig. 11.1c). Sector **c** comprises a large area of MTDs with variable degrees of disruption, which spreads down to ~4,900 m water depth (sector **c** in Figs. 11.2b, 11.3 and 11.4b, c). The MTDs together make a total area of ~150,000 km^2 (Fig. 11.1c).

11.3.2 Seismic Architecture and Depositional Features

Failed masses move downslope along multiple flat-lying bed-parallel basal shear surfaces, indicating the occurrence of translational slide masses (Figs. 11.2b and 11.3). However, MTDs in different locations involve shear surfaces that follow

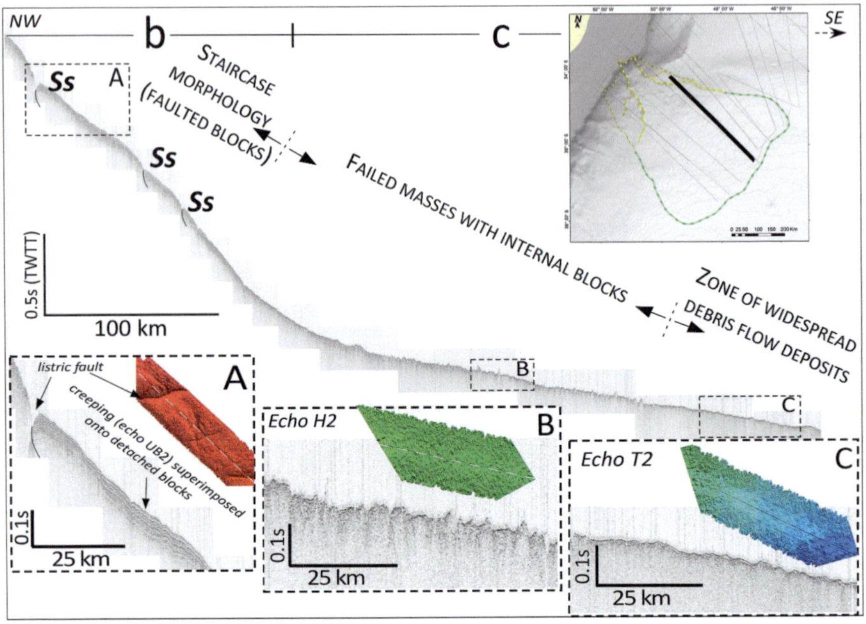

Fig. 11.4 Dip-oriented 3.5 kHz profile, coupled with extracts of along-track swath batymetry, across the Chuí Megaslide Complex, showing the variable nature of mass-transport deposits as indicated by varying echofacies across the morphological sector *a*. *Ss* Secondary scarps; *a* and *c* are morphological sectors across of the submarine landslide complex

different stratigraphic horizons. Besides that, shear surfaces often step down the stratigraphy to create a series of staircase-like geometries, pointing to the creation of new décollements at depth and to incremental incorporation of new material during the emplacement of sediment slides (Fig. 11.3). MTDs are stacked either directly one on the top of the other, or separated from each other by continuous units of strong plan-parallel reflectors, acoustically interpreted as a combination of turbiditic and hemipelagic deposition. Failed masses can also be interbedded with well stratified uniform lens- or wedge-shaped strata, previously interpreted in the area as contourite deposition (Pereira et al. 2010). These architectural elements evidence the occurrence of multiple and complexly-interbedded landslide events and/or failed masses, all of them sourced from the main failure scar (sector ***a***) and spreading for hundreds of kilometers downslope (Figs. 11.1c, 11.3 and 11.4b, c). For these reasons, all MTDs were grouped under the designation of the Chuí Megaslide Complex (CMC), which consists of a stack of 8–12 large-scale landslides, depending on the area considered and on the resolution of the available data, cutting into Pliocene-Quaternary sedimentary successions (Figs. 11.2 and 11.3). The MTDs were classified as megaslides due to their extension (hundreds of km large) and thickness (individual thickness up to *c*. 400 m; considering a value of stacking layer velocity of 2,000 m/s).

On both seismic and 3.5 kHz lines, the shallow stratigraphy across sector **b** appears as uniformly-bedded truncated layers, comprising large extensional zones of closely-spaced faulted-detached blocks with small runouts, replaced downslope by failed masses (Figs. 11.2b and 11.4a). Such structural features allow us to interpret morphological sector **b** as consisting of zones of mainly retrogressive slide experiencing ongoing seabed disruption to give rise to downslope mass-transport processes.

However, dip-oriented seismic and 3.5 kHz profiles across depositional area **c** can show variable seismic facies and features depending on the region considered, as well as on the seismic resolution involved. Seismic facies vary from disrupted masses containing large internal blocks (up to several kilometers across) to dominant cohesive mass flows. Owing to the relatively limited seismic resolution, failed masses are seen mostly as transparent to chaotic in seismic profiles, with individual blocks resolved only by their rough morphology (Fig. 11.3). On the other hand, on 3.5 kHz lines and on along-track swath bathymetric profiles two distinctive seafloor textures can be seen within the landslide deposits (Fig. 11.4): (i) a very rough sea-floor texture depicts rather irregular hyperbolic echoes (our echo type H2 in Fig. 11.4b) indicating the occurrence of large zones of failed masses containing internal preserved blocks, extending in downslope continuity to sector **b** down to *c.* 4,600 m water depth; this area being followed by (ii) a smoother seafloor dominated by continuous transparent layers (our echo type T2 in Fig. 11.4c), interpreted as a distal fringe of dominant debris flow deposits, according to its dominant transparent seismic facies that can be traced further down to a water depth of ~4,900 m (Fig. 11.1c). The mechanism involved in the downslope physical transformation of the failed masses remain unclear in the area.

11.3.3 Possible Preconditioning Parameters and Triggering Mechanisms

Slope angle may have an important impact on slope instabilities but it is not always considered as a good indicator of susceptibility to landsliding (e.g., Migeon et al. 2006). Accordingly, slope angle alone does not seem to have played a significant role in inducing slope failure on the South Brazilian margin. The MTDs that compose the CMC were all developed along the southernmost margin where the slope is inclined at ~1,6°, but similar gradients occur across the continental slope of the entire southern margin (profiles a, b and c in Fig. 11.1a). Nonetheless, we do not exclude the action of contour current erosion in contributing to the local over-steepening of the slope around the outreach of the Rio Grande Cone.

On the other hand, structural features and the stratigraphy provide hints at other preconditioning parameters and triggering mechanisms that may have played a most significant role in the area, such as: (i) structures of thin-skinned tectonics affecting the Rio Grande Cone (Zalán 2005) seem to have played a major role in the initiation of slope failures. The high angle fault plane across scarp **Is$_2$** points to a clear causal relation between block slump-off and the gravity-related listric faults (Fig. 11.2a). Actually, mass-wasting induced by gravity tectonics has also been identified in other segments of the Brazilian margin (e.g., the Amazon fan; Reis et al. 2010); (ii) the main slide scar (sector **a**) flanks the thickest Early Miocene-Quaternary depocenter of the entire southern Brazilian margin, which extends across the shelf edge into the Rio Grande Cone (Fig. 11.1b). Higher sedimentation rates related to the deposition of the Rio Grande Cone would possibly have induced pore pressure elevation in the upslope area of the megaslides occurrence (as required to initiate landslides). Equally, high sedimentary influx into the area has also provided enough sediment to partially refill the failure scar and to the successive reconstruction of the shelf edge-upper slope. Actually, the occurrence of landslides on the flanks of major submarine fans has been recurrent on various continental margins, as for instance the Amazon fan (Silva et al. 2010).

11.4 Conclusions

The combination of a diversified geophysical dataset provides insights into the slope failure processes that occur in the Southern Brazilian Margin. Results lead us to the following main conclusions:

1. The Chuí Megaslide Complex stands out as a regional-scale megaslide complex within the context of a margin largely dominated by sediment drifts;
2. The megaslide complex, composed of 8–12 large MTDs, affects significantly both the margin morphology and sedimentation across a large area (~150,000 km^2), extending from the shelf-break ~650 km downslope to ~4,900 m water depth;
3. Active gravity tectonics has to be assumed as the probable major mechanism inducing slope failures, since the main upslope slide scar seems likely to have been initiated by movements of gravity-related structures;
4. Preconditioning for the submarine mass-wasting process may also have been favoured by rapid sediment accumulation in the head area of the slide complex, together with the erosive action of contour currents around the Rio Grande Cone.

Acknowledgements The authors wish especially to thank the *Executive Committee of the LEPLAC PROGRAMME* for permission to publish the acquired geophysical data. We would also thank the Brazilian National Oil and Gas Agency (ANP) for supplying extra seismic and well data. We are also very grateful to reviewers Sébastien Migeon and Nathalie Babonneau for their constructive suggestions while revising the paper. Special thanks are due to Seismic Micro-Technology Inc. (SMT) for the use of the educational licenses of the software Kingdom Suite®. Finally, we thank the Brazilian National Research Agency (CNPq) for providing research grants for the first and second authors.

References

Bassetto M, Alkmim F, Szatmari P, Mohriak WU (2000) The oceanic segment of the southern Brazilian margin: morpho-structural domains and their tectonic significance. In: Mohriak WU, Talwani M (eds) Atlantic rifts and continental margins, geophysical monograph 115. American Geophysical Union, Washington, pp 235–259

Migeon S, Cattaneo A, Hassoun V et al (2006) Morphology, distribution and origin of recent submarine landslides of the Ligurian Margin (north-western Mediterranean): some insights into geohazard assessment. Mar Geophys Res. doi:10.1007/s11001-011-9123-3

Pereira F, Reis, AT, Silva C (2010) Contourites drifts – the main Cenozoic architectural element of the central Pelotas Basin – southernmost Brazilian Margin. In: Abstracts of the AGU Meeting of the Americas, 08–12 August 2010, Foz do Iguassu, American Geophysical Union, Brazil

Preu B, Hernández-Molina FJ, Violante R et al (2013) Morphosedimentary and hydrographic features of the northern Argentine margin: the interplay between erosive, depositional and gravitational processes and its conceptual implications. Deep-Sea Res I 75:157–174. doi:10.1016/j.dsr.2012.12.013

Reis AT, Perovano R, Silva CG, Vendeville BC et al (2010) Two-scale gravitational collapse in the Amazon Fan: a coupled system of gravity tectonics and mass-transport processes. J Geol Soc (Lond) 167(3):593–604. doi:10.1144/0016-76492009-035

Silva CG, Araújo EFS, Reis AT et al (2010) Megaslides in the Foz do Amazonas Basin, Brazilian Equatorial Margin. In: Mosher DC, Shipp RC, Moscardelli L et al (eds) Submarine mass movements and their consequences, advances in natural and technological hazards research, vol 28. Springer, New York, pp 581–591

Stramma L, England M (1999) On the water masses and mean circulation of the south Atlantic Ocean. J Geophys Res 104(9):863–883

Zalán PV (2005) End members of gravitational fold and thrust belts (GTFBs) in deep waters of Brazil. In: Shaw JH, Connors C, Suppe J (eds) Seismic interpretation of contractional fault-related folds, Seismic Atlas, Studies in Geology, vol 53. AAPG, Tulsa, pp 147–153

Chapter 12
Submarine Landslides and Incised Canyons of the Southeast Queensland Continental Margin

Thomas Hubble, Jody Webster, Phyllis Yu, Melissa Fletcher, David Voelker, David Airey, Samantha Clarke, Angel Puga-Bernabeu, David Mitchell, Floyd Howard, Stephen Gallagher, and Tara Martin

Abstract An investigation conducted aboard the *RV Southern Surveyor* (*SS2013-V01*) in January 2013 offshore east Australia collected regional bathymetric data for the continental margin of southern Queensland between Noosa Heads in the south and Indian Head, Fraser Island in the north. This newly mapped area presents a particularly steep portion of continental slope (5–10°) that presents numerous submarine landslides, including two 'whole-of-slope' features (the Wide Bay Canyon, and Inskip Slides). The slope is also dissected by three large submarine canyons offshore northern Fraser Island, Wide Bay, and Noosa Heads (i.e. the Fraser Canyons, the Wide Bay Canyon and the Noosa Canyon). Dredge and core samples were collected from slide scars in the northern, central, and southern areas of the bathymetric survey area. The initial examination of the area's bathymetry, the core and dredge sample sedimentology, and determination of biostratigraphic ages for these sediment samples indicates that the larger submarine slides present in this study area have probably been shed from the slope since the late Pliocene and that canyon incision is currently active on this portion of the slope. In one case, canyon incision is partly responsible for generating slides due to undercutting and

T. Hubble (✉) • J. Webster • P. Yu • M. Fletcher • D. Airey • S. Clarke • D. Mitchell
School of Geosciences, The University of Sydney, Sydney, NSW, Australia
e-mail: tom.hubble@sydney.edu.au

D. Voelker
MARUM, University of Bremen, Bremen, Germany

A. Puga-Bernabeu
Department of Geology, University of Granada, Granada, Spain

F. Howard
Geoscience Australia, Canberra, Australia

S. Gallagher
School of Earth Sciences, University of Melbourne, Melbourne, VIC, Australia

T. Martin
Marine National Facility, CSIRO, Hobart, TAS, Australia

G. Lamarche et al. (eds.), *Submarine Mass Movements and their Consequences*, Advances in Natural and Technological Hazards Research 41,
DOI 10.1007/978-3-319-20979-1_12

removal of the toe of the slope. Slope sediments are dominantly comprised of hemipelagic muds but also include grain-flows and turbidites comprised of shelf-derived sands and upper slope sediment that have abraided the slope muds. The results confirm previous work that indicates that this margin is in an active phase of deconstruction dominated by mass failure.

12.1 Introduction

Contrary to common expectation, it is not the seismically active continental margins that generate the largest submarine landslides. Instead it is the low seismicity, passive margins such as northern Europe, Northwest Africa, and South Africa that present these enormous mass failure structures (c.f. Dingle 1983; Hjelstuen et al. 2005; Georgiopoulou et al. 2010). Similarly, low rates of sedimentation and seismicity combined with geomechanically strong sediment (high shear strengths and high friction angles), characteristics common on the eastern Australian continental margin, are also generally considered to reduce the likelihood of mass failures on continental slopes. Consequently, the marine geological community were somewhat surprised when recent work conducted on the RV Southern Surveyor (*SS2006-V10* and *SS2008-V12*; c.f. Boyd et al. 2010; Clarke et al. 2012; Hubble et al. 2012) demonstrated that the continental slope of eastern Australia's passive continental margin (EAPCM) presents abundant examples of small to large, upper slope and mid slope submarine landslides including a substantial number that were potentially tsunamogenic (Clarke et al. 2014). Here we report initial bathymetric and sedimentologic data as well biostratigraphic dates for material collected during a third investigation of the EAPCM conducted offshore Fraser Island, southern Queensland, onboard the *RV Southern Surveyor* in 2013 (*SS2013-V01*). These results extend the area of the EAPCM known to affected by the phenomenon of submarine landsliding, including two large slides that are apparently whole-of-slope features, and will assist in constraining the causes and history of landsliding in this area.

12.2 Study Area Location and Bathymetric Features

The study area is located along the east Australian continental margin, approximately 30–60 km offshore the coast of southern Queensland between Indian Head on Fraser Island in the north and Noosa Heads in the south. Water depths range between 200 m at the continental shelf edge and 4,000 m at the abyssal plain (Figs. 12.1, 12.2 and 12.3). High-resolution multibeam data of the local and regional bathymetry were collected using a Kongsberg EM300 multibeam echosounder (Hubble 2013). Fledermaus V7.3.3b software (http://www.qps.nl/) was used to process and merge the raw bathymetric data to produce a single 50 m gridded digital elevation model

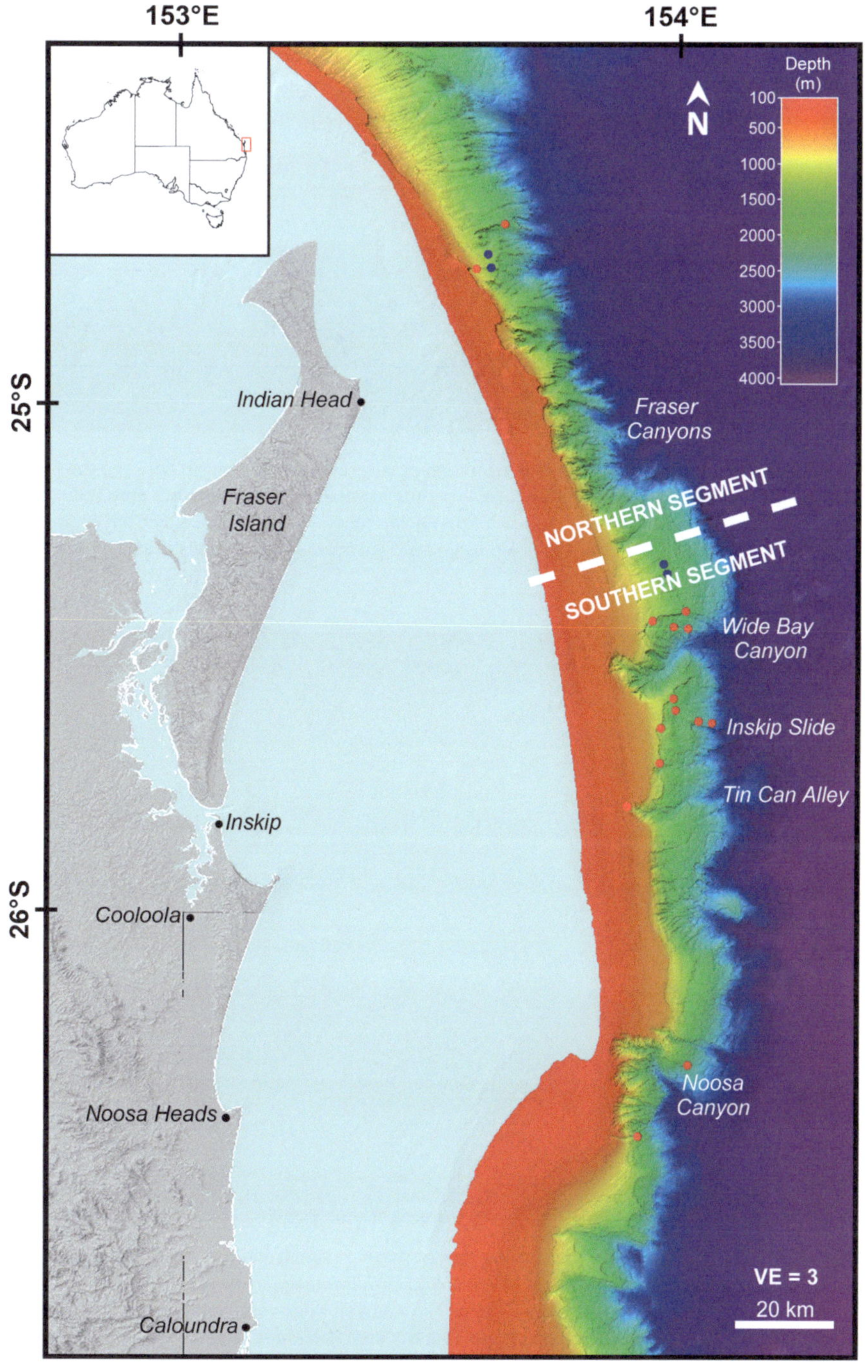

Fig. 12.1 Location map showing digital elevation model for the continental slope bathymetry, major features referred to in the text, as well as dredge (*red dots*) and core (*blue dots*) site locations

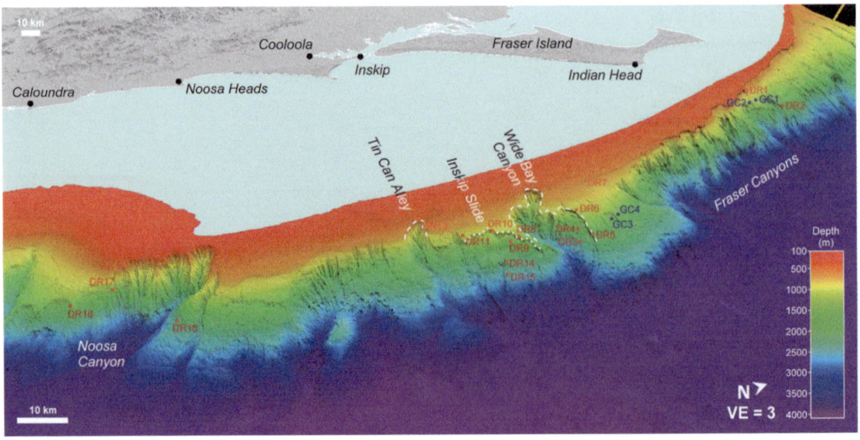

Fig. 12.2 Oblique view of the eastern Australian passive continental margin between the Noosa Canyon and Indian Head, Fraser Island, showing the location of several gravity core (GC; *blue*) and dredge (DR; *red*) sites. *White dashed lines* indicate canyon heads (Wide Bay Canyon and Tin Can Alley) and *white dotted lines* indicate submarine landslide (Inskip Slide) headscarps

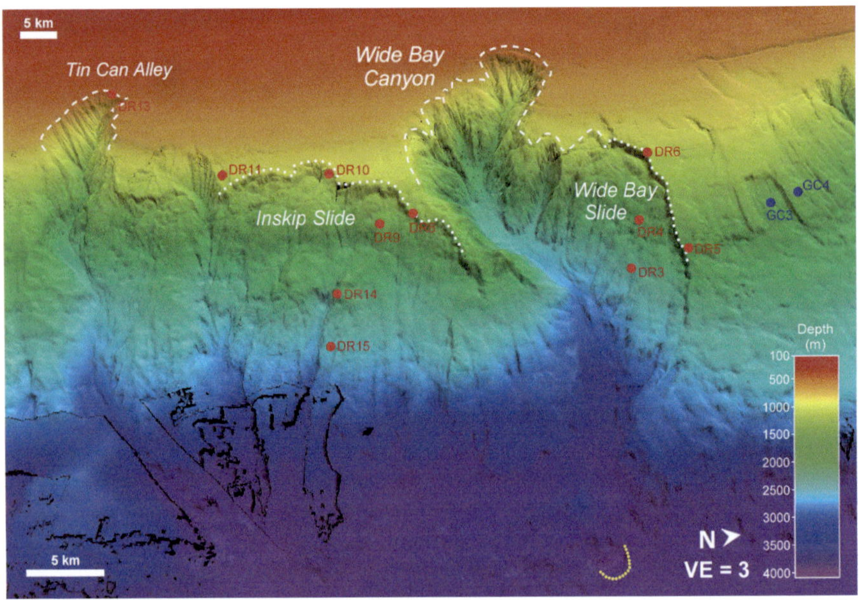

Fig. 12.3 Oblique view of the eastern Australian passive continental margin showing the characteristics of the slope and the location of several features discussed in the text including: the Wide Bay Canyon, Tin Can Alley Canyon, the Wide Bay Canyon Slide (northern canyon wall) and the Inskip Slide as well as the locations of several gravity core (GC; *blue*) and dredge (DR; *red*) sites. *White dashed lines* indicate canyon heads (Wide Bay Canyon and Tin Can Alley) and *white dotted lines* indicate submarine landslide (Wide Bay Slide, Inskip Slide) headscarps. *Yellow dotted line* shows the perimeter of the elliptical depression at the mouth of the Wide Bay Canyon

(DEM) covering the region investigated. The 50 m *RV Southern Surveyor* data was then displayed using Fledermaus as part of the merged gbr100 dataset (http://www.deepreef.org/), a high-resolution bathymetry and Digital Elevation Model (DEM) covering the Great Barrier Reef, Coral Sea, and neighbouring Queensland coastline. The aerial extent of the high-resolution multibeam coverage used in this study is shown in Fig. 12.1.

The continental slope offshore Fraser Island exhibits a spectacular morphology with a distinct change in the morphology of the transition between the shelf and slope at around 25°10′S (Fig. 12.2). North of that latitude (northern segment), the seafloor drops from a well-defined shelf edge at 280–320 m water depth to a steep upper slope (20–30°) and presents a moderately steep lower slope (~10°), that terminates in the abyssal plain at around 3,800 m water depth. The slope is incised by deep gullies. The shelf edge is affected by a series of partly overlapping arcuate indentations. These indentations have a width of ~3–6 km and they cut ~1–2 km into the shelf edge. These intricate, arcuate indentations of the shelf edge and the slope morphology in the northern segment indicate intense modification of the slope by mass failure processes and scouring by the downslope motion of dense fluids giving this segment a "mature" character that resembles dendritic, bedrock-eroded river terrains. Segments of the upper slope from a shelf depth of ~300 m down to the middle slope have failed as individual blocks and moved downslope. The products of this process appear to have disintegrated and spilled into the abyssal plain as fluidized material because no deposits of such debris have been identified on the abyssal plain.

South of 25°10′S (southern segment Fig. 12.1) the transition from shelf to slope is generally much more subtle except where the slope is dissected by the spectacular Wide Bay and Tin Can Alley Canyons. South of 25°10′S the shelf to slope transition also steps seawards by a distance of 2 km at the shelf edge and by 5 km at the foot of the slope. This portion of the study area presents an upper slope morphology that is typically smooth with an average slope gradient of around 5°, although the upper slope also presents a set of radially disposed incisions in the upper slope at 26° S. The lowermost slope is significantly steeper (23–40°) and shows a number of landslide headscarps, and lower slope canyons (Figs. 12.2 and 12.3).

The Wide Bay Canyon system is developed at 25°30′S. It extends from the shelf to the abyssal plain at 3,800 m, and significantly modifies the shelf edge (Fig. 12.2a). It is up to 8 km wide, sinuous and has a flat 1–2.5 km wide bottom. A 45 km long escarpment of 150 m height towers above the northern wall of the canyon and forms a geomorphic boundary between the bathymetrically rugged canyon to the south and relatively smooth continental slope to the north. A slide at least 80 km^2 is situated below this headscarp and it is evident that a slab of slope sediment 150 m thick was removed in a single slide or a succession of translational slides that effectively removed the entire slope above the Canyon axis. We name this slide 'The Wide Bay Canyon Slide' and infer that the slide material collapsed into the canyon and was then transported out onto the abyssal plain. It is likely that the canyon incision played a major role in the dislocation of the slide by removing the basal support of the slope over a large distance. An intriguing characteristic of the easterly canyon exit is the absence of a distributary sediment fan. Instead, an elliptical, 150 m deep, 3 × 1 km wide depression is present at the canyon's mouth that is surrounded by an almost-enclosed, ring-shaped, 70 m high wall. We infer that the canyon must have been actively delivering sediment to this site during past periods and it follows that such sediment would probably be deposited within the depression due to the decrease in slope that marks the entry of the canyon onto the abyssal plain. The identified depression should therefore be filled up by sediment discharged from the canyon unless this material is removed by vigorous bottom water flow. A plausible explanation for this extraordinary morphological feature is that it has been formed and maintained by vigorous bottom-water currents that flow north along the toe of the slope as suggested for this margin by Hubble et al. (2012).

A second whole-of-slope slide, the Inskip Slide, is situated beneath the prominent headscarp feature that separates the smooth upper slope from the steep middle slope immediately to the south of the Wide Bay Canyon (Fig. 12.3). The south of the study area is bounded by the deeply incised Noosa Canyon which presents a dendritic pattern of incisions that resembles the morphology commonly associated with steep, stream-incised, mountainous terrains. This morphology is interpreted to have been formed during the Late Quaternary by cascades of shelf sands driven over the shelf-break by the vigorous, southerly flowing East Australian Current which commonly achieves speeds in excess of four knots (c.f. Boyd et al. 2004).

12.3 Sediment Sample Characteristics and Ages

Unconsolidated and compacted but unlithified sediments were collected from 11 - mid-slope locations (4 gravity cores and 7 dredge hauls; Fig. 12.4) in the study area. Physical and geotechnical parameters have been determined for representative samples and will be reported in detail elsewhere but the samples are dominantly hemipelagic muds identical to those described in Clarke et al. (2012) and Hubble et al. (2012) except for a 20 cm thick, coarse, normally graded sand unit (a grainflow) deposited between a moderately consolidated mud (below) and unconsolidated mud (above) which is evident in a gravity core *SS2013-V01* GC01 (Fig. 12.4).

12.3.1 Dredge Sample Ages

At the time of writing five samples have been dated on the basis of their contained foraminifera and microfossils. These samples were prepared by initial disaggregation and then wet-sieving. The coarsest fractions (63–150 mm, 150–1 mm, >1 mm) were sent for biostratigraphic analysis (Yu et al. 2014; Fletcher et al. 2014). These results indicate that the compacted, hemipelagic muds recovered in dredge hauls of material exposed on the continental slope immediately to the south of Wide Bay

Fig. 12.4 Images of *SS2013-V01* dredge samples (**a**) DR11 and (**b**) DR15, and core sample (**c**) GC01. Locations of DR11 and DR15 are indicated in Fig. 12.3 and location of GC01 is indicated in Fig. 12.2. Radiocarbon sample locations are shown on the GC01 core image (**c**). Two samples are radiocarbon dead, meaning that no dates could be obtained

Canyon (sites DR14 and DR15; Fig. 12.4) is of Late Miocene to Mid Pliocene age and was deposited approximately 4 million years ago. The material exposed in the head scarp of this slope (site DR08) is Mid-Miocene in age and was deposited approximately 15 million years ago.

12.3.2 Core Sample Ages

The biostratigraphy available for core GC01 (Fletcher et al. 2014) indicates that the oldest material recovered in the base of the core which is a compacted hemipelagic mud is Early Pleistocene in age. This material was deposited between 2 and 2.5 Ma.

The moderately consolidated hemipelagic material immediately overlying this basal core material is Late Pleistocene in age with its contained microfossils yielding a depositional age of 0.45 Ma. Four radiocarbon ages were used to establish minimum age constraints for material recovered within gravity core (GC01; Fletcher et al. 2014). Radiocarbon ages were determined at the CHRONO Centre, Queen's University, Belfast UK and are indicated on Fig. 12.4. This indicates that the sand layer is comprised of shell fragments that are 31 ka. The sand layer presents an erosional contact with the underlying consolidated mud; the immediately overlying unconsolidated mud was deposited 14.6 ka just before the start of the Holocene This date is penecontemporaneous with that determined for a slide dated by Clarke et al. (2012) that is located 450 km to the south offshore Byron Bay (15.8 ka).

12.4 Discussion and Conclusions

The bathymetric, lithologic, and age data presented here indicates that the slope instability and erosional processes that have been documented for the continental slope south of the Noosa Canyon in northern New South Wales (Boyd et al. 2010; Hubble et al. 2012; Clarke et al. 2012) have also affected the steeper section of continental slope that is present between the Noosa Canyon and the termination of the Tasman Sea Basin offshore northern Fraser Island. The dominant processes are (1) canyon incision (Wide Bay, Tin Can Alley, and Noosa Canyon); (2) large, infinite slope type failures which we suspect present as whole-of-slope events and were probably pre-conditioned by canyon incision undercutting the toe of the slope in the case of the northern wall of the Wide Bay Canyon and by bottom current erosion removing sediment from the toe of the slope in the case of the large slope slide identified immediately to the south of the Wide Bay Canyon; and (3) translational and/or rotational failure of large masses from the middle and upper slope in the northern half of the study area (North of 25°10′S). In addition to these large scale slope failure and slope erosion processes, the identification of the layer of graded sand as a grainflow and the age relationships above and below this layer provides strong evidence for an interpretation of the ongoing and perhaps 'continual removal' of the slope sediment which has accumulated on this slope during the Pleistocene either due to mass failure or by abrasion; it is likely that sand-waves driven over the shelf-break by the action of the East-Australian Current (c.f. Boyd et al. 2004) generated grainflows that have cascaded down, and abraided the slope, removing the hemipelagic slope muds and transporting them downslope.

We have also noted the absence of depositional features such as sediment fans at the mouth of the Wide Bay and Tin Can Alley canyons, and the apparent absence of slide debris or sediment mounds at the foot of the slope in the study area. This suggests that any material that is delivered to these locations is either moving with sufficient energy that it is transported out quite some distance onto the abyssal plain or that material deposited at these sites is removed by the action of strong bottom water currents as posited for this margin in Hubble et al. (2012). The data presented here is certainly consistent with both these interpretations.

The Late Pliocene age determined for the compacted hemipelagic muds dredged from the continental slope failure surface located immediately to the south of the Wide Bay Canyon, and the Early Pleistocene age determine for the material at the base of GC01 indicates that relatively large slope failures along this portion of the EAPCM have occurred during relatively recent geologic time, probably since the latest Pliocene (Wide Bay) and Middle Pleistocene (Northern Fraser Island). This finding confirms the assertions of Boyd et al. (2010) and Hubble et al. (2012) that further submarine landsliding on this margin should be expected, considered to be common at geological time scales, and are probably an ongoing phenomenon.

Acknowledgements We would like to acknowledge the P&O crew and the CSIRO scientific support team who assisted us on the *RV Southern Surveyor* voyage SS2013-V01 as well as the provision of the CSIRO managed Marine National Facility that is funded by the Australian government. This paper benefitted from thoughtful and constructive reviews by Professor Neville Exon and Dr Asrar Talukdar.

References

Boyd R, Ruming K, Roberts JJ (2004) Geomorphology and surficial sediments on the southeast Australian continental margin. Aust J Earth Sci 51:743–764

Boyd R, Keene J, Hubble T, Gardner J, Glenn K, Ruming K, Exon N (2010) Southeast Australia: a cenozoic continental margin dominated by mass transport. In: Mosher DC, Moscardelli L, Baxter CDP et al (eds) Submarine mass movements and their consequences, vol 28, Advances in natural and technological hazards research. Springer, Dordrecht, pp 491–502. doi:10.1007/978-90-481-3071-9_40

Clarke S, Hubble T, Airey D, Yu P, Boyd R, Keene J, Exon N, Gardner J (2012) Submarine landslides on the upper Southeast Australian passive continental margin – preliminary findings. In: Yamada Y, Kawamura K, Ikehara K et al (eds) Submarine mass movements and their consequences, vol 31, Advances in natural and technological hazards research. Springer, Netherlands, pp 55–66. doi:10.1007/978-94-007-2162-3_5

Clarke S, Hubble T, Airey D, Yu P, Boyd R, Keene J, Exon N, Gardner J, Ward S (2014) Morphology of Australia's Eastern continental slope and related Tsunami hazard. In: Krastel S, Behrmann J, Volker D, et al. (eds.) Submarine mass movements and their consequences, advances in natural and technological hazards research. Springer International Publishing, Switzerland, 37:529–538

Dingle R (1983) Slump structures on the outer continental margin of southwestern Africa. Seismic Expression of Structural Styles. Am Assoc Pet Geol Stud Geol 15:1.2

Fletcher M, Hubble T, Clarke S, Airey D, Yu P (2014) Submarine landslide morphology of Box slides present on the continental slope offshore Fraser Island, Queensland, Australia. Spring Meeting, EGU, Vienna

Georgiopoulou A, Masson D, Wynn R, Krastel S (2010) Sahara slide: age, initiation, and processes of a giant submarine slide. Geochem Geophys Geosyst 11, Q07014. http://dx.doi.org/10.1029/2010GC003066

Hjelstuen BO, Sejrup HP, Haflidason H, Nygård A, Ceramicola S, Bryn P (2005) Late Cenozoic glacial history and evolution of the Storegga Slide area and adjacent slide flank regions, Norwegian continental margin. Mar Pet Geol 22(1):57–69

Hubble TCT (2013) SS 2013-V01 Voyage Summary: Submarine landslides offshore northern New South Wales and southern Queensland; their geomechanical characteristics, timing and triggers. CSIRO Hobart, p 34

Hubble T, Yu P, Airey D, Clarke S, Boyd R, Keene J, Exon N, Gardner J (2012) Physical properties and age of continental slope sediments dredged from the eastern Australian continental margin – implications for timing of slope failure. In: Yamada Y, Kawamura K, Ikehara K et al (eds) Submarine mass movements and their consequences, vol 31, Advances in natural and technological hazards research. Springer, Dordrecht, pp 43–54

Yu P, Hubble T, Airey D, Gallagher S, Clarke S, Keene J, Gardner J (2014) The Wide Bay Canyon system: a case study of canyon morphology on the east Australian continental margin. In: Fall Meeting, AGU, San Francisco, CA, 15–19 Dec. 2014

Chapter 13
Novel Method to Map the Morphology of Submarine Landslide Headwall Scarps Using Remotely Operated Vehicles

Veerle A.I. Huvenne, Aggeliki Georgiopoulou, Leo Chaumillon, Claudio Lo Iacono, and Russell B. Wynn

Abstract Submarine landslides are one of the main marine geohazards worldwide. In order to better estimate their risk and develop mitigation measures, a better understanding of the failure mechanisms is needed. However, observing landslides in real time is near to impossible, hence careful study of both the failed sequence and the deposit is still the main source of information. Until recently, studies of the headwall scarps of submarine landslides were mainly based on shipboard acoustics and the descriptions of analogues on land. The increasing availability of Remotely Operated Vehicles (ROVs) now allows in-situ, close-up investigations in the marine environment. In this paper we present a novel methodology to obtain a detailed map of the headwall morphology of submarine landslides, including detail on vertical cliffs, overhanging strata and biological colonisation. Using a high-resolution multibeam system mounted on the front of a working-class ROV, rather than in a traditional downward looking configuration, we mapped three sections of a scallop-shaped headwall scarp which is part of the Rockall Bank Slide Complex, west of Ireland, as part of the SORBEH cruise funded by the Irish Government and the ERC CODEMAP project. The resulting 3D model provides insights in the build-up of the failed sequence, allowing advances in the understanding of rupture mechanisms. It can be combined with close-up video ground-truthing and carefully collected samples to create a complete picture of the headwall scarp.

V.A.I. Huvenne (✉) • C. Lo Iacono • R.B. Wynn
Marine Geoscience, National Oceanography Centre, University of Southampton Waterfront Campus, European Way, Southampton, UK
e-mail: vaih@noc.ac.uk

A. Georgiopoulou
UCD School of Geological Sciences, University College Dublin, Dublin, Ireland

UCD Earth Institute, University College Dublin, Dublin, Ireland

L. Chaumillon
Marine Geoscience, National Oceanography Centre, University of Southampton Waterfront Campus, European Way, Southampton, UK

INTECHMER, Cherbourg, France

G. Lamarche et al. (eds.), *Submarine Mass Movements and their Consequences*, Advances in Natural and Technological Hazards Research 41,
DOI 10.1007/978-3-319-20979-1_13

13.1 Introduction

Submarine landslides can be found along all of the World's ocean margins (e.g. Hühnerbach et al. 2004). Although research into the field has made significant progress since the first discoveries (e.g. Heezen and Ewing 1952; Krastel et al. 2014), the processes behind submarine slope failures are still only partly understood. The inaccessibility of the marine environment makes detailed studies challenging and expensive. However, the development of new technologies, especially based on robotic systems, now makes this possible, also in the deep sea.

One aspect of submarine landslides that is always difficult to study, is the full morphology and nature of the headwall scarps. Micallef et al. (2008), for example, carried out a detailed geomorphometric study of 115 mass movement headwalls within the Storegga Slide scar. Their shape was indicative of the principal formative geological processes of the failures (spreading vs. debris flow formation). However, the analysis, although based on full-coverage bathymetric grids, was only carried out in plan view. More detailed information about the 3D morphology of headwalls may lead to better insights into the mechanics of the rupture, and into the nature of the failed sequence. The aim of this paper is to present a novel method to obtain a high-resolution (sub-metre pixel size) morphological model of the headwall scarp of a submarine landslide, including internal stratification, overhanging morphologies and biological colonisation.

13.2 Rockall Bank

The headwall scarp mapped in this study is located on the SE flank of Rockall Bank, W of Ireland (Fig. 13.1), in an area characterised by a large number of landslide scars and mass flow deposits, collectively known as the Rockall Bank Mass Flow complex (Elliot et al. 2010) or more recently, the Rockall Bank Slide Complex (RBSC, Georgiopoulou et al. 2013).

Rockall Bank is a structural high that forms the steep western margin of the Rockall Trough, a sediment-starved, elongated deep-water basin that formed in Mesozoic times, during the opening of the North Atlantic (Shannon 1991). Bottom currents in Rockall Trough flow in an anticlockwise direction, impinging on Rockall nticlockwise direction, impinging on Rockall Bank with a south-westward flow that created a number of contourite deposits (Stoker et al. 2005). It is here that the RBSC developed as a result of multiple phases of slope collapse. A large along-slope, channelised moat or 'erosional zone' divides the area at ca. 1350–1650 m water depth, and may have contributed to some of the slope failures through differential sedimentation and erosion (Elliot et al. 2010). The moat is also characterised by the occurrence of several pinnacles of unknown origin. Based on the recovery of coral-rich sands and live coral specimens from one of those, it has been suggested they may either be cold-water coral carbonate mounds or volcanic cones capped by a coral cover (Øvrebø et al. 2005; Georgiopoulou et al. 2013). The

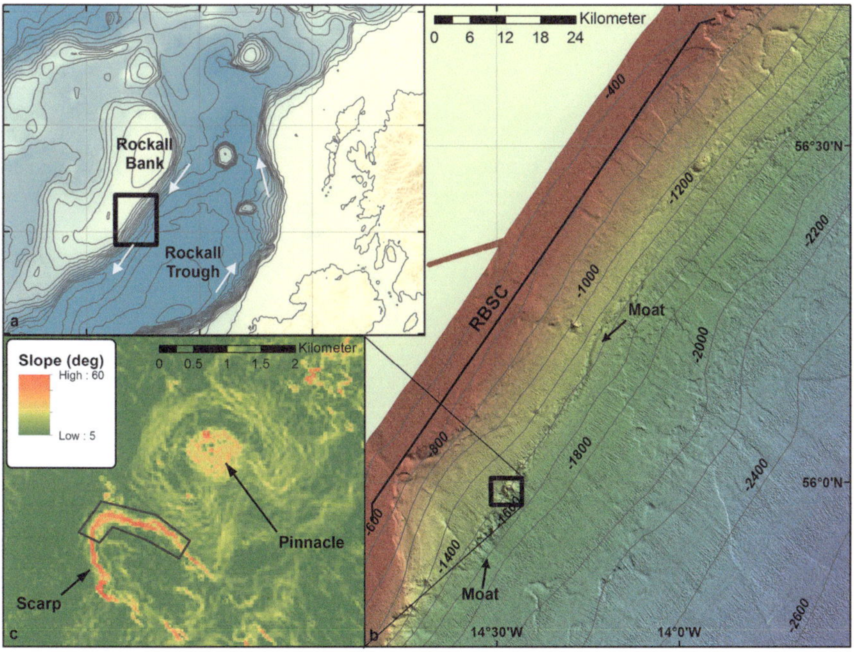

Fig. 13.1 Location map: (**a**) Regional map illustrating the setting of Rockall Bank and Rockall Trough, W of Ireland. *White arrows* indicate the general direction of bottom currents in the area. (**b**) Bathymetric map of SE Rockall Bank, showing the numerous scarps and seabed features making up the Rockall Bank Slide Complex (RBSC, Georgiopoulou et al. 2013). (**c**) Slope gradient map of the working area, illustrating both the landslide scarp under study and the nearby pinnacle to the NE of it. The section of the headwall and sidewall mapped with the ROV-based forward technique is shown in the *black polygon*

small headwall scarp presented here is formed on the upslope flank of this moat, and is located immediately SW of a pinnacle at ca. 1400 m water depth (Fig.13.1c). Shipboard multibeam bathymetry data provided by the Geological Survey of Ireland, INFOMAR programme (www.infomar.ie) shows that the scarp is typically horse-shoe shaped, up to 175 m high and 1200 m wide. Reconnaissance ROV video transects on the walls of the scarp, carried out immediately prior to the mapping exercise, demonstrated the occurrence of rich cold-water coral and clam communities.

13.3 Method

13.3.1 Data Collection

The data for this study were collected during the 2014 expedition CE14011 (‘SORBEH’) on board the *RV Celtic Explorer*, and form part of the ERC Starting

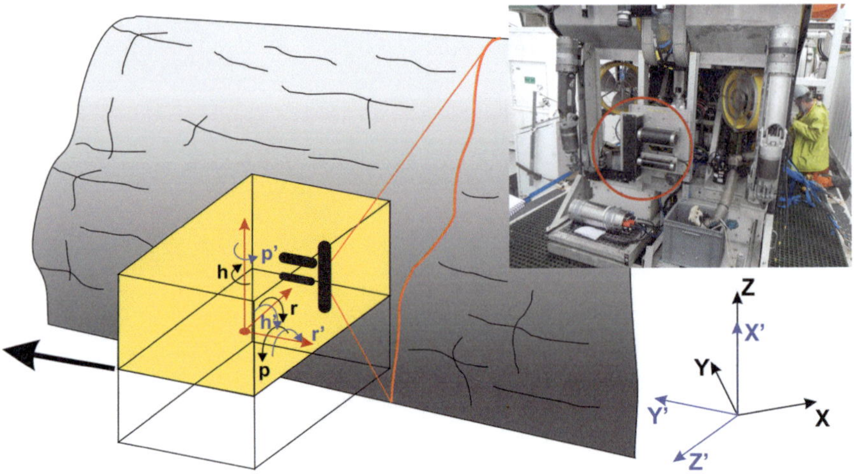

Fig. 13.2 Schematic representation of the ROV forward mapping set-up, illustrating the coordinate systems and attitude conventions (pitch, roll, heading) of the original (*black*) and rotated framework (*blue*). Inset: *Holland I* ROV with Reson 7125 multibeam system mounted on the front

Grant project CODEMAP. The Irish scientific ROV *Holland I* was equipped with a Reson 7125 dual frequency multibeam echosounder (MBES). Instead of using a traditional downward-facing configuration, the system was mounted on the front of the vehicle, in a forward-looking set-up (Fig. 13.2). Using the 400 kHz mode, the data were collected by moving the ROV sideways, keeping the vehicle heading constant and approximately perpendicular to the overall strike of each cliff section under investigation. Due to the curved nature of the scarp (horse-shoe shape), the cliff was surveyed in three separate sections. Data recording took place using the 7K and PDS2000 software packages. Navigation was based on Ultra Short Base Line recordings (USBL, Sonardyne Ranger), smoothed in real time using a Kalman filter steered by a Doppler Velocity Log (DVL). Unfortunately the low-frequency (300 kHz) Doppler system of the ROV caused interference with the MBES acoustics, hence only the high-frequency (1200 kHz; RDI) system could be used, limiting the operation altitude of the ROV to ca. 25 m above the seafloor. Vehicle attitude was recorded using an Ixsea Octans 3000 gyro, and sound velocity at the transducers (and by extension at the water depth where the data were recorded) was calculated in real time from the ROV CTD readings.

13.3.2 Data Processing

No standard software packages are available to carry out MBES data processing for this type of forward-looking configuration. Hence we applied a double coordinate transformation to the navigation and attitude data (rotation around a horizontal and vertical axis), projecting them to a local coordinate system where the X′ and Y′ axes define a plane parallel to the overall strike of the wall, which could be interpreted as the 'imaginary sea surface' for the recorded MBES data. This way the vehicle depth became the new X′ coordinate, while the new Z′-axis was oriented in the direction opposite to the average ROV heading. In this configuration, the recorded ROV pitch represented the (inverse of the) MBES roll, the recorded ROV roll represented the variation in heading, and the deviations of the ROV heading from the average for the section represented the (inverse) pitch (Fig. 13.2). Using these parameters, the individual sections were processed in the standard package Caraibes (developed by IFREMER). The resulting grids were then exported as ascii files and re-projected to their actual coordinates using Matlab. Finally, they were merged and visualised in the Fledermaus and Surfer software packages.

Each section was mapped twice. The first run was recorded at ca. 40 m distance from the cliff, to create a 3D grid with more extensive coverage, and to obtain a feel for the complex terrain (ROV risk mitigation). Subsequently, a second run was carried out at a distance of ca. 15–20 m, to obtain a grid of 10 × 10 cm resolution. Due to the limited Doppler range, no parallel lines at different depths could be acquired.

13.4 Results

A range of images, showing the processed grids and the final point cloud model, are presented in Fig. 13.3. Although this method of ROV-based 'forward mapping' was tested once before along a straight wall in the Whittard submarine canyon (Huvenne et al. 2011), the data presented here represent the first application of the technique on an irregular feature, requiring the mapping to be carried out in several sections. The resulting sections, once back-projected, match seamlessly, correctly representing the curved nature of the headwall scarp and the complex terrain morphology in a true 3D model.

The model integrates well with the regional shipboard bathymetry (Fig. 13.3a), but also underlines the shortcomings of that broadscale dataset: while the maximum slope values for the headwall scarp, calculated from the shipboard bathymetry, reach 60–65°, the forward mapped data show that some cliff sections are actually vertical, or even overhanging. The effect of apparent terrain smoothing as a result of mapping at lower resolution is well known (Wilson et al. 2007), and could be avoided by using higher resolution equipment, potentially deployed closer to the seabed (e.g. using Autonomous Underwater Vehicles, Wynn et al. (2014)). However, the ability of mapping terrain morphology under overhangs can never be achieved by traditional, downward-looking equipment, which would always produce acoustic shadowing at whatever achieved resolution.

The dataset is affected by one major artefact, however: excessive noise in the navigation data caused across-track striping which manifests itself as vertical

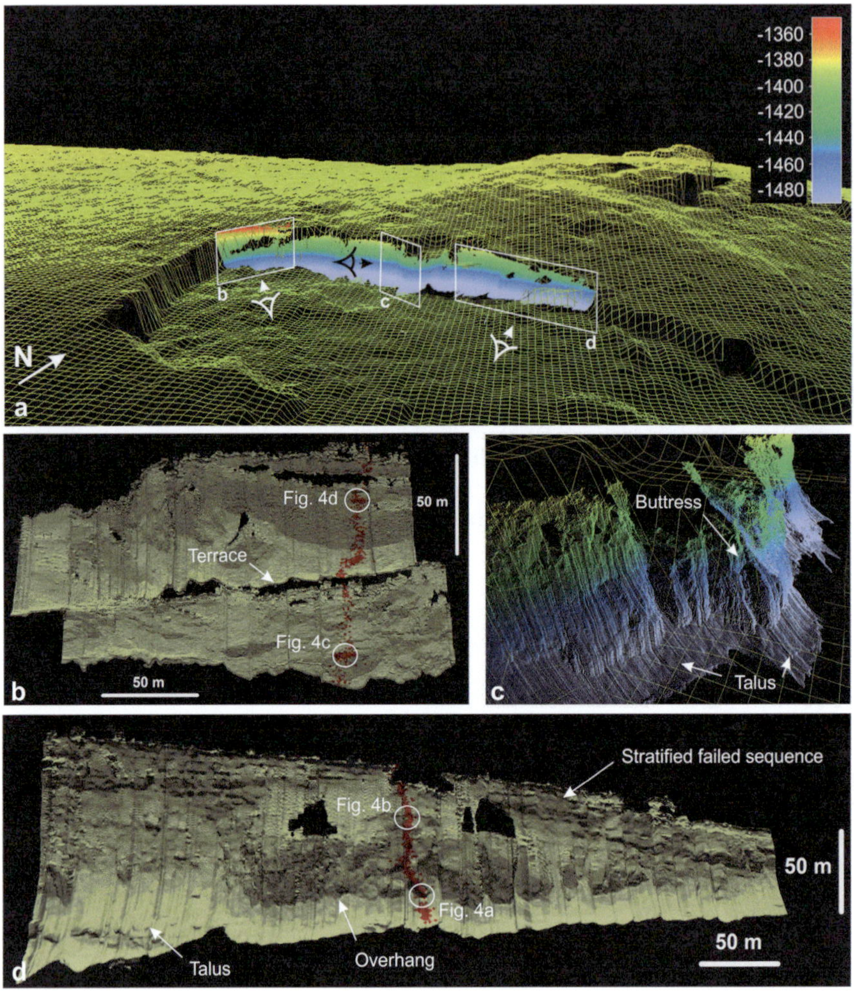

Fig. 13.3 (**a**) 3D view of the ROV-based forward-mapped headwall scarp (colour scale representing depth, 0.4 × 0.4 m resolution point cloud) embedded within the shipboard bathymetry (*yellow*, 20 × 20 m mesh). No vertical exaggeration. Locations of individual details are indicated by *white boxes*. (b–d) Detailed views of individual sections of the headwall scarp, including the location of 2 ROV video transects (*red*) carried out in the area. Section (**b**) and (**d**) illustrate gridding results in the vertical plane in which the data were processed (i.e. the X′-Y′ plane, Fig. 13.2)

striping in the 3D model. The USBL navigation reached a nominal accuracy of ~0.5° of the slant range. At 1400 m water depth, this resulted in potential errors in the order of ~12 m, orders of magnitude above the desired pixel resolution. Carrying out the surveys directly adjacent to a highly reflective, vertical cliff may also have increased the noise level in the USBL data. Even Kalman-filtered with the help of the DVL, the navigation tracks still contained considerable irregularities. Further smoothing during post-processing (before projection) visibly improved the results, but some of the effects remain. Although the human eye easily compensates for this, and the data can be interpreted without too much difficulty, the effect will interfere with any potential automated analysis. The artefact is known from ROV-based and ultra-high resolution bathymetric surveys: as the desired grid resolution reduces to sub-metre pixel sizes, the accuracy of navigation and attitude sensors often cannot reach similar standards, typically causing across-track striping (Hughes Clarke 2003).

13.5 Interpretation and Discussion

The new, ultra-high resolution 3D model clearly shows the overall morphology of the scarp, which is irregular with major blocks protruding as buttresses from the wall (Fig. 13.3c). They may represent remnant sections that resisted rupture, or could include individual blocks that partly slid down the wall during a subsequent phase of collapse. The entire mapped portion of the scarp is characterised by a talus at its foot (>1470 m depth), followed by steep, vertical or even overhanging sections (1470–1430 m depth), and finally a receding section that abruptly ends at the edge with the overall (near-horizontal) seafloor. The apex of the headwall scarp shows a two-step morphology, with a low-relief terrace in between (Fig. 13.3b).

Correlation with the two ROV video transects carried out in the area shows different rock types for the different sections of the scarp (Fig. 13.4). At the vertical and overhanging walls, outcrops can be found which appear to be of extrusive volcanic origin (Fig. 13.4a), supporting the hypothesis that the pinnacle to the NW could be a volcanic cone (Georgiopoulou et al. 2013). Above that, where the cliff starts to recede, sedimentary rocks are present with distinctive layering and apparently coarse-grained texture. Based on the erosional pattern, we suggest a potential volcaniclastic origin (Fig. 13.4b, c).

Finally, the upper cliff at the headwall apex is formed in a different type of lithology, more homogeneous, probably calcareous, and etched by the burrows and excavations of different benthic species, which suggest it is softer and finer-grained (Fig. 13.4d). The megafauna communities observed throughout both video transects, although varying with depth and substrate type, are dominated by a stony coral, most probably the scleractinian *Solenosmilia variabilis*, and the limid clam *Acesta excavata*. Talus deposits, developed at the base of each steep cliff section, are composed of a mixture of fossil coral rubble and other faunal debris (e.g., shells) in addition to sediments and small rock fragments that most probably originated from the erosion of the outcrops above (Fig. 13.4d).

The overall morphology of this small landslide scarp points to a translational slide mechanism associated with extension and spreading: the fairly high and steep walls, which illustrate the cohesive strength of the failed material, their irregular morphology, and the approximately planar surface are all coherent with such

Fig. 13.4 Example photographs illustrating the rock types and biological communities found at different locations on the headwall scarp. Scale bars are 20 cm, locations of the photographs are indicated in Fig. 13.3. (**a**) Rock outcrop, potentially of volcanic origin. (**b**) Stratified, potentially volcaniclastic rock with irregular erosion pattern. (**c**) Stratified, potentially volcaniclastic rock showing cross-bedding. (**d**) More homogeneous outcropping rock type. Note talus of coral rubble and finer sediment at the bottom. The dominant species visible in all photographs are the stony coral *Solenosmilia variabilis* and limid bivalve *Acesta excavata*. In addition there are various (encrusting) sponges, hydroids, brittle stars etc

interpretation (Micallef et al. 2007; Masson et al. 2010). However, although some remnant blocks were found in the centre of the slide scar (identified on ROV video), the failed material seems to be nearly completely evacuated, which would suggest rapid disintegration and transport down the slope (Hogan et al. 2013). Talus material observed on the ROV video data appears to have been formed in situ, through gradual erosion of the cliff faces, subsequent to the failure event.

The location of the slide, on the flank of the moat, may have contributed to the pre-conditioning factors. Undercutting by the slope-parallel current may have reduced the stability of this slope section (Elliot et al. 2010). Deflection of this current around the volcanic cone may have altered the flow from potentially laminar to locally turbulent, enhancing the erosive action. However, the role of the heterogeneous material in the area (volcaniclastic, contouritic, hemipelagic) with different geotechnical properties will need to be investigated further in order to fully understand the failure mechanism that created this small landslide.

13.6 Conclusions and Further Work

The method described above provides a new and unique way to look at submarine headwall scarps, and has the potential to initiate similar scientific breakthroughs as the development of LIDAR technology has done in the terrestrial domain (e.g. Glenn et al. 2006; Lissak et al. 2014). Especially when combined with further ROV-based video surveys and sampling, this will allow for fully integrated studies of headwall scarp morphology and lithology, leading to better insights into the processes behind headwall rupture, and to the potential identification of unstable parts and incipient scarps. It also gives unprecedented access to the failed sequence – which otherwise can only be sampled through expensive deep coring or drilling. The next step in the development of the method will include processing and visualisation of the multibeam backscatter within the 3D model, resulting in acoustic images of cliff reflectivity, containing information on substrate, rock types, sediment cover and potential biological communities.

Acknowledgements The SORBEH expedition ('Slope collapses On Rockall Bank and Escarpment Habitats') was supported by the Irish Marine Institute and was funded under the Marine Research Sub-Programme by the Irish Government. VH is holder of the ERC Starting Grant CODEMAP ('Complex Deep-sea Environments: Mapping habitat heterogeneity As Proxy for biodiversity', Grant number 258482), which co-funded the vertical mapping exercise. The authors would like to thank the cruise-organising team in the Marine Institute, the captain and crew of the RV Celtic Explorer, and the ROV engineering team, especially Patrick O'Driscoll and Will Handley for making this work possible. We are very grateful to Terje Thorsnes and John Hughes Clarke for their helpful and supportive reviews.

References

Elliot GM, Shannon PM, Haughton PDW, Øvrebø LK (2010) The Rockall Bank Mass Flow: collapse of a moated contourite drift onlapping the eastern flank of Rockall Bank, west of Ireland. Mar Pet Geol 27:92–107

Georgiopoulou A, Shannon PM, Sacchetti F, Haughton PDW, Benetti S (2013) Basement-controlled multiple slope collapses, Rockall Bank Slide Complex, NE Atlantic. Mar Geol 336:198–214

Glenn NF, Streutker DR, Chadwick DJ, Thackray GD, Dorsch SJ (2006) Analysis of LiDAR-derived topographic information for characterising and differentiating landslide morphology and activity. Geomorphology 73:131–148

Heezen BC, Ewing M (1952) Turbidity currents and submarine slumps, and the 1929 Grand Banks earthquake. Am J Sci 250:849–873

Hogan KA, Dowdeswell JA, Mienert J (2013) New insights into the slide processes and seafloor geology revealed by side-scan imagery of the massive Hinlopen Slide, Arctic Ocean margin. Geo-Mar Lett 33:325–345

Hughes Clarke JE (2003) Dynamic motion residuals in swath sonar data: ironing out the creases. Int Hydrogr Rev 4:6–23

Hühnerbach V, Masson DG, project potC (2004) Landslides in the North Atlantic and its adjacent seas: an analysis of their morphology, setting and behaviour. Mar Geol 213:343–362

Huvenne VAI, Tyler PA, Masson DG, Fisher EH, Hauton C, Hühnerbach V, Le Bas TP, Wolff GA (2011) A picture on the wall: innovative mapping reveals cold-water coral refuge in submarine canyon. PLoS One 6:e28755

Krastel S, Behrmann J-H, Völker D, Stipp M, Berndt C, Urgeles R, Chaytor J, Huhn K, Strasser M, Harbitz CB (2014) Submarine mass movements and their consequences. Springer, Switzerland

Lissak C, Maquaire O, Malet J-P, Bitri A, Samyn K, Grandjean G, Bourdean C, Reiffsteck P, Davidson R (2014) Airborne and ground-based data sources for characterizing the morpho-structure of a coastal landslide. Geomorphology 2017:140–151

Masson DG, Wynn RB, Talling PJ (2010) Large landslides on passive continental margins: processes, hypotheses and outstanding questions. In: Mosher DC, Shipp RC, Moscardelli L, Chaytor JD, Baxter CDP, Lee HJ, Urgeles R (eds) Submarine mass movements and their consequences. Springer, Heidelberg, pp 153–165

Micallef A, Masson DG, Berndt C, Stow DAV (2007) Morphology and mechanics of submarine spreading: a case study from the Storegga Slide. J Geophys Res Earth Surf 112:F03023

Micallef A, Berndt C, Masson DG, Stow DAV (2008) Scale invariant characteristics of the Storegga Slide and implications for large-scale submarine mass movements. Mar Geol 247:46–60

Øvrebø LK, Haughton PDW, Shannon PM (2005) Temporal and spatial variations in late quaternary slope sedimentation along the undersupplied margins of the Rockall Trough, offshore west Ireland. Nor J Geol 85:279–294

Shannon PM (1991) The development of Irish offshore sedimentary basins. J Geol Soc Lond 148:181–189

Stoker MS, Praeg D, Hjelstuen BO, Laberg JS, Nielsen T, Shannon PM (2005) Neogene stratigraphy and the sedimentary and oceanographic development of the NW European Atlantic margin. Mar Pet Geol 22:977–1005. doi:10.1016/j.marpetgeo.2004.11.007

Wilson MFJ, O'Connell B, Brown C, Guinan JC, Grehan AJ (2007) Multiscale terrain analysis of multibeam bathymetry data for habitat mapping on the continental slope. Mar Geod 30:3–35

Wynn RB, Huvenne VAI, Le Bas TP, Murton BJ, Connelly DP, Bett BJ, Ruhl HA, Morris KJ, Peakall J, Parsons DR, Sumner EJ, Darby SE, Dorrell RM, Hunt JE (2014) Autonomous Underwater Vehicles (AUVs): their past, presence and future contributions to the advancement of marine geoscience. Mar Geol 352:451–468

Chapter 14
Flow Behaviour of a Giant Landslide and Debris Flow Entering Agadir Canyon, NW Africa

Sebastian Krastel, Russell B. Wynn, Peter Feldens, Anke Schürer, Christoph Böttner, C. Stevenson, Matthieu J.B. Cartigny, Veit Hühnerbach, and Daniel Unverricht

Abstract Agadir Canyon is one of the largest submarine canyons in the World, supplying giant submarine sediment gravity flows to the Agadir Basin and the wider Moroccan Turbidite System. While the Moroccan Turbidite System is extremely well investigated, almost no data from the source region, i.e. the Agadir Canyon, are available. New acoustic and sedimentological data of the Agadir Canyon area were collected during RV Maria S. Merian Cruise 32 in autumn 2013. The data show a prominent headwall area around 200 km south of the head of Agadir Canyon. The failure occurred along a pronounced weak layer in a sediment wave field. The slab-type failure rapidly disintegrated and transformed into a debris flow, which entered Agadir Canyon at 2500 m water depth. Interestingly, the debris flow did not disintegrate into a turbidity current when it entered the canyon despite a significant increase in slope angle. Instead, the material was transported as debrite for at least another 200 km down the canyon. It is unlikely that this giant debris flow significantly contributed to the deposits in the wider Moroccan Turbidite System.

S. Krastel (✉)
Institute of Geosciences, Christian-Albrechts-University at Kiel, Kiel, Germany
e-mail: skrastel@geophysik.uni-kiel.de

R.B. Wynn • M.J.B. Cartigny
National Oceanography Centre, Southampton, UK

P. Feldens • A. Schürer • C. Böttner • D. Unverricht
Institute for Geosciences, Kiel University, Kiel, Germany

C. Stevenson
School of Earth and Environment, University of Leeds, Leeds, UK

V. Hühnerbach
GEOMAR Helmholtz Centre for Ocean Research Kiel, Kiel, Germany

G. Lamarche et al. (eds.), *Submarine Mass Movements and their Consequences*,
Advances in Natural and Technological Hazards Research 41,
DOI 10.1007/978-3-319-20979-1_14

14.1 Introduction

Submarine canyons are major incisions at continental margins around the world, which are usually formed by downslope traveling sediment-laden gravity flows, so-called turbidity currents (Daly 1936; Clark and Pickering 1996; Puig et al. 2014). They represent important pathways for the transfer of sediments from the shelf and upper continental slope into the deep sea. Agadir Canyon (Fig. 14.1) is one of the largest submarine canyons in the World; it is 450 km long, up to 30 km wide and 1250 m deep. It incises the Moroccan Shelf at 200 m water depth and terminates on the floor of Agadir Basin at 4300 m water depth. The Agadir Canyon supplies some of the World's largest submarine sediment-gravity flows (>100 km^3 volume and up to 2000 km run-out) to Agadir Basin and the wider Moroccan Turbidite System. The Moroccan Turbidite System has been intensively studied for almost 30 years based on acoustic data and more than 200 shallow sediment cores (e.g., Wynn et al. 2002; Frenz et al. 2009; Hunt et al 2013). Coring and drilling of basin-floor sediments has revealed a long sequence of turbidites, mostly sourced from the Moroccan continental margin and the volcanic Canary Islands. The largest individual flow deposits in the Moroccan Turbidite System contain sediment volumes >100 km^3, although these large-scale events are relatively infrequent with a recurrence interval of ~10,000 years (over the last 200,000 years). The largest siliciclastic flow in the last 200,000 years was the 'Bed 5 event' (Wynn

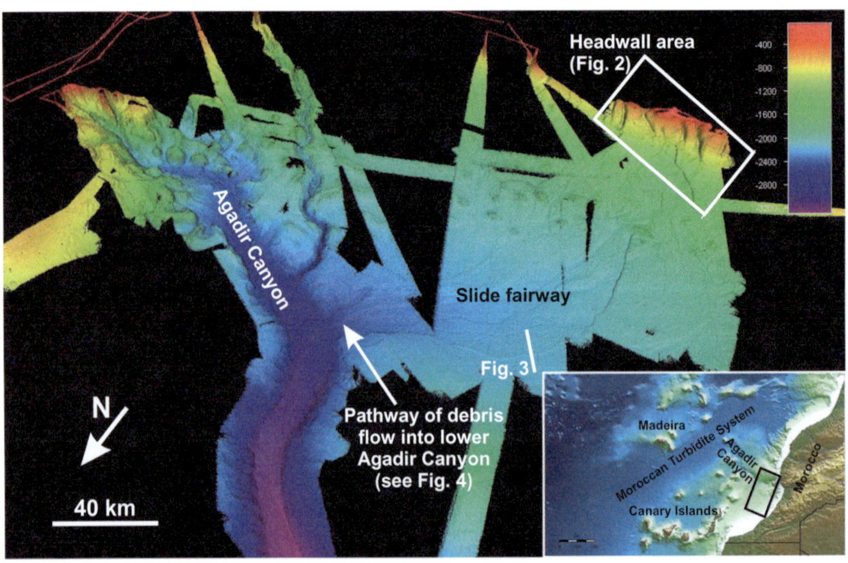

Fig. 14.1 3D perspective view of shipborne multibeam bathymetry data from the survey area, showing Agadir Canyon and a pronounced landslide headwall area south of the canyon. The slide fairway can be traced from the headwall area to lower Agadir Canyon. *Inset* shows general location off NW Africa

et al. 2010), which transported 160 km^3 of sediment up to 2000 km from Agadir Canyon to the southwest Madeira Abyssal Plain. The flow is dated at 60 ka and originated on the upper slope offshore Atlantic Morocco in the vicinity of Agadir Canyon. Two further siliciclastic flows have reached the Madeira Abyssal Plain in the last 200,000 years, both transporting >100 km^3 of sediment (Wynn et al. 2002; Frenz et al. 2009). These large flow volumes strongly suggest a landslide source. Therefore, it is likely that at least three different giant landslides have occurred in and around upper Agadir Canyon over this 200,000-year period. However, the exact source locations of these landslides are yet unknown.

In order to investigate the source area of these large-volume flows and determine whether the landslide mass transformed to a large-volume turbidity current and contributed to the sedimentary succession of the wider Moroccan Turbidite System, a combined acoustic (bathymetric multibeam sounder, sediment echo sounder, airgun seismic) and core data set was collected from RV Merian in the upper Agadir Canyon region in autumn 2013. While no obvious major landslide scarps (headwall height >20 m, diameter of scar area >1 km) were identified in the canyon area itself but a very prominent failure area was identified about 200 km south of the canyon head (Fig. 14.1). The objective of this manuscript is to provide the first description of this giant landslide, and investigate flow behaviour along its pathway from the upper continental margin into Agadir Canyon.

14.2 Methods

Data were acquired during RV Maria S. Merian Cruise MSM32 in autumn 2013. Bathymetric data were collected with a hull-mounted Kongsberg Simrad EM122 system. An Atlas Parasound parametric system was used for collecting sediment echo sounder data, with 4 kHz selected as secondary low frequency. A digital 88-channel Geometrics GeoEel streamer (channel distance of 1.56 m) and a 1.7 l standard GI-gun were used to acquire high-resolution multichannel seismic data. Processing of the airgun seismic data included binning at 2 m bin distance, frequency filtering (40–400 Hz), velocity analysis, normal moveout correction, stacking, and post-stack time migration. Vertical resolution is in the range of a few meters.

14.3 Results

Agadir Canyon (Fig. 14.1) is a deeply incised canyon, which can be traced for about 450 km into the Agadir Basin. The canyon head area comprises several tributary canyons, but no large landslide scarps are visible. In contrast, a pronounced failure area is imaged ~200 km south of Agadir Canyon (Fig. 14.1). This landslide is marked on some previously published maps of the NW African margin (e.g., Wynn et al. 2000; Weaver et al. 2000; Masson et al. 2006) but, to our knowledge, it has not been described. We will therefore use the name ‘Agadir Slide’ for this feature. An obvious ‘fairway’ of landslide material is clearly imaged beneath the failure area; this fairway enters lower Agadir Canyon at ~2500 m water depth. The individual sections of the Agadir Slide are described below.

14.3.1 The Headwall Area and the Slide Fairway

The headwall area of the Agadir Slide (Fig. 14.2) shows a complex pattern of individual landslide scarps, incised canyons and sediment waves. The main headwall cuts back to 500 m water depth, where it intersects a deeply incised canyon. Headwall heights are in the range of 60–100 m. The uppermost section of the evacuated area is very narrow (~2 km) but it quickly widens to ~10 km at ~1300 m water depth. Well-developed sidewalls up to 90 m high are imaged on both sides of the scar area at this water depth. The area with missing sediments (negative relief) covers ~600 km^2; the volume of missing sediment equates to ~40 km^3. No landslide deposits, such as major blocks, are visible on bathymetric data in the headwall region. Additional smaller landslide scarps can be identified on bathymetric and sediment echo sounder data.

Sediment waves are especially well developed on the slopes adjacent to and downstream of canyon mouths. Sediment wave heights are 10–60 m and wavelengths are 1–3 km. They are located in 330–1000 m water depth, which corresponds to the main failure area of the Agadir Slide. A dip-oriented seismic section crossing the sediment waves and the Agadir Slide headwall is shown on Fig. 14.2b. The sediment waves developed above a basal reflector, which is found at ~300 ms Two-way-Travel Time (TWT, corresponding to ~250 m at 1700 m/s assuming water-rich hemipelagic sediments). The height of the sediment waves increases with decreasing subsurface depth, which in combination with a shallow dip angle of ~3–14° between individual waves, clearly characterizes them as sediment waves and not extensional fault features (Wynn and Stow 2002). Correlation of acoustic layers between individual waves is possible (Fig. 14.2b), revealing that the near-surface succession above the Agadir Slide headwall is 60 ms TWT thicker than that below the headwall, which is in good agreement with the headwall height mentioned above. On the seismic profile shown in Fig. 14.2b it is difficult to distinguish between the relatively steep seaward-facing flanks of individual sediment waves and the slide headwall, but the headwall is clearly visible as a morphological step (cutting the waves obliquely) in planform bathymetric data (Fig. 14.2a). It is interesting to note that the sediment waves are not destroyed by the landslide beneath the headwall; they are fully intact, except for the fact that ~60 ms (TWT) of sediment above the glide plane has been removed by the landslide.

Downslope of the headwall, the landslide morphology is more typical of a normal depositional debris flow. Debrite deposits up to ~50 m thick are imaged at ~1600 m water depth at a distance of 30 km from the headwall. Typical slope gradients in this region are $<1°$, compared to average slope gradients in the failure

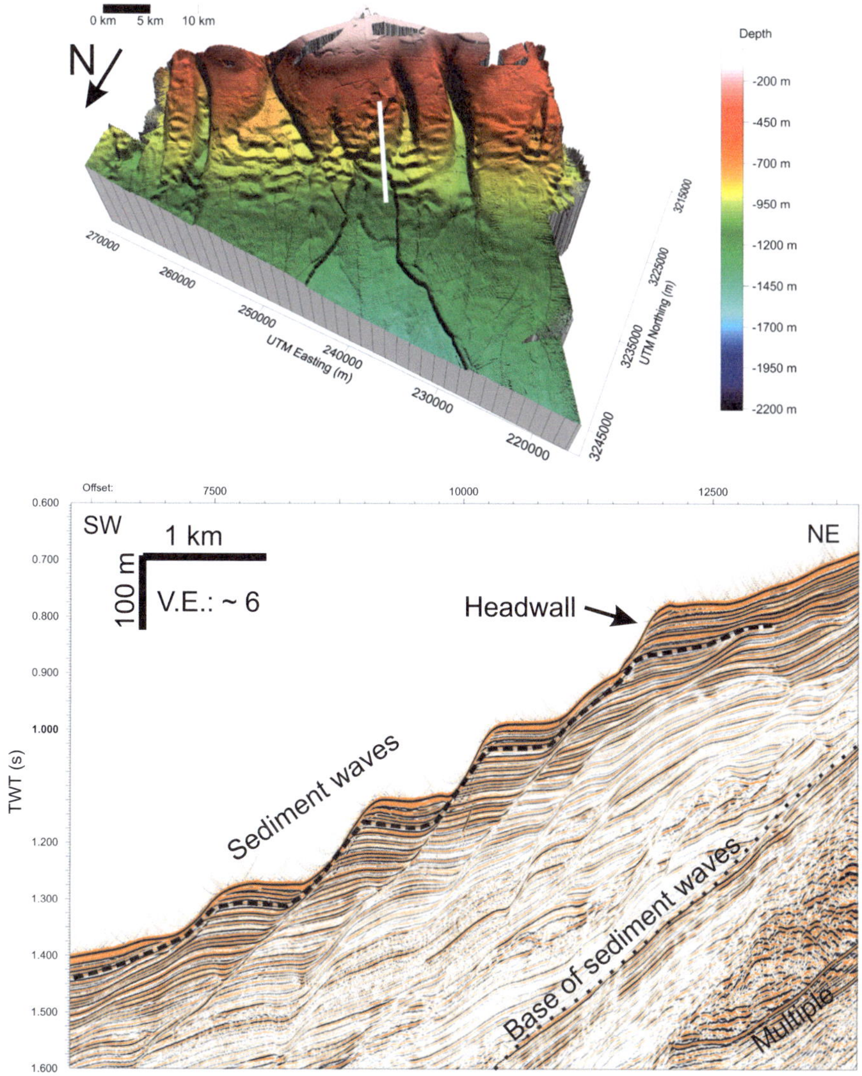

Fig. 14.2 *Upper* panel: 3D perspective view of the headwall area of the Agadir Slide. See Fig. 14.1 for location. *Lower panel*: Seismic profile crossing the headwall and the sediment waves beneath the headwall. Sediment thickness above the dashed reflector is significantly greater for the area above the headwall, suggesting removal of a substantial volume of sediment by the landslide. The location of the profile is marked as a *solid white line* on Fig. 14.2a

area of ~4°. The debris flow pathway is here termed the fairway, and is clearly visible on bathymetric data (Fig. 14.1). Debrite deposits display positive relief in this area, and are characterized by a relatively rough surface and transparent seismic facies on sediment echo sounder data (Fig. 14.3). Cores taken in this area (Fig. 14.3) recover a sheared and contorted debrite beneath 470 cm of undisturbed hemipelagic

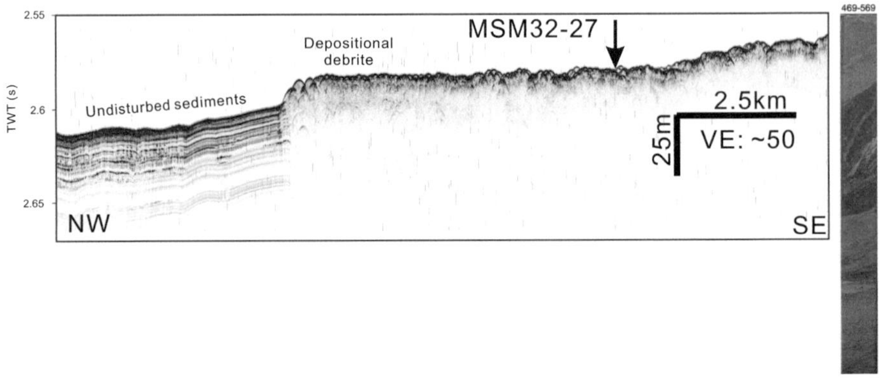

Fig. 14.3 Sediment echo sounder profile crossing the margin of Agadir Slide in the fairway area. For location see Fig. 14.1. The photograph of the core shows a section of core MSM32-27 from 469 to 569 cm sub-sea floor depth (location marked on profile). A sheared and contorted debrite was sampled beneath ~470 cm of undisturbed background sediments

'background' sediments. The estimated age of the slide deposits is ~140 ka based on litho-stratigraphic correlation of the undisturbed units with a dated core outside the slide area. These numbers result in a long-term sedimentation rate of 3.5 cm/ 1000 years, which is in god agreement with sediment accumulation rates reported for this area (Kuhlmann et al. 2004).

14.3.2 Slide Fairway into Lower Agadir Canyon

The Agadir Slide fairway can be tracked into lower Agadir Canyon at ~2500 m water depth (Fig. 14.4). The slope angle of the fairway immediately south of the canyon is only ~0.3°, but increases to >2° on the canyon margin. The fairway has negative relief where it enters the canyon, especially on its western margin (Figs. 14.4 and 14.5a). Slope sediments on either side of the fairway are well stratified and mainly undisturbed. A relatively small and shallow secondary failure is visible on the western margin of the fairway (Figs. 14.4 and 14.5a). Debrite deposits of the youngest landslide event are clearly visible within the fairway as an up to 50 m-thick and ~16 km-wide transparent to chaotic unit. The sedimentary succession beneath the youngest slide deposit shows a variable and disturbed seismic pattern with abundant mass transport deposits of varying sizes down to 700 ms TWT. The disturbed zone narrows to about 10 km width at ~250 ms TWT subsurface depth (Fig. 14.5a). This narrowing coincides with a slight increase of seismic amplitudes, but the general disturbed pattern remains the same. The seismic data suggest a formerly deeply incised slide fairway, which was subsequently filled with mass transport deposits, turbidites, and hemipelagic background sediments. This suggests the fairway has been active for a long time, and therefore represents an important sediment transport pathway into Agadir Canyon.

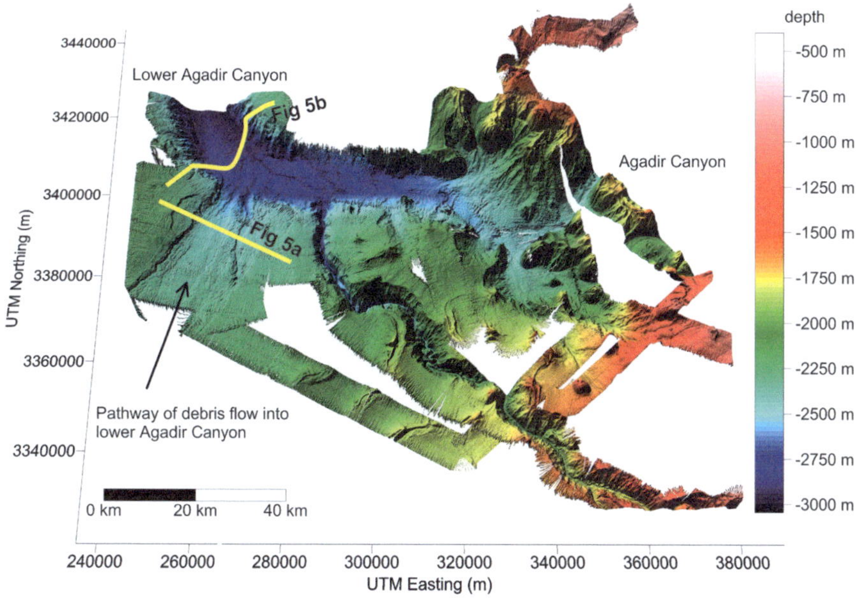

Fig. 14.4 3D perspective view of Agadir Canyon, showing where the Agadir Slide fairway joins the canyon from the south. See Fig. 14.1 for general location

Figure 14.5b shows a seismic profile across Agadir Canyon slightly downslope of the location where the fairway enters the canyon. Agadir Canyon is more than 500 m deep at this location and deeply incised into well-stratified sediments. Canyon width is almost 20 km. The canyon floor is relatively flat but slightly inclined to the SE. At least three thick (>80 m) transparent units are imaged beneath the present canyon floor, which are interpreted to be thick debrite deposits that entered the canyon via the fairway (they are absent on seismic profiles crossing the canyon further upslope). Gravity coring of the uppermost transparent unit confirmed the presence of sheared and contorted debrite. A similar succession of stacked debrites is found on seismic profiles crossing Agadir Canyon further downslope, some 400 km from the canyon head. This suggests that debris flows did not undergo further disintegration upon entering the canyon, although more detailed analysis of sediment core data (including dating) will be required to confirm this result.

14.4 Discussion and Conclusion

The Agadir Slide forms a major mass transport deposit, which can be traced from its failure area at ~500 m water depth down to the lower Agadir Canyon (Fig. 14.1). The total run-out distance is >400 km. The depth of the failure area and the height of the headwall are typical for the NW African margin (Hühnerbach et al. 2004;

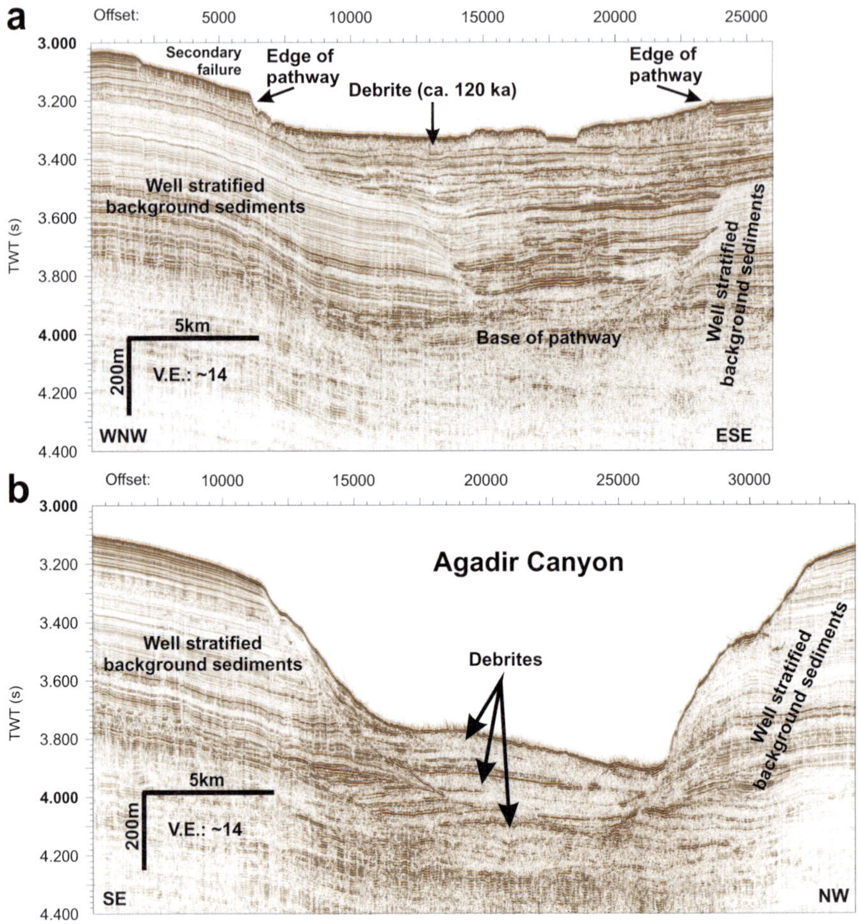

Fig. 14.5 (**a**) Seismic profile crossing the fairway, where it enters Agadir Canyon. (**b**) Seismic profile crossing Agadir Canyon slightly downslope of the location where the fairway enters the canyon. Details are given in the text. See Fig. 14.4 for location of profiles

Krastel et al. 2012). The morphology of the headwall suggests a kind of slab-type failure along a pronounced weak layer where sediment waves beneath the Agadir headwall are intact and not destroyed by the landslide (Fig. 14.2). This points to a modified slab failure because slab-type failures usually occur on more planar, seaward dipping features (e.g., Masson et al. 2006). However, the preservation of the sediment waves beneath the headwall of the Agadir Slide strongly supports the importance of a weak layer for this failure, as suggested for many landslides (e.g., Masson et al. 2006). On possible explanation for the preservation of the sediment waves may be their subdued morphology and low aspect ratio.

We can only speculate about the nature of the weak layer because we do not have cores penetrating through this layer. However, the general arrangement of (i) canyon mouths, (ii) sediment waves, and (iii) slope sediments with good acoustic penetration around the landslide headwall (Fig. 14.2) suggests that a sequence of well-sorted turbidite sand bases interbedded with fine-grained turbiditic and hemipelagic sediments is likely. Such a setting may result in (i) good grain-size sorting of turbidite sands making the sediments weaker than poorly sorted sediments; this may facilitate liquefaction in response to cyclic loading during, for example, earthquake shaking, and (ii) high water contents and underconsolidation leading to reduced shear strengths (Laberg and Camerlengi 2008). Hence, it is likely to have weak layers in the sedimentary succession and the landslide occurred along one of these weak layers but the exact preconditioning factors and the final trigger of the Agadir Slide failure remain speculative. Morphological data indicate that the initial failure rapidly disintegrated and transformed into a debris flow. Thick debrite deposits (~50 m) are found all along the slide fairway. The decreasing slope angle from ~1° in the uppermost depositional area at ~1600 m water depth to about 0.3° before the landslide enters Agadir Canyon does not have an obvious significant effect on the flow behaviour of the Agadir Slide.

A significant increase of the slope angle to ~2° is found where the Agadir Slide enters Agadir Canyon. Such changes in slope angle are potential locations for flow transformations and the development of an associated turbidity current. For example, Talling et al. (2007) observed such transformation in sediment flows crossing very gentle slope breaks in the Moroccan Turbidite System. The seismic and core data presented in this study, however, suggest the Agadir Slide continued to move as debris flow when entering the canyon and it continued to move as debris flow for another 200 km within the canyon.

The new seismic data show that the Agadir Slide fairway has been a persistent conduit for debris flows entering Agadir Canyon. We do not know if older landslides entering the canyon along the fairway contributed to the sedimentary succession of the Moroccan Turbidite System, but we do know that some of the major flows in the Moroccan Turbidite System are significantly younger than the ~140 ka-old Agadir Slide. For example, the gigantic 'Bed 5 event' with a volume of ~160 km^3 is dated at ~60 ka (Wynn et al. 2010). This flow must have originated from another source away from the Agadir Slide, because the ~140 ka-old Agadir Slide is the youngest major landslide originating from the headwall area identified here, ~200 km south of Agadir Canyon. We therefore hypothesise that the most likely source area for some of the major flow deposits in the Moroccan Turbidite System is the head region of Agadir Canyon, despite the fact that no clear landslide scarps are visible on the bathymetric data. Future work will test this hypothesis through detailed provenance and dating analysis of flow deposits in cores recovered from Agadir Canyon and adjacent slopes.

Acknowledgments We thank all scientists and crew of RV Merian Cruise MSM32 in autumn 2013 for their support during data collection. The authors are thankful to Jason Chaytor and David Völker for their reviews and constructive comments. Financial support was provided by the Deutsche Forschungsgemeinschaft and the UK Natural Environment Research Council (NERC grant NE/J012955/1).

References

Clark JD, Pickering KT (1996) Submarine channels; processes and architecture. Vallis Press, London

Daly RA (1936) Origin of submarine canyons. Am J Sci Fifth Ser XXXI 186:401

Frenz M, Wynn RB, Georgiopoulou A, Bender VB, Hough G, Masson DG, Talling PJ, Cronin BT (2009) Provenance and pathways of Late Quaternary turbidites in the deep-water Agadir Basin, northwest African margin. Int J Earth Sci 98:721–733

Hühnerbach V, Masson DG, Project Partners (2004) An analysis of submarine landslide dynamics and processes in the North Atlantic. Mar Geol 213:343–362

Hunt JE, Wynn RB, Talling PJ, Masson DG, Douglas G (2013) Frequency and timing of landslide-triggered turbidity currents within the Agadir Basin, offshore NW Africa: are there associations with climate change, sea level change and slope sedimentation rates? Mar Geol 346:274–291

Krastel S, Wynn RB, Georgiopoulou A, Geersen J, Henrich R, Meyer M, Schwenk T et al (2012) Large scale mass wasting at the NW-African continental margin: some general implications for mass wasting at passive continental margins. In: Yamada Y (ed) Submarine mass movements and their consequences, vol 31, Advances in natural and technological hazards research. Springer, Berlin, pp 189–199

Kuhlmann H, Freudenthal T, Helmke P, Meggers H (2004) Reconstruction of paleoceanography off NW Africa during the last 40,000 years: influence of local and regional factors on sediment accumulation. Mar Geol 207:209–224

Laberg JS, Camerlenghi A (2008) The significance of contourites for submarine slope stability. In: Rebesco M, Camerlenghi A (eds) Contourites. Developments in sedimentology 60. Elsevier, Amsterdam, pp 537–556

Masson DG, Harbitz CB, Wynn RB, Pedersen G, Løvholt F (2006) Submarine landslides: processes, triggers and hazard prediction. Philos Trans Soc A 364:2009–2039

Puig P, Palanques A, Martín J (2014) Contemporary sediment-transport processes in submarine canyons. Annu Rev Mar Sci 6:5.1–5.25

Talling PJ, Wynn RB, Masson DG, Frenz M, Cronin BT, Schiebel R, Akhmetzhanov AM, Dallmeier-Tiessen S, Benetti S, Weaver PPE, Georgiopoulou A, Zühlsdorff C, Amy LA (2007) Onset of submarine debris flow deposition far from original giant landslide. Nature 450:541–544

Weaver PPE, Wynn RB, Kenyon NH, Evans J (2000) Continental margin sedimentation with special reference to the Northeast Atlantic margin. Sedimentology 47:239–256

Wynn RB, Stow D (2002) Classification and characterisation of deep-water sediment waves. Mar Geol 192:7–22

Wynn RB, Masson DG, Stow DAV, Weaver PPE (2000) The Northwest African slope apron: a modern analogue for deep-water systems with complex seafloor topography. Mar Petrol Geol 17:253–265

Wynn RB, Weaver PPE, Masson DG, Stow DAV (2002) Turbidite depositional architecture across three interconnected deep-water basins on the northwest African margin. Sedimentology 49:669–695

Wynn RB, Talling PJ, Masson DG, Stevenson CJ, Cronin BT, Le Bas TP et al (2010) Investigating the timing, processes and deposits of one the world's largest submarine gravity flows: the 'Bed 5 event' off northwest Africa. In: Mosher DC (ed) Submarine mass movements and their consequences, vol 28, Advances in natural and technological hazards research. Springer, Berlin, pp 463–474

Chapter 15
Fine-Scale Morphology of Tubeworm Slump, Monterey Canyon

Charles K. Paull, Krystle Anderson, David W. Caress, Eve Lundsten, and Roberto Gwiazda

Abstract Multibeam bathymetry and chirp seismic reflection profiles collected using an autonomous underwater vehicle reveal the morphology and shallow seafloor structure of Tubeworm Slump on the flank of Monterey Canyon at an unprecedented resolution. The data show smaller subsidiary deformation above the headwall, on the headwall, within the sediment drape that covers the sole of the slide, and on the sidewall of Monterey Canyon below Tubeworm Slump. The AUV data indicate that the existing slump scar represents a composite of gravity-driven deformation generated by multiple failure events.

15.1 Introduction

Multibeam bathymetric data is the primary tool for identifying and characterising geomorphic evidence for submarine landslides in the marine environment. Accurate determination of their geometry and shallow structure is important for understanding the history of individual scars and their failure mechanisms. Autonomous underwater vehicles (AUV) are increasingly being used to map selected areas of the seafloor (Caress et al. 2008). Significantly higher resolution is achieved with AUVs compared to ship mounted multibeam sonars operating in deep waters ($\geq$500 m) because they operate close to the seafloor (minimising the issues associated with spherical spreading), in an acoustically quiet environment, and they are isolated from surface waves.

Tubeworm Slump (Fig. 15.1) is a slide scar in 2.1 km water depth with a diameter of ~3.2 km and a distinctive ~150 m high arcuate-shaped headwall scarp, perched at the edge of Smooth Ridge on the northern rim of Monterey Canyon, offshore California (Greene et al. 2002; Greene and Ward 2003). The downslope end of Tubeworm Slump is in 2,350–2,500 m water depths where the slope abruptly increases to ~20° along the flank of Monterey Canyon.

C.K. Paull (✉) • K. Anderson • D.W. Caress • E. Lundsten • R. Gwiazda
Monterey Bay Aquarium Research Institute, Moss Landing, CA, USA
e-mail: paull@mbari.org

G. Lamarche et al. (eds.), *Submarine Mass Movements and their Consequences*, Advances in Natural and Technological Hazards Research 41,
DOI 10.1007/978-3-319-20979-1_15

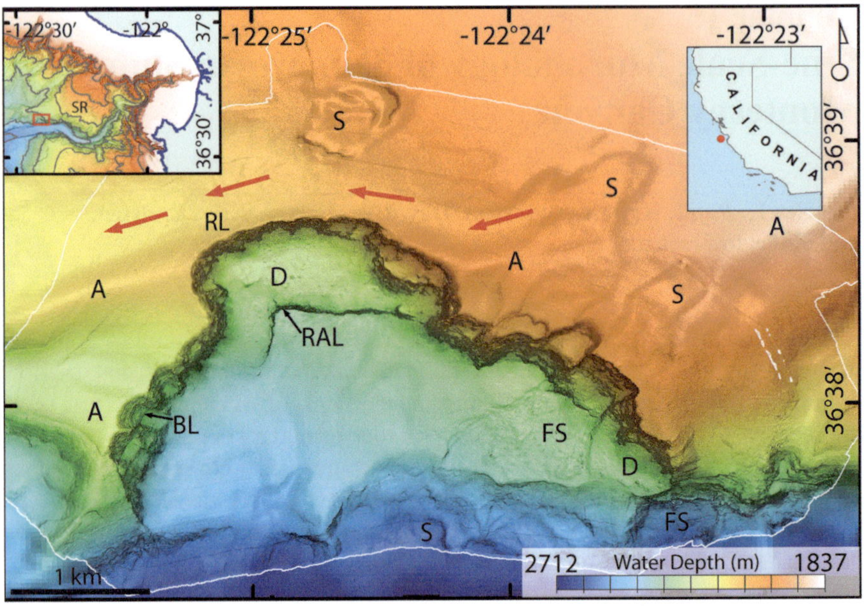

Fig. 15.1 Map showing AUV bathymetry (*circled* with *white line*) of Tubeworm Slump overlain on 30 kHz regional survey data (Greene et al. 2002). Inset maps shows location of Tubeworm. Slump (*red boxes*) and Smooth Ridge (SR). *Red arrows* indicate run-out path of slides up-slope of the raised lip on the main headwall. (*A* Anticline, *S* Slide scar, *FS* Fresh slide scar, *RL* Raised lip, *BL* Bedding lineations, *D* Debris, *RAL* Right angle ledge)

Tubeworm Slump was originally identified in regional multibeam data (Greene et al. 2002). This 30 kHz multibeam system had a 2° by 2° degree beam resolution. Therefore, the resolution based on the beam footprint diameter at nadir in 2,500 m water depths was 87 m.

Tubeworm Slump was named by Naehr et al. (2000) because Vestimentiferan tubeworms were discovered during an exploratory remotely operated vehicle (ROV) dive within this slide scar. Subsequent ROV dives showed that numerous chemosynthetic biological communities occur along the western headwall of this scar, consistent with the presence of methane oozing from the relatively fresh exposures on the slide scar surface (Paull et al. 2005). Models of Greene and Ward (2003) indicated that this slide potentially generated an up to 2 m high tsunami within Monterey Bay. This paper re-examines Tubeworm Slump utilising the increased resolution enabled by AUV technology.

15.2 Methods

Two AUV dives were conducted to map Tubeworm Slump (Figs. 15.1 and 15.2) by the Monterey Bay Aquarium Research Institute (MBARI). The AUV carried a Reson 7125 200-kHz multibeam sonar and an Edgetech 2- to 16-kHz chirp

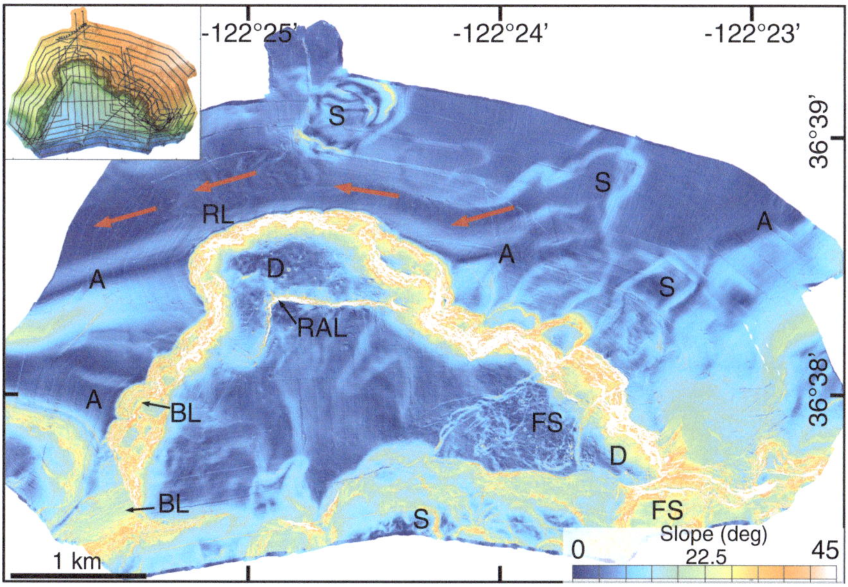

Fig. 15.2 Slope map for Tubeworm Slump covering the same area shown in Fig. 15.1. Inset map shows track lines ran during AUV surveys. *Red arrows* indicate run-out path of slides upslope of the raised lip on the main headwall. (*A* Anticline, *S* Slide scar, *FS* Fresh slide scar, *RL* Raised lip, *BL* Bedding lineations, *D* Debris, *RAL* Right angle ledge)

sub-bottom profiler (Caress et al. 2008). The AUV run up to 18 h long missions at a speed of 3 knots while maintaining an altitude of ~50 m off the seafloor. Track lines were laid out to avoid going into the steeper slopes, while trying to obtain overlapping multibeam bathymetric coverage (Fig. 15.2; inset map). In this mode, bathymetry (0.15 m vertical resolution and 1 m horizontal footprint) and chirp seismic-reflection profiles (~0.11 m vertical resolution) were collected simultaneously. The 200 kHz Reson 7125 has a 1° by 1° degree beam resolution, so at 50 m altitude the beam footprint diameter at nadir is 0.87 m. Given that the beam footprints stretch out away from nadir, the stated resolution is 1 m.

15.3 Results

15.3.1 Surface of Smooth Ridge Surrounding Tubeworm Slump

The surface of Smooth Ridge (Fig. 15.1; inset) outside the main headwall scarp of Tubeworm Slump contains subtle ridges and troughs. Most obvious is a 400 m wide, 15 m high ENE-WSW oriented ridge on the western side of Tubeworm Slump (Figs. 15.1 and 15.2). Chirp profiles perpendicular to the ridge show that

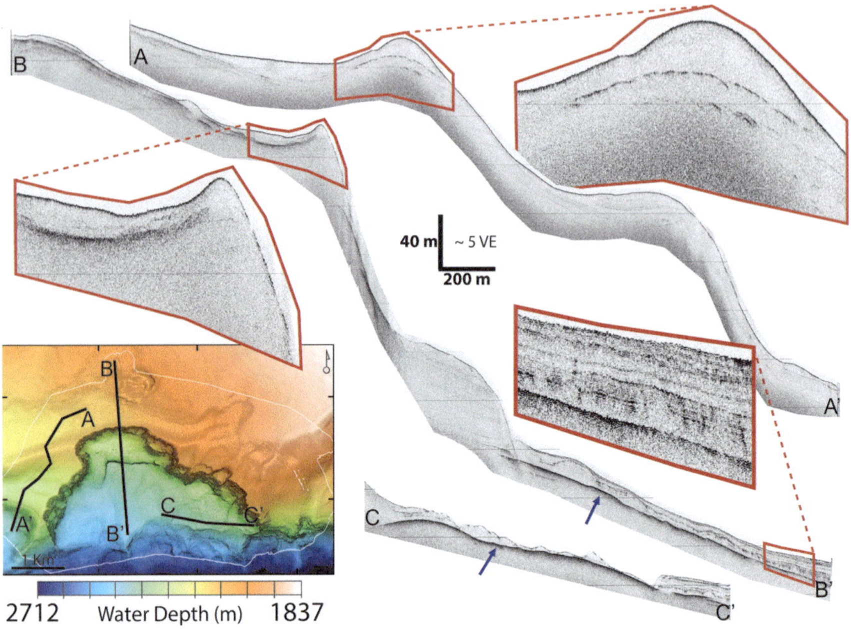

Fig. 15.3 Illustrative sections of chirp profiles. Inset map shows line locations. A-A′ is a profile on Smooth Ridge crossing the anticlines. Expanded section shows reflectors on both flanks dipping in opposite directions. B-B′ is a profile crossing the headwall and extending over the sole of the slide. Expanded sections show strata on top of the main headwall scarp dipping to the north and closely spaced small en echelon disruptions occurring within the lower half of the sediment cover on the slide scar. C-C′ is a profile crossing a secondary slide scar. *Blue arrow* shows prominent reflector interpreted to be main failure surface

reflectors on its flanks dip in opposite directions (Fig. 15.3A-A'), indicating that the ridge is the crest of an anticline (A in Figs. 15.1 and 15.2). The headwall scarp cuts through the ridge (Fig. 15.2), and the northern limb of the anticline forms a raised lip rimming the northern edge of the main headwall scarp (Figs. 15.1, 15.2, and 15.3). Chirp profiles across the headwall confirm that the raised lip is the remaining limb of the truncated anticline (Fig. 15.3B-B') because the strata underlying the lip dip to the north. The anticline remains intact to the east of the main headwall scarp. A similar sized anticline, also truncated by the slide, occurs ~900 m further down slope at the western flank of the slide scar.

Subtle slide scars are also seen above the main headwall scarp (S in Figs. 15.1 and 15.2). Although these scars have U-shaped headwalls that are up to 450 m across with $\leq$10 m of relief, they are comparatively subdued features as their headwalls and sidewall slopes are $<$10°. The run-out path for the scars on the southern side of the truncated anticline extend SW to the main headwall scarp of Tubeworm Slump. In contrast, the run-out path for the scars to the north of the truncated anticline are redirected to the WSW by the anticlinal ridge, away from main headwall scarp (Figs. 15.1 and 15.2).

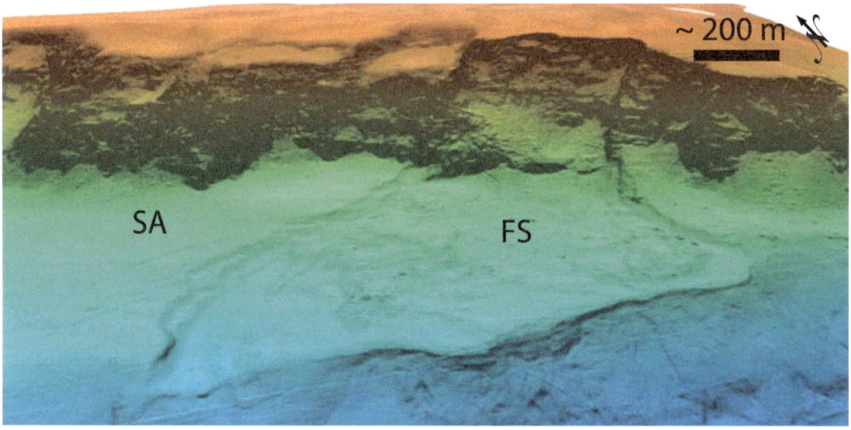

Fig. 15.4 Perspective view of fresh scar within Tubeworm Slump. No vertical exaggeration. (FS Fresh slide scar, *SA* Sediment apron)

15.3.2 Main Headwall Scarp

The main headwall scarp is a composite of about twenty smaller secondary arcuate scarps, each 50–350 m across (Figs. 15.1 and 15.2). Aprons of sediment drape lap-up onto most of these secondary scarps. Neither the sediment drape, nor the bedded strata exposed on the headwalls are crossed by gullies (Fig. 15.4). The sediment drape within some of the secondary headwall scarps is smooth, while in others, numerous blocks of debris stick up through the aprons. A sediment drape covers the sole of the slide below the headwall, but partly buried debris blocks that are up to 20 m long protrude up to 3 m above the sediment cover.

15.3.3 Sole of Slide Scar

A prominent strong reflector occurs in chirp profiles collected over the main slide scar's sole at up to 20 m sub-bottom depth. This reflector consistently slopes at 5° to the SSW into the canyon. The few weak reflectors that are resolved below this surface are parallel to it. In places this prominent reflector steps up or down suggesting shifts to other bedding planes. For example, a large shift in the level of this prominent strong reflector occurs along a 40 m high scarp (RAL in Figs. 15.1 and 15.2) This reflector is interpreted to be the main slide's failure plane, which developed along several bedding surfaces (Fig. 15.3B-B').

Chirp profiles show considerable internal deformation in the sediment cover overlying the prominent strong reflector on the sole of the slide scar. The lower half of this sediment cover contains closely spaced small en échelon breaks and lobes of apparently distorted sediments (Fig. 15.3B-B') that do not offset the surface of the strong reflector.

A distinctive secondary slide scar crosses the eastern side of the main slide scar (Figs. 15.1, 15.2, and 15.4). This secondary scar has well defined sidewall scarps with 5–14 m of relief, and forms a trough in the sediment drape of the main slide scar that widens down slope to ~850 m width at the edge of the main scar. On its upslope side the trough connects on the face of the main headwall with a distinct 360 m wide scallop scarp. This scarp appears to be a relatively fresh feature: it has a rough surface texture, contains lineations suggesting exposure of nearly horizontal beds and lacks substantial sediment drape (Figs. 15.3C-C' and 15.4). The upper edge of the headwall scarp abruptly transitions at its edge to the run-out trail of one of the subtle U-shaped scars observed above the main slide.

Chirp profiles show that the prominent strong reflector extends under the secondary slide scar without substantial interruption (Fig. 15.3C-C'). However, the sediment cover over this surface is of variable thickness, ranging from 0 to 15 m and lacking internal layering. No drape of post-failure sediment is identified within this slide scar, suggesting it was produced by a comparatively recent event.

15.4 Discussion and Conclusions

The AUV data show that the existing slide scar morphology of Tubeworm Slump is the result of several mass failures rather than a single catastrophic event. However, a prominent reflector dipping ~5° to the SW is evidence of a major failure surface within the main scar (Fig. 15.3B-B' and C-C'). This failure surface developed along various bedding planes. Because these dipping bedding planes occur at the edge of Smooth Ridge on the flanks of Monterey Canyon, they lack down slope support, and are inherently susceptible to down slope creep and pre-conditioned for slope failure. Down slope movement above the main reflector is inferred to have deformed the lower sediment accumulations within the main slide scar. This slip was accommodated along the surface associated with the prominent strong reflector. In contrast, the upper half of the sediment drape consistently shows laterally continuous reflectors that are parallel to the present seafloor with no internal deformation.

The field relationships between different geomorphological features provide a relative chronology of events, and some clues about the mechanisms for the development of the slide. Because the failure event that created the main headwall scarp truncates the anticlines, the formation of the main headwall scarp is either synchronous with or postdates the formation of these anticlines. The formation of the anticline also pre-dates the formation of the slide scars above the headwall because it has redirected the run-out from these slides.

The origin of these anticlines needs to be considered. Similar anticlines suggestive of regional tectonic forces have not been identified in the surrounding region. Instead these anticlines appear to be unique to the unsupported flank of Smooth

Ridge. No evidence exists to suggest the near surface anticlines that are clearly developed on both sides of the main scar exist in the deeper strata that still underlie the main slide scar (Fig. 15.3B-B'). Downslope slip along bedding surfaces dipping into the canyon may have lead to the deformation that shaped the anticlines, which is consistent with the gravity driven mass wasting events that generated the slide scar.

The main failure surface within the main slide scar is now largely covered with a $\geq$20 m thick sediment drape (Fig. 15.3B-B' and C-C'). While sedimentation rates in this area are poorly constrained, accumulation of such thickness of layered sediments typically requires tens of thousands of years. Thus, the failures that generated the main slide scar are not recent. Additional deformation within only the lower half of the sediment fill (Fig. 15.3B-B') indicates deformation took place while the drape was accumulating. The lack of deformation on the upper half of the sediment fill indicates that this deformation has not continued.

The most recent failure events can be identified in the AUV surveys by the rough morphology and lack of sediment drape on the modified landscape. The most significant is the distinctive fresh secondary slide scar (Fig. 15.4) cutting across the main scarp, which was not detected in the 30 kHz surface ship multibeam data (Greene et al. 2002). This failure may have included material from up-slope of the main headwall as well (Figs. 15.1 and 15.2). Numerous chemosynthetic biological communities have been found within the headwall of the fresh scar (Paull et al. 2005). Methane oozing from the relatively fresh exposures are believed to be supporting chemosyntheic biological communities.

The anticlinal ridges upslope of the main headwall have deflected most sediment from being transported over the main headwall scar. Consistent with this observation, the absence of gullies on the face of Tubeworm Slump (Fig. 15.4) substantiates that down slope channeling of sediment has not played a major role in modifying the scalloped faced of the scarp.

Another similar-sized deep-water slope failure scar (Eel Slump) was surveyed with the same MBARI AUV (Gwiazda et al. 2015). The geomorphic evolutions of these two slide scars differ considerably. Recurring sediment failure is the primary process sculpting the detailed surface morphology of Tubeworm Slump. In contrast, Eel Slump is notably lacking in evidence for continuing mass wasting deformation; and has a surface morphology that is primarily attributed to erosion, sediment ponding and fluid venting. Apparently continuing slope failure and gravity deformation is the dominant processes in altering the morphology of some slide scars, while the gravity failure ceases to be a significant factor in the morphologic evolution in other slide scars.

Acknowledgements The David and Lucile Packard Foundation provided support. Thanks are given to *R/V Zephyr* crew, AUV team, and reviews by Danny Brothers and Sara Benetti.

References

Caress DW, Thomas H, Kirkwood W, et al (2008) High-resolution multibeam, sidescan, and subbottom surveys using the MBARI AUV D. Allan. B., In: Reynolds JR, Greene HG (eds) Marine habitat mapping technology for Alaska: Alaska Sea Grant College Program, University of Alaska Fairbanks, doi:10.4027/mhmta.2008.04

Greene HG, Ward SN (2003) Mass movement along the central California Marine and their consequences for tsunami generation. In: Locat J, Mienert J (eds) Submarine mass movements and their consequences. Kluwer Academic Publishers, Dordrecht, pp 343–356

Greene HG, Maher N, Paull CK (2002) Physiography of the Monterey Bay National Marine Sanctuary and implications about continental margin development. Mar Geol 181:55–62

Gwiazda R, Paull CK, Caress DW, et al (2015) Eel Canyon slump scar and associated fluid venting. In: Lamarche G, Mountjoy J (eds) Submarine mass movements and their consequences. Springer, Dordrecht, pp 411–418

Naehr TH, Stakes DS, Moore WS (2000) Mass wasting, ephemeral fluid flow, and barite deposition on the California continental margin. Geology 28:315–318

Paull CK, Ussler W III, Greene HG et al (2005) Bioerosion by chemosynthetic biologic communities on Holocene submarine slide scars. Geo-Mar Let 25:11–19

Chapter 16
Submarine Slide Topography and the Distribution of Vulnerable Marine Ecosystems: A Case Study in the Ionian Sea (Eastern Mediterranean)

Alessandra Savini, Fabio Marchese, Giuseppe Verdicchio, and Agostina Vertino

Abstract In this work, we sought to document how submarine mass-movements influence the submarine landscape and associated habitat distributions on the upper portion of the northern Ionian Margin (eastern Mediterranean Sea) between 200 m to greater than 1,000 m in water depth (w.d.). In this area, mass-wasting processes have created unique morphological forms that, in turn, have generated high diversity for edaphic and hydrogeologic conditions; and these areas are marked by the patchy occurrence of varying natural benthic habitats. Surficial or sub-surficial Mass-Transport Deposits (MTDs) were documented by seismic and high-resolution morpho-bathymetric data and displayed dense aggregation for detached blocks spread over 1,200 km^2 between 400 and 1,000 m in w.d.. Living Cold-Water Coral (CWC) communities populate the blocky region and form coral topped mounds. These habitats are important Vulnerable Marine Ecosystems (VMEs) that are exposed to human pressure in the deep sea. Through production of a detailed geomorphological map and an examination of published data on the extent and distribution of CWC communities in the area, we sought to document how comprehensive research into submarine slide topography should also take into account the peculiar characteristics of their biotopes.

16.1 Introduction

Over the past 10 years, important United Nations General Assembly (UNGA) resolutions (i.e.: 59/25 in 2004 and 61/105 in 2010) have demanded that high seas fishing nations and regional fisheries management organizations (RFMOs)

A. Savini (✉) • F. Marchese • A. Vertino
Department of Earth and Environmental Sciences, Milano-Bicocca University, Milan, Italy
e-mail: alessandra.savini@unimib.it

G. Verdicchio
Impresub Diving and Marine Contractor, Trento, Italy

G. Lamarche et al. (eds.), *Submarine Mass Movements and their Consequences*,
Advances in Natural and Technological Hazards Research 41,
DOI 10.1007/978-3-319-20979-1_16

protect Vulnerable Marine Ecosystems (VMEs) from destructive fishing practices. As a result of these resolutions, improved knowledge regarding the distribution of deep-sea benthic habitats is required in order to promote specific conservation measures so that VMEs can be protected. In UNGA Resolution 61/105, VMEs are vaguely defined and can include seamounts, hydrothermal vents, sponge fields, and Cold-Water Corals (CWCs). CWCs are ecosystem engineers and provide habitat for rich and diverse fisheries and invertebrate communities. CWCs are filter feeders that are able to flourish in deep waters where a complex interplay between geologic, oceanographic, and geochemical factors promotes their growth (Roberts et al. 2006). CWCs are widely distributed along oceanic and Mediterranean continental margins and display preferential occurrence in areas with a complex seabed topography (Rengstorf et al. 2013; Savini et al. 2014; Lo Iacono et al. 2014a) where the topographically-induced acceleration of near-bed currents enhances the food supply (Mortensen et al. 2001; Thiem et al. 2006). Submarine slides are morpho-sedimentary features capable of producing a complex seabed topography that strongly disrupts the original homogeneity of the seafloor, and, indeed, provides suitable substrates for developing CWC provinces (Savini and Corselli 2010; Lo Iacono et al. 2014b). In 2007, De Mol et al. (2007) illustrated the first case for which submarine mass-movements played a critical role in creating favorable conditions for CWC settlement and growth (i.e.: offshore Mauritania and at the headwall of the Storegga Slide).

Hereafter, in particular, we consider how mass-movements shaped the submarine landscape of the northern Ionian Margin and created unique morphological forms marked by the patchy occurrence of varying natural benthic habitats, including CWCs. Our work was based on the analysis of a consistent dataset collected over the past 10 years that consisted of bathymetric, seismic (Savini and Corselli 2010), sedimentologic (Malinverno et al. 2010; Rosso et al. 2010), hydrologic (Budillon et al. 2010), and video data (Etiope et al. 2010; Vertino et al. 2010; Savini et al. 2014). All data were imported into a GIS environment (ESRI ArcGIS™). A digital terrain model (DTM) at a 40×40 m grid cell size (obtained from multibeam bathymetric data) was analyzed using basic geomorphometric techniques, including the Bathymetric Position Index (BPI), which was computed in order to emphasize relationships between mass-transport generated topography and the distribution of CWC habitats.

16.2 Submarine Slide Topography on the Ionian Margin

Extensive Mass-Transport Deposits (MTDs) dominate the north-central/eastern sector of the study area, downslope of a series of arcuate-shaped scars that mark the upper slope at the transition zone of the continental shelf (Savini and Corselli 2010) (Fig. 16.1a). Seismic profiles indicated MTDs as multiple and often coalescent sediment bodies exposed on the seafloor or partially/totally buried by recent

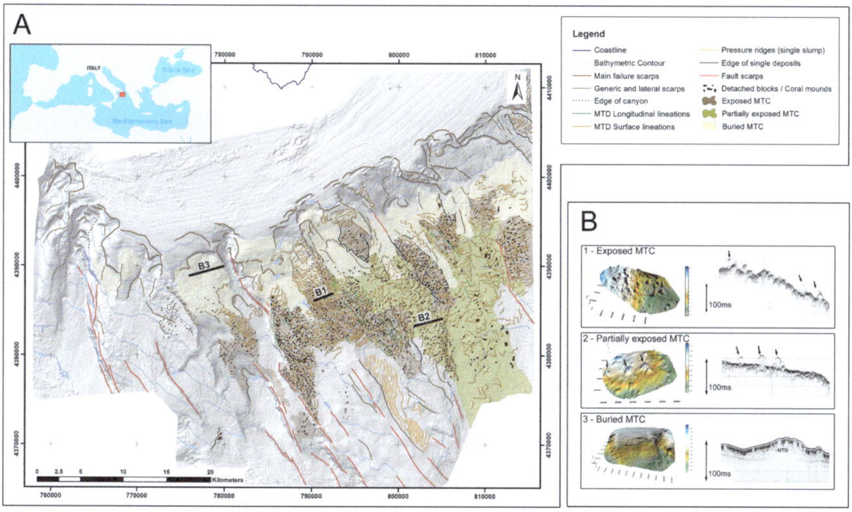

Fig. 16.1 (**a**) A geomorphological map of the study area (northern Ionian Sea, eastern Mediterranean). (**b**) Three main echotypes characterizing the entire Leuca MTC associated with a 3D view (obtained from multibeam data) of the acquired area (modified from Savini and Corselli 2010). The precise location is indicated in (**a**). The *red line* indicates the ROV dive that documented CWC occurrences (Etiope et al. 2010)

sediments (Fig. 16.1b). Hereafter, the term Leuca Mass-Transport Complex (MTC) is used to indicate the eastern sector of the explored area (Fig. 16.1a) that extends over ca. 1,200 km^2 (Savini et al. 2013). Where exposed or partially exposed, the Leuca MTC exhibits very complex morphological structures, including extensional and compressional ridges and numerous separated blocks of sediment (Fig. 16.1a) up to a few hundreds of meters in diameter and between 10 and 20 m in height (Fig. 16.1b; Savini and Corselli 2010). Sediment blocks form characteristic mound-like features that are definitely more pronounced and aggregated where the MTC is fully exposed on the seafloor (Fig. 16.2). The Leuca MTC is particularly well exposed (Fig. 16.1) on the top of large-scale, up-thrown faulted regional blocks (i.e., horsts) of the Apulian Plateau (Etiope et al. 2010) between 500 and 900 m in w.d. (Fig. 16.1). At the same depth range, where down-dropped, hanging-wall basins occur, and in some of the deeper areas, the MTC is only partially exposed (Fig. 16.1) and blocks display significantly more spacing in relation to one another (Fig. 16.2). Total or partial slide exposure is particularly evident at the same depth range (i.e.: 500–1,000 m) in which Adriatic Dense Water (ADW) flows, resulting in enhanced bottom currents (Budillon et al. 2010). The MTC is, instead, totally buried at a shallower depth from the slide scar area down to 400/500 m of water depth (Fig. 16.1).

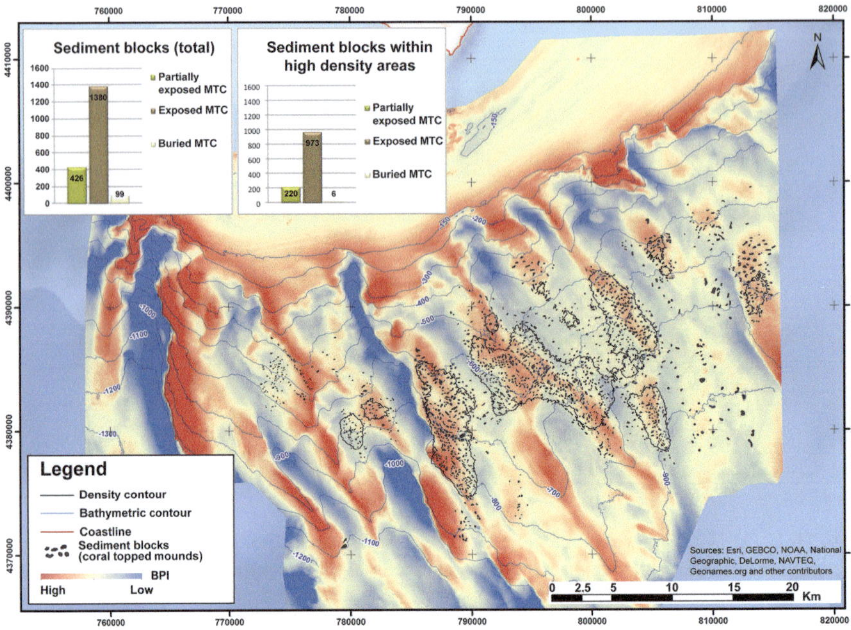

Fig. 16.2 The Bathymetric Positioning Index computed for the study area. Elevated areas are shown in *red*, whereas depressed areas are shown in *blue*. *Black polygons* indicate areas in which the density of sediment blocks (mapped in *dark gray* on the map) is greater than 5 per 1 km-radius circle

16.3 Deep-Sea Habitats of the Ionian Margin and Relationships with Landslide Morphologies

In 2000, the discovery of a CWC province dominated by frame-building scleractinians, located a few miles off Santa Maria di Leuca (SML), led to significant ecology and biodiversity research within the northern Ionian Margin (Corselli 2010, and references therein). Thirteen distinct macro-habitats (sensu Greene et al. 1999), including the *L. pertusa* and *M. oculata* coral frameworks, and hard-ground crusts and boulders colonized by the antipatharian species *L. glaberrima*, were identified based on acoustic backscatter, video, and grab surveys obtained from two mound-like features within the SML CWC province (Vertino et al. 2010; Savini et al. 2014). A total of ten mounds were explored using underwater cameras (Freiwald et al. 2009; Etiope et al. 2010) in the Leuca MTC and all indicated dense occurrences for CWC habitats on their tops and northeastern flanks (Vertino et al. 2010). Within the entire sector of the explored margin, Savini et al. (2014) isolated 5,820 mounds (i.e., sediment blocks) particularly aggregated over 600 km^2 (i.e., the blocky region of the exposed Leuca MTC). Using habitat predictive modelling, 82 % of the mounds were identified as populated by CWC macro-habitats, or a total of 1,902 coral-topped mounds (Savini et al. 2014). Use of the term "coral-topped mounds" is consistent with results obtained from sediment core analyses that only revealed the occurrence of coral fragments within the uppermost portion of the various sampled mounds (up to a core depth varying from 97 to 145 cm), with the underneath portion composed of stiff mud and lithogenic silt (Malinverno et al. 2010; Fink et al. 2012). The complex, small-scale, and irregular topography created by mass-wasting phenomena has been hypothesized to have exposed older and harder strata, thereby providing necessary substrates for coral colonization. Following colonization (which according to chronostratigraphic information from a coral-topped mound began around 13.96 ± 18 ky B.P., as indicated by the oldest dated coral in Malinverno et al. 2010), subsequent coral growth enhanced the original elevation and the morphology of failed blocks (Savini and Corselli 2010; Vertino et al. 2010). Two hundred and twenty-two species were identified within coral-topped mound habitats (Mastrototaro et al. 2010). A heterogeneous distribution of pennatulaceans was also documented along inter-mound areas (Vertino et al. 2010; Mastrototaro et al. 2013), approximately between 400 and 500 m of w.d. where the MTC appears buried by a roughly 20 ms thick drape of sediments (Fig. 16.1b).

Different authors noted that species diversity and the composition of macro-habitats, within and outside the mounds of the study area, changed across space, especially due to interactions between changes in seabed terrain and current speed (Rosso et al. 2010; Vertino et al. 2010; Mastrototaro et al. 2013). Bottom currents crossing the study area likely had a significant role in preventing corals from being buried and in providing food, promoting their growth and controlling their distribution at a variety of spatial scales. Over a broader scale, we determined that coral-topped mounds are particularly aggregated at the top of the regional blocks (Fig. 16.3) that dissect the Apulian Plateau through a series of NNW-SSE oriented normal faults (Etiope et al. 2010). Such configurations have been interpreted to be a result of a recent margin evolution that underwent regional uplift during the Late-Middle Pleistocene (Doglioni et al. 1994). Uplift reduced the accommodation space and created a physiographic context where large areas became swept due to local bottom-currents, especially on the tops of large-scale, up-thrown, faulted regional blocks due to their elevated position. Such areas include most of the portions of the coral-topped mounds mapped within the Leuca MTC (Fig. 16.3). At a finer scale, the preferential location of live scleractinian coral colonies corresponded to the tops and northeastern flanks of mounds, whereas the dominance of fine-grained sediments corresponded to the southwestern flank (Vertino et al. 2010; Savini et al. 2014). Coral growth at elevated positions, where nutrient inputs were supplied by rich NE bottom currents (Budillon et al. 2010), emphasized seafloor roughness originally generated by mass-movements that occurred over time, especially along the regional topographic high of the margin (Fig. 16.3).

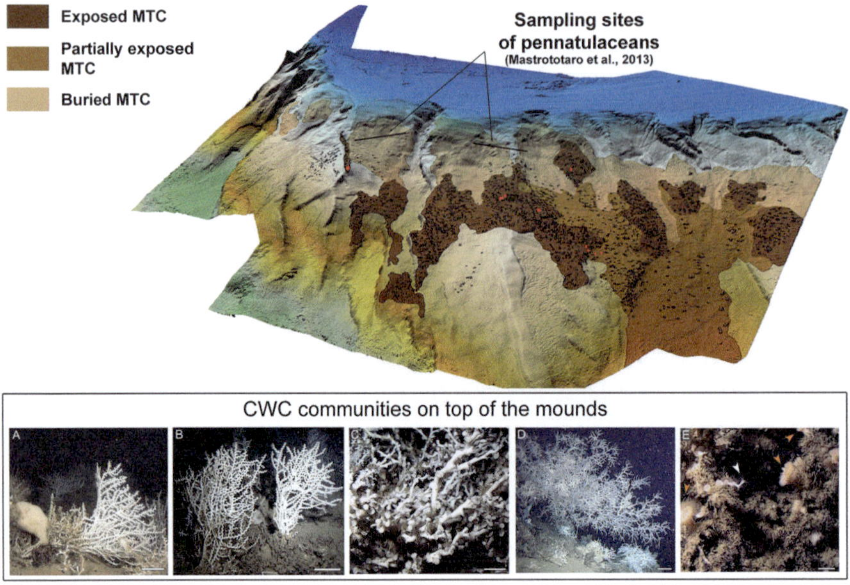

Fig. 16.3 A 3D view of the study area. Exposed, partially exposed, and buried sectors of the Leuca MTC are mapped along with the sediment blocks. Sampled coral-topped mounds are indicated in *red* (From Etiope et al. 2010; Vertino et al. 2010; Savini et al. 2014). *Track lines* over which pennatulaceans were sampled are also indicated (Mastrototaro et al. 2013). Representative Pluto-ROV seafloor images of coral-dominated macro-habitats are showed in the *black rectangular* frame: (**a**) Live colony of *Madrepora oculata* and the sponge *Poecillastra compressa* (attached to the death basal portion of the coral – modified after Vertino et al. 2010); (**b**) Live planar colonies of *M. oculata* settled on black-coated coral framework; (**c**) Detail of live *Lophelia pertusa*, on the *left upper* corner some tiny branches of *M. oculata* are visible; (**d**) Large colony of the antipatharian coral *Leiopathes glaberrima* colonising, together with scleractinians, dead coral frameworks and hardground crusts interspersed with mud; (**e**) Dead and black-coated scleractinian coral branches densely colonised by hydrozoans, sponges and scleractinians (note the white tiny branch of *M. oculata*, white arrow, and the pale orange solitary *Desmophyllum dianthus*). Scale bar: 10 cm (**a**, **b**, **d**), 3 cm (**c**), 2 cm (**e**)

16.4 Economic Significance of Submarine Landslide Areas

We determined that the peculiar biotopes that make the Ionian Margin a relevant hotspot of biodiversity within the eastern Mediterranean Basin result from specific environmental conditions strictly controlled by the interplay between landslide seafloor morphologies (generated by past multiple failure events), tectonic activity, and local hydrographic patterns.

The importance of submarine mass-movements in supporting deep-sea biodiversity has often been referred to as a negative effect produced by mass wasting phenomena that causes a decrease in species abundance and richness or that totally obliterates soft sediment benthic fauna (Glover et al. 2001). Nevertheless, since rough topography, nature of substrata and nutrient supply affect biodiversity on local scales (Levin et al. 2001), here, we considered how Pleistocene submarine mass-movements along the northern Ionian Margin played a critical role in generating complex seafloor topographies and small-scale changes in substrate distributions, yielding long-term, suitable environmental conditions (hard substrates and increased bottom currents) for the settling of specific Vulnerable Marine Ecosystems (VMEs).

Submarine mass-wasting deposits can not only be considered a geo-hazard due to their potential instability, but also as favorable locations for protected habitats in need of conservation measures. Therefore, increased efforts for mapping mass wasting features could bring important advances in recognizing VMEs and better designing procedures for natural resource management.

Acknowledgments The authors' work benefited from EU FP7 project CoralFish (Grant agreement number: 213144 – www.eu-fp7-coralfish.net), the National Italian project MAGIC, the RITMARE project, funded by the Italian Ministry of University and Research (MIUR) and the ESF COCARDE Network. ROV video footage were collected within the framework of the CoNISMa project RBAU 01RKC7_004 APLABES (2004-2006FIRB program). The authors sincerely thank the reviewers (Claudio Lo Iacono and Jean-Francois Bourillet) for their valuable suggestions and constructive comments.

References

Budillon G, Lo Bue N, Siena S, Spezie G (2010) Hydrographic characteristics of water masses in the northern Ionian Sea. Deep-Sea Res II 57(5–6):441–457

Corselli C (2010) Introduction: cold-water coral communities in the Mediterranean Sea. Deep-Sea Res II 57(5,6):345–359

De Mol B, Kozachenko M, Wheeler A, Alvares H, Henriet JP, Olu-Le RK (2007) Thérèse Mound: a case study of coral bank development in the Belgica Mound Province, Porcupine Seabight. Int J Earth Sci 96(1):103–120

Doglioni C, Mongelli F, Pieri P (1994) The Puglia uplift SE Italy: an anomaly in the foreland of the Apenninic subduction due to buckling of a thick continental lithosphere. Tectonics 13(5): 1309–1321

Etiope G, Savini A, Lo Bue N, Favali P, Corselli C (2010) Deep sea survey for the detection of methane at the "Santa Maria di Leuca" cold-water coral mounds (Ionian Sea, South Italy). Deep-Sea Res II 57(5–6):431–440

Fink HG, Wienberg C, Hebbeln D, McGregor HV, Schmiedl G, Taviani M, Freiwald A (2012) Oxygen control on Holocene cold-water coral development in the eastern Mediterranean Sea. Deep-Sea Res I 62:89–96

Freiwald A, Beuck L, Ruggeberg A, Taviani M, Hebbeln D (2009) The white coral community in the central Mediterranean Sea revealed by ROV surveys. Oceanography 22(1):58–74

Glover AG, Paterson G, Bett B, Gage J, Sibuet M, Sheader M, Hawkins L (2001) Patterns of polychaete abundance and diversity from the Madeira Abyssal Plain, northeast Atlantic. Deep-Sea Res I 48:217–236

Greene HG, Yoklavich MM, Starr RM, O'Connell VM, Wakefield WW et al (1999) A classification scheme for deep seafloor habitats. Oceanol Acta 22(6):663–678

Levin LA et al (2001) Environmental influences on regional deepsea species diversity. Annu Rev Ecol Syst 132:51–93

Lo Iacono C, Robert K, Gonzalez R, Gori A, Orejas C, Gili JM (2014a) Predictive ensemble maps for cold-water coral distributions in the Cap de Creus Canyon (NW Mediterranean). Advantages and limitations. Dissertation, GeoHab Conference – Marine Geological and Biological Habitat Mapping. Lorne, Australia. May 2014. http://geohab.org/conferences/2014-lorne-australia/

Lo Iacono C, Gracia E, Ranero C, Emelianov M, Huvenne VAI, Bartolome R, Booth-Rea G, Prades J, MELCOR Cruise Party (2014b) The West Melilla cold water coral mounds, Eastern Alboran Sea: morphological characterization and environmental context. Deep-Sea Res II 99:316–326

Malinverno E, Taviani M, Rosso A, Violanti D, Villa I et al (2010) Stratigraphic framework of the Apulian deep-water coral province, Ionian Sea. Deep-Sea Res II 57(5–6):345–359

Mastrototaro F, D'Onghia G, Corriero G, Matarrese A, Maiorano P et al (2010) Biodiversity of the white coral bank off Cape Santa Maria di Leuca (Mediterranean Sea): an update. Deep-Sea Res II 57:412–430

Mastrototaro F, Maiorano P, Vertino A, Battista D, Indennidate A et al (2013) A Kophobelemnon (Cnidaria, Octocorallia) facies from Santa Maria di Leuca coral province (Mediterranean Sea). Mar Ecol. doi:10.1111/maec.12017

Mortensen PB, Hovland MT, Fossa JH, Furevik DM (2001) Distribution, abundance and size of Lophelia pertusa coral reefs in mid-Norway in relation to seabed characteristics. J Mar Biol Assoc UK 81:581–597

Rengstorf AM, Yesson C, Brown C, Grehan AJ (2013) High-resolution habitat suitability modelling can improve conservation of vulnerable marine ecosystems in the deep sea. J Biogeogr 40:1702–1714

Roberts JM, Wheeler AJ, Freiwald A (2006) Reefs of the deep: the biology and geology of cold-water coral ecosystems. Science 312:543–547

Rosso A, Vertino A, Di Geronimo I, Sanfilippo R, Sciuto F et al (2010) Hard versus soft-bottom thanatofacies from the Santa Maria di Leuca deep-water coral mound province, recent Mediterranean. Deep-Sea Res II 57(5–6):360–379

Savini A, Corselli C (2010) High resolution bathymetry and acoustic geophysical data from Santa Maria di Leuca cold water coral province (northern Ionian Sea–Apulian continental slope). Deep-Sea Res II 57(5–6):326–344

Savini A, Verdicchio G, Marchese F (2013) Geomorphological mapping of complex submarine slides (Apulian ridge, eastern Mediterranean Sea): implication for geohazard assessments. Dissertation, 8th international conference (IAG) on Geomorphology, Paris

Savini A, Vertino A, Marchese F, Beuck L, Freiwald A (2014) Mapping cold-water coral habitats at different scales within the northern Ionian Sea (central Mediterranean): an assessment of coral coverage and associated vulnerability. PLoS One 9(1):e87108. doi:10.1371/journal.pone.0087108

Thiem RE, Fossa JH, Berntsen J (2006) Food supply mechanisms for cold-water corals along a continental shelf edge. J Mar Syst 26:1481–1495

Vertino A, Savini A, Rosso A, Di Geronimo I, Mastrototaro F et al (2010) Benthic habitat characterization and distribution from two representative sites of the deep-water SML Coral Mound Province (Mediterranean). Deep-Sea Res II 57:380–396

Part III
Geotechnical Aspects of Mass Movement

Chapter 17
Shear Strength of Siliciclastic Sediments from Passive and Active Margins (0–100 m Below Seafloor): Insights into Seismic Strengthening

Joshua R. DeVore and Derek E. Sawyer

Abstract Submarine geohazards threaten coastal communities and global economies. Submarine debris flows are the largest mass-wasting events observed on the Earth's surface, comprising of up to 50 % of basin fill. Further insight can be gained into these important processes by understanding in-situ preconditioning factors that lead to slope destabilization. We examine two locations from the International Ocean Discovery Program data archive to determine how external effects on sediment properties compare between passive margins and active margins. We select representative passive margin (Amazon Fan) and active margin sites (Nankai Trough), and analyse peak shear strength, void ratio, and composition from the uppermost 100 m below seafloor. This depth corresponds to a depth range in which most submarine mass movements originate. However, it is not appropriate to directly compare shear strength and void ratio of samples from different settings due to differing stress histories, sedimentary composition, and consolidation properties. We focus on ideal locations on both margin types that have solely undergone one-dimensional burial, no diagenesis/cementation, and no unroofing. We find that active margin sediments exhibit an increase in shear strength when compared to their passive margin counterparts, while void ratio tends to be higher on active margins. We are currently conducting a focused lab program to better understand compositional effects and determine the intrinsic properties of each site to more definitively normalize the in-situ sediment profiles. Our results suggest a potential link between shear strength and margin seismicity.

J.R. DeVore (✉) • D.E. Sawyer
School of Earth Sciences, The Ohio State University, 125 Oval Drive South, Columbus, OH 43210, USA
e-mail: devore.87@osu.edu

G. Lamarche et al. (eds.), *Submarine Mass Movements and their Consequences*, Advances in Natural and Technological Hazards Research 41,
DOI 10.1007/978-3-319-20979-1_17

17.1 Background and Significance

Earthquakes can be an effective triggering mechanism of underwater landslides and slide-generated tsunami because they impose large and sudden horizontal and vertical loads on the slope that induce excess pore pressure, reducing shear strength. When a submarine landslide occurs, it disrupts the overlying water column and can generate a slide-induced tsunami. Submarine debris flows are the largest mass-wasting events observed on the Earth's surface, yet we know little about in-situ preconditioning factors that lead to slope destabilization in the submarine environment. Our ability to model and predict landslides and tsunami is greatly affected by an insufficient understanding of the physical properties of seafloor sediments and how they evolve with depth, time, and location.

Most (~80 %) of submarine mass movements fail within 100 m below the seafloor (mbsf) (Hampton 1996; McAdoo et al. 2000, 2004; Moscardelli 2007). Not all earthquakes generate submarine landslides; however, all earthquakes impart seismic energy to slope sediments as a shear stress. When a shear stress is applied to saturated, fine-grained sediment during an earthquake, an instantaneous increase in pore pressure is induced. This pore pressure increase weakens the sediment, reducing its shear strength temporarily. If this temporary weakening does not cause failure, pore pressure will slowly diffuse over time allowing the soil to compact into an overconsolidated state. Thus, an earthquake that does not induce a submarine landslide could have a strengthening effect on the slope, increasing its resistance to future failure. This process has been termed 'seismic strengthening' (Locat et al. 2002; Lee et al. 2004) and is proposed to be a fundamental process that controls shear strength evolution on continental margins.

Boulanger (1999), Locat et al. (2002), and Lee et al. (2004) performed small-scale geotechnical tests on reconstituted marine sediments. These studies exposed samples to high-frequency simulated earthquakes and measured the reduction of pore water with subsequent periods of drainage, correlating void ratio reduction to an increase in shear strength (Fig. 17.1). Bench-top experiments are performed under different timescales and conditions than natural field-scale settings. In order to compliment these successful pilot studies, we analyse shear strength of margin sediments from the International Ocean Discovery Program (IODP) archive. We aim to explore the rationale that, if seismic strengthening is a fundamental process, near-surface sediment on active margins will exhibit higher shear strength values versus passive margins.

17.2 Global Shear Strength Trends

We acquired datasets from IODP, including peak shear strength (measured with an automated vane shear device), moisture and density, and compositional characteristics. We only selected sites from active and passive margins in siliciclastic

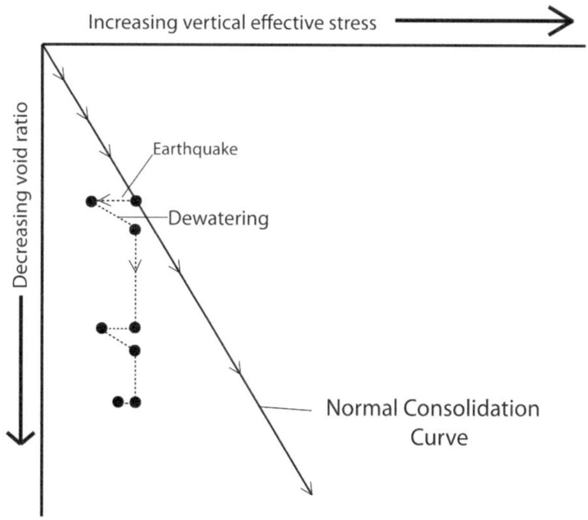

Fig. 17.1 Example of "seismic strengthening" produced with direct shear tests on reconstituted sediment (Adapted from Lee et al. 2004). Through four seismic cycles, the sample becomes progressively overconsolidated. Vertical effective stress is reduced as pore pressure increases and returns to its natural hydrostatic state following drainage of excess pore fluid

sediments. Active margins used in this study are from the Cascadian subduction zone, the Nankai Trough, the Japan Trench, the Lapulapu Ridge; and east Taiwan. Passive margin sites are West Africa, Amazon Fan, the eastern United States, and the Gulf of Mexico.

A preliminary survey of shear strength from active and passive margins shows a tantalizing result with a simple plot of shear strength for the upper 100 mbsf (Fig. 17.2). Higher shear strength on active margins is clearly observed both with depth and frequency of shear strength values. Global shear strength data suggests that there is an apparent strengthening signal exhibited on active margins (Fig. 17.2).

It is not appropriate to directly compare shear strength and void ratio of samples from different settings due to the fact that each setting will have different sedimentary composition and consolidation properties. These data must first be normalized. The challenge is to find ideal type locations on both margin types that have undergone exclusively one-dimensional burial, no diagenesis/cementation, and no unroofing of overburden.

17.3 Ideal Type Sites

We have identified one passive margin site and one active margin site from the global database that have hydrostatic pore pressure conditions, continuous burial histories, and are both composed of fine-grained siliciclastics,: Amazon Fan (Site 942) and South Japan (Site C0001E) (Fig. 17.3). South Japan site C0001E from IODP bleg 315 is representative of siliciclastic sedimentation on an active margin that is composed dominantly of silty clays and clayey silts with rare interbedded ash

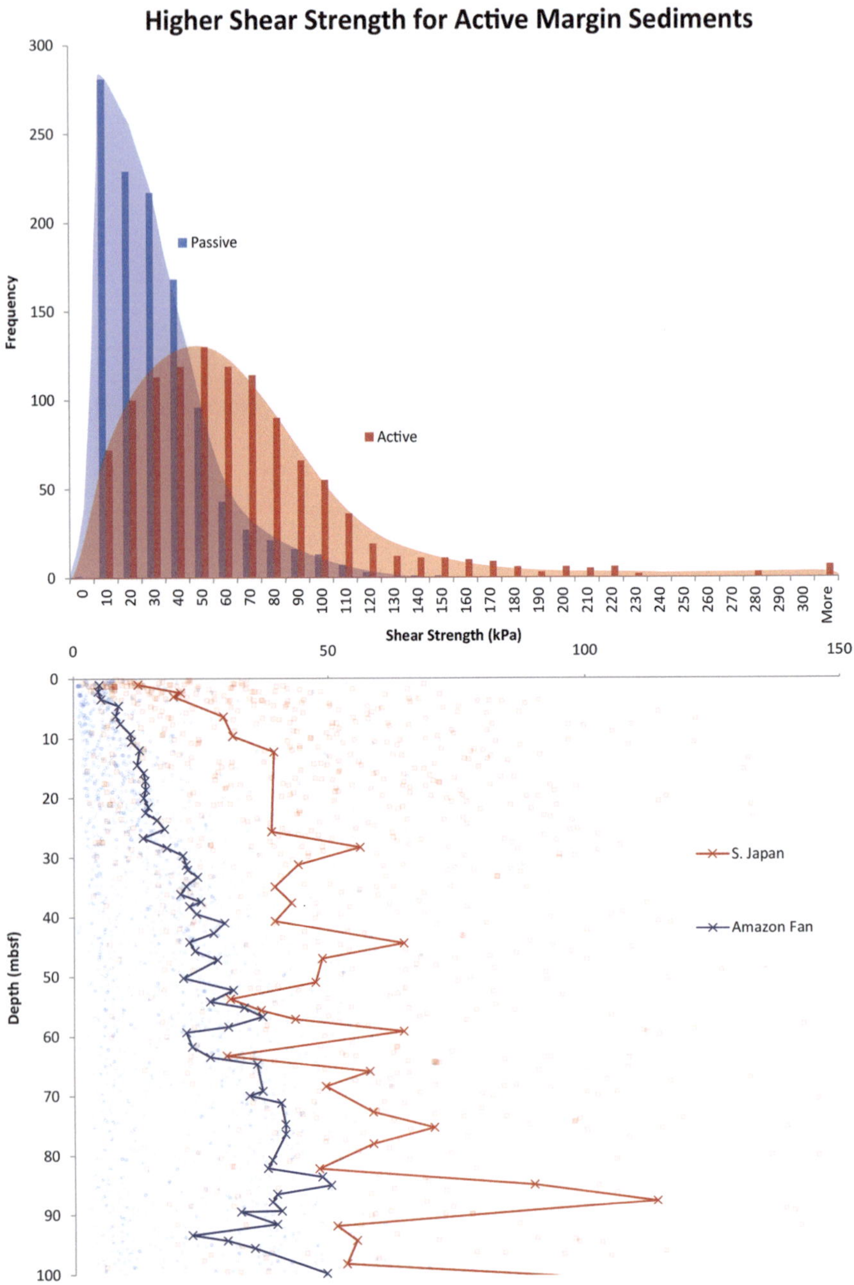

Fig. 17.2 Active margin sediments exhibit higher shear strength than passive margin sediments. (**a**) Frequency and distribution of shear strength of 83 sites from siliciclastic margins. All data points (>10,000) are measured by the automatic vane shear system from the IODP data archive. (**b**) Shear strength plotted versus depth. The *blue* and *red* translucent background data are the entire dataset that is shown in 1A. It is observed that active margins (*red*) generally have higher shear strength values with depth. Given that shear strength is a function of many variables, we have identified one 'pristine' site on each margin. 'Pristine sites" are those that have no evidence for

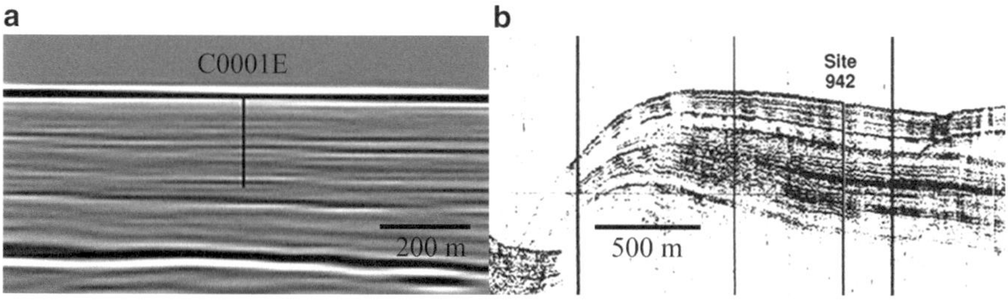

Fig. 17.3 Seismic surveys of South Japan site C0001E (**a**) and Amazon Fan site 942 (**b**). These sites represent pristine depositional conditions displaying horizontal sediment layers in the upper 100 mbsf and are free from any noticeable deformation. Note that the transparent layer at site 942 may be a debris flow, however, this layer is well below 100 mbsf and does not affect our layer of interest. The *black vertical bar* represents the core locations

deposits, and has experienced relatively low sedimentation rates (~0.1 mm/year) over the Quaternary. Amazon Fan site 942 from leg 155 of the Ocean Drilling Program (ODP) was chosen as a paleoceanographic site to provide an expanded hemiplegic sedimentation record for the Holocene.

This site has experienced slower sedimentation rates, ~0.7 mm/year over the Holocene due to its distal location from the main sediment source for the region. This site is an example of a siliciclastic system on a passive margin and is characterized by beds of clayey silt with interbedded laminae to thin beds of silts, with rare laminae of fine silty sands and nannofossil-foraminifer rich clays. These sites represent the most pristine sedimentation with no evidence of soft sediment deformation, slumping or landsliding, and have undergone relatively low rates of sedimentation (0.1–0.7 mm/year). Selecting sites with similar grain distribution, stress histories, and low sedimentation rates will help to normalize any differences that may exist between the sites, having undergone only uniform one-dimensional vertical compression. After selection of these ideal sites, the active margin site still has higher shear strength than the passive margin site margins at equivalent depths (Fig. 17.2).

Fig. 17.2 (continued) diagenesis, soft sediment deformation, landsliding, or gas hydrate presence. S. Japan (C0001E) (*red*) and Amazon (Site 942) (*blue*) are these sites. Following this screening process, the active margin profile is stronger than the passive margin and appears to trend toward a convergence with depth

17.4 Hydrostatic Pore Pressure Conditions at Type Sites

Shear strength is directly impacted by the vertical effective stress that the sediment experiences and an increase in pore pressure can drive slope failure. We estimate in-situ pore pressure to determine if it could be responsible for the observed shear strength differences.

Skempton (1970) demonstrates that clays undergoing sedimentation rates of less than 2 mm/year are at least 95 % consolidated and considered to be normally pressured to depths of 50 m. To further constrain that the type sites are at or near vertical effective stresses corresponding to hydrostatic pore pressures to depths of 100 mbsf, we apply a Gibson time factor equation to each of our sites. Gibson (1958) presents the theoretical solutions for pore pressure during one-dimensional consolidation of clay (Eq. 17.1).

$$u = \gamma' mt - \gamma' (\pi C_v t)^{-1/2} \exp\left(\frac{-x^2}{4C_v t}\right) \int_0^\infty \xi \tanh \frac{m\xi}{2C_v} \cosh \frac{x\xi}{2C_v t} \exp - \frac{\xi^2}{4C_v t} d\xi \quad (17.1)$$

Where u is pore pressure in excess of hydrostatic pressure, γ' is the bulk density of the sediment, m is the sedimentation rate, C_v is the coefficient of consolidation, t is time, and ξ is the depth of interest within the sediment column divided by the total sediment column height as a function of time $\left(\frac{x}{h(t)}\right)$. When the curves of $\frac{u}{\gamma' h}$ are plotted against $\frac{x}{h}$ one develops curves that represent the time factor (T), where:

$$T = \frac{m^2 t}{C_v} \quad (17.2)$$

Variables m and t are assessed from biostratigraphic analyses and regional correlations developed by the scientific party onboard each IODP leg and C_v values are taken from similar sediments in the region. Using these parameters, Table 17.1 represents the calculated T for each site.

Pore pressure will range from 0 to 1 (hydrostatic to lithostatic pressure) and indicates the possible overpressure percentage (Gibson 1958). The maximum possible pore pressure at site C0001E is in excess of 1 %, while site 942 possibly has 0.5 % excess pore pressure. We use this analysis to show that the sites are very close to hydrostatically pressured and that we can assume normal vertical effective

Table 17.1 Gibson time factor analysis for target sites

Site	Location	m (m/year)	t (year)	C_v (m^2/year)	T
942	Amazon Fan	0.0007	140,000	13.25	0.0051
C0001E	Japan	0.0001	1,240,000	1.19	0.0104

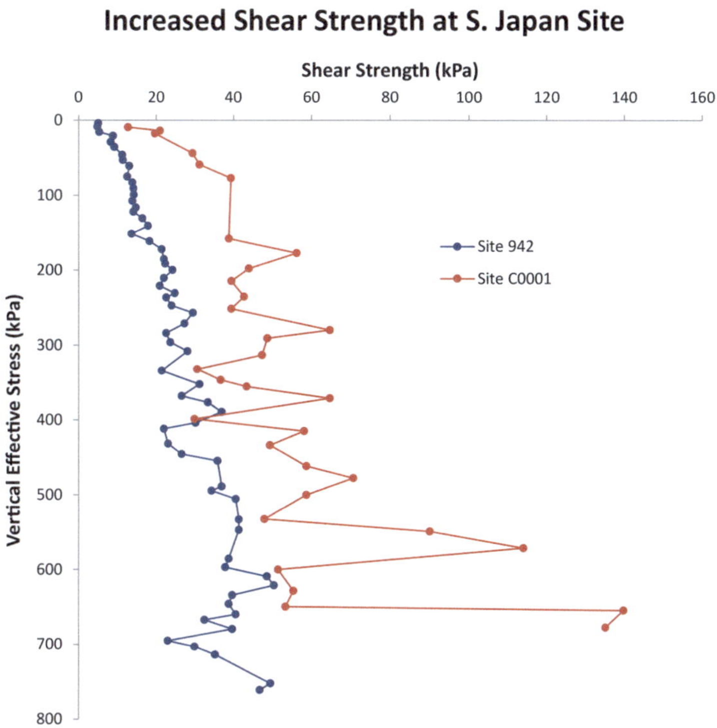

Fig. 17.4 Shear strength with vertical effective stress for isolated 'pristine' sites. Active margin site C0001 still exhibits increased shear strength following the correction for density and pore pressure contrasts between the two sites

stress distributions at our type locations. In this example, shear strengths are greater on active margins than passive margins at the same vertical effective stress (Fig. 17.4).

17.5 Continental Margin Sediment Shear Strength

The coring process can facilitate a reduction in sediment shear strength and is difficult to avoid. If there were no disturbance we might expect shear strength to be approximately 20 % of vertical effective stress (Lambe and Whitman 1969). Standard practices in measuring shear strength are to measure from the interior of the core once it is halved, reducing the likelihood of shear strength alteration. We assume any reduction in shear strength due to coring disturbance to be similar at both sites.

In the preceding discussion, we have analysed sites from south Japan (Site C0001E) and Amazon Fan (Site 942) that exhibit hydrostatic pressure, undisturbed sedimentation, and no evidence for diagenesis or hydrates. Following this screening, the active margin site still exhibits greater shear strength (Fig. 17.4). However, the final step in normalization is to account for potential lithological differences between Japan and Amazon. Seismic strengthening may be a fundamental, margin-wide process affecting shallow (0–100 mbsf) sediment on active margins, increasing resistance to failure and compaction behavior of the continental shelves and slopes. The shear strength trend on active margins appears to move toward a possible convergence with the passive margin trend as vertical effective stress increases. This convergence may be analogous to the boundary between contractive and dilative soil behavior, indicating that only contractive, and therefore shallow sediment, is capable of becoming strengthened by seismicity and requires future work. We are currently conducting a detailed study on the sediment composition and intrinsic physical properties at these sites to assist in isolating the effects of seismicity on shear strength development and void ratio. This study will enhance our understanding of the physical properties of seafloor sediments in response to seismicity.

Acknowledgments The authors gratefully acknowledge the pioneering investigations into seismic strengthening, helping lay the foundation for this research, and the helpful comments and insights provided by Homa Lee and Jacques Locat.

References

Boulanger E (1999) Cyclic behavior of sediments of the continental margin of the Eel River: a possible explanation for the small superficial submarine landslides in this Region. University of Laval, Quebec, Quebec City, Canada

Gibson R (1958) The progress of consolidation in a clay layer increasing in thickness with time. Geotechnique 8(4):171–182

Hampton MA (1996) Submarine landslides. Rev Geophys 34(1):33–59

Lambe TW, Whitman RV (1969) Soil mechanics. Wiley, New York

Lee HJ, Locat J, Boulanger E, Konrad JM (2004) Seismic strengthening, a conditioning factor influencing submarine lanslide development. Paper presented at the 57th Canadian Geotechnical Conference; 5th Joint CGS/IAH-CNC Conference, Quebec City, Quebec, Canada, October 24–27, 2004

Locat J, Lee H, Kayen R, Israel K, Savoie M-C, Boulanger É (2002) Shear strength development with burial in Eel River margin slope sediments. Mar Georesour Geotechnol 20(2):111–135. doi:10.1080/03608860290051 831

McAdoo BG, Pratson LF, Orange DL (2000) Submarine landslide geomorphology, US continental slope. Mar Geol 169(1–2):103–136, http://dx.doi.org/10.1016/S0025-3227(00)00050-5

McAdoo BG, Capone MK, Minder J (2004) Seafloor geomorphology of convergent margins: implications for Cascadia seismic hazard. Tectonics 23(6):TC6008. doi:10.1029/2003TC001570

Moscardelli LG (2007) Mass transport processes and deposits in offshore Trinidad and Venezuela, and their role in continental margin development. Dissertation, The University of Texas at Austin

Skempton AW (1970) The consolidation of clays by gravitational compaction. Q J Geol Soc Lond 125, Part 3. Geological Society of London, London

Chapter 18
A Small Volume Calibration Chamber for Cone Penetration Testing (CPT) on Submarine Soils

Matthias Fleischer, Stefan Kreiter, Tobias Mörz, and Marc Huhndorf

Abstract A new small volume CPT calibration chamber with dynamically controlled boundary conditions has been built to improve the correlation between in-situ data and soil parameters. The sample volume in the new CPT calibration chamber has a diameter of 30 cm and a height of 54.5 cm. Therefore, it is possible to use reconstituted samples of limited quantity, e.g. from boreholes. The chamber is able to simulate large overburden stresses and overconsolidation ratios (OCR) up to 5 MPa. Horizontal, vertical and pore pressures are independently applied via syringe pumps while recording volume changes. All pressures are dynamically controlled and allowing stress, strain and mixed stress-strain stress boundary conditions BC1–BC5 to be enforced. The sample deformation is measured by circumferential laser triangulation sensors. In a first series of tests using Cuxhaven Sand, a 12 mm cone and BC1 conditions, CPT tip resistances reach steady state about a length of at least 10 cm. The corrected tip resistances and the inferred relative densities for Cuxhaven Sand differ substantially from previously established correlations, confirming the need for more advanced correction factors and relationships between CPT data and in-situ soil properties.

18.1 Introduction

Cone penetration testing (CPT) is a common method to measure in situ soil mechanical properties in the marine realm (Lunne 2010). Applications include the investigation of submarine slopes to understand mass movement processes and soil investigation for marine constructions (Lunne 2010). Crucial parameters inferred from CPT measurements include friction angle, undrained shear strength and relative density.

M. Fleischer (✉) • S. Kreiter • T. Mörz • M. Huhndorf
MARUM – Center for Marine Environmental Sciences, University of Bremen, Leobener Strasse, 28359 Bremen, Germany
e-mail: mfleischer@marum.de

G. Lamarche et al. (eds.), *Submarine Mass Movements and their Consequences*, Advances in Natural and Technological Hazards Research 41,
DOI 10.1007/978-3-319-20979-1_18

Table 18.1 Boundary conditions for CPT calibration chambers (Huang and Hsu 2005)

Boundary condition	Vertical stress	Vertical strain	Horizontal stress	Horizontal strain
BC1	Constant	–	Constant	–
BC2	–	0	–	0
BC3	Constant	–	–	0
BC4	–	0	Constant	–
BC5	Constant	–	Controlled	–

A long history of CPT measurements has led to a wide range of methods to correlate CPT measurements with soil mechanical properties (Baldi et al. 1986; Kulhawy and Mayne 1990; Lunne et al. 1997; Jamiolkowski et al. 2003; Robertson and Cabal 2014) The available empirical correlations are not widely applicable for a range of soil types (Kulhawy and Mayne 1990). Therefore, more specific correlations may improve the reliability of CPT interpretation.

A robust and reliable technique to empirically derive correlation factors is the use of CPT calibration chambers (Jamiolkowski et al. 2003). The advantages of laboratory testing are the repeatability of test conditions, specimen uniformity, known/controlled boundary conditions and the ability to apply a specific stress history (Hsu and Huang 1998). Usually, four different boundary conditions are applied (BC1 to BC4 – Table 18.1). However, because of the finite dimensions of calibration chambers all these boundary conditions cause artificial boundary effects, as documented in various studies (Baldi et al. 1982; Bolton et al. 1999; Salgado 1993; Salgado et al. 1998; Jamiolkowski et al. 2003). For dense sands the measured tip resistance appears to match the field value, if the diameter ratio of sample diameter to cone diameter is higher than 60 (Kulhawy and Mayne 1990; Ghionna and Jamiolkowski 1991). For loose sands, tip resistances are less dependent on boundary effects (Parkin 1988). Use of large calibration chambers with diameters greater than 1 m reduces boundary effects (Ghionna and Jamiolkowski 1991), however a disadvantage of large chambers is the need for large sample volumes. Small chambers are more economic, practical and are necessary in cases where samples of limited quantity are tested, for example material from marine cores. Testing in small chambers utilizes correspondingly small Mini-CPT cones. Generally, boundary effects need to be either accounted for by correction factors specific to small diameter ratios (Mayne and Kulhawy 1991; Kulhawy and Mayne 1990; Baldi et al. 1982; Salgado 1993) or compensated for by the dynamic regulation of the horizontal pressure in order to simulate an infinite surrounding soil, as first proposed by Ghionna and Jamiolkowski (1991). The latter was adapted by other studies (Hsu and Huang 1998; Bałachowski 2006) and is called BC5 boundary condition.

Here, we describe a new small CPT calibration chamber which has been developed at MARUM – Center for Marine Environmental Sciences, University of Bremen and referred to as the MARUM Calibration Chamber (MARCC). This chamber combines controlled horizontal stress, small sample dimensions and the ability to simulate overburden stresses up to 5 MPa.

18.2 Methods

18.2.1 New MARCC Calibration Chamber Design

The essential feature of this chamber is its small sample volume, which is cylindrical with a height of 54.5 cm and a diameter of 30 cm (Fig. 18.1). The sample is laterally confined by a latex membrane and the horizontal stress is applied via water pressure. Vertical stress is applied by a water-filled latex cushion from the bottom of the chamber. Vertical and circumferential pressures are isolated by a small annular part of the bottom plate and controlled independently. The pore water is connected via a porous geotextile on the water cushion from below and via a sintered metal filter from above. Thus the horizontal and lower sample boundaries are flexible while the upper boundary is rigid. The horizontal, vertical, and pore fluid pressures are independently operated either by controlled stress or controlled volume.

There are three sets of windows at three different heights in the chamber cell through which the lateral sample deformation is monitored by laser triangulation with a resolution of 0.3 μm. The windows are made of high pressure resistant borosilicate glass. The entire chamber is rated for a pressure of up to 5 MPa. An interchangeable CPT-cone inlet piece is centred on top of the small lid which allows usage of various cone diameters.

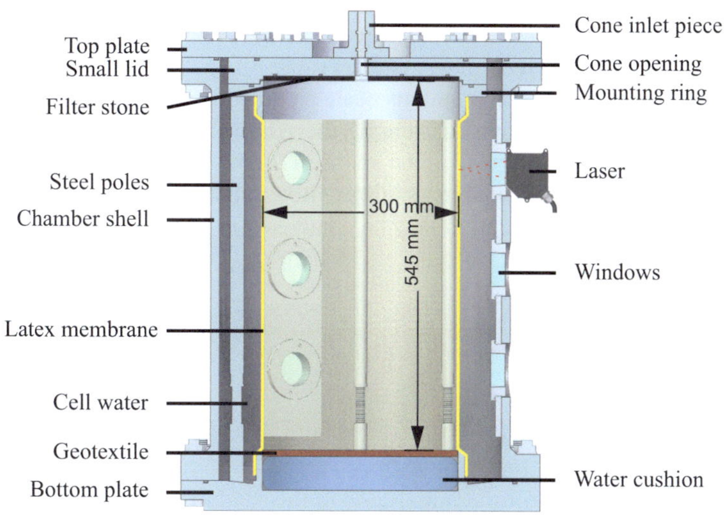

Fig. 18.1 Cross section of the MARCC calibration chamber

Table 18.2 Input data set for the real time system

No.	Input data	Control	Recording
1	Inductive position of hydraulic cylinder (penetration speed and depth)	x	x
5–9	Laser triangulation sensors (radial displacement)		x
3	Vertical, horizontal and pore pressures	x	x
3	Position of syringe pumps (volume changes)		x
1	External penetration force		x
1	Mini CPT tip resistance		x

18.2.2 Sensors, Control and Measurement Devices

The stress – strain boundary conditions BC1–BC5 (Table 18.1) and the pore pressure are controlled by three 1 l high pressure syringe pumps with a capability of 5 MPa and a volume resolution of 1 ml.

To apply BC2, BC3 and BC5 conditions, the horizontal strain of the sample during the penetration process is measured with up to nine laser triangulation sensors in horizontal direction. In addition, the vertical and horizontal strains are determined independently via fluid volume measurement.

The mini-cone used here has a diameter of 12 mm, resulting in a sample-to-cone diameter ratio of 25. The mini-cone is equipped with four strain gauges (DMS half bridges) to measure the tip force. The cone is pushed via a hydraulic cylinder into the sample.

Up to 18 sensors, gather data from measurements in and outside the chamber via a real-time control and recording system (Table 18.2). The data input from all sensors during experiments is simultaneously recorded at a maximum frequency of 5 kHz to gain high resolution data during the relatively short penetration test of about 20 s with a standard penetration speed of 2 cm/s. Penetration speed and all applied pressures are controlled in real time with an update rate of 5 kHz. An entire test including sample preparation may be completed in 10 h.

18.3 Results

18.3.1 Specimen Preparation and Cuxhaven Test Sand

Cuxhaven Sand is used for the initial tests in this study (Table 18.3). The loosest state of this sand has a void ratio of $e_{max} = 0.82$ and at its densest state $e_{min} = 0.48$, determined following DIN 18126 (1996). The samples are prepared in a mould consisting of two half cylinders, which are removed after sample preparation. The

Table 18.3 Index properties of test sands

Sand	Ticino sand (Baldi et al. 1986)	Ticino sand (Jamiolkowski et al. 2003)	Hokksund sand	Toyoura sand	Cuxhaven sand (This study)
γ_{max} (max. dry density)	17.00–17.05	16.67	17.24	16.13	16.88
γ_{min} (min. dry density)	13.91–13.98	13.64	14.10	13.09	14.21
Specific density [–]	–	2.68	2.72	2.65	2.64
Uniformity coefficient [–]	–	1.3	1.91	1.31	3.09
D_{50} [mm]	–	0.60	0.45	0.22	0.25
Quartz [%]	30	30	35	90	95

sand is prepared by dry pluviation with a 12.5 cm long handheld sand curtain directly in the sample mould at a constant falling height of 60 cm. The void ratio e of the tested sample is controlled by changing the width of the sand curtain. The smaller the flow area and the sand flow rate, the denser the reconstituted sand sample (Lagioia et al. 2006). After the pluviation the top of the sample is trimmed, vacuum stabilized, vacuum saturated, consolidated and as a last step the state of sample saturation is tested with the Skempton-B coefficient: the ratio of change in pore-water pressure to change in cell pressure (Skempton 1954). The initial relative density D_R of the sample is calculated as:

$$D_R = (e_{max} - e)/(e_{max} - e_{min}) \tag{18.1}$$

18.3.2 Laboratory CPT Experiments

Figure 18.2 illustrates the first series of CPT depth-profiles obtained with the MARCC at different relative densities. The tests were conducted under BC1 conditions at an isotropic effective stress of 100 kPa and the saturation tests resulted in Skempton-B coefficients >0.97 (Skempton 1954). The tip resistance q_c of all profiles increases rapidly within the first 5 cm of cone penetration. Samples with densities $D_R = 60$ and 65 % reach a steady state tip resistance earlier than the samples with D_R >90 %. Yet, less dense samples show more q_c fluctuations along the profile. For calibration purposes the steady state tip resistance is considered. As expected the tip resistance increases with increasing relative density. Below 30–35 cm the tip resistance slightly decreases.

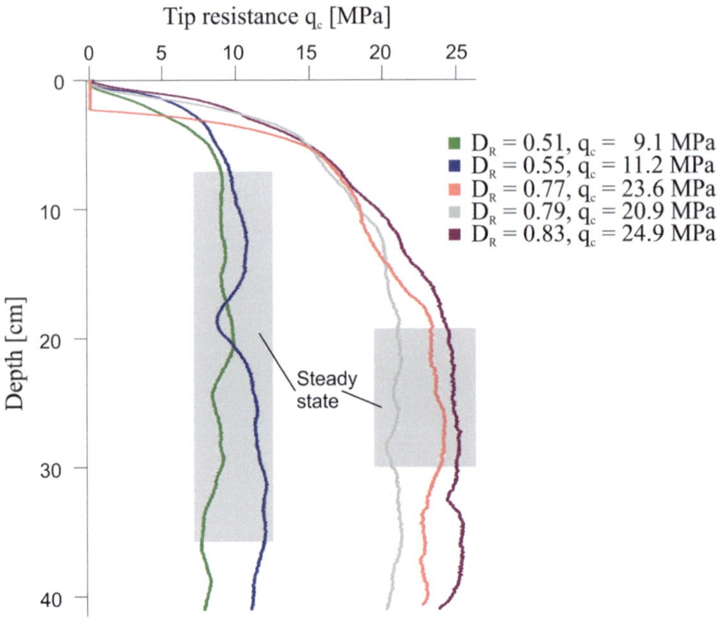

Fig. 18.2 CPT of Cuxhaven sand with different relative densities under B1 conditions at an isotropic effective stress of 100 kPa. The used tip resistance q_c [MPa] is the mean value of the steady state

The measured tip resistance is corrected for boundary effects and diameter ratio (here $R_D = 25$) by:

$$q_c(\text{corrected}) = q_c(\text{measured}) \left[(R_D - 1)/70\right]^{-0.5\, D_R} \tag{18.2}$$

after Kulhawy and Mayne (1990).

In a first attempt to correlate the tip resistance q_c to the relative density D_R we compared our data with correlations from Baldi et al. (1986) and Jamiolkowski et al. (2003) for normally and overconsolidated reconstituted siliceous sands with a mean effective stress σ'_{mo} of 100 kPa (Fig. 18.3). The correlations are based on Ticino Sand in Baldi et al. (1986) and Ticino, Hokksund and Toyoura Sands in Jamiolkowski et al. (2003) (Table 18.3). The earth pressure at rest K_0 is 0.4–1.4 in Baldi et al. (1986) and 0.4–1.0 in Jamiolkowski et al. (2003).

The corrected tip resistances for Cuxhaven Sand do not fit the functions of Baldi et al. (1986) and Jamiolkowski et al. (2003). The specific correlation function through the corrected tip resistance of Cuxhaven Sand is:

$$D_R = 0.28\ \ln(q_c) - 0.20 \tag{18.3}$$

with $\sigma'_{mo} = 100$ kPa and $K_0 = 1$, although the fit is based on few data and should be considered preliminary.

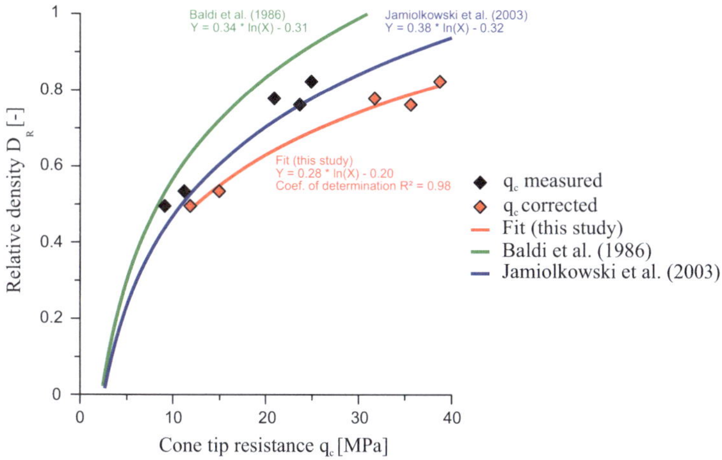

Fig. 18.3 Correlation of relative density D_R vs tip resistance q_c for fine to medium grain size, uncemented, unaged siliceous sands at a mean effective stress of 100 kPa

18.4 Discussion

Fluctuations in the q_c depth-profiles might be attributed to non-uniform sample preparation (Salgado et al. 1998). It seems to be more difficult to achieve homogeneous samples of lower density using the handheld pluviation device. The decreasing tip resistance in the lower part of the sample is probably the effect of the flexible lower boundary (Bolton and Gui 1993). However, most of the observed fluctuations are small and do not affect the general pattern of increasing steady-state tip resistance as a function of sample density.

Because the sample-to-cone diameter ratio of 25 is far below the suggested value of 60, measured tip resistances are underestimated applying BC1 (Kulhawy and Mayne 1990; Ghionna and Jamiolkowski 1991). A correction for boundary effects is essential to compare measured and field CPT data. The correction factor increases with increasing relative density (Kulhawy and Mayne 1990). The correction is empirically derived using various fine and medium sands yet it might be of limited applicability for this study.

The data for corrected tip resistance as a function of relative density for Cuxhaven Sand does not fit the available empirical correlations for reconstituted siliceous sands. Most likely this is due to the use of different sands for the correlations. Different soil properties, foremost the grain compressibility controlled by grain size, grain shape and mineralogy, exhibit different tip resistances (Kulhawy and Mayne 1990; Robertson and Cabal 2014). Cuxhaven Sand has a low compressibility because of its high quartz content. This property leads to higher tip resistances in comparison to the other sands (Robertson and Cabal 2014). The tip resistance is also highly dependent on the effective stress configuration (Tümay

1976). All three correlations used in this study are based on the mean effective stresses. However, it is known from calibration chamber tests that the horizontal stress has a higher influence on the tip resistance than the vertical effective stress (Baldi et al. 1986; Jamiolkowski et al. 2003). Thus the tip resistance also depends on the earth pressure at rest K_0. Furthermore, differences in sample preparation methods, including degree of sample saturation and determination of loosest and densest state to calculate relative density might affect the tip resistance measurements (Tümay 1976; Kulhawy and Mayne 1990; Robertson and Cabal 2014).

18.5 Conclusion and Outlook

A new calibration chamber is presented for efficient testing and small sample volumes. A whole testing cycle takes approximately 10 h. The results show that despite the small dimensions a steady state in tip resistance is reached, which depends on the sample density. The difference between the data and previously proposed empirical correlations show the importance of establishing sample-specific correlation functions to properly interpret in situ data for civil engineering and research purposes. Therefore, a specific correlation function was established for Cuxhaven Sand.

Acknowledgments The authors acknowledge funding by the Federal Ministry for Economic Affairs and Energy on the basis of a decision by the German Bundestag. Valuable support on the chamber design was provided by Tom Lunne and An-Bing Huang. We appreciate the effort of Matt Ikari and Max Kluger for proof reading our manuscript. We thank Wout Broere and Diego Lo Presti for their reviews and helpful comments.

References

Bałachowski L (2006) Penetration resistance of Lubiatowo Sand in calibration chamber tests. Arch Hydro-Eng Environ Mech 53(4):311–329

Baldi G, Bellotti R, Ghionna VN, Jamiolkowski M, E P (1982) Design parameters for sands from CPT. In: Proceedings of the Second European Symposium on Penetration Testing, Amsterdam

Baldi G, Bellotti R, Ghionna V, Jamiolkowski M, Pasqualini E (1986) Interpretation of CPTs and CPTUs; 2nd part: drained penetration of sands. In: Proceedings of the 4th International Geotechnical Seminar, 1986. p 143–156

Bolton MD, Gui MW (1993) The study of relative density and boundary effects for cone penetration tests in Centrifuge. Engineering Department, Cambridge University, Cambridge

Bolton M, Gui M, Garnier J, Corte J, Bagge G, Laue J, Renzi R (1999) Centrifuge cone penetration tests in sand. Géotechnique 49(4):543–552

DIN 18126 (1996) Bestimmung der Dichte nichtbindiger Böden bei lockerster und dichtester Lagerung. Baugrund, Untersuchung von Bodenproben, vol DIN 18126:1996-11. Deutsches Institut für Normung, Berlin

Ghionna V, Jamiolkowski M (1991) A critical appraisal of calibration chamber testing of sands. In: Huang A-B (ed) Proceedings of the first International Symposium on Calibration Chamber Testing (ISOCCT1), Potsdam, New York, 1991. Elsevier, New York, pp 13–40
Hsu HH, Huang AB (1998) Development of an axisymmetric field simulator for cone penetration tests in sand. ASTM Geotech Test J 21(4):348–355
Huang AB, Hsu HH (2005) Cone penetration tests under simulated field conditions. Géotechnique 55(5):345–354
Jamiolkowski M, Lo Presti D, Manassero M (2003) Evaluation of relative density and shear strength of sands from CPT and DMT. In: Soil Behavior and Soft Ground Construction, 2003. ASCE, pp 201–238
Kulhawy FH, Mayne PW (1990) Manual on estimating soil properties for foundation design. Geotechnical Engineering Group, Cornell University, Ithaca
Lagioia R, Sanzeni A, Coleselli F (2006) Air, water and vacuum pluviation of sand specimens for the triaxial apparatus. Soils Found 46(1):67
Lunne T (2010) The CPT in offshore soil investigations: a historic perspective. Paper presented at the 2nd International Symposium on Cone Penetration, Huntington Beach, CA, May 9–11, 2010
Lunne T, Robertson PK, Powell JJM (1997) Cone penetration testing in geotechnical practice. Blackie Academic & Professional/Routledge, New York
Mayne PW, Kulhawy FH (1991) Calibration chamber database and boundary effects correction for CPT data. In: Huang A-B (ed) Proceedings of the First International Symposium on Calibration Chamber Testing (ISOCCT1), Potsdam, New York, 1991. Elsevier, New York, pp 257–264
Parkin A (1988) Calibration of cone penetrometers. In: Procedures of the 1st International Symposium on Penetration Testing (ISOPT-1), Orlando, Florida, 1988. vol 1. Pergamon, p 221–243
Robertson PK, Cabal KL (2014) Guide to cone penetration testing for geotechnical engineering, 6th edn. Gregg Drilling and Testing, Signal Hill
Salgado R (1993) Analysis of penetration resistance in sands. Rivista Italiana di Geotecnica 4/98
Salgado R, Mitchell J, Jamiolkowski M (1998) Calibration chamber size effects on penetration resistance in sand. J Geotech Geoenviron Eng 124(9):878–888
Skempton AW (1954) The pore-pressure coefficients A and B. Géotechnique 4:143–147
Tümay M (1976) Realtive density versus CPT cone bearing in cohesionless soils. Dissertation, Bogazici University, Istanbul

Chapter 19
Underwater Mass Movements in Lake Mjøsa, Norway

Carl Fredrik Forsberg, Håkon Heyerdahl, and Anders Solheim

Abstract This study was initiated because underwater mass movements in Lake Mjøsa, Norway had caused utility pipeline breakages. Multibeam bathymetry, sub bottom profiler data and samples were acquired to allow morphological interpretations of the processes leading to the mass movements. The underwater slopes of the lake generally show gradients of 15–20°, but exceed 30° in places. The sediment thickness above acoustic basement interpreted from sub bottom profiler data show that the accumulation rate in the central lake basin is about 2 mm/year. Numerous channels are seen on the bathymetry as well as several slide scarps that are about 2 m high. The channels are interpreted to be caused by dense water cascading, probably induced by winter cooling. Calculations based on geotechnical tests of samples indicate that sediment layers in excess of 2 m have a factor of safety less than 1.5 on slopes above 30°. Triggering of slides may thus be spontaneous due to sedimentation, but may also be triggered by oversteepening due to erosional channeling.

19.1 Introduction

Mass movements in lakes have not been studied as frequently as those in the marine environment even though lakes are often associated with human infrastructure that is vulnerable to the effects of such movements. Of course, not all lakes have a bathymetry conducive to the development of mass movements. A prerequisite for mass movements is in general accumulation of sediments and a steep slope where they can be triggered. Lake Mjøsa, Norway's largest lake (Fig. 19.1), fulfils this requirement with numerous slopes with inclinations of 15–30°. The lake has water depths up to 449 m. It is an overdeepened glacial trough and forms an extension of the Gudbrandsdalen Valley. It has a U-shaped cross section, a length of 117 km and a maximum width of 9 km. In the vicinity of the study area, it is about 2 km wide (Fig. 19.1). As such it is similar to Lake Lucern, Lake Zurich and Lake Windermere which have been the subject of several investigations related to subaquatic mass

C.F. Forsberg (✉) • H. Heyerdahl • A. Solheim
Norwegian Geotechnical Institute (NGI), 3930, Ullevål Stadion,
NO-0806 Oslo, Norway
e-mail: carl.fredrik.forsberg@ngi.no

G. Lamarche et al. (eds.), *Submarine Mass Movements and their Consequences*,
Advances in Natural and Technological Hazards Research 41,
DOI 10.1007/978-3-319-20979-1_19

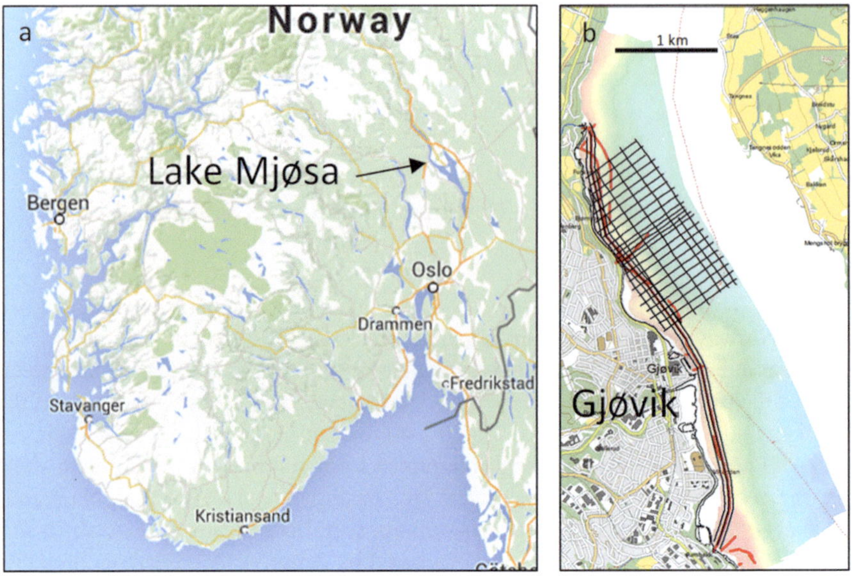

Fig. 19.1 (**a**) Location of Lake Mjøsa (Map data © 2015 GeoBasis-DE/BKG (©2009), Google). (**b**) SBP tracks (*black lines*) and multibeam coverage (*coloured area*) in the lake close to the municipality of Gjøvik. *Red lines* are pipelines (Background map © Kartverket)

movements (Vardy et al. 2010; Lowag et al. 2012; Pinson et al. 2013; Strasser et al. 2011, 2013; Stegmann et al. 2007).

The current study was initiated after ROV inspections revealed that breakages of utility pipelines in the lake outside the municipality of Gjøvik (Figs. 19.1 and 19.2) were caused by mass movements. The breakages were not associated with any noticeable earthquake activity, nor were any lake level fluctuations registered.

The purpose of the initial study was to evaluate the slope stability and mass movement mechanisms in the lake as an aid to routing and siting of future utilities on the lake bed. However, vital to such evaluations is an understanding of the preconditioning and triggering mechanisms related to sliding and other slope processes. We present here the morphology and the interpretations of the processes that contribute to the slope instabilities in the area and discuss these in relation to conditions in other lakes.

19.2 Methods

The bathymetry available prior to this study was based on hand leadline measurements from 1906 (NVE 1984), which were not suitable for a detailed study. Parker Maritime Ltd. was therefore contracted to acquire multibeam bathymetry and sub bottom profiles in Lake Mjøsa offshore from Gjøvik (Fig. 19.1) on behalf of NGI

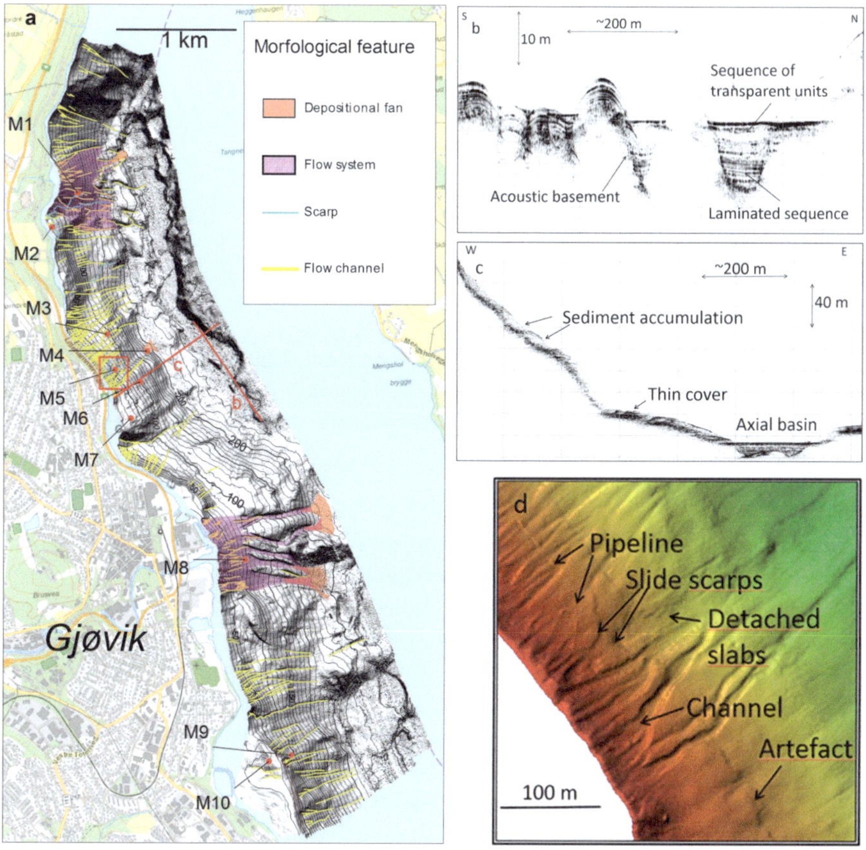

Fig. 19.2 (**a**) Overview of morphological interpretations. The flow systems are broad areas with down slope flow in front of rivers. M1 to M10 are sample locations. Slopes have a gradient that is generally between 15 and 25°, but may exceed 30° in some places. *Red lines* show location of SBP profiles in (**b**) and (**c**). *Red rectangle* is area with pipeline breakages shown on (**d**). *Bold contours* are at 50 m intervals (Background map © Kartverket). (**b**) Axial SBP showing that the fill above acoustic basement is up to 14 m thick. *Horizontal grid line* spacing is 2 m. (**c**) SBP section passing down slope into axial basin. Transparent sediment cover has accumulated on terraces in acoustic basement. *Horizontal grid line* spacing is 10 m. (**d**) Detailed view of area where pipeline breakages occurred. The pipeline position is shown after its broken end was temporarily relocated onshore. Also shown are the slide scarps and some detached slabs. Acquisition artefacts are also marked

(Norwegian Geotechnical Institute). The data was collected using a 7.9 m long boat as a platform for a Kongsberg Maritime EM 3002 multibeam and an Innomar SES 2000 parametric echo sounder for sub bottom profiler (SBP) records. Navigation was performed using differential GPS and registered on a Seapath 200 RTK system that performed real time correction of the data for the boat's movements.

While preliminary bathymetry was compiled using Olex software, final bathymetric grids were generated using the Generic Mapping Tools (Wessel et al. 2013)

Table 19.1 Overview of sediment coring with comments from ROV inspection

Sample [no.]	Recovery [m]	Comments from ROV inspection
M1	1.5	Stones. Sampler fell over. Sample after second attempt
M2	1.8	Soft lake bed
M3	1.4	Soft lake bed
M4	1.0	Stones. Sampler fell over. New sample after two attempts
M5	No sample	Stones. Sampler fell over. No sample even after two additional attempts
M6	1.5	Soft lake bed
M7	1.5	Soft lake bed
M8	0.4	Stones. Sampler fell over. New sample after two attempts
M9	2.0	Soft lake bed. Clay. Full recovery
M10	1.2	Sampler fell over. New sample after two attempts. Very soft material

suite of tools. Bathymetric raw data prior to processing had a spacing of 0.1–0.2 m whereas the gridded bathymetry has a resolution of 1×1 m.

The results of the bathymetric survey varied with water depth, however all data were satisfactory.

The steep and uneven slopes in the lake meant that the SBP profiles were challenging to interpret due to interference from side echoes. Additionally, very little signal penetration was achieved in many areas. No systematic interpretation of the data was therefore attempted. Instead, profiles were used to evaluate the stratigraphy near the lake bed at selected sites in conjunction with morphological analyses and sampling.

Morphological interpretations were performed using shaded relief bathymetric maps derived from the bathymetric grids with additional input from the SBP profiles.

Following inspection of the bathymetric and geophysical data, nine gravity cores were recovered from ten selected sites using a 1-ton, 2-m barrel coring device deployed from a barge with a frame and winch (Table 19.1, Fig. 19.1).

The laboratory programme for the cores consisted of measurements of grain size, water content, density, liquid limit, plastic limit, humus content and fall cone tests of undisturbed and remoulded undrained shear strengths.

19.2.1 *Morphology*

The bathymetric data were inspected using a combination of Global Mapper and Geographix GeoAtlas software. Emphasis was put on identifying features associated with mass transport phenomena such as slide scars, channelling and depositional lobes (Fig. 19.2).

The main features identified are small slide scarps, a metre or two high (Fig. 19.2), numerous channels indicating down slope density flows and broader areas with down slope sediment transport in front of rivers entering the lake. No slide debris has been identified downslope of the headwalls although some detached slabs have been identified in the slide scar associated with the pipeline breakages (Fig. 19.2). The bulk of the slide masses are therefore thought to have continued into the lake basin. On SBP data from the lake basin, the acoustic basement is covered by a laminated sequence on top of which are several transparent units separated by relatively prominent reflectors (Fig. 19.2). Areas with rougher topography are interpreted as bedrock outcrops or regions where the acoustic basement has a thin cover of sediments only. These areas are mostly found surrounding the deep lake basin and near the shore line.

Several slide scarps about 1.5–2 m high form a "terraced" slide scar in the area where the pipeline breakages occurred (Fig. 19.2); some detached slabs are found in the scar area. We believe that this is the slide that caused the breakages. The site is also crossed by some of the more prominent mapped channels. There is a small depositional fan at about 200 m water depth, at a break in the slope close to sample M4; the seafloor is characterised by stones in this area (Table 19.1, Fig. 19.2).

19.2.2 Slope Stability

The slopes in the study area are quite steep, frequently in the range of 20–30°, and quite extended compared to the height of the slide scarps (headwalls) identified during the morphological interpretations. The slides have therefore probably been translational and infinite slope analyses are therefore appropriate for slope stability calculations.

Undrained shear strengths and unit weight are needed as input in the calculations and were measured on the retrieved samples.

A conservative "design" profile for undrained shear strength (s_{ud}) was used for the stability calculations (Fig. 19.3):

$$s_{ud} = 3z + 0,5\ [\text{kPa}], \text{where } z \text{ is the vertical depth.}$$

The measured unit weights varied from 13.1 to 15.3 kN/m^3, with a value of 15 kN/m^3 used for the calculations.

The factor of safety for infinite slope analyses (FS) is the ratio between the resisting forces (S) from the undrained shear strength and the driving forces (τ). S is given by the undrained shear strength multiplied by the area of the slip plane:

$$S = s_{ud}\ a/\cos\alpha, \text{ for a slip plane with horizontal extent a.}$$

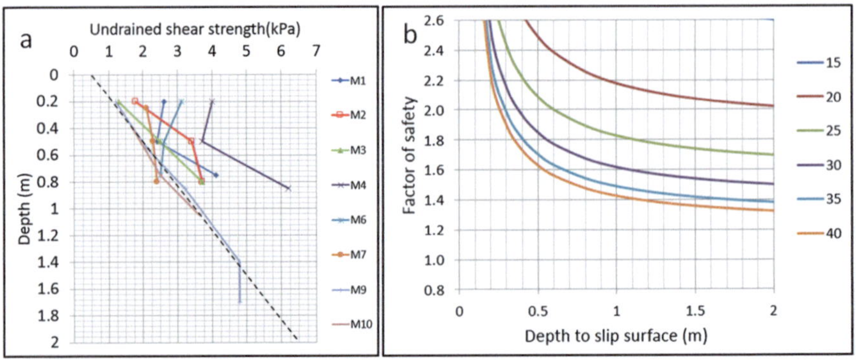

Fig. 19.3 (**a**) Undrained shear strengths vs. depth. The *dashed line* shows the undrained shear strength vs depth used for the stability calculations. (**b**) Factor of safety for various slopes (in degrees) vs depth to slip surface

The driving force τ along the failure plane is given by:

$$\tau = a\gamma' z \sin\alpha,$$

where γ' is the submerged unit weight and α the slope angle.

Combining these two gives the factor of safety (FS):

$$FS = s_{ud}/(\gamma' z \quad \sin\alpha\cos\alpha)$$

Slopes in excess of about 30° will have a safety factor below about 1.5 if the soft sediment thickness exceeds 2 m (Fig. 19.3) using the relationships above.

19.2.3 Discussion

Slide scarps, channeling and flow systems are the three main morphological features related to down slope movements mapped in the lake. They are all found on the slopes. The basin itself has a planar lake bed without topographic features indicative of mass movements.

The seismic stratigraphy of the axial basin defined two main facies above the acoustic basement. A basal laminated sequence (~8 m thick) overlain by a series of transparent units (~6 m). The acoustic basement is most likely comprised of bedrock covered by basal till from the last glaciation. The laminated sequence therefore represents the first deposits after glacial retreat, maybe a proglacial meltwater or glaciolacustrine deposit similar to that proposed for the SSSII unit in Lake Windermere (Lowag et al. 2012; Pinson et al. 2013) or in Swiss lakes (Strasser et al. 2013). The documented Swiss deposits are however up to 140 m thick (Strasser et al. 2013), i.e. significantly thicker than those in Lake Mjøsa. The uppermost sequence of transparent units is interpreted to represent the post glacial Holocene sediments. We suggest their transparent nature is due to the influence and accumulation of mass transport deposits that originate on the slope, the bounding reflectors representing periods of normal lacustrine sedimentation. Debris flow deposits may have formed the depositional lobes found at a break in the slope upslope from the axial basin. In general, only turbidites or dense water flows may therefore presently pass towards the axial basin. The horizontal basin floor supports this concept suggesting that lacustrine sedimentation combined with ponding of turbiditic flows dominate accumulation at present.

The total thickness of the axial deposits of Lake Mjøsa is 14 m above acoustic basement. This means that the accumulation rate in Lake Mjøsa is low, on average below about 2 mm/year for the post glacial period and including the influence of mass transport processes.

The channels on the slope are erosional features that are probably related to density flows from a process analogous to dense shelf water cascading (DSWC) seen in the marine environment (Canals et al. 2006; Puig et al. 2008). The density of the flows is unlikely to be due to suspended sediments as the basin accumulation rate is low. We therefore attribute the densification to cooling of surface water during cold winter weather and believe that the shallow water near the shore may be especially prone to cooling because of the added effect of waves washing across the cold rocks at the water's edge. For Lake Mjøsa the concentration of channels is greater where the slope to shoreline distance is small (Fig. 19.2). We suggest that the short distance to the slope will not allow the dense water to pond before descending the slope. However, where the distance from the shore is somewhat larger, some ponding occurs and the dense water arrives at the slope at points with a greater separation than where the "shelf" is narrow. An additional unknown factor is the contribution from numerous small streams that do not appear on maps.

The flow systems /areas in front of the rivers exhibit evidence of sediment accumulations at depth in the form of depositional fans. The systems probably represent an amalgamation of many channels that are fed by hyperpycnal flows associated with the rivers. The flows would depend on the relative densities of the river water and the lake water with both temperature and suspended sediments contributing to the differences.

The results of the slope stability analyses rely heavily on the undrained shear strength profile used. Considerably different profiles may seem equally appropriate when compared to the measured data. We explore this a bit further. The slope stability analyses use a "design" profile for the undrained shear strength which is based on the lower bound for the measured data. This lower bound (Fig. 19.3) has a gradient (3 kPa/m) somewhat higher than usually expected for normally consolidated sediments (~1.3 kPa/m) with a unit weight of 15 kNm3. This may be due to unreliable results for some gravity core samples of the softest sediments as a result of sample disturbance. The design profile uses a seafloor strength of 0.5 kPa. Figure 19.3 shows that a less conservative strength at the seafloor intercept may be around 2 kPa. Using this intercept with the 1.3 kPa/m gradient gives better stability for sediment sheets thinner than 2 m, but decreased stability for thicker sheets. Thus the main result still holds; i.e. that sediment sheets in excess of 2–3 m in thickness on gradients in excess of 30° will be prone to sliding. The difference between undrained shear strength as measured for compressional loading by the fall cone method and horizontal shearing occurring during a translational slide may warrant a reduction in the FS. The preconditioning for sliding is therefore thought to be the sedimentation draping the lake bed in the area. Triggering may be spontaneous, but channel erosion by the density flows may set off a sliding event by removing toe support. Secondary mass flows may be set off by impact from an initial sliding event further up slope. This contrasts to the Swiss lake slides, many of which have been synchronously triggered by earthquakes.

Once sliding has taken place, the slope at the site will remain stable until a fresh 2–3 m thick layer has accumulated. However, neighbouring regions with intact sediments and equivalent slopes will still be prone to sliding. Thus mass flows in Lake Mjøsa will occur regularly and sheet like landslides will occur when sediment thicknesses reach critical values.

Compared to the Swiss lakes studied by Lowag et al. (2012), Stegmann et al. (2007), and Strasser et al. (2011, 2013) the slides presented here are minor (2–3 m thick and in the order of 100 × 100 m or around 30,000 m^3). The slope stability analyses demonstrate that thick accumulations of sediment are unlikely on the slopes in this part of Lake Mjøsa and consequently large slides equivalent to the Swiss slides are unlikely to occur. However, although the slides in Lake Mjøsa were minor, they caused damages that cost 55 million NOKs according to a local newspaper, and so understanding the risk from such events is important.

References

Canals M, Puig P, de Madron XD, Heussner S, Palanques A, Fabres J (2006) Flushing submarine canyons. Nature 444(7117):354–357

Lowag J, Bull JM, Vardy ME, Miller H, Pinson LJW (2012) High-resolution seismic imaging of a Younger Dryas and Holocene mass movement complex in glacial lake Windermere, UK. Geomorphology 171:42–57

NVE (Norges Vassdrags og Elektrisitestvesen), Vassdragsdirektoratet (1984) http://gis3.nve.no/metadata/tema/DKBok1984/Dybdekart_1984.htm. Accessed Jan 2015

Pinson LJW, Vardy ME, Dix JK, Henstock TJ, Bull JM, Maclachlan SE (2013) Deglacial history of glacial lake Windermere, UK: implications for the central British and Irish Ice Sheet. J Quat Sci 28(1):83–94

Puig P, Palanques A, Orange DL, Lastras G, Canals M (2008) Dense shelf water cascades and sedimentary furrow formation in the Cap de Creus Canyon, northwestern Mediterranean Sea. Cont Shelf Res 28(15):2017–2030

Stegmann S, Strasser M, Anselmetti F, Kopf A (2007) Geotechnical in situ characterization of subaquatic slopes: the role of pore pressure transients versus frictional strength in landslide initiation. Geophys Res Lett 34(7):L07607

Strasser M, Hilbe M, Anselmetti FS (2011) Mapping basin-wide subaquatic slope failure susceptibility as a tool to assess regional seismic and tsunami hazards. Mar Geophys Res 32(1–2):331–347

Strasser M, Monecke K, Schnellmann M, Anselmetti FS (2013) Lake sediments as natural seismographs: a compiled record of late quaternary earthquakes in Central Switzerland and its implication for Alpine deformation. Sedimentology 60(1):319–341

Vardy ME, Pinson LJW, Bull JM, Dix JK, Henstock TJ, Davis JW, Gutowski M (2010) 3D seismic imaging of buried Younger Dryas mass movement flows: Lake Windermere, UK. Geomorphology 118(1–2):176–187

Wessel P, Smith WHF, Scharoo R, Luis J, Wobba F (2013) New, improved version of the generic mapping tools improved version released. Eos Trans AGU 94(45):409–410

Chapter 20
In Situ Cyclic Softening of Marine Silts by Vibratory CPTU at Orkdalsfjord Test Site, Mid Norway

Max Oke Kluger, Stefan Kreiter, Jean-Sebastien L'Heureux, Sylvia Stegmann, Vicki Moon, and Tobias Mörz

Abstract Earthquake induced cyclic loading has the potential to destabilize submarine slopes either by liquefaction in coarse-grained deposits or by cyclic softening in cohesive sediments. Vibratory cone penetration tests (VCPTU) represent a new approach for the evaluation of cyclic softening in fine grained sediments. In the past, VPCTU were utilized to evaluate liquefaction potential of sands, but cyclic softening of fine-grained marine sediments has not yet been tested with VCPTU in situ. At the study site in Orkdalsfjord, mid Norway marine clayey silt deposits are interbedded with coarse silt and clay layers. Static and vibratory CPTU were performed down to 19 m penetration depth using the Geotechnical Offshore Seabed Tool (GOST) and in addition, two gravity cores were taken for cyclic triaxial testing and geotechnical index tests. From static and vibratory CPTU a number of coarse silt layers with a distinct drop in cyclic cone resistance were identified. Compared to surrounding finer sediments the coarse silt layers exhibited a higher potential for cyclic softening. This assumption is supported by cyclic triaxial tests on very coarse and surrounding medium-coarse silts, respectively, revealing a strong loss of cyclic shear strength in a controlled and documented stress-strain regime. This study highlights the potential for VCPTU as a promising tool to qualitatively evaluate the vulnerability of marine silts to cyclic softening. In combination with advanced laboratory tests these results are envisioned to help better identifying submarine slopes subjected to failure during earthquakes.

M.O. Kluger (✉) • S. Kreiter • S. Stegmann • T. Mörz
MARUM – Center for Marine Environmental Sciences, University of Bremen, Klagenfurter Strasse, 28359 Bremen, Germany
e-mail: mkluger@marum.de

J.-S. L'Heureux
Norwegian Geotechnical Institute, Trondheim, Norway

V. Moon
Department of Earth and Ocean Sciences, University of Waikato, Hamilton, New Zealand

G. Lamarche et al. (eds.), *Submarine Mass Movements and their Consequences*, Advances in Natural and Technological Hazards Research 41,
DOI 10.1007/978-3-319-20979-1_20

20.1 Introduction

Submarine and near-shore slope failures often result from earthquakes or other forms of cyclic loading such as waves, tides or from human activities like blasting or machine vibrations. In order to assess the liquefaction potential in sands (Ishihara 1985; Youd et al. 2001) or the cyclic softening behaviour in cohesive soils (Boulanger and Idris 2006, 2007), cone penetrometer (CPTU) data and empirical methods are commonly used in practice (Robertson and Wride 1998; Seed et al. 1983). Alternatively, vibratory cone penetration tests (VCPTU) may have potential to directly determine the cyclic behaviour of in situ soils as they mechanically induce cyclic loads into the sediment (Sasaki et al. 1984; Wise et al. 1999; Tokimatsu 1988; Jorat et al. 2014). When pushing a vibratory CPTU tip in a liquefiable soil, the cone resistance (q_{cv}) is substantially reduced compared to that from a static test (q_{cs}). In order to relate q_{cs} and q_{cv}, Sasaki et al. (1984) proposed the reduction ratio (RR):

$$RR = 1 - q_{cv}/\, q_{cs} \qquad (20.1)$$

Following Tokimatsu (1988), sediments with RR values of more than 0.8 exhibit high liquefaction potential. In addition to high RR values, liquefiable sediment layers often show an increase in the induced pore pressure (e.g. Mitchell 1988; Bonita 2000). To our knowledge, most VCPTU studies found in the literature have evaluated the liquefaction potential of terrestrial and marine sands only. Here we evaluate the cyclic behaviour of fine-grained marine soils from Orkdalsfjord, mid Norway, based on both CPTU and VCPTU profiles obtained with the Geotechnical Offshore Seabed Tool (GOST). The in situ data set is combined with soil index properties and compared to the results of cyclic undrained triaxial tests.

20.2 Geological Setting

Orkdalsfjord is a 25–30 km long branch of the Trondheimsfjord located 25 km southwest of Trondheim, mid Norway. The sediment strata consists of layered loose fine sand, silt and silty clay deposits originating from the Orkla river (Bjerrum 1971). In May 1930 a submarine retrogressive landslide occurred along the fjord initiating a 15 m high tsunami wave that caused damage to harbour installations and killed one person (L'Heureux et al. 2014). The volume of sediment that evacuated downslope of the 8–12 m high and 3 km long headwall is estimated to $18.5{\cdot}10^6$ m^3 (Fig. 20.1, L'Heureux et al. 2014).

20.3 Material and Methods

The study site GeoB18623 is located in 38 m water depth at the intact seafloor above the headwall scar of the 1930 landslide (Fig. 20.1). A static and a vibratory CPTU profile as well as two up to 4.6 m long gravity cores were retrieved during

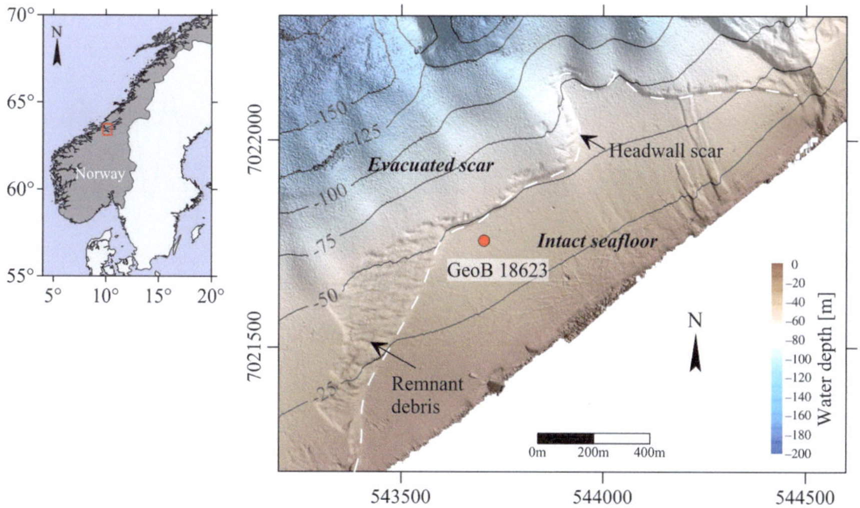

Fig. 20.1 Orkdalsfjord study area (Bathymetry from L'Heureux et al. (2014) used with courtesy from the Geological survey of Norway, NGU)

R/V Poseidon expedition POS472 in 2014. For better comparability the VCPTU depth was correlated with the CPTU profile resulting in maximal vertical offsets of 20 cm. The core GeoB18623-1 was split for sediment description and geotechnical index tests, while GeoB18623-2 was kept closed for subsequent triaxial testing. Undrained shear strength s_u was determined directly after splitting using vane shear and fall cone test DIN 4094-4:2002-01 (2002). Atterberg limits were measured in every lithology following ASTM D4318-10 (n.d). Grain size distribution was determined every 20 cm with a Beckman Coulter laser diffraction particle size analyser (LS 13320) adopting Udden (1914) and Wentworth (1922) classification.

20.3.1 CPTU

The static and vibratory CPTU were conducted using the GOST system (Steiner et al. 2014; Moon et al. 2013; Jorat et al. 2014). The system uses a 5 cm^2 high-resolution piezovibrocone, which measures cone resistance q_c, sleeve friction f_s and pore pressure u_2, which, in static mode, is pushed at a rate of 2 cm/s down to maximal 40 mbsf. As originally developed for static deep sea operations, GOST has a limited hydraulic energy supply restricting penetration speed of vibratory mode to 1.35 cm/s. The CPTU results from cohesive sediments are generally strain-rate dependent and correction procedures for different penetration speeds exist that are also validated for Norwegian fjord sediments (Steiner et al. 2014). The resulting correction for the used setup in q_c is approximately 3 % and was therefore neglected. Vertical laser displacement measurements of the vibratory mode show

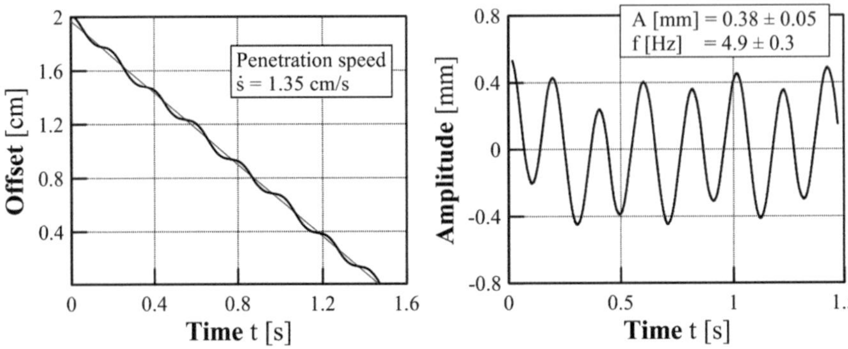

Fig. 20.2 Cyclic signal of VCPTU using GOST

a steady sinus-shaped cyclic amplitude of 0.38 mm at 4.9 Hz during penetration (Fig. 20.2). Note that the movement of the rod is hydraulically driven from the seafloor and regulated by external displacement sensors. Therefore the actual cyclic movement of the cone may vary with depth.

20.3.2 Triaxial Laboratory Testing

Cyclic laboratory experiments were conducted using the MARUM dynamic triaxial testing device (DTTD) (Kreiter et al. 2010a, b). The axial stress is measured inside the cell directly below the soil specimen, whereas the stress control is using an external force sensor. The test specimens have an area of 10 cm^2 and a length of 7 cm, following the test procedure of DIN 18137-2:2011-04 (2011). Bulk specimens are tested with undrained isotropically consolidated cyclic single-stress level experiments (e.g. Ishihara 1985; Singh 1996). The applied confining pressure of $\sigma_c' = 100$ kPa was higher than the in situ stress conditions to overcome friction loss at low stress cycles with the external force control. A medium cyclic stress ratio of $CSR = (\sigma_1' - \sigma_3') / 2\sigma_c' = 0.1$ at a frequency $f = 0.5$ Hz was applied.

20.4 Results

20.4.1 Geotechnical Characterization of Silt Layers

The sediments of gravity core GeoB18623-1 are layered fine to very coarse silt deposits with water contents near the liquid limit w_L (Fig. 20.3). The sediments have plasticity indices between 12 and 16 % and are therefore expected to show clay-like behaviour upon cyclic loading (Boulanger and Idriss 2006, 2007). Gas escape fractures are common throughout the core indicating a considerable amount

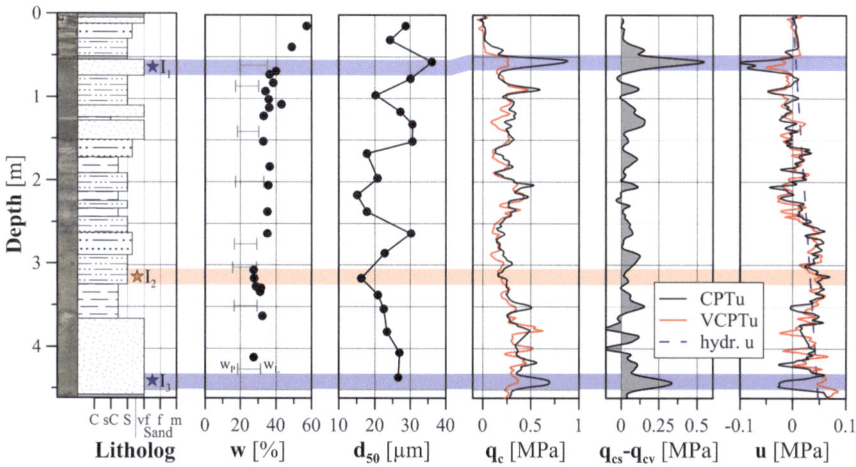

Fig. 20.3 Sediment properties and in situ CPTU at study site GeoB18623. Lithology includes core photos and qualitative grain size description. *C*: Clay, *sC*: Clayey silt, *S*: Silt, *vf*: very fine sand, *f*: fine sand, *m*: medium sand. I_1–I_3: samples for cyclic triaxial tests, *Blue*: (Very) coarse silt intervals, *Pink*: medium-coarse silt interval

of free gas in the in situ fjord sediments. The undrained shear strength from lab vane and fall cone is more or less constant and varies between $15 \leq s_u \leq 25$ kPa.

Two layers, I_1 and I_3, consisting of very coarse and coarse silt ($25 \leq d_{50} \leq 36$ μm, Fig. 20.4) with high static cone resistances and distinctly lower cyclic cone resistance are detected at depths of 0.55 and 5.5 m, respectively (Fig. 20.3). The decrease in cone resistances is 540 and 340 kPa for layers I_1 and I_3, respectively. Both silt layers have RR-values (Eq. 20.1) between 0.5 and 0.7 and are therefore classified to exhibit medium liquefaction potential (Tokimatsu 1988). This corresponds to cone resistance reductions of 68 % and 45 %, respectively, which is quite dramatic. In the pore pressure profile intervals with negative u-values are identified. This extreme pore pressure drop is especially well pronounced in the (very) coarse silt layers I_1 and I_3 during static CPTU, while the pore pressure drop is less for vibratory CPTU.

To differentiate the cyclic softening behaviour of the very coarse silt layers with surrounding finer sediments a reference interval I_2 was selected consisting of medium-coarse silt with $d_{50} = 16$ μm (Fig. 20.4). In contrast to I_1 and I_3 the reference interval is characterized by low static cone resistance and low RR-values.

20.4.2 Cyclic Triaxial Response of Silt Layers

In order to evaluate the in situ VCPTU observations, cyclic triaxial tests were conducted on silt layers I_1 and I_2 as their difference in grain size is higher compared

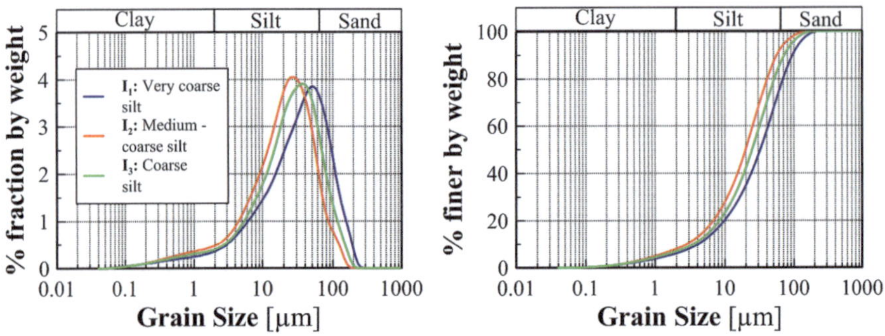

Fig. 20.4 Grain size distribution of samples I_1, I_2 and I_3 tested in DTTD

to I_3 (Fig. 20.5). The cyclic softening is derived from axial strain ε_1 and normalized excess pore pressure $u_n = u / \sigma_c'$ at the same number of cycles (Fig. 20.5a, b). The very coarse silt I_1 softens five times faster than the medium coarse silt I_2. The effective mean pressure p' is decreasing in both samples with each cycle due to excess pore pressure build-up, however, the initial incremental loss in p' is larger in I_1 (Fig. 20.5c). In total: The very coarse silt appears to be more vulnerable to cyclic softening than the medium-coarse silt.

20.5 Discussion and Conclusion

In this pilot study cyclic geotechnical tests were conducted in situ and under laboratory conditions allowing insight into the cyclic softening behaviour of marine silts from Orkdalsfjord. The very coarse silt layers lose up to 68 % cone resistance during VCPTU and exhibit moderate RR-values of maximal 0.7 making it more vulnerable against cyclic softening compared to surrounding finer silts with RR-values around zero. This assumption is supported by DTTD tests stating that cyclic softening potential of very coarse silt is significantly higher than the softening potential of the medium coarse silt. The in situ silts generate negative excess pore pressures during CPTU. However, this phenomenon was also observed near Finneidfjord with GOST (Steiner et al. 2014), in the Norwegian Sea and offshore Africa (Lunne 2010) and in the North Sea (Bayne and Tjelta 1987) with different CPTU tools. Negative excess pore pressure seems to be characteristic for some cohesive materials with thin interbedded coarser layers. During VCPTU the very coarse silt seems to be less dilative as a response to the cyclically-induced pore water pressures, however the effect may partly be influenced by the difference in penetration speed.

The comparison of VCPTU results with DTTD data is considered with caution as both analyses differ in terms of stress state. VCPTU was conducted in unconsolidated silts subjected to natural anisotropic stress conditions. In DTTD experiments a higher isotropic stress of 100 kPa was used in order to overcome friction

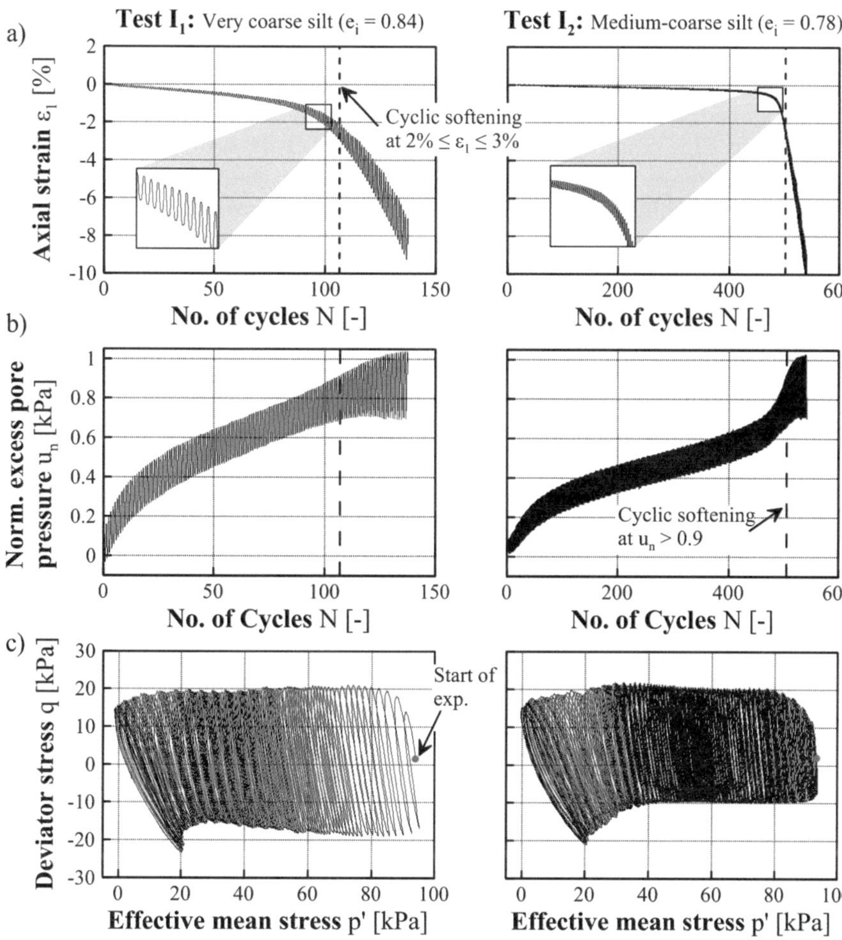

Fig. 20.5 Undrained cyclic triaxial tests on samples I_1 and I_2 with limits for cyclic softening after Kramer (1996) and Boulanger and Idriss (2006). Test conditions: $\sigma_c' = 100$ kPa, CSR = 0.1, $0.98 \leq B \leq 0.99$

loss at low stress cycles. In addition, material properties such as relative density, stress strain history, cementation and the fabric of the soil directly affect the cyclic softening potential (Kramer 1996). The latter likely varies between the in situ and lab as the lab specimens may have been disturbed through coring, gas escape, transport and preparation. Therefore only a qualitative comparison is possible. A general problem comparing static and vibratory CPTU is the mandatory perfect correlation of the two test sites, in this case the lateral offset is around 3–5 m. Imperfect correlation or lateral variations of the sediment would cause artefacts in the difference of cone resistances and the reduction ratio RR, however the correlation at GeoB18623 is convincing.

The failure mechanism leading to the 1930 submarine landslide in Orkdalsfjord was comprehensively investigated by L'Heureux et al. (2014). The authors showed that the slide plane most likely involved a clay-rich layer at 8–10 m below the original seabed. Results presented in the actual study show that the silt deposits overlying the slide prone layer are vulnerable against cyclic softening and that only small vibrations are necessary to considerably reduce their shear strength. This softening process might have been important in the retrogressive landslide process in order for the landslide debris to completely evacuate the landslide scar.

The present study showed that cyclic softening is a considerable phenomenon in marine silts from Orkdalsfjord and that it might be under certain circumstances an important factor facilitating failure of submarine landslides. We further showed that VCPTU may be a helpful tool to qualitatively characterize in situ cyclic softening potential of marine silts. However, to comprehensively understand the effect of cyclic loading on the soil strength cyclic laboratory tests are indispensable.

Acknowledgments The authors acknowledge funding by the Integrated Coastal Zone and Shelf Sea Research Training Group INTERCOAST and the MARUM Centre for Marine Environmental Science at the University of Bremen. The Geological Survey of Norway is gratefully acknowledged for the access to the bathymetric data. We thank Wolfgang Schunn from the University of Bremen who operated GOST and improved data acquisition on board the R/V Poseidon. We thank Dr Nicole Baeten, Alexander Rösner and Robert Roskoden for invaluable assistance during data acquisition and discussion. Marc Huhndorf from the University of Bremen is acknowledged for his efforts regarding laser measurements of the cyclic signal of GOST. We thank both research students David Seibel and Nele Lamping for their sophisticated geotechnical laboratory work. Dr Daniel A. Hepp and Lotta C. Kluger are acknowledged for helpful discussions on the paper.

References

ASTM D4318-10 (n.d.) Standard test methods for liquid limit, plastic limit, and plasticity index of soils

Bayne JM, Tjelta TI (1987) Advanced cone penetrometer development for in-situ testing at Gullfaks C. In: Offshore technology conference. Offshore Technology Conference

Bjerrum L (1971) Subaqueous slope failures in Norwegian fjords. Norw Geot Inst Publ (88)

Bonita JA (2000) The effects of vibration on the penetration resistance and pore water pressure in sands. PhD thesis, Virginia Polytechnic Institute and State University

Boulanger RW, Idriss IM (2006) Liquefaction susceptibility criteria for silts and clays. J Geotech Geoenviron Eng 132(11):1413–1426

Boulanger RW, Idriss IM (2007) Evaluation of cyclic softening in silts and clays. J Geotech Geoenviron Eng 133(6):641–652

DIN 18137-2:2011-04 (2011) Soil investigation and testing – Determination of shear strength – part 2: triaxial test

DIN 4094-4:2002-01 (2002) Subsoil – field testing – part 4: field vane test

Ishihara K (1985) Stability of natural deposits during earthquakes. In: Proc 11th Int Con Soil Mech Found Eng 1(1), 321–375

Jorat ME, Kreiter S, Mörz T, Moon V, de Lange W (2014) Utilizing cone penetration tests for landslide evaluation. In: Krastel S et al (eds) Submarine mass movements and their consequences. Springer, Heidelberg, pp 55–71

Kramer SL (1996) Geotechnical earthquake engineering. In: Prentice–Hall international series in civil engineering and engineering mechanics

Kreiter S, Moerz T, Strasser M, Lange M, Schunn W, Schlue BF, Otto D, Kopf A (2010a) Advanced dynamic soil testing—introducing the new MARUM dynamic triaxial testing device. In: Submarine mass movements and their consequences. 4th international symposium, Austin, Springer, Dordrecht, pp 31–41

Kreiter S, Hepp DA, Ossig B, Hebig J, Otto D, Schlue BF, Mörz T, Kopf A (2010b) Cyclic soil parameters of complex offshore soils: a practical engineering perspective. In: Triantafyllidis T (ed) Workshop Gründung von Offshore-Windenergieanlagen. Karlsruher Institut für Technologie, Karlsruhe, pp 165–181

L'Heureux JS, Longva O, Hansen L, Vanneste M (2014) The 1930 landslide in Orkdalsfjorden: morphology and failure mechanism. In: Krastel S et al (eds) Submarine mass movements and their consequences. Springer, Heidelberg, pp 239–247

Lunne T (2010) The CPT in offshore soil investigations-a historic perspective. In: 2nd international symposium on cone penetration testing (CPT'10). Huntington Beach

Mitchell JK (1988) New developments in penetration tests and equipment. Proc 1st Int Symp Penetr Testing, Orlando, FL, vol 1 AA Balkema, 245–261

Moon V, Cunningham MJ, Wyatt JB, Lowe DJ, Mörz T, Jorat ME (2013) Landslides in sensitive soils, Tauranga, New Zealand. Proc 19th NZGS Geotech Symp

Robertson PK, Wride CE (1998) Evaluating cyclic liquefaction potential using the cone penetration test. Can Geotech J 35(3):442–459

Sasaki Y, Itoh Y, Shimazu T (1984) A Study of the relationship between the results of vibratory cone penetration tests and earthquake induced settlement of embankments. In: Proc 19th Annu Mtg of JSSMFE

Seed HB, Idriss IM, Arango I (1983) Evaluation of liquefaction potential using field performance data. J Geotech Eng 109(3):458–482

Singh S (1996) Liquefaction characteristics of silts. Geotech Geol Eng 14(1):1–19

Steiner A, Kopf AJ, L'Heureux JS, Kreiter S, Stegmann S, Haflidason H, Mörz T (2014) In situ dynamic piezocone penetrometer tests in natural clayey soils—a reappraisal of strain-rate corrections. Can Geotech J 51(3):272–288

Tokimatsu K (1988) Penetration tests for dynamic problems. In: Proc 1st

Udden JA (1914) Mechanical composition of clastic sediments. Bull Geol Soc Am 25:655–744

Wentworth CK (1922) A scale of grade and class terms for clastic sediments. J Geol 30:377–392

Wise CM, Mayne PW, Schneider JA (1999) Prototype piezovibrocone for evaluating soil liquefaction susceptibility. Geotech Earthq Eng 2:537–542

Youd TL, Idriss IM, Andrus RD et al (2001) Liquefaction resistance of soils: summary report from the 1996 NCEER and 1998 NCEER/NSF workshops on evaluation of liquefaction resistance of soils. J Geotech Geoenviron Eng 127(10):817–833

Chapter 21
First Results of the Geotechnical In Situ Investigation for Soil Characterisation Along the Upper Slope Off Vesterålen: Northern Norway

Sylvia Stegmann, Stefan Kreiter, Jean-Sebastien L'Heureux, Maarten Vanneste, David Völker, Nicole Jeanne Baeten, Siren Knudsen, Leif Rise, Oddvar Longva, Jo Brendryen, Haflidi Haflidason, Shyam Chand, Tobias Mörz, and Achim Kopf

Abstract High-resolution geophysical data reveal the presence of several spatially-isolated, small-scale landslides along the gently dipping (~3–4°) upper slope off Vesterålen, Northern Norway. Dynamic slope stability analysis suggests that seismicity may be largely responsible for the occurrence of these slope failures. The landslides are clustered in two groups, with one group of parallel features with their headwalls in ~500 m water depths. The second group is found in ~800 m water depths.

We present first results of geotechnical *in situ* Cone Penetration Test (CPTU) data and TOPAS sub-bottom profiles collected during two cruises in summer 2013/ 2014. We obtained a total of six static CPTU profiles penetrating the top 20 m of soil. Three of these were taken across one of the landslide complexes (SL3) from the slide scar down to the depositional area. The other three are reference sites in the adjacent undisturbed areas.

The combination of geophysical and geotechnical data allows us to divide the well-stratified glacio-marine slope deposits into three different sediment-mechanical units, and reveals the occurrence of mechanically weaker zones (MWZ). These zones are interbedded by coarser layers with high values of cone

---

S. Stegmann (✉) • S. Kreiter • D. Völker • T. Mörz • A. Kopf
MARUM – Center for Marine and Environmental Sciences, University of Bremen, Bremen, Germany
e-mail: sstegmann@marum.de

J.-S. L'Heureux • M. Vanneste • S. Knudsen
Norwegian Geotechnical Institute (NGI), Oslo/Trondheim, Norway

N.J. Baeten • L. Rise • O. Longva • S. Chand
Geological Survey of Norway (NGU), Trondheim, Norway

J. Brendryen • H. Haflidason
Department of Earth Science, University of Bergen, Bergen, Norway

G. Lamarche et al. (eds.), *Submarine Mass Movements and their Consequences*, Advances in Natural and Technological Hazards Research 41,
DOI 10.1007/978-3-319-20979-1_21

tip resistance. The occurrence of sensitive fine-grained material may be responsible for the loss of strength in the deeper portion.

One-dimensional pseudo-static stability analysis attests that the Vesterålen slope is stable except for exceptionally large earthquakes, that induce a peak-ground acceleration (PGA) of 0.224 g or larger to the MWZ. The depth levels of the MWZ correspond well with the slide planes of the landslides.

21.1 Landslides Along the Slope Off Vesterålen

Landslide processes off the Norwegian margin are abundant, with the Storegga Slide being the most prominent example (e.g. Haflidason et al. 2004). In this study, we focus on clusters of small-scale landslide features on the gently dipping slope (~3–4°) off Vesterålen, Northern Norway (Fig. 21.1). The landslides in one cluster

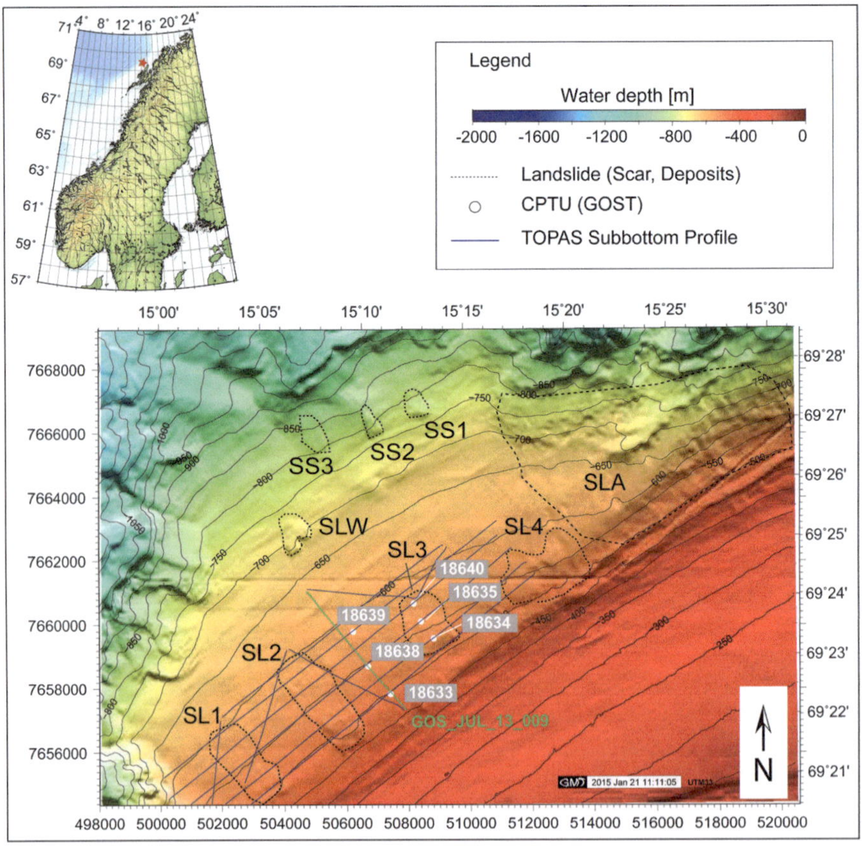

Fig. 21.1 Bathymetric map of the study area on the gently dipping slope off Vesterålen (www.mareano.no). Isolated slope failure features are aligned parallel along 500 and 800 isobath

have their headwalls at around the 500 m isobaths, whereas the second cluster is located in slightly deeper water areas. The spatially-isolated landslides occur in an area essentially devoid of topographic relief and disconnected from both canyon systems and the glacial termini further upslope (Laberg et al. 2007). The volume of failed material varies between $8.5{\cdot}10^6$ and $16.0{\cdot}10^6$ m^3. The run-out distances of these landslides do not exceed 3350 m (L'Heureux et al. 2013). The landslides (SL1-SL4, SLA, SLW) cut the slope-parallel stratified glacio-marine silty to sandy clays on two well-defined slide planes (Rise et al. 2012). The sliding surfaces coincide with spatially continuous high-amplitude reflections (Rise et al. 2012; L'Heureux et al. 2013, Fig. 21.3, horizons a and c). It is unclear, however, whether the sliding planes of the slides are different in composition or have happened at the same time or not. Based on an already existing multidisciplinary database (see Vanneste et al. 2014), the soils consist of predominantly normally consolidated clays to silty clays with intermediate plasticity. Triaxial test results show intervals with strain-softening behaviour. A comprehensive slope stability assessment by L'Heureux et al. (2013) (i) attests that the slope is *per se* stable under static conditions and (ii) indicates that external seismically induced load and post-seismic pore pressure accumulation are the most relevant triggers for landslide and deformation processes.

21.2 Methods

21.2.1 Sub-bottom Mapping

In addition to the first sub-bottom mapping in 2010 (Rise et al. 2012), a complementary survey in summer 2013 was acquired using a parametric sub-bottom profiler system (TOPAS PS18) operated from R/V G.O. Sars (Univ. Bergen). Focus of the survey was set on the mapping of the landslide complexes SL1–SL4 (Fig. 21.1; Haflidason et al. 2010).

21.2.2 CPTU Investigation of Slope Sediments

During the cruise POS472 (MARUM/Univ. Bremen) with R/V Poseidon, geotechnical *in situ* data were collected using the high-resolution static Geotechnical Offshore Seabed Tool (GOST) developed at MARUM. The system is equipped with a 5 cm^2 electrical piezocone with a pore pressure port (CPTU). After landing the GOST on the seafloor, the probe is pushed hydraulically with a constant velocity of 2 cm/s into the sediments. Cone resistance q_c, sleeve friction f_s and excess pore water pressure response Δu_2 are recorded with a sampling frequency of ~30 Hz, which results in a vertical resolution of ~2 mm. Excess pore water pressure is measured differentially against the water filled rod directly behind the cone in u_2 location. For further details, see Jorat et al. (2014).

Processing of the CPTU data and the derivation of strength properties was accomplished according to standard practice summarized in Lunne et al. (1997). Undrained shear strength s_u was calculated using the static cone factor N_{kt} =12, 15 and 17 for normally consolidated clayey sediments (Lunne et al. 1997). The normalised undrained shear strength (s_u/σ'_{v0}) can be used to determine the state of consolidation. For normally-consolidated soil, this ratio typically falls in the range of 0.2–0.3 (Karlsson and Viberg 1967). A ratio exceeding 0.3 may indicate overconsolidation.

Identification of soil types was carried out following the soil behaviour type classification by Ramsey (2002; SBT_{Ramsey}) and Robertson (1990; $SBT_{Robertson}$) using the normalised cone resistance Q_t [$(q_t - \sigma_{vo})/\sigma'_{vo}$] and the pore pressure parameter B_q [$\Delta u_2/(q_t - \sigma_{vo})$] with the vertical total stress (σ_{vo}) and the effective vertical stress (σ'_{vo}) (Lunne et al. 1997).

21.2.3 Pseudo-static Factor of Safety (FoS)

The 1D undrained, infinite slope stability analysis by Morgenstern and Price (1965) defines the pseudo-static factor of safety (FoS) – the ratio of resisting forces to driving forces during earthquake shaking. Seismic-induced shear stresses are considered to be constant over the period of shaking (Hampton et al. 1996). If FoS >1, the sediments are assumed to be stable. FoS <1 indicates permanent deformation and/or failure.

FoS is defined as

$$FoS = \frac{s_u}{\sigma'_{v0}\left(\sin(\alpha)\cos(\alpha) + k\left(\frac{\gamma}{\gamma'}\right)\cos^2(\alpha)\right)}$$

with α = the slope angle, γ' = the effective unit weight and the seismic coefficient k. Given the fact that soil slopes are not rigid and the peak ground acceleration (PGA) lasts for a very short period of time, seismic coefficients used in practice correspond to lower PGA values (Kramer 1996). Here, k was calculated with 0.3 · PGA [g] recommended by Marcuson (1981) and following the proposition by Hynes-Griffin and Franklin (1984): $k = 0.5 \cdot \text{PGA}$ [g]. The latter criteria is suggested to be appropriate for most slopes by Kramer (1996).

We run the analysis using a PGA of 0.051 g and 0.224 g. These values correspond to potential seismic events offshore Vesterålen with a 90 % probability of no exceedance for 475 years and 10,000 years recurrence period, respectively (NORSAR 1998).

21.3 Results

Geophysical as well as *in situ* geotechnical CPTU data presented here (Figs. 21.2 and 21.3) focus on the characterization of the uppermost strata of the slope in terms of sediment-mechanical properties and slope stability (Fig. 21.1).

The TOPAS data illustrate that the sediments are well stratified with continuous strata packages of different reflectivity (Fig. 21.2). The scars of the different slide complexes clearly cut the slope sediments and terminate at different levels, typically on spatially-continuous reflections. Since the new TOPAS data closely resemble those presented by L'Heureux et al. (2013), we refrain from showing them here.

The CPTU penetrated between 7 and 19.8 m and profiled (i) intact material (GeoB18633, −38, −39) (Fig. 21.2) and (ii) different morphological parts of SL3 (GeoB18634, −35, −40) (Fig. 21.1) The CPTU located upslope (GeoB18633, −34) were stopped at 7 m and 9 m, when q_c reached 20 MPa and 44 MPa, respectively to avoid bending or breaking of the rods. Availability of reliable f_s data is limited as the sleeve friction jacket was blocked during some measurements.

21.3.1 *Sedimentological and Geotechnical Characterisation of Slope Sediments*

The correlation of geophysical and *in situ* CPTU data identifies three different sediment-mechanical units in the upper 20 m of slope sediments (Fig. 21.3).

Unit I has a well pronounced slope-parallel, distinct reflection pattern, in which corrected cone resistance q_t reaches peaks up to 0.93 MPa. Excess pore pressure Δu_2 rises up to 0.12 MPa. q_t of the background sediment increases linearly to

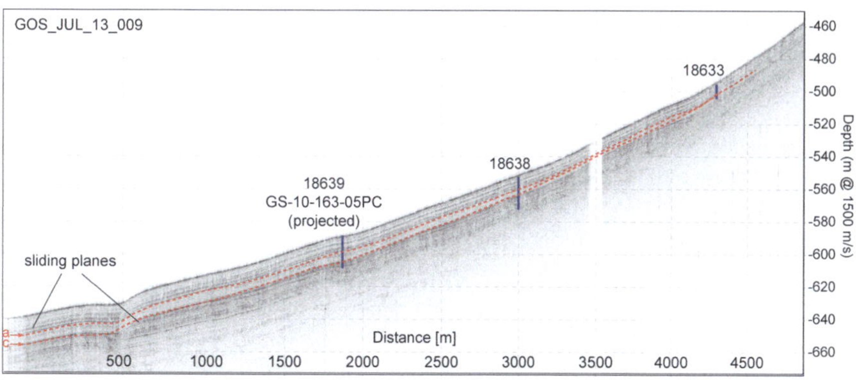

Fig. 21.2 Sub-bottom TOPAS profile crossing the location of the CPTU transect along the intact portion of the slope. Location of the seismic profile is shown in Fig. 21.1. Position as well as penetration of three CPTU are illustrated by *blue lines*. *Dashed red lines* represent the sliding planes according to Rise et al. (2012) and L'Heureux et al. (2013)

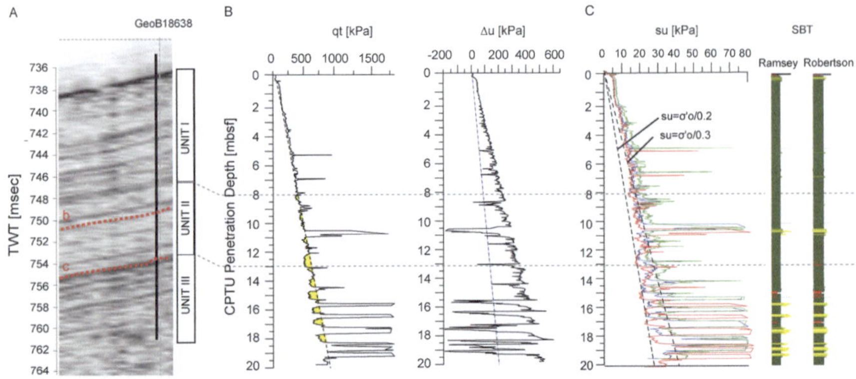

Fig. 21.3 Sediment-mechanical characterisation of intact slope sediments: (**a**) Position and penetration depth of GeoB18638 on the TOPAS line GOS_13_JUL_009; (**b**). *Yellow-shaded* areas indicate mechanically weaker zones (MWZ). (**c**) Undrained shear strength s_u was calculated with $N_{kt} = 12$ (*green*),15 (*black*) and 17 (*red*). *Dashed lines* are indicative of the state of consolidation based on s_u/σ'_{vo}. Soil behaviour types were determined by Robertson (1990) and Ramsey (2002). Colour-coding of the soil types: *green* = Clay, NC (Ramsey), silty clay (Robertson), *yellow*: silty sand to sand, *red*: sensitive fine-grained material. Please note that the CPTU and s_u data are clipped for better illustration

0.35 MPa, which corresponds to s_u of 18 kPa (Fig. 21.3c). With the normalized undrained shear strength up to 0.45 (based on s_u with $N_{kt} = 15$), the sediments appear slightly overconsolidated.

The reflections within Unit II appear less coherent than those of Unit I. A locally distributed coarse layer is detected between 10.5 m and 11 m with a peak in q_t reaching 12 MPa. A peak of that order does not appear in GeoB18639. The linear trend of s_u is decreasing marginally and approaches values typical for normally consolidated sediments (Fig. 21.3c). Immediately underneath the distinct peaks of q_t we find intervals with exceptional low q_t. These mechanically weaker zones (MWZ) are shaded in yellow in Fig. 21.3b.

Unit III is characterized by the alteration of coarse layers (high q_t and low Δu_2 values) with relative weaker zones (q_t values = 0.58 MPa, Fig. 21.3b). In this unit, again the geophysical data reveal clearly accentuated slope-parallel layering, s_u of MWZ decreases to 23 kPa minimum. The material is normally consolidated (Fig. 21.3c).

Excess pore pressure response Δu_2 increases linearly with depth to 330 kPa. Δu_2 lows coincide with q_t-peaks, which corresponds to dilative behaviour induced by shearing of the penetrating cone in coarser-grained material.

Two different types of soil type behaviour classification were used (SBT_{Ramsey}, $SBT_{Robertson}$) to characterize the thin-layered strata with q_t-values >1500 kPa as sandy layers. The background sediment is characterized by SBT_{Ramsey} as normally consolidated clay and by $SBT_{Robertson}$ as silty clay. Based on SBT the MWZ in Unit II and III correspond partly to sensitive fine-grained material (Fig. 21.3c).

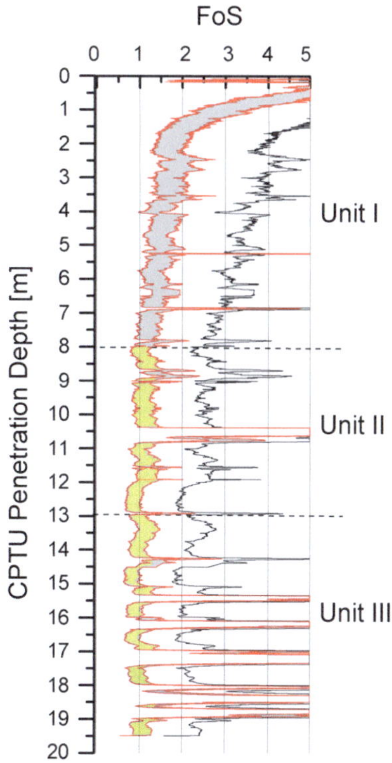

Fig. 21.4 FoS was calculated for the CPTU profile GeoB18638. The *lower* boundary represents the FoS calculated with k = 0.5 * PGA [g], the *upper* boundary relates to k = 0.3 * PGA [g]. Slope angle α is 3°. s_u derived with $N_{kt} = 15$ was used for the calculation of FoS with a PGA of 0.224 g (10,000-years recurrence period). The *black line* represents the FoS related to a PGA = 0.051 g (475-years recurrence period). *Yellow-shaded* portions correlate to MWZ (see Fig. 21.3)

21.3.2 Pseudo-static Slope Stability Analysis

Results of the FoS analysis show that the PGA of the seismic event with 475 years-recurrence time (PGA = 0.051 g) is not sufficient to destabilize the slope sediment in any depth level (FoS >2; Fig. 21.4). When assuming an earthquake with a PGA of 0.224 g (i.e. 10,000 year return period event) the FoS comes close to 1 (meta-stable) or falls below unity within the sediment-mechanical Unit II and III for cases where s_u is estimated with $N_{kt} = 15$, respectively. These failure depths correlate to the MWZ identified by means of CPTU tests in Fig.21.3b.

21.4 Discussion and Outlook

Along the stable slope offshore Vesterålen, static CPTU data reveal the occurrence of mechanically weaker zones (MWZ) by the drop of q_t-values (Fig. 21.3b). Within the sediment-mechanical Unit III, some MWZ coincide with layers where sensitive fine-grained sediments were classified by $SBT_{Robertson}$ and SBT_{Ramsey} (Fig. 21.3; Units II, III).

The 1D infinite, undrained pseudo-static stability calculation presents several critical slope layers (meta-stable to unstable) in Unit II and III, indicating permanent deformation or failure in case additional dynamic load is induced by an earthquake (Fig. 21.4). According to L'Heureux et al. (2013), a 10,000 year-seismic event with a PGA of 0.224 g is necessary to produce failure or permanent deformation along the slope. Such seismic events, which historically had magnitude M5.7 may have been frequent due to the glacio-isostatic uplift after the last glaciation along the slope off Vesteralen (Bungum et al. 1991; Olesen et al. 2013).

Given the fact that the stability analysis does not account for earthquake-induced excess pore pressure and inherent shear strength reduction due to accumulation of deformation, this undrained pseudo-static analysis is only a first-order approximation providing a relative stability index for the upper slope sediments. The seismic reflectors b and c (Figs. 21.2 and 21.3a) represent the sliding surfaces for SL1-SL4 and SLA, respectively (L'Heureux et al. 2013; Rise et al. 2012). The results of our analysis reveals the occurrence of critical layers, which correspond to reflector b (~10 mbsf) and c (~13 mbsf) (Fig. 21.4). The analysis attests that the mechanically weaker zones (MWZ) correlate well with the portions where FoS ≤ 1 (Fig. 21.4). Thus, the mechanically weak layers can be considered as one important pre-condition for landslide activity along the gently dipping slope. External trigger (e.g. large magnitude seismic event), however, is necessary to generate landslide activity along the slope off Vesterålen.

In conclusion, the integration of *in situ* CPTU data with geophysical data improves soil characterization and hence foster a better understanding of the pre-conditioning factors for slope instability at the upper continental slope off Vesterålen.

Based on these preliminary *in situ* CPTU data, further (2D) analysis is necessary to understand i) the liquefaction potential of the coarse-grained layers and ii) the influence of the stratified fine-grained – sandy sequence in Unit III on the dissipation behaviour of seismic-induced excess pore pressure, and iii) the factors, which control the stability of the RWZ in deeper portion (13 m–18 m, Unit III; Fig. 21.4). Risk assessment for the present-day slope requires these studies for the slope off Vesterålen, where the opening for offshore oil and gas exploration is controversially debated (Misund and Olsen 2013).

Acknowledgments The study was founded by the DFG Research Center/Cluster of Excellence 'MARUM — The Ocean in the Earth System'. We thank Wolfgang Schunn (Univ. Bremen), who operated GOST and improved data acquisition onboard and the crew of R/V POSEIDON for excellent effort during the GOST operations. Sub-bottom data were collected as part of the Norwegian SEABED project C-DOG. NGI's internal research funding scheme is also acknowledged. We appreciate both reviewers Gabriela Dan and Sara Lafuerza for their constructive comments – merci bien!

References

Bungum H, Alsaker A, Kvamme LB, Hansen RA (1991) Seismicity and seismotectonics of Norway and surrounding continental shelf areas. J Geophys Res 96:2249–2265

Haflidason H et al (2004) The Storegga Slide: architecture, geometry and slide-development. Mar Geol 213:201–234

Haflidason H et al (2010) Marine geological cruise report from mid-Norwegian continental shelf, Ranafjorden-Sørfjorden and Lofoten-Vesterålen margin. Report No. 100-02/10, Department of Earth Science, University of Bergen, Bergen, 39 pp

Hampton MA, Lee HJ, Locat J (1996) Submarine landslides. Rev Geophys 34(1):33–59

Hynes-Griffin ME, Franklin AG (1984) Rationalizing the seismic coefficient method. Miscellaneous Paper GL-84-13, US Army Corps of Engineers, Vicksburg

Jorat ME, Moerz T, Schunn W, and Kreiter S (2014) Geotechnical offshore seabed tool (GOST): a new cone penetrometer, In: Robertson PK, Cabal KL (eds) CPT'14. procs. 3rd intern. symp. on cone penetration testing, CPT'14 Organizing Comittee, Las Vegas, pp 207–215

Karlsson R, Viberg L (1967) Ratio of c/p in relation to liquide limit and plasticity index, with special reference to Swedish clays. In: Proceedings, conference on shear strength properties of natural soils and rocks, vol 1. Oslo

Kramer SL (1996) Geotechnical earthquake engineering. Prentice Hall, Upper Saddle River

L'Heureux JS et al (2013) Stability, mobility and failure mechanism for landslide at the upper continental slope off Vesteralen, Norway. Mar Geol 346:192–207

Laberg JS, Guidard S, Mienert J, Vorren TO, Haflidason H, Nygård A (2007) Morphology and morphogenesis of a high-latitude canyon area; the Andøya Canyon, Norwegian Sea. Mar Geol 246:68–85

Lunne T, Robertson PK, Powell JJM (1997) Cone penetration testing in geotechnical practice. Spon Press, London

Marcuson WF (1981) Modeator's report for session on "Earth dams and stability of slopes under dynamic loads". Proccs, Intern Conference on Recent Advances in Geotechnical Earthquake Engineering and Soil Dynamics, St. Louis, Vol 3, 1175

Misund OA, Olsen E (2013) Lofoten–Vesteralen: for cod and cod fisheries, but not foroil? ICES J Mar Sci. doi:10.1093/icesjms/fst086

Morgenstern NR, Price VE (1965) Analysis of stability of general slip surfaces. Geotechnique 15(1):79–93

NORSAR (1998) Development of a seismic zonation for Norway. Norwegian Council for Building Standardization (NBR), p 162

Olesen O, Bungum H, Dehls J, Lindholm C, Pascal C, Roberts D (2013) Neotectonics, seismicity and contemporary stress field in Norway – mechanisms and implications. In: Olsen L, Fredin O, Olesen O (eds) Quaternary geology of Norway, vol 13, Geological survey of Norway special publication., pp 145–174

Ramsey N (2002) A calibrated model for the interpretation of cone penetration tests (Cpts) in north sea quaternary soils. Offshore site investigation and geotechnics 'Diversity and Sustainability'; Proceedings of an international conference, 26–28 Nov, London

Rise L et al (2012) Investigations of slides at the upper continental slope off Vesterålen, North Norway. In: Yamada Y, Kawamura K, Ikehara K, Ogawa Y, Urgeles R, Mosher DC, Chaytor J, Strasser M (eds) Submarine mass movements and their consequences (5th international symposium). Springer Netherlands, pp 167–176

Robertson PK (1990) Soil classification using the cone penetration test. Can Geotech J J 27(1):151–158

Vanneste M et al (2014) Seafloor instabilities and sediment deformation processes: the need for integrated, multidisciplinary investigations. Mar Geol. doi:10.1016/j.margeo.2014.01.005

Chapter 22
A Novel Micro-shear Tester for Failure Analysis of Fine and Cohesive Granular Matter

Lutz Torbahn, Stefan Strege, and Arno Kwade

Abstract We apply a novel micro-shear tester (μST) to investigate the shear flow behaviour of very small granular volumes (15 μl). We compare the results to standard ring shear devices which confirm that these small volumes are sufficient for granular flow analysis, and hence are helpful to identify the stable and unstable areas of very small granular quantities. Within the laboratory experiments we realised shear tests by using a torsional shear movement which was applied to a cohesive granular calcium carbonate. For the analysis of the shear behaviour the yield loci are used which were derived from the bulk shear and normal stresses. The comparison of the yield loci determined by μST and ring shear device shows only a minor deviation. A particle characterization of the calcium carbonate material according to the particle size distribution with a mean grain size of 5 μm using laser diffraction and determining cohesion forces on single particles with 34 nN using an atomic force microscope (AFM) is part of this study. Finally, we combine a X-ray computed tomography (XCT) with the μST to reveal changes in the microstructure and notice shear bands caused by the shearing process.

22.1 Introduction

Commonly submarine mass movements are caused by zones of failure which range over large areas. The physical behaviour along these failure zones is important and influences the dynamics of mass movements. Many studies suggest that the development of failure zones occur especially in weak layers caused by trigger mechanisms which are crucial for the mass movement initiation (Hampton et al. 1996; Locat and Lee 2002). Diverse trigger mechanisms are discussed in that context (e.g. Wright and Rathje 2003; Sultan et al. 2004). Most of them are linked to

L. Torbahn (✉) • A. Kwade
Institute for Particle Technology (iPAT), TU Braunschweig, Braunschweig, Germany
e-mail: l.torbahn@tu-braunschweig.de

S. Strege
BASF SE, Ludwigshafen, Germany

G. Lamarche et al. (eds.), *Submarine Mass Movements and their Consequences*,
Advances in Natural and Technological Hazards Research 41,
DOI 10.1007/978-3-319-20979-1_22

physical parameters such as shear strength based on friction and cohesion. These parameters describe the macroscopic material behaviour of a bulk which is the sum of many single processes on a particulate level influenced by, e.g. particle friction and cohesion forces. In laboratory experiments it is often tried to reproduce the natural macroscopic failure behaviour. Therefore, different shear devices are applied to analyse the macroscopic granular failure behaviour (Marone 1998; Schulze 2008). The required volumes for the standard shear devices are between 1 and 1000 ml. For fine granular matter such large quantities make it difficult to investigate on a particle level. Hence, the use of smaller volumes enables the investigation on the mesoscale and builds a bridge between the macro- and micro-scale. For single spherical silica particles cohesion forces and friction were already determined on a particle level by atomic force microscopy (AFM) by others (e.g. Fuchs et al. 2014; Mader-Arndt et al. 2014). Moreover, first mesoscale experiments also used spherical silica particles as a bulk and apply compressional forces for compaction analysis (Strege et al. 2014; Strege 2014).

So far, mainly numerical simulations are used to analyse shear bands during shear deformation for particle sizes of ~30 μm. In that context shear plane development of non-spherical particles were analysed based on numerical simulations regarding friction and changes in pore volumes (e.g. Torbahn and Huhn 2014).

In contrast to this the aim of the current study is to investigate the behaviour of irregular shaped calcium carbonate particles in a mesoscale environment regarding shear band evolution during a rotary shear process. Therefore, the bulk material was investigated in mesoscale with a combination of new developed micro-shear tester and X-ray computed tomography (XCT).

22.2 Characterization of the Calcium Carbonate Sample

Here fine granular calcium carbonate with irregular shaped particles (Fig. 22.1) was used and characterized according to the particle size distribution and cohesion forces.

The shear behaviour of the granular calcium carbonate depends on the particle sizes and its distribution. Thus, a determination of the particle size distribution by laser diffraction method using Helos (Sympatec, Germany) was realized. For this purpose the calcium carbonate particles were initially dispersed within deionised water and ultrasonic was applied for 30 s. A mean value of $x_{50,3} = 4.8$ μm was found for the calcium carbonate particles (cumulative and frequency distribution see Fig. 22.2a). Particles of that size usually have non-negligible cohesion forces. Therefore, the cohesion forces between single particles were determined.

An AFM XE 100 (Park Systems, Korea) was used for this analyses which is able to detect forces down to a few nN. The cohesion was determined according to the Colloidal Probe Method (Butt et al. 2005), where one particle is fixed at the apex using glue strengthened by ultraviolet light.

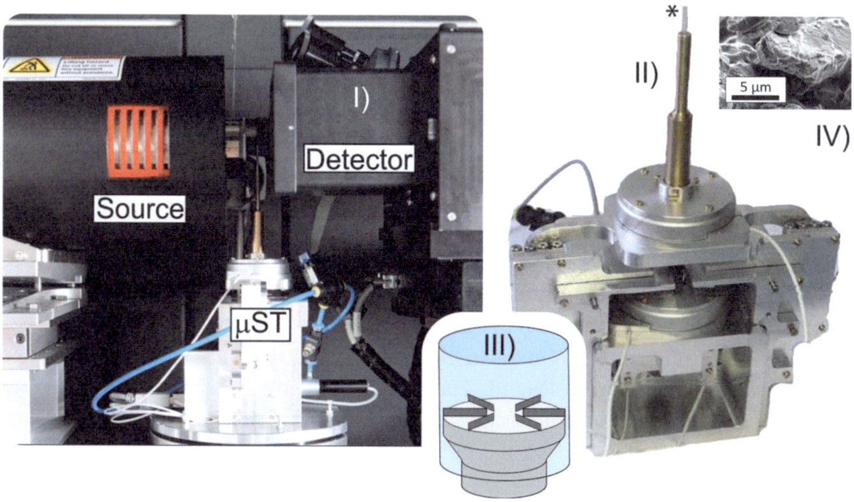

Fig. 22.1 (**a**) Setup of μST between source and detector of the XCT, (**b**) μST with sample chamber*, (**c**) Schematic view of glass capillary as shear cell boundary and lower piston as the lower boundary, and (**d**) SEM image of irregular shaped calcium carbonate particles

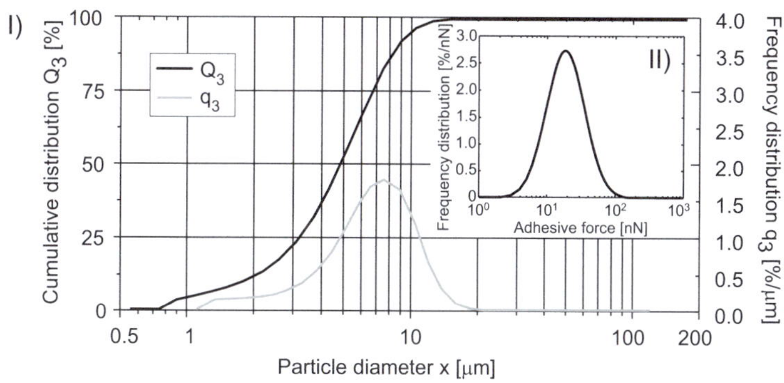

Fig. 22.2 (**a**) Particle-size-distribution of calcium carbonate with laser diffraction method, (**b**) Frequency distribution of cohesive forces with AFM

Another particle is fixed on an object slide. After the particles are brought to contact the force changes to break that contact again are logged. Because of strong fluctuations within the measurement results, at least ten iterations of the experiment are necessary (Fig. 22.2b). Here, ten different particles were fixed on cantilevers and for each particle the cohesion forces to ten different particles were determined. Often and also in our case the frequency distribution of the maximum cohesion force has a log-normal distribution (Wargenau and Kwade 2010). The force between calcium carbonate particles has a mean value of 34 nN with a standard deviation of ~25 nN.

22.3 Micro Shear Tester and X-ray Computed Tomography

22.3.1 Micro Shear Tester

The new μST was designed for the implementation within a XCT setup, and therefore, developed at the iPAT, Braunschweig (Strege 2014). It uses torsional shear geometry and differs regarding a ring shear cell only by using the whole cell radius, and hence, is also not limited in the shear distance. In contrast to other well-known shear devices, e.g., ring shear cell (Schulze 1994) and Jenike shear cell (Jenike 1964) with sample volumes in the range of 4–1000 ml the μST is well suited for very small volumes of a few μl. The optimal sample volume for XCT investigations is in the range between 6 μl (2 mm sample diameter) and 16 μl (2.7 mm sample diameter) depending on the radius of the sample chamber and thus much smaller than the established shear devices. The sample chamber consists of a very fine borosilicate glass capillary with a wall thickness of 50 μm. On top and bottom the sample is confined either by flat pistons for compressional sample analyses or structured pistons grabbing into sample for analysing shear processes. The pistons are driven by high accurate piezo engines with positioning in single μm scale. The advantage of μST is that forces and torque are independently determined and decoupled from each other. Moreover, a method for frictionless bearing of the sample was implemented by using an air bearing. By doing this a disturbing frictional noise could be excluded or at least minimized. Additionally, the force transmission is realized frictionless via magnetic spring. In this way the force transmission is absolutely contactless. All this features results in very accurate measurements. As a consequence, the μST is able to resolve precise positions of less than 1 μm, forces that are less than 1 mN and torques less than 1 nNm. This is realized by a compensation unit using a laser spot – CCD-camera combination to detect tiny displacements during loading processes which are used to determine the stresses and torque.

For the analyses of compression and shear processes normal and shear stresses in the range of 0.1–20 kPa can be applied, equivalent to ~4 m depth in the marine environment.

For the determination of the calcium carbonate yield loci a sequence of pre-shear and shear processes was applied. The purpose of the pre-shear step is to prepare the sample again and again with the same loading conditions until the steady state flow is achieved. The sample is loaded with larger normal force than in the shear state, and hence, causes an over-consolidation. During the shear process different normal forces were applied and the shear forces were logged. The peaks of the shear stresses of the over-consolidated matter caused by the pre-shear process are summarized as the points of incipient flow and belong to the yield loci. The yield loci for the different loadings are illustrated in Fig. 22.3a. We observe an already known correlation between increasing normal and shear stresses (e.g. Schulze 2008). A comparison of the yield loci at 10 kPa measured with the μST and with a ring

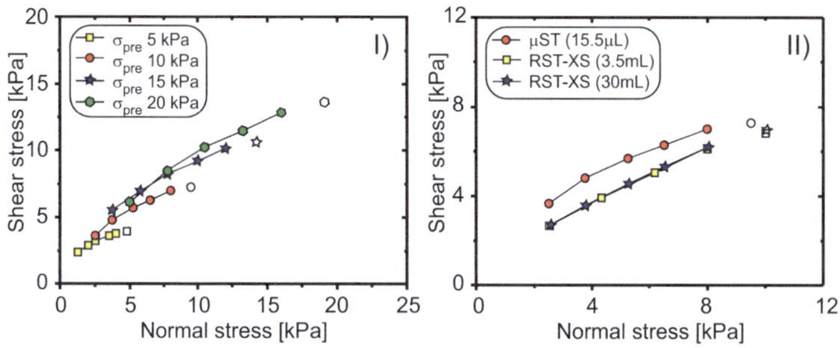

Fig. 22.3 (**a**) Yield loci of calcium carbonate at different applied normal loading, blank symbols indicate the pre-shear process and signify the steady state flow (Schulze 2008) whereas filled symbols belongs to the shear process, (**b**) Comparison of yield loci results for normal load of 10 kPa according to μST and RST-XS

shear device RST-XS according to sample volumes of 3.5 and 30 ml is given in Fig. 22.3b. The data comparison shows only a small shift between both devices. Shifts are a common phenomenon comparing different shear devices because of the variability in geometry as an important factor (Schulze 2008).

22.3.2 X-ray Computed Tomography (XCT)

The X-ray computed tomography is a non-destructive imaging technique that allows for scanning samples within the μST during deformation. In this study the MicroXCT-400 (Xradia. Inc., USA), device was applied which has a well-defined contrast behaviour. For the experimental setup the following parameters were applied: source voltage = 40 keV, no optical filter, strength of electric current 200 mA.

22.3.3 Combination of μST and XCT

There are many advantages of combining the μST and the XCT, e.g., on the fly observation of deformation processes or detailed 3D image reconstructions of distinct time steps. For the experiments the sample was arranged between the source and the detector so that the sample was full-sized in the image-scans (Fig. 22.1). In these experiments glass capillaries with a diameter of 2.7 mm were used. In order to achieve a full-size image for reconstruction an optical magnification of 4x, 6 μm/voxel, binning 4 and 800 projections was applied. During the reconstruction process improvements to the scanned data were applied such as a binning of 2, ring removal, centre shift and beam hardening (Xradia 2011).

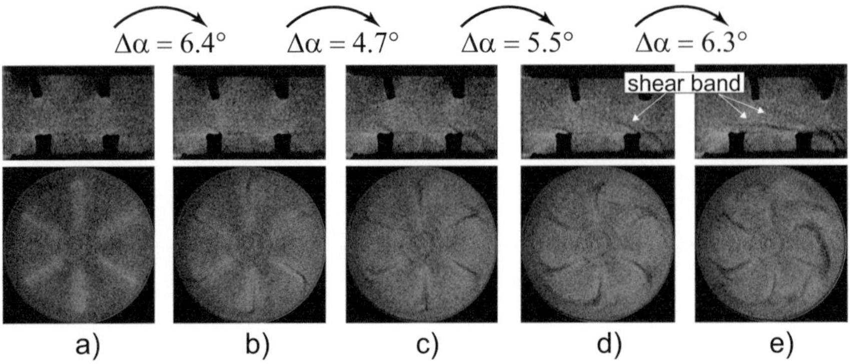

Fig. 22.4 Different states of XCT scans during shear deformation. The sample diameter is 2.7 mm. Development of shear bands during states (d) and (e)

To prepare the sample the calcium carbonate was sieved to exclude large agglomerates formed by cohesion forces with a fill height of 2.7 mm according to the diameter of the glass capillary. Afterwards, the sample was compressed with a pre-shear load and a sequence of pre-shear and shear processes were performed.

In Fig. 22.4 a sequence of five shear state images illustrated each by a horizontal and vertical slice are shown for different shear angles with steps of $\Delta\alpha \sim 5°$. The upper pictures show the vertical slices close to two third of the shear cell radius. However, the lower pictures show the horizontal slices 50 μm above the lower piston. In the first one (a) the sample is intact. The second image (b) shows the initiation of a shear band at the piston (horizontal slice), which evolves until the last image (e). Further the last image show two shear bands in the vertical slice. In this image sequence we show the structural changes within the mesoscopic shear band initiation and evolution during the applied shear for the very fine and cohesive calcium carbonate.

22.4 Conclusion

The new developed μST device works with fine granular matter and with very small sample volumes. The experimentally determined shear behaviour is comparable to established ring shear cells with significant larger volumes. There is only a small shift between the compared data. The combination of the μST and the XCT extend the possibilities for investigations and merge results of structural data with material response behaviour as shown in this study. This helps to understand the shear behaviour of granular matter. This is also applicable to submarine mass movement processes where the influence of physical parameter behaviour and changes in the microstructure are relevant.

So far only materials which are interesting for process engineering issues of bulk solids and their flow behaviour were investigated. But according to the sample size of the μST geo materials on a particle or mesoscale such as silt, volcanic ashes etc. could be investigated as well regarding shear band initiation and evolution during shear deformation.

Acknowledgments We wish to thank, Stefan Kreiter and Jürgen Tomas for very constructive reviews. We also would like to thank the DFG for financial support within SPP 1486.

References

Butt H-J, Cappella B, Kappl M (2005) Force measurements with the atomic force microscope: technique, interpretation and applications. Surf Sci Rep 59(1–6):1–152. doi:10.1016/j.surfrep.2005.08.003

Fuchs R, Weinhart T, Meyer J, Zhuang H, Staedler T, Jiang X, Luding S (2014) Rolling, sliding and torsion of micron-sized silica particles: experimental, numerical and theoretical analysis. Granul Matter 16(3):281–297. doi:10.1007/s10035-014-0481-9

Hampton MA, Lee HJ, Locat J (1996) Submarine landslides. Rev Geophys 34:33–59

Jenike AW (1964) Storage and flow of solids, Bulletin of the University of Utah. University of Utah – Utah Engineering Experiment Station, Salt Lake City

Locat J, Lee HJ (2002) Submarine landslides: advances and challenges. Can Geotech J 39(1):193–212. doi:10.1139/t01-089

Mader-Arndt K, Kutelova Z, Fuchs R, Meyer J, Staedler T, Hintz W, Tomas J (2014) Single particle contact versus particle packing behavior: model based analysis of chemically modified glass particles. Granul Matter 16(3):359–375. doi:10.1007/s10035-013-0478-9

Marone C (1998) Laboratory-derived friction laws and their application to seismic faulting. Annu Rev Earth Planet Sci 26:643–696

Schulze D (1994) Development and application of a novel ring shear tester. Aufbereitungstechnik 35(10):524–535

Schulze D (2008) Powders and bulk solids. Springer, Berlin/Heidelberg. doi:10.1007/978-3-540-73768-1

Strege S (2014) Röntgenmikrotomographische Analyse der Verdichtung und Scherung feiner und kohäsiver Pulver. Dissertation, TU Braunschweig

Strege S, Weuster A, Zetzener H, Brendel L, Kwade A, Wolf DE (2014) Approach to structural anisotropy in compacted cohesive powder. Granul Matter 16(3):401–409. doi:10.1007/s10035-013-0454-4

Sultan N, Cochonat P, Canals M, Cattaneo A, Dennielou B, Haflidason H, Laberg JS, Long D, Mienert J, Trincardi F, Urgeles R, Vorren TO, Wilson C (2004) Triggering mechanisms of slope instability processes and sediment failures on continental margins: a geotechnical approach. Mar Geol 213(1–4):291–321. doi:10.1016/j.margeo.2004.10.011

Torbahn L, Huhn K (2014) Interrelationship between sediment fabric, pore volume variations as indicator for pore pressure changes, and sediment shear strength. In: Krastel S, Behrmann J-H, Völker D et al (eds) Submarine mass movements and their consequences, vol 37, Advances in natural and technological hazards research. Springer, The Netherlands, pp 85–94. doi:10.1007/978-3-319-00972-8_8

Wargenau A, Kwade A (2010) Determination of adhesion between single aspergillus Niger spores in aqueous solutions using an atomic force microscope. Langmuir 26(13):11071–11076. doi:10.1021/la100653c

Wright SG, Rathje EM (2003) Triggering mechanisms of slope instability and their relationship to earthquakes and tsunamis. Pure Appl Geophys 160(10-11):1865–1877. doi:10.1007/s00024-003-2410-4

Xradia (2011) MicroXCT-200 and microXCT-400, user's guide, vol version 8

Chapter 23
Knickpoint Migration Induced by Landslide: Evidence from Laboratory to Field Observations in Wabush Lake

Dominique Turmel, Jacques Locat, Gary Parker, and Jean-Marie Konrad

Abstract Wabush Lake is characterized by a nearly constant sediment input, resulting from the deposition of mine tailings. Five bathymetric surveys were conducted in this lake, in order to understand the sedimentation pattern over more than 12 years. Of the morphologies studied, submarine channel and knickpoints were traced out and documented. A physical model of Wabush Lake was also constructed in order to understand specific morphologies, such as the knickpoints.

As part of this study, it was observed that knickpoints migration could be explained, at least in the laboratory, by two mechanisms: landslide and erosion, and not only erosion as previously thought. This previous conclusion will be applied to some knickpoints found in Wabush Lake. Two cases are analysed: (1) a knickpoint where a tension crack is present and (2) another knickpoint that shows no sign of instability. It is found that, when an excess of pore water pressure is present due to rapid sedimentation, static liquefaction may occur at the head of a knickpoint leading to a localized slope failure.

23.1 Introduction

Knickpoints are topographic depressions, characterized by a steep scarp slope and a flatter profile downslope. This morphology is found both in rivers in the subaerial environment and in submarine environments such as in canyons (Mitchell 2006) or

D. Turmel (✉) • J. Locat
Laboratoire d'études sur les risques naturels, Département de géologie et de génie géologique, Université Laval, G1V 0A6 Québec, QC, Canada
e-mail: dominique.turmel.1@ulaval.ca

G. Parker
Ven Te Chow Hydrosystems Lab, Department of Civil and Environmental Engineering and Department of Geology, College of Engineering, University of Illinois at Urbana-Champaign, Urbana, IL 61801, USA

J.-M. Konrad
Département de génie civil et de génie des eaux, Université Laval, G1V 0A6 Québec, QC, Canada

G. Lamarche et al. (eds.), *Submarine Mass Movements and their Consequences*, Advances in Natural and Technological Hazards Research 41,
DOI 10.1007/978-3-319-20979-1_23

in submarine channels (Heiniö and Davies 2007). In the submarine environment, these morphologies are formed by the interaction between turbidity current and an erodible bed (e.g. Mitchell 2006). These morphologies migrate upwards towards the shoreline, leading to the formation of submarine or subaqueous channels. Previous studies on the migration of knickpoints (e.g. Toniolo and Cantelli 2007) concluded that this migration is solely due to the erosion by turbidity currents.

In their studies for the Wabush Lake situation, Turmel et al. (2012) concluded that, at least in the laboratory environment, the migration of knickpoints could also be due to landslides. Their experiments were aimed to reproduce the physical settings of a lake which is not influenced by any allogenic forcing. Because of this singularity, they were able to demonstrate that coupling of fluid and soils mechanics has to be considered for a full understanding of knickpoint migration. A remaining aspect here is to answer the following question: are field conditions in Wabush Lake favourable for the development of slope instabilities related to knickpoint migration?

The materials involved in Wabush Lake are tailings, and below the lake level they have the grain size mostly in the range of silts (Turmel et al. 2015). Previous studies show that tailings are materials prone to static (Wagener et al. 1997; Dawson et al. 1998) and dynamic (Dobry and Alvarez 1967) liquefaction. Furthermore, in the area studied, the sedimentation rates are very high. The sedimentation rate is an important factor to consider in such an analysis since for low hydraulic conductivity soils, the consolidation process when subjected to very rapid sedimentation rate will generate excess of pore water pressure, which facilitates slope failure.

The objective of this paper is therefore to use the observations made in the laboratory for a better understanding of the field environment in order to demonstrate that landslides may be part of knickpoint progression.

23.2 Wabush Lake

Wabush Lake is an N-S oriented lake located near Labrador city, New-Foundland, Canada. This 15 km long and 3.5 wide lake (Fig. 23.1) is subjected to a continuous hyperpycnal flow since 1964, resulting from the discharge of iron tailings into the lake, with a deposition rate of approximately 20 million tons per year. This mean rate increased to 25 million tons per year between 2008 and 2011. Those tailings are discharged onshore, in the southern portion of the lake, thus creating a hyperpycnal delta. The location of the outlet is moved according to the evolution of the delta (Turmel et al. 2014).

In order to follow the evolution of the lake floor, bathymetric surveys of the lake were carried out in 1999, 2004, 2006, 2008 and 2011. The morphological evolution of the lake is possible to follow in time since all these surveys use the same horizontal and vertical datum, with an horizontal precision of about 20 cm and a vertical resolution of about 10 cm. It was also possible to approximate the lake bathymetry prior to the mine operations, by the interpretation of seismic survey data

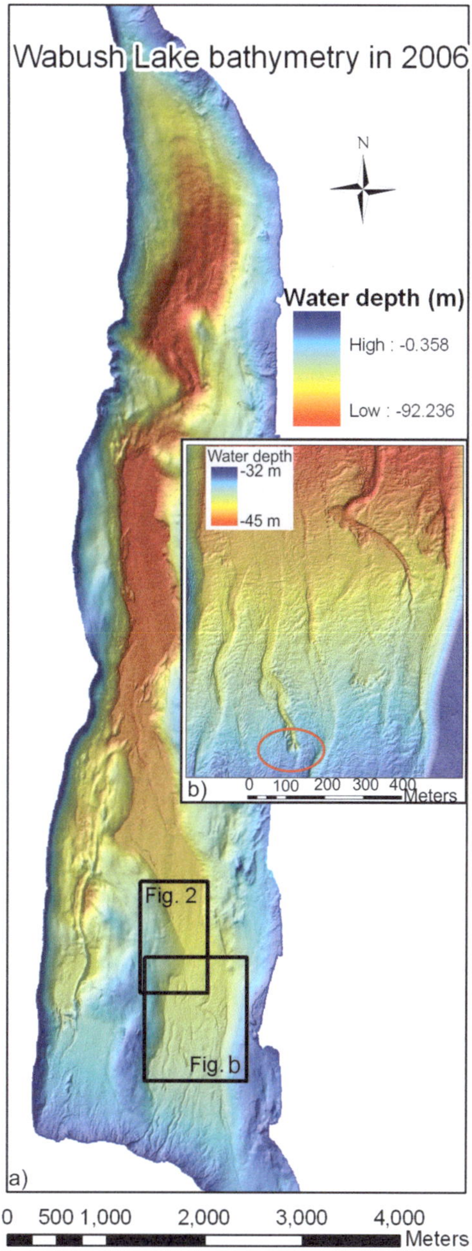

Fig. 23.1 (**a**) Bathymetric data from 2006 survey. In (**b**) is *circled* the location of one of the knickpoints studied

acquired with a Seistec boomer in 2006 (Cauchon-Voyer and Locat 2007). For the seismic interpretations, a constant seismic velocity of 1500 m/s was used. More information on Wabush Lake morphology and evolution as well as the methodology used in the multibeam and seismic surveys are presented in Turmel et al. (2010, 2012, 2014, 2015).

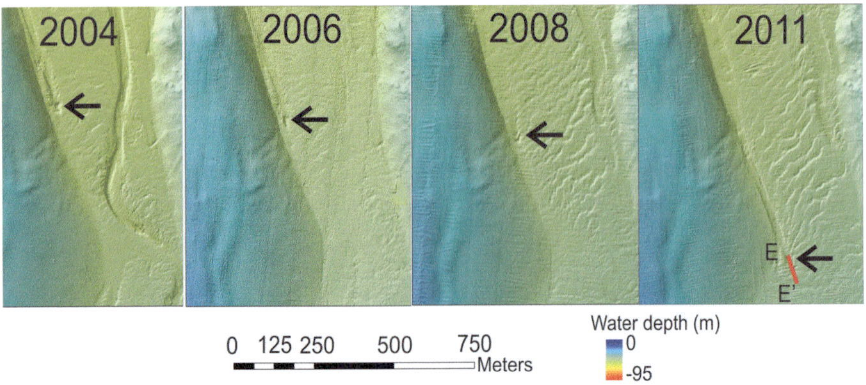

Fig. 23.2 Evolution of the second studied knickpoint between 2004 and 2011. *Arrows* indicate the head of the knickpoint at each year. The location of the 2011 profile for the stability analysis (Fig. 23.4b) is marked in *red* as *E-E'*

Two sites were chosen for stability analysis. The first site is at 35 m water depth (Fig. 23.1b). This 3.5 m high knickpoint was chosen because, in the 2006 bathymetry, there are signs of an incipient failure revealed by a small depression 8 m upslope of the knickpoint. The slope angle of the modelled knickpoint is 16°. In this area, the water depth estimated before mine operations is 66 m, 41 m in 1999, 38 m in 2004, and 35 m in 2006.

The second site is at 45 m water depth (Figs. 23.1 and 23.2). In this section of the lake, a 3 m high knickpoint and a channel created by its migration are visible on all surveys between 2004 and 2011 (Fig. 23.2). The slope angle of this knickpoint is 11°. The lake depth here was 61 m before mine operations, 51 m in 1999, 50 m in 2004, 48 m in 2006, 46.5 m in 2008 and 45.7 m in 2011. On Fig. 23.2, another channel is visible in 2004 and was buried between 2004 and 2006, illustrating the highly dynamic environment in this area. Sediment waves are also visible in 2004, 2008 and 2011.

23.3 Methodology

Four main steps were taken in order to analyse the stability of the knickpoints. First of all, remolded surface sediments were sampled in the Lake using an Ekman grab sampler and samples were also taken at the outlet of the mine. They were then analysed in the laboratory to obtain geotechnical properties of the tailings. An evaluation of the excess pore water pressure through numerical analysis was then made and slope stability analysis undertaken.

23.3.1 Geotechnical Properties

Triaxial tests (Table 23.1) were performed on reconstituted sediments from the raw tailings, i.e. tailings taken directly from the pipes. Ten triaxial tests were conducted

Table 23.1 Characteristics and summary of the triaxial tests. In this table, P means pluviation method, MT moist tamping method, w_i is the initial water content, e_i the initial void radio, e_c the void ratio after saturation and consolidation to p'_0, p'_0 the effective mean stress, p'_{ss} the effective mean stress at steady state and q_{ss} is the deviator stress at steady state

Sample #	Sample preparation	w_i	e_i	e_c	p'_0	p'_{ss}	q_{ss}
1	P	24.8	0.746	0.679	51.2	763	1,242
3	P	31.3	0.64	0.597	450	780	1,242
5	MT	4	0.9	0.8	371	15.2	18.8
6	MT	4	0.9	0.8	269	0.2	1.6
7	MT	4	0.9	0.84	171	1.5	6.4
8	MT	5	0.65	0.63	320	350	544
9	MT	5	0.72	0.69	500	479	742
10	MT	5	0.72	0.7	513	299	492
11	MT	5	0.78	0.695	506	55.9	104
13	MT	5	0.78	0.738	500	18.9	34.2

(Konrad 2004), among which two were prepared in a dense state using the pluviation method and the remaining were prepared in a loose state using moist tempering method (e.g. Ladd 1974). The loose specimens were premixed at an initial water content of 4 or 5 % and saturated after the specimen preparation. All the tests were performed in undrained condition after having been isotropically consolidated. The specimens were axially strained at 1 %/min.

A Sedimentation Consolidation (SedCon) test (e.g. Locat 1982) was done, using sediments taken between 30 and 50 m water depth. In a large Plexiglas cell, a pre-weighted and pre-mixed amount of tailings and water is delivered inside the cell. The first part of the test (sedimentation) is to follow the vertical displacement of the water-suspension interface, until a constant level is reached. The second phase of the test (consolidation) consists in loading the sample via weights added on a piston placed over the sediments. The load was gradually increased with a factor of 50 %, when the strain rate reached values lower than $10^{-7}s^{-1}$. The range of vertical stress varied between 0.02 and 207 kPa. During the consolidation of the material, the drainage at the base can be opened or not. The drainage at the upper boundary is always possible. Sensors at the base of the soil column allow monitoring of the evolution of the pore water pressure at the base of the soil column. When the drainage at the bottom is closed it is possible to calculate the hydraulic conductivity of the soil.

23.3.2 Excess Pore Pressure and Stability Analysis

In order to calculate the stability of the slopes, some steps had to be taken to try to establish the hydraulic conditions of the soil studied. Since sedimentation rates are high in all those areas, and hydraulic conductivity of the tailings is low (values are presented in the results section), an excess of pore water pressure may be present due to consolidation process. To estimate the excess of pore water pressure, the software Basin (Kvalstad and Kaynia 2001) has been selected for this study. The main geotechnical inputs needed in Basin are the evolution of the void ratio as a function of the effective stress and the evolution of the permeability as a function of the void ratio. These relationships are available from the SedCon tests. The sedimentation rate values determined from seismic analysis were used in the Basin model.

The stability analysis was done using Slope/W and Seep/W from the GeoStudio 2004 package. Slope stability analyses were done in terms of effective stress. Friction angle and stability line angle were used in the analysis. Hanzawa (1980) observed that peak shear stress points in undrained triaxial tests on loose sands can be defined, in effective stress, by the stability line angle. The use of the instability line in the liquefaction analysis was used to explain some submarine landslides that occurred via liquefaction, such as the Nerlerk liquefaction failure (Sladen et al. 1985), or flowslides in coal mine waste dumps (Dawson et al. 1998).

23.4 Results

23.4.1 Geotechnical Tests

Results obtained from the triaxial tests are shown in Fig. 23.3a in a graphic showing the deviatoric stress versus mean normal stress. In this Figure, results from both dense and loose specimens are shown. The critical state line as well as the instability line is also traced on this Figure. The instability line is defined by the peak points of the undrained stress path (Fig. 23.3) (Lade 1993). The instability line is not unique but varies with the void ratio of the sample. In this case, the instability line was traced for the samples of experiments 5, 6, and 7, i.e. samples with the highest void ratio (0.8, 0.8 and 0.84 respectively).

Analysis of the triaxial tests results shows that, for the specimens that were mounted at a low effective mean stress and are in a loose state, a strain-softening behaviour is present. Tests 6 and 7 even achieved static liquefaction. At a higher effective mean stress (samples 8, 9, 10) or samples mounted in a dense state (samples 1 and 3), triaxial results shows a strain hardening behaviour.

The critical state line, having a $M_{cs} = 1.60$, corresponds to a friction angle of 40°. The instability line has a $M_{IL} = 0.7$ slope, corresponding to a mobilized friction angle at the instability line of 18°.

The SedCon test started with an initial water content of 95.5 %, which represents a void ratio of 3.15. After 3 days, the settlement is completed and the void ratio is 1.9. Then, the consolidation portion was initiated. Figure 23.3b shows the evolution of the void ratio as a function of the effective vertical stress. The final void ratio reached, at an effective vertical stress of 207 kPa, is 1.02. The mean compression index calculated from the SedCon tests is of 0.21. On Fig. 23.3b is also shown the function incorporated in the basin model for this SedCon test. Hydraulic

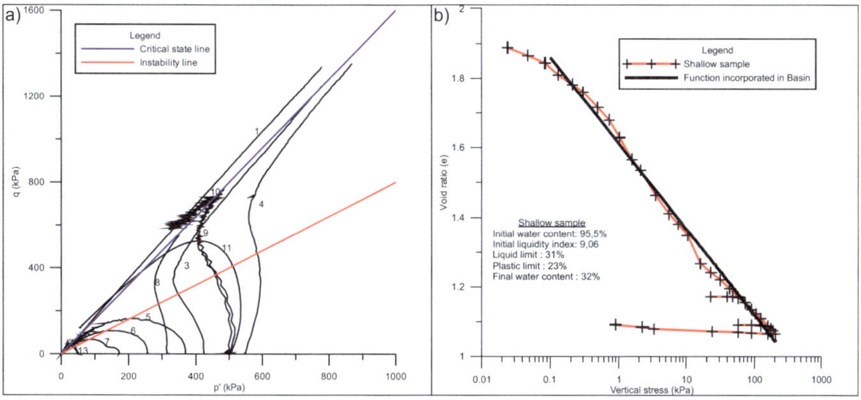

Fig. 23.3 (**a**) Results from the triaxial tests, showing the instability line and the critical state line. (**b**) Results from the consolidation part of the SedCon test. Shown in *black* is the function incorporated in Basin

conductivity was also calculated, using the decay in pore water pressure information. The hydraulic conductivity was found to vary between 3e-8 m/s for a void ratio of 1.46–1.23e-9 m/s for a void ratio of 1.1.

23.4.2 Excess Pore Water Pressure and Stability Analysis

Using the Basin software and the sediments characteristics obtained from the SedCon tests, as well as the sedimentation rates computed from the analysis of the bathymetric and seismic data, the excess of pore water pressure can be calculated for both sites. For site 1, with a mean sedimentation rate of 0.8 m/year between 1964 and 2006, the excess pore water pressure is 14 kPa at the elevation of the rupture surface (3.5 m). For the second site, that shows a lower sedimentation rate (0.34 m/year between 1964 and 2007), the excess pore water pressure is approximately 1.5 kPa at 1 m deep and 5 kPa at 3 m. These parameters were incorporated in the slope stability analysis using Seep/W by imposing the results from Basin as boundary conditions.

The possibility of a landslide occurring at the head of the first knickpoint was modelled with one major constraint: since the tension crack is approximately 8 m behind the slope crest, the surface of rupture position behind the crest was set in Slope/W to be 8 m behind the slope crest (Fig. 23.4a). The slope modelled was the slope of the channel walls adjacent to the knickpoint, which shows a slope of 18°, i.e. 14 m long and 4.6 m high. The soil modelled is cohesionless and properties were kept constant for the whole layer. In hydrostatic conditions, the factor of safety found is approximately 3.8, and a mobilized friction angle of 13° is necessary to obtain a factor of safety of one. When considering the excess pore water pressure, the lowest safety factor found is 2.7, using a friction angle of 40° which was

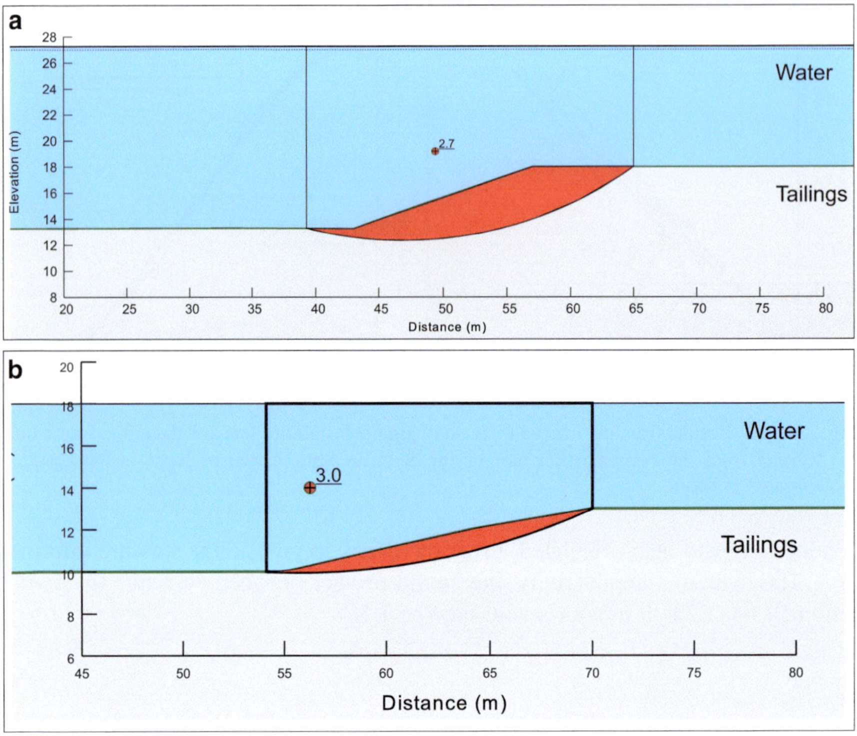

Fig. 23.4 Geometry of the two knickpoints modeled and their slope surface for the case where excess pore water pressure are taken into account

obtained from the triaxial tests. As this is loose silt, the friction angle of the material should be lower and a safety factor of 1 or less is achieved if the friction angle is 18° or less.

The second knickpoint studied shows no apparent sign of instability, at least at the time of the survey. The slope modelled, i.e. the knickpoint head, is 3 m high, for a length of 15 m, i.e. 11° (Fig. 23.4b). The only constraints on the surface of rupture imposed to the model were that the entrance of the failure surface have to extent at least from the top of the slope to somewhere at the bottom of the slope. Stability analysis with hydrostatic conditions brings a factor of safety of 3.4. In these conditions, the slope would be stable, and the mobilised friction angle at rupture is around 16°. When considering the excess pore water pressure calculated with Basin, the factor of safety lowers to 3.0, and the mobilised friction angle at rupture is 17°.

23.5 Discussion

These two examples of knickpoint scarp stability show that landslides may, in certain circumstances, migrate upwards via landsliding processes, for both knickpoints, the mobilised friction angles at failure are approximately match the instability line. Some differences exist between the two knickpoints studied. The first one is located in an area of higher sedimentation, which influences the stability of the soil by enhancing the excess of pore water pressure. In this case, the presence of a tension crack behind the slope scarp shows that a landslide was already initiated at the head of this knickpoint, or that liquefaction occurred in only part of the slope. The other knickpoint, located in an area of lower sedimentation, shows no sign of instability at the time of the survey. However, it was shown in the laboratory (Turmel 2012), that knickpoints do not evolve in a linear way. The slope angle at the head of the knickpoints is always evolving, and a difference of more than 20° for slope angles were noted in the laboratory. Since the knickpoint is always influenced by erosion, and the head scarp of the knickpoint is always evolving, it can be argued that this scarp, even if stable at the time of the survey, is not always stable throughout the lifetime of the knickpoint.

To reach failure at the knickpoint head scarp, one must keep in mind that the mobilized friction angle is lower than the friction angle at steady state. In the case of Wabush Lake, the materials tested in the triaxial apparatus showed static liquefaction when they were in a loose state and at low confining pressure which corresponds to the conditions present in Wabush Lake. In order to have a static liquefaction, the instability must take place in an undrained manner (e.g. Kramer 1988). This may be the case in Wabush Lake since all the knickpoints are influenced by very rapid and intense erosion caused by turbidity currents and because the hydraulic conductivity of the tailings is low. It is shown that, for the two knickpoints, the mobilized friction angle at rupture is between 16 and 18°, which matches the angle of the instability line found in the triaxial tests (18°). That means that a part of the slope may be subjected to liquefaction if an undrained loading or unloading is applied to the slope. Therefore, contrary to the conclusion of Toniolo and Cantelli (2007), static liquefaction may occur and may lead to a localized landslide helping the knickpoint to migrate upward. Other geotechnical factors must be taken into account in the stability analysis and not only the fact that the angle of the knickpoint is lower than the friction angle at steady state.

23.6 Conclusion

We have tested the hypothesis that in the subaqueous environment knickpoint migration can be caused by landslides. The geotechnical and hydrological properties of the soils were derived from laboratory tests on the sediments. Using the information on the sedimentation rate, derived from analysis of seismic lines and bathymetric surveys, the excess of pore water pressure was calculated using the software Basin. Values found were then incorporated in the slope stability model. The triaxial tests showed that, at low confining pressure and when the materials were in a loose state, the stability line was lower than the steady state line. In some samples, static liquefaction appeared.

Using the strength parameters derived from the triaxial and hydraulic parameters from SedCon tests, the stability of two knickpoints were analysed. One of the knickpoints indicates incipient failure and required a friction angle in the order of 18° in order to obtain a factor of safety of 1. This friction angle is equivalent to the instability line when using excess of pore water pressure derived from the basin model. The second knickpoint shows no sign of previous instability, and the factor of safety was over 1 even with a friction angle of 18°. However, since the friction angle for a safety factor of 1 is calculated to be 17° we conclude that parts of the slope may be subjected to liquefaction during undrained conditions.

This study demonstrates that landslides can occur at the head of knickpoints when, due to erosion of the knickpoint by turbidity currents, the scarp of the knickpoint is steepened and undrained conditions occur. We conclude that landslides in fine grained sediments in areas of high deposition are a fundamental mechanism for knickpoint retreat and thus channel development.

Acknowledgments The authors are grateful to F. Gilbert and C. Juneau for their help with the experiments. We also thank Rio Tinto and the National Science and Engineering Research Council of Canada for their financial support. Both reviewers, Mylène Sansoucy and Jean-Sébastien L'Heureux, are acknowledged. Their comments greatly helped to improve the paper. We would finally like to thanks the NGI for allowing us to use their software Basin.

References

Cauchon-Voyer G, Locat J (2007) Spatio-temporal accumulation of tailings in Wabush Lake. Université Laval, Québec, Internal report, report LERN-2007-Wabush-03

Dawson RF, Morgenstern NR, Stokes AW (1998) Liquefaction flowslides in rocky mountain coal mine waste dumps. Can Geotech J 35:328–343

Dobry R, Alvarez L (1967) Seismic failures of Chilean tailings dams. J Soil Mech Found Div ASCE 93(6):237–260

Hanzawa H (1980) Undrained strength and stability analysis for a quick sand. Soils Found 20(2):17–29

Heiniö P, Davies RJ (2007) Knickpoint migration in submarine channels in response to fold growth, western Niger Delta. Mar Pet Geol 24(6–9):434–449

Konrad J-M (2004) Laboratory triaxial test results IOC mine tailings. Université Laval, Laval

Kramer SL (1988) Initiation of soil liquefaction under static loading conditions. J Geotech Eng 114(4):412–430

Kvalstad TJ, Kaynia AM (2001) Ormen lange slope stability assessment, BASIN – a program for one-dimensional large-strain consolidation of non-homogeneous multi-layered sedimentation. NGI, Oslo

Ladd RS (1974) Specimen preparation and liquefaction of sands. J Geotech Geoenviron 100:1180–1184

Lade PV (1993) Initiation of static instability in the submarine Nerlerk berm. Can Geotech J 30(6):895–904

Locat J (1982) Contribution à l'étude de l'origine de la structuration des argiles sensibles de l'est du Canada. Université de Sherbrooke, Sherbrooke

Mitchell N (2006) Morphologies of knickpoints in submarine canyons. Bull Geol Soc Am 118(5–6):589–605

Sladen JA, D'Hollander RD, Krahn J, Mitchell D (1985) Back analysis of the Nerlerk berm liquefaction slides. Can Geotech J 22(4):579–588

Toniolo H, Cantelli A (2007) Experiments on upstream-migrating submarine knickpoints. J Sediment Res 77:772–783. doi:10.2110/jsr.2007.067

Turmel D (2012) Évolution des knickpoints initiés par un courant hyperpycnal continu: le cas du lac Wabush, *Thèse de doctorat,* Département de géologie et de génie géologique, Université Laval, Québec

Turmel D, Locat J, Parker G (2012) Upstream migration of knickpoints: geotechnical considerations. In: Yamada Y, Kawamura K, Ikehara K, Ogawa Y, Urgeles R, Mosher D, Chaytor J, Strasser M (ed) Submarine mass movements and their consequences. Springer, Dordrecht, pp 123–132

Turmel D, Locat J, Parker G (2014) Evolution of a lakebed resulting from subaqueous tailings discharge: coupling multibeam surveys and airborne LiDAR. CIM J 5(2):135–149

Turmel D, Locat J, Parker G (2015) Morphological evolution of a well-constrained, subaerial-subaqueous source to sink system: Wabush Lake. Sedimentology. doi:10.1111/sed.12197

Turmel D, Locat J, Cauchon-Voyer G, Lavoie C, Simpkin P, Parker G, Lauzière P (2010) Morphodynamic and slope instability observations at Wabush Lake, labrador. In: Mosher DC, Shipp RC, Moscardelli L, Chaytor JD, Baxter CDP, Lee HJ, Urgeles R (ed) Submarine mass movements and their consequences. Springer, Dordrecht, pp 435–446

Wagener F, Strydom K, Craig H, Blight G (1997) The tailings dam flow failure at Merriespruit, South Africa: causes and consequences. In: Proceedings of the conference on tailings and mine waste, vol 97, pp 657–666

Chapter 24
Multiple Flow Slide Experiment in the Westerschelde Estuary, The Netherlands

Dick Mastbergen, Geeralt van den Ham, Matthieu Cartigny, André Koelewijn, Marco de Kleine, Mike Clare, Jamie Hizzett, Maria Azpiroz, and Age Vellinga

Abstract Flow slides form a major threat to flood defences along coastlines and riverbanks in the Netherlands. Due to the uncertainties with respect to the process in combination with the severity of the consequences and costs for prevention measures, there is a need to improve existing models for prediction or occurrence of, and damage by, flow slides. One of the key questions to be answered is whether slope failure by a flow slide is caused by up-slope migrating breaches or by static liquefaction. Although fundamentally different mechanisms, both result in a flowing sand-water mixture or turbidity current that eventually redeposits on a gentle slope. Over the last decades numerical models have been developed for both mechanisms, based on flume experiments. Upscaling these experiments is complex, as scaling rules are different for the various processes involved. To evaluate the failure mechanism on a natural scale, validate numerical models and test new technology to monitor the occurrence of flow slides, a large, controlled field test was performed.

The test site was situated in the Westerschelde estuary, in the south-western part of the Netherlands (Fig. 24.1). Several flow slides of 10^5–10^6 m^3 have occurred in this area in the past. In advance of the experiment, cone penetration tests and boreholes were performed on the test location. Pore water pressure sensors were installed in the sand. Triaxial tests and grain size distribution measurements were performed on collected soil samples.

The flow slides were initiated by means of steepening of the slope by dredging. Eventually several autonomously retrogressing flow slides were observed, running several hours over a maximum distance of about 100 m and resulting in a total displaced volume of several 10^3 m^3 of sand. During the test the evolution of the

D. Mastbergen (✉) • G. van den Ham • A. Koelewijn • M. de Kleine
Deltares, Delft, The Netherlands
e-mail: Dick.Mastbergen@deltares.nl

M. Cartigny • M. Clare • J. Hizzett • M. Azpiroz • A. Vellinga
National Oceanography Centre Southampton, Southampton, UK

G. Lamarche et al. (eds.), *Submarine Mass Movements and their Consequences*, Advances in Natural and Technological Hazards Research 41,
DOI 10.1007/978-3-319-20979-1_24

slope topography was monitored continuously by three multibeam survey vessels. This resulted in a full multibeam survey of the area almost every quarter of an hour and, assisted with other advanced instruments, enabling us to witness the early development of a flow slide.

24.1 Introduction

Flow slides are a mainly subaqueous type of landslide in sandy and silty slopes along coasts, estuaries and rivers, which transform steep slopes into more gentle slopes. Flow slides can result in much longer, partly subaerial retrogression lengths in banks or foreshore, compared to other geotechnical slope failure mechanisms and can cause significant damage when occurring in the vicinity of dikes, levees and river banks or hydraulic structures (Van den Ham et al. 2014). Risk indications are oversteepened slopes due to erosion at the toe or deposition at the top, lack of protection of the surface to waves or currents, presence of loosely packed and recently deposited fine sand mixed with silt or clay layers, extreme low water, heavy rainfall and human interventions like dredging.

There is a 200 year history of flow slide observations in the Netherlands, especially in the Schelde estuary (Wilderom 1979), however, the events were generally observed only after the event had occurred. This means that final geometries of flow slides are well known, but the actual development of the slides over time was seldom witnessed, let alone measured. Over 40 years, many research programmes have been focussed on flow slides worldwide and field observations were described (De Groot et al. 1987). Elements of the flow slide process, such as liquefaction (resulting in slope deformations in very loosely packed sand), or the retrogression of a steep wall in densely packed sand during dredging (breaching), have been reproduced in flume tests and described with numerical models (Van den Ham et al. 2014). However, even in large scale laboratory tests it has never been possible to reproduce the full process of a flow slide autonomously retrogressing into a sandy slope and depositing sand over a large distance as observed in the field.

Combining geotechnical investigations, probabilistic methods, statistics of characteristics of past flow slides and physics-based models, can help assess the probability of occurrence of, and damage due to a flow slide. For instance slope height and steepness, relative density and grain size distribution are generally considered as soil mechanical criteria for a flow slide hazard (Van Duinen et al. 2014). However, the actual nature of the failure mechanism, required for sound numerical modelling, has never been determined in full scale flow slides so far.

Two types of flow slide failure mechanisms are generally considered in literature: static liquefaction followed by a shear slide, and retrogressive breaching, generating a sustained turbidity current downslope of the location of initiation. Theoretically, both types of flow slide failure will result in a similar slope profile, although the evolution over time is very different. Static liquefaction is related to loosely packed sand layers in the subsurface that lose their strength instantaneously due to excess pore water pressure, whereas breaching is related to more densely packed fine sand layers at the soil-water interface and may slowly retrogress over several hours. Both result in a flowing sand-water mixture or turbidity current that eventually redeposits on a gentle slope.

In September 2014 a field test was performed in The Netherlands as part of the flood defence policy framework (Van den Ham et al. 2015). The main reasons to perform this field test to induce and monitor a flow slide were:

- Monitoring the 'birth' and development of a flow slide;
- Testing sensor and monitoring technology;
- Produce field-scale benchmark data for validation of numerical models.

24.2 Field Test Set-Up

The test site should have favourable natural conditions for flow slides, yet not carry a risk of damaging a dike or any other civil construction. Such a test site was found in the Westerschelde estuary on the slopes of the Plaat van Walsoorden tidal flat (see Fig. 24.1). The Westerschelde is a highly active morpho-dynamic system with

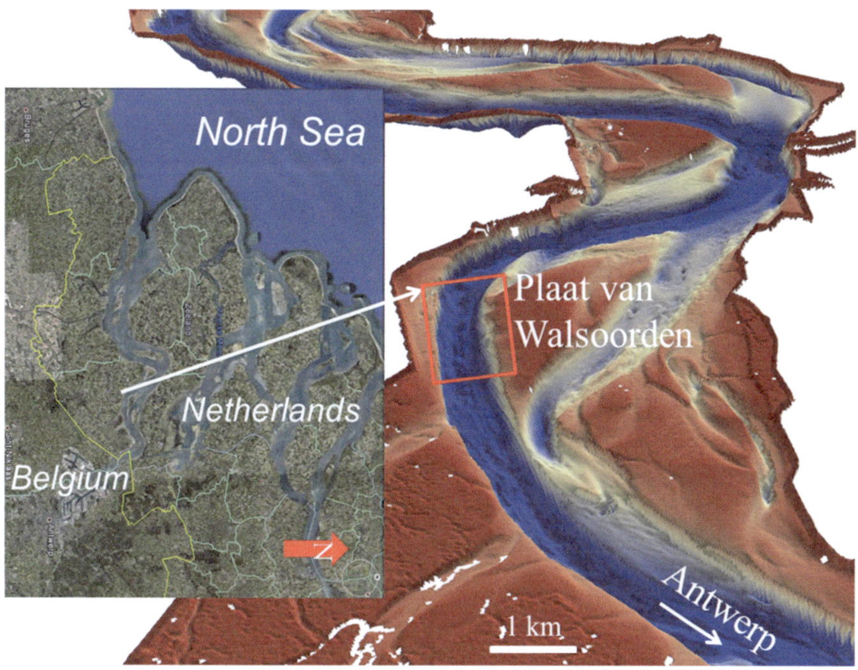

Fig. 24.1 Westerschelde estuary (inset shows map of location), seen downstream in western direction with test area indicated in *red square*, see Figs. 24.2 and 24.3

Fig. 24.2 View towards the South from Plaat van Walsoorden sand flat at location of 22nd of July 2014 flow slide. Inset aerial view. Scar of flow slide clearly visible. *Blue dot* is position of photo. Test area indicated by *red square*, see also Figs. 24.1 and 24.3

strong tidal currents (over 1.5 m/s), almost 5 m tidal range and an abundance of sediments, mainly fine sand, resulting in continuous sedimentation on the tidal flats. Natural flow slides are reported here once every couple of years.

Most recently a large flow slide (almost 1 Million m^3) occurred on 22nd of July 2014 (see Figs. 24.2 and 24.3), exactly at the originally proposed test site. After a week, on 28th of July 2014, further activity was observed on the shoreline, where gullies incised the tidal flat. Vertical slabs of sand were observed to gradually fall into the water. The duration of the event was at least several hours, but probably lasted for over 24 h (Van den Ham et al. 2015). The eroded sand was deposited into the shipping channel, leading to a decrease in the water depth of more than 5 m. As a result, the minimum navigation depth needed to be restored by dredging to the depth required by the Antwerp Port authorities. The bulk of the sand is still present after 6 months and will erode by reworking through tidal currents. The scarp that was created forms a natural bay (see Fig. 24.2) and is being filled in with loosely packed fine sediments. Given the occurrence of this recent failure, the test site was relocated 400 m to the east (see Fig. 24.3), where almost identical pre-failure conditions were found.

A soil investigation programme was performed at the site before and after the experiment. Before the test, borings were made, both regular and seismic cone penetration tests (CPT) were taken (Van Duinen et al. 2014) and geophysical

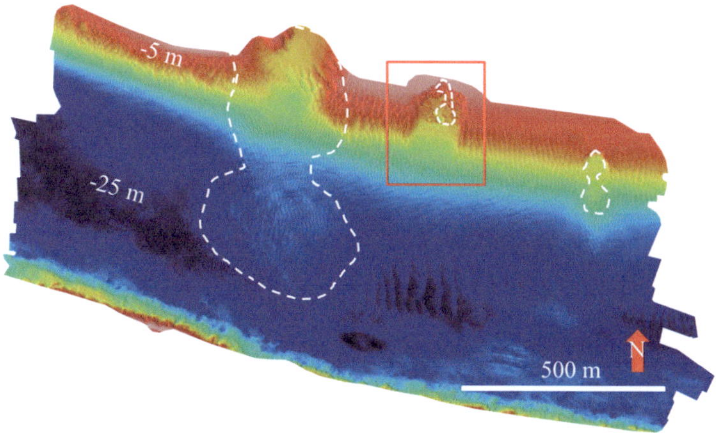

Fig. 24.3 Multibeam survey of complete test area (*red square* in Figs. 24.1 and 24.2): *left* flow slide 22nd of July 2014, *middle* dredging location (in *red square*, see Fig. 24.4), *right* dumping location

surveys (multibeam and sub-bottom profiling) were performed. Pore-water pressure meters were installed in the boreholes at about 5 m depth below the surface for continuous monitoring during the test. After the test vibrocore drillings and additional CPT's were taken. The soil investigation showed a generally uniform fine sand (d_{50} about 140 μm, d_{60}/d_{10} about 1.5) with low clay content and some loosely packed layers at about 10 m depth below the surface. In the tidal zone, clay layers and organic material were found.

To create the conditions for a flow slide in the available test period, continuous excavation of the slope was performed by dredging with a 750 m^3 trailing suction hopper. This aimed to steepen the slope over a certain height, approaching natural oversteepening by erosion as close as possible and to excavate the slope to a depth where loosely packed sand layers were expected, based on the CPT data. It was anticipated that the triggering mechanism would be the dredging itself, creating a small breach or shear slide or inducing liquefaction in a loosely packed sand layer. It was predicted from experience that a flow slide could develop spontaneously in fine sand if the slope steepness was at least 1:3 over a height of at least 5 m with a total slope height of at least 15 m. Continuous monitoring of the seabed with survey vessels was required during the dredging period, as the moment and location of initiation remained uncertain.

24.3 Applied Instrumentation

While the dredger progressively steepened the 1:12 slope over a width of about 100 m with the lower base at about −13 m, two vessels equipped with multibeam sensors for seabed mapping surveyed the area continuously, resulting in a combined

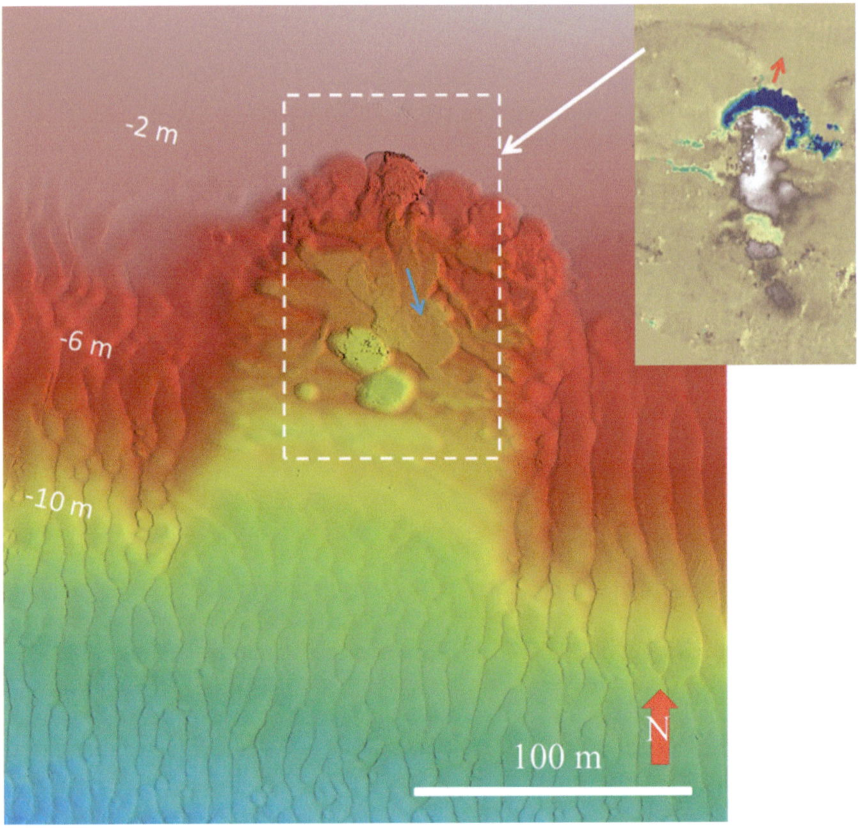

Fig. 24.4 Multibeam detail dredging area (*red square* in Fig. 24.3) at 1-10-2014-19:30 UTC with active retrogressing breach and turbidity current (see also Fig. 25.5). Inset differential plot with 15′ interval (*blue* = erosion, *yellow/white* = deposition)

bathymetry map about every 15 min (see Fig. 24.4). Near-real time differential bathymetry plots and cross-sections clearly showed the evolution of the profile (see Fig. 25.5). Since the dredger itself partly blocked the full view on the seabed, the dredged area could be surveyed in detail during the periods that the dredger disposed of the sand in two dedicated areas 300 m up- and downstream of the test site. These dump sites were also surveyed in detail (see Fig. 24.3). Moving sand ripples and dunes could be observed clearly. A third larger vessel served as the test control centre and moved slowly alongside the dredger. This third vessel was equipped with three Acoustic Doppler Current Profilers (ADCP) and three mutually differently oriented 2D imaging multibeam sonars to image the flow slide and monitor the resulting turbidity currents. The pore water pressure meters were registered with a 10 Hz sampling rate during the total test period and could be read on-line on the survey vessel and on a land-based control centre. In this way, the very beginning of the event could be detected. As soon as any flow slide activity

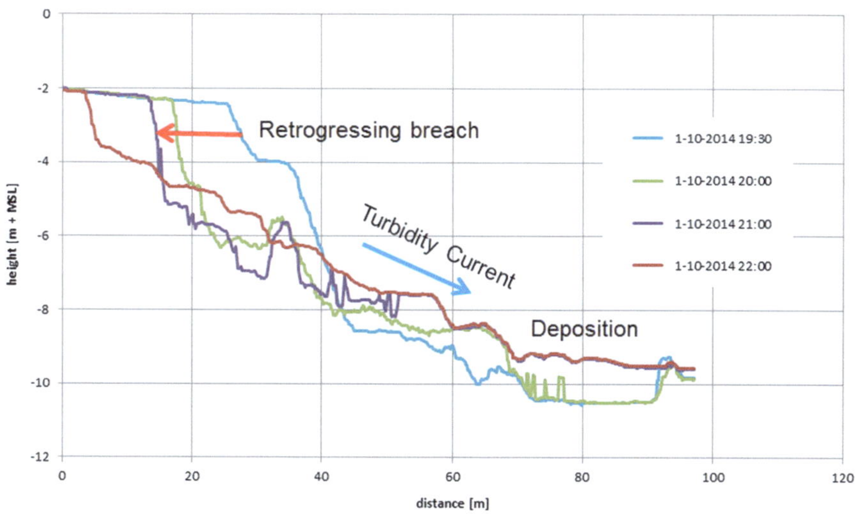

Fig. 25.5 Cross-section flow slide of Fig. 24.4. Breach retrogression speed about 20 m in 2 h. Active breach height 6–8 m. Flow length 60–100 m

was observed, the dredger was moved off the location, only to resume if any activity ceased.

At low tides the subaerial part of the flat was surveyed by a laser-altimeter installed on the survey vessel operating in the shallow water zone. On two survey vessels a low-frequency echo-sounder (sub-bottom profiler) was running, to be able to penetrate into the turbidity current and to image changes in subsurface conditions.

The larger survey vessel was equipped with three ADCP's of different frequencies (300, 600, 1200 kHz) to primarily measure the flow velocity of turbidity currents that could be triggered by a flow slide. The advantage of having three different frequencies is that it gives the possibility to not only measure the flow velocity, but also grain size and sediment concentration of the turbidity current, as these different frequencies respond differently to variation in grain sizes and concentrations (Kostaschuk et al. 2005). Three 2D imaging multibeam sonars were also installed on the other side of the survey vessel. These imaging sonars have the advantage of giving much higher resolution images of the turbidity currents in both plan and cross-section view, as demonstrated in (Hughes Clark et al. 2012).

24.4 Results and Conclusions

The test took place from 23rd of September until 2nd of October 2014 in generally calm weather conditions. After about 5 days the development of several steep, amphitheatre-shaped breaches started (see Fig. 24.4). These breaches retrogressed upslope above the dredged zone and generated turbidity currents that depleted quickly downslope (see Fig. 25.5). A number of these small flow slides retrogressed into the shallow water zone of the tidal flat, north to the dredging area. However, the flow slides did not retrogress far enough to appear on the sand flat above the low water zone, as was predicted and was seen a few months before, only a few hundred metres to the west. Nevertheless, in total seven flow slide events were observed in detail during the test, including a steep active breach of up to 6 m height and 30 m width, which retrogressed over a couple of hours and generated a turbidity current that ran out over about 100 m (Van den Ham et al. 2015). The autonomous retrogression speed of the active breaches was between 10 and 20 m/h. These observations are well in line with a breaching flow slide as defined in (Van den Ham et al. 2014) and explain the historical observations in the Netherlands (Van Duinen et al. 2014), but also fit with recently published field observations in Australia, where retrogressing breach events occur in sand beaches in the coastal area of Queensland (Beinssen et al. 2014). No positive or negative excess pore water pressures, that could be an indication of static liquefaction or breaching respectively, were measured, but since the flow slides did not get close to the sensors (over 200 m), it is still not clear if any excess pore water pressures occurred near the flow slide. Water levels and even ship waves were monitored accurately however.

It is still uncertain what condition is responsible for the upslope breaching process to continue or to stop, but it may be related to the downslope turbidity current. Since the slope of the dredged area was quite gentle (see Fig. 25.5), the sand may have settled quickly and blocked the flow downslope, resulting in a decrease of the breach height and halting of the flow. This will be further analysed by means of simulation of the observed flow slides with numerical models (Van den Ham et al. 2015) and more detailed data analysis.

It is likely that the flow slide of 22nd of July 2014 started in a similar way, with a small retrogressing breach, but it continued retrogressing for many hours into the subaerial zone of the tidal flat, finally resulting in the observed large amphitheatre shaped bay and depositing the eroded sand in the shipping channel. Seemingly, the turbidity current extended over a much longer distance (over 750 m). It may have been catalysed by the effects of draining of pore water at low tide in combination with flushing of fresh rain water in gullies (see Fig. 24.2), since, unlike our test, this natural flow slide was preceded by heavy rainfall.

The test has resulted in a large quantity of data that still has to be analysed in detail and will be published in subsequent papers, especially the data of pore water pressure meters, ADCP and multibeam imaging sonar, detailed soil investigations, future developments of the site and numerical analyses.

Acknowledgements The test was made possible by collaboration of research institutes, government authorities and various companies from the Netherlands, Belgium and the UK. The experiment was performed by the Stichting FloodControl IJkdijk and funded by the Ministry of Economic Affairs of The Netherlands, STOWA and Rijkswaterstaat. We want to thank our reviewers J.H. van den Berg and J. Eggenhuisen for their helpful advice.

References

Beinssen K, Neil DT, Mastbergen DR (2014) Field observations of retrogressive breach failures at two tidal inlets in Queensland, Australia. Aust Geomech 49(3):55–63

De Groot MB, Silvis F, van Rossum H, Koster MJ (1987) Liquefied sand flowing over a gentle slope. Groundwater effects in geotechnical engineering, IXth European conference on soil mechanics and foundation engineering, Dublin, pp 595–598

Hughes-Clarke JE, Brucker S, Muggah J, Church I, Cartwright D, Kuus P, Hamilton T, Pratomo D, Eisan B (2012) The squamish proDelta: monitoring active landslides and turbidity currents, new and emerging technology. Canadian hydrographic conference, Niagara Falls, 15 pp

Kostaschuk R, Best J, Villard P, Peakall J, Franklin M (2005) Measuring flow velocity and sediment transport with an acoustic doppler current profiler. Geomorphology 68(1):25–37

Van den Ham GA, de Groot MB, Mastbergen DR (2014) A semi-empirical method to assess flow slide probability. In: Submarine mass movements and their consequences, 6th Int. Symp., Kiel, Springer, pp 213–223

Van den Ham GA, Mastbergen DR, Koelewijn AR, ter Brake CKE, Zomer WS (2015) Eindrapport validatie-experiment zettingsvloeiing, Meten aan zettingsvloeiing. Amersfoort (The Netherlands): STOWA/FloodControl IJkdijk

Van Duinen A, Bezuijen A, van den Ham GA, Hopman V (2014) Field measurements to investigate submerged slope failures. In: Submarine mass movements and their consequences, 6th Int. Symp., Kiel. Springer, pp 13–21

Wilderom NH (1979) Resultaten van het vooroeveronderzoek langs de Zeeuwse stromen. Rijkswaterstaat, Nota 75.2, Vlissingen (The Netherlands)

Part IV
Multidisciplinary Case Studies

Chapter 25
Submarine Mass Wasting on Hovgaard Ridge, Fram Strait, European Arctic

Matthias Forwick, Jan Sverre Laberg, Katrine Husum, and Jenny A. Gales

Abstract Hovgaard Ridge is an 1800 m high bathymetric high in the Fram Strait, the only deep-water gateway between the Arctic Ocean and the other World's oceans. The slopes of the ridge provide evidence of various types of sediment reworking, including (1) up to 12 km wide single and merged slide scars with maximum ~30 m high headwalls and some secondary escarpments; (2) maximum 3 km wide and 130 m deep slide scars with irregular internal morphology, partly narrowing towards the foot of the slope; (3) up to 130 m deep, 1.5 km wide and maximum 8 km long channels/gullies originating from areas of increasing slope angle at the margins of a plateau on top of the ridge. Most slide scars result presumably from retrogressive failure related to weak layers in contourites or ash. The most likely trigger mechanism is seismicity related to tectonic activity within the nearby mid-ocean fracture zone. Gully/channel formation is suggested to result from cascading water masses and/or from sediment gravity flows originating from failure at the slope break after winnowing on the plateau of the ridge.

25.1 Introduction

Slope instabilities and a variety of sediment reworking processes of various extents have affected many slopes at high northern latitudes (e.g. Vorren et al. 1998; Haflidason et al. 2004; Vanneste et al. 2006; Baeten et al. 2013, 2014; Rise et al. 2013; Elger et al. 2015). Slope failures occurred frequently during peak glaciation when grounded ice sheets extended to shelf break and the mouths of

M. Forwick (✉) • J.S. Laberg
Department of Geology, University of Tromsø – The Arctic University of Norway, NO-9037 Tromsø, Norway
e-mail: Matthias.Forwick@uit.no; Jan.Laberg@uit.no

K. Husum
Norwegian Polar Institute, NO-9296 Tromsø, Norway

J.A. Gales
National Oceanography Centre, University of Southampton Waterfront Campus, European Way, Southampton SO14 3ZH, UK

G. Lamarche et al. (eds.), *Submarine Mass Movements and their Consequences*, Advances in Natural and Technological Hazards Research 41,
DOI 10.1007/978-3-319-20979-1_25

glacial shelf troughs (e.g. Laberg and Vorren 1995; Vorren et al. 1998). However, the rapid deposition of glacigenic debris lobes on fine-grained contouritic deposits with high water content may lead to overpressure in the latter, facilitating the development of large-scale submarine landslides during interglacials (e.g. Haflidason et al. 2004; Solheim et al. 2005; Laberg and Camerlenghi 2008). Submarine slope failures on continental margins can also occur in areas unaffected by rapid deposition of glacigenic debris lobes (e.g. Baeten et al. 2013, 2014; Elger et al. 2015). Triggering factors of failures at high latitudes can include tectonic/glacitectonic processes or processes related to gas-hydrate dissociation (e.g. Vanneste et al. 2006; Winkelmann and Stein 2007; Elger et al. 2015).

The upper slopes of high-latitude continental margins are often incised by gullies/channels of various sizes and shapes (Vorren et al. 1998; Gales et al. 2013a, b). Gullies are most probably formed by hyperpycnal flows carrying sediment-laden subglacial meltwater emanating from the margins of grounded ice sheets at the shelf edge during glacial maxima (Gales et al. 2013b). However, the processes leading to the formation of gullies remain poorly understood.

The documentation of large-scale sediment reworking in areas of the Arctic that are detached from continental margins remains sparse. However, it has been suggested that sediment reworking on the Lomonosov Ridge, central Arctic Ocean, resulted from iceberg erosion on the ridge crest, seismicity or an impact of an extraterrestrial object in the Arctic Ocean (Kristoffersen et al. 2004, 2007).

In this study, we present swath-bathymetry and high-resolution penetration echo sounder (chirp) data revealing repeated mass wasting on Hovgaard Ridge, central Fram Strait. We discuss potential preconditions and trigger mechanisms in an area that has not been covered with a grounded ice sheet, and that is located in a hydrographically complex setting close to a mid-ocean fracture zone.

25.2 Study Area

Hovgaard Ridge is located at the boundary of the Fram Strait and the Boreas Basin, i.e. within the only deep-water gateway between the Arctic Ocean and the Atlantic Ocean (Fig. 25.1). The ridge is approx. 110 km long and rises up to ~1800 m above the surrounding sea floor. Its width decreases from ~20 km in the SE to <4 km in the NW. The dip of its southern slope is ~10–40° and the gradients of the western and eastern slopes decrease gradually from up to 30° in the south to <15° in the north. The ridge top is an up to 10 km wide, northward-narrowing, "convex plateau" between ~1150 and 1600 m water depth, with surface gradients typically below 5°. It remains debated whether the Hovgaard Ridge is composed of oceanic or continental crust split off from the Svalbard Platform (Myhre and Eldholm 1988; Engen et al. 2008).

Six iceberg ploughmarks between ~1180 and 1210 m on the "convex plateau" provide evidence of grounding of giant, southward drifting icebergs during one or several glaciations (Arndt et al. 2014; Arndt and Forwick accepted).

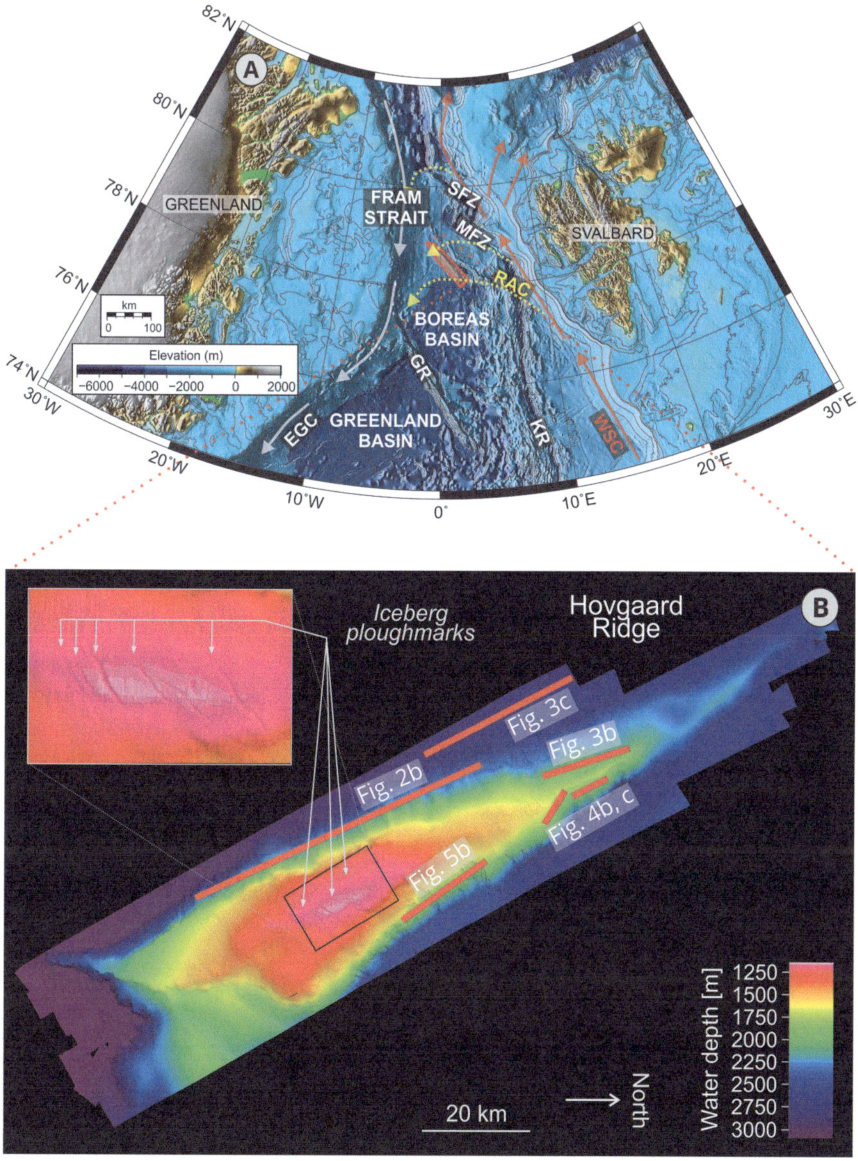

Fig. 25.1 Overview maps. (**a**) Map (IBCAO 3.0; Jakobsson et al. 2012) of the northern Nordic Seas and the Fram Strait. EGC/*white arrows*: East Greenland Current, WSC/*red arrows*: West Spitsbergen Current, RAC/*yellow*, *dashed arrows*: Return Atlantic Current (After Gascard et al. 1995). *GR* Greenland Ridge, *KR* Knipovich Ridge, *MFZ* Molloy Fracture Zone, *SFZ* Spitsbergen Fracture Zone. (**b**) Bathymetry map of Hovgaard Ridge. Locations of bathymetry and chirp profiles in Figs. 25.2, 25.3, 25.4, and 25.5 are indicated. The iceberg ploughmarks on top of the ridge are described by Arndt et al. (2014) and Arndt and Forwick (accepted)

The northward flowing West Spitsbergen Current (WSC) and southward flowing East Greenland Current (EGC) characterize the large-scale hydrography in Fram Strait (Fig. 25.1a). However, in the vicinity of Hovgaard Ridge, the hydrography is more complex with a general weak southward flow of deep-water masses, but also the formation of gigantic meanders or eddies, as well as the formation of the Return Atlantic Current (RAC; e.g. Gascard et al. 1995; Fahrbach et al. 2001; Schauer et al. 2004).

The upper ~340 m of the sediment column on top of Hovgaard Ridge are divided into two lithological units (Shipboard Scientific Party 1995). Unit I (0–185 m below sea floor) consists almost entirely of Pliocene-Quaternary muddy sediment with varying amounts of dropstones. Exceptions are layers of foraminifera-bearing clayey or silty mud close to the seafloor (maximum 30 cm thick), as well as 3 up to 66 cm thick altered ash layers close to the base of the unit. Unit II is composed of late Oligocene biosilica-bearing or biosiceous mud (Shipboard Scientific Party 1995).

25.3 Material and Methods

Swath-bathymetry data were collected in July 2011 with a hull-mounted Kongsberg Maritime EM 300 multibeam echo sounder installed on R/V *Helmer Hanssen* owned by the University of Tromsø – The Arctic University of Norway. A sound-velocity profile was calculated from a CTD cast (conductivity-temperature-depth) prior to data acquisition. The collected data were processed using the software *Neptune* (grid cell size of 25 m by 25 m). They are visualized using *Fledermaus*.

Sub-bottom chirp data were collected with a hull-mounted Edgetech 3300-HM sub-bottom profiler synchronously with the acquisition of the swath-bathymetry data using a pulse mode 2–5 kHz and a shot rate of ~0.2 Hz. All acoustic data were acquired at a ship's speed of approx. 10 kn.

25.4 Results

25.4.1 Western Slope

Numerous channels/gullies occur in the southern and central parts of the western slope (Fig. 25.2). The largest incisions are approx. 130 m deep, up to 1.5 km wide and maximum 8 km long. They originate in areas of increasing slope angle at the margins of the plateau, below 1300 m water depth. Depositional lobes are typically absent.

The northern parts of the western slope contain multiple slide scars with up to ~30 m high headwalls (Fig. 25.3). Single scars are up to 5 km wide. Along approx.

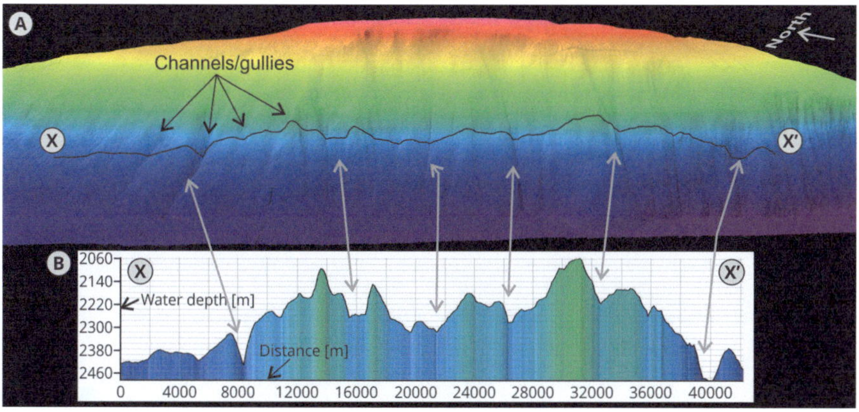

Fig. 25.2 (**a**) Swath bathymetry from the central and southern parts of the western slope. (**b**) Profile from the mid-slope showing up to ~130 m deep channels/gullies

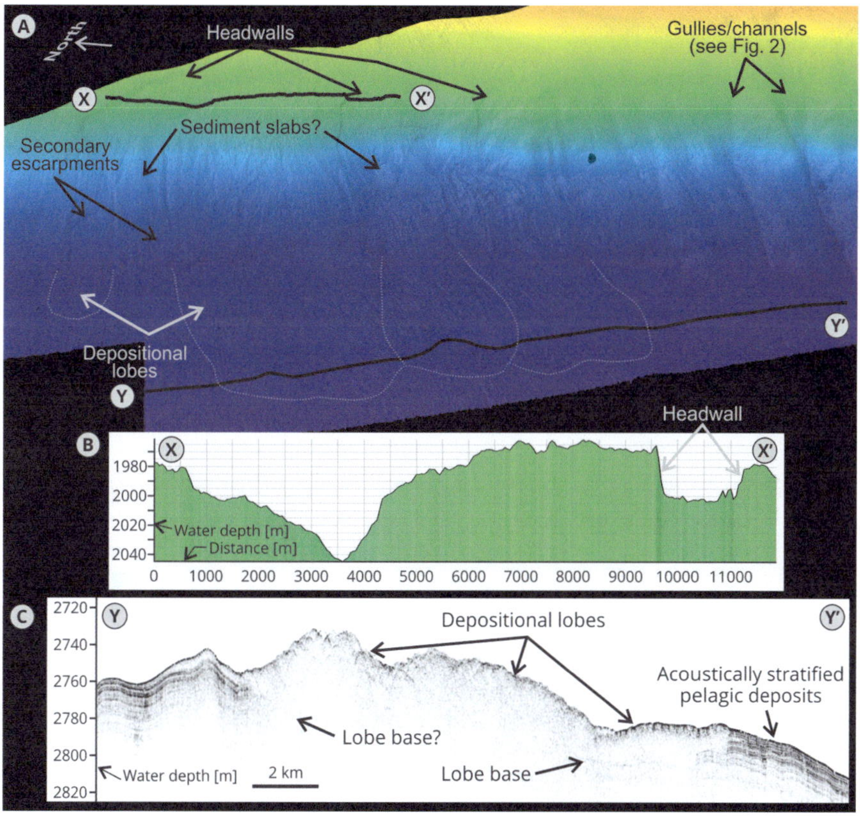

Fig. 25.3 (**a**) Swath bathymetry from the northern parts of the western slope. (**b**) Profile from the upper slope showing up to ~30 m high headwalls. (**c**) Chirp penetration echo sounder profile across the distal parts of multiple depositional lobes off the ridge foot

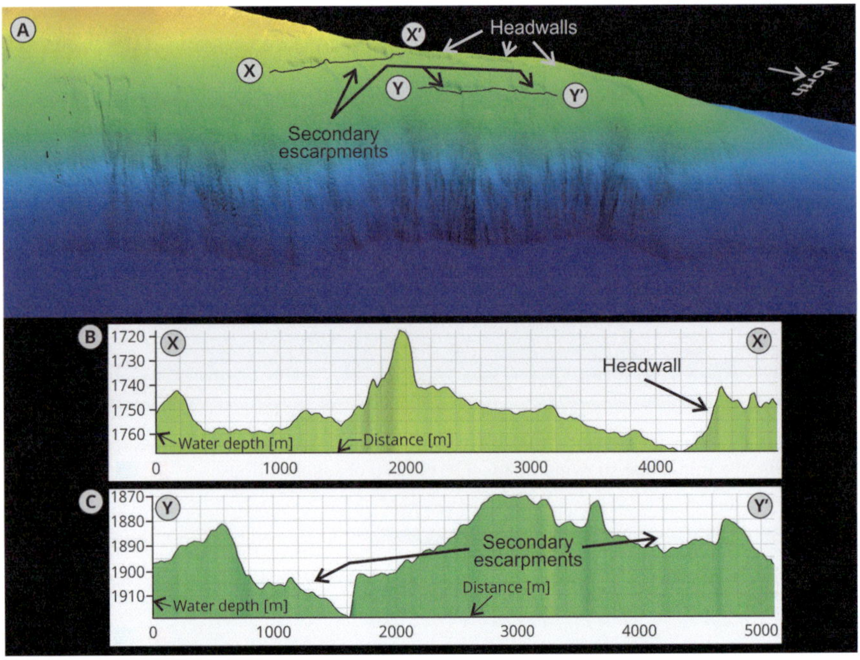

Fig. 25.4 (**a**) Swath bathymetry from the northern parts of the eastern slope. (**b**) and (**c**) Profiles from the upper slope showing up to ~25 m high headwalls

14 km of the foot of the ridge, up to 50 m thick, single and stacked depositional lobes extent up to 6.5 km into the Boreas Basin. They partly erode into and cover the acoustically stratified pelagic deposits in the northernmost Boreas Basin (Fig. 25.3c).

25.4.2 *Eastern Slope*

The northern parts of the eastern slope contain multiple slide scars with up to ~25 m high headwalls (Fig. 25.4). Single scars are up to 5 km wide, and the maximum lateral extent of merged multiple scars is 12 km. Multiple secondary escarpments can be identified. Depositional lobes are absent.

Up to 130 m deep and 3 km wide slide scars with rough morphology occur in the central parts of the eastern slope (Fig. 25.5). The width of the slide scars is either constant or it narrows towards the foot of the slope. Few depositional lobes measuring maximum 3 km width, 2.5 km length and 20 m height extend beyond the foot of the ridge.

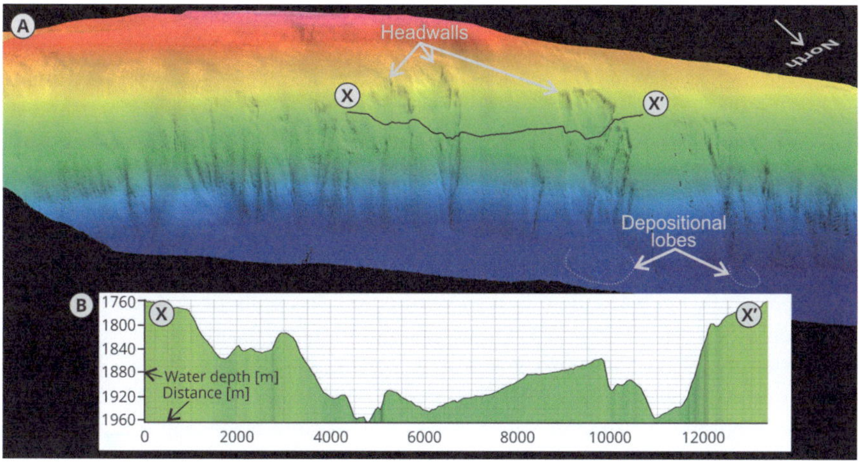

Fig. 25.5 (**a**) Swath bathymetry from the central parts of the eastern slope. (**b**) Profile from the upper/mid-slope revealing up to ~130 m slide scars

25.5 Discussion

Whereas the presence of slide scars on the northern slopes and in the central parts of the eastern slope provide evidence of relatively voluminous sediment evacuation by submarine landslides, the channels/gullies on the central and southern parts of the western slope indicate smaller-scale sediment evacuation, as well as predominantly sediment bypass.

The general absence of secondary escarpments within the slide scars on the western slope of northern Hovgaard Ridge (Fig. 25.3) is interpreted to be due to failures limited to one stratigraphic level. A staircase-like morphology on the eastern slope (Figs. 25.4 and 25.5) suggests slope failures affecting weak layers occurring at different stratigraphic levels, probably reflecting retrogressive failure. The relatively high width-to-depth ratios of the scars on either side of the northern ridge probably indicate failure of sediment slabs. However, the downward narrowing of scars in the central parts of the eastern ridge (Fig. 25.5) may indicate that evacuation of failed sediments from the slide scars was incomplete (compare with Laberg et al. 2014).

The channels/gullies in the central and southern parts of the ridge (Fig. 25.2) are suggested to reflect predominantly erosion and sediment bypass. These incisions might be eroded by gravity flows that are fed with winnowed sediment from the plateau of the ridge (Kaminski et al. 2015), or with sediment pushed over the break of the top of the plateau by grounding giant icebergs (Kristoffersen et al. 2004, 2007; Arndt et al. 2014; Arndt and Forwick accepted). Furthermore, smaller failures and sediment evacuation from the gullies/channels may have occurred due to slope undercutting (e.g. Micallef et al. 2012). The incisions may also have been influenced by cascading water masses within southward-flowing currents from the Arctic Ocean or the RAC (Fig. 25.1).

Systematic studies of the sedimentary processes on Hovgaard Ridge remain sparse, but the seismic stratigraphy from the ridge crest (Myhre et al. 1993) suggests that contouritic deposition predominates. Lithological variations and/or variations in physical properties within contourites can lead to the development of weak layers that may either fail due to sediment load or an external trigger (e.g. Kvalstad et al. 2005; Laberg and Camerlenghi 2008). We regard such conditions as a potential mechanisms facilitating slope failure on the ridge.

Three ash layers were identified ~180 m below the seafloor on top of the ridge (Shipboard Scientific Party 1995). As local variations in sediment thickness occur (Myhre et al. 1993), these ash layers may appear at shallower depths below the seafloor on the slopes of the ridge. They may have acted as weak layers for some of the deep-eroding slides on the central parts of the eastern slope (Fig. 25.5; compare with Harders et al. 2010).

Slide scars on the eastern parts of Hovgaard Ridge are generally more abundant and larger than on the western side. This might either be the result of different lithological compositions of the sediments on either side of the ridge or that the eastern slope is located closer to the transform-fault system in the east where tectonic processes can lead to earthquakes (compare with Läderach et al. 2011).

The depositional lobes off the northern parts of the western slope occurring at very shallow stratigraphic levels or even at the seafloor (Fig. 25.3) provide evidence of relatively young slope failures. These failures could e.g. have been triggered by high-magnitude earthquakes in the Spitsbergen Fracture Zone (for location see Fig. 25.1) which occasionally occur also at present times (e.g. M_b 6.6 event on March 6th 2009; Läderbach et al. 2011).

Sediment cores are required to date slope failures reliably. The absence of sediment lobes in several parts around Hovgaard Ridge suggests that (1) the failures are relatively old and that the reworked sediments are buried by younger contouritic and/or glacimarine, hemipelagic sediments or (2) lobe deposition occurred beyond the extent of the data set.

25.6 Conclusions

- The slopes of Hovgaard Ridge provide evidence of various types of sediment reworking including (1) single and merged slide scars with some secondary escarpments, (2) slide scars with irregular morphology, partly narrowing downslope and (3) gullies/channels.
- Large-scale mass wasting may have been retrogressive originating from failure in weak layers in contourites or in ash. The most probable triggering mechanism was earthquakes originating along transform faults east of the study area.

- Gullies/channels have probably formed from sediment gravity flows fed by winnowing on the plateau on the ridge top, by sediment pushed over the break of the plateau by grounding icebergs and/or from cascading water masses.
- Mass wasting appears to have taken place over a long time. Depositional lobes occurring at very shallow stratigraphic levels or even at the seafloor provide evidence of relatively young slope failures.

Acknowledgements The data were collected during the teaching cruise GEO-8144/3144 "Past and present geological processes and climate: from fjords to the deep sea" arranged by the Ph.D. Trainee School in Arctic Marine Geology and Geophysics (AMGG) at the Department of Geology, University of Tromsø – The Arctic University of Norway. We thank the captain and crew of R/V Helmer Hanssen, as well as Steinar Iversen, Jan P. Holm and the participating students for their support during and after the cruise. The regional bathymetry in Fig. 25.1a was visualized using Generic Mapping Tools (GMT; Wessel and Smith 1998). Finally, we thank Reidulv Bøe and Maarten Vanneste for their constructive reviews of the manuscript, as well as co-editor Sebastian Krastel for his support.

References

Arndt JE, Forwick M (accepted) Deep-water iceberg ploughmarks on Hovgaard Ridge, Fram Strait. In: Dowdeswell JA et al (eds) Atlas of submarine glacial landforms: modern, quaternary and ancient. Geological Society, London, Memoirs

Arndt JE, Niessen F, Jokat W, Dorschel B (2014) Deep water paleo-iceberg scouring on top of Hovgaard Ridge – Arctic Ocean. Geophys Res Lett. doi:10.1002/2014GL060267

Baeten NJ, Laberg JS, Forwick M, Vorren TO, Vanneste M, Forsberg CF, Kvalstad TJ, Ivanov M (2013) Morphology and origin of smaller-scale mass movements on the continental slope off northern Norway. Geomorphology 187:122–134

Baeten NJ, Laberg JS, Vanneste M, Forsberg CF, Kvalstad TJ, Forwick M, Vorren TO, Haflidason H (2014) Origin of shallow submarine mass movements and their glide planes – sedimentological and geotechnical analyses from the continental slope off northern Norway. J Geophys Res B (Earth Surf) 119:2335–2360. doi:10.1002/2013/JF003068

Elger J, Berndt C, Krastel S, Piper DJW, Gross F, Spielhagen RF, Meyer S (2015) The Fram Slide off Svalbard: a submarine landslide on a low-sedimentation-rate glacial continental margin. J Geol Soc 172:153–156

Engen Ø, Faleide JI, Dyreng TK (2008) Opening of the Fram Strait gateway: a review of plate tectonic contraints. Tectonophysics 450:51–69

Fahrbach E, Meincke J, Østerhus S, Rohardt G, Schauer U, Tverberg V, Verduin J (2001) Direct measurements of volume transports through Fram Strait. Polar Res 20:217–224

Gales JA, Forwick M, Laberg JS, Vorren TO, Larter RD, Graham AGC, Baeten NJ, Amundsen HB (2013a) Arctic and antarctic submarine gullies – a comparison of high latitude continental margins. Geomorphology 201:449–461

Gales JA, Larter RD, Mitchell NC, Dowdeswell JA (2013b) Geomorphic signature of antarctic submarine gullies: implications for continental slope processes. Mar Geol 337:112–124

Gascard JC, Richez C, Rouault C (1995) New insights on large-scale oceanography in Fram Strait: the West Spitsbergen Current. In: Smith W, Grebmeier J (eds) Arctic oceanography: marginal ice zones and continental shelves. Coastal Estuarine Studies, AGU, Washington, DC, pp 131–182

Haflidason H, Sejrup HP, Nygård A, Mienert J, Bryn P, Lien R, Forsberg CF, Berg K, Masson D (2004) The Storegga Slide: architecture, geometry and slide development. Mar Geol 213:201–234

Harders R, Kutterolf S, Hensen C, Moerz T, Brueckmann W (2010) Tephra layers: a controlling factor on submarine translational sliding? Geochem Geophys Geosyst. doi:10.1029/2009GC002844

Jakobsson M et al (2012) The International Bathymetric Chart of the Arctic Ocean (IBCAO) Version 3.0. Geophys Res Lett. doi:10.1029/2012GL052219

Kaminski M, Niessen F, PS87 Shipboard Geoscience Party (2015) Agglutinated foraminifera from the Hovgård Ridge, Fram Strait, west of Spitsbergen: evidence for a deep bottom current. J Geol Soc Pol 85:309–320

Kristoffersen Y, Coakley B, Jokat W, Edwards M, Brekke H, Gjengedal J (2004) Seadbed erosion on the Lomonosov Ridge, central Arctic Ocean: a tale of deep draft icebergs in the Eurasia Basin and the influence of Atlantic water inflow on iceberg motion. Paleoceanography. doi:10.1029/2003PA000985

Kristoffersen Y, Coakley BJ, Hall JK, Edwards M (2007) Mass wasting on the submarine Lomonosov Ridge, central Arctic Ocean. Mar Geol 243:132–142

Kvalstad TJ, Andresen L, Forsberg CF, Berg K, Bryn P, Wangen M (2005) The Storegga slide: evaluation of triggering sources and slide mechanisms. Mar Pet Geol 22:245–256

Laberg JS, Camerlenghi A (2008) The significance of contourites for submarine slope stability. In: Rebesco M, Camerlenghi A (eds) Developments in sedimentology 60: contourites. Elsevier, Amsterdam, pp 537–556

Laberg JS, Vorren TO (1995) Late Weichselian submarine debris flow deposits on the Bear Island Trough Mouth Fan. Mar Geol 127:45–72

Laberg JS, Kawamura K, Amundsen H, Baeten N, Forwick M, Rydningen TA, Vorren TO (2014) A submarine landslide complex affecting the Jan Mayen Ridge, Norwegian-Greenland Sea: slide-scar morphology and processes of sediment evacuation. Geo-Mar Lett 34:51–58

Läderach C, Schlindwein V, Schenke HW, Jokat W (2011) Seismicity and active tectonic processes in the ultra-slow spreading Lena Trough, Arctic Ocean. Geophys J Int 184:1354–1370

Micallef A, Mountjoy JJ, Canals M, Lastras G (2012) Deep-seated bedrock landslides and submarine canyon evolution in an active tectonic margin: cook strait, New Zealand. In: Yamada Y et al (eds) Advances in natural and technological hazards research, vol 31. Springer, Dordrecht

Myhre AM, Eldholm O (1988) The Western Svalbard margin (74°–80°N). Mar Pet Geol 5:134–156

Myhre AM, Thiede J, Firth J (1993) Ocean Drilling Program Leg 151 Scientific Prospectus: North Atlantic Arctic Gateways. Ocean Drilling Program, College Station, 79 pp

Rise L, Bøe R, Riis F, Bellec VK, Laberg JS, Eidvin T, Elvenes S, Thorsnes T (2013) The Lofoten-Vesterålen continental margin, North Norway: canyons and mass-movement activity. Mar Pet Geol 45:134–149

Schauer U, Fahrbach E, Østerhus S, Rohardt G (2004) Arctic warming through the Fram Strait: oceanic heat transport from 3 years of measurements. J Geophys Res. doi:10.1029/2003JC001823

Shipboard Scientific Party (1995) Site 908. In: Myhre AM, Thiede T, Firth JV et al (eds) Proc. ODP init. repts, vol 151. Ocean Drilling Program, College Station, pp 113–158

Solheim A, Berg K, Forsberg CF, Bryn P (2005) The Storegga Slide complex: repetitive large scale sliding with similar cause and development. Mar Pet Geol 22:97–107

Vanneste M, Mienert J, Bünz S (2006) The Hinlopen Slide: a giant, submarine slope failure on the northern Svalbard margin, Arctic Ocean. Earth Planet Sci Lett 245:373–388

Vorren TO, Laberg JS, Blaume F, Dowdeswell JA, Kenyon NH, Mienert J, Rumohr J, Werner F (1998) The Norwegian-Greenland Sea continental margins: morphology and Late Quaternary sedimentary processes and environment. Quat Sci Rev 17:273–302

Wessel P, Smith WHF (1998) New, improved version of generic mapping tools realeased. EOS Trans Am Geophys Union. doi:10.1029/98EO00426

Winkelmann D, Stein R (2007) Triggering of the Hinlopen/Yermak Megaslide in relation to paleoceanography and climate history of the continental margin north of Spitsbergen. Geochem Geophys Geosyst. doi:10.1029/2006GC001485

Chapter 26
3D Seismic Investigations of Pleistocene Mass Transport Deposits and Glacigenic Debris Flows on the North Sea Fan, NE Atlantic Margin

Berit Oline Hjelstuen and Siv Grinde

Abstract Mass movements are a frequent sedimentary process in the marine realm, affecting both glaciated and non-glaciated continental margins. Here a 3D seismic data set from the North Sea Fan, NE Atlantic margin, is used to study internal architecture, external geometry and surface geomorphology of different types of buried sediment transport. We identify three mass transport deposits, at a depth of ca 100–1000 m below seabed, corresponding to the previously mapped Tampen (~130 ka) and Møre (~300 ka) slides and the Stad (~400 ka) Slide, identified in this study. These slides all eroded underlying sedimentary units and their surfaces include curvilinear ridges up to 20 m high and 10–15 km long. Locally, the slide surfaces also include rafted slide blocks, up to 200 m wide, 300 m long and 30 m high. We relate the curvilinear ridges and the slide blocks to submarine spreading. Intervals of glacigenic debris flows are identified between the mass transport deposits. These lens-shaped bodies are seen in plan to be flows that widen downslope and which are fed from 10 to 20 m deep, 50–200 m wide and >2 km long transport channels that extend from the paleo-shelf edge. Analyses of time slices suggest that such flows may have been operating on the North Sea Fan for the last ~1.1 million years.

26.1 Introduction

During the Quaternary time period the NE Atlantic continental shelves have been dominated by a glacial regime, and throughout the last 0.5 million years the Fennoscandian and British-Irish ice sheets have extended repeatedly to the shelf

B.O. Hjelstuen (✉)
Department of Earth Science, University of Bergen, Bergen, Norway
e-mail: berit.hjelstuen@uib.no

S. Grinde
Department of Earth Science, University of Bergen, Bergen, Norway

Statoil ASA, Bergen, Norway

G. Lamarche et al. (eds.), *Submarine Mass Movements and their Consequences*,
Advances in Natural and Technological Hazards Research 41,
DOI 10.1007/978-3-319-20979-1_26

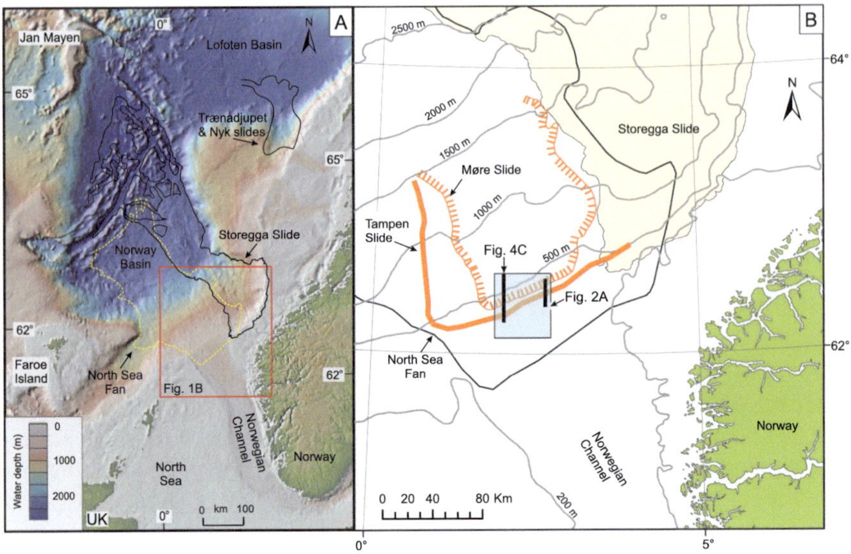

Fig. 26.1 (**a**)Bathymetric map of the Norwegian continental margin. Outline of North Sea Fan from Nygård et al. (2005). Outline of Nyk and Trænadjupet slides from Laberg et al. (2002). (**b**) Location of 3D-cube (*blue box*) analyzed in this study. Outline of North Sea Fan, Møre and Tampen slides from Nygård et al. (2005)

edge (e.g. Sejrup et al. 2005). During these glacial periods huge trough mouth fans were locally constructed on the upper continental slopes. These fans are dominantly composed of mass transport deposits (MTDs) and glacigenic debris flows (GDFs) (e.g. Vorren et al. 1998). Here we focus on the Quaternary North Sea Fan, deposited in front of the Norwegian Channel (Fig. 26.1a). Due to numerous studies over the last couple of decades the overall development of this fan complex is rather well known (e.g. King et al. 1998; Nygård et al. 2005). However, previous investigations mainly relied upon analyses of 2D seismic profiles and seabed data. In this study, access to a 3D seismic cube, from the uppermost North Sea Fan (Fig. 26.1), affords new insights into the geometries, internal architecture and flow mechanisms of MTDs and GDFs.

26.2 Geological Background

The North Sea Fan reaches a maximum thickness of about 2 km near the shelf edge, covers an area of about 110×10^3 km^2 and extends for a distance of nearly 500 km, from the shelf edge into the central part of the Norway Basin (Fig. 26.1a). The study area has been impacted by shelf-edge glaciations during Marine Isotope Stages (MIS) 12, 10, 8, 6 and 2, inferred to have involved an active ice stream in the Norwegian Channel (Fig. 26.1) (e.g. Sejrup et al. 2005). Previous studies have shown that the North Sea Fan is dissected by submarine slides (Fig. 26.1b), resulting in deposition of thick MTDs. The previously mapped Møre and Tampen

MTDs are assumed to be of MIS 9 and MIS 6 age, respectively (Nygård et al. 2005). During glacial maximums GDFs were fed from deforming till beneath the Norwegian Channel Ice Stream (NCIS), building the North Sea Fan. The individual GDFs are 2–40 km wide and up to 60 m thick, and can be followed for distances of >100 km down-slope (King et al. 1998; Nygård et al. 2002).

26.3 Data and Methods

The 3D seismic cube (Fig. 26.1b), analyzed in this study, covers an area of 1540 km^2 of the upper North Sea Fan and has been made available by Petroleum Geo-Services. The vertical and horizontal resolution for the upper 2.2 s of the 3D cube, i.e. the time interval in focus in the study, is ~6 m (Grinde 2012). The seismic interpretation software Petrel (v 2010), from Schlumberger AS, is used to analyze the seismic data. In addition to interpretation of sequence boundaries and units in vertical profiles (Fig. 26.2) we have generated surfaces of interpreted horizons, analyzed horizontal time slices and performed RMS attribute analyses (Figs. 26.3 and 26.4). A sediment velocity of 1800 m/s (Grinde 2012) is utilized in order to estimate vertical dimensions of mapped seismic features.

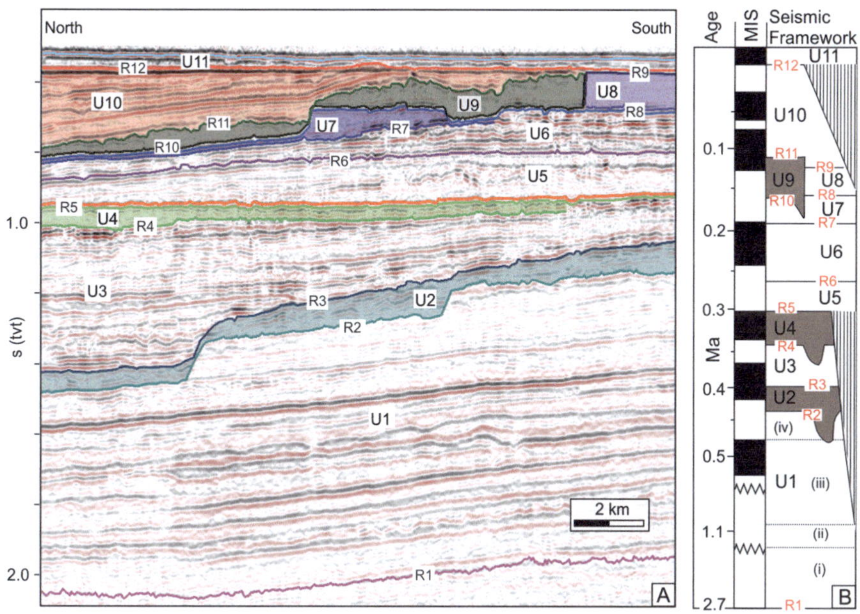

Fig. 26.2 (**a**)Seismic example, location in Fig. 26.1b, showing identified sequence boundaries (R1–R12), seismic units (U1–U11) and seismic facies characteristics in the study area. (**b**)Seismo- and chronostratigraphic framework for the Quaternary North Sea Fan. MIS:Marine Isotope Stage; U1–U11: Identified seismic units; R1–R12: Identified seismic sequence boundaries. Units with a *gray* colour represent MTDs, whereas hatched regions represent erosional unconformities

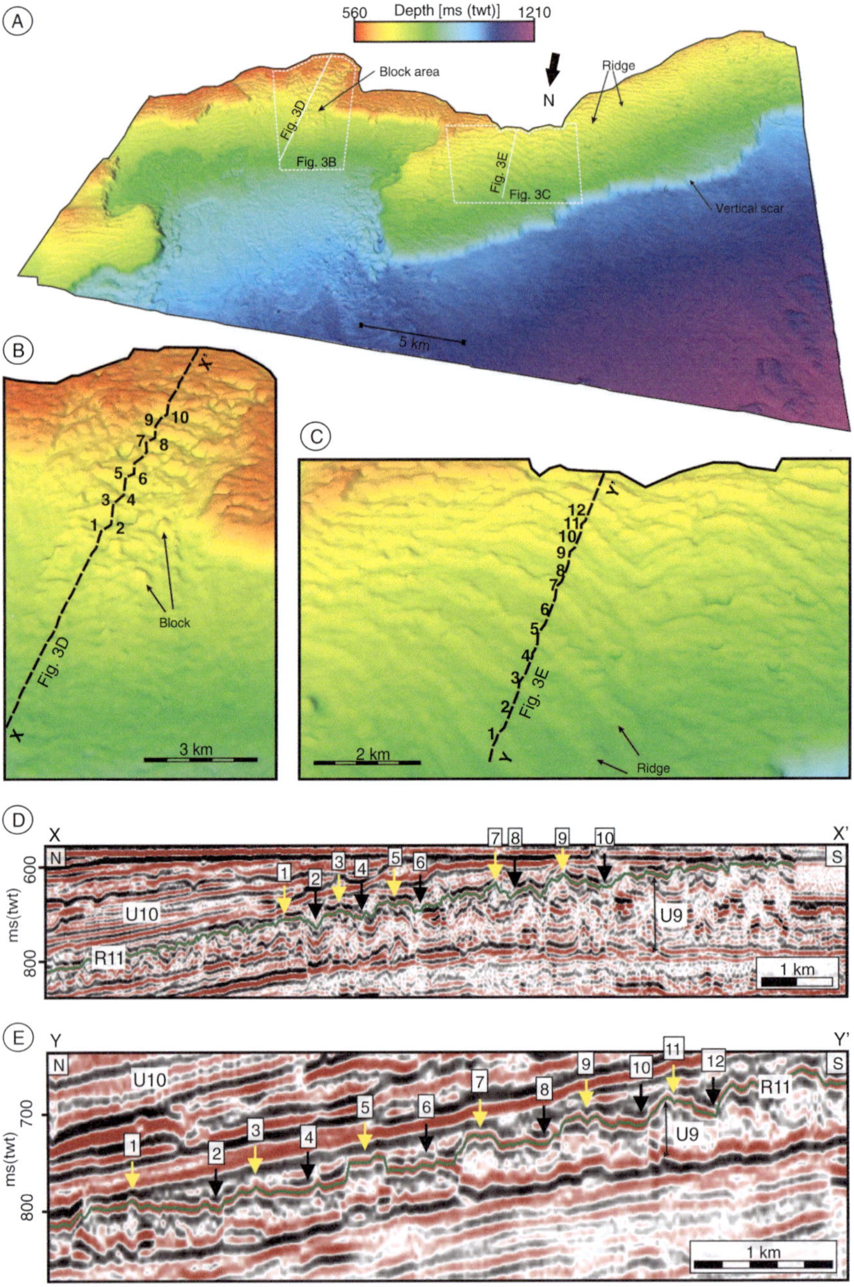

Fig. 26.3 (**a**)Generated surface of sequence boundary R11, i.e. the top of the Tampen MTD (Unit U9). (**b**) Block-like features observed on the R11 surface. The identified *blocks* are numbered 1–10. Figure location in (**a**). (**c**) Ridge-like features, observed at R11 surface. The identified curvilinear features are numbered 1–12. Figure location in (**a**). (**d**) Seismic profile crossing the block structures in (**b**). The annotated numbers refer to the numbered (1–10) block-like features in (**b**). R11: interpreted sequence boundary; U9 and U10: identified seismic units. (**e**) Seismic profile crossing the ridge structures in (**c**). The annotated numbers refer to the numbered (1–12) curvilinear ridges in (**c**). R11: interpreted sequence boundary; U9 and U10: identified seismic units

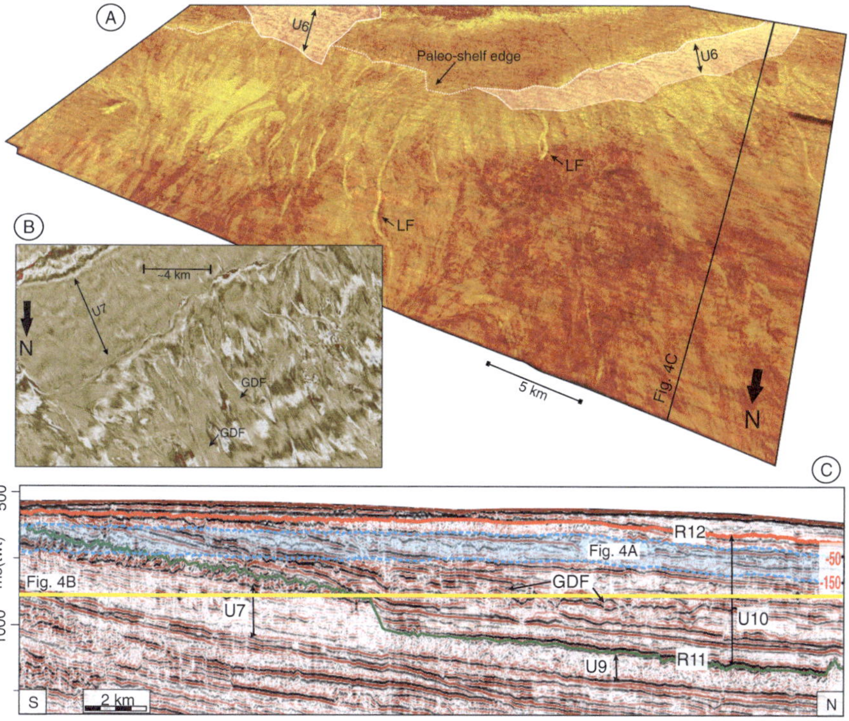

Fig. 26.4 (**a**) RMS attribute surface of a 100 ms time volume (location in **c**) in Unit U10. Linear furrows (*LF*) represent transport channels for GDFs. U6: Identified seismic units. (**b**)Time slice at 900 ms (twt) below the sea level (location in **c**), showing GDFs within unit U10, seaward of unit U7. (**c**) Seismic profile (location in **a**) showing typical seismic facies in the study area. Location of RMS attribute volume, shown in (**a**), and time slice, shown in (**b**), are indicated. R11 and R12: Identified sequence boundaries; U7, U9 and U10: Identified seismic units

26.4 Results

The Quaternary succession of the North Sea Fan sedimentary succession is divided into 11 units, U1 (oldest) – U11, by the identification of 12 sequence boundaries, R1 (oldest)–R12 (Fig. 26.2). The oldest unit has been divided into four sub-units, U1 (i)–UI(iv). To date the sequence boundaries and the units we adapt the chronostratigraphy established for the North Sea Fan by Nygård et al. (2005). Sequence boundary R1 represents the initiation of Northern Hemisphere glaciation at 2.74 Ma, while R2 marks the change to repetitive shelf edge glaciations. Three of the identified units, U2, U4 and U9 (Figs. 26.2 and 26.3d), display a chaotic internal seismic facies. Unit U11 is dominated by an acoustic laminated seismic facies. The oldest unit, U1, together with units U3, U5–U8 and U10 (Figs. 26.3d, e and 26.4b), all show a subparallel acoustic laminated facies which also include wavy reflections (lens-shaped facies).

Sequence boundaries R3, R5 and R11 (Fig. 26.2) all represent the top of chaotic sediment packages and for each of these surfaces have been generated. The three surfaces all display a comparable assemblage of morphological features including curvilinear ridges, isolated blocks and vertical scars (R11 shown in Fig. 26.3a). The slope-parallel ridges are up to 20–30 m high, 200–300 m wide and can be followed continuously for 5–10 km (Fig. 26.3c, e). The isolated blocks (Fig. 26.3b, d) show maximum lengths, widths and heights of about 200 m, 300 m and 50 m, respectively. The vertical scars (Fig. 26.3a) are in places more than 125 m in height and can be followed for distances of up to 50 km, i.e. across the entire 3D-cube.

Analyses of time slices across units U1, U3, U5–U8 and U10 show that the lens-shaped intervals observed in vertical profiles correspond to linear flow-like bodies that widen down-slope (Fig. 26.4b). The flows reach a maximum width of 500–700 m within the study area. RMS attribute analyses within unit U10, furthermore, show that the flow-like features, near the paleo-shelf edge, are associated with narrow curvilinear furrows, 10–20 m deep and 50–200 m wide, that can be followed for more than 2 km (Fig. 26.4a).

26.5 Discussion

Acoustically chaotic seismic units, often observed in association with vertical scars, are commonly interpreted to represent MTDs (e.g. Solheim et al. 2005). We observe three units, up to 100–150 m in thickness, that we interpret to be MTDs. The two youngest units correlate to the previously identified Tampen (U9 in Fig. 26.2) and Møre (U4 in Fig. 26.2) MTDs, whereas the oldest (U2 in Fig. 26.2) is deposited during a mass failure, not previously mapped, which we name the Stad Slide. We suggest that the Stad Slide occurred around 0.4 Ma (Fig. 26.2). The sediments set into motion during the repetitive failure events were transported down-slope, creating curvilinear ridges (Fig. 26.3c). Similar ridge-like features have been observed in bathymetric records and in side scan sonar data from the nearby seabed-exposed Storegga, Trænadjupet and Nyk slides (Fig. 26.1a) (e.g. Canals et al. 2004; Lindberg et al. 2004). We note that a 3D seismic study of the Tampen MTD (Gaferia et al. 2010), at the North Sea Fan-Storegga Slide boundary in a water depth of 1300–1800 m, indicate that the ridge-like nature of the MTD are present across the entire length of the slide.

We relate the curvilinear ridges to submarine spreading (Kvalstad et al. 2005; Micallef et al. 2007). Submarine spreading is controlled by factors such as pore pressure and friction causing extension in the mobilized sediments forming the characteristic curvilinear ridges with associated troughs. It has been inferred that spreading can occur over large areas, an observation also supported by our study. For the Tampen MTD we now connect spreading features at the uppermost slope to those observed by Gaferia et al. (2010) in a water depth of >1300 m.

We infer the isolated block structures (Fig. 26.3b), observed within the MTDs, to be rafted slide blocks related to submarine spreading. In the Møre, Tampen and

Stad slides it seems that the slide blocks were set in motion after the main failure events, modifying the slide scars (Fig. 26.3a). Slide blocks have commonly been reported from seabed-exposed slides. In the ~3000 km^3 Storegga Slide (Fig. 26.1), blocks with dimensions of 1.8 0.2 km and a height of 70 m are observed at seabed, whereas in the 0.2 km^2 Afen Slide, in the Faeroe-Shetland Channel, slide blocks with dimensions of up to 400 400 m and a height of about 6 m are identified (Canals et al. 2004). Thus, the dimensions of the slide blocks may depend on the size of the slide.

Previous studies from the North Sea Fan (e.g. Nygård et al. 2002) have shown that the observed lens-shaped bodies represent GDFs. Time slices and RMS attribute analyses now document that the GDFs seem to be associated to narrow, and deep, transport channels upslope (Fig. 26.4a, b). GDFs are assumed related to ice stream activity (King et al. 1998), which is inferred to have been initiated in the Norwegian Channel at ~1.1 Ma (Sejrup et al. 2000). Our time slice analyses show that the first occurrence of GDFs on the North Sea Fan is observed in sub-unit U1 (ii) (Fig. 26.2), allowing us to suggest that they are associated with the ~1.1 Ma NCIS advance to the shelf edge. It has previously not been observed GDFs on the North Sea Fan during this shelf edge ice advance.

26.6 Conclusions

A 3D seismic cube from the North Sea Fan, NE Atlantic margin, provides new insight into the geometries, internal architecture and flow mechanisms of buried Pleistocene MTDs and GDFs. The main conclusions are:

- Three large-scale failure events, the Tampen (~130 ka), Møre (~300 ka) and Stad (~400 ka), are mapped in the study area. The Stad failure event is identified for the first time in this study.
- The sediment mobilized during these slide failures did not fully disintegrate but moved down-slope in a ridge-like pattern, with sediment blocks in certain areas. We relate the formation of these features to submarine spreading.
- The GDFs have been funneled to the North Sea Fan through narrow transport channels on the uppermost continental slope. These channels are 10–20 m deep, 50–200 m wide and >2 km in length.
- Sediment transport on the North Sea Fan, related to GDFs, has taken place since ~1.1 Ma. i.e. since the first occurrence of an ice stream in the Norwegian Channel.

Acknowledgements We acknowledge Petroleum Geo-Services for access to the 3D seismic data set. We thank Dr Nicole Baeten and Dr Daniel Praeg for reviewing the paper.

References

Canals M, Lastras G, Urgeles R, Casamor JL, Mienert J, Cattaneo A, De Batist M, Haflidason H, Imbo Y, Laberg JS, Locat J, Long D, Longva O, Masson DG, Sultan N, Trincardi F, Bryn P (2004) Slope failure dynamics and impacts from seafloor and shallow sub-seafloor geophysical data: case studies from the COSTA project. Mar Geol 213:9–72

Gaferia J, Long D, Sutton R, Evans D (2010) 3D seismic investigation of the internal structure within the Tampen Slide on the North Sea Fan: are chaotic deposits that chaotic? J Geol Soc 167:605–616

Grinde S (2012) Studie av glasiale sedimentære prosesser på Nordsjøviften med hjelp av 3D-seismiske data. Unpub MSc thesis, Dep of Earth Science, University of Bergen, 200 pp

King EL, Haflidason H, Sejrup HP, Løvlie R (1998) Glacigenic debris flows on the North Sea trough mouth fan during ice stream maxima. Mar Geol 152:217–246

Kvalstad TJ, Andersen L, Forsberg CF, Berg K, Bryn P, Wangen M (2005) The Storegga Slide; evaluation of trigger sources and slide mechanics. Mar Pet Geol 22:245–256

Laberg JS, Vorren TO, Mienert J, Bryn P, Lien R (2002) The Trænadjupet Slide: a large slope failure affecting the continental margin of Norway 4,000 years ago. Geo-Mar Lett 22:19–24

Lindberg B, Laberg JS, Vorren TO (2004) The Nyk Slide – morphology, progression, and age of a partly buried submarine slide offshore northern Norway. Mar Geol 213:277–289

Micallef A, Masson DG, Berndt C, Stow DAW (2007) Morphology and mechanisms of submarine spreading: a case study from the Storegga Slide. J Geophys Res 112(21):F03023. doi:10.1029/2006JF000739

Nygård A, Sejrup HP, Haflidason H, King EL (2002) Geometry and genesis of glacigenic debris flows on the North Sea Fan: TOBI imagery and deep-tow boomer evidence. Mar Geol 188:15–33

Nygård A, Sejrup HP, Haflidason H, Bryn P (2005) The glacial North Sea Fan, southern Norwegian Margin: architecture and evolution from the continental slope to the deep-sea basin. Mar Pet Geol 22:71–84

Sejrup HP, Aarseth I, Haflidason H, Løvlie R, Bratten Å, Tjøstheim G, Forsberg CF, Ellingsen KI (2000) Quaternary of the Norwegian Channel: glaciation history and palaeoceanography. Nor J Geol 75:65–87

Sejrup HP, Hjelstuen BO, Dahlgren KIT, Haflidason H, Kuijpers A, Nygård A, Praeg D, Stoker MS, Vorren TO (2005) Pleistocene glacial history of the NW European continental margin. Mar Pet Geol 22:1111–1129

Solheim A, Berg K, Forsberg CF, Bryn P (2005) The Storegga Slide complex: repetitive large scale sliding with similar cause and development. Mar Pet Geol 22:97–107

Vorren TO, Laberg JS, Blaume F, Dowdeswell JA, Kenyon NH, Mienert J, Rumohr J, Werner F (1998) The Norwegian-Greenland Sea continental margins: morphology and late Quaternary sedimentary processes and environment. Q Sci Rev 17:273–302

Chapter 27
Do Embedded Volcanoclastic Layers Serve as Potential Glide Planes?: An Integrated Analysis from the Gela Basin Offshore Southern Sicily

Jannis Kuhlmann, Katrin Huhn, and Matt J. Ikari

Abstract The NE portion of the Gela Basin (Strait of Sicily) shows evidence of multiple mass wasting events of predominantly translational character. In this context, recent investigations implicate volcanoclastic layers as key stratigraphic surfaces acting as preferential planes of failure. We present an integrated analysis of a representative sedimentary transition from overlying homogeneous background sedimentation of silty clay to a volcanoclastic layer. A high-resolution CT scan and three drained direct-shear laboratory experiments from a 20 cm whole-round section (~28.2 mbsf) allow the delineation of the role of this volcanoclastic layer in the framework of slope stability and failure initiation. The mechanical results indicate a general strengthening of the material with increased volcanoclastic content. Tendency for failure is expected to be highest within the silty clay due to relatively lower shear strength and strain-weakening behaviour, which promotes progressive sediment failure. In contrast with recent findings, this suggests that volcanoclastic sediment would not act as a weak layer. However, the volcanoclastic layer exhibits significant mesoporosity (i.e., fracturing) and may therefore host large volumes of fluid. Temporarily undrained conditions, for example during seismic activity, could transiently elevate fluid pressures and thus reduce the material shear strength below that of the surrounding silty clay. Such a weak layer may preferentially form along the interface of fractured volcanoclastic material and relatively impermeable silty clay, where differences in material strengths are lowest.

J. Kuhlmann (✉) • K. Huhn • M.J. Ikari
MARUM – Center for Marine Environmental Sciences, Universität of Bremen, Leobener Str., 28359 Bremen, Germany
e-mail: jkuhlmann@marum.de

G. Lamarche et al. (eds.), *Submarine Mass Movements and their Consequences*, Advances in Natural and Technological Hazards Research 41,
DOI 10.1007/978-3-319-20979-1_27

27.1 Introduction

The destabilization of continental slopes in form of bedding-parallel translational slides is frequently related to discrete *weak layers* and their material properties (e.g., Hampton et al. 1996; Masson et al. 2006). In this context, embedded volcanoclastic layers have been proposed to act as potential detachment planes promoting translational sliding, although the exact mechanisms remain speculative (e.g., Harders et al. 2010).

The NE region of the Gela Basin within the Strait of Sicily, Mediterranean Sea, features a continental slope that is characterized by multiple failure events of predominantly translational character (Fig. 27.1; Minisini et al. 2007; Minisini and Trincardi 2009; Kuhlmann et al. 2014). Along with sequence-stratigraphic boundaries, sub-horizontal volcanoclastic layers have been hypothesized as key stratigraphic surfaces that act as preferential planes of failure in this homogeneous muddy shelf-edge setting (Minisini et al. 2007). Drilled core samples from one such layer provide an excellent opportunity to test this hypothesis by evaluating the internal structure and mechanical behaviour of the volcanoclastic material and neighbouring sediment.

This study presents new evidence from geochemically identified marine marker tephra Y-7 (Kuhlmann et al. 2015), recovered at site GeoB14403 from undisturbed sediments upslope of the *Twin Slide* complex in the Gela Basin (Fig. 27.1c). We introduce high resolution computer tomography (CT) scans as well as drained

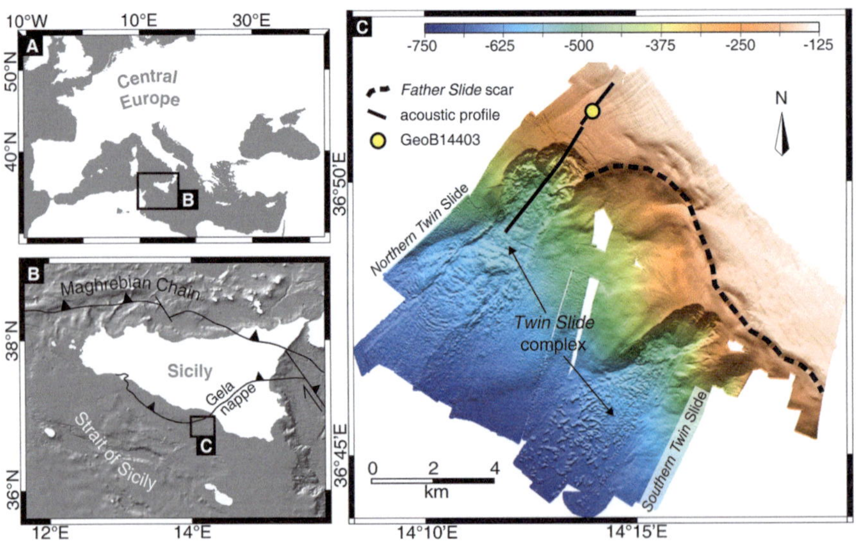

Fig. 27.1 (**a**) Overview of the study area in the Mediterranean Sea, (**b**) Location of the study area within the Strait of Sicily and major regional tectonic features, (**c**) Multibeam shaded relief image of the NE Gela Basin illustrating the position of drill site GeoB14403 and the acoustic profile over the *Twin Slide* complex presented in Fig. 27.2 (See Kuhlmann et al. 2014)

direct-shear laboratory tests from a 20 cm whole-round (WR) core sample (GeoB14403-8 5P-2 WR) from the lithologic transition from overlying homogeneous silty clay to volcanoclastic material at a depth of ~28.2 mbsf. Additional information is provided by petrophysical Multi-Sensor Core Logger (MSCL) and geochemical X-Ray Fluorescence (XRF) data on neighbouring split core samples.

27.2 Material and Methods

The principal data set for this study is based on a 54.6-m-long sediment succession at site GeoB14403, recovered with the Bremen seafloor drill rig MeBo (Freudenthal and Wefer 2007, 2013) during RV Maria S. Merian cruise MSM15/3 in 2010 (Fig. 27.2a; Kuhlmann et al. 2014, 2015). Acoustic data was obtained from a parametric sediment echosounder with dm-scale vertical resolution operating at 4 kHz (Atlas Parasound) and two bathymetric multibeam echosounder operating at 12 kHz (Kongsberg Simrad EM120) and 95 kHz (EM1002).

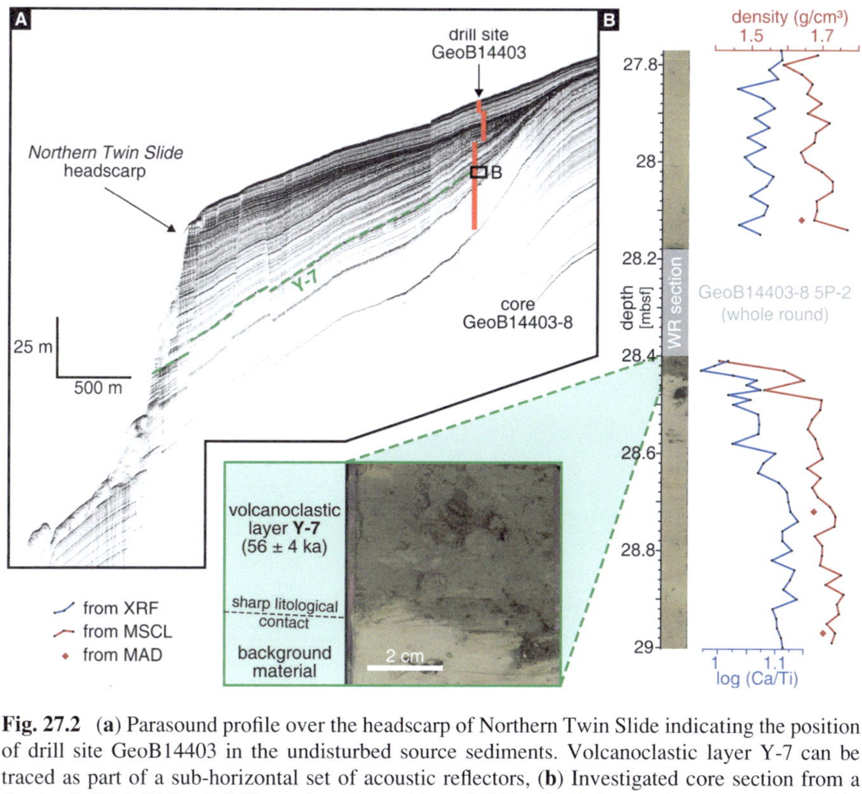

Fig. 27.2 (**a**) Parasound profile over the headscarp of Northern Twin Slide indicating the position of drill site GeoB14403 in the undisturbed source sediments. Volcanoclastic layer Y-7 can be traced as part of a sub-horizontal set of acoustic reflectors, (**b**) Investigated core section from a depth of ~27.8–29.0 mbsf illustrating the position of Y-7 and whole-round section GeoB14403-8 5P-2. Petrophysical and geochemical measurements indicate a significant drop in sediment density and a relative enrichment of Ti for the volcanoclastic layer compared to the background material

Non-destructive measurements of core-physical properties on split core sections surrounding the analysed WR section (Fig. 27.2b) were obtained with a GEOTEK Ltd. Multi-Sensor Core Logger (MSCL), while geochemical logging for light elements (Al to Fe) was performed with an Avaatech II core scanner using a generator setting of 10 kV, 0.2 mA and a sampling time of 20 s.

The 20 cm WR section GeoB14403-8 5P-2 was scanned by a Toshiba Aquilion 64 computer tomograph at the hospital Klinikum Bremen-Mitte with an exposure time of 35 s per rotation (~14 s per scan) at an X-ray source voltage of 120 kV and a tube current of 600 mA per rotation/210 mAs. Scan resolution was 0.35 mm in x–y direction and 0.5 mm in z direction and images were reconstructed at a 0.3 mm interval. The resulting X-ray attenuation data (radiodensity) was quantified according to the Hounsfield scale (HU = Hounsfield units (Hounsfield 1973)) in order to visualize local differences in sediment densities and to extract measurements of mesoporosity (i.e., fractures; Fig. 27.3).

Drained direct-shear experiments on three intact samples were performed on cylindrical WR samples of 56 mm diameter and ~25 mm height using a GIESA RS5

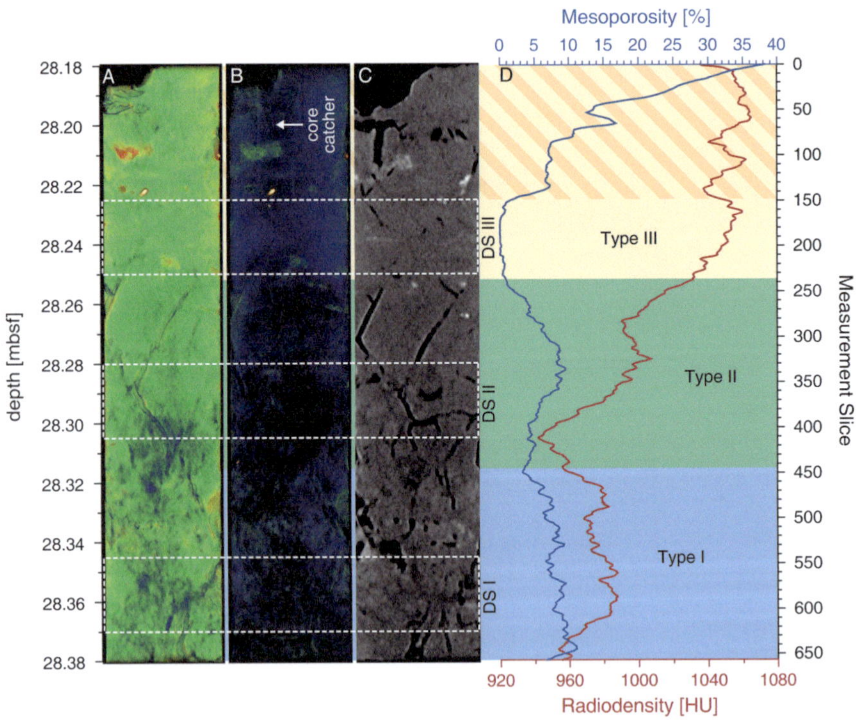

Fig. 27.3 Processed images and extracted information from CT data of GeoB14403-8 5P-2 WR, (**a**) Sliced volume image cut along depth-axis (*blue colours* indicate low relative density, *green colours* high relative density; *orange/red colours* show presence of very high density (i.e., shell fragments)), (**b**) Full volume image (*black/blue/green* = low/high/very high relative density), (**c**) Thin orthoslice revealing mesoporosity, (**d**) Radiodensity and fraction of mesoporosity

device (see Ikari and Kopf 2011 for details). The effective normal stress (σ_n') applied to the experiments was controlled at a constant 192 kPa to match the in situ effective vertical stress (σ_v') at ~28 mbsf, which was calculated from shipboard bulk density measurements (Ai et al. 2014). The shear displacement for each experiment was 12 mm, at a constant displacement rate of 0.008 mm/min to allow sample drainage and complete pore pressure dissipation (in accordance with DIN 18137-3 (Deutsches Institut für Normung 2002) and following plasticity indices presented by Ai et al. (2014)). During the test, shear stress (τ) as well as vertical and horizontal displacement were constantly measured at a frequency of 0.1 Hz.

In this study we also report an apparent coefficient of friction (μa) that is calculated from the effective normal stress and the measured shear stress, which includes both frictional and cohesive strengths:

$$\mu_a = \frac{\tau}{\sigma_n'} \tag{27.1}$$

27.3 Material Characteristics and Internal Structure

The investigated depth interval exhibits a homogeneous background sedimentary section with a dominant lithology of nannofossil silty clay (Fig. 27.2b) (Kuhlmann et al. 2014). Shipboard moisture and density (MAD) and MSCL density values average ~1.7 g/cm^3 and XRF scans reveal logarithmic Ca/Ti ratios well over the value of 1. At 28.45 mbsf the base of tephra layer Y-7 intercalates this succession, visually distinguishable by its relatively darker colour indicating increased presence of volcanoclastic material. The sharp lower contact of this layer is reflected in both physical and chemical properties, which show that density values and logarithmic Ca/Ti ratios drop significantly. The latter indicates a relative enrichment of Ti, a characteristic element in this volcanoclastic layer.

The WR section samples the upper part of Y-7 and its diffuse transition into the background material, as observed in the CT data (Fig. 27.3). Variability of radiodensities as well as internal fractures within this sample are observable by colour-coded reproductions of: (A) a sliced volume cut along the depth-axis, (B) the full volume data, and (C) a thin orthoslice (for information on the colour scheme please refer to the according figure legend). Additional information as to the mesoporosity (i.e., fractures) as well as the mean radiodensity within a single measurement slice (0.3 mm vertical resolution) is presented in Fig. 27.3d. Generally, the WR section may be subdivided into three sediment domains, each of which was sampled for subsequent direct-shear laboratory experiments as listed below (Fig. 27.3d):

DS I	Upper section of volcanoclastic layer Y-7 (low relative radiodensity – high degree of mesoporosity)
DS II	Transitional zone (upwards decreasing mesoporosity – respective rise in radiodensity)
DS III	Background material (highest relative radiodensity – virtually no mesoporosity)

The CT data confirms the negative density anomaly of the volcanoclastic layer compared to the background material and reveals a significantly higher degree of fracturing in this layer. Note that the mesoporosity values for the upper ~4 cm of the WR section are unreliable, as they are biased by the cylindrical void left by the core catcher (Fig. 27.3b).

27.4 Drained Shear Strength

The experiments performed at in situ effective normal stresses reveal markedly different stress-strain behaviour for the investigated materials (Fig. 27.4). For the background sediment (DS III) a pronounced peak in shear strength ($\tau_{peak} = 102.5$ kPa) followed by a decay toward a lower residual value ($\tau_{peak} = 69.0$ kPa) can be observed. In contrast, the volcanoclastic layer (DS I) maintains a relatively constant shear strength once a maximum of $\tau_{peak} = 135.0$ kPa is reached and only slightly decreases toward a steady-state value of $\tau_{peak} = 129.1$ kPa. The stress-strain curve of sample DS II exhibits a peak strength intermediate to DS I and DS III ($\tau_{peak} = 119.5$ kPa), but with a displacement-weakening trend similar to DS III. However, since this decay does not reach a steady value by the end of the experiment (~12 mm displacement), the residual value of $\tau_{peak} = 101.2$ kPA should be considered an upper estimate.

Correspondingly, the calculated apparent coefficients of friction (Eq. 27.1) are lowest in case of the background material (DS III, $\mu_{peak} = 0.53$ and and $\mu_{res} = 0.36$), reflecting the predominantly clayey nature of the sediment. With increasing volcanoclastic content these increase to a maximum of $\mu_{peak} = 0.7$ and $\mu_{res} = 0.67$ in sample DS I.

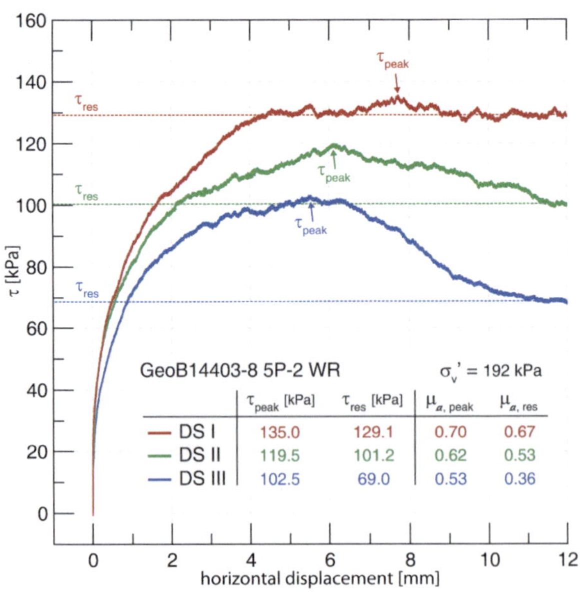

Fig. 27.4 Stress-strain diagram from drained direct shear experiments on samples from each sediment type of the WR section. Additionally, peak and residual values for shear strength and apparent coefficient of friction are reported. The data reveals a strain-softening behaviour of the background sediment (DS III) and significantly higher strength of the volcanoclastic layer (DS I)

27.5 Discussion and Conclusions

The investigated core interval, which samples the transition from homogeneous background sediments to marine marker tephra Y-7, reveals fundamental differences between the two end members in terms of internal structure and shear strength and suggests that failure would require two distinct failure mechanisms.

The background sediment displays the lowest peak shear strength of all three investigated samples under drained conditions and may be considered as the most susceptible to failure. It follows a stress-strain path indicating strain-softening behaviour typical for the reorientation of clay particles parallel to the direction of shear – a characteristic that favours the phenomenon of progressive failure and hence the development of failure surfaces once shear stresses locally reach the material peak strength (e.g., Leroueil 2001). This material behaviour is absent within the ash-bearing layer and appears to correlate with diminished volcanoclastic content.

While mechanical investigations provide evidence of preferential failure within the background material, mesoscale observations suggest a distinct failure mechanism that relates to the volcanoclastic layer, which works on much shorter timescales: excess pore pressure generation in response to external transient events. Existing fractures within the volcanoclastic layer may host significant volumes of fluid, which could generate transiently elevated pressures in response to temporarily undrained conditions. In order for the volcanoclastic layer to act as a failure plane it must become weaker than the surrounding silty clay. Assuming a constant apparent coefficient of friction, this would require a reduction in effective normal stress via an increase in pore water pressure of $\geq$45.6 kPa for domain DSI (Eq. 27.1). With diminished volcanoclastic content this value decreases to $\geq$26.7 kPa for domain DSII and approaches 0 near the volcanoclastic/silty clay boundary, suggesting preferential failure along this interface. This mechanism is similar to the weak layer generation through rearrangement of ash-particles as previously proposed by Harders et al. (2010), though it works on a different spatial scale.

Considering the high frequency of failure, we propose a seismic trigger (see Minisini et al. 2007; Ai et al. 2014) to repeatedly invoke the above introduced mechanism. We suggest that transiently increased fluid pressures due to earthquake shaking may promote instantaneous translational sliding along the interface between embedded volcanoclastic material and overlying regular sediments – thus impeding failure initiation within the background sediments, which are weaker over longer timescales. Evidence of volcanoclastic layers at the base of two drilled mass-transport deposits further supports this theory. In the framework of submarine slope stability in muddy shelf-edge settings, our results indicate that given the occurrence of suitable transient external events (i.e., earthquakes) at the site of investigation, slope failure depends on a competition between mechanical sediment properties and hydrologic effects via mesoscale structures such as fracturing, as frequently found in marine volcanoclastic layers.

Acknowledgments This work has been funded by the DFG through MARUM – Universität Bremen. We gratefully acknowledge the constructive reviews by Dr. Marzia Rovere and Dr. Rieka Harders and wish to particularly thank Dr. Gauvain Wiemer for his help with the shear box experiments as well as Dr. Jürgen Titschack for support with the CT data. MSCL and XRF core scanning were carried out at the lab facilities of MARUM.

References

Ai F, Kuhlmann J, Huhn K et al (2014) Submarine slope stability assessment of the central Mediterranean continental margin: the Gela Basin. In: Krastel S, Behrmann J-H, V€olker D et al (eds) Submarine mass movements and their consequences. Springer, Heidelberg, pp 225–236

Freudenthal T, Wefer G (2007) Scientific drilling with the sea floor drill rig MeBo. Sci Drill 5:63–66. doi:10.5194/sd-5-63-2007

Freudenthal T, Wefer G (2013) Drilling cores on the sea floor with the remote-controlled sea floor drilling rig MeBo. Geosci Instrum Methods Data Syst 2(2):329–337. doi:10.5194/gi-2-329-2013

Hampton MA, Lee HJ, Locat J (1996) Submarine landslides. Rev Geophys 34(1):33–59. doi:10.1029/96RG03287

Harders R, Kutterolf S, Hensen C (2010) Tephra layers: a controlling factor on submarine translational sliding? Geochem Geophys Geosyst 11(5). doi:10.1029/2009GC002844

Hounsfield GN (1973) Computerized transverse axial scanning (tomography). Part I. Description of system. Br J Radiol 46(552):1016–1022. doi:10.1259/0007-1285-46-552-1016

Ikari MJ, Kopf AJ (2011) Cohesive strength of clay-rich sediment. Geophys Res Lett 38(16). doi:10.1029/2011GL047918

Kuhlmann J, Asioli A, Strasser M (2014) Integrated stratigraphic and morphological investigation of the Twin Slide complex offshore southern Sicily. In: Krastel S, Behrmann J-H, V€olker D et al (eds) Submarine mass movements and their consequences. Springer, Heidelberg, pp 583–594

Kuhlmann J, Asioli A, Trincardi F et al (2015) Sedimentary response to Milankovitch-type climatic oscillations and formation of sediment undulations: evidence from a shallow-shelf setting at Gela Basin on the Sicilian continental margin. Quat Sci Rev 108:76–94. doi:10.1016/j.quascirev.2014.10.030

Leroueil S (2001) Natural slopes and cuts: movement and failure mechanisms. Geotechnique 51(3):197–243. doi:10.1680/geot.2001.51.3.197

Masson DG, Harbitz CB, Wynn RB et al (2006) Submarine landslides: processes, triggers and hazard prediction. Philos Trans R Soc Lond A: Math, Phys Eng Sci 364(1845):2009–2039. doi:10-1098/rsta.2006.1810

Minisini D, Trincardi F (2009) Frequent failure of the continental slope: the Gela Basin (Sicily Channel). J Geophys Res 114(F3). doi:10.1029/2008JF001037

Minisini D, Trincardi F, Asioli A et al (2007) Morphologic variability of exposed mass-transport deposits on the eastern slope of Gela Basin (Sicily channel). Basin Res 19(2):217–240. doi:10.1111/j.1365-2117.2007.00324.x

Deutsches Institut fu¨r Normung (2002) Soil, investigation and testing – determination of shear strength – part 3: direct shear test. Rep DIN 18137–3, Berlin

Chapter 28
Sediment Failure Affecting Muddy Contourites on the Continental Slope Offshore Northern Norway: Lessons Learned and Some Outstanding Issues

Jan Sverre Laberg, Nicole J. Baeten, Maarten Vanneste, Carl Fredrik Forsberg, Matthias Forwick, and Haflidi Haflidason

Abstract The formerly glaciated continental margin off Norway has experienced a relatively large number of submarine landslides of varying sizes, volumes, and ages originating from contourites deposited on the continental slope. We review: (i) the origin and occurrence of weak layers involved, (ii) sediment disintegration and initial flow, and (iii) sediment run-out and resulting deposits. The following major knowledge gaps, critical for further progress in this field are identified: information on lithology and sediment properties of material recovered from below the depth of conventional coring, and in situ measurements of sediment physical properties at the depth of weak layers.

28.1 Introduction

The continental slopes of formerly glaciated continental margins commonly include: (i) trough-mouth fans and prograding wedges comprising mainly glacigenic debris flows (GDFs), (ii) contourite drifts containing ice-rafted debris, and (iii) canyons and/or channel systems affected by turbidity currents. The clay-to-sand ratio (Elverhøi et al. 2010) and/or the gradient of the continental slope

J.S. Laberg (✉) • M. Forwick
Department of Geology, University of Tromsø – The Arctic University of Norway, 9037 Tromsø, Norway
e-mail: jan.laberg@uit.no

N.J. Baeten
Geological Survey of Norway, 7491 Trondheim, Norway

M. Vanneste • C.F. Forsberg
Norwegian Geotechnical Institute & International Centre for Geohazards (ICG), 0806 Oslo, Norway

H. Haflidason
Department of Earth Science, University of Bergen, 5020 Bergen, Norway

G. Lamarche et al. (eds.), *Submarine Mass Movements and their Consequences*, Advances in Natural and Technological Hazards Research 41,
DOI 10.1007/978-3-319-20979-1_28

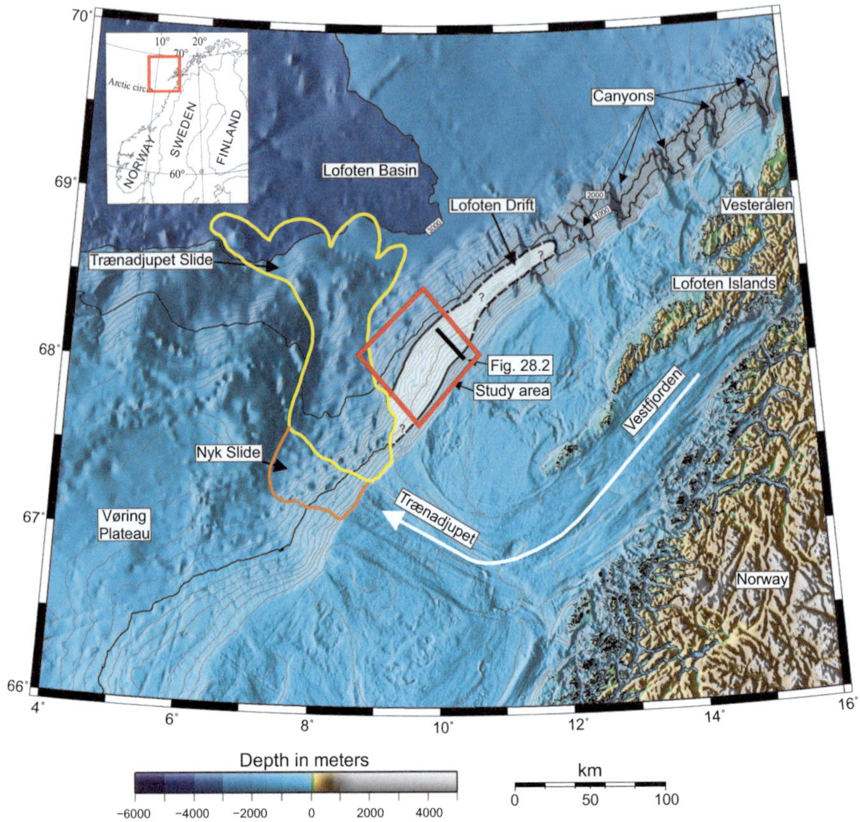

Fig. 28.1 Bathymetric map showing the location of the study area, the outline of the Lofoten Drift (*white shading*), the Trænadjupet Slide (*yellow outline*), and part of the upper area of the Nyk Slide (*orange outline*, most of this event was later remobilized by the Trænadjupet Slide). The location of Fig. 28.2 is indicated. The index map locates the map offshore Norway. The bathymetry is from Jakobsson et al. (2012) displayed by the GMT software (Wessel and Smith 1998)

(Piper and Normark 2009) exerts a fundamental control on sediment flow dynamics in submarine failures derived from such systems. Examples from offshore Norway include the giant Storegga Slide Complex (Bryn et al. 2005), as well as a number of buried larger (Laberg et al. 2001; Rise et al. 2006) and smaller slides (Baeten et al. 2013; L'Heureux et al. 2013). In this review our main focus lies on the Lofoten area (Fig. 28.1), where we conducted a multi-disciplinary approach to study the smaller submarine landslides within the Lofoten Contourite Drift during the LOSLOPE project. LOSLOPE built on previous projects that have focused on the larger-sized submarine slides offshore Norway (Haflidason et al. 2004; Bryn et al. 2005; Vanneste et al. 2006; Hjelstuen et al. 2007).

Swath bathymetry, side-scan sonar, high-resolution seismic data, geological and geotechnical laboratory analyses from a giant piston-core sample complemented with numerical modelling provided the data base for the LOSLOPE project (Baeten et al. 2013, 2014; Laberg et al. 2013; Vanneste et al. 2012). Here, we review and elaborate on some of the key results from the project, including: (i) the occurrence and origin of weak layers (the definition of a weak layer is after Locat et al. 2014), (ii) sediment disintegration and initial flow and (iii) sediment run-out and resulting deposits. Remaining knowledge gaps are highlighted subsequently.

28.2 Background

Exploration of oil and gas is now moving into the high north and the deep-water areas (e.g. Rise et al. 2013). Therefore, a proper evaluation of the stability of the continental slope succession of these areas is of paramount importance, including the origin and nature of submarine landslides in the past. This review focuses on new results from offshore northern Norway where the up to 360 m-thick, about 150 km-long (along drift axis) Lofoten Contourite Drift, a mounded, giant, elongated drift (sensu Faugères et al. 1999) is located. Drift formation has taken place since the mid-Miocene (Laberg et al. 1999; Rise et al. 2013). A layered, continuous, parallel to slightly divergent seismic signature of medium amplitude reflections characterizes the drift (Fig. 28.2). As the drift deposit comprises up to 80–90 % mud, this review concerns the composition and properties of sediments involved in muddy submarine landslides.

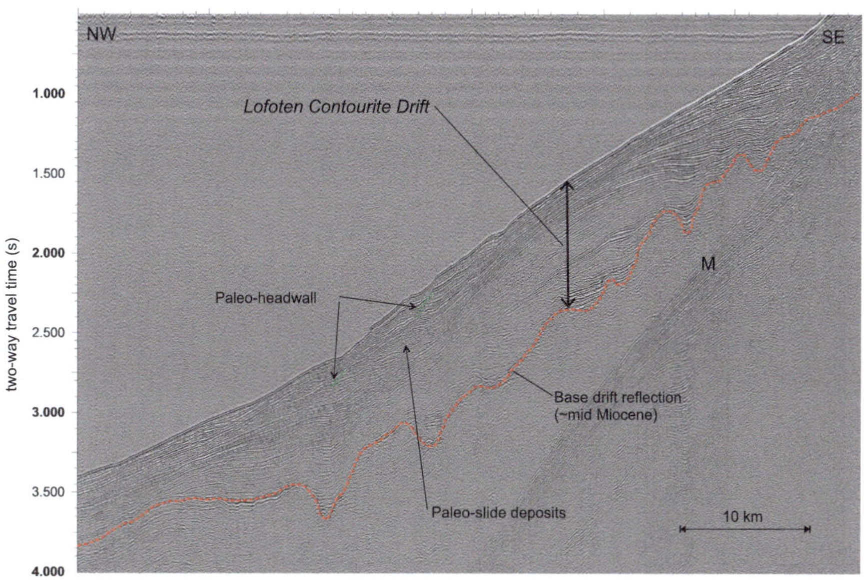

Fig. 28.2 *Above*: Seismic profile across the Lofoten Contourite Drift. Buried areas of sediment disturbance (paleo-headwall, paleo slide deposits) are seen within a mostly undisturbed part of the drift. The base drift reflection has been suggested to be of ~mid-Miocene age (Laberg et al. 1999). *M* multiple. See Fig. 28.1 for location

28.3 Lessons Learned and Some Outstanding Issues

1. *Contouritic sediments: importance for failure at high-latitude margins*
   Investigations of the Storegga (Bryn et al. 2005) and Trænadjupet slides (Laberg et al. 2003) revealed that basal glide planes were located within contouritic deposits (see Laberg and Camerlenghi (2008) for a review). Geotechnical analyses concluded that the contouritic sediments were weaker (contractive, stain softening behavior) compared to the GDF deposits (dilatant behavior) Therefore, contouritic deposits were suggested to be the weak layer that originally failed (see also Vanneste et al. (2014)).

   Later studies of the Lofoten Contourite Drift have shown that various styles of failures affect drift deposits, paleo-slides (buried) as well as more recent events (Figs. 28.2 and 28.3). The basal glide planes are parallel to the original layering of the contourites, implying that lithology and/or physical properties control the initial failure, subsequent sediment disintegration and flow. Relatively recent failures in restricted geographical areas may utilize glide planes from ~13 to ~100 m below the sea-floor. Whereas some slides had their glide plane mainly on one level, others involved multiple glide planes (Baeten et al. 2013). The reasons for these differences remain unknown.
2. *Contouritic sediments: their lithology, origin and distribution*
   Baeten et al. (2014) identified the following lithofacies in a sediment core from the Lofoten Contourite Drift: homogeneous mud, bioturbated mud, bioturbated sandy mud, laminated mud, laminated sandy mud and laminated gravelly, sandy

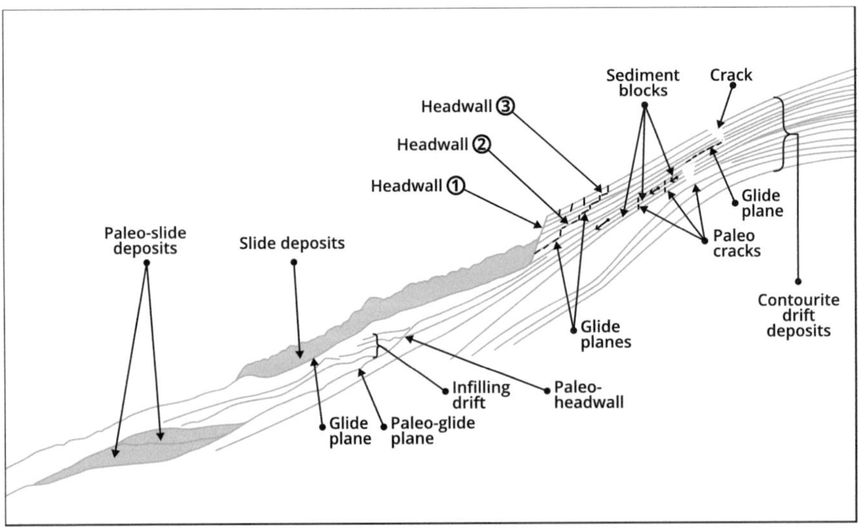

Fig. 28.3 Line drawing summarizing features diagnostic of mass wasting affecting the Lofoten contourite drift on the continental slope offshore Norway

mud. Based on sediment structures, as well as geotechnical testing they inferred that deformation occurred in a laminated interval at the base of a sediment block.

Lamination is often the result of variations in grain size (faint clay – silty clay alternations) occasionally exclusively visible on X-rays. However, in other cases it is more clearly expressed through clay – sandy mud/sand lamina. At least two hypothesis have been proposed for the formation of lamination within the continental slope deposits: (1) meltwater outbursts from ice sheets at or near the shelf break, i.e. "plumites" (Hesse et al. 1997; Lekens et al. 2005), and (2) fluctuations in ocean-current flow velocity affecting contourite drift deposits (Martin-Chivelet et al. 2008). Differentiating between these two processes remains challenging, but larger variation in grain size are probably related to plumites as these vary more distinctly and rapidly than ocean currents. Also, plumite intervals may have a sharp lower boundary, defined by a colour change while contourites may have more gradual lower boundaries.

Plumites are deposited beyond the termini of warm-based ice streams releasing sediment-laden melt-water into the ocean. These deposits have a local or regional lateral distribution controlled by the nature and location of the meltwater discharge and may also be influenced by ocean currents (Lekens et al. 2005). On the other hand, laminated contouritic sediments related to ocean current erosion, transportation and deposition may be deposited over larger parts of continental slopes at any latitude. So far, the level of sedimentological information from high-latitude continental slopes is restricted to the upper sediment succession reached by conventional coring. Offshore Norway, one of the best studied high-latitude continental margins, very little information exists about sediment composition and physical properties deeper than 15 m below the sea-floor in areas of sediment failures. This also includes the areas of the Storegga and Trænadjupet slides.

3. *Weak layers and/or weak zones*?

Due to their stress-strain behaviour, contouritic sediments may act as weak layers in which the initial deformation occurs and glide plains develop (Fig. 28.4a). But is this the only way to form glide planes in this type of deposits? Alternatively, could it also be that in settings where uniform contouritic sediments are interbedded with geotechnically stronger glacigenic sediments (GDFs), it is at the zone of rapid in-crease (or decrease) in strength, i.e. at the contourite – glacigenic sediment interface (Fig. 28.4b, c) where the glide plane develops (analogous to an icy layer acting as a glide plane of a snow avalanche)? If correct, it may not only be the properties of a specific layer but also the magnitude of property change and the interval over which it occurs that is of importance. As shown by Baeten et al. (2014), the thickness of layers of glacigenic sediments within a contouritic succession may be below seismic resolution and thus difficult to detect without access to sediment samples. Also, how do weak layer sediments initially deform, due to liquefaction or is it related to their clay content as the frictional resistance of clay is low compared to coarser sediments (Leroueil 2001; Wiemer and Kopf 2015)?

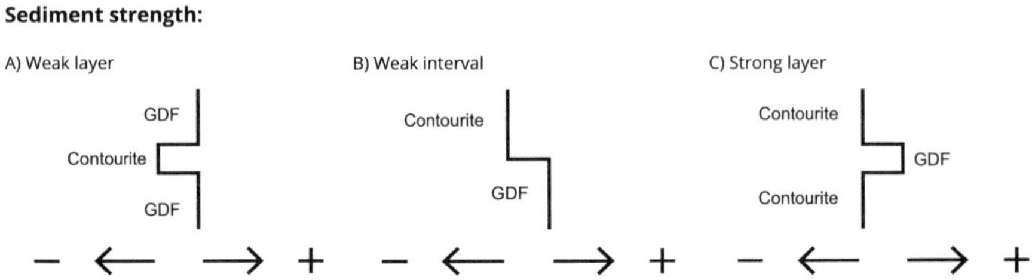

Fig. 28.4 Line drawing summarizing the sediment strength contrast (increasing: +, decreasing: –) that may occur within contouritic sediments; (**a**) contouritic sediments within a GDF succession represent a weak layer, (**b**) and/or a weak zone occurs at the contourite – GDF interface, or (**c**) on top of (or at the base) of a GDF within a contouritic succession

4. *Pore pressure in contouritic sediments*
   Several studies infer that excess pore pressure is one of the factors playing a role in order for contouritic sediments to fail (e.g. Laberg and Camerlenghi 2008). But under what conditions is excess pore pressure likely to develop? Which lithofacies is most likely associated with excess pore pressure build-up? Is it related to periods of high sedimentation rate? Is excess pore pressure build-up a prerequisite for these sediments to liquefy? So far, excess pore pressure has only been reported from the headwall area of the Storegga Slide (Strout and Tjelta 2005). In situ measurements are vitally needed to determine if this is a wide spread phenomenon.
5. *Sediment disintegration and flow*
   Studies from the headwall areas of submarine landslides affecting the Lofoten Contourite Drift revealed that initial sediment disintegration may lead to the formation of blocks and slope-parallel ridges with glide planes at ~13 and ~100 m below sea floor, respectively (Baeten et al. 2013). Most blocks and ridges disintegrate rapidly. Block formation and disintegration of most blocks occurred also in the nearby Trænadjupet Slide (Laberg and Vorren 2000).

   Large debris-flow deposits dominate the lower parts of the Trænadjupet Slide (Laberg et al. 2006). They contain multiple large sediment blocks implying that some failed sediment did not disintegrate entirely, even though they may have experienced a run-out distance of up to 200 km. This is possible when sediment deformation mainly occur in a basal layer due to shear deformation, thus, not affecting the upper parts of the sediments (De Blasio et al. 2004). The volume of failed sediments transforming into turbidity currents (Laberg et al. 2006) remains to be identified.
6. *The importance of the paleo-relief on the location of the failure*
   The larger submarine landslides found offshore Norway are located near the shelf break or they occasionally form the shelf break (Fig. 28.3). This has been ascribed to a retrogressive slide behavior that terminates when the failures reach overconsolidated subglacial deposits of the continental shelf deposited beneath grounded ice sheets during repeated glacials (Bryn et al. 2005). However, the headwall on the Lofoten Contourite Drift slides is located on the mid-slope where slope gradients are highest whereas the shallower, less-steep and mounded parts remain unaffected (Baeten et al. 2013).

28.4 Outstanding Issues

The following knowledge gaps can be deduced from the discussion presented above:

- Information on sediment composition and geotechnical properties (strength behaviour) beyond the depth reached by conventional coring offshore Norway is of paramount importance for a better understanding of the preconditions leading to some of the world's largest submarine landslides.
- In situ measurements of the physical and geotechnical properties of the sediments at the depth of weaker layers are presently not available.

Other relevant aspects for a better understanding of the occurrence of submarine landslides at high-latitude continental margins include:

- In situ monitoring of sediment physical properties for a better understanding of their natural variation and controlling factors.
- Availability of high-quality samples for geotechnical testing in order to better characterize and model the slide dynamics and flow velocity.
- Studies of the microfabric of contouritic sediments and its potential influence on the stability of the sediments.

28.5 Concluding Remarks

- Lithological and physical properties of contouritic sediments are vital parameters for the stability of the slopes of formerly glaciated continental margins.
- Within contouritic sediments recovered by conventional coring, the sediment deformation is located within laminated sediments. The precise origin of these sediments is often not known. Beyond the depth of conventional coring, little information is available.
- Internal variations of lithology and physical properties can lead to the formation of glide planes within contourites. Whether or not glide planes also develop at the interface of contourites – GDFs (zones of rapid changes in shear strength) remains unknown.
- Knowledge about conditions leading to excess pore pressure within contouritic sediments and about preferred lithofacies remains sparse.
- Some failed sediment blocks can disintegrate within a few km whereas other blocks move up to 200 km. Such long movements are related to deformation limited to a basal layer of the block. The volume of sediments transforming into turbidity currents is presently not known.

Acknowledgement This is a contribution to the LOSLOPE project funded by the Research Council of Norway. We acknowledge the skills, guidance and inspiration from Profs. T.O. Vorren and M. Ivanov, both contributed to the initiation of this work, but sadly passed away before the project came to conclusion. We are grateful for constructive reviews from C. Berndt and B.O. Hjelstuen.

References

Baeten NJ et al (2013) Morphology and origin of smaller-scale mass movements on the continental slope off northern Norway. Geomorphology 187:122–134

Baeten NJ et al (2014) Origin of shallow submarine mass movements and their glide planes—Sedimentological and geotechnical analyses from the continental slope off northern Norway. J Geophys Res: Earth Surf 119, doi:10.1002/2013JF003068

Bryn P, Berg K, Forsberg CF, Solheim A, Kvalstad TJ (2005) Explaining the Storegga Slide. Mar Pet Geol 22:11–19

De Blasio FV, Elverhøi A, Issler D, Harbitz CB, Bryn P, Lien R (2004) Flow models of natural debris flows originating from overconsolidated clay materials. Mar Geol 213:439–455

Elverhøi A, Breien H, De Blasio FV, Harbitz CB, Pagliardi M (2010) Submarine landslides and the importance of the initial sediment composition for run-out length and final deposit. Ocean Dyn 60:1027–1046

Faugères J-C, Stow DAV, Imberg P, Viana A (1999) Seismic features diagnostic of contourite drifts. Mar Geol 162:1–38

Haflidason H et al (2004) The Storegga Slide: architecture, geometry and slide development. Mar Geol 213:201–234

Hesse R, Khodabakhsh S, Klaucke I, Ryan WBF (1997) Asymmetrical turbid surface-plume deposition near ice-outlets of the Plesitocene ice sheet in the Labrador Sea. Geo-Mar Lett 17:179–187

Hjelstuen BO, Eldholm O, Faleide JI (2007) Recurrent Pleistocene mega-failures on the SW Barents Sea margin. Earth Planet Sci Lett 258:605–618

Jakobsson M et al (2012) The International Bathymetric Chart of the Arctic Ocean (IBCAO) version 3.0. Geophys Res Lett 39:L12609. doi:10.1029/2012GL052219

L'Heureux J-S, Vanneste M, Rise L, Brendryen J, Forsberg CF, Nadim F, Longva O, Chand S, Kvalstad TJ, Haflidason H (2013) Stability, mobility and failure mechanism for landslides at the upper continental slope off Vesterålen, Norway. Mar Geol 346:192–207

Laberg JS, Camerlenghi A (2008) The significance of contourites for submarine slope stability. Dev Sedimentol 60:537–556, Elsevier

Laberg JS, Vorren TO (2000) The Trænadjupet Slide, offshore Norway – morphology, evacuation and triggering mechanisms. Mar Geol 171:95–114

Laberg JS, Vorren TO, Knutsen S-M (1999) The Lofoten contourite drift off Norway. Mar Geol 159:1–6

Laberg JS, Dahlgren T, Vorren TO, Haflidason H, Bryn P (2001) Seismic analyses of Cenozoic contourite drift development in the Northern Norwegian Sea. Mar Geophys Res 22:401–416

Laberg JS, Vorren TO, Mienert J, Haflidason H, Bryn P, Lien R (2003) Preconditions leading to the Holocene Trænadjupet Slide offshore Norway. In: Locat J, Mienert J (eds) Submarine mass movements and their consequences. Kluwer Academic, Dordrecht, pp 247–254

Laberg JS, Vorren TO, Kenyon NH, Ivanov M (2006) Frequency and triggering mechanisms of submarine landslides of the North Norwegian continental margin. Nor J Geol 86:155–161

Laberg JS, Baeten NJ, Lågstad P, Forwick M, Vorren TO (2013) Formation of a large submarine crack during the final stage of retrogressive mass wasting on the continental slope offshore northern Norway. Mar Geol 346:73–78

Lekens WAH, Sejrup HP, Haflidason H, Petersen GØ, Hjelstuen B, Knorr G (2005) Laminated sediments preceding Heinrich event 1 in the northern North Sea and southern Norwegian Sea: origin, processes and regional linkage. Mar Geol 216:27–50

Leroueil S (2001) Natural slopes and cuts: movement and failure mechanisms. Geotechnique 51(171):197–243

Locat J, Leroueil S, Locat A, Lee H (2014) Weak layers: their definition and classification from a geotechnical perspective. In: Krastel S et al (eds) Submarine mass movements and their consequences, vol 37, Advances in natural and technological hazards research. Springer International Publishing, Cham. doi:10.1007/978-3-319-00972-8 1

Martin-Chivelet J, Fregenat-Martinez MA, Chacón B (2008) Traction structures in contourites. Dev Sedimentol 60:159–182, Elsevier

Piper DAW, Normark WR (2009) Processes that initiate turbidity currents and their influence on turbidites: a marine geology perspective. J Sediment Res 79:347–362

Rise L, Ottesen D, Longva O, Solheim A, Andersen ES, Ayers S (2006) The Sklinnadjupet slide and its relation to the Elsterian glaciation on the mid-Norwegian margin. Mar Pet Geol 23:569–583

Rise L, Bøe R, Riis F, Bellec VK, Laberg JS, Eidvin T, Elvenes S, Thorsnes T (2013) The Lofoten-Vesterålen continental margin, North Norway: canyons and mass-movement activity. Mar Pet Geol 45:134–149

Strout JM, Tjelta T (2005) In situ pore pressures: what is their significance and how can they be reliably measured? Mar Pet Geol 22:275–285

Vanneste M, Mienert J, Bünz S (2006) The Hinlopen Slide: a giant, submarine slope failure on the northern Svalbard margin, Arctic Ocean. Earth Planet Sci Lett 245:373–388

Vanneste M et al (2012) Shallow landslides and their dynamics in coastal and deepwater environments, Norway. In: Yamada Y et al (eds) Submarine mass movements and their consequences (5th International Symposium). Springer, p 29–41

Vanneste M, Sultan N, Garziglia S, Forsberg CF, L'Heureux J-S (2014) Seafloor instabilities and sediment deformation processes: the need for integrated, multi-disciplinary investigations. Mar Geol 352:183–214

Wessel P, Smith WHF (1998) Improved version of the generic mapping tools released. Eos Trans AGU 79:579

Wiemer G, Kopf A (2015) Altered marine tephra deposits as potential slope failure planes? Geo-Mar Lett 35(4):305–314

Chapter 29
Mass Wasting History Within Lake Ohrid Basin (Albania/Macedonia) Over the Last 600 ka

Katja Lindhorst, Sebastian Krastel, and Henrike Baumgarten

Abstract Lake Ohrid (LO), a transboundary lake shared by Macedonia and Albania on the Balkan Peninsula, is not only considered to be the oldest lake in Europe (~2 Ma) but has a long and continuous sedimentary history. An advantage at LO is the availability of hydroacoustic data sets of good quality covering the entire lake basin. The tectonically formed basin is filled with thick undisturbed sediments. However, the overall internal structure of LO is characterized by numerous faults, clinoform structures, and several Mass Transport Deposits (MTDs). By using a seismic chronology model (SCM) correlating seismic reflector packages with Marine Isotope Stages (MIS) we estimate the occurrence of the deepest MTD detected in the southern basin at the transition of MIS9 to MIS8 (~300 ka) defining the onset of the sliding history in LO that is still ongoing today. In general, MTDs are widespread within the basin but they do cluster at active faults. Two large MTDs occurred in the early MIS7 (~230 ka, ~220 ka) and after a quiesence period of about ~70 ka two additional large MTDs have been detected in the late penultimate glacial period MIS6 (~150 ka, 130 ka). MIS5 seemed to be another quiet period with respect to mass wasting. In the younger sedimentary history mass movement is a common process with several large and mid-sized deposits mapped at all stratigraphic levels. The youngest slide deposits are estimated to occur within the last 2,000 years. The main outcome of this paper is a model for the spatial and temporal distribution of mass wasting for Lake Ohrid.

K. Lindhorst (✉) • S. Krastel
Institute of Geosciences, Christian-Albrechts-University at Kiel, Kiel, Germany
e-mail: klindhorst@geophysik.uni-kiel.de

H. Baumgarten
Leibniz- Institute for Applied Geophysics (LIAG), Stilleweg 2, 30655 Hannover, Germany

G. Lamarche et al. (eds.), *Submarine Mass Movements and their Consequences*,
Advances in Natural and Technological Hazards Research 41,
DOI 10.1007/978-3-319-20979-1_29

29.1 Lake Ohrid

Lake Ohrid (LO) is located on the Balkan Peninsula within a mountain chain that formed in a late phase of the alpine orogeny (Dumurdzanov et al. 2005; Burchfiel et al. 2008, Fig. 29.1). LO, a transboundary lake shared by Macedonia and Albania, is 30 km long and 15 km wide, covers an area of 358 km^2 and contains a water volume of about 50.7 km^3 with a maximum water depth of 293 m (Popovska and Bonacci 2007; Lindhorst et al. 2010). It is considered to be the oldest continuously existing lake in Europe (~1.2 Ma, Wagner et al. 2014). With respect to its size, LO has a high density of endemic species with more than 212 of endemic species described, which makes it a hotspot of biodiversity (Albrecht and Wilke 2008).

The tectonic and sedimentary evolution of LO started as a pull-apart basin, with subsequent E-W extension that is still active today, as evidenced by active faults partly visible at the lake floor (Lindhorst et al. 2015, Fig. 29.1). Intensive studies of the sedimentary and tectonic evolution of LO by interpreting hydro-acoustic and seismic data revealed that the basin infill is a valuable and continuous sedimentological archive favorable to study the relationship between biodiversity and

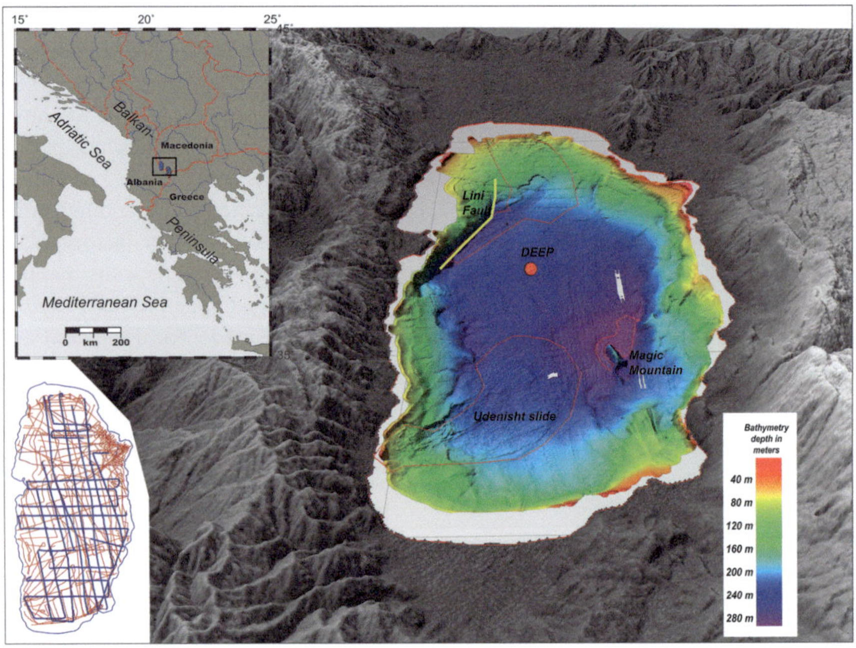

Fig. 29.1 Overview map showing the topography and bathymetry of Lake Ohrid Basin and its surrounding area. *Red dot* marks the main drill site of the ICDP SCOPSCO (Scientific Collaboration on Past Speciation Conditions in Lake Ohrid) drilling campaign in spring 2013. The Lini Fault (*yellow line*), and Magic Mountain a prominent basement high in the southeast are labeled. Rough topography indicating recent slide deposits are outlined in *red*. A track map (*lower left* corner) shows the dense grid of sediment echo-sounder lines (*red*) and multichannel seismic lines (*blue*)

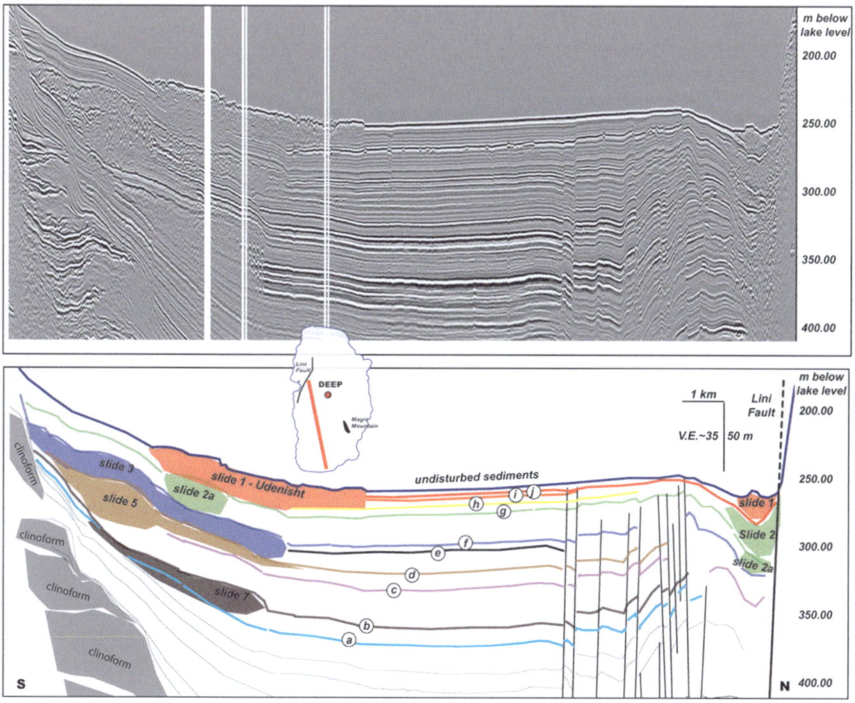

Fig. 29.2 Seismic cross section illustrates the internal structure of Lake Ohrid. A prominent fault (Lini Fault) at the northern end, slide deposits on its hanging wall, and antithetic faulting can be detected. The central basin is characterized by thick undisturbed sediments whereas several slide deposits are clearly visible as transparent units in the southern area overlying continuous reflectors and clinoform structures. Slides are color-coded according to their time of occurrence and important reflectors (‘*a*’ to ‘*j*’) used for the slide chronology are marked

climatic and/or short and long-term environmental changes within the northern Mediterranean Region (Wagner et al. 2008, 2014). Beside a complex internal structure characterized by clinoforms, active faulting, as well as Mass Transport Deposits (MTDs); thick undisturbed sediment succession in the central part are also present without unconformities or erosional features (Fig. 29.2); for this reason it was selected as the key site of a drilling campaign realized in summer 2013 in order to establish a continuous age model for the basin (Wagner et al. 2014, see Fig. 29.1 for location of drill site).

The aim of this paper is a detailed mapping of subaquatic slide deposits by means of hydro-acoustic and seismic data in order to analyze potential factors controlling the spatial and temporal occurrence of subaquatic landslides. We used a seismic chronological model to define and trace ten stratigraphic horizons relevant in terms of MTDs (Lindhorst et al. 2015). Multichannel seismic (MCS) data provide information of the onset of sliding in LO as well as recurrence intervals in its early sliding history. In addition, we used a dense grid of sediment echosounder (SES) data to give a more detailed description of MTDs occurring in the uppermost sediments (<50 m).

29.2 Data Sets and Interpretation

We used high resolution bathymetric and seismic data that were recorded between 2004 and 2009 during several site surveys at LO. A high resolution bathymetric map, a dense grid of MCS data (~500 km) and SES (>900 km) are available (Figs. 29.1, 29.2, and 29.3). In addition, a Vertical Seismic Profiling (VSP) was carried out during the deep drilling campaign within the frame of the International Continental Drilling Program (ICDP) in spring 2013 at site DEEP in the central part of the basin (Fig. 29.1, Wagner et al. 2014). Picking the first break arrivals of the VSP data results in a time-depth conversion chart and subsequently allows us to convert Two-Way-Travel-Time (TWTT) into sediment depth of all seismic lines. Reflectors are traced by using the Interpretation Software IHS Kingdom Suite 8.8 starting with a starting point at site DEEP. A selection of reflectors was traced on all seismic lines throughout the entire basin. Due to the fact that great offsets of active faults hamper a direct tracing into some areas, we were not able to correlate reflectors in the very shallow parts, e.g. in the northern area, and along the margins, because these areas are still tectonically active.

In order to evaluate the spatial and temporal distribution of MTDs we chose ten reflectors (a–j, Figs. 29.2 and 29.3) each of them fulfilling two criteria: (1) they must be traceable throughout the basin and (2) they can be regarded as sediment cover of transparent bodies imaged in seismic reflection profiles (reflector ‘a’ only

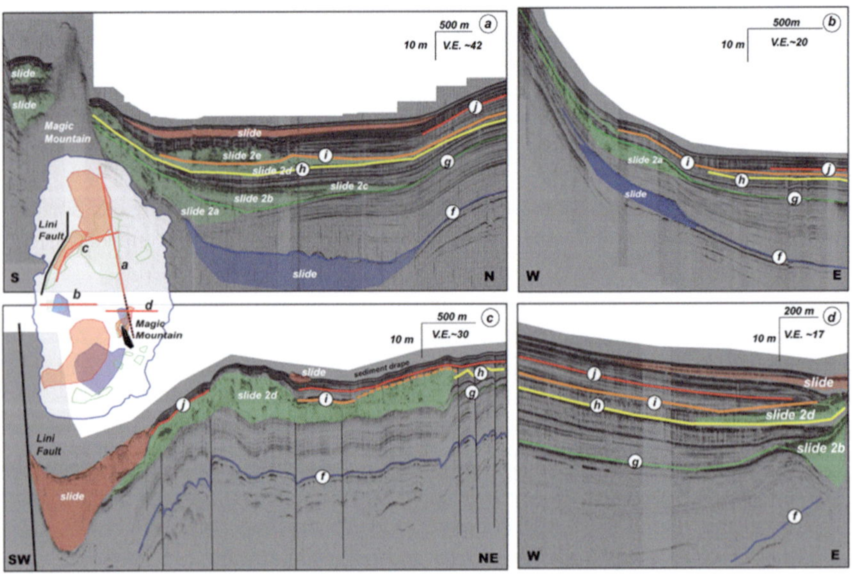

Fig. 29.3 Sediment echo-sounder profiles of the uppermost sediments (<50 m) imaging the younger sedimentary evolution of Lake Ohrid. Four areas with extensive mass transport deposits are shown: (**a**) and (**d**) around Magic Mountain, (**b**) along the western margin, (**c**) and offshore Lini Peninsula in an area of active faulting

fulfills the first criteria and thus marks the lower limit of sliding). In order to estimate an exact timing of slides, it would be favorable to pick the eroded base and its respective top reflector adjacent to the toe of a MTD (Schnellmann et al. 2002). Although MCS lines are covering the entire lake, the resulting grid of seismic lines is not dense enough in order to detect a toe for each individual slides. Nevertheless we could define a top reflector for each MTD; giving us a minimum age. If two MTDs are covered by the same reflector we assume that they have occurred at the same time.

For the establishment of a chronology of subaquatic slides in LO we apply an age model (here called SCM) recently developed solely based on seismic data (Lindhorst et al. 2015). An alternating pattern of high and low amplitudes within seismic packages were interpreted to reflect a climate induced signal. In summary, the authors argued that during a warm period the lake was more productive, vegetation was dense and forest-dominated, stabilizing the catchment area, leading to sedimentation within the basin that was mainly biogenic (low seismic amplitudes). In contrast during Glacials, the catchment area was less stable because vegetation was only steppe-like and subsequently more clastic material was transported into the lake (high seismic amplitudes). For a more detailed description of the seismic chronological model we refer to Lindhorst et al. (2015). Ages for reflectors used here are ‘a’: ~340 ka (MIS10 to MIS9), ‘b’: ~300 ka (MIS9 to MIS8), ‘c’: ~230 ka, ‘d’: ~220 ka, ‘e’: ~150 ka, ‘f’: ~130 ka (MIS6 to MIS5), ‘g’: ~80 ka (MIS5 to MIS4), ‘h’: ~60 ka (MIS4 to MIS3), ‘i’: ~30 ka, and ‘j’: ~12 ka (start Holocene), see Fig. 29.4.

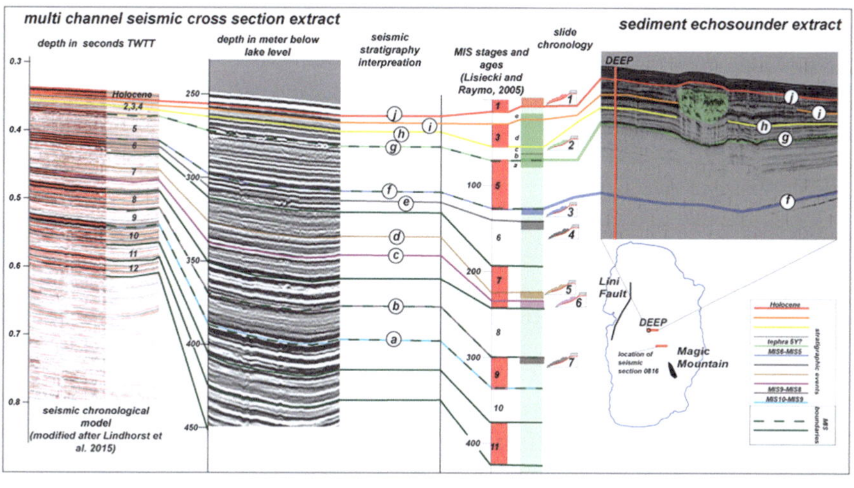

Fig. 29.4 (**a**) Seismic section used for developing seismic chronology model (*red-to-white*), (**b**) respective depth converted seismic section (*black* and *white*). (**c**) MIS boundaries (*green lines*) and color coded reflectors defined for slide chronology are illustrated (‘*a*’ to ‘*j*’, see legend for further explanation). (**d**) All reflectors are assigned to times applying the seismic chronology model resulting in a slide chronology (categories 1–7). (**e**) Sediment echosounder data crossing site DEEP (*upper right* corner) and important strong reflectors (‘*f*’ to ‘*j*’). For location see inlet map

29.3 Results

29.3.1 Temporal Distribution

All MTDs are detected above reflector 'a', consequently it marks the onset of the sliding history at the transition of MIS10 to MIS9. Therefore, all subaquatic slides in LO are younger than ~340 ka (Figs. 29.2 and 29.4). The oldest MTDs are covered by reflector 'b' that is interpreted to represent sediments deposits around the transition of MIS9 to MIS8 (~300 ka). Unfortunately, the vertical resolution of our seismic data does not allow better precision.

We could not detect any evidence for mass wasting within MIS8. The next two younger sliding events occurred within the early MIS7 (~230 ka, 220 ka, Fig. 29.4). Another quiescence interval without detectable MTDs are the time period in between ~220 and ~150 ka. The next occurrence of MTDs happened at the end of the penultimate glacial interval MIS6 (two events at ~150 ka, 130 ka, Fig. 29.4). The warm period MIS5 was again quiet with respect to mass wasting and only toward its end we detected significant MTDs covered by reflector 'f' (Figs. 29.2 and 29.3).

Within the uppermost sediments, due to the availability of a dense grid of high resolution SES data we were able to detect many MTDs. A subdivision was made within respective subaquatic slides that belong to category 2 (Figs. 29.3 and 29.4). Although individual strong reflectors between MTDs of category 2 are traceable, a correlation with sediment cores is necessary to obtain precise ages. Here, we will only give relative timings of five individual MTDs occuring between the transition of MIS5 to MIS4 and the Holocene (category 2a–e, 4).

Ages are only estimates within the range of the seismic resolution. Once a detailed age model based on detailed analysis of sediment cores will be available, it will be possible to extract more precise ages for slides deposited since the penultimate glacial period by correlating strong seismic reflectors with sediment physical parameters.

A rough topography and transparent units with only a thin sediment cover point to an ongoing process of mass wasting within the basin during the Holocene. One example is the Udenisht slide complex that has been interpreted to be younger than 2000 years (Lindhorst et al. 2012) as well as areas in the north (Fig. 29.1).

29.3.2 Spatial Distribution

Mass wasting is a widespread phenomenon in LO, with transparent bodies present along lake margins but not in the central basin (Figs. 29.2 and 29.3). In general, three areas show the most frequent occurrence of MTD: (1) on the hanging-wall of the Lini Fault (LF, Figs. 29.2 and 29.3), (2) around Magic Mountain (MM), a basement high in the southeastern part of LO (Fig. 29.3), and (3) the southern area (Fig. 29.2).

A MCS cross section of LO (N-S) gives an overview impression of the internal structure of the basin and its spatial distribution pattern of MTDs (Fig. 29.2). A large offset at its northern end marks the LF, forming a half graben with associated faults further to the south (Fig. 29.2). Several MTDs are present on the hanging wall filling the depression adjacent to the most active fault (LF) in LO (Figs. 29.2 and 29.3c). The middle part of the seismic line nicely illustrates undisturbed sediments where we did not find any evidence for slide deposits (Fig. 29.2). However, further south, several large MTDs are stacked on top of each other up to a sediment depth of ~100 m (Fig. 29.2). Reflectors below are onlapping on the southern margin overlying several clinoform structures (Fig. 29.2).

Spatial distribution pattern of MTDs of the younger (up to a sediment depth of ~50 m) sedimentary history of LO are illustrated in Fig. 29.3. Several MTDs can be found in the area around MM (Fig. 29.3a, d). Smaller MTDs have been detected along the western margin (Fig. 29.3b). The northern shelf area is characterized by a rough topography and slide deposits directly present at the lake floor (Figs. 29.1 and 29.3c).

29.4 Discussion

All MTDs are most likely younger than ~340 ka because we did not find any transparent bodies below reflector ‘a’. Although it is very likely that seismic shaking occurred frequently even long before the onset of sliding due to the fact that the lake basin was affected by E-W extension since its existence as a water filled basin. However, around ~340 ka ago sediments deposited along steep margin were destabilised for the first time and mass wasting occurred. Analysis of sediment cores will provide essential information to constrain causes for the changes in sediment deposition and/or depositional processes leading to an onset of sliding. No sliding occur during three time intervals: within glacial period MIS8, between ~220 and ~150 ka, and during interglacial MIS5 implying that the basin was in a stable condition probably with a reduce sediment input and/or without oversteepening along the margins. It may also point to a time of quiesence with respect to seismic shaking in the area.

Overall, we do not see a direct correlation between climate and recurrence of sliding as they occur at transitions of warm and cold periods as well as right within MISs (Fig. 29.4). The lack of slides during a cold period (MIS8) and a warm period (MIS5) further suggest that there seem to be no link to climate conditions but other factors are controlling the appearance of MTDs. Although it has to be stated that given ages are only approximations because the age model we used is preliminary and was developed with several assumption (Lindhorst et al. 2015). In addition, the mapping of MTDs is based solely on seismic data inhibiting a limited vertical resolution, especially in great sediment depth.

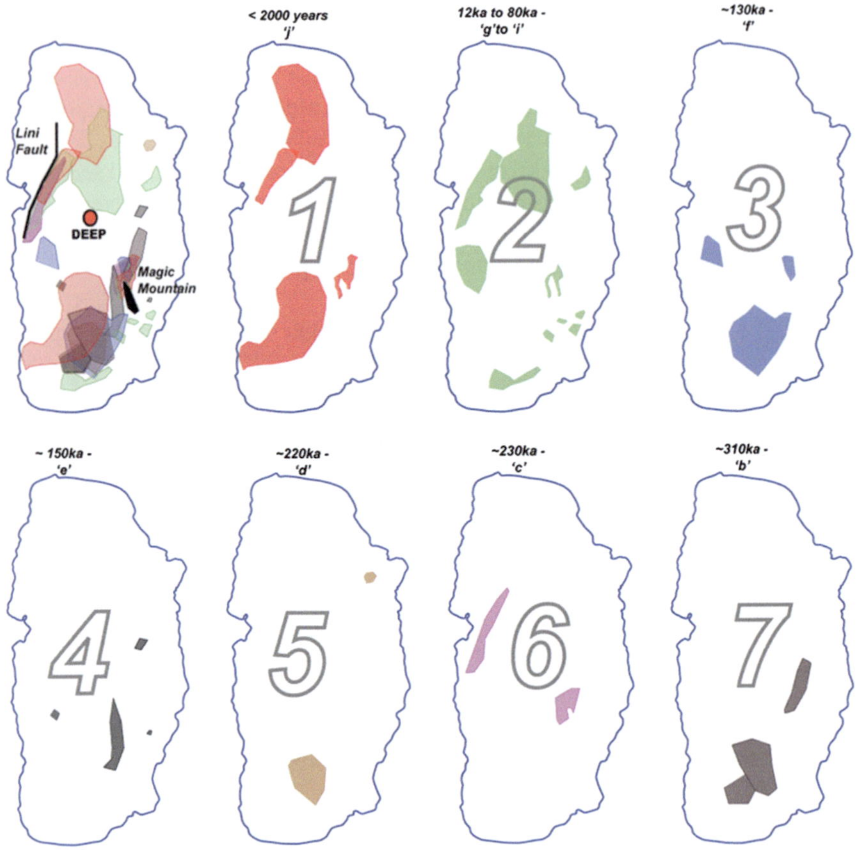

Fig. 29.5 Slide distribution maps of Lake Ohrid illustrating spatial distribution of slides within individual time intervals (1–7). Only the central basin has not been affected by subaquatic sliding as imaged in the first map with all mass movements mapped in Lake Ohrid

Nevertheless, it can be said that MTDs are widespread within the LO sedimentary fill. We found several large MTDs at greater depth mainly in the south, but also many small deposits along the margins and in the northern area indicating that mass wasting was an important sedimentation process in the sedimentary history of LO. In general, MTDs cluster around major faults (e.g. LF, around Magic Mountain), deposited in small depressions associated with hanging-walls (Figs. 29.2, 29.3, and 29.5). Those MTDs are interpreted as the results of fault activity along the respective faults. Undisturbed sediments in between point to time intervals without fault ruptures hence are promising features in order to study paleoseismicity of distinctive faults within LO (Reicherter et al. 2011, Figs. 29.4 and 29.5).

Recent sliding (<2000 years) mainly occurred around Magic Mountain, along the southwestern margin and in the northern area where MTDs are transported southwards into the basin filling up small depression adjacent to the LF (Fig. 29.4).

A rough topography along the northern slope area implies that mass wasting is still ongoing and thus MTDs are younger than 2000 years. Such young ages were suggested also for the Udenisht (Lindhorst et al. 2012) and Lini slide (Wagner et al. 2012). Triggers of both slides could be linked to a early 6th AD earthquake indicating that further investigation of mass wasting in LO have the potential to reconstruct the paleoseismicity within the surrounding area.

29.5 Conclusion

MTDs are widespread features that can be found within the entire lake basin at different stratigraphic levels of the Lake Orid (LO) sedimentary fill. Subaquatic sliding started around ~340 ka ago and is still an ongoing process. One exception is the central basin characterized by thick undisturbed sediments; here we find no indications for large MTDs. In general, large MTDs are found in the southern area whereas small MTDs are clustered within sub-basins on top of hanging-walls of major active faults such as the LF and around faults associated with Magic Mountain. Such clusters along faults zones indicate a link to the seismic activity along respective faults. Intervals of seismic sequences of undisturbed lacustrine sediments interlaced with MTD suggest that these areas are promising locations for studying past activities of these faults.

The main outcome of this study is an improvement of the sedimentary history of LO with special emphasis on the mass wasting history that started within MIS9. We assume that seismic shaking caused by an E-W extension within the surrounding area always hit the region and can be regarded as a likely trigger for subaquatic sliding. Nevertheless, sediments seemed to be stable before ~340 ka and at some point got unstable and subsequently failure occurred. In the early slide history, material was transported mainly into the southern basin. We identified five large MTDs within the time interval ~300 to ~150 ka implying long recurrence times. In contrast, in the recent most history of the basin (up to a sediment depth of 50 m) mass wasting seems to be a dominant sedimentation process in the entire basin as evidenced by a high number of large and mid-sized MTDs distributed throughout the basin. Sliding in LO is still ongoing evidenced by the presence of MTDs on the modern lake bathymetry that have either a small drape of sediment, e.g. Udenisht slide, or even no sediment cover at all as observed in the northern area.

Acknowledgments We thank the reviewers Alina Polonia (ISMAR, Bologne, Italy) and Flavio Anselmetti (University Bern, Switzerland) as well as the editor Joshu Mountjoy (NIWA, New Zealand). We are grateful for the support of our Macedonian Colleagues as well as all members of the scientific community of the SCOPSCO project. Special thanks to Bernd Wagner from University of Cologne.

References

Albrecht C, Wilke T (2008) Ancient Lake Ohrid: biodiversity and evolution. Hydrobiologia 615:103–140. doi:10.1007/s10750-008-9558-y

Burchfiel B, Nakov R, Dumurdzanov N, Papanikolaou D, Tzankov T, Serafimovski T, King R, Kotzev V, Todosov A, Nurce B (2008) Evolution and dynamics of the Cenozoic tectonics of the South Balkan extensional system. Geosphere 4:918

Dumurdzanov N, Serafimovski T, Burchfiel B (2005) Cenozoic tectonics of Macedonia and its relation to the South Balkan extensional regime. Geosphere 1(1):1–22. doi:10.1130/GES00006.1

Lindhorst K, Vogel H, Krastel S, Wagner B, Hilgers A, Zander A, Schwenk T, Wessels M, Daut G (2010) Stratigraphic analysis of lake level fluctuations in Lake Ohrid: an integration of high resolution hydro-acoustic data and sediment cores. Biogeosciences 7 (in: Evolutionary and geological history of Balkan lakes Ohrid and Prespa. (eds: B. Wagner, T. Wilke, and V. Brovkin)): Springer, The Netherlands, pp 3531–3548. doi:10.5194/bgd-7-3651-2010

Lindhorst K, Gruen M, Krastel S, Schwenk T (2012) Hydroacoustic analysis of mass wasting deposits in Lake Ohrid (Former Yugoslavian Republic of Macedonia/Albania). In: Yamada Y, Kawamura K, Ikehara K et al (eds) Submarine mass movements and their consequences, vol Advances in Natural and Technological Hazards Research 31. Springer, The Netherlands, pp 245–253

Lindhorst K, Krastel S, Reicherter K, Stipp M, Wagner B, Schwenk T (2015) Tectonic and sedimentary evolution of Lake Ohrid (Albania/Macedonia). Basin Res 27(1):84–101

Lisiecki LE, Raymo ME (2005) A Pliocene-Pleistocene stack of 57 globally distributed benthic δ18O records. Paleoceanography 20(1):1–17

Popovska C, Bonacci O (2007) Basic data on the hydrology of Lakes Ohrid and Prespa. Hydrol Process 21(5):658–664

Reicherter K, Hoffmann N, Lindhorst K, Krastel S, Fernandez-Steeger T, Grutzner C, Wiatr T (2011) Active basins and neotectonics: morphotectonics of the Lake Ohrid Basin (FYROM and Albania). Z Dtsch Ges Geowiss 162(2):217–234

Schnellmann M, Anselmetti F, Giardini D, McKenzie J, Ward S (2002) Prehistoric earthquake history revealed by lacustrine slump deposits. Geology 30(12):1131–1134

Wagner B, Francke A, Sulpizio R, Zanchetta G, Lindhorst K, Krastel S, Vogel H, Daut G, Grazhdani A, Lushaj B (2012) Seismic and sedimentological evidence of an early 6th century AD earthquake at Lake Ohrid (Macedonia/Albania). Clim Past 8:2069–2078

Wagner B, Reicherter K, Daut G, Wessels M, Matzinger A, Schwalb A, Spirkovski Z, Sanxhaku M (2008) The potential of Lake Ohrid for long-term palaeoenvironmental reconstructions. Palaeogeog Palaeoclimatol Palaeoecol 259(2–3):341–356

Wagner B, Wilke T, Krastel S, Zanchetta G, Sulpizio R, Reicherter K, Leng M, Grazhdani A, Trajanovski S, Francke A, Lindhorst K, Levkov Z, Cvetkoska A, Reed J, Zhang X, Lacey J, Wonik T, Baumgarten H, Vogel H (2014) The SCOPSCO drilling project recovers more than 1.2 million years of history from Lake Ohrid. Sci Drill 17:19–29

Chapter 30
Implications of Sediment Dynamics in Mass Transport along the Pianosa Ridge (Northern Tyrrhenian Sea)

Elda Miramontes, Antonio Cattaneo, Gwenael Jouet, and Sebastien Garziglia

Abstract The Pianosa Ridge forms the eastern flank of the Corsica Trough in the Northern Tyrrhenian Sea: it is the site of preferential accumulation of contourites and Mass Transport Deposits (MTDs). Along the Pianosa Ridge, 11 MTDs with a total volume of 6.5 km^3 were identified. These MTDs are distributed in three areas: (A) one small MTD associated to canyon flank destabilisation in the northern part of the study area; (B) six intermediate size MTDs in the central area; (C) four MTDs of larger size (up to 2.62 km^3) to the south, including the Pianosa Slump, which is the most recent MTD in this area (aged at 42–50 kyr BP) and analysed in more detail. The main factor controlling the formation of MTDs in areas A and B seems to be steep slopes associated to erosion and heterogeneous sedimentation caused by bottom currents, respectively. In contrast, multiple factors may control slope instability in the zone where the largest MTDs took place (area C): the incision generated by contour currents, the presence of coarser layers in contourite drifts that may accumulate gas and the location of normal faults near the headwall.

30.1 Introduction

The Corsica Trough is located between Corsica and the Tuscan Shelf in the Northern Tyrrhenian Sea. It has a maximum depth of 900 m and connects the Tyrrhenian and Ligurian Seas. The eastern flank of the Corsica Trough is formed by the Pianosa Ridge, a tectonic structure with typical slopes between 3 and 10°, reaching 20° locally (Fig. 30.1). This area is characterised by several submarine landslides and contourite deposits (Cattaneo et al. 2014; Fig. 30.1). Contourite drifts are formed by the Levantine Intermediate Water (LIW), flowing northwards (Artale and Gasparini 1990) and generating stronger bottom currents on the Pianosa

E. Miramontes (✉) • A. Cattaneo • G. Jouet • S. Garziglia
IFREMER, Géosciences Marines-EDROME, Centre de Brest, BP70, CS10070, 29280 Plouzané, France
e-mail: elda.miramontes.garcia@ifremer.fr

G. Lamarche et al. (eds.), *Submarine Mass Movements and their Consequences*, Advances in Natural and Technological Hazards Research 41,
DOI 10.1007/978-3-319-20979-1_30

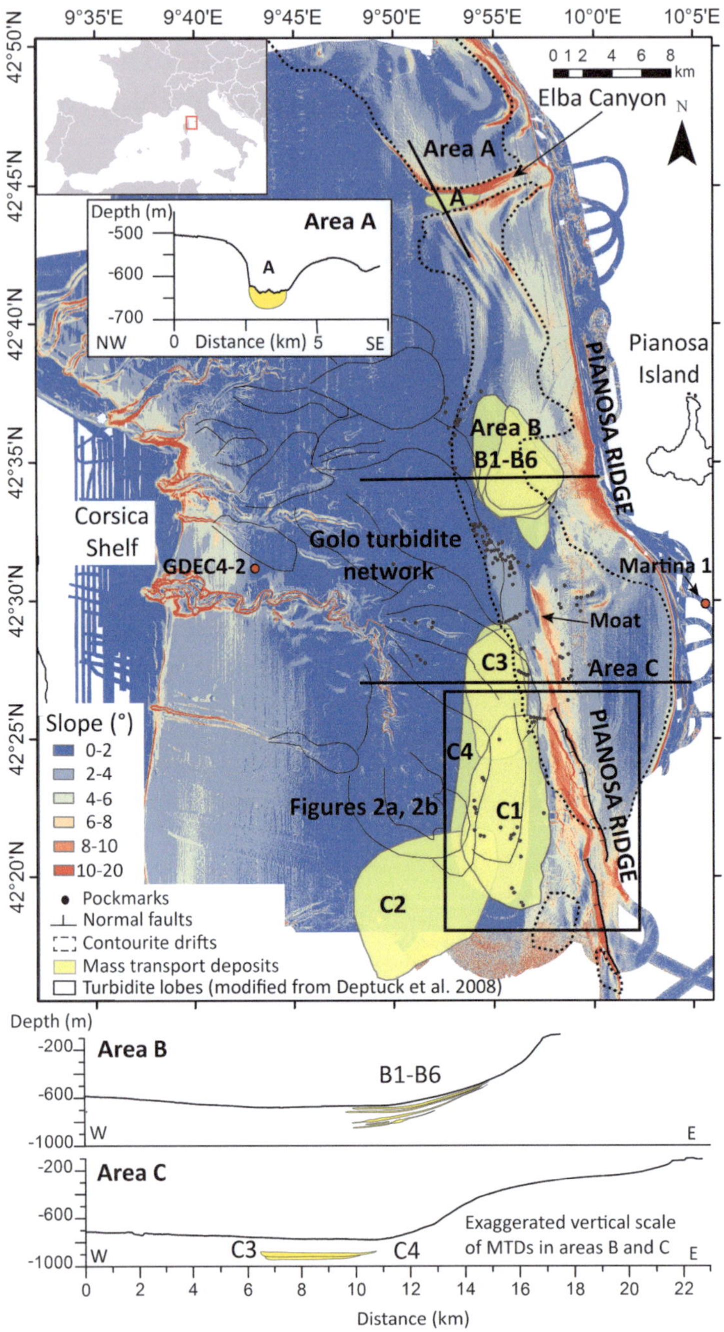

Fig. 30.1 Seafloor gradient map of the Corsica Trough with superposed location of morpho-sedimentary features: Mass Transport Deposits (MTDs), contourite drifts and turbidite lobes (see legend). Three bathymetric cross-sections across areas A, B and C show the location of MTDs

Ridge due to the Coriolis force. The sediment input from the Tuscan Shelf is low since sediment from Italian rivers is mostly trapped in the shelf and cannot reach the Corsica Trough (Roveri 2002). Most of the sediment deposited in the Corsica Trough is transported from the Corsica shelf by the Golo turbidite network (Deptuck et al. 2008; Fig. 30.1).

The aims of this study are to present the size and the stratigraphic distribution of Mass Transport Deposits (MTDs) along the Pianosa Ridge, and to analyse their relationship with the pre-existing seabed morphology.

30.2 Material and Methods

In 2013 two cruises, PRISME2 and PRISME3 onboard the R/V Atalante and R/V Pourquoi pas?, respectively, took place along the Pianosa Ridge. Multibeam bathymetry, 72-channel high-resolution mini GI gun (50–250 Hz) and CHIRP (1800–5300 Hz) seismic reflection profiles, and Calypso piston cores were collected along the Pianosa Ridge. P-wave velocity was measured on whole core sections using a Geotek Multi-Sensor Core Logger (MSCL). The bulk sediment semi-quantitative geochemical composition was measured with an Avaatech X-ray fluorescence (XRF) core scanner on split cores. Radiocarbon dating was obtained from bulk planktonic foraminifera at the Poznan radiocarbon laboratory. Radiocarbon ages were corrected for a marine reservoir effect of 400 years and calibrated using Calib 7.0.2 radiocarbon calibration software and the IntCal13 calibration curve (Reimer et al. 2013). MTDs were identified by seismic facies validated by the analysis of equivalent coring samples. MTD volume was calculated using the Kingdom Suite ® software by quantifying the time interval between the top and the bottom of the MTD from the seismic profiles. The velocity applied for time-depth conversion (1550–1560 m/s) was obtained from the velocity analysis calculated with the ProMAX ® Seismic Processing Software.

30.3 Results

Along the Pianosa Ridge 11 MTDs were identified in the Quaternary sediment record. Differences in location, volume and morphology were used to distinguish three areas of MTDs: A, B and C (Table 30.1). Seismic correlation with the borehole GDEC4-2 (Fig. 30.1; Angue Minto'o 2014) provided information about the stratigraphy in the region.

Area A shows a small submarine landslide associated to the destabilisation of the northern flank of the Elba Canyon (west of the Elba Island), where the slope is 8–20°. The headwall scarp and the deformation features of the MTD are clear in the bathymetry (Fig. 30.1). MTD A is of Holocene age, since it affects a regional reflector dated at 12 kyr.

Table 30.1 Characterisation of mass transport deposits in areas A, B and C

Area	Name	Volume (km^3)	Maximum runout (km)	Maximum length (km)	Maximum width (km)	Width/ length ratio	Age (kyr BP)
A	A	0.10	1.5	1.2	3.6	3.0	<12
B	B1	0.19	7.0	5.7	7.6	1.3	42–59
	B2	0.09	8.0	5.6	4.6	0.8	125–137
	B3	0.26	7.5	5.9	7.2	1.2	420–437
	B4	0.15	4.2	2.5	7.4	3.0	452->542
	B5	0.05	6.0	3.4	4.4	1.3	>542
	B6	0.04	6.2	3.6	4.4	1.2	>542
C	C1	2.62	9.0	6.0	13.5	2.3	42–50
	C2	0.47	16.0	10.0	7.8	0.8	125–160
	C3	0.67	6.0	4.0	8.7	2.2	265–321
	C4	1.86	10.4	5.3	18.6	3.5	321–380

Six small to intermediate size MTDs (B1 to B6) are observed in area B (Table 30.1). They are located west of the Pianosa Island, where the Pianosa Ridge forms an edge, generating in this zone high slope gradients (8–10°). The MTDs have an oval shape in plan view. B1, B2, B3 and B4 have a mean thickness of about 10 m; B5 and B6 mean thickness is 5 m. The continental slope in area B is starved during Pliocene and Quaternary. Pre-Messinian deposits (age deduced by correlation with the well Martina 1; Fig. 30.1; Pascucci et al. 1999) crop out in patches on the upper slope. Despite the reduced size of these MTDs, their thickness corresponds to about 30 % of the Pliocene-Quaternary deposits in the lower continental slope. B6 and B5 MTDs are older than 542 kyr, while the sediment covering B4 was deposited at 452 kyr, based on correlation to the bottom of borehole GDEC4-2. B3 is close in time to B4 but separated by 300 kyr from B2 (Table 30.1). B1 is the most recent MTD (42–59 kyr) in area B, it took place about 80 kyr after B2. The frequency of MTDs in area B is irregular and it was higher during the Middle Pleistocene than during the Upper Pleistocene.

Four MTDs (C1 to C4) are located in area C, in the southern zone of the study area. Their volume ranges from 0.67 to 2.62 km^3 and they are characterised by a high width/length ratio, except for C2 that has a ratio lower than 1 (Table 30.1). Between 265 and 380 kyr BP C3 and C4 were formed, followed by C2 at 125–160 kyr (Table 30.1). The Pianosa Slump (C1) is the most recent MTD of area C and it was analysed in more detail than the other MTDs since it was sampled by the Calypso piston core PSM3-CS022 (Figs. 30.2 and 30.3). The top of C1 (i.e. the bottom of the post-slide hemipelagic sediment; Fig. 30.3) is dated at 42,326 cal year BP and Cattaneo et al. (2014) report that the MTD disrupted sediments as old as 50,000 cal year BP. C1 is overlain by about 17–20 m of sediment (Fig. 30.3). However, various morphological features related to MTD C1 can still be appreciated at the present seafloor (Fig. 30.2).

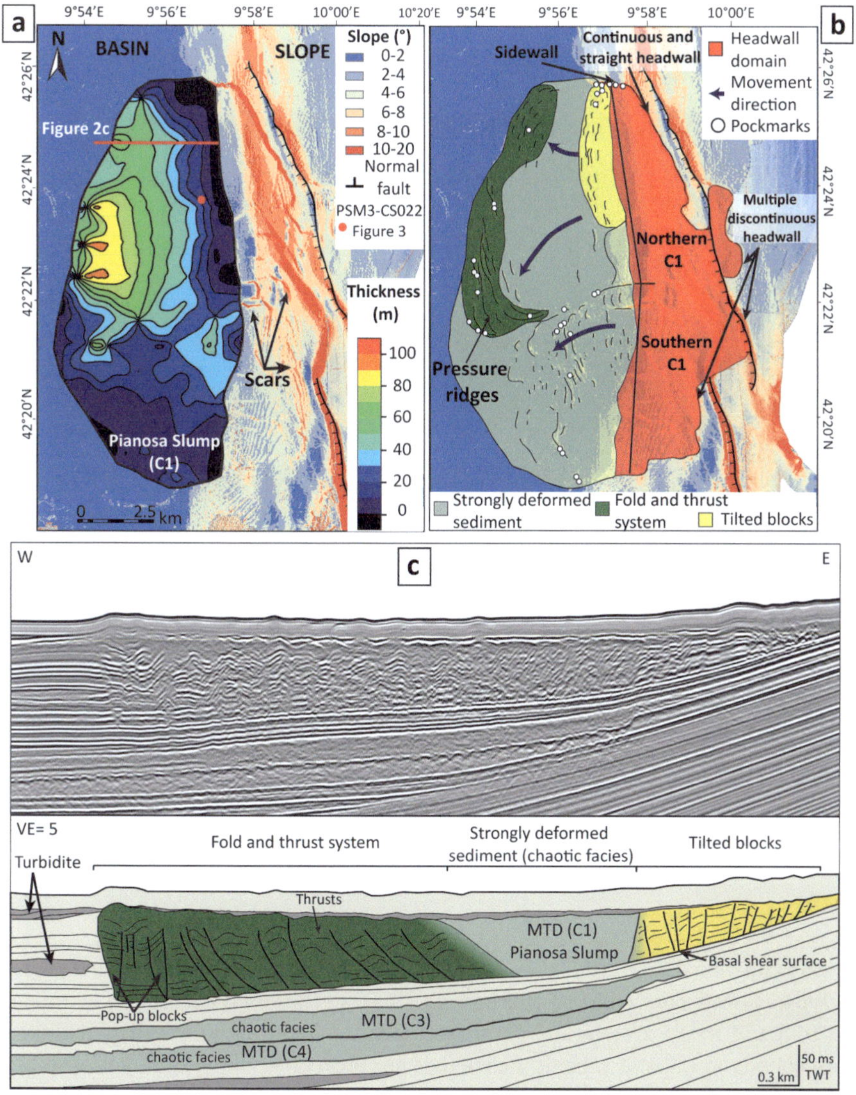

Fig. 30.2 (**a**) Isopach map of the Pianosa Slump (C1) overlaying the present seafloor gradient map with the location of Calypso piston core PSM3-CS022. (**b**) Morphological interpretation of C1. (**c**) Multichannel high-resolution mini GI gun seismic reflection profile across MTDs C1, C3 and C4 and line drawing with the interpretation

The headwall domain of C1 has the same lateral extent as the Pianosa Slump deposit. However, the headwall morphology is rather irregular. In the northern part, the headwall is continuous, parallel to a normal fault (Fig. 30.2b). It is connected to the MTD by a sidewall that is 30 m high at the present seafloor. South of this zone, the slope is cut by multiple discontinuous headwalls and scarps. C1 is composed of

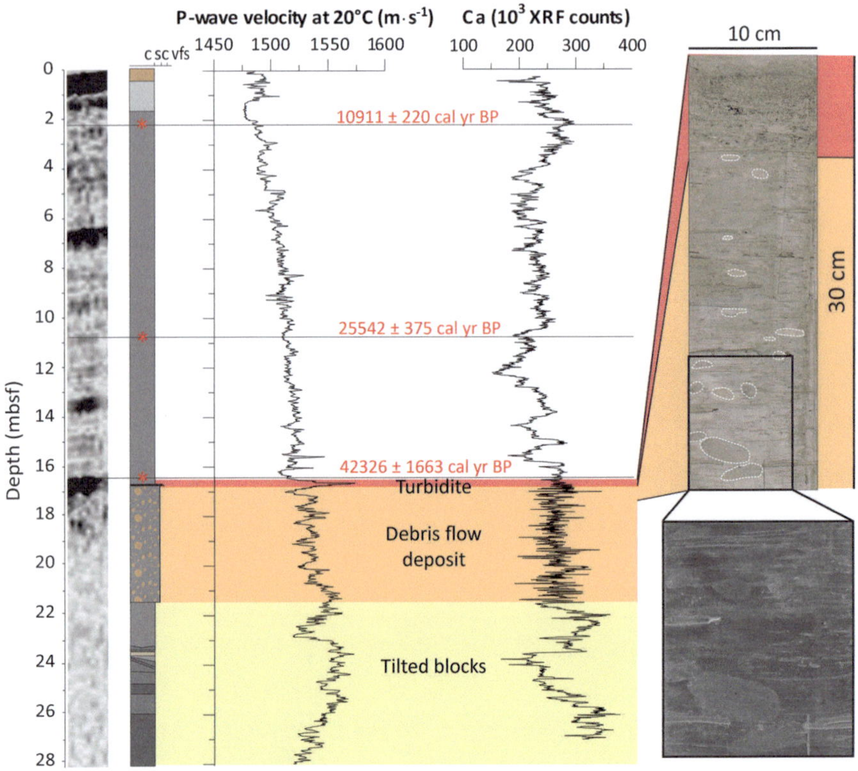

Fig. 30.3 From *left* to *right*: CHIRP seismic facies, core log, P-wave velocity and Calcium curves of Calypso piston core PSM3-CS022. Photo of the core and zoom with inverted colour. The *red* area shows a turbidite (peak in P-wave velocity), the *orange* area a debris flow deposit and the *yellow* area tilted blocks. Three calibrated radiocarbon ages are represented with red stars

three main parts (Fig. 30.2b, c) deduced from seismic profiles, bathymetry and Calypso piston core PSM3-CS022 (Fig. 30.3): (1) strongly deformed sediment, (2) fold and thrust system and (3) tilted blocks. The strongly deformed sediment corresponds to chaotic seismic facies that are present in the central northern part and in the southern part of C1, where the headwall is discontinuous. In this area the surface morphology shows lobate shapes and crests that indicate a movement towards the west that turned slightly to the southwest, in the direction of increasing depth. The MTD thickness of the southern part shows that there is a deposit at the foot of the slope up to 50 m thick and a thicker zone (50–70 m) in the central distal part.

In the northern part of C1 the internal structure of the deposit is more complex than in the southern part. In this area the toe domain of the MTD is formed by a fold and thrust system. The frontal part of C1 is composed of hemipelagites and turbidites that are deformed and compressed forming a system of thrusts, folds and pressure ridges, with individual pop-up blocks apparent in the frontal part of the

MTD (Fig. 30.2). The greatest thickness of C1 deposits (up to 100 m) is found in the central zone of C1 where the pressure ridges are higher (up to 10 m high) and larger (up to 2 km long) (Fig. 30.2a). The proximal part consists of tilted blocks (Fig. 30.2b) that become more deformed towards the south. Tilted blocks are identified in seismic profiles where reflectors dip in different directions. Moreover, tilted blocks form crests that are apparent in the seafloor. Groundtruthing evidence of the presence of tilted blocks comes from the bottom of Calypso piston core PSM3-CS022 (Fig. 30.3). This core shows that a 4.7-m-thick debris flow deposit composed of mud with clay pebbles lies on top of the tilted blocks. These distinct seismic and lithological facies correspond to different degrees of deformation, which can be identified in the Calcium curve. When the degree of deformation is very high (debris flow deposit), the Calcium signal is very noisy (Fig. 30.3). Between the compressed frontal zone and the zone of tilted blocks, the sediment is strongly deformed, characterised by chaotic seismic facies and by a relatively smooth upper surface (Fig. 30.2). Since the pressure ridges and the crests are perpendicular to the mass movement, they provide information on the direction of the transport. The northern part of the Pianosa Slump comes from the area of the straight headwall parallel to the fault. The northernmost part moved slightly towards the northwest, while most of the sediment was transported in a southwest direction, towards the deepest part of the basin.

Multibeam bathymetry reveals the presence of many pockmarks located in the contourite drifts between areas B and C and close to MTDs in zone C (Fig. 30.1). The fluid escape paths over the Pianosa Slump are mainly found in the pressure ridges, in the crests and in the sidewall (Fig. 30.2b).

30.4 Discussion and Conclusions

Slope instability has been a recurrent process along the Pianosa Ridge, although the origin and the control factors could vary according to the location of the submarine landslides. The differences in size and volume between MTDs in area B and C are linked to an uneven sedimentation along the Pianosa Ridge related to the basement morphology and to bottom currents. Sediment accumulation on the slope in area C is much higher compared to area B. Therefore, more sediment could be potentially mobilised by mass transport during a single event, generating larger submarine landslides. In area B bottom currents are accelerated due to the higher slope gradient, preventing sediment deposition.

Group B and C present some similarities in the time intervals of slope failure, since B1-C1 and B2-C2 are in similar ranges of age. In contrast, the other MTDs are not synchronous. When C3 and C4 were formed, area B was in stable conditions (Table 30.1).

The main predisposing factor in areas A and B seems to be the steep slopes (8–20° and 8–10°, respectively) (Fig. 30.1). Within the Elba Canyon, due to the Coriolis effect, bottom currents could produce a stronger erosion on the northern flank of the canyon, generating the observed canyon asymmetry with a steeper northern flank (Fig. 30.1). Incisions and oversteepening on the northern flank might have contributed to the formation of MTD A. In area B, six MTDs are stacked, and represent a relevant percentage (30 %) of the Pliocene-Quaternary deposit in the lower slope, suggesting that an important proportion of the sediment accumulated on the middle and upper slope was affected by mass transport. The formation of submarine landslides in area C (Fig. 30.1) seems to be much more complex. Multiple factors may predispose slope instability in this area. Bottom currents may have an important role since they provide sediment and generate incisions with higher slopes, corridors of non deposition or slight erosion (moat with slope of 15°; Fig. 30.1). In addition, layers of coarser sediment in contourite drifts seem to influence the location of failure planes (Miramontes et al. 2014) and could potentially stock more gas than the overlying mud. Numerous pockmarks affect the present day seafloor on top of the MTDs and contourite drifts (Figs. 30.1 and 30.2). Some of these pockmarks are clearly rooted on buried MTDs and turbidite lobes (Fig. 30.1). Such a configuration suggests that the emplacement of these sedimentary bodies might have influenced subsequent instability.

In the Pianosa Ridge the present-day seismicity is considered to be very low (Scisciani and Calamita 2009), while the area entered in a post-rift phase from the late Messinian (Pascucci et al. 1999). However, the location of the headwall of the Pianosa Slump next to a normal fault that affects the Pliocene-Quaternary record (Fig. 30.3) raises questions about the possible relationship between faulting and slope instability.

Acknowledgments We thank the Captain and the crew of the PRISME2 and PRISME3 cruises (2013) onboard R/V Atalante and R/V Pourquoi pas?, respectively. We are grateful to reviewers M. Rovere and C. Campbell for their valuable suggestions that improved the manuscript. The thesis of Elda Miramontes is co-funded by TOTAL and Ifremer as part of the scientific project TOTAL-Ifremer PAMELA. We thank projects LabexMER ANR-10-LABX-19-01 Axis 4 for lab support.

References

Angue Minto'o CM (2014) Enregistrements sédimentaires des changements climatiques et environnementaux pendant le quaternaire terminal sur la marge Est-Corse. PhD Thesis, Université de Perpignan, Perpignan, 309

Artale M, Gasparini GP (1990) Simultaneous temperature and velocity measurements of the internal wave field in the Corsican Channel (Eastern Ligurian Sea). J Geophys Res 95(C2):1635–1645

Cattaneo A, Jouet G, Charrier S, Thereau E, Riboulot V (2014) Submarine landslides and contourite drifts along the Pianosa Ridge (Corsica Trough, Mediterranean Sea). In: Submarine Mass Movements and Their Consequences: 6th International Symposium 37:435–445

Deptuck ME, Piper DJW, Savoye B, Gervais A (2008) Dimensions and architecture of late Pleistocene submarine lobes off the northern margin of East Corsica. Sedimentology 55:869–898

Miramontes E, Cattaneo A, Jouet G, Garziglia S, Thereau E, Gaillot A, Roubi A, Rovere M (2014) The Pianosa Contourite Depositional System (Corsica Trough, North Tyrrhenian Sea): stratigraphic evolution and possible role in slope instability. In: Van Rooij D, Rüggeberg A (eds). 2014. Book of abstracts. 2nd deep-water circulation congress: the contourite log-book. Ghent, Belgium, 10–12 Sept 2014. VLIZ Special Publication 69:15–16

Pascucci V, Merlini S, Martini IP (1999) Seismic stratigraphy of the Miocene-Pleistocene sedimentary basins of the Northern Tyrrhenian Sea and Western Tuscany (Italy). Basin Res 11:337–356

Reimer PJ et al (2013) IntCal13 and Marine13 radiocarbon age calibration curves 0–50,000 years cal BP. Radiocarbon 55:1869–1887

Roveri M (2002) Sediment drifts of the Corsica Channel, northern Tyrrhenian sea. In: Stow DAV, Pudsey CJ, Howe JA, Fauge'res J-C, Viana A (eds) Deep-water contourite systems: modern drifts and ancient series, seismic and sedimentary characteristics, vol 22. Geological Society of London, London, Memoirs, pp 191–208

Scisciani V, Calamita F (2009) Active intraplate deformation within Adria: examples from the Adriatic region. Tectonophysics 476:57–72

Chapter 31
Late-Holocene Mass Movements in High Arctic East Lake, Melville Island (Western Canadian Arctic Archipelago)

Alexandre Normandeau, Gabriel Joyal, Patrick Lajeunesse, Pierre Francus, Scott Lamoureux, and François Lapointe

Abstract East Lake, located at Cape Bounty (Melville Island, Canadian High Arctic), was mapped using a high-resolution swath bathymetric sonar and a 12 kHz sub-bottom profiler, allowing for the first time the imaging of widespread occurrence of mass movement deposits (MMDs) in a Canadian High Arctic Lake. Mass movements occurred mostly on steep slopes located away from deltaic sedimentation. The marine to lacustrine transition in the sediment favours the generation of mass movements where the underlying massive mud appears to act as a gliding surface for the overlying varved deposits. Based on acoustic stratigraphy, we have identified at least two distinct events that triggered failures in the lake during the last 2000 years. The synchronicity of multiple failures and their widespread distribution suggest a seismic origin that could be related to the nearby Gustaf-Lougheed Arch seismic zone. Further sedimentological investigations on the MMDs are however required to confirm their age and origin.

A. Normandeau (✉)
Centre d'études nordiques, GEOTOP and Département de Géographie, Université Laval, Québec, QC, Canada G1V 0A6

Department of Geography, Queen's University, Kingston, ON, Canada K7L 3N6
e-mail: alexandre.normandeau@queensu.ca; alexandre.normandeau.1@ulaval.ca

G. Joyal • P. Lajeunesse
Centre d'études nordiques, GEOTOP and Département de Géographie, Université Laval, Québec, QC, Canada G1V 0A6

P. Francus • F. Lapointe
Institut national de la recherche scientifique, Centre Eau, Terre et Environnement and GEOTOP, Québec, QC, Canada G1K 9A9

S. Lamoureux
Department of Geography, Queen's University, Kingston, ON, Canada K7L 3N6

G. Lamarche et al. (eds.), *Submarine Mass Movements and their Consequences*, Advances in Natural and Technological Hazards Research 41,
DOI 10.1007/978-3-319-20979-1_31

31.1 Introduction

Sediment archives from Arctic lakes are commonly used for paleoenvironmental and paleoclimatic reconstructions (e.g., Lapointe et al. 2012). Lake basins in the Arctic can accumulate large amounts of sediment derived from landscape runoff and in some cases preserve varves (e.g., Lapointe et al. 2012). Because this polar region is highly sensitive to climate change via feedback processes, high-resolution sedimentary records are particularly beneficial for assessing natural climate variability and for validating climate models (Kaufman 2009). However, an increasing number of studies demonstrate that lacustrine sedimentation is also influenced by mass movements, which can be unrelated to climate variability (e.g., Beck 2009; Waldmann et al. 2011). Understanding sedimentary processes within lakes is thus of major interest for interpreting past climate and predicting future climatic change. Studying such processes in Arctic lakes is challenging because they are often inaccessible with an ice cover for up to 10 months per year. Further, lake studies in the Arctic rely solely on the interpretation of sediment cores while the geological, stratigraphic and geomorphological context in which sediment deposition takes place are often left undocumented.

In this paper, we report and describe for the first time sub-lacustrine landslides and sedimentary processes that occurred in a Canadian High Arctic Lake (East Lake, Melville Island) from the analysis of high-resolution swath bathymetry and sub-bottom profiling. This paper also aims at documenting pre-conditioning factors and the approximate timing of mass movement deposits in order to identify possible triggers for extreme events recorded in previously analyzed sedimentary records from the lake (Cuven et al. 2011; Lapointe et al. 2012).

31.2 Regional Setting

East Lake (unofficial name, 74° 53′ N, 109° 32′ W) is located on the south-central coast of Melville Island, in the western Canadian Arctic Archipelago (Fig. 31.1a). It is a small-sized and ~30-m deep lake (1 × 2 km) located 5 m above sea-level (asl; Fig. 31.1b). The area was entirely covered by the Laurentide Ice Sheet (LIS) during the Late-Wisconsinan (Nixon et al. 2013). By 14 ka cal BP, the retreat of the LIS margin led to the marine invasion, which reached areas today located at ~75 m asl. Following emergence from the sea due to glacio-isostatic rebound, an estuary and then a lake formed by 2195 BC and 243 AD respectively, providing a favorable environment for varve formation (Cuven et al. 2011).

The watershed of East Lake (and the adjacent West Lake) has been monitored since 2003 for hydrological, limnological and sediment transport and deposition studies, making it the longest comprehensive hydrological-limnological monitoring program in the Canadian High Arctic. For example, Cuven et al. (2011) and Lapointe et al. (2012) have linked hydroclimatic variability to the physical and geochemical properties of the varves. Lapointe et al. (2012) documented a recent

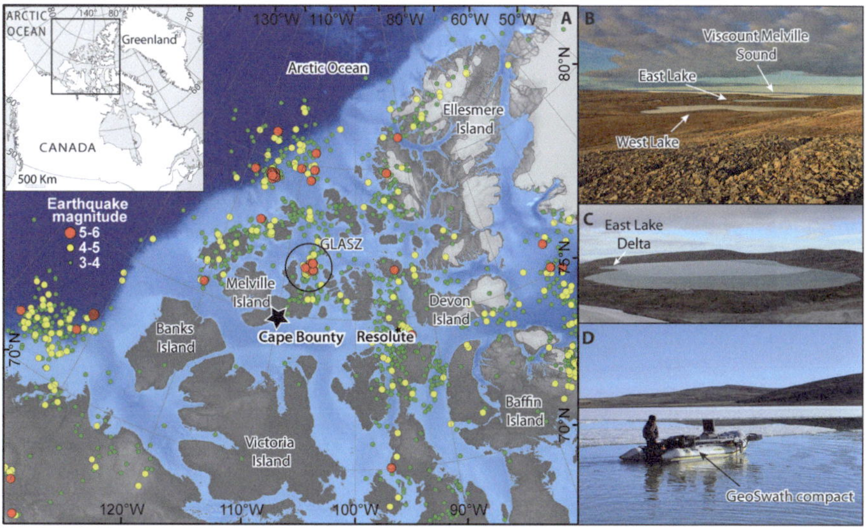

Fig. 31.1 (**a**) Location of Cape Bounty (Melville Island, Canadian High Arctic) and of recent earthquakes (GLASZ = Gustaf-Lougheed Arch Seismic Zone); (**b**) Overview of East and West Lake in early August 2013; (**c**) View of East Lake and its delta. Note the presence of ice, which limited the coverage of the hydroacoustic surveys over the entire lake; (**d**) GeoAcoustics GeoSwath echosounder mounted on an inflatable boat on East Lake

increase in rainfall events since 1920, with unprecedented levels occurring late in the twentieth century that are likely linked with the ongoing warming in the Arctic.

East Lake is located ~150–200 km south-west from the Gustaf-Lougheed Arch Seismic Zone (GLASZ) (Fig. 31.1a). An earthquake swarm (65 locatable earthquakes in 55 days) with four major (M>5) earthquakes occurred in the region during November-December 1972 (Hasegawa 1977). An earthquake of M~5.6 also occurred in 2001.

31.3 Methods

A high-resolution bathymetric map of East Lake was produced from a hydroacoustic survey undertaken in 2013. A GeoAcoustics GeoSwath Plus compact interferometric bathymetric sonar (250 kHz) was deployed on a 2.5 m inflatable boat (Fig. 31.1d). A total of 12.5 km of subsurface data was acquired on the northern half of the lake using a 12 kHz Knudsen 3212 echosounder from the same boat and interpreted using The Kingdom Suite® and SonarWiz 5.0® softwares. Due to the presence of residual lake ice during the survey in early August 2013 (Figs. 31.1c, d), swath bathymetry data was collected over 75 % of the lake while sub-bottom profiles were collected only over the northern half of the lake area. The survey capabilities were highly dependent on daily wind directions which pushed the seasonal lake-ice cover from one end of the lake to the other.

31.4 Results

31.4.1 High-Resolution Bathymetry

East Lake forms a 30 m-deep overall round basin. The greatest depths are located in the central part of the lake, near the East River delta (Fig. 31.2). The slopes are steep on all sides of the basin, reaching $\geq 10°$ (Fig. 31.3a). The slopes are especially steep

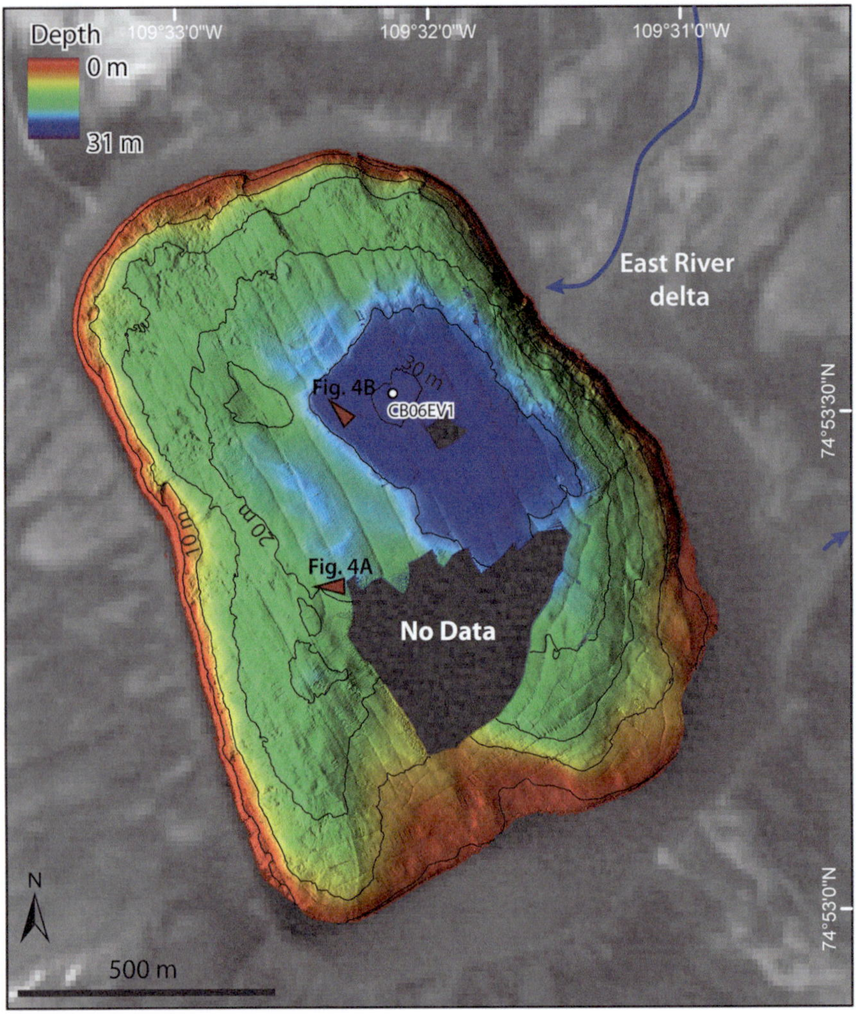

Fig. 31.2 Swath bathymetry map of East Lake (Cape Bounty, Melville Island). Regions of no data are due to the presence of ice during the 2013 survey. *Red arrows* indicate the viewpoints of Fig. 31.4

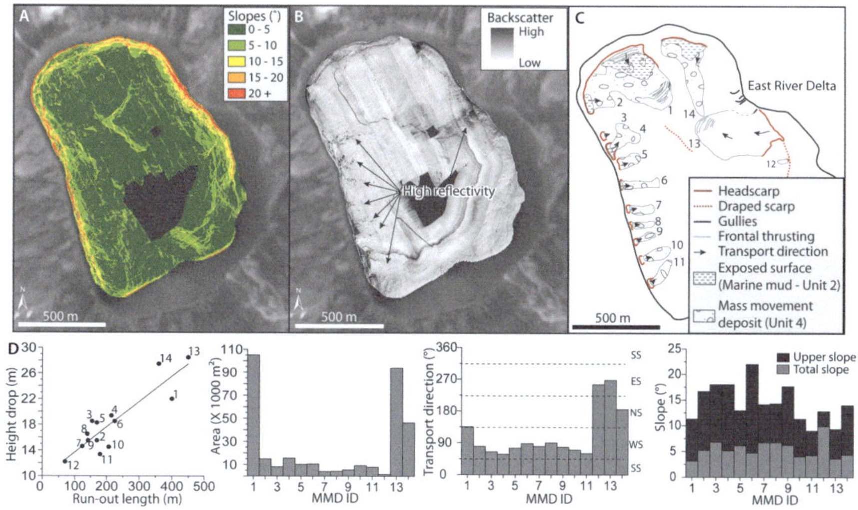

Fig. 31.3 Morphological characteristics of East Lake mass movements: (**a**) Slope map; (**b**) Reflectivity map; (**c**) Geomorphological map of mass movement's morphologies of the lake floor; (**d**) Morphological properties of 14 mass movements identified in (**c**)

on the eastern, northern and western margins of the lake, where they occasionally reach ≥15°. The southern portion has a gentle slope that rarely reaches ≥5°.

Two main types of lake-floor morphologies are observed in the lake: (1) gullies at the delta front; and (2) mass movement deposits (MMDs) throughout the lake, mostly on the northern and western slopes (Fig. 31.3c). Erosional gullies are ~50 m long and less than 10 m wide and are located at the mouth of the East River (Fig. 31.2). Fourteen MMDs were identified from surface morphology alone (Fig. 31.3c). The sizes of the MMDs vary along the slopes of the lake, but most have height drops of 14–20 m and run-out distances of 150–250 m. Three MMDs are much larger (No. 1, 13, 14; Fig. 31.3c), with run-out distances and height drops of >350 m and >22 m, respectively (Figs. 31.3c, d). The scar gradient of each MMD is generally ≥10°. The mean area of the deposits is 10,000 m^2, excluding the two largest that are ~100,000 m^2. One of the two largest MMDs is found near the delta front, and the other one on the northern slope. Compressional ridges are located at the front of these two larger MMDs while extensional ridges are observed near the headscarp of the northernmost MMD (Fig. 31.4b). The smaller MMDs consist mainly of boulder-size debris (Fig. 31.4a).

The backscatter map reveals a general low-intensity backscatter of the lake floor (Fig. 31.3b). This low intensity is interpreted to represent the undisturbed and normally deposited lacustrine sediment. The highest intensities observed along the western slope of the lake floor illustrate the debris nature of the MMDs (Figs. 31.3b and 31.4a). The lowest intensities are observed on the northern and southern parts of the lake, where mass movements cannot be identified from

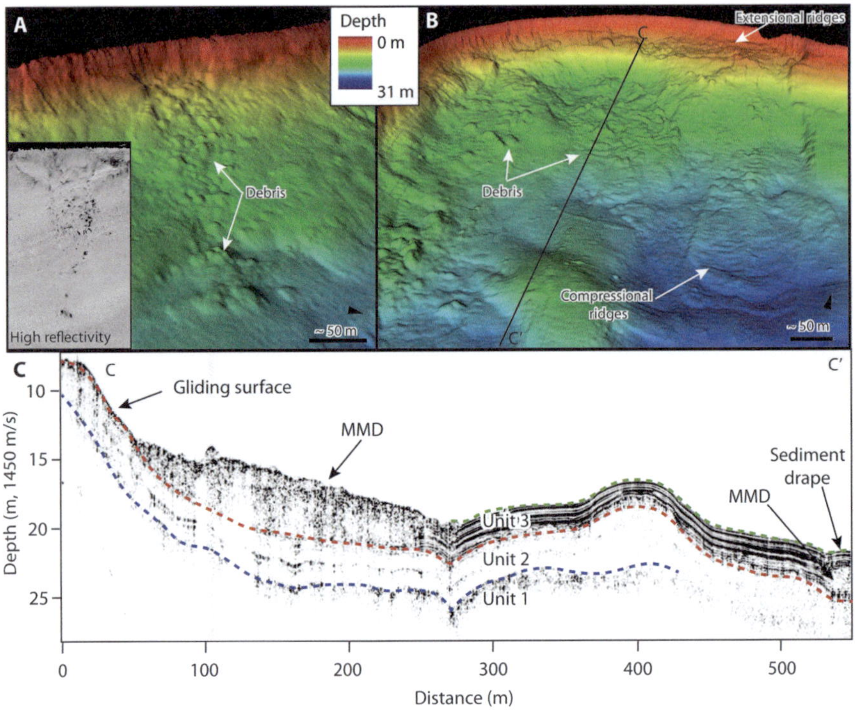

Fig. 31.4 3D view of mass movements 6 (**a**) and 1 (**b**) identified in East Lake. Location of images is provided in Fig. 31.2; (**c**) Sub-bottom profile illustrating the three acoustic units and MMDs. Location of profile provided in (**b**)

backscatter intensities. The delta front is also characterized by higher intensities, reflecting the coarser nature of the sediment discharged from the East River.

31.4.2 *Acoustic Stratigraphy*

Three acoustic stratigraphic units were identified in East Lake (Figs. 31.4c and 31.5). Lowermost Unit 1 is characterized by an absence of penetration and represents the acoustic basement. Unit 2 overlies Unit 1 and is acoustically transparent with few low-amplitude reflections. It appears to be present throughout the lake. Based on the depth of this unit from the lake bottom ($\leq$4 m), its uppermost part appears to have been reached by a 7 m-long core (Cuven et al. 2011) and represents massive compacted mud, interpreted as marine sediments deposited prior to the lacustrine phase $\geq$4 ka BP. Unit 3 consists of high-amplitude reflections and drapes the underlying units. It was previously cored and consists of annually deposited silt and sand rythmites (varves) (Cuven et al. 2011; Lapointe et al. 2012).

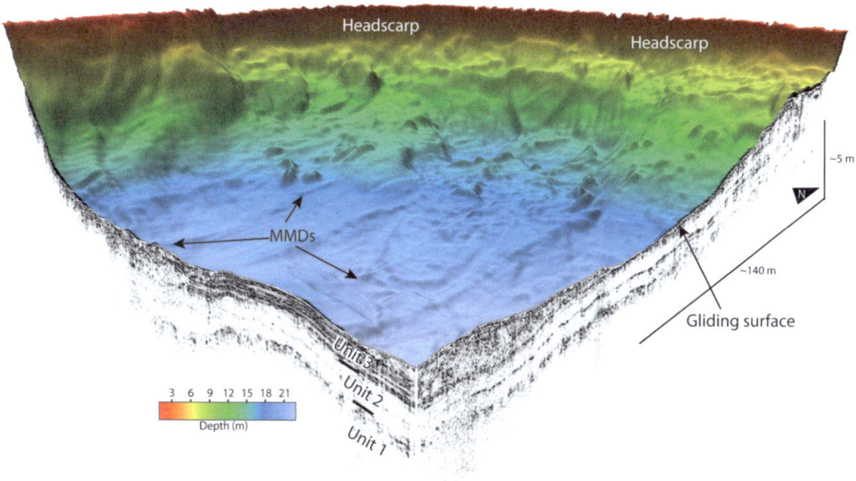

Fig. 31.5 3D view of swath bathymetry on sub-bottom profiles illustrating units 1–3 present in East Lake

Unit 3, however, is not present throughout the entire lake subsurface (Figs. 31.4c and 31.5). It is absent on steep sub-lacustrine slopes but can reach 4 m in thickness in the deepest part of the lake. At the top or within Unit 3, transparent to chaotic bodies (Fig. 31.4c) are interpreted as localized MMDs and are mostly observed at the base of steep slopes. The massive mud of Unit 2 is often exposed where the mass movements eroded the overlying Unit 3 (Figs. 31.4c and 31.5), illustrating that this unit probably acts as a gliding surface for mass movements.

31.5 Discussion

31.5.1 Factors Pre-conditioning Failures

High-resolution bathymetric data and sub-bottom profiles collected in East Lake show widespread evidence of MMDs. These MMDs are mostly located on steep slopes of the western and northern sectors of the lake. On the eastern side, steep slopes are also observed but gullies indicate that turbidity currents erode the lake floor, particularly at the delta front. Sub-bottom profiles reveal that the varved sediments (Unit 3) slid over the underlying massive mud (Unit 2), suggesting a stratigraphic control in the generation of mass movements where the top of Unit 2 acts as a gliding surface for the overlying silty-to-sandy laminated sediments. This abrupt change in type of sediment (e.g., marine to lacustrine) reduces slope stability and favours the development of a weak layer or weak interface that can subsequently lead to failure (e.g., Strasser et al. 2007; Baeten et al. 2014). Therefore, the highly stratified varved sediments, the marine to lacustrine transition and the steepness of the slopes surrounding the lake are pre-conditioning factors favoring the triggering of mass movements in East Lake.

31.5.2 Recent Sediment Failures

The number of individual events responsible for the occurrence of MMDs in East Lake is difficult to quantify due to the absence of sediment cores associated with each mass movement. Based on the available sub-bottom profiles, at least two different episodes of sediment failures appear to have occurred during the last ~2000 years. ~1 m of varved sediment covers two MMDs, one of them visible on Fig. 31.4c. Moreover, core CB06EV1 (Fig. 31.2) has a prominent coarse layer at a depth of 1.3 m that was deposited at 1300 AD (Lapointe et al. 2012). This coarse layer (turbidite?) is probably the distal result of the lowermost MMD located on the eastern margin (MMD 13) thus suggesting a date of ~1300 AD for this first mass movement episode (Lapointe et al. 2012).

Core CB06EV1 recorded many coarse layers and turbidites, which suggests an influence from river floods in the accumulation of sediments in the lake (Lapointe et al. 2012). The presence of gullies reveals the occurrence of turbidity currents that could have been triggered by flood discharge and associated high suspended sediment loads. Hyperpycnal currents are frequent in freshwaters (e.g., Simonneau et al. 2013) and occur in East Lake since suspended sediment concentrations are especially high during snowmelt (Cockburn and Lamoureux 2008). Presently, the differentiation between river-generated turbidites (flood event layers) and mass movement induced turbidites has not been accomplished because previous studies in East Lake have not identified the distinct sedimentological signature of both deposits.

Thirteen other MMDs appear to have a similar and more recent age since they are located in the upper acoustic stratigraphy sequence (Fig. 31.4c). One of the largest turbidites analyzed in previous studies in the lake was dated with varve counting, ^{210}Pb and ^{137}Cs dating at 1972–1973 AD by Cuven et al. (2011) and at 1971 AD by Lapointe et al. (2012). This turbidite could be related to hyperpycnal flows but we rather interpret it as the distal result of a mass movement that reached the deeper basin (e.g., MMD 1 and/or 14). Based on this interpretation, the second mass movement event observed in the lake could have been triggered between 1971 and 1973.

The synchronicity of multiple MMDs and their widespread distribution in the lake suggest that mass movements were triggered by seismicity. An earthquake swarm occurred in the GLASZ during November-December 1972, the strongest event reaching M~5.7, at a focal depth of 9–14 km (Fig. 31.1a) (Hasegawa 1977). The distance of the earthquakes (200 km) would attenuate the intensity felt at East Lake. However, accepting Cuven et al. (2011) core chronology, the recent MMDs and the turbidite observed in core CB06EV1 could be related to seismic activity in the GLASZ since it is dated to the winter of 1972–1973. This hypothesis is also supported by the presence of the MMDs on the northern slope, far from the East River inflow where sediment loading would be minimal. Earthquake shaking causes significant strength reduction in stratified sediments (Cauchon-Voyer et al. 2008) and could have caused slipping over Unit 2 and the failure of the steep lake margins. However, more sediment cores are needed to constrain the chronology of MMDs in East Lake and infer their exact triggers through time.

31.6 Conclusion

Our survey in East Lake allowed for the first time the imaging of sediment instabilities in a Canadian High Arctic Lake. These results show that mass movements are ubiquitous in East Lake (Melville Island) and that their deposits cover a large part of the lake floor. Data suggest that steep slopes and the stratified package of lacustrine sediments on top of compacted marine deposits are favouring factors for the triggering of mass movements. Swath bathymetry and acoustic stratigraphy indicates that at least two different mass movement events occurred during the last 2000 years. A tentative timing for the events is proposed based on previous sediment core analysis (Cuven et al 2011; Lapointe et al. 2012) and acoustic stratigraphy. A first event appears to have occurred near ~1300 AD while more recent events are tentatively dated to 1970–1973 based on a prominent turbidite present in a core record. These recent MMDs could be related to an earthquake swarm that occurred in the GLASZ in the winter of 1972. Collection of cores from MMDs will lead to a better dating of the events and allow reconstructing Holocene paleo-seismicity or environmental change leading to sediment failures in this High Arctic Lake. Although the exact processes responsible for MMDs in East Lake are not adequately explained for the moment, we have demonstrated that the interpretation of sediment cores from high-latitude lacustrine environments for paleo-environmental reconstructions needs to be supported by a high-resolution analysis of the geomorphology and stratigraphy of the lake floor.

Acknowledgements This study was supported by NSERC through Discovery grants to P.L., P.F., and S.L., NSERC Northern supplement grants to P.F., the Canadian Foundation for Innovation and the Ministère de l'Éducation du Québec through equipment grants to P.L., and the Polar Continental Shelf Program through grants to P.F. and S.L. Flavio Anselmetti, Marc DeBatist and Editor Sebastian Krastel provided helpful comments that improved the quality of this paper.

References

Baeten NJ, Laberg JS, Vanneste M, Forsberg CF, Kvalstad TJ, Forwick M, Vorren TO, Haflidason H (2014) Origin of shallow submarine mass movements and their glide planes – sedimentological and geotechnical analyses from the continental slope off northern Norway. J Geophys Res Earth Surf 119:2335–2360

Beck C (2009) Late-Quaternary lacustrine paleo-seismic archives in north-western Alps: examples of earthquake-origin assessment of sedimentary disturbances. Earth Sci Rev 96:327–344

Cauchon-Voyer G, Locat J, St-Onge G (2008) Late-Quaternary morpho-sedimentology and submarine mass movements of the Betsiamites area, Lower St. Lawrence Estuary, Quebec, Canada. Mar Geol 251:233–252

Cockburn JMH, Lamoureux SF (2008) Hydroclimate controls over seasonal sediment yield in two adjacent high Arctic watersheds. Hydrol Process 22:2013–2027

Cuven S, Francus P, Lamoureux S (2011) Mid to Late Holocene hydroclimatic and geochemical records from the varved sediments of East Lake, Cape Bounty, Canadian High Arctic. Quat Sci Rev 30:2651–2665

Hasegawa HS (1977) Focal parameters of four Sverdrup Basin, Arctic Canada, earthquakes in November and December of 1972. Can J Earth Sci 14:2481–2494

Kaufman DS (2009) An overview of late Holocene climate and environmental change inferred from Arctic lake sediment. J Paleolimnol 41:1–6

Lapointe F, Francus P, Lamoureux SF, Saïd M, Cuven S (2012) 1750 years of large rainfall events inferred from particle size at East Lake, Cape Bounty, Melville Island, Canada. J Paleolimnol 48:159–173

Nixon FC, England JH, Lajeunesse P, Hanson MA (2013) Deciphering patterns of postglacial sea level at the junction of the Laurentide and Innuitian Ice Sheets, western Canadian High Arctic. Quat Sci Rev 91:165–183

Simonneau A, Chapron E, Vannière B, Wirth SB, Gilli A, Di Giovanni C, Anselmetti FS, Desmet M, Magny M (2013) Mass-movement and flood-induced deposits in Lake Ledro, southern Alps, Italy: implications for Holocene palaeohydrology and natural hazards. Clim Past 9:825–840

Strasser M, Stegmann S, Bussmann F, Anselmetti FS, Rick B, Kopf A (2007) Quantifying subaqueous slope stability during seismic shaking: Lake Lucerne as model for ocean margins. Mar Geol 240:77–97

Waldmann N, Anselmetti FS, Ariztegui D, Austin JA Jr, Pirouz M, Moy CM, Dunbar R (2011) Holocene mass-wasting events in Lago Fagnano, Tierra del Fuego (54°S): implications for paleoseismicity of the Magallanes-Fagnano transform fault. Basin Res 23:171–190

Chapter 32
Pleistocene Mass Transport Deposits Off Barbados Accretionary Prism (Lesser Antilles)

Thibaud Pichot, Sara Lafuerza, Martin Patriat, and Walter Roest

Abstract Two Pleistocene mass transport deposits (MTDs), with volumes of thousands of km^3, have been identified from multi-channel seismic data in the abyssal plain at the front of the Barbados accretionary prism. Estimated sediment volumes for these MTDs are likely underestimated due to limited seismic coverage. In this work, we suggest that these MTDs are comparable in size to large submarine landslides as reported in the literature. These MTDs lie on the vicinity of two major oceanic ridges, the Barracuda Ridge and the Tiburon Rise. It is also suggested in this work that the MTDs come from seismicity associated with the formation of the Barracuda Ridge or the Barbados accretionary prism; however, triggering mechanisms involved in their formation remain uncertain. The present study discusses the potential causal factors accounting for the formation of these MTDs.

32.1 Introduction

The Lesser Antilles volcanic arc and the Barbados accretionary prism have been created by the subduction of the oceanic lithosphere of the North America and South America plates beneath the Caribbean plate (Fig. 32.1). In the forearc domain, numerous localised submarine landslides have been described mostly in the southern part of the Barbados accretionary prism (Sumner and Westbrook 2001; Deville et al. 2015). At the latitude of Anguilla and Martinique islands, the

T. Pichot (✉)
Ifremer, Centre de Brest, Géosciences Marines, ZI de la Pointe du Diable, CS 10070, 29280 Plouzané, France

Beicip-Franlab, 232 avenue Napoleon Bonaparte, BP 213, 92502 Rueil- Malmaison, France
e-mail: thibaud.pichot@gmail.com

S. Lafuerza
Fugro Geoconsulting SAS, 39 rue des Peupliers, 92752 Nanterre, France

M. Patriat • W. Roest
Ifremer, Centre de Brest, Géosciences Marines, ZI de la Pointe du Diable, CS 10070, 29280 Plouzané, France

G. Lamarche et al. (eds.), *Submarine Mass Movements and their Consequences*, Advances in Natural and Technological Hazards Research 41,
DOI 10.1007/978-3-319-20979-1_32

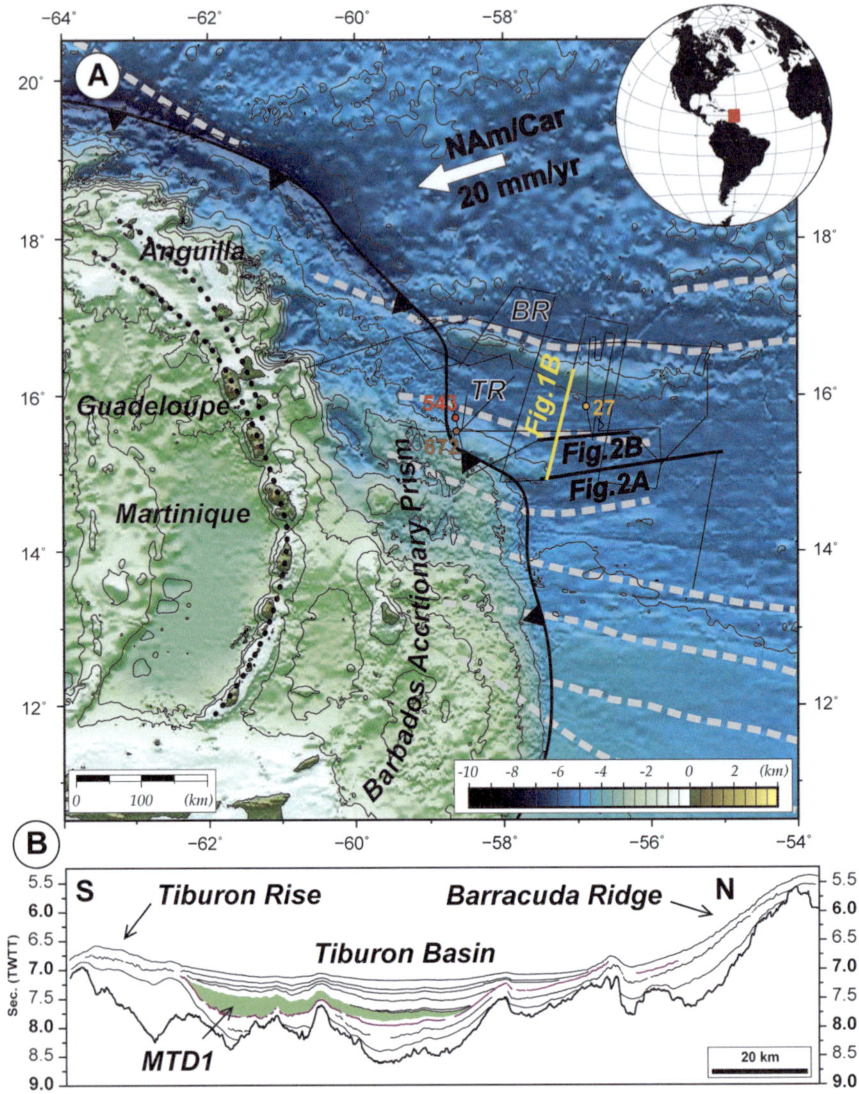

Fig. 32.1 (**a**) Bathymetric map (Smith and Sandwell 1997). *Grey dashed lines*; fracture zones (From Pichot et al. 2012). *Black dots*; volcanic arc. ANTIPLAC seismic profiles are indicated by *black lines*. *TR* Tiburon Rise, *BR* Barracuda Ridge. The DSDP-ODP drilling sites 27, 543 and 672 are shown by *colored dots*. NAm/Car; North America – Caribbean relative plate motion. (**b**) Interpretation of seismic profile crossing the Tiburon Basin. *Thick black* reflector; acoustic basement. *Purple* reflector; unconformity at the base of the quaternary sequence (H02)

Barbados accretionary prism is located above the diffuse plate boundary (between North America and South America). As a result of local readjustments of the oceanic lithospheres, two ridges are derived from oceanic fracture zones that were uplifted in the Middle-Late Miocene (Tiburon Rise) and the early Pleistocene

(Barracuda Ridge) (Patriat et al. 2011; Pichot et al. 2012). Of these two ridges, the Tiburon Rise is currently subducting and creating roughness in the downgoing plate beneath the accretionary prism.

This study uses seismic reflection data, multibeam bathymetry data and published borehole data to evaluate the origin of two giant MTDs. We present an age estimation for these MTDs and we also discuss the potential triggering mechanisms that could be associated with their emplacement.

32.2 Data and Methods

32.2.1 Geophysical Data

The data used in this study were acquired during the ANTIPLAC Cruise conducted by *Ifremer* in 2007 in offshore Martinique and Guadeloupe. The survey consisted of 4,160 km of multi-channel seismic reflection and multibeam bathymetry data (Fig. 32.1). Additional multibeam bathymetry data compiled from different sources for the ANTIPLAC program (Pichot et al. 2012) has been also used. The seismic data experiment was comprised of a 24-channel streamer and 2 generator injector air guns with a total volume of 300 in^3. Two additional seismic lines provided by the NOC (courtesy of Alan Evans), have also been used for this study (lines DM127 and EP129, Fig. 32.1).

32.2.2 Seismic Calibration Based on Borehole Data

The Demerara Abyssal Plain adjacent to the Barbados accretionary front has been subject of several drilling legs in the past, among them: DSDP leg 4 Site 27 (1969), leg 78A Site 543 (1981) and ODP legs 110 Site 672 (1986) (Fig. 32.1). The ANTIPLAC cruise was designed to cross these DSDP-ODP drilling sites. Because of the highest amount of quaternary sediments recovered at Site 27 (Bader et al. 1970) this site is of particular importance to study recent sedimentation processes occurring seaward of the Barbados accretionary prism. Considering velocity laws from previous studies in the area (Bangs et al. 1990; Dolman 2000), borehole depth and stratigraphic boundaries reported at Site 27 were converted into time (Patriat et al. 2011; Pichot et al. 2012;). This allowed us to calibrate borehole data into de seismic profiles to estimate the age of the MTDs (Fig. 32.2d).

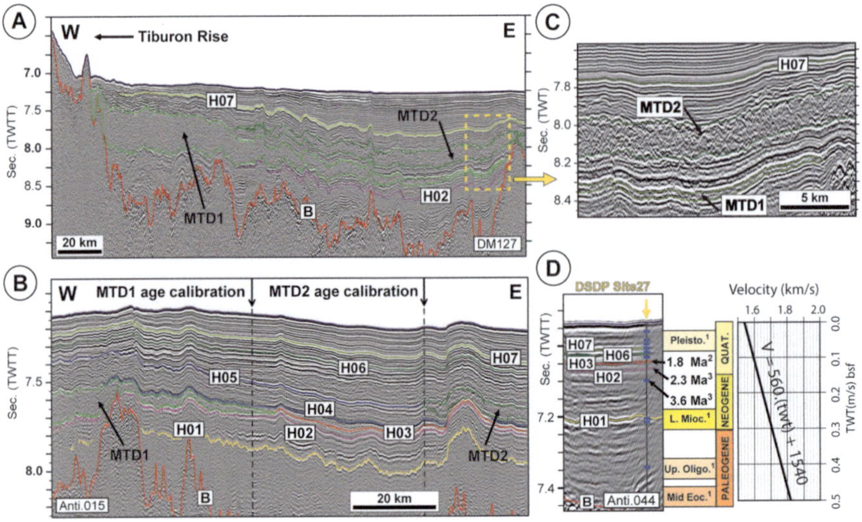

Fig. 32.2 (**a**) Seismic profile DM127 showing MTD1 and MTD2. (**b**) Seismic section Anti-015 showing quaternary horizons discussed in the text. The *black arrows* indicate the positions where the age calibration has been estimated. The *red* reflector refers as the acoustic basement. (**c**) Zoom of the MTDs seismic facies. (**d**) Correlation of section Anti-044 with the DSDP site 27 borehole. The ages are from Bader et al. (1970; superscripts 1) except for H03 (This study; superscripts 2), H02 and earlier marker (Patriat et al. 2011; superscripts 3). *Blue marks* is the top of cored intervals at Site 27. The velocity low used for time to depth conversions is also indicated

32.3 Results

The available seismic data shows a Quaternary sequence characterized by continuous, well-stratified seismic reflectors of weak and strong amplitude (Fig. 32.2). According to the stratigraphic record reported at Site 27 (Bader et al. 1970) and other regional studies (Wright 1984; Dolman 2000), the quaternary succession generally comprises interbedded distal turbidites within hemipelagic sediments. The Quaternary sequence onlaps against a progressive unconformity (horizon H02) (Fig. 32.1), which marks the onset of the Barracuda ridge uplift (Patriat et al. 2011; Pichot et al. 2012). Two bodies with chaotic seismic facies have been identified south of the Barracuda Ridge and east of the Tiburon Rise (Fig. 32.2). MTDs are named here, from oldest to youngest, as MTD1 and MTD2. Both bodies show erosional basal surfaces and are interpreted as MTDs (Fig. 32.2c). It is not known if these MTDs were multistage or emplaced as single events.

The relative ages of these MTDs were estimated by considering the timing of seismic horizons that define their bases and tops. This approach considers that MTD1 is bounded by horizons H02 and H03 (Fig. 32.2b). According to Patriat et al. (2011), H02 is 2.3 ± 0.2 Ma at Site 27, using the velocity law shown in Fig. 32.2c and considering a sedimentation rate of 4 cm/ka (Sigurdsson et al. 1980). The same sedimentation rate has been assumed for the interval comprised between

H02 and H03 (at line Anti-015, Fig. 32.2b). Accordingly, we estimate the age of H03 as 1.8 ± 0.2 Ma at Site 27. The base of MTD1 is approximately located 0.022 s (TWT) above H02 (see arrow in Fig. 32.2b), which corresponds to 24 m according to the velocity profile that was used. Considering a minimum sedimentation rate of 4 cm/ka, it is inferred then that the base of MTD1 is younger than 2.0 ± 0.2 Ma.

Sedimentation rates of 4 cm/ka, used at Site 27, cannot be applied for the quaternary sediments above seismic horizon H03 (Line Anti-015, Fig. 32.2b, d). For the interval between the seafloor and H03, where MTD2 is located, a mean sedimentation rate of 28 ± 4 cm/ka was used (uncertainty calculated from the vertical resolution of the multi-channel seismic data of 10 m). Sediment thickness between H03 and the base of the MTD2 is ~114 m (~0.10 s TWT, Fig. 32.2b), which provides an age of the MTD2 base of 1.4 ± 0.3 Ma. If the thickness between H03 and the top of the MTD2 is ~142 m (~0.14 s TWT, Fig. 32.2b) the resulting age of the MTD2 top is 1.3 ± 0.3 Ma. The time of formation of MTD2 can then be bracketed between 1.3 and 1.4 ± 0.3 Ma. The uncertainties provided here are too large to precise if these MTDs were deposited as the result of a multiphase event or as part of one single episode of mass wasting.

The available seismic data does not cover the entire extent of these MTDs and therefore only minimum estimates of their areal distribution could be provided as part of this work (Fig. 32.3c). According to these estimates MTD1 and MTD2 cover minimum areas of $15 \cdot 10^3$ and $19.6 \cdot 10^3$ km^2 respectively. Both MTDs thin northwards reaching maximum thicknesses of 0.5 s (MTD1) and 0.4 s TWT (MTD2) (Fig. 32.3c). Erosional surfaces defining the base of these MTDs are more significant in the south whereas these units lie conformably above older sediments toward the north (Pichot et al. 2012). Taking into consideration the relative location of the Barracuda Ridge and the Tiburon Rise, and the partial thickness trends and areal distribution observed in the seismic data, we hypothesized that the source area for MTD1 might be located in the eastern edge of the Tiburon Rise. On the other hand, the source area for MTD2 remains still speculative even though it is our assumption that the source for this unit is located toward the south. Minimum estimated sediment volumes for MTD1 and MTD2 are 1,800 and 2,600 km^3 respectively. Sediment compaction curves were not taken into consideration when these calculations were performed.

32.4 Discussion

32.4.1 Slope Instability at Oceanic Ridges

Submarine landslides generated along slopes of oceanic highs have already been described by a multitude of authors (Pimm et al. 1972; Kristoffersen et al. 2007; Rodriguez et al. 2012). The fact that MTD1 is located at the edge of the eastern flank of the Tiburon Rise would point to the uplift of this rise as a potential trigger.

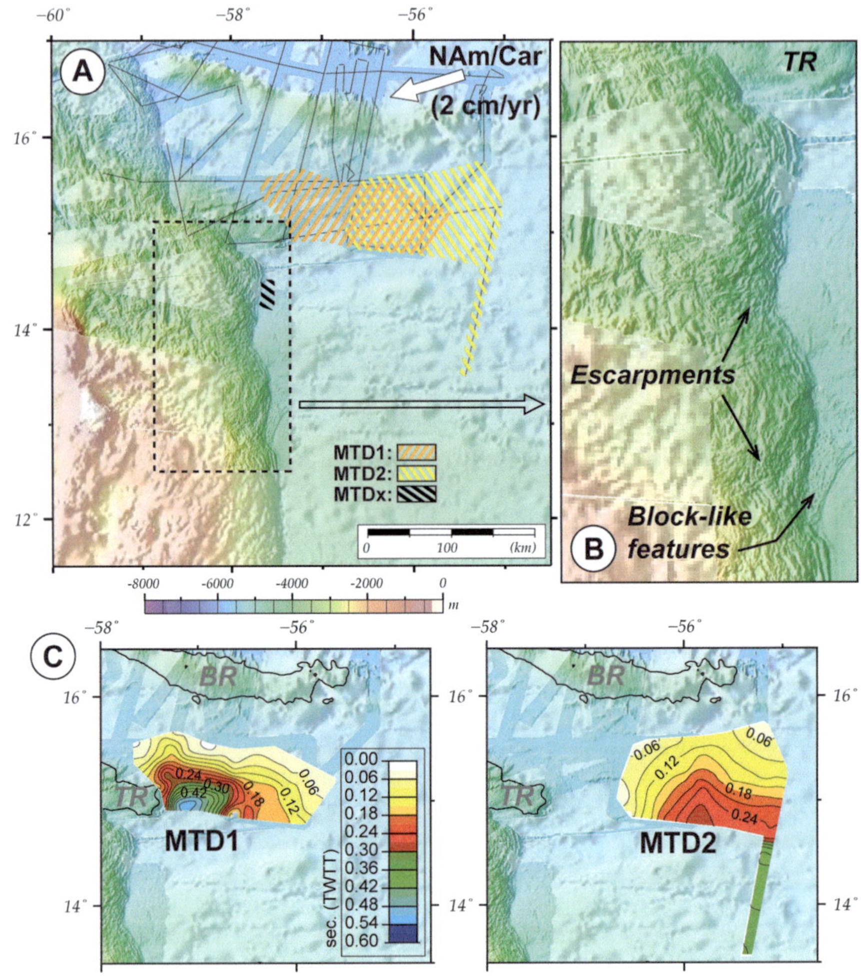

Fig. 32.3 (**a**) Combined multibeam bathymetric and satellite altimetry (Smith and Sandwell 1997) map showing the area covered by MTD 1 and MTD2. MTDx refers to the *slump deposit* described by Sumner and Westbrook (2001). (**b**) Zoom of the Barbados accretionary front showing destabilization features. (**c**) Isopach maps of each MTD showing their spatial distribution. Note that the thickness is in two way travel time (TWTT)

However, the uplift of the Tiburon Rise occurred in the Middle-Late Miocene long before the estimated emplacement of MTD1 (Pichot et al. 2012). This indicates that MTD1 was not generated during the main phase of uplift associated with the Tiburon Rise; instead a more recent seismic event could have destabilized the flanks of the Tiburon Rise triggering the formation of MTD1. It is interesting to notice that the timing of both MTDs is close to the time of uplift of the Barracuda Ridge at 2.3 Ma (Pichot et al. 2012); however, seismic and bathymetry data do not suggest that the source area for these units is linked to this structure (no headwall escarpments or other erosional features indicates that mass wasting affected the Barracuda Ridge). Even though no clear evidence exists to state that these MTDs were sourced from the Barracuda Ridge, it is possible that seismicity associated with the emplacement of this structural high could have affected the stability of the adjacent Tiburon Rise triggering the formation of these units.

32.4.2 Slope Instability at Barbados Accretionary Prism

Sumner and Westbrook (2001) described a deposit located to the east of the Barbados accretionary front that presents seismic facies that are similar to those observed within the studied MTDs (Fig. 32.3a). This deposit, referred as *slump deposit*, was dated between 1.8 and 1.1 Ma. Both, the seismic facies and the age of this slump deposit coincide well with the MTDs we described further north. A direct link between them is not clear because of the lack of seismic data between the two areas. It is unlikely that all these events were part of a single unit but it is possible that these MTDs were formed during the same time period. The suggestion is that broad slope instability processes at the beginning of the Quaternary affected the Barbados accretionary prism. Additional multibeam bathymetry data available for the ANTIPLAC program shows spectacular features of destabilisation of the front of the prism between 13 and 14°N (Fig. 32.3b). Adjacent to these escarpments, the seafloor presents numerous block-like features that can be attributed to MTDs running toward the northeast. The presence of these deposits strengthens the hypothesis that slope instabilities affected the Barbados accretionary prism. The geometry of the distal part of the studied MTDs reveals that the frontal part of the MTDs thins progressively towards the north and lies between parallel-stratified reflections. No compressional features such as reverse faults have been observed in the frontal part of these MTDs. They seem to be translated in an unconfined way over the seafloor, which could suggest relatively long travelled distances. It seems plausible therefore that the Barbados accretionary prism might also constitute a source area for MTD1 and MTD2 and we offer this hypothesis as an alternative interpretation.

32.5 Conclusions

Two MTDs (MTD1 and MTD2) were interpreted near the Barbados accretionary prism using seismic reflection and multibeam bathymetry data. The minimum areal extension of these MTDs ranges from $15 \cdot 10^3$ to $19.6 \cdot 10^3$ km^2 and their minimum estimated sediment volumes range from 1,800 to 2,600 km^3. Borehole data allowed us to bracket the age of the MTD1 and MTD2 between 2.0–1.8 $\pm$ 0.2 Ma and 1.4–1.3 $\pm$ 0.3 Ma, respectively. Determination of exact source areas for these events remains unclear due to lack of proper seismic coverage. However, we developed two hypotheses that explore potential triggering mechanisms and source areas for these units. Hypothesis 1 suggests that MTD1 was derived from the Tiburon Rise as suggested by stratigraphic relationships observed in one of the seismic lines (Fig. 32.2a); however, potential triggering mechanisms for MTD1 are associated with seismicity related to the later uplift of the Barracuda Ridge. Hypothesis 2 suggests that both MTDs were sourced from the Barbados accretionary prism to the south where similar looking MTDs have been imaged using multibeam bathymetry and satellite altimetry data. Hypothesis 2 also implies that triggering mechanisms are associated with seismicity associated with the formation of the Barbados accretionary prism. These working hypotheses need to be confirmed by the acquisition and examination of more extensive data sets that will allow for the proper mapping of these units.

Acknowledgments The authors sincerely thank Benoit Loubrieu for multibeam processing and Alan Evans for providing additional seismic profiles. We thank the government of Barbados and notably its ministry of Foreign Affairs to have given permission to use their seismic reflection data in this paper. Alan Evens of the National Oceanographic Centre, Southampton, facilitated the access to this data set. We strongly thank Jean-Yves Collot, Christian Beck and Lorena Moscardelli for their useful suggestions, which helped to improve the manuscript.

References

Bader RG, Gerard RD, Benson WE, Bolli HM, Hay WW, Thomas Rothwell W Jr, Ruef MH, Riedel WR, Sayles FL (1970) Site 27. Initial Rep Deep Sea Drill Proj 4:93–123

Bangs NLB, Westbrook GK, Ladd JW, Buhl P (1990) Seismic velocities from the Barbados Ridge complex: indicators of high pore fluid pressures in an accretionary complex. J Geophys Res 95:8767–8782

Deville E, Mascle A, Callec Y, Huyghe P, Lallement S, Lerat O, Mathieu X, Padron de Carillo C, Patriat M, Pichot T, Loubrieu B, Granjeon D (2015) Tectonics and sedimentation interactions in the east Caribbean subduction zone: an overview from the Orinoco delta and the Barbados accretionary prism. Mar Pet Geol 64:76–103

Dolman R (2000) Physical properties derived from seismic modeling at the toe of the Barbados accretionary complex. PhD thesis, University of Birmingham, UK

Kristoffersen Y, Coakley BJ, Hall JK, Edwards M (2007) Mass wasting on the submarine Lomonosov Ridge, central Arctic Ocean. Mar Geol 243:132–142

Patriat M, Pichot T, Westbrook GK, Umber M, Deville E, Bénard F, Roest WR, Loubrieu B, The ANTIPLAC Cruise Party (2011) Evidence for quaternary convergence between the North American and South American plates, east of the Lesser Antilles. Geology 39(10):979–982

Pichot T, Patriat M, Westbrook GK, Nalpas T, Gutscher MA, Roest WR, Deville E, Moulin M, Aslanian D, Rabineau M (2012) The Cenozoic tectono-stratigraphic evolution of the Barracuda Ridge and Tiburon Rise, at the western end of the North America-South America plate boundary zone. Mar Geol 303–306:154–171

Pimm AC, Burroughs RH, Bunce ET (1972) Oligocene sediments near chain ridge, northwest Indian Ocean: structural implications. Mar Geol 13:14–18

Rodriguez M, Fournier M, Chamot-Rooke N, Huchon P, Zaragosi S, Rabaute A (2012) Mass wasting processes along the Owen Ridge (NW Indian Ocean). Mar Geol 326–328:80–100

Sigurdsson H, Sparks RJS, Carey S, Huang TC (1980) Volcanogenic sedimentation in the Lesser Antilles arc. J Geol 88:523–540

Sumner R, Westbrook GK (2001) Mud diapirism in front of the Barbados accretionary wedge: the influence of fracture zones and North America–South America plate motions. Mar Pet Geol 18:591–613

Wright A (1984) Sediment distribution and depositional processes operating in the Lesser Antilles intraoceanic island arc, eastern Caribbean. In: Biju-Duval B, Moore JC et al (eds) Initial reports of the deep sea drilling project (DSDP), vol 78A. US Government Printing Office, Washington, DC, pp 301–324

Chapter 33
Exploring the Influence of Deepwater Currents as Potential Triggers for Slope Instability

Maria I. Prieto, Lorena Moscardelli, and Lesli J. Wood

Abstract Gravity-induced sediment transport processes and associated deposits in deepwater settings have been investigated for decades. However, the role of deepwater bottom currents as a preconditioning factor for mass-transport deposits (MTDs), as well as their capacity to redistribute sediments, is poorly understood. MTDs form an important component of the stratigraphic column within continental margins, and these units are often found in association with current-generated sediment waves (CGSWs). Our analysis of geophysical data from the Gulf of Mexico (GOM) indicates that there are different types of stratigraphic associations between CGSWs and MTDs in this region. Comparison of our data with existing studies where the relationship between MTDs and CGSWs has also been reported seems to suggest that there might be causal mechanisms and pre-conditioning factors that could explain this coupling. In this study, we discuss these relationships, taking into account the geologic and paleo-environmental conditions under which these units were deposited. Two main factors appear to control the stratigraphic coupling of MTDs and CGSWs in the study area: (1) fine-grained CGSWs act as shear basal surfaces that apparently precondition the slope for mass wasting events, and (2) contour-following bottom currents erode the lowermost slope, destabilizing its base and increase the likelihood of mass-wasting events.

M.I. Prieto (✉)
Jackson School of Geosciences, The University of Texas at Austin, Austin, TX, USA
e-mail: miprieto@utexas.edu

L. Moscardelli
Bureau of Economic Geology, Austin, TX, USA now at Statoil RDI, Austin, TX, USA

L.J. Wood
Department of Geology and Geological Engineering, Colorado School of Mines, Golden, CO, USA

G. Lamarche et al. (eds.), *Submarine Mass Movements and their Consequences*, Advances in Natural and Technological Hazards Research 41,
DOI 10.1007/978-3-319-20979-1_33

33.1 Introduction

The generally accepted term *contourite* is defined as "sediments deposited or substantially reworked by the persistent action of bottom currents" (Rebesco et al. 2014). This term was introduced nearly 50 years ago when Heezen et al. 1966 observed that geostrophic contour-following bottom currents controlled the formation of these deposits; the term itself implied a contour-conforming orientation of the deposits. To date, a broad continuum of sediments deposited by bottom currents is found in the literature. Deepwater sediment waves are defined as "large-scale undulating sedimentary bedforms" (tens of meters to kilometers in wavelength and several meters in wave height) that develop beneath deepwater currents (Wynn et al. 2000a; Wynn and Stow 2002). These features can develop through a variety of processes that include the action of episodic turbidity currents and of bottom currents (Wynn et al. 2000a; Wynn and Stow 2002). Deepwater sediment waves have been widely documented around the world using modern analogs; a series of criteria such as grain size, depositional environment, and morphometric parameters have been used to classify these features (Wynn and Stow 2002). The wave-crest orientation of the deposit is indicative of the flow direction. The majority of bottom-current-generated sediment waves are contour conforming, and the non-contour-conforming character of a sediment wave field most likely implies a turbidity-current origin. For purposes of this paper, deepwater current-generated sediment waves (CGSWs) will be defined as contour conforming when associated with bottom currents and as non–contour conforming when associated with sediment gravity flows (turbidity currents).

In this work, we present interpretations from a geophysical dataset that was acquired in the ultra-deepwater region of the Gulf of Mexico (GOM). In addition, we incorporate into our analysis analogs from previous case studies regarding the occurrence of deepwater CGSWs and mass-transport deposits (MTDs) that have been documented in different geologic settings (Fig. 33.1). Our main objective is to briefly discuss potential relationships associated with the nature of the coupling between deepwater CGSWs and MTDs. It is clear that examination of these relationships requires an in-depth analysis of core data so that a proper description of composition, grain-size distribution, state of consolidation, permeability variations, and estimations of sedimentation rates and pore-pressure profiles can be performed for both CGSWs and MTDs. Access to rock data is generally difficult when working in the subsurface; however, seismic-imaging techniques allow us to obtain a view of the main stratigraphic and geormorphological relationships that exist between CGSWs and MTDs within these domains. Description of these architectural relationships is also crucial to trying to unravel the nature of the relationship between CGSWs and MTDs; this work tackles this component of the problem. We seek to understand the geological context in which CGSWs and MTDs occur together; however, given data limitations, we explore these relationships by posing hypotheses that will need further confirmation.

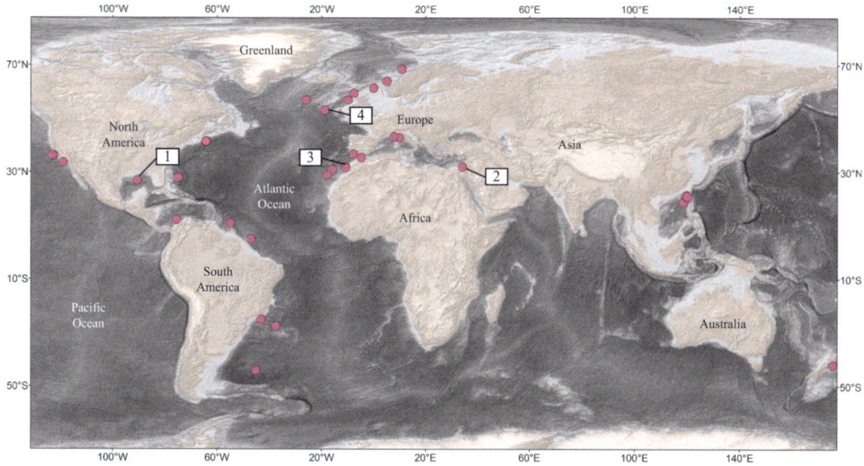

Fig. 33.1 Bathymetric map showing location of documented examples of deepwater CGSWs (*dots* in map). Numbered *white boxes* refer to the following areas: (1) Deepwater Gulf of Mexico (this paper); (2) Eastern Mediterranean (Frey-Martínez et al. 2006); (3) Morocco's Atlantic margin (Dunlap et al. 2010, 2013); (4) Rockall Bank, offshore western Ireland (Georgiopoulou et al. 2013)

33.2 MTDs and Current-Generated Sediment Waves

The relationship between MTDs and CGSWs has been noted previously by several authors (Bryn et al. 2005; Solheim et al. 2005; Frey-Martínez et al. 2006; Laberg and Camerlenghi 2008; Dunlap et al. 2010, 2013; Georgiopoulou et al. 2013; and many others). Deepwater CGSWs are found either above or below MTDs; preliminary observations suggest that these stratigraphic relationships can have genetic connotations. In this section, we explore the nature of the relationships that exist between MTDs and underlying CGSWs, using observations derived from geophysical data from the GOM and discussing potential analogies with other documented case studies.

33.2.1 CGSW and MTD Association in the Holstein Minibasin

High-resolution sub-bottom profiles acquired along the southern rim of the Holstein minibasin in deepwater central GOM show intercalations of parallel continuous reflections and chaotic semitransparent units interpreted as MTDs (Fig. 33.2a). Sub-bottom profiles also show the presence of undulating features below an MTD (Fig. 33.2b), which, based on their morphology, are interpreted as CGSWs. These features have wavelengths on the order of 200 m, wave heights of about 7 m, and a wave-crest orientation that is parallel to the local slope. The bedforms are

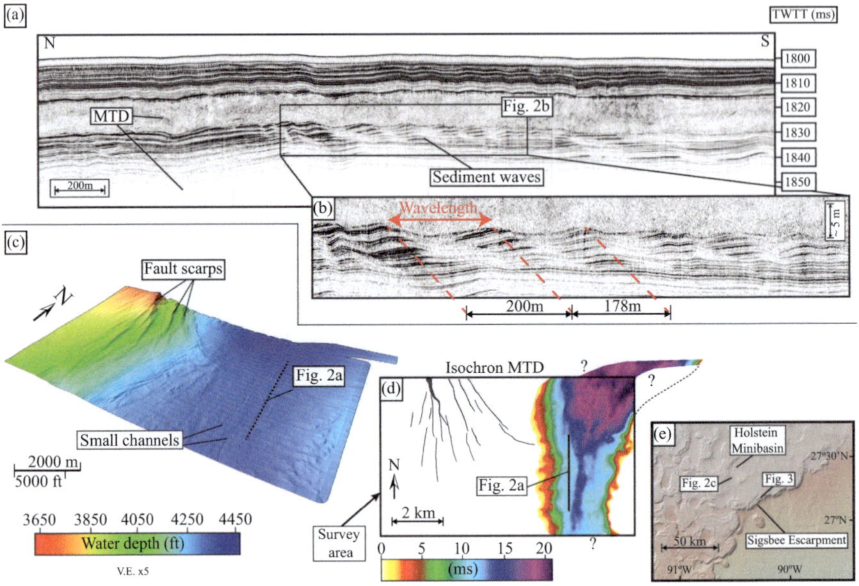

Fig. 33.2 (**a**) Subbottom profile showing CGSWs below an MTD, Holstein minibasin, deepwater Gulf of Mexico. (**b**) Crest-to-crest measurement methodology of sediment wave-train morphometrics. (**c**) Bathymetry at the southern edge of Holstein minibasin showing an open rim and concentric faults that define the edge of the minibasin. (**d**) Isochron map of MTD on Figure a. The unit shows a depocenter to the north and a narrow conduit to the south where it wedges to both west and east edges of the open rim. (**e**) Bathymetry map showing location of Holstein minibasin, deepwater Gulf of Mexico

asymmetrical, and present thicker beds in the upslope/upcurrent direction, which is indicative of upslope migration in the opposite direction of the dominant flow (Wynn et al. 2000b). The sub-bottom profiles show light and dark intervals, suggesting that the deposits are alternating temporally between high- and low-reflectivity material, possibly between coarser, low-reflectivity sands and finer, high-reflectivity muds (Young et al. 2003). The morphology, non-contour-conforming orientation, and scale of these deepwater CGSWs suggest that these features were most likely formed by turbidity currents flowing through the minibasin. The bedforms are not present throughout the entire study area but are seen only between two bathymetric highs (Fig. 33.2c). It is possible that this change in topographic configuration (narrower cross-sectional area and/or difference in topographic height moving from the depocenter toward the southern rim) (Fig. 33.2c) could have favored hydrodynamic conditions in which the flow became supercritical and promoted the formation of antidunes (Wynn et al. 2000a).

Subsequent deposition of an MTD is linked to gravitational instabilities associated with salt movement within the flanks of the minibasin. This MTD is a locally derived unit that is 20-ms thick (~15 m, assuming a constant acoustic velocity of 1500 m/s) and occupies ~20 km^2 in the study area (Fig. 33.2). However, this unit shows slight thinning between the two bathymetric highs, where thickness is ~15 ms (~11.5 m). This area seems to act as a narrow sedimentary conduit, which is also coincident with the area of underlying CGSWs. It is common for MTDs to have erosive bases; however, in this particular case the MTD does not appear to have eroded since the overall topography of the wave-set surface is retained. The overlying deposit appears to have 'slid' over the top of the local sediment wave field as it makes its way out from the confines of the minibasin, while being laterally constrained by the bathymetric configuration. The total aereal extent of the MTD is not imaged within the survey, and it is possible the incision rim is found upslope toward the depocenter.

33.2.2 *Nature of the CGSW and MTD Association in the Holstein Minibasin*

Shallow core data from this area was not available to the authors. However, it is possible to establish some analogies and derive hypotheses in terms of potential preconditioning factors that could be associated with the coupling between CGSWs and MTDs in the Holstein area. Multiple studies depicting CGSWs have reported that the majority of these units are composed of fine-grained lithologies (Wynn and Stow 2002). Even though not all deepwater muds are transformed into mechanically weak layers, a percentage of these muddy intervals develop conditions that are favorable for the formation of basal shear surfaces (Laberg et al. 2003; Ogiesoba and Hammes 2012; Locat et al. 2014). This is the case for an upper-Pliocene, frontally confined MTD off the shore of Israel that overlies contour-conforming deposits thought to have served as a detachment surface (see Fig. 9 of Frey-Martinez et al. 2006). The Israel slump complex in the eastern Mediterranean was deposited during a period of high clastic input. Some regions associated with the basal surface of the slump have been clearly identified as contour-conforming CGSWs and as potential detachment surfaces. Well logs confirmed that the CGSWs intervals are composed of fine-grained clastic sediments that are susceptible to overpressure owing to sediment loading (Frey-Martínez et al. 2006). The Mediterranean, like the GOM, is an area of important and complex water circulation; the occurrence of deepwater CGSWs within the Mediterranean is therefore common (Marani et al. 1993). Based on the character of the lithology and location of CGSWs, it is possible to hypothesize that these units can develop into MTD detachment surfaces if the right set of conditions is present to transform these underlying muddy intervals into mechanically weak layers. In order to confirm the validity of this hypothesis, it is necessary to secure core data penetrating these intervals and establish a pore-pressure profile with depth among other rock property profiles.

33.3 Bottom Currents and MTD Association in the Mad Dog Area (Sigsbee Escarpment, GOM)

In the Mad Dog area, located within the Sigsbee Escarpment of the GOM, steep slopes with scalloped morphologies and abundant slumps and slide scars are ubiquitous (Fig. 33.3). In the landward segment of the escarpment, the location and geometry of sedimentary pathways are controlled by underlying extensional grabens associated with deflated salt walls and minibasins (Fig. 33.3a). Debris flows and furrows characterize the front and toe of the escarpment. The presence of large, furrowed erosional features is a good indicator that bottom currents running parallel to the slope managed to rework the lower part of the escarpment into a furrowed topography. These furrows are evident both in the bathymetry and side-scan images (Fig. 33.3). Furrows have a width between 10 and 30 m and are 1–3 m deep. The furrowed seafloor extends for ~2 km off the toe of the escarpment in a basinward direction and has been mapped for several kilometers along the escarpment (Bryant and Bean 2000; Niedoroda et al. 2003).

The clear presence of furrows in close proximity to the toe of the slope within the Sigsbee Escarpment strongly suggests that bottom currents could have played an important role as destabilizing agents linked to MTD generation in this region. This observation is not unique to the GOM; it has been argued that a change in bottom-current intensity during the Cretaceous time in offshore Morocco promoted cannibalization of the lower continental rise, causing widespread slope instability and deposition of MTDs (Dunlap et al. 2010, 2013). The morphometrics of the CGSWs in offshore Morocco (wave height of 35 m and wavelengths of 2–15 km), along with their wave-crest orientation relative to the paleo-slope and the similarity of

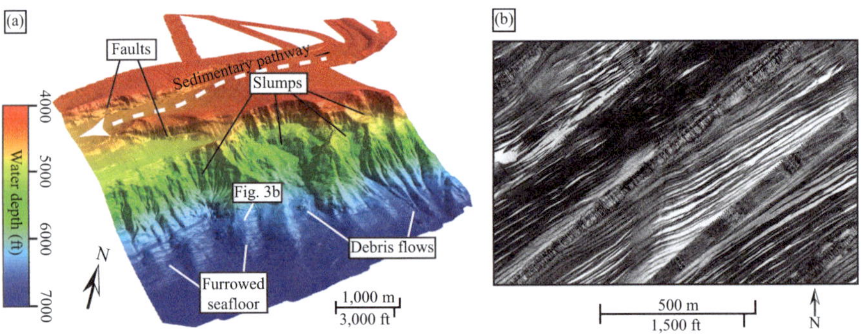

Fig. 33.3 (**a**) Bathymetry at the Mad Dog area, Sigsbee Escarpment, deepwater Gulf of Mexico. The escarpment is characterized by steep slopes, a scalloped morphology, and abundant slumps and slide scars. In the region landward of the escarpment, sedimentary pathways are defined by the planform trackways of extensional grabens associated with deflated salt walls. Debris flows and erosional furrows characterize the toe of the escarpment. *Dashed white lines* mark the axis of bathymetric lows, along which sediments are funneled to the escarpment slope and down to the abyssal plain. (**b**) Side-scan sonar images at the base of the escarpment showing erosional furrows trending parallel to the slope. These features have average widths of 10–30 m

their seismic expression to that of modern bottom-current sediment waves documented at other continental margins, support the view that these bedforms are fine grained and bottom-current generated (see Figs. 14 and 15 of Dunlap et al. 2013).

A similar mechanism of formation has been suggested for the Rockall Bank Slide Complex of offshore Ireland (Georgiopoulou et al. 2013). By investigating oceanographic processes controlling thermohaline circulation patterns and their relationship with gravity-induced deposits, it was concluded that the development of the Feni Drift was associated with the genesis of the Rockall Bank Mass Flow (Georgiopoulou et al. 2013). These observations debunk previous ideas suggesting that bottom-water circulation cannot affect the lower part of the slope and that these processes were mostly active in the abyssal plain. Based on these observations and analogies, it is reasonable to hypothesize that the action of bottom currents along continental margins has the capacity to destabilize the lower slope and make it prone to mass wasting events. Current data, as presented in this work, strongly supports this idea.

33.4 Discussion

Although the coupling of MTDs and CGSWs is common along several continental margins, there is no single link that governs the occurrence of this pairing. Controlling factors include (1) the cannibalization of the slope by bottom currents remobilizing sediments and triggering MTDs, and (2) the fine-grained lithologic nature of the blanketing wave fields defining basal detachment surfaces that facilitate subsequent mass wasting. Climatic conditions can favor bottom-current competence, appeasing downslope gravity processes. In some instances, however, local structural controls affecting bathymetry (such as diapirism) define a complex topographic configuration that can amplify and enhance bottom-current activity and result in the presence of local contour-conforming sediment waves. In the case of fields of non-contour-conforming sediment waves, where turbidity currents are the main forming agents and gravity-driven sedimentary processes are dominant, the resulting stratigraphic pair of CGSWs and MTDs is a consequence of high sediment input and/or local structural controls (e.g., diapirism) that induce sediment gravity- flow processes. The underlying non-contour-conforming sediment waves also play an important role as a preconditioning factor (detachment surfaces).

Acknowledgements This work was made possible thanks to the donation of data by BP America and through the generous members of the Quantitative Clastics Laboratory (QCL) consortia, the Jackson School of Geosciences, the GCSSEPM Ed Picou Fellowship Grant and the GSA Graduate Student Research grant. Constructive and insightful reviews by Dr. David Voelker and Dr. James Goff significantly improved the final manuscript and are gratefully acknowledged. The University of Texas at Austin acknowledges support of this research by Landmark Graphics Corporation via the Landmark University Grant Program. Publication authorized by the Director, Bureau of Economic Geology, Jackson School of Geosciences, The University of Texas at Austin.

References

Bryant W, Bean D (2000) Massive bed-forms, mega-furrows, on the continental rise at the base of the Sigsbee Escarpment, northwest Gulf of Mexico. Gulf Coast Assoc Geol Soc Trans L 2000

Bryn P, Berg K, Stoker MS et al (2005) Contourites and their relevance for mass wasting along the mid-Norwegian margin. Mar Pet Geol 22:85–96. doi:10.1016/j.marpetgeo.2004.10.012

Dunlap DB, Wood LJ, Weisenberger C, Jabour H (2010) Seismic geomorphology of offshore Morocco's east margin, Safi Haute Mer area. Am Assoc Pet Geol Bull 94:615–642. doi:10.1306/10270909055

Dunlap DB, Wood LJ, Moscardelli L (2013) Seismic geomorphology of early North Atlantic sediment waves, offshore northwest Africa. Interpretation 1:75–91

Frey-Martínez J, Cartwright J, James D (2006) Frontally confined versus frontally emergent submarine landslides: a 3D seismic characterisation. Mar Pet Geol 23:585–604. doi:10.1016/j.marpetgeo.2006.04.002

Georgiopoulou A, Shannon PM, Sacchetti F et al (2013) Basement-controlled multiple slope collapses, Rockall Bank Slide Complex, NE Atlantic. Mar Geol 336:198–214. doi:10.1016/j.margeo.2012.12.003

Heezen B, Hollister C, Ruddiman W (1966) Shaping of the continental rise by deep geostrophic contour currents. Science 152:502–508. doi:10.1126/science.152.3721.502

Laberg JS, Camerlenghi A (2008) The significance of contourites for submarine slope stability. In: Rebesco M, Camerlenghi A (eds) Developments in sedimentology. Elsevier, Amsterdam, pp 537–556

Laberg JS, Vorren TO, Mienert J et al (2003) Preconditions leading to the holocene Traenadjupet slide offshore Norway. In: Locat J, Mienert J (eds) Submarine mass movements their consequences. Kluwer Academic Publishers, Dordrecht, pp. 247–254

Locat J, Leroueil S, Locat A, Lee H (2014) Weak layers: their definition and classification from a geotechnical perspective. In: Krastel S, Behrmann J.-H, Völker D, Stipp M, Berndt C, Urgeles R, Chaytor J, Huhn K, Strasser M, Harbitz CB (eds) Submarine mass movements their consequences. Springers, Dordrecht, pp. 3–12

Marani M, Argnani A, Roveri M, Trincardi F (1993) Sediment drifts and erosional surfaces in the central Mediterranean: seismic evidence of bottom-current activity. Sediment Geol 82:207–220. doi:10.1016/0037-0738(93)90122-L

Niedoroda A, Reed C, Hatchett L (2003) Bottom currents, deep sea furrows, erosion rates, and dating slope failure-induced debris flows along the Sigsbee escarpment in the deep Gulf of Mexico. Offshore Technol Conf 0–5

Ogiesoba O, Hammes U (2012) Seismic interpretation of mass-transport deposits within the upper Oligocene Frio formation, south Texas Gulf Coast. Am Assoc Pet Geol Bull 96:845–868. doi:10.1306/09191110205

Rebesco M, Hernández-Molina FJ, Van Rooij D, Wåhlin A (2014) Contourites and associated sediments controlled by deep-water circulation processes: state-of-the-art and future considerations. Mar Geol 352:111–154. doi:10.1016/j.margeo.2014.03.011

Solheim A, Berg K, Forsberg C, Bryn P (2005) The Storegga Slide complex: repetitive large scale sliding with similar cause and development. Mar Pet Geol 22:97–107. doi:10.1016/j.marpetgeo.2004.10.013

Wynn R, Stow DAV (2002) Classification and characterisation of deep-water sediment waves. Mar Geol 192:7–22. doi:10.1016/S0025-3227(02)00547-9

Wynn R, Weaver P, Ercilla G (2000a) Sedimentary processes in the Selvage sediment-wave field, NE Atlantic: new insights into the formation of sediment waves by turbidity currents. Sedimentology 47:1181–1197. doi:10.1046/j.1365-3091.2000.00348.x

Wynn RB, Masson DG, Stow DAV, Weaver PP (2000b) Turbidity current sediment waves on the submarine slopes of the western Canary Islands. Mar Geol 163:185–198. doi:10.1016/S0025-3227(99)00101-2

Young A, Bryant D, Slowey D (2003) Age dating of past slope failures of the Sigsbee Escarpment within Atlantis and Mad Dog developments. Proceedings of OTC 2001 International Conference, p 24

Part V
Tectonics and Mass Movements

Chapter 34
French Alpine Foreland Holocene Paleoseismicity Revealed by Coeval Mass Wasting Deposits in Glacial Lakes

Emmanuel Chapron, Anaëlle Simonneau, Grégoire Ledoux, Fabien Arnaud, Patrick Lajeunesse, and Patrick Albéric

Abstract The French alpine foreland area has been struck by several earthquakes with magnitudes above 5 on Richter scale in recent history. In this paper we document the regional impact of historical and Holocene earthquakes based on the identification of mass wasting deposits in glacial lakes at different settings. Lake Le Bourget and Lake Paladru are situated at low elevations (respectively 231 m–492 m) and Lake Blanc Huez is located at 2500 m altitude. Through the integration of high-resolution acoustic profiles and accurately dated sediment samples from cores, recent coeval mass wasting deposits in each lake were correlated with nearby historical earthquakes, whereas coeval mass wasting deposits around 5200 cal BP and 9550 cal BP in these three lakes were correlated to regional earthquakes events.

E. Chapron (✉)
GEODE UMR 5602 CNRS-Université Toulouse Jean Jaurès, 5 Allée A. Machado, 31058 Toulouse cedex 9, France

Institut des Sciences de la Terre d'Orléans, UMR 7327 CNRS-Université d'Orléans, 1A rue de la Férollerie, 45071 Orléans cedex 2, France
e-mail: emmanuel.chapron@univ-tlse2.fr

A. Simonneau • P. Albéric
Institut des Sciences de la Terre d'Orléans, UMR 7327 CNRS-Université d'Orléans, 1A rue de la Férollerie, 45071 Orléans cedex 2, France

G. Ledoux • P. Lajeunesse
Centre d'études nordiques, Département de Géographie, Université Laval, Québec, QC G1VOA6, Canada

F. Arnaud
EDYTEM UMR 5204 CNRS-Université de Savoie, 73376 Le Bourget du Lac, France

G. Lamarche et al. (eds.), *Submarine Mass Movements and their Consequences*, Advances in Natural and Technological Hazards Research 41,
DOI 10.1007/978-3-319-20979-1_34

34.1 General Setting and Data

The French Alpine foreland area (Fig. 34.1) was covered by ice during the last glaciation. Following deglaciation, thick lacustrine sequences were deposited in numerous over-deepened basins (Van Rensbergen et al. 1999). Over the last 300 years, several earthquakes with magnitudes above 5 on Richter scale struck this part of the alpine range (Nomade et al. 2005). Seismic activity on active faults (Thouvenot et al. 2003) in this populated and industrial region is a hazard because

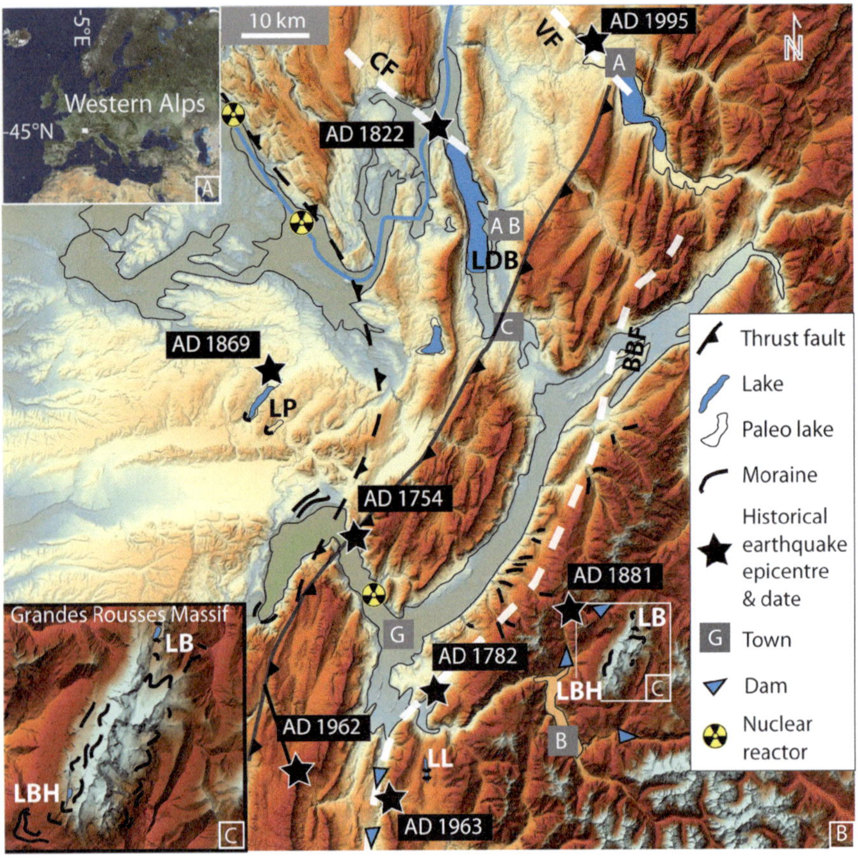

Fig. 34.1 General location of the study area (**a**) and geomorphological settings (**b** and **c**) of investigated lakes in the western Alps (*LP* Lake Paladru, *LDB* Lake Le Bourget, *LBH* Lake Blanc Huez), the main paleo lakes, the Rhone River (*blue line*), outstanding moraine ridges and historical earthquake epicentres (*black stars*). The location of major regional thrusts and the active strike slip Belledonne Border Fault (*BBF*), Culoz Fault (*CF*) and Vuache Fault (*VF*) are also illustrated. Lakes Bramant (*LB*) and Laffrey (*LL*) together with regional cities (*G* Grenoble, *C* Chambéry, *AB* Aix-les-Bains, *A* Annecy and *B* Bourg d'Oisans), nuclear reactors and hydroelectricity dams are also located

of seismic wave amplification in lacustrine deposits within valleys (site effect) or in mountain ranges (topographic effect). In order to better understand the occurrence of these natural hazards, it is necessary to extend the current earthquake record, based on instrumental and historical data, back in time. It has been demonstrated that lake deposits are natural earthquake archives. Lacustrine sedimentation is highly sensitive to regional earthquake activity and exposed to subaquatic slope failures, violent waves and seiche effects (Siegenthaler et al. 1987; Chapron et al. 1999; Hilbe and Anselmetti 2014). In the present paper, we use high-resolution seismic profiling and well-dated sediment cores to report regional, coeval Holocene mass-wasting deposits (MWDs) in three lakes, at variable altitudes. We discuss their sensitivities to earthquake activity according to their setting and dominating sedimentary processes.

Lake Le Bourget is a long and narrow over-deepened basin of glacial origin located at 231.5 m above sea level (a.s.l.) located in a syncline at the inner Jura mountains along active strike-slip transfer faults (Fig. 34.1). The area was struck by the largest historical earthquake of the Western Alps: the AD 1822 Chautagne event had a Medvedev-Sponheuer-Kamik (MSK) intensity VII-VIII and a 5.5–6 equivalent magnitude (Chapron et al. 1999). Lake Le Bourget was part of a larger Rhone valley lake system but progradation of the Rhone fan delta in the Chautagne area isolated the present lake from its main tributary around 10,060 ± 150 cal BP (Arnaud et al. 2012). Today, the lake has two active delta systems build by its present-day tributaries; the Leysse and the Sierroz rivers (Fig. 34.2). Its basin-fill has been intensively studied over the last decades (Arnaud et al. 2012). For this study we used high-resolution (vertical resolution <1 m) bathymetry (Ledoux et al. 2010) and unpublished 12 kHz Chirp subbottom profiles. The recent basin-fill consists of a major slope failure deposit (the Hautecombe Disturbed Unit, HDU), which extends over the entire northern part of Lake Le Bourget (Chapron et al. 2005; Debret et al. 2010). The HDU was dated to the Early Holocene (9550 ± 150 cal BP). After this event, Holocene lacustrine sedimentation is dominated by autochthonous carbonates and formed an up to 15 m thick lacustrine drape.

Lake Paladru is a small basin of glacial origin in the alpine foreland (Fig. 34.1) damned at 492 m a.s.l. by Pleistocene frontal moraines. It has two tributaries (Surand and Courbon rivers) that developed small deltas. Its basin-fill is exclusively documented by 12 kHz chirp profiling, a piston core (PAL09) and seven short gravity cores (Fig. 34.3). A detailed correlation of the cores is based on the identification of distinct lithological facies, aided by samples analysis with a Scanning Electron Microscope (SEM) and laser grain-size measurements on the upper part of PAL09. Since ca. 14,000 years cal BP, sedimentation of autochthonous carbonates formed a lacustrine drape of up to 5 m thick (Simonneau et al. 2013).

Lake Blanc Huez is a small narrow proglacial lake located at 2543 m a.s.l. at the south western side of the Grandes Rousses Massif (Fig. 34.1). As further detailed in Chapron et al. (2007) and Simonneau et al. (2014), this lake deglaciated before ca. 9700 cal BP. Glacier activity within its catchment area (3.2 km^2) was very limited from the Early to the Mid Holocene but increased since the onset of the

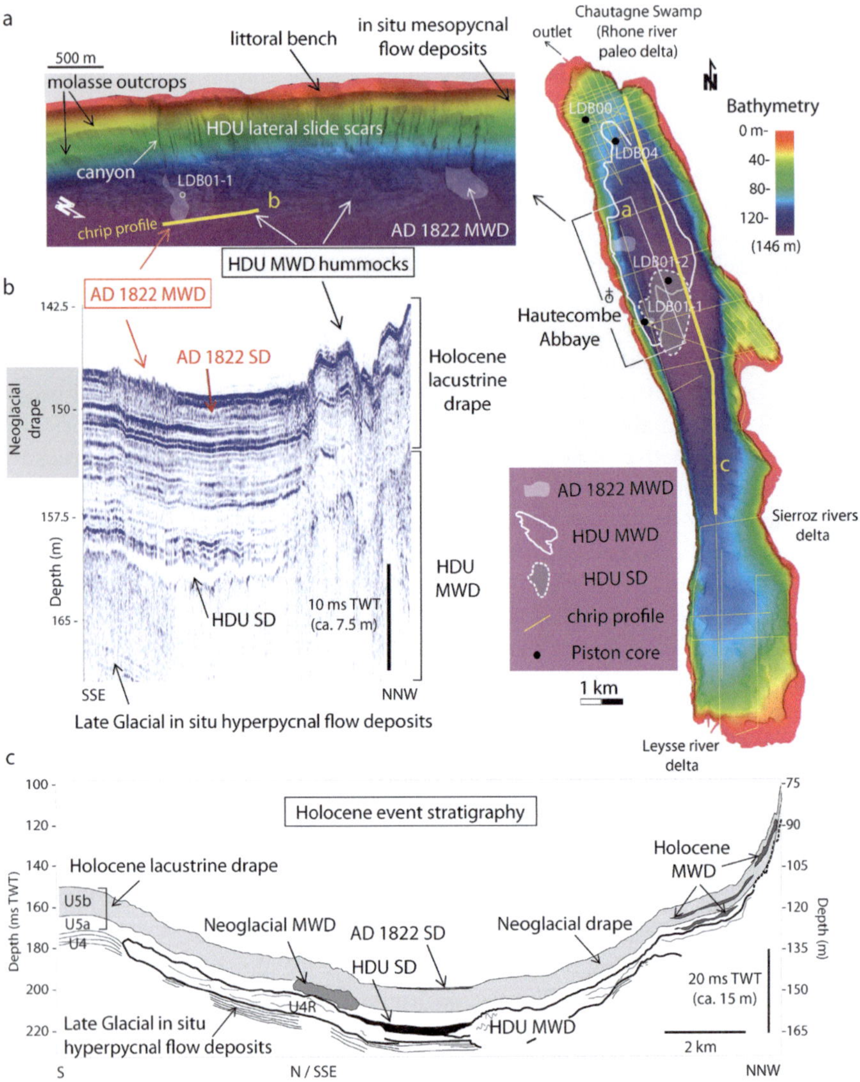

Fig. 34.2 Illustration of Lake Le Bourget geomorphology and stratigraphic characteristics of earthquake-induced mass wasting deposits (*MWD*) and seiche deposits (*SD*) developed in AD 1822, at the onset of the Neoglacial period and in the early Holocene (*HDU* Hautecombe Disturbed Unit) according to 12 kHz Chirp sub-bottom profiles (*yellow lines*), multibeam bathymetric data and sediment cores (*black dots*). A zoom of the bathymetric map (**a**) localize chrip profile given in (**b**) and illustrate the main slide scars and MWDs related to the AD 1822 and the HDU events. Chirp profile (**b**) illustrates the acoustic facies of the lacustrine drape, MWDs and SDs discussed in the text. The interpretation of a chirp profile along the main basin (**c**) is locating the Neoglacial MWD and synthetizing the Holocene event stratigraphy based on seismic and core data

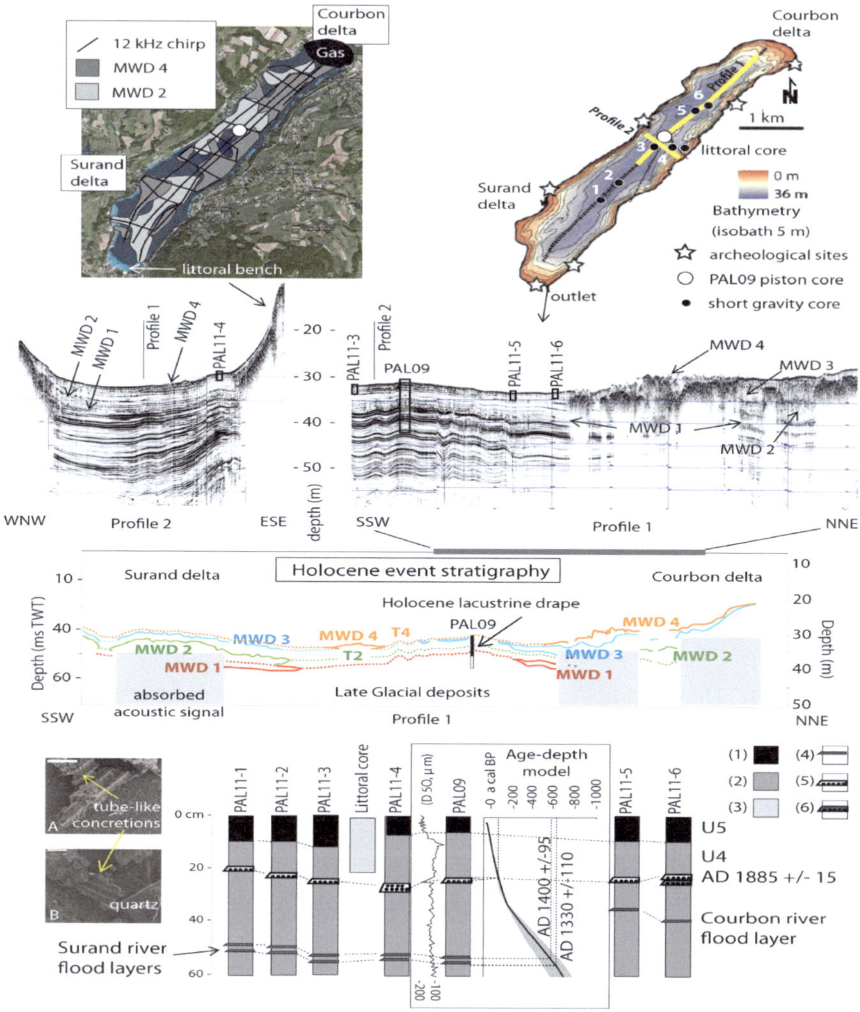

Fig. 34.3 Illustration of Lake Paladru geomorphology and stratigraphic characteristics of mass wasting deposits (*MWD*) and turbidites (*T*) according to 12 kHz Chirp sub-bottom profiles (*black lines*) and well-dated sediment cores (*dots*) lithologies as detailed in the text and in the *lower panel*: organic rich lacustrine marls (*1*); lacustrine marls (*2*); littoral marls rich in carbonate concretions (*3*); tributary flood layer (*4*); carbonated turbidite (*5*) and minerogenic turbidite (*6*). Up to four generations of slope instabilities are identified and dated based on seismic stratigraphy, together with PAL09 age-depth model, grain size measurements and scanning electron microscopy images (*white* bar = 100 µm) from selected sediment samples in the littoral gravity core (**a**) and PAL09 piston core (**b**)

Neoglacial period (ca. 5400 cal BP), which favored the construction of a steep prodelta in the northern part of the lake.

Historical sedimentary events (SEs) associated with MWD's or turbidites in the central basin were dated to AD 1963 (SE11), 1881 (SE10) and either AD 1782 or

Table 34.1 Records of historical and Holocene earthquakes in lakes Le Bourget (LDB), Paladru (LP), Blanc Huez (LBH), Bramant (LB) and Laffrey (LL) reported in this study and in (1) Simonneau et al. (2014), (2) Guyard et al. (2007), and (3) Nomade et al. (2005) by Mass Wasting Deposits (*MWDs*) ranging from slump (*S*), debris flow (*DF*), turibidite (*T*) and seiche deposit (*SD*)

LDB	LP	LBH (1)	LB (2)	LL (3)	Historical earthquake
		SE11 (S)		DF1/DF2	AD 1962/63
	MWD4 + T4 ?	SE10 (T)	S	DF4	AD 1881
	MWD4 + T4			DF4 ?	AD 1869
MWD+SD		SE9(T)	T		AD 1822
				DF5	AD 1782
	MWD3			DF6	AD 1754
	(AD 1,660 +/– 100				
					Holocene earthquake
Neogl.MWD	MWD2+T2	SE4			
(5,020 +/– 200)	(5,190 +/– 50)	(5,375 +/– 175)			ca. 5,200 cal BP
HDU MWD+SD	MWD1	SE1			
(9,550 +/– 150)	(9,010 +/– 600)	(9,680 +/– 140)			ca. 9,550 cal BP

1822 (SE9). They highlight the sensitivity of this lake to regional historical earthquakes (Fig. 34.1 and Table 34.1). Similar Holocene sedimentary events (Simonneau et al. 2014) are dated to 5375 ± 175 cal BP (SE4) and 9680 ± 140 cal BP (SE1).

34.2 Results

In Lake Le Bourget, multiple MWDs coeval to the impact of the AD 1822 Chautagne earthquake remolded mesopycnal distal flood deposits from the Rhone River and carbonate sandy silt from the littoral bench down to 130 m water depths (Chapron et al. 1999, 2005). A canyon and a related carbonate MWD at its downslope termination are clearly imaged on bathymetry and 12 kHz Chirp profiles (Fig. 34.2). The MWD has been sampled in core LDB01-1 and interpreted as a slump deposit (Arnaud et al. 2005). This slump is laterally evolving into a pluridecimetric seiche deposit. The seiche deposit, located in the deepest part of the central basin, is characterized by a transparent acoustic facies on Chirp data (Fig. 34.2) and corresponds to a fine-grained homogeneous sedimentary facies in gravity cores (Chapron et al. 1999, 2005; Jenny et al. 2014). The Holocene lacustrine drape in Lake Le Bourget is subdivided in two acoustic facies (Fig. 34.2): a transparent basal facies (U5a) matching Early Holocene lacustrine marls in piston cores (Arnaud et al. 2005, 2012; Debret et al. 2010); and a upper stratified facies (U5b) characterized by several high-amplitude and continuous reflections resulting from more frequent Rhone River flood events since the Neoglacial period. The base of the so-called "Neoglacial drape" is thus a key chrono-stratigraphic marker horizon on seismic profiles and allows dating a MWD to ca. 5000 ± 200 cal BP using LDB01-1, LDB01-2 and LDB04 piston cores (Fig. 34.2) age-depth models (Chapron et al. 2005; Arnaud et al. 2012). This MWD is a chaotic to transparent lens-shaped body at the transition from facies U5a to U5b, located at the eastern edge the eastern central basin. The Early Holocene HDU event forms a MWD consisting in a decametric chaotic to transparent lens-shaped body with a hummocky topography. It is evolving southward into a plurimetric turbidite characterized by a transparent acoustic facies ponded in the deepest part of the basin. The paleo-topography of associated MWDs at the foot of northern and western slopes is draped by Holocene sediments and still clearly visible on bathymetric data (Fig. 34.2).

In Lake Paladru, up to four generations of MWDs are identified within its lacustrine drape on 12 kHz profiles (Fig. 34.3). They are stacked in the front of Surand and Courbon deltas and developing lens-shaped bodies with chaotic to transparent facies that are locally absorbing the acoustic signal. Only the upper (MWD4) and the intermediate (MWD2) slope failure events can be clearly mapped. These two events consist in coeval slope failures both impacting deltaic environments and littoral benches. MWD4 is laterally associated with an outstanding light-colored sandy turbidite, rich in carbonate concretions (T4) dated to AD 1885 ± 15 and identified on a transect of six cores across the basin. In core PAL11-6, this key horizon is, in addition, covering a dark colored minerogenic sandy turbidite. The age of MWD3 is estimated to 290 ± 100 cal BP (AD 1660 ± 100) based on seismic stratigraphy (i.e. tracing across the basin the low amplitude but continuous reflection that drape MWD3) and on PAL09 age-depth model (Simonneau et al. 2013). Dating uncertainty is derived by assuming a vertical resolution on seismic profiles of 7 cm and considering a mean sedimentation rate of 0.7 mm/year on PAL09 at the related core depth. The age of MWD2 is similarly based on seismic stratigraphy and is further supported by the identification of a striking light colored sandy layer (T2) at 275 cm in core PAL09 precisely dated to 5190 ± 50 cal BP. Finally, the age of MWD1 is estimated to 9010 ± 600 cal BP by seismic stratigraphy, PAL09 age-depth model and taking into consideration the vertical resolution of seismic data with a mean sedimentation rate of 0.3 mm/year.

34.3 Discussion

The sedimentary record of historical earthquakes in the area (Fig. 34.1 and Table 34.1) documented by Nomade et al. (2005), Guyard et al. (2007), and Simonneau et al. (2014) are summarized in this study. Due to dating uncertainties, MWD4 and MWD3 in Lake Paladru may be, contemporary to the AD

1881 Allemond and the AD 1754 Voreppe events, respectively. But MWD4 may also result from the combination of a locally documented earthquake (the AD 1869 Charavine event) and an abrupt lake level change in AD 1870 described by Dutey Grinet (1873) and reported in Colardelle and Verdel (1993). This last event triggered numerous slope failures along its littoral benches in AD 1870. This interpretation is further supported by the identification on SEM images of similar tube-like carbonate concretions both in T4 and in the littoral core (Fig. 34.3), that are typical from the edge of a littoral bench (Magny et al. 2008). The siliciclastic turbidite deposited just below T4 in PAL11-6 may thus be earthquake triggered, while T4 and several MWDs may result from the following lake-level changes in AD 1870. A similar combination of ground motion and lake-level changes may also explain the development of MWD2 and T2 at ca. 5200 cal BP. The same regional earthquake may have triggered SE4 in Lake Blanc Huez and the Neoglacial MWD in Lake Le Bourget. Both of these slope failure events were likely facilitated by changes in sedimentation modes due to enhanced glacier activities in their catchment areas (Simonneau et al. 2014) and clastic supplies in deltaic areas. Similarly, a regional earthquake at ca. 9550 cal BP occurring during the Boreal period that was associated with important environmental changes (in vegetation, glacier activity and lake-level) in Western Europe (Magny et al. 2014), may have triggered MWD1 in Lake Paladru, SE1 in Lake Blanc Huez and the HDU in Lake Le Bourget. The larger amplitude of the coeval HDU event is here clearly resulting from its isolation from the Rhone River and a drastic drop in siliciclastic sediment supply. These observations suggest that the sensitivities of lake sediments to earthquakes are not only related to the distance and intensity of earthquakes, but also to the geomorphology and sedimentation patterns in the lakes.

34.4 Conclusion

Coeval regional MWDs in lake sediments from the French alpine foreland highlight the impact of historical earthquakes in AD 1822, 1881 and 1962. Similarly, and within radiocarbon and seismic stratigraphic dating uncertainties, Holocene regional earthquakes potentially triggered coeval MWDs in lakes Le Bourget, Paladru and Blanc Huez at ca. 5200 and 9550 cal BP. This study suggest that lakes sensitivity to ground motion was enhanced by changes in sedimentation modes following environmental changes associated with the Little Ice Age, Neoglacial and Boreal periods.

Acknowledgments The authors wish to thank G. Vargas and P. Van Rensbergen for their constructive reviews and G. Lamarche for his contribution to the improvement of this manuscript.

References

Arnaud F, Revel M, Chapron E et al (2005) 7200 years of Rhone river flooding activity in Lake Le Bourget, France: a high-resolution sediment record of NW Alps hydrology. The Holocene 15:420–428

Arnaud F, Révillon S, Debret M et al (2012) Lake Bourget regional erosion patterns reconstruction reveals Holocene NW European Alps soil evolution and paleohydrology. Quat Sci Rev 51:81–92

Chapron E, Beck C, Pourchet M, Deconinck J-F (1999) 1822 earthquake-triggered homogenite in Lake Le Bouregt (NW Alps). Terra Nova 11:86–92

Chapron E, Arnaud F, Noël H et al (2005) Rhone River flood deposits in Lake Le Bourget: a proxy for Holocene environmental changes in the NW Alps, France. Boreas 34:404–416

Chapron E, Faïn X, Magand O et al (2007) Reconstructing recent environmental changes from proglacial lake sediments in the Western Alps (Lake Blanc Huez, 2543 m a.s.l., Grandes Rousses Massif, France). Palaeogeogr Palaeoclimatol Palaeoecol 252:586–600

Colardelle M, Verdel E (1993) Les habitats du lac de Paladru dans leur environnement: la formation d'un terroir au XIe siècle. Documents d'archéologie française, Edition de la Maison des Sciences de l'Homme

Debret M, Chapron E, Desmet M et al (2010) North western Alps Holocene paleohydrology recorded by flooding activity in Lake Le Bourget, France. Quat Sci Rev 29:2185–2200

Guyard H, Chapron E, St-Onge G et al (2007), High-altitude varve records of abrupt environmental changes and mining activity over the last 4000 years in the Western French Alps (Lake Bramant, Grandes Rousses Massif). Quat Sci Rev 26(19):2644–2660

Hilbe M, Anselmetti F (2014) Mass movement-induced tsunami hazard on perialpine Lake Lucerne (Switzerland): scenarios and numerical experiments. Pure Appl Geophys. doi:10.1007/s00024-014-0907-7

Jenny J-P, Wilhelm B, Arnaud F et al (2014) A 4D sedimentological approach to reconstructing the flood frequency and intensity of the Rhone River (Lake Bourget, NW European Alps). J Paleolimnol 51:469–483

Ledoux G, Lajeunesse P, Chapron E, St-Onge G (2010) Multibeam bathymetry investigations of mass movements in Lake Le Bourget (NW Alps, France) using a portable platform. In: Mosher DC et al (eds) Submarine mass movements and their consequences, advances in natural and technological hazards research. Springer, The Netherdlands, vol 28, pp 423–433

Magny M, Peyron O, Gauthier E et al (2008) Quantitative reconstruction of climatic variations during the Bronze and early Iron ages based on pollen and lake-level data in the NW Alps France. Quat Int. doi:10.1016/j.quaint.2008.03.011

Magny M, Combourieu-Nebout N, de Beaulieu J-L et al (2014) North-south palaeohydrological contrast in the central Mediterranean during the Holocene: tentative synthesis and working hypotheses. Clim Past 9:2043–2071

Nomade J, Chapron E, Desmet M et al (2005) Reconstructing historical seismicity from lake sediments (Lake Laffrey, Western Alps, France). Terra Nova 17:350–357

Siegenthaler C, Finger W, Kelts K, Wanf W (1987) Earthquake and seiche deposits in Lake Lucerne, Switzerland. Eclogae Geol Helv 80:241–260

Simonneau A, Doyen E, Chapron E et al (2013) Holocene land-use evolution and associated soil erosion in the French Prealps inferred from Lake Paladru sediments and archaeological evidences. J Archaeol Sci 40:1636–1645

Simonneau A, Chapron E, Garçon M et al (2014) Tracking Holocene glacial and high-altitude alpine environments fluctuations from minerogenic and organic markers in proglacial lake sediments (Lake Blanc Huez, Western French Alps). Quat Sci Rev 89:27–43

Thouvenot F, Fréchet J, Jenatton L, Gamond J-F (2003) The Belledonne border fault: identification of an active seismic strike-slip fault in the western Alps. Geophys J Int 155:174–192

Van Rensbergen P, de Batist M, Beck C, Chapron E (1999) High-resolution seismic stratigraphy of glacial to interglacial fill of a deep glacigenic lake: Lake Le Bourget, Northwestern Alps, France. Sediment Geol 128:99–129

Chapter 35
Spatial and Temporal Relation of Submarine Landslides and Faults Along the Israeli Continental Slope, Eastern Mediterranean

Oded Katz, Einav Reuven, Yonatan Elfassi, Anner Paldor, Zohar Gvirtzman, and Einat Aharonov

Abstract A new study of the Israeli Mediterranean continental slope provides an understanding of the interaction between submarine landslides, fault scarps, and subsurface evaporites. Faults and landslides interact in the northern part of the studied continental slope where fault scarps rupture the seabed. In this area landslides are thought to be triggered by over-steepened fault-scarps and are observed to cover older fault scarps, or be cut by younger faults. These variable cross-cutting relationships indicate a multi-phase history in which landsliding and faulting both post-date and pre-date one another. Isopach maps of the Messinian evaporites further reveal that fault scarps are mainly found along a slope-parallel belt where the underlying salt layer is 150–500 m thick. We suggest that this rather thin sequence of the Messinian evaporites associated with faulting serves as a localized detachment zone for the overlaying strata. We argue that the multi-phase and spatially variable association of landslides and faults reveal a highly dynamic continental slope, which may be still active in the present day.

35.1 Introduction

The present work analyses hundreds of recently mapped submarine landslides exposed on the surface of the Levant continental slope (SE Mediterranean), off central Israel, hereafter called the study area (Fig. 35.1). These landslides spatially overlap and interact with scarps of growth faults rupturing the sea floor. This work

O. Katz (✉) • Z. Gvirtzman
Geological Survey of Israel, Jerusalem, Israel
e-mail: odedk@gsi.gov.il

E. Reuven • Y. Elfassi • A. Paldor
Geological Survey of Israel, Jerusalem, Israel

Institute of Earth Sciences, Hebrew University, Jerusalem, Israel

E. Aharonov
Institute of Earth Sciences, Hebrew University, Jerusalem, Israel

G. Lamarche et al. (eds.), *Submarine Mass Movements and their Consequences*, Advances in Natural and Technological Hazards Research 41,
DOI 10.1007/978-3-319-20979-1_35

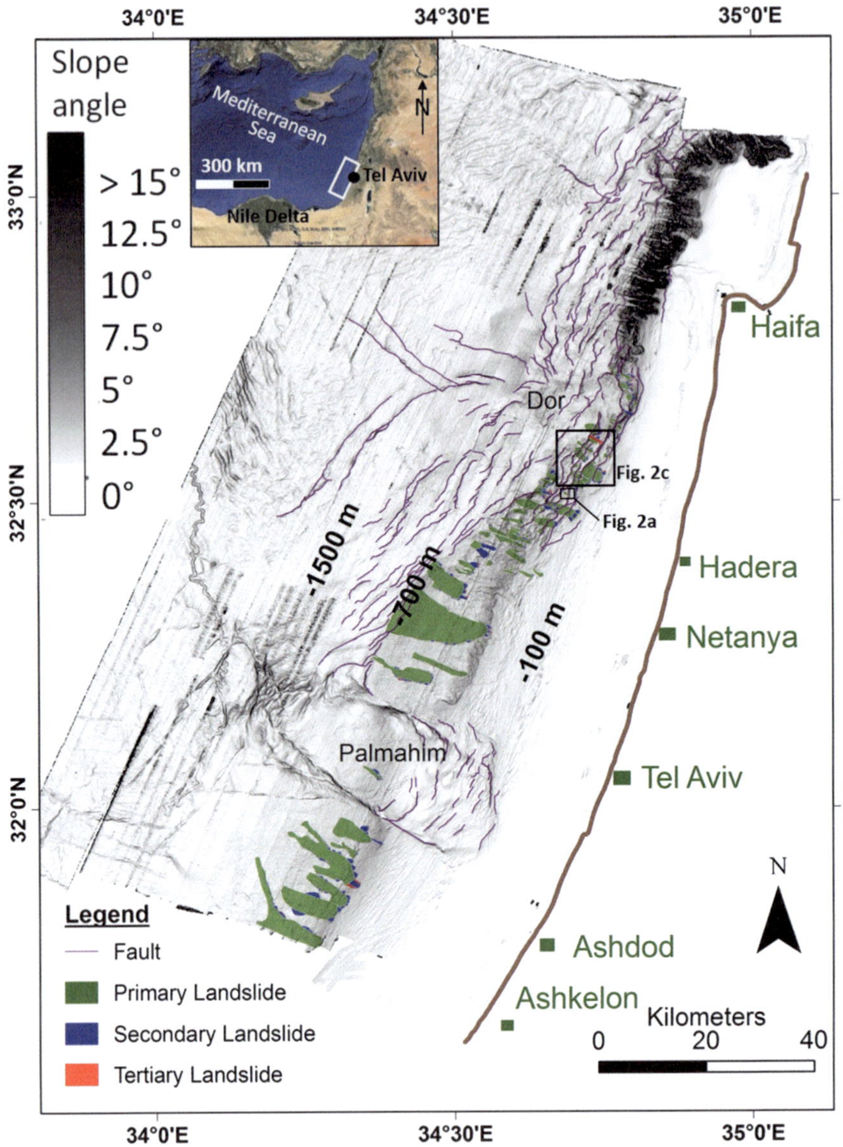

Fig. 35.1 Landslides and fault scarps mapped in the studied area are shown on top of slope map. *Gray* colouring is the slope angle calculated from 50 m bathymetric grid (Tibor et al. 2013). The study area (*open* slope province) is south of 'Dor' disturbance and the canyon province is to its north

investigates the spatial, temporal, and mechanical relations between landslides, fault scarps, and the underlying sequence of Messinian evaporites.

The Levant coast is controlled by northward along-shore transport of sediments from the Nile Delta (Stanley et al. 1998). During Pliocene-Quaternary times, 1–2 km of Nile-derived clay-rich siliciclastic sediments accumulated in the eastern

Mediterranean over sandy turbidites and Messinian evaporites (e.g. Gvirtzman and Buchbinder 1978).

The current submarine offshore morphology consists of a shelf that is up to 20 km wide in the south and narrows northwards. The shelf is shallow, with slopes not exceeding 2° at water depths below 200 m, and further to the west, a continental slope with up to 4° (and locally higher gradients), at water depths of 200 m to more than 1,000 m. The continental slope consists of two lateral provinces (Fig. 35.1): The submarine canyons province in its northern part (north of Latitude 32.70°N) and the open slope province further south. Two main types of mass transport deposits (MTD) are observed along the open slope: The first type consists of very large slumps, over 1 km thick, with their primary sliding-plane located within the Messinian evaporites (Almagor 1984). The internal deformation within these landslides, also known as, disturbances' (Garfunkel 1984), consists of landward block rotation on listric growth faults rooted in the underlying evaporites. Two sites across the studied continental slope are believed to reveal this kind of very large gravitational collapse: the Palmahim and Dor Disturbances (Mart et al. 1978; Fig. 35.1). The second type of submarine MTD along the open slope, which forms the subject of the current work, consist of 447 small to medium landslides, which were identified previously (Almagor and Garfunkel 1979; Frey-Martinez et al. 2005) and recently comprehensibly mapped by Katz et al. 2015. These landslides are found at water depths ranging between 130 and 1,100 m, where slopes exceed a critical gradient of around 4°. Their surface areas range from 0.0024 to 91 km^2. The landslides show a hierarchical pattern, resulting from sequential slope-failure events where secondary landslides developed in the steep scars of the primary ones (Fig. 35.1).

In addition the studied area hosts north-south elongated morphological step structures, recognized as surficial scarps of growth faults that are a result of subsurface salt tectonic related to the buried Messinian evaporitic sequence that results in extensional deformation of the overlaying Pliocene-Quaternary sequence (Mart and Ryan 2007; Cartwright et al. 2012; Gvirtzman et al. 2015 and references there in).

In this work we document the spatial relations between landslides and fault scarps exposed on the sea floor, as well as the spatial relations between the surficial landslides and fault scarps, and the sequence of the Messinian Evaporites, buried more than a kilometer below the surface of the continental slope. We then analyse the temporal and possible mechanical relations between the landslides and faults, and between the buried evaporites and the surficial features.

35.2 Methods

Fifty time-migrated N-S and E-W orientated 2D seismic reflection sections, acquired for oil and gas exploration by TGS-NOPEC Geophysical Company L.P were analysed. Upper and lower bounding of the Messinian evaporite sequence were mapped and time structure maps produced (using HIS's Kingdom software). These maps were then converted to the depth domain (using velocities of 1,500 m\s for sea water, 1,800 m\s for the Plio-Quarternary sediments and 4,200 m\s for the evaporites) and the difference calculated to produce an isopach map, illustrating the thickness variation in the Messinian sequence (Fig. 35.3).

35.3 Results

35.3.1 Spatial Relations of Landslides and Faults Scarps

Landslides and fault scarps spatially overlap and interact only in the northern part of the study area (Fig. 35.1). Farther south, surficial faults are identified west of the continental slope whereas landslides are located upslope (Fig. 35.1). South of the Palmahim disturbance, faulting is absent (Fig. 35.1).

Eighty-four out of the 447 mapped landslides reveal three main types of cross-cutting relations with faults scarps: (a) 55 landslides were found to initiate from fault scarps, with triggering thought to be promoted by the over steepness of the scarps (Fig. 35.2a, b). In this case landsliding and faulting are mechanically related, although landsliding postdates the faulting. (b) In five cases, landslide scars are located upslope of faults, which are buried by material derived from the landslides (Fig. 35.2c, d). Here, landsliding postdates the faulting but is not mechanically affected by it. (c) In 15 cases landslide deposits are cut by fault scarps. In this case faulting postdates the landsliding (Fig. 35.2c, e). In an additional seven cases landslides are both cut by younger faults and also cover older faults with their deposits. Two further landslides were observed to initiate from a fault scarp, where one of them is cut by, and the other covers a different fault scarp further down slope.

35.3.2 Spatial Relation of Landslides, Faults Scarps and Messinian Evaporites

The isopach map of the Messinian Evaporites (Fig. 35.3) indicates that the sequence thickens westwards. Close to the coast, the evaporites are thinner than 100 m increasing to 400 m under the continental slope, >500 m under the foot of the slope, and >1,000 m further west (Fig. 35.3). In addition, a shore-line parallel thickness variation is observed. The evaporite sequence is significantly thicker under the Palmahim disturbance due to the pre-Messinian Hanna canyon (Steinberg et al. 2011) and is somewhat thicker under the Dor disturbance area (labeled, Fig. 35.1). The spatial relations between the landslides and the fault scarps exposed on the sea floor have been analysed along with the thickness of the underlying Messinian evaporites. Where exposed on the sea floor (north of Palmahim

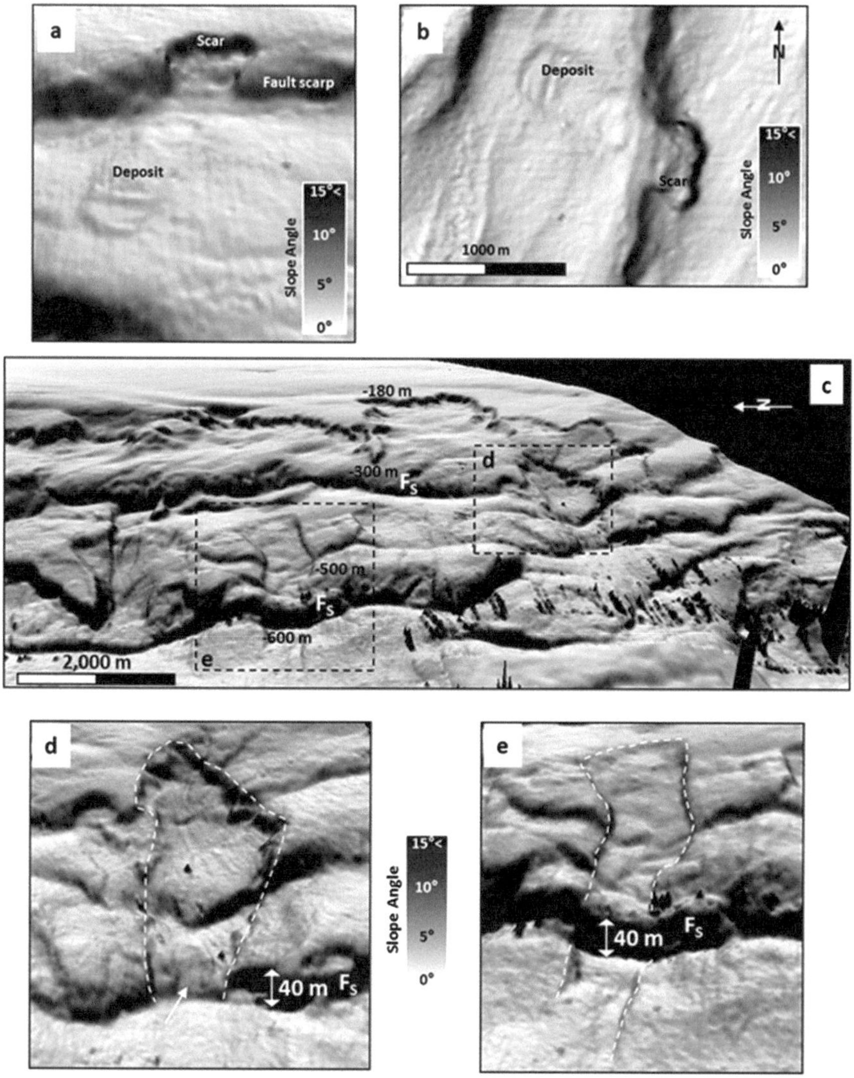

Fig. 35.2 Types of crosscutting relationships between landslides and fault scarps shown on a slope map (calculated from a 50 m bathymetric grid), (**a**) Perspective view of an example of a landslide that initiates from a fault scarp, (**b**) Map view of the same example, (**c**) General perspective view of an area of the continental slope. *Fs* fault scarp, (**d**) Zoomed in perspective view of a landslide whose derived material covers a fault scarp, (**e**) Zoomed in perspective view of an example of a landslide cut by a fault scarp

disturbance), submarine fault scarps coincide in most cases (80 % of their total length) with evaporites thickness of 150–500 m (Fig. 35.3). Along most of the studied continental slope, the evaporites attain this thickness at the slope foot. At two sites, the Palmahim and Dor disturbances, evaporites are up to c.150 m thick at the upper part of the continental slope, close to the shelf break. It is noted that in this

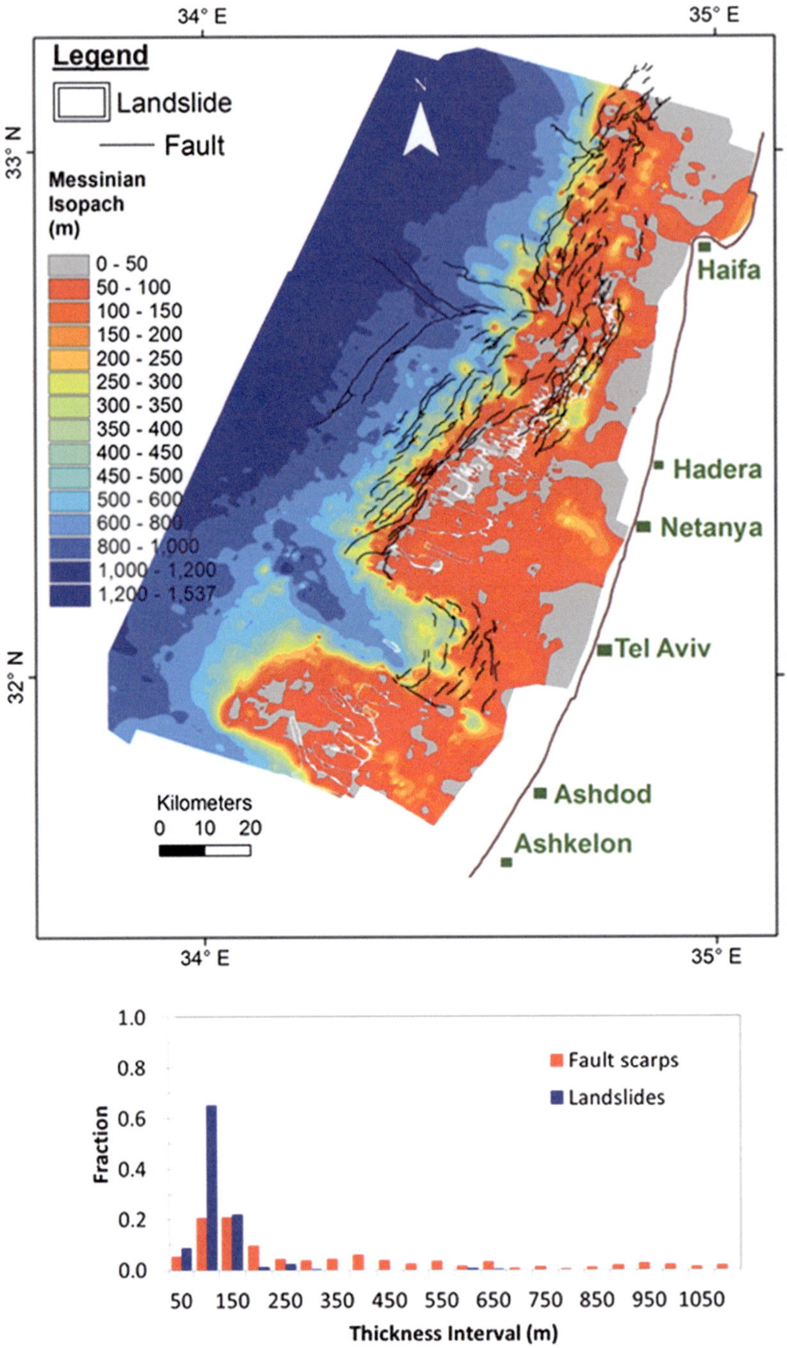

Fig. 35.3 Isopach map of the Messinian sequence in the studied area. *Top*: Landslides and fault scarps mapped in the studied area are shown on *top* of the isopach map. *Bottom*: Histograms of the evaporites-thickness undelaying the fault scarps and landslides

area, fault scarps appear at generally shallower water depths. South of Palmahim disturbance, the evaporites sequence appears thinner, and thickens towards the basin at a slower rate than is observed to the north (Fig. 35.3). It is noted that, for reasons that are not totally clear, faults are absent in this area (Katz et al. 2015).

In the vicinity of the Dor disturbance, landslides spatially overlap with faults, and coincide with evaporites thickness exceeding 150 m. Further south, landslides do not initiate from fault scarps and correlate in general with evaporite thicknesses of c.100 m.

35.4 Discussion

35.4.1 The Relationship Between Evaporites and Faulting

The fault scarps observed are the surface expression of growth faults rooted in the Messinian sequence (Garfunkel 1984). A spatial relation between fault scarps and the thickness of the evaporite sequence (i.e. 150–500 m) is observed. Faults infrequently appear above where the evaporite sequence is thicker or thinner than this range (Fig. 35.3). Following Hudec and Jackson (2007), we suggest that this rather thin sequence of the Messinian evaporites associated with faulting serves as a localized detachment zone for the overlaying strata hosting this extensional deformation. Further seawards, where the evaporite sequence is over 500 m thick, decreasing slope gradients do not enable similar styles of extensional deformation within the post Messinian sequence. Gradients in evaporite thickness are apparently also important, which may explain why in the south no faults appear in the vicinity of 150–500 m thick evaporites, since the rate of basin-ward thickening is smaller. This additional controlling factor is subject to further analysis.

35.4.2 The Relationship Between Landslides, Faults Scarps, and Evaporites

In general, the distribution of submarine landslides does not correlate with any particular thickness of Messinian evaporites. Landslides appear over areas of negligible thickness in some cases, and in areas where evaporates are more than 150 m thick as well. Thus, we suggest that the spatial distribution of landslides in the study area is not directly affected by underlying salt-tectonic processes. It seems that the largest controlling factor on the spatial distribution of the submarine landslides is the local gradient of the continental slope (Fig. 35.1). In the southern part of the study area, head scars of the landslides appear at a constant depth (ca. 350–400 m) along a 80 km coast parallel line (Katz et al. 2015). The slope angle along this trend exceeds 4°, which apparently is the critical angle for slope instability in the area. Similar critical slope angles for submarine failure have been noted elsewhere (Masson et al. 2006; Brothers et al. 2014). Where both faults and landslides are present in the northern part of the study area, slope angle exceeds this critical value of 4° within fault scarps, thus making such sites favourable for the initiation of landsliding. Since these fault scarps occur in association with a particular thickness of Messinian evaporites (>150 m), the location of landslides in the north part of the study area is interpreted to be indirectly controlled by the spatial variation of the underlying evaporite sequence.

In the north, variable cross cutting relationships of landslides and fault scarps appear. These cross-cutting relations attest to a history of landsliding and active fault rupturing, which both post- and pre-date one another. Landslides in the southern part of the study area do not appear to overlap with faults (Fig. 35.1). Since the surface expression of landslides (e.g. scar height and slope) in the south is not significantly different than the northern landslides the northern and southern landslides are believed to be contemporaneous.

Sedimentation rates in the study have been calculated at 0.25–1.30 m/ky (Schilman et al. 2001; Almogi-Labin et al. 2009). With landslide scars and fault scarps in the study area in the order of tens meters high, it follows that they will likely be covered by sedimentation within tens of thousands of years following their formation. As such we conclude that the observed landslides and faults are younger than c.50,000 years, and possibly still active. Similar age constrains are suggested by Frey Martinez et al. (2005), for landslides in the south of the study area.

35.5 Conclusion

Landslides and faults scarps exposed along the continental slope off the Israeli coast in southeastern Mediterranean have been analysed in order to better understand the interrelation between surficial mass wasting processes and faults. Faults and landslides interact in the northern part of the study area where the fault scarps rupture the continental slope sea bed. Here, landslides are thought to be triggered by over-steepened fault scarps, and are observed to cover older fault scarps, or be cut by younger faults. These variable cross-cutting relations testify to a multi-phase history where landsliding and surface rupturing by active faults both post- and pre-date one another. Fault scarps are found along a slope-parallel belt where the thickness of underlying Messinian evaporites is 150–500 m thick. We suggest that this thickness range is an important factor encouraging extensional deformation within the overlaying Pliocene-Quaternary sequence. We argue that the multi-temporal and spatially variable relationship between landslides and faults reveal a highly dynamic continental slope, which may be still active in the present day.

Acknowledgments This work was funded by the Ministry of Science and Technology, Israel. Reviews by U. ten-Brink and Y. Mart significantly contributed to the quality of this work. Editorial work by S. Bull is much appreciated.

References

Almagor G (1984) Salt-controlled slumping on the Mediterranean slope of central Israel. Mar Geophys Res Lett 6:227–243

Almagor G, Garfunkel Z (1979) Submarine slumping on the continental margin of Israel and northern Sinai. Bull Am Assoc Pet Geol 63:324–340

Almogi-Labin A, Bar-Matthews M, Shriki D, Kolosovsky E, Paterne M (2009) Climatic variability during the last w90 ka of the southern and northern Levantine Basin as evident from marine records and speleothems. Q Sci Rev 28:1–15

Brothers DS, Ruppel C, Kluesner JW, ten-Brink US, Chaytor JD, Hill JC, Andrews BD, Flores C (2014) Seabed fluid expulsion along the upper slope and outer shelf of the US Atlantic continental margin. Geo Res Lett 41:96–101

Cartwright J, Jackson M, Dooley T, Higgins S (2012) Strain partitioning in gravity-driven shortening of a thick, multilayered evaporite sequence. In: Alsop GI, Archer SG, Hartley AJ, Grant NT, Hodgkinson R (eds) Salt tectonics, sediments and prospectivity. Geological Society, London, pp 449–470 (Special Publications)

Frey-Martinez J, Cartwright J, Hall B (2005) 3D seismic interpretation of slump complexes: examples from the continental margin of Israel. Basin Res 17:83–108

Garfunkel Z (1984) Large-scale submarine rotational slumps and growth faults in the eastern Mediterranean. Mar Geol 55:305–324

Gvirtzman G, Buchbinder B et al (1978) The late tertiary of the coastal plain and continental shelf of Israel and its bearing on the history of the eastern Mediterranean. In: Ross DA (ed) Initial Rep. of the Deep Sea Drilling Project, vol 42A. U.S. Government Printing, Washington, DC, pp 1195–1222

Gvirtzman Z, Reshef M, Buch-Leviatan O, Groves-Gidney G, Karcz Z, Makovsky Y, Ben-Avraham Z (2015) Bathymetry of the Levant basin: interaction of salt-tectonics and surficial mass movements. Mar Geol 360:25–39

Hudec MR, Jackson MPA (2007) Terra infirma: understanding salt tectonics. Earth Sci Rev 82:1–28

Katz O, Reuven E, Aharonov E (2015) Submarine landslides and fault scarps along the eastern Mediterranean Israeli continental-slope. Mar Geol (under review). doi:10.1016/j.margeo.2015.08.006

Mart Y, Ryan W (2007) The levant slumps and the phoenician structures: collapse features along the continental margin of the southeastern Mediterranean Sea. Mar Geophys Res 28:297–307

Mart Y, Eisin B, Folkman Y (1978) The Palmahim structure — A model of continuous tectonic activity since the Upper Miocene in the Southeastern Mediterranean off Israel. Earth Planet Sci Lett 39:328–334

Masson DG, Harbitz CB, Wynn RB, Pedersen G, Lovholt F (2006) Submarine landslides: processes, triggers and hazard prediction. Phil Trans R Soc A 364:2009–2039

Schilman B, Bar-Matthews M, Almogi-Labin A, Luz B (2001) Global climate instability reflected by Eastern Mediterranean marine records during the late Holocene. Palaeogeogr Palaeoclimatol Palaeoecol 176:157–176

Stanley DJ, Nir Y, Galili E (1998) Clay mineral distributions to interpret Nile cell provenance and dispersal: III. Offshore margin between Nile Delta and Northern Israel. J Coast Res 14:196–217

Steinberg J, Gvirtzman Z, Folkman Y, Garfunkel Z (2011) The origin and nature of the rapid late tertiary filling of the Levant Basin. Geology 39:355–358

Tibor G, Sade R, Sade H, Hall JK (2013). Data collection and processing of multibeam data from the deep water offshore Israel. IOLR report H-31/2013

Chapter 36
Earthquake Induced Landslides in Lake Éternité, Québec, Canada

Jacques Locat, Dominique Turmel, Marion Habersetzer, Annie-Pier Trottier, Patrick Lajeunesse, and Guillaume St-Onge

Abstract Lake Éternité, located between the Upper Saguenay Fjord and the St. Lawrence River has registered many submarine slides caused by at least one earthquake. Landslides are mostly rooted in the gyttja (Holocene sediments). Mapping of landslides revealed a total of 128 scars over an area of only 3.2 km^2. A larger proportion of the landslide scars are located on the SE and NW facing slope which may support an epicentre location for the strongest earthquake (1663?), to the NW or NE of the lake. The preliminary numerical analysis of the site effects caused by topography on local preferred seismic amplification is not conclusive enough to support the observations made for landslides. Associating landslides to specific earthquakes will only be possible with further investigations, including coring of various features including rupture surfaces. The study also revealed interesting slide morphologies developed in homogeneous sediments, providing excellent examples for future modelling of similar events.

36.1 Introduction

Recent studies have attempted to investigate the morphology and sedimentology of lakes and fjords in Québec and Ontario in order to better define the postglacial seismic records (e.g., Shilts 1984; Doig 1990; Shilts and Clague 1992; Levesque et al. 2006; Lajeunesse et al. 2008; Normandeau et al. 2013; Doughty et al. 2014). Central to this research is the February 5th 1663 earthquake (Lamontagne 1987),

---

J. Locat (✉) • D. Turmel • M. Habersetzer
Laboratoire d'études sur les risques naturels, Département de géologie et de génie géologique, Université Laval, Québec, QC G1V 0A6, Canada
e-mail: jacques.locat@ggl.ulaval.ca

A.-P. Trottier • P. Lajeunesse
Laboratoire de géosciences marines, Centre d'études nordiques, Université Laval, Québec, QC G1V 0A6, Canada

G. St-Onge
Institut des Sciences de la mer, Université du Québec à Rimouski (ISMER), Rimouski, QC G5L 3A1, Canada

G. Lamarche et al. (eds.), *Submarine Mass Movements and their Consequences*, Advances in Natural and Technological Hazards Research 41,
DOI 10.1007/978-3-319-20979-1_36

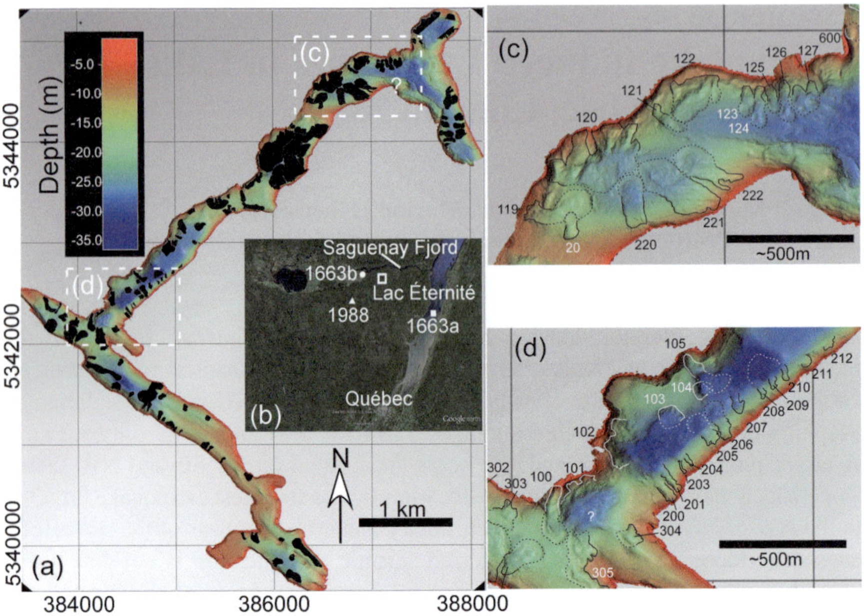

Fig. 36.1 (**a**) Lake Éternité and the landslide coverage in black, note the almost complete disturbance of the lake floor near the central portion of the NE arm; (**b**) Insert showing the location of the Lake Éternité and the position of the reference points referred to in the text (**c**) and (**d**) are examples of basin, located in (**a**), offering accommodation space for landslides. The *dashed line* contours the debris area. *Black* and *white* numbers in '**c**' and '**d**' are slide identification

one of the largest historical earthquake to occur in Eastern Canada with a magnitude Mw of about 7.5, likely the largest for the last 7200 ycal BP (St-Onge et al. 2004), resulting in many aerial and subaqueous landslides (Locat 2011). As part of the review of the 1663 Charlevoix earthquake, there is still some debate about the location of its epicentre (either 1663a or 1663b in Fig. 36.1b) which has significant implications on seismic hazard maps (actual location shown as 1663a in Fig. 36.1b). In his analysis of mass movement signatures in lakes, based on core analysis and shallow seismic survey (no sidescan survey), Doig (1998) suggested that their concentration on SE and SW facing slopes could imply that the epicentre of the earthquake responsible for these mass movements may be located to the NW or NE of the lake.

Therefore, to continue our search for the location of the epicentre of the 1663 earthquake, the work of Doig (1998) will be re-visited in light of more recent research development in this region since 1998, in addition to this 2014 investigation. By doing so, we will also challenge his conclusion that the events noted in Lake Éternité were not related to the 1663 earthquake. At this point, we present our initial analysis of Lake Éternité using two approaches. The first one is based on a morphological analysis of detailed bathymetry map acquired in September 2014 using a GeoSwath interferometric sidescan sonar at a frequency of 250 kHz (1 m resolution) and explored with a 12 kHz shallow seismic echo-sounder. The second approach is a numerical analysis using SPECFEM3D Cartesian software which uses the continuous Galerkin spectral-element method to simulate elastic wave propagation caused by earthquakes (Komatitsch and Tromp 1999) to look at the effect that source location and topography of the region on the distribution of landslides. It is known that surface topography can significantly affect earthquake ground motion (e.g. Lee et al. 2009a), and SPECFEM3D Cartesian may simulate this effect on a regional scale (e.g. Komatitsch 2004; Lee et al. 2009a, b).

36.2 Geological Setting and Sedimentological Observations

Lake Éternité is located 160 km North of Québec City (Fig. 36.1a). It has a 'T' shape with its longest arm (~5 km) oriented NE-SW. The hills surrounding the lake reach a maximum elevation of 200 m above lake level which is at an altitude of 256 m. The area of the lake covered by the bathymetric survey is 3.8 km^2 (~90 % of the lake total area) with an average width of 0.3 km. The bathymetry is variable with small basins, the deepest one being at a depth of 38 m. The lake geometry is controlled by the underlying bedrock consisting of metamorphic rocks (mostly gneiss) of the Grenville Province that are intensively fractured and faulted. These fractures and faults were later carved by the successive glaciations (Lajeunesse 2014) which are responsible for the 'U' shape valleys in the region, including the Saguenay Fjord. The area was just south of the retreating Wisconsinian ice front about 10,400 years ago (Occhietti et al. 2011).

From the morphology and shallow seismic surveys, the stratigraphic sequence of Quaternary sediments found in lake Éternité consists of till or glacio-fluvial sand and gravel, overlain by glacio-lacustrine sediments which were later covered by a Holocene mud (gyttja). Over the last 2000 years the sedimentation rate in Lake Éternité is about 0.6 mm/year (Doig 1998). The distribution of the glacio-lacustrine sediments may have been controlled by the presence of stagnant ice around which the sediments could accumulate leaving a sort of kettle-like morphology (small basins). So upon the final melting of the stagnant ice, the lake was left with few small basins (e.g. dark blue in Fig. 36.2) often separated by flat lying glacio-lacustrine clays or bedrock sills. In the SE portion of NW-SE arm of the lake, glacio-lacustrine deposition is of less than 1 m and is restricted to shallow basins. As we will see later, most of the small basins provided accommodation space for the accumulation of slide debris resulting from earthquakes.

36.3 Landslide Morphology and Distribution

The distribution of landslides is shown in Fig. 36.1a. A total of 128 landslides have been mapped in the lake. From our knowledge of the region they are believed to have been triggered by earthquakes (Levesque et al. 2006; Dougthy et al. 2014). In most cases, sediments involved consists 2–5 m thick gyttja, seen in seismic profiles

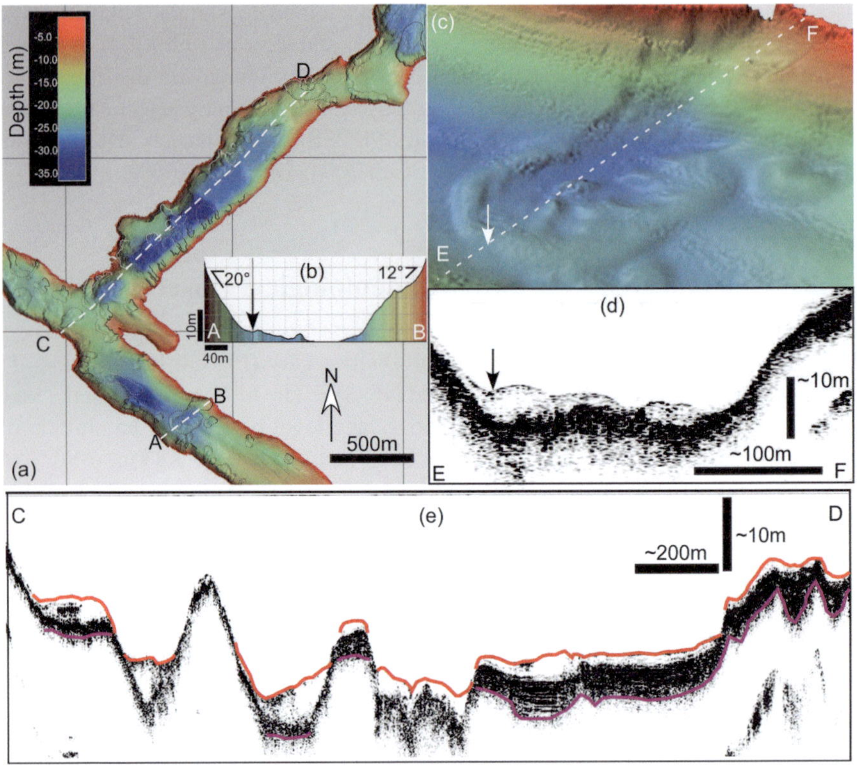

Fig. 36.2 (**a**) A portion of Lake Éternité showing the distribution of the landslides on part of the lake floor; (**b**) A bathymetry profile along *dashed line A–B*; (**c**) 3D view of a landslide with *line E–F* giving the position of the seismic line shown in (**d**), this line is almost at the same position than *line AB* in (**a**); (**d**) a seismic line across the landslide in (**c**); (**d**) An interpreted seismic line along *line C–D* shown in (**a**). The *red line* shows the top of the gyttja transparent layer and the magenta the bottom of the underlying glacio-lacustrine layer. The *arrow* in (**c**) and (**d**) shows the run out distance of the debris of the slide

as a transparent layer draping most of the lake floor (Fig. 36.2e). As shown in Figs. 36.1c, d and 36.2c, almost all slides present a similar morphology: a steep failure surface (10–20°) in the starting zone with an accumulation of mud showing arcuate lobes with a small depression between the back slope and the frontal debris. These debris lobes can be symmetric when they are not in contact with other slides (Fig. 36.2c). For the slide shown in Fig. 36.2b, the head scarp is very close to the shoreline and its height is difficult to define. The rupture surface angle, in the main portion, is about 10° and the run out distance is 290 m. The debris lobe reaches a maximum thickness of about 2.5 m. The estimated volume is 14,000 m^3. In general, scarps are typically less than 2 m high while lateral scarp can reach 3–4 m where the initial morphology had a convex slope near the bottom (as seen in the NW facing slopes along the NE arm). Interestingly, in most basins, gyttja sediments present a hummocky morphology. To that effect, Shilts and Clague (1992) noted that except for 1 or 2 of the 150 lakes they surveyed in Canada, such a morphology likely resulted from seismic shaking.

Regarding the spatial distribution of landslides, amongst the 128 slides mapped, 71 are on either SE or SW facing slopes while 57 are on NE or NW facing slopes (Fig. 36.1a). The SE facing slopes of the NE arm are more dissected than on the other side of the arm except in a small area where the slopes are flatter or where the underlying bedrock is close to the lake floor.

36.4 Effect of Earthquake Source Location

In order to create a regional Digital Elevation Model, the isohypse (10 m interval) were extrapolated on a 40 m grid, and bathymetric data were added on this map. The mesh used for the numerical simulation, in the preliminary results showed here, have a lower resolution mesh: a grid of 190 m was used. The total simulation area (black square in Fig. 36.3a) consisted in a grid of 216 × 216 elements, and the domain had a depth of 30 km. As a simplification, velocities for the whole domain were set 4500 and 3500 m/s for primary and secondary wave velocity. The S-wave value is consistent with values determined by Leblanc and Buchbinder (1977).

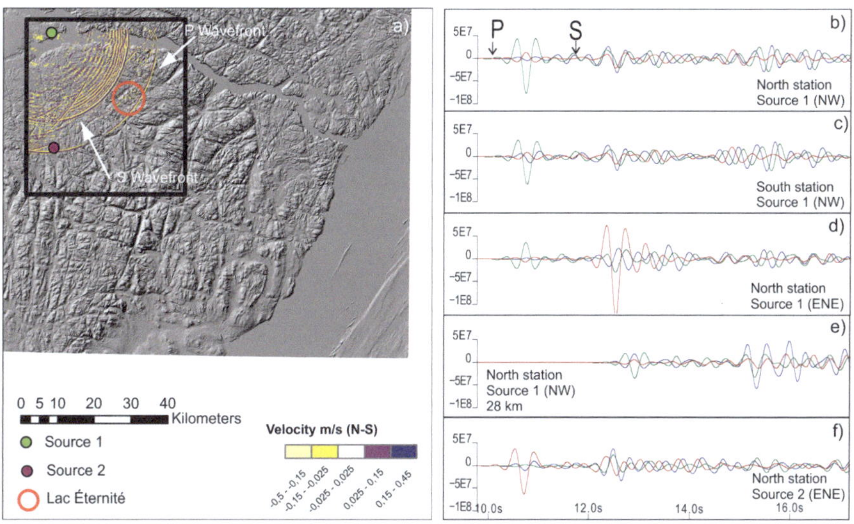

Fig. 36.3 (**a**) Example of seismic wave propagation from source 1; (**b–f**) Synthetic seismograms showing the accelerations on the three components (*green* represent the components perpendicular to the shoreline (N315), *red* the component along the shoreline (N45) and *blue* the vertical component) for site located north and south of the main arm of Lake Éternité and for the two sources shown in (**a**). Scale units are not important in this analysis, but all simulations were scaled the same way

However, in order to obtain a more robust model at lower period, the P-wave value was set lower than the standard value that should be around 6000 m/s. Mesh definition in Specfem3D do not allow to model very superficial layers such as the gytjja. Furthermore, since the distance between the source and the receivers is short and authors only want to look at the relative effect of the source location and topography, the rock quality factor was set to 0 (no attenuation), even if the rock is intensely fractured. Two point source areas for the earthquake were simulated (Fig. 36.3a). The first one is located in the Baie des Ha!Ha! (Source 1, Fig. 36.3a), about 25 km NW from the site, the second one is located near the Lake Ha!Ha! (Source 2, Fig. 36.3a), 25 km SW from the site. These two sites were chosen because they are in line with the two main branches of the Lake and because a fault trace visible on aerial photographs that has the same orientation than the presume orientation of the 1988 Chicoutimi earthquake at source 1 was suggested by Locat (2011) to be the fault associated with the 1663 earthquake. Two directions of the source were simulated for source 1: one with the direction of the fault which is approximately to the NW, and another direction consistent with the stress field derived from most focal mechanism of earthquakes elsewhere in eastern Canada, i.e. N60E (Adams and Basham 1989). For source 2, the direction of the source was set to be N60E. The point source simulated has a force of 1×10^{21} N and the hypocenter was at a depth of 12 km. This depth was chosen as most earthquakes in this area (more specifically in the Charlevoix-Kamouraska zone) have their hypocenter at a depth between 5 and 15 km (Boore and Atkinson 1992) with a median depth of 12 km (Lamontagne 1999). One simulation was also made with a depth of 28 km, which was the depth of the 1988 earthquake, the biggest earthquake that was recorded in this area. Figure 36.3a shows an example of the propagation of the wave for a simulation made with the epicenter at the Baie des Ha!Ha!. This figure shows the N-S speed component at the surface in m/s.

Results from five different simulations will be presented. The first three simulations have their source in the Baie des Ha!Ha!, but for simulation (1), the direction of the source is set to be the direction of the fault, simulation (2) uses the N60E source direction, and simulation (3) uses the same setting as (1) but with a depth of 28 km. The fourth simulation used the same setting as simulation (2), but the source was moved to the Lake des Ha!Ha!.

Synthetic seismograms were calculated in the model on both sides of the Northern arm of Lake Éternité (Fig. 36.2a). The seismograms were filtered with a Butterworth filter, where the period between 0.35 and 10 s was kept. The smaller period correspond to the shortest period where the model was found stable. Figure 36.3 shows the results of four different simulations. In (b) and (c) are shown the results for the first simulation from the north side and the south side of the northern branch. Figure 36.3d shows the results for the north side of the lake, but for the second simulation. Figure 36.3e shows the results for the north side for the third simulation, and Fig. 36.3f shows the results, again for the north side, but when the source is moved to the Lake des Ha!Ha!.

36.5 Discussion

The distribution of landslides around the lake could support the effect of the orientation of the lake relative to the seismic source at the origin of these slides. We could make the hypothesis that they were triggered by the largest events since 7200 ycal BP, i.e. the 1663 Charlevoix earthquake. Doig (1998), using short sediment cores conclude that he could see re-suspended sediments form the 1988 Saguenay earthquake but not for the 1663 one, although he could see the record of an older one, about 1200 years ago. Doig (1998) paper could not take into account the many studies carried out since that time which showed that there are many very large landslides events, related to the 1663 earthquake, in the Upper Saguenay Fjord area (Locat 2011). In addition, the position of the cores used by Doig (1998) is not known and considering the very disturbed nature of the lake floor this is very important to know. In his 1990 and 1998 papers, Doig indicate that his method would better in lakes were there are no landslides, which is not the case. With our current detailed bathymetry, it would now be much easier to select a site with minimum interference with nearby slope movements. We should also keep in mind the results of the work of Levesque et al. (2006) who showed that for the Saguenay Fjord the initial hypothesis that all slide scars were caused by the 1663 earthquake was only valid for 6 of 17 sites investigated. As it was done by Levesque et al. (2006) future work on Lake Éternité will focus on coring at various landslides sites (on the rupture surface and in undisturbed areas) to establish a reliable chronology considering that major earthquakes occurred also in 1791 (*M*6), 1860 (*M*6), 1871 (*M*6.5), 1925 (*M*6.2) and 1988 (*M*5.9) (Levesque et al. 2006).

The effects of the topography on site response were observed at various locations, such as in California (Spudich et al. 1996), where a site, 15 m high, 500 m long and 130 m wide, instrumented with multiple geophones, was subjected to an earthquake. They showed that at the top of the hill amplifications were more than three to four times higher than at the base of the hill. In the simulations presented here, there was no noticeable difference between the seismograms presented with or without topography, as well as between seismograms on the two sides of the Lake, at the same elevation.

Based on Fig. 36.3b, c, showing the difference between the acceleration on the northern side and the southern side of the lake, we can conclude that, at least with the resolution of the mesh that was used here, the effect of topography between the north and south shores, is present but negligible. Furthermore, the acceleration along the coastline direction (red) is approximately the same as the acceleration perpendicular to the coastline direction (green), except for the P waves, where the acceleration perpendicular to the coastline is much higher than along the coastline. For both seismograms, the highest acceleration is caused by the arrival of P waves, and the acceleration on the north station is slightly higher than at the south station. The main reason that may explain this small difference, may be the resolution of the mesh used in the simulation that is too coarse to capture greater amplification. With a minimum period resolved of 0.35 s and a S-wave speed of 3500 m/s, minimum wavelength resolved of the S-wave will be of 1225 m. Topographic effects would probably not be resolved for height difference of less than a quarter of this wavelength.

Three other factors must be taken into account: the source location, its depth and the direction of the source. As for the source directivity, Raghukanth et al. (2012), in their regional study of landslides associated with the 2011 Sikkim Earthquake near the border of Nepal and the Indian state of Sikkim, found correlations between the simulated ground motion obtained with simulations made with SPECFEM3D and landslide locations. Furthermore, they showed a correlation between the displacement and the directivity effect due to the fault orientation and rupture direction at specific sites.

The direction of the source was taken into account by simulations 1 and 2 where the source was kept constant, but the direction of the source was moved from the NW to the ENE (Fig. 36.3b, d). Major differences may be seen for both the P wave and the S wave. For the P wave, the acceleration, when the source is in the ENE direction, is about half the acceleration of the simulation 1. However, when the source is in the ENE direction (Fig. 36.3d), the amplitude of the acceleration in the direction perpendicular to the shoreline as well as in the vertical direction, for the P waves will be slightly higher than when the source is in the NW direction. The alongshore component will be three to four times higher when the source is to the ENE.

Scenarios 1 and 3, i.e. Fig. 36.3b, e, show the influence of the depth of the source. When the source is deeper, the horizontal components will be slightly lower; however, the vertical component will be higher.

The effect of source location is well seen in the simulations. Comparison of Fig. 36.3d, f, i.e. seismograms for the same station location but with a different source, shows that the signal is also dependant on the source location. On Fig. 36.3f, for the P wave, the component along the coastline is the major component, which is the opposite of what is seen in Fig. 36.3e. For the P waves, for source 2, the amplitudes in all three directions are about the same, which is very different than at source 1 where the component along the coastline direction is way higher. In all cases, the maximum acceleration amplitudes seen are lower when the source is near the Lake des Ha!Ha!.

36.6 Concluding Remarks

The following remarks can be made from our initial analysis of geomorphological and geophysical data at Lake Éternité and a regional wave propagation numerical modeling:

1. The great disturbance of the gytjja by the earthquakes(s) resulted in more than 128 landslide scars with volume from 3000 to 15,000 m^3.

2. There are more slide scars on the SE and NW facing slopes. Dating these slides and their associated earthquakes could be used to map their relative spatial occurrence and validate or not the hypothesis that the epicentre of the 1663 earthquake could be located to the NW or NE of the lake, i.e. towards the Saguenay Graben.
3. Preliminary numerical simulations suggest that, at least at the resolution of the model, the topography does not appears to play a significant role as a seismic amplification factor. However, only regional simulations were run and the resolution of the model had to be kept coarse. Simulations with a smaller domain but with a higher resolution will need to be run. On the other hand, the numerical simulations clearly illustrate the effect of the directivity of the source and of its position. For a position of the source based on Locat (2011), for both direction of the source, the acceleration for all components are higher than when the source is near the Lake des Ha!Ha!. More numerical analysis along these lines will be carried in a near future. Modification of the source position to be at the position of the 1988 earthquake, 50 km from the Lake des Ha!Ha!, as well as more local simulations, will need to be done.

Acknowledgments The authors would like to thank the Association of Lac Éternité for providing access to the lake and to NSERC for their funding support. We also thank Nabil Sultan and Didier Perret for their constructive review of the manuscript.

References

Adams J, Basham P (1989) The seismicity and seismotectonics of Canada east of the Cordillera. J Geol Assoc Can 16(1):3–16

Boore DM, Atkinson GM (1992) Source spectra for the 1988 Saguenay, Quebec, earthquakes. Bull Seismol Soc Am 82(2):683–719

Doig R (1990) 2300 yr history of seismicity from silting events in Lake Tadoussac, Charlevoix, Quebec. Geology 18:820–823

Doig R (1998) 3000-year paleoseismological recods from the region of the 1988 Saguenay, Québec, earthquake. Bull Seismol Soc Am 88(5):1198–1203

Dougthy M, Eyles N, Eyles CH, Wallace K, Boyce JI (2014) Lake sediments as natural seismographs: earthquake-related deformation (seismites) in Canadian lakes. Sediment Geol 313:45–67

Komatitsch D (2004) Simulations of ground motion in the Los Angeles Basin based upon the spectral-element method. Bull Seismol Soc Am 94(1):187–206

Komatitsch D, Tromp J (1999) Introduction to the spectral element method for three-dimensional seismic wave propagation. Geophys J Int 139(3):806–822

Lajeunesse P (2014) Buried preglacial fluvial gorges and valleys preserved through Quaternary glaciations beneath the eastern Laurentide Ice Sheet. Geol Soc Am Bull 126:447–458

Lajeunesse P, St-Onge G, Randall K, Moreau-Labrecque A (2008) Mouvements de masse subaquatiques postglaciaires au lac Jacques-Cartier, Réserve faunique des Laurentides (Québec): résultats préliminaires. Comptes rendus de la 4e Conférence canadienne sur les géorisques: des causes à la gestion. Presse de l'Université Laval, Québec 313–321

Lamontagne M (1987) Seismic activity and structural features in the Charlevoix region, Quebec. Can J Earth Sci 24:2118–2129

Lamontagne M (1999) Rheological and geological constraints on the earthquake distribution in the Charlevoix Seismic Zone, PhD. thesis, Carleton University, Ottawa, 353 pp

Leblanc G, Buchbinder GGR (1977) Second micro-earthquake survey of the St. Lawrence Valley near La Malbaie, Quebec, Canada. Can J Earth Sci 14:2778–2789

Lee SJ, Chan YC, Komatitsch D, Huang BS, Tromp J (2009a) Effects of realistic surface topography on seismic ground motion in the Yangminshan Region of Taiwan based upon the spectral-element method and LiDAR DTM. Bull Seismol Soc Am 99(2A):681–693

Lee SJ, Komatitsch D, Huang BS, Tromp J (2009b) Effects of topography on seismic-wave propagation: an example from Northern Taiwan. Bull Seismol Soc Am 99(1):314–325

Levesque CL, Locat J, Leroueil S (2006) Dating submarine mass movements triggered by earthquakes in the upper Saguenay Fjord, Quebec, Canada. Nor J Geol 86:231–242

Locat J (2011) La localisation et la magnitude du séisme du 5 février 1663 (Charlevoix) revues à l'aide des mouvements de terrain. Rev Can Géotech 48:1266–1286

Normandeau A, Lajeunesse P, Philibert G (2013) Late-Quaternary morphostratigraphy of Lake St-Joseph (southeastern Canadian shield): evolution from semi-enclosed glacimarine basin to a postglacial lake. Sediment Geol 295:38–52

Occhietti S, Parent M, Lajeunesse P, Robert F, Govare E (2011) Late Pleistocene-early Holocene decay of the Laurentide ice sheet in Québec-Labrador. In: Ehlers J, Gibbard PL, Hughes PD (eds). Elsevier, Netherlands, Developments in Quaternary science, vol 15. pp 601–630

Raghukanth STG, Lakshmi K, Kavitha B (2012) Estimation of ground motion during the 18th September 2011 Sikkim Earthquake, earthquake. Geomatics Risk 3(1):9–34

Shilts WW (1984) Sonar evidence for postglacial tectonic instability of the Canadian Shield and Appalachians. Curr Res Pt A Geol Surv Can Pap 84-1A:567–579

Shilts WW, Clague JJ (1992) Documentation of earthquake induced disturbance of lake sediments using sub-bottom acoustic profiling. Can J Earth Sci 29:1018–1042

Spudich P, Hellweg M, Lee WHK (1996) Directional topographic site response at Tarzana observed in aftershocks of the 1994 Northridge, California, Earthquake: implications for mainshock motions. Bull Seismol Soc Am 86(1b):S193–S208

St-Onge G, Mulder T, Piper DJW, Hillaire-Marcel C, Stoner S (2004) Earthquake and flood-induced turbidites in the Saguenay Fjord (Québec): a Holocene paleoseismicity record. Quat Sci Rev 23(3–4):283–294. doi:10.1016/j.quascirev.2003.03.001

Chapter 37
Large Mass Transport Deposits in Kumano Basin, Nankai Trough, Japan

Gregory F. Moore and Michael Strasser

Abstract Large-scale landsliding is a common process in the Kumano Forearc Basin of the Nankai Trough accretionary prism. We use a 3D seismic data volume to map the seafloor reflection, which shows that there are two surficial landslides, one rotational slump ~3.4 km wide, 1.8 km long and 150 m thick and one disintegrative slide that has left a seafloor scar ~ >3.65 km wide, 2.6 km long and ~200 m deep. We see no evidence for any deposits related to the latter in our data, so the entire mass must have been transported as debris flows/turbidites outside the area covered by 3D data. The slump failures occurred along a bedding plane that dips ~5–7° landward, but the disintegrative landslide has a gently-dipping base and is associated with steep normal fault scarps. Several large subsurface mass-transport deposits (MTD)s are mapped in the 3D seismic data – all have slid along single landward-dipping bedding planes. Their bases range in depth from 140 to 700 m below sea floor (mbsf). The thickest MTD is ~6.5 $km^2 \times$ 155 m thick, encompassing a volume of ~1.0 km^3. The three other large MTDs range from 0.3 to 0.6 km^3 in volume. The toes of the MTDs are imbricated, and the imbricate structure, as imaged in continuity displays, is aligned parallel to the slope. Many less extensive, thinner (<20 m thick) MTDs are also present in the Kumano Basin. Regional seismic-stratigraphy and age-constraints on MTD-correlative seismic reflections drilled at IODP drill Sites C0009 and C0002 reveal that four of the investigated MTDs are younger than 0.3–0.44 Ma, three are 0.44–0.9 Ma, and three others are between ~0.9 and 1.24 Ma.

37.1 Introduction

Submarine landslides are common on active margins because of the interaction between tectonic uplift that creates steep slopes and frequent earthquakes that serve as landslide triggers. The landslides are capable of generating destructive tsunamis

---

G.F. Moore (✉)
Department of Geology & Geophysics, University of Hawaii, Honolulu, HI, USA
e-mail: gmoore@hawaii.edu

M. Strasser
Geological Institute, ETH Zurich, Zurich, Switzerland

G. Lamarche et al. (eds.), *Submarine Mass Movements and their Consequences*, Advances in Natural and Technological Hazards Research 41,
DOI 10.1007/978-3-319-20979-1_37

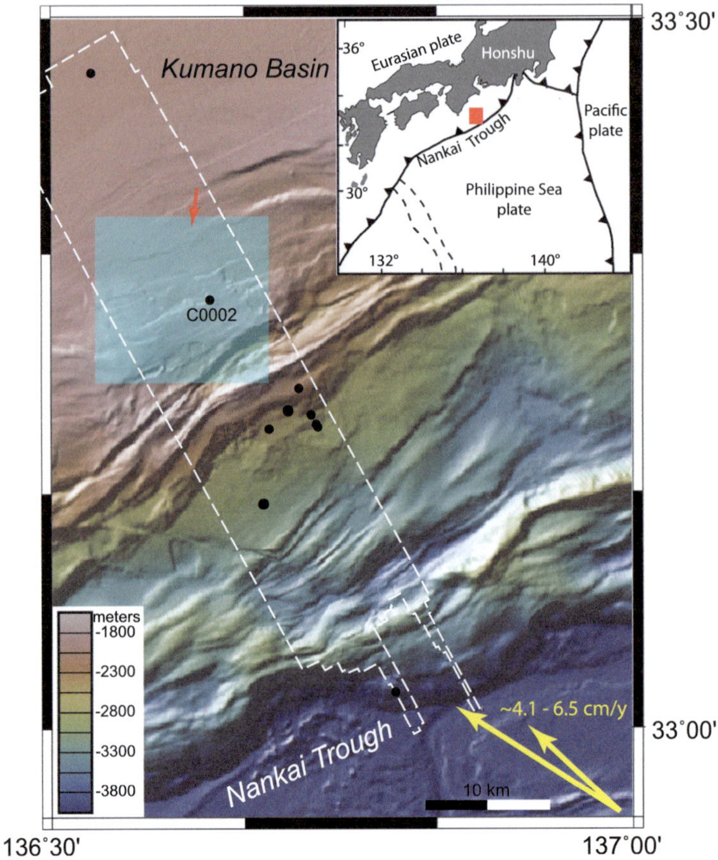

Fig. 37.1 Location map with regional bathymetry showing location of 3D seismic survey (*dashed white box*), location of Fig. 37.2 (*green box*) and IODP drill sites (*black dots*). *Red arrow* indicates look direction for Fig. 37.2. *Yellow arrows* indicate convergence vectors for Philippine Sea plate with respect to Japan (Seno et al. 1993; Miyazaki and Heki 2001; Heki 2007). Inset in upper right is a regional tectonics map showing the setting of the Nankai Trough study area. *Red box* shows location of main map

that can devastate nearby coastal communities (e.g., Bardet et al 2003; Satake 2012). Quaternary landslides on the Nankai Trough outer accretionary wedge have occurred at toe (Kawamura et al. 2012) and in trench slope basins (Strasser et al. 2011, 2012).

In this study, we describe several Quaternary submarine landslides and their deposits in the Kumano Basin of the Nankai Trough inner accretionary prism. We restrict our analysis to the seaward part of the basin (blue area in Fig. 37.1) because the recent slide scars and most of the large buried mass-transport deposits (MTDs) are restricted to this region.

37.1.1 Geologic Setting

The Nankai Trough is the locus of subduction of the Philippine Sea plate beneath southern Japan (Fig. 37.1). Plate convergence occurs at a rate of ~4.0–6.5 cm/year at an azimuth of ~300–315 (Seno et al. 1993; Miyazaki and Heki 2001). Accretion of a thick section (>1 km) of terrigenous sediment has built a wide accretionary prism (Aoki et al. 1982). Landward of the actively deforming outer prism is the less active inner prism (Wang and Hu 2006) that is capped by the Kumano forearc basin (Okino and Kato 1995). Kumano Basin has formed since ~1.95–2.0 Ma behind a ridge that is believed to have been uplifted by movement along the mega-splay fault (Gulick et al. 2010; Moore et al. 2015). The basin strata are cut by a series of normal faults that cut the seafloor, indicating that they were very recently active (Moore et al. 2013).

37.1.2 3D Seismic Data

We use a three-dimensional (3D) seismic reflection data set collected across the outer part of Kumano Basin and the Nankai prism in 2006 by Petroleum Geo-Services (Moore et al. 2009). The 12-km-wide, 56-km-long survey was acquired with four 4.5 km-long hydrophone streamers and dual 3,090 in.^{3} (50.6 l) airgun source arrays. Basic processing through pre-stack time migration was carried out by Compagnie Générale de Géophysique (CGG) and full 3D pre-stack depth migration (3D PSDM) was completed by the Japan Agency for Marine Earth Science and Technology (JAMSTEC). See details of processing procedures in Moore et al. (2009). The interval between inlines (oriented NW-SE) and cross lines (oriented SW-NE) of the resulting dataset is 18.75 m and 12.5 m, respectively. The vertical resolution for the interval of interest in this study is ~5–7 m near the seafloor and ~10–20 m at depths near 1,400 mbsf (Moore et al. 2009). We extracted a coherency volume from the final 3D PSDM (Bahorich and Farmer 1995; Marfurt et al 1998), which facilitated recognition of the boundaries and internal structure of the MTDs.

The evolution of Kumano Basin is documented by seismic stratigraphic analysis and ocean drilling (Gulick et al 2010; Underwood and Moore 2012; Moore et al. 2015). The Quaternary section is divided into 12 landward-dipping seismic stratigraphic units (K1-K12 after Gulick et al. 2010). We concentrate on the section younger than ~0.9 Ma (sequences K4-K1).

37.2 Characteristics of Mass Wasting in Kumano Basin

Two basic types of submarine landslides and their deposits occur in the Kumano forearc basin: Surficial slumps and older, buried MTDs.

37.2.1 Surficial Landslides

A large rotational slump (slide #1) occurs at the seafloor along the NE margin of our survey area (Figs. 37.2, 37.3 and 37.4). The basal sliding surface is developed along a bedding plane that dips landward ~7°. The slump has a well-developed headwall,

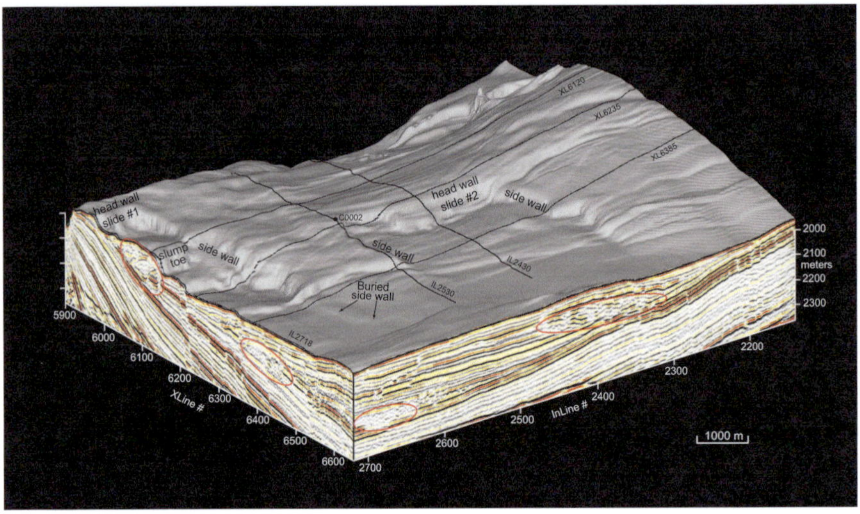

Fig. 37.2 Perspective view of 3D seismic volume with surface bathymetry showing prominent slide scars #1 and #2. *Black lines* on surface = locations of in-lines (IL) and cross-lines (XL) displayed in Figs. 37.3 and 37.4. Major MTDs circled in *red*. Location shown as *green* box in Fig. 37.1. View is looking to the SSW (*red arrow* in Fig. 37.1)

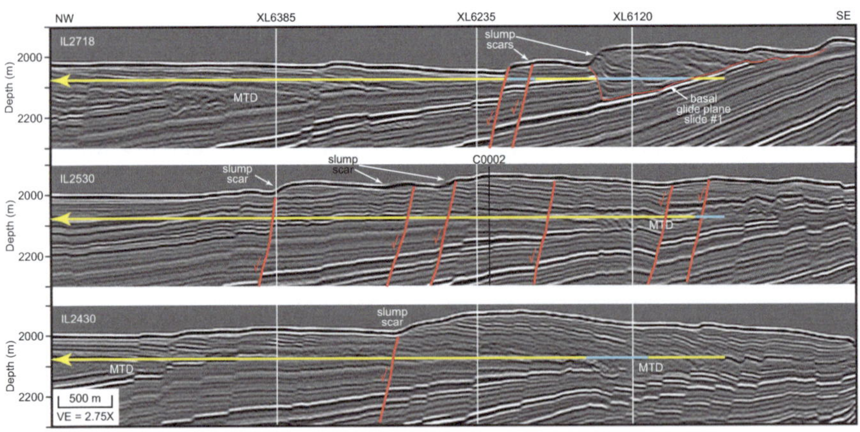

Fig. 37.3 Seismic in-lines (IL) extracted from 3D volume. The upper panel crosses slide #1 and the lower two panels cross slide #2. A few normal faults (*red lines*) that are associated with surficial slumps are annotated. *Yellow line* shows depth and extent of map in Fig. 37.5; *blue lines* show extent of slumps on Fig. 37.5

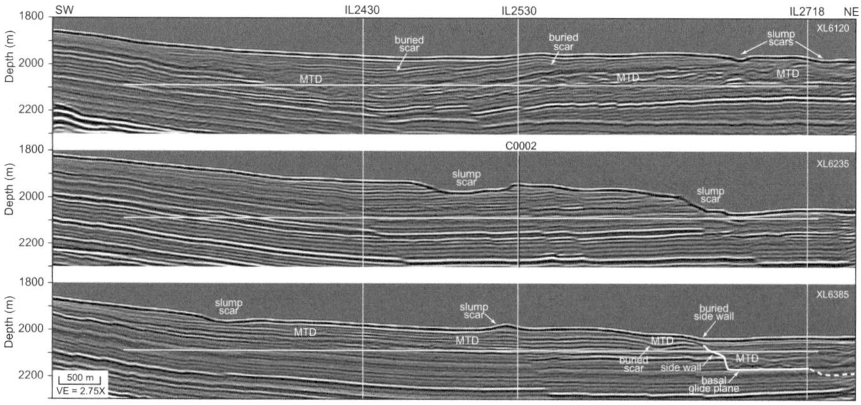

Fig. 37.4 Seismic cross-lines (XL) extracted from 3D volume. Several recent as well as buried MTDs and a buried slide scar are imaged. A buried side wall scarp has been covered by younger strata in the lower panel. *Yellow line* shows depth and extent of map in Fig. 37.5; *blue lines* show extent of slumps on Fig. 37.5

SW sidewall and slump toe. The NE sidewall is beyond the edge of our survey; the slump is imaged on a 2D seismic line (KR0413-D5) 2 km to the east, so the slump is at least 3.4 km wide × 1.8 km long and 0.15 km thick, and thus covers at least 6 km^2 area and has a volume of at least 0.9 km^3. The seismic cross-line that crosses the lower reaches of the slide (Fig. 37.4, XL6385) shows that the area is covered by younger strata that have buried the side wall of the slide.

A second surficial landslide (slide #2) in the central part of the area is 3.65 km wide × 2.6 km long and 0.2 km thick, thus covering ~9.5 km^2 area with a volume of ~1.9 km^3. Most of that material has been completely removed from the region and was probably deposited as debris flows/turbidites in the deeper part of the basin. The slump scars all dip to the NW, and the underlying strata dip 1–2° either landward or seaward. The slump scars are closely associated with normal faults that have offsets of 20–30 m (Fig. 37.3).

37.2.2 *Buried MTDs*

The seismic data image two large buried MTDs (Figs. 37.5 and 37.6). The younger deposit (MTD-U) was a slump that formed a positive topographic feature on the seafloor, as indicated by the onlap relations of the overlying strata (Fig. 37.6). Its area is 6.5 km^2 and its volume is ~1.0 km^3. It has a well-defined basal glide plane and its internal structure is characterized by coherent blocks with small-offset thrust faults between them. These internal pressure ridges are displayed well in the coherency plot (Fig. 37.5) that shows the orientation of the blocks. The blocks form arc-like structures on coherence plot. The base of MTD-U is less than 60 m

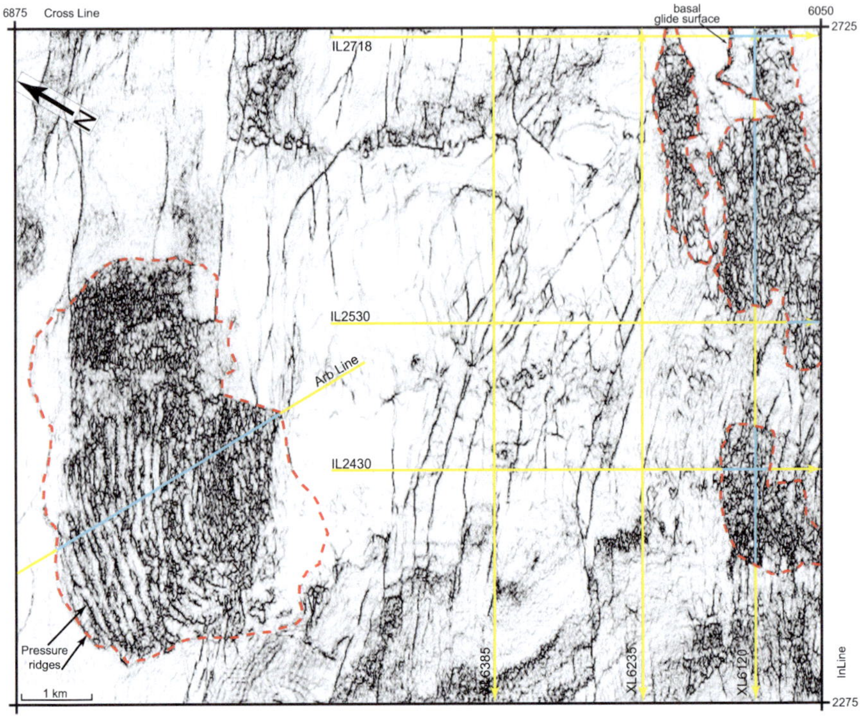

Fig. 37.5 Depth slice at 2078 m through 3D coherency volume. Darker masses outlined by *red dashed lines* are MTDs. *Yellow lines* show extent of seismic in-lines (IL) and cross-lines (XL) shown in Figs. 37.3 and 37.4; *blue lines* show extent of slumps on Figs. 37.3 and 37.4; *blue lines* show extent of slumps on Figs. 37.3 and 37.4. "Arb Line" = shows location of arbitrary line extracted from the seismic volume and shown in Fig. 37.6. Most *dark* lineations are normal faults

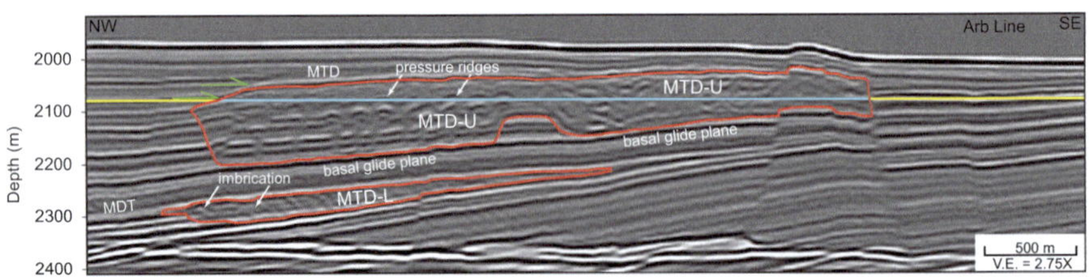

Fig. 37.6 Arbitrary seismic line extracted from 3D seismic volume orientated approximately perpendicular to the direction of structural trends within MTD-U. *Green arrows* show onlap onto MTD. Note imbricate structure within MTD-L. Thinner MTDs overlie both MTD-U and MTD-L. *Yellow line* shows location of depth slice shown in Fig. 37.5

above the top of seismic stratigraphic unit K4 of Gulick et al. (2010), which has an age of 0.3–0.44 Ma at IODP Site C0002 (Exp 315 Scientists 2009; Moore et al. 2014).

The older deposit (MTD-L) is lens-shaped and pinches out in the seaward (SE) direction. It is thicker in the north and thins gradually to the south. It also has a well-defined basal glide plane, and the toe of the slide is imbricated. The base of MTD-L is ~8 km wide, 3.5 km long and ~100 m thick (area ~28 km^2; volume ~2.8 km^3) and is less than 50 m above seismic unit K5 (<0.9 Ma).

There are many more buried MTDs that range from a few metres to a few 10's of metres in thickness (Figs. 37.2, 37.3, 37.4, and 37.6). Most are recognised by their chaotic internal reflection character. Both MTD-L and MTD-U are overlain by thinner MTDs.

37.3 Discussion

As expected in this tectonically active setting, submarine landsliding has been an active process in the Kumano Basin during the Quaternary. Four of the slumps/ slides are younger than seismic sequence K4 (~0.3–0.44 Ma), three MTDs are within sequence K4 (~0.44–0.9 Ma) and there is one MTD each in sequences K5 (~0.9–1.04 Ma), K6 (~1.04–1.07 Ma) and K7 (~1.07–1.24 Ma). Thus, the approximate recurrence interval is ~0.05–0.10 Ma. This is much less frequent than the recurrence rate for Nankai Trough great earthquakes (~100–200 year, Ando 1975). Thus, although the slides may ultimately be triggered by earthquake shaking, the region must be pre-conditioned in order to be susceptible to failure during the quakes, as suggested for slides in the slope basins seaward of Kumano Basin (Strasser et al. 2012). One recent slide occurred along a landward-dipping slope of ~5–7°, but other slides are closely associated with regional normal faults that have 20–30 m of vertical slip. It is thus likely that the slides occurred along the exposed fault scarps, so this is one type of potential pre-conditioning in Kumano Basin. It is also possible that gas hydrate dissociation may play a role in slope destabilization (e.g., Bangs et al. 2010), or potentially even bottom current activity. The timing of slide activity in Kumano Basin is roughly similar to activity seaward of the basin (Strasser et al 2012), so it will be important for future studies to more precisely date the Kumano MTDs.

Although there are many MTDs in the basin, all were derived locally. That is, all of the sediment transported in the MTDs came from the flanks of the basin, a process that simply redistributes the existing sediment.

Compared to submarine slides generated on the Nankai frontal prism, the larger volume of the Kumano Basin MTDs, their coherent mass, the shallower water depth, closer distance to the shoreline and, most importantly, the landward-facing transport direction are all important ingredients to increasing their hazard potential with respect to tsunami generated from such slides (e.g., Satake 2012).

37.4 Concluding Remarks

Submarine landslides are ubiquitous in the Quaternary strata of Kumano Basin. One surficial slide and at least one buried MTD are coherent slumps that accumulated as positive features on the seafloor, but most other slides evolved downslope into debris flows and spread out into the deeper parts of the basin.

The Kumano Basin MTDs were all generated by local submarine landslides. Thus, although there are many MTDs in the basin, this sliding process only displaces local sediment into deeper parts of the basin and is not responsible for supplying shelf sediment to the basin as in most passive continental margins.

The number of Quaternary submarine slides in Kumano Basin is much less than the number of major earthquakes that occurred during this time, so earthquake shaking might have been the final trigger for these slides, but the regions were preconditioned to fail prior to the quakes. For instance, many of the surficial slides were generated as failures along normal fault scarps.

Acknowledgments This work was supported by grants from the US National Science Foundation (OCE-0451790), the US Science Support Program and the Swiss National Science Foundation (Grant # 133481). We thank Paradigm Geophysical for making their software available to us. Derek Sawyer, Yasuhiro Yamada and Aaron Micallef are acknowledged for constructive reviews. SOEST Contribution #9301.

References

Ando M (1975) Source mechanisms and tectonic significance of historical earthquakes along the Nankai Trough, Japan. Tectonophysics 27:119–140

Aoki Y, Tamano T, Kato S (1982) Detailed structure of the Nankai Trough from migrated seismic sections. In: Watkins, JS, Drake, CL, (eds) Studies in continental margin geology, Am Assoc Petrol Geol Mem 34:309–322

Bahorich M, Farmer S (1995) The coherence cube. Lead Edge 14:1053–1058

Bangs NL, Hornbach MJ, Moore GF, Park JO (2010) Massive methane release triggered by seafloor erosion offshore southwestern Japan. Geology 38(11):1019–1022. doi:10.1130/g31491.1

Bardet JP, Okal EA, Synolakis CE, Davies HL, Imamura F (2003) Landslide tsunamis: recent findings and research directions. Pure Appl Geophys 160(10–11):1793–1809

Expedition 315 Scientists (2009) Expedition 315 Site C0002. In: Kinoshita M, Tobin H, Ashi J et al. (eds) Proc. IODP 314/315/316. Integrated Ocean Drilling Program Management Int., Washington, DC. doi:10.2204/iodp.proc.314315316.124.2009

Gulick SPS, Bangs NLB, Moore GF, Ashi J, Martin KM, Sawyer DS, Tobin HJ, Kuramoto S, Taira A (2010) Rapid forearc basin uplift and megasplay fault development from 3D seismic images of Nankai Margin off Kii Peninsula, Japan. Earth Planet Sci Lett 300(1–2):55–62. doi:10.1016/j.epsl.2010.09.034

Heki K (2007) Secular, transient and seasonal crustal movements in Japan from a dense GPS array: implications for plate dynamics in convergent boundaries. In: Dixon T, Moore JC (eds) The seismogenic zone of subduction thrust faults. Columbia University Press, New York, pp 512–539

Kawamura K, Sakaguchi A, Strasser M, Anma R, Ikeda H (2012) Detailed observation of topography and geologic architecture of a submarine landslide scar in a toe of an accretionary prism. In: Yamada Y. et al (eds) Submarine mass movements and their consequences. Springer, Dordrecht/Heidelberg/London/New York pp 301–309, 10.1007/978-94-007-2162-3_27

Marfurt KJ, Kirlin RL, Farmer SL, Bahorich MS (1998) 3-D seismic attributes using a semblance-based coherency algorithm. Geophysics 63:1150–1165

Miyazaki SI, Heki K (2001) Crustal velocity field of southwest Japan: subduction and arc-arc collision. J Geophys Res 106(3):4305–4326

Moore GF, Park J-O, Bangs NL, Gulick SP, Tobin HJ, Nakamura Y, Sato S, Tsuji T, Yoro T, Tanaka H, Uraki S, Kido Y, Sanada Y, Kuramoto S, Taira A (2009). Structural and seismic stratigraphic framework of the NanTroSEIZE Stage 1 transect. In: Kinoshita M, Tobin H, Ashi J, Kimura G, Lallement S, Screaton EJ, Curewitz D, Masago H, Moe KT, The Expedition 314/315/316 Scientists (ed) Proceedings of the IODP, 314/315/316. Integrated Ocean Drilling Program Management International, Inc., Washington, DC. doi:10.2204/iodp.proc.314315316.102.2009

Moore GF, Boston BB, Sacks AF, Saffer DM (2013) Analysis of normal fault populations in the Kumano Forearc Basin, Nankai Trough, Japan: 1. Multiple orientations and generations of faults from 3-D coherency mapping. Geochem Geophys Geosyst 14(6):1989–2002. doi:10.1002/ggge.20119

Moore GF, Kanagawa K, Strasser M, Dugan B, Maeda L, Toczko S, Sci. Party IODP Exp. 338 (2014) IODP Expedition 338: NanTroSEIZE Stage 3: NanTroSEIZE plate boundary deep riser 2. Sci Drill 17:1–12. doi:10.5194/sd-17-1-2014

Moore, GF, Boston, BB, Strasser, M, Underwood, MB, Ratliff, RA (2015) Evolution of tectono-sedimentary systems in the Kumano Basin, Nankai Trough forearc. Mar Petrol Geol 67:604–616. doi:10.1016/j.marpetgeo.2015.05.032

Okino K, Kato Y (1995) Geomorphological study on a clastic accretionary prism: the Nankai Trough. Island Arc 4:182–198

Satake K (2012) Tsunamis generated by submarine landslides. In: Yamada Y (ed) Submarine mass movements and their consequences. Advances in natural and technological hazards research, vol 31. Springer, Dordrecht/Heidelberg/London/New York pp 475–484. doi:10.1007/978-94-007-2162-3_42

Seno T, Stein S, Gripp AE (1993) A model for the motion of the Philippine Sea plate consistent with NUVEL-1 and geological data. J Geophys Res 98:17, 941–917, 948

Strasser M, Moore GF, Kimura G, Kopf AJ, Underwood MB, Guo J, Screaton EJ (2011) Slumping and mass transport deposition in the Nankai fore arc: evidence from IODP drilling and 3-D reflection seismic data. Geochem Geophys Geosyst 12. doi:10.1029/2010gc003431

Strasser M, Henry P, Kanamatsu T, Thu MK, Moore GF, IODP Expedition 333 scientists (2012) Scientific drilling of mass-transport deposits in the Nankai accretionary wedge: first results from IODP Exp 333. In: Yamada Y et al (eds) Submarine mass movements and their consequences. Advances in natural and technological hazards research, vol 31. Springer, Dordrecht/Heidelberg/London/New York pp 671–681. doi:10.1007/978-94-007-2162-3_60

Underwood MB, Moore GF (2012) Evolution of sedimentary environments in the subduction zone of southwest Japan: recent results from the NanTroSEIZE Kumano transect. In: Busby CJ, Azor AP (eds) Tectonics of sedimentary basins: recent advances. Wiley-Blackwell, New York, pp 310–326

Wang K, Hu Y (2006) Accretionary prisms in subduction earthquake cycles: the theory of dynamic Coulomb wedge. J Geophys Res 111(B6):B06410. doi:10.1029/2005jb004094

Chapter 38
Insights into Effectiveness of Simplified Seismic Displacement Procedures to Evaluate Earthquake Behavior of a Deepwater Slope

Aurelian C. Trandafir and Mihail E. Popescu

Abstract This paper employs numerical modeling to investigate the ability of simplified procedures based on sliding block methodology to provide a reasonable characterization of the seismic displacements of a deepwater slope. Earthquake-induced permanent shear displacements obtained from dynamic finite-element analyses of a deepwater slope subjected to various input base excitations are presented and compared with the seismic displacements predicted by two relatively recent simplified procedures available in the literature. The numerical outcomes indicate that the simplified procedures may be overly conservative in evaluating the earthquake-induced permanent shear displacements along the sliding surface of deepwater slopes. Based on the limited displacement data set developed under the present study, correlations aiming at improving the predictive capability of the selected simplified procedures in respect to evaluation of seismic performance of deepwater slopes are provided.

38.1 Introduction

Characterization of seismic behavior of submarine slopes represents a major component of a geohazard assessment program associated with offshore oil and gas infrastructure development in earthquake prone areas. Published simplified displacement evaluation procedures based on the sliding block methodology represent and attractive screening option to evaluate the earthquake induced displacements of submarine slopes especially in the planning stage of field development. However, a major drawback associated with the application of various published simplified displacement prediction relationships in evaluations of earthquake behavior of

A.C. Trandafir (✉)
Fugro GeoConsulting, Inc., 6100 Hillcroft Ave., Houston, TX 77081, USA
e-mail: atrandafir@yahoo.com; atrandafir@fugro.com

M.E. Popescu
Illinois Institute of Technology, 3201 South, Dearborn Street, Chicago, IL 60616, USA

G. Lamarche et al. (eds.), *Submarine Mass Movements and their Consequences*, Advances in Natural and Technological Hazards Research 41,
DOI 10.1007/978-3-319-20979-1_38

submarine slopes is the fact that these relationships have only been verified against onshore case histories of earthquake induced permanent slope displacements.

In this context, the present study employs numerical modeling to investigate the ability of simplified displacement evaluation procedures to provide a reasonable characterization of the seismic performance of a deepwater slope. Permanent seismic shear displacements obtained from two-dimensional total stress undrained dynamic finite-element analyses employing a cyclic nonlinear constitutive soil model are compared with the displacements predicted by selected published simplified procedures employing the sliding block methodology. The outcomes of the numerical investigation are presented and discussed in the light of likely conservatism associated with the application of simplified displacement procedures for various earthquake excitations and depths of the potential sliding surface. The generic deepwater sediment geotechnical properties and slope geometry utilized in the analysis have been selected based on experience with deepwater slopes offshore East Africa.

It is noteworthy that the finite-element based numerical scheme used in this study has been verified against observed earthquake-induced deformations of clayey slopes in laboratory centrifuge tests. However, further research will be needed to calibrate the numerical model against well documented case histories of earthquake-induced displacements of deepwater slopes when such information will become available.

38.2 Finite-Element Model and Input Geotechnical Parameters

Figure 38.1 shows the finite-element mesh of the analyzed submarine slope characterized by a slope angle of about 8° along segment GF, and a free face along segment AG making an angle of about 14° with the horizontal. The finite-element model consists of 712 eight – noded isoparametric quadrilateral elements, and has a total horizontal length of 900 m. The bottom boundary BCD, assumed to represent a high acoustic impedance contrast boundary between the overlying clay sediments and a much stronger partially-lithified sedimentary unit, occurs at a depth of 64 m below the seafloor measured along vertical "X" in Fig. 38.1. The sliding surface of the model is located in a 1 m – thick weak clayey interlayer. Two depths of the sliding surface of 11 m and 24 m (measured to the top of the interlayer along vertical "X"), corresponding to sliding surface 1 and sliding surface 2 in Fig. 38.1, respectively, are considered in the present numerical study.

The boundary conditions of the finite-element model for the dynamic analysis involved restrained horizontal and vertical relative displacements along the bottom boundary BCD where the input ground motion was applied, and absorbing (viscous) boundaries along the vertical edges AB and DE to ensure appropriate dissipation of the outward propagating seismic waves. The dynamic finite-element

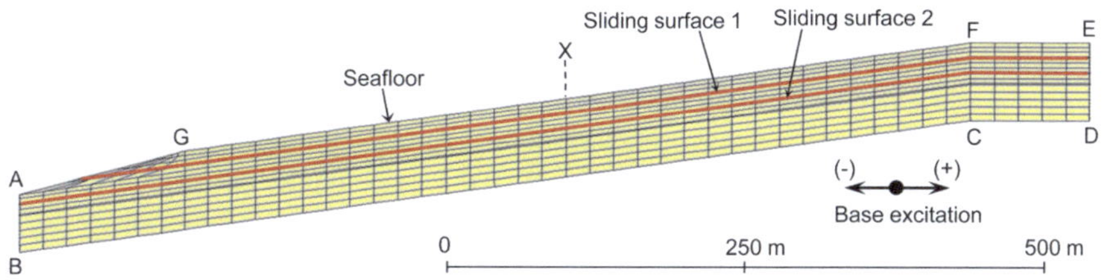

Fig. 38.1 Finite-element model of the analyzed deepwater slope

Fig. 38.2 Submerged unit weight (*SUW*) and undrained shear strength (S_u) profiles considered in the analysis

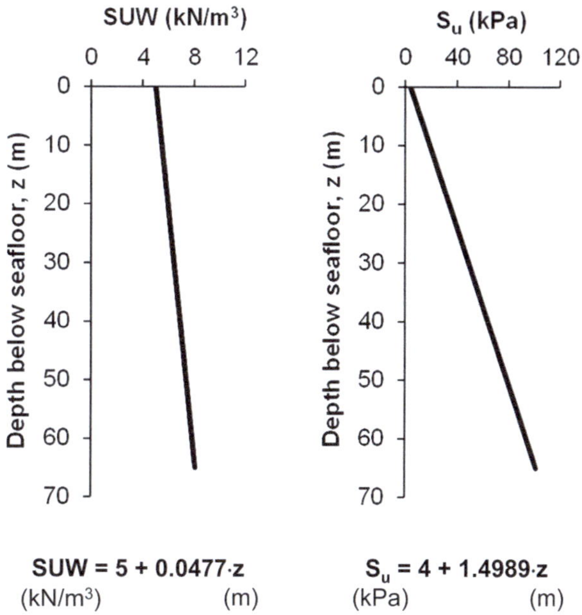

analysis employed the initial stresses obtained from a static finite-element analysis that involved activation of the gravitational loads in the system.

Figure 38.2 presents the submerged unit weight (*SUW*) and dynamic undrained shear strength (S_u) profiles characterizing the deepwater clay sediments of the analyzed slope. The dynamic undrained shear strength of the weak interlayer (S_u^{weak}) was taken as 70 % of the dynamic undrained shear strength S_u of the clay sediment at the depth of the interlayer.

The dynamic stress-strain behavior of both clay sediments and weak interlayer was modeled using a cyclic nonlinear constitutive soil model developed by Wakai and Ugai (2004) for finite-element based characterization of dynamic deformation of slopes during earthquakes, referred herein as Wakai-Ugai model. Similar to previous constitutive formulations employing a phenomenological approach to describe experimental stress – strain relationships observed in laboratory cyclic shear tests, such as modified Hardin-Drnevich model (Hardin and Drnevich 1972) and modified Ramberg-Osgood model (Jennings 1964), the selected constitutive model is based on mathematical equations describing, in the shear stress (τ) – shear strain (γ) domain, a backbone curve that represents the locus of unloading/reloading points and a hysteresis loop characterizing the unloading/reloading behavior departing from the backbone curve. The Wakai-Ugai model offers the advantage of a more flexible hysteresis loop formulation translating into a better characterization of observed experimental damping ratio versus shear strain relationship for clayey soils compared to the modified Hardin-Drnevich model that typically provides unrealistically high damping ratios at shear strain levels in excess of about 1 % (Wakai and Ugai 2004) or even lower.

A brief introduction of the basic equations describing the Wakai-Ugai constitutive stress-strain formulation is provided herein. The backbone curve of the constitutive model is defined by the following hyperbolic relationship originally proposed by Kondner and Zelasko (1963):

$$\tau = \frac{G_0\gamma}{1 + \frac{G_0\gamma}{\tau_f}} \tag{38.1}$$

where G_0 = initial shear modulus; and τ_f = soil shear strength (= S_u for the current total stress dynamic finite-element analysis).

The equation describing the hysteresis loop in $\tau-\gamma$ domain between two points of coordinates ($-\gamma_a$, $-\tau_a$) and ($+\gamma_a$, $+\tau_a$) located on the backbone curve (Fig. 38.3) is as follows (Wakai and Ugai 2004)

$$\widetilde{\tau} = \frac{a\widetilde{\gamma}^n + G_0\widetilde{\gamma}}{1 + b\widetilde{\gamma}} \tag{38.2}$$

where $\widetilde{\tau} = |\tau - \tau_a|$; $\widetilde{\gamma} = |\gamma - \gamma_a|$; b, n = constants; and a = parameter provided by the following equation:

$$a = \frac{1}{(2\gamma_a)^n}\{2\tau_a(1 + 2b\gamma_a) - 2G_0\gamma_a\} \tag{38.3}$$

An important feature of the Wakai-Ugai constitutive formulation is the ability of the backbone curve to translate along the horizontal (γ) axis in order to intercept the last point of strain reversal for irregular cyclic loading conditions. It is noteworthy that the Wakai-Ugai constitutive model along with the associated finite-element based numerical scheme developed by the two authors and employed in the present dynamic analysis, were successfully validated against observed earthquake-induced deformations of clayey slopes in laboratory centrifuge tests (Wakai and Ugai 2004).

The following Wakai-Ugai constitutive model parameters, describing both the clay sediments and the weak interlayer, were selected for the present finite-element study: $G_0 = 1{,}000 \cdot S_u$; $b = 1{,}450$; $n = 2$. Figure 38.3a shows, in the normalized shear

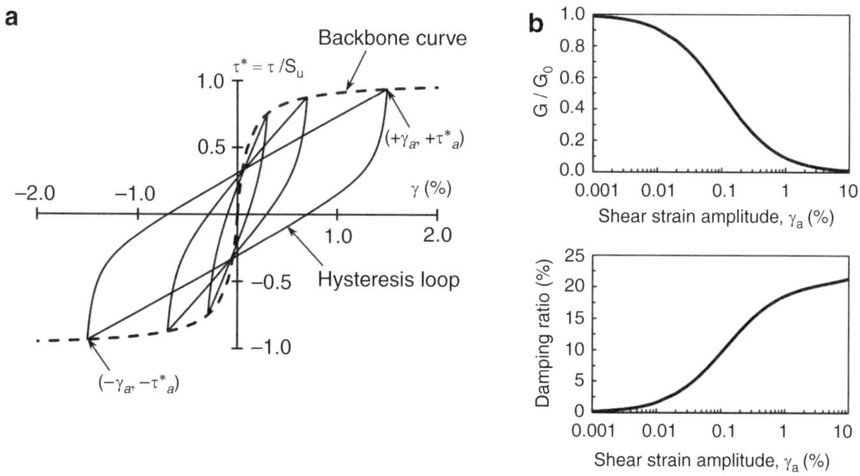

Fig. 38.3 (**a**) Modeled stress-strain behavior in cyclic simple shear and (**b**) computed normalized shear modulus (G/G_0) and damping ratio of the clay sediments in relation to cyclic shear strain amplitude (γ_a) based on selected constitutive soil model parameters

stress (τ/S_u) versus shear strain (γ) space, the predicted backbone curve and hysteresis loops at various cyclic shear strain amplitudes for uniform cyclic simple shear conditions with no initial (static) shear stress, based on the selected constitutive model parameters. For illustrative purposes, model based predictions of typical dynamic soil properties consisting of normalized shear modulus (G/G_O) and damping ratio in relation to cyclic shear strain amplitude (γ_a), based on the selected constitutive model parameters are presented in Fig. 38.3b since such information is familiar to the geotechnical engineering practice.

38.3 Numerical Results

The earthquake response analysis of the deepwater slope shown in Fig. 38.1 was performed using the finite-element code UWLC developed by Forum8 Inc. (2006) which employs the finite-element based numerical scheme developed by Wakai and Ugai (2004). Finite-element computed displacements presented in this paper correspond to a representative point on the slope at the seafloor located along vertical "X" in Fig. 38.1. In addition to the damping associated with the modeled nonlinear stress-strain behavior of sediments on the slope, Rayleigh damping was also introduced in the finite-element model and maintained at levels not greater than about 3 %.

Eleven horizontal ground motion records with the characteristics presented in Table 38.1 were used as base excitation in the present dynamic finite-element

Table 38.1 Horizontal ground motions used as input base excitation in finite-element analysis

Earthquake event	Moment magnitude	Record ID	PGA[a](g)	T_m[b](s)	$D_{5-95\ \%}$[c](s)
Chi-Chi, Taiwan (1999)	7.6	HWA 034-000	0.142	0.56	20.1
Chi-Chi, Taiwan (1999)	7.6	NST 090-000	0.309	0.35	10.2
Chi-Chi, Taiwan (1999)	7.6	TCU 078-090	0.444	0.43	25.9
Chi-Chi, Taiwan (1999)	7.6	TCU 079-090	0.743	0.49	24.2
Loma Prieta, CA (1989)	6.9	CAP-090	0.443	0.49	13.2
Loma Prieta, CA (1989)	6.9	HVR-000	0.134	0.68	16.2
Northridge, CA (1994)	6.7	CYP-143	0.149	0.38	9.9
Northridge, CA (1994)	6.7	LOS-000	0.410	0.59	6.3
Northridge, CA (1994)	6.7	CWC-180	0.298	0.54	15.0
Whittier Narrows, CA (1987)	6.0	GLP-177	0.296	0.34	6.5
Whittier Narrows, CA (1987)	6.0	CYP-143	0.137	0.49	9.8

[a]Peak acceleration of the base input ground motion
[b]Mean period of the base input ground motion as defined by Rathje et al. (2004)
[c]Significant duration of the base input ground motion as defined by Dobry et al. (1978)

analysis. The records were extracted from the ground motion database available in the computer program SLAMMER developed by the U.S. Geological Survey (Jibson et al. 2013). An example of computed seismic response of the analyzed deepwater slope along sliding surface 1 for the Loma-Prieta-CAP-090 ground motion record (Table 38.1) is shown in Fig. 38.4. The shear displacement time history in Fig. 38.4 corresponds to the shear strains developed solely along the sliding surface, thus disregarding yielding of sediments below and above the sliding surface.

The permanent shear displacement (D_{FEA}) values obtained from the finite-element analyses were utilized in evaluations of the predictive capability of simplified displacement procedures in order to be consistent with the mechanism of simplified sliding block models which allow for plastic yielding to occur only along the sliding surface of the model whereas the sliding mass is treated as a rigid or elastic system. As seen in Fig. 38.4, the computed permanent shear displacement at the end of the earthquake (29 cm) is significantly smaller than the permanent displacement of the slope at the seafloor (76 cm), with the latter incorporating also the permanent deformations associated with yielding of sediments below and above the sliding surface. Finite-element computed parameters required in evaluation of the selected simplified displacement procedures included permanent shear displacement (D_{FEA}), peak seafloor acceleration (a_{max}), and 5 % damped elastic spectral acceleration (S_a) of the site below the sliding mass at a degraded period of $1.5T_s$. The analyzed slope is characterized by an initial (small strain) fundamental period (T_s) of 1.2 s which was estimated from the free vibration of the system at the end of the earthquake shaking.

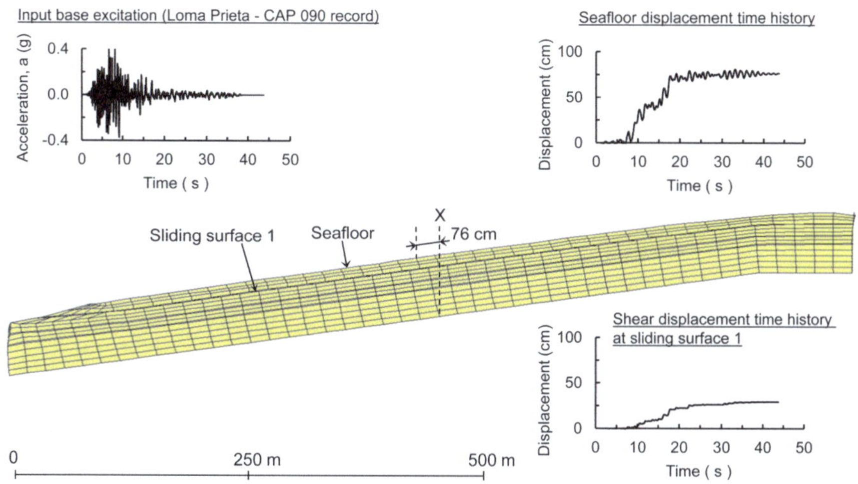

Fig. 38.4 Example of finite-element computed seismic slope performance along sliding surface 1, in terms of deformed finite-element mesh at the end of the earthquake (30× magnification) and time histories of shear displacement and displacement at the seafloor along vertical X

38.4 Evaluation of Simplified Displacement Procedures

Several simplified procedures for evaluation of seismic slope displacements based on the sliding block methodology are currently available in the literature. A thorough review of these procedures can be found in Meehan and Vahedifard (2013). The recent simplified procedures of Bray and Travasarou (2007) and Jibson (2007) have been selected for evaluation in the present study. Bray and Travasarou (B-T) procedure is gaining wide acceptance in seismic slope design practice for onshore projects, whereas Jibson (J) procedure may represent an attractive option as one of the recently developed procedures employing ground motion parameters readily available from the basic information accompanying an earthquake record.

The yield coefficient (k_y) of the sliding mass required in seismic displacement (D_{SB}) assessment based on simplified sliding block models was estimated from a limit-equilibrium based pseudo-static slope stability analysis using the infinite-slope theory. The yield acceleration, defined as the earthquake acceleration required to bring the sliding mass to the limit equilibrium condition, corresponding to a safety factor of 1.0, is obtained as the yield coefficient (k_y) multiplied by the gravitational acceleration (g). k_y is a key input parameter required by the sliding block methodology which assumes that sliding is activated when the ground shaking acceleration exceeds the yield acceleration of the sliding mass. k_y values of 0.038 and 0.026 characterize the sliding mass along sliding surface 1 and sliding surface 2, respectively (Fig. 38.1).

Mean seismic displacements ($D_{SB(B\text{-}T)}$) based on B-T procedure were estimated using the functional form for the amount of nonzero seismic displacement provided in Bray and Travasarou (2007). Mean seismic displacements ($D_{SB(J)}$) based on J

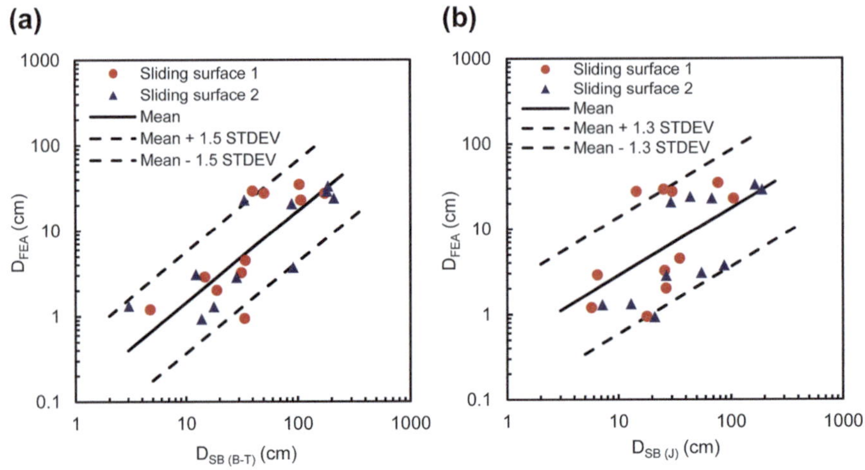

Fig. 38.5 Finite-element computed permanent shear displacements (D_{FEA}) versus displacements $D_{SB(B\text{-}T)}$ and $D_{SB(J)}$ predicted by the simplified Bray and Travasarou (2007) and Jibson (2007) procedures, respectively

procedure were obtained using the functional form that correlates the seismic displacement with the earthquake moment magnitude (M) and peak horizontal ground acceleration (a_{max}) provided in Jibson (2007).

Figure 38.5 shows plots of predicted seismic displacements, based on B-T ($D_{SB\ (B\text{-}T)}$) and J ($D_{SB(J)}$) procedures, versus permanent shear displacements computed from finite-element analyses (D_{FEA}). The results indicate that the above mentioned simplified displacement procedures may significantly overestimate the earthquake-induced permanent shear displacement of a deepwater slope when compared to the finite-element computed permanent shear displacement. Eighty-nine percent of the $D_{SB(B\text{-}T)}$ values are 2–15 times larger than D_{FEA}, and 78 % of the $D_{SB(J)}$ values are 2–24 times larger than D_{FEA}.

In an attempt to improve the predictive capability of the considered simplified displacement procedures, the following regression equations describing the correlation between computed D_{FEA} and D_{SB} values (Fig. 38.5) have been developed:

$$\ln D_{FEA} = 1.0699\ \ln D_{SB(B-T)} - 2.0977 \pm 1.5STDEV_{(B-T)} \qquad (38.4)$$

$$\ln D_{FEA} = 0.7876\ \ln D_{SB(J)} - 0.7683 \pm 1.3STDEV_{(J)} \qquad (38.5)$$

The *STDEV* term in Eqs. 38.4 and 38.5 represents the standard deviation of the regression model, and the displacements are expressed in cm. For the D_{SB} values derived based on B-T and J simplified procedures, $STDEV_{(B\text{-}T)} = 0.9112$ and $STDEV_{(J)} = 1.2115$, respectively. Equations 38.4 and 38.5 above bracket the limited data set of displacements presented in Fig. 38.5. The mean relationships obtained by removing the *STDEV* related term in Eqs. 38.4 and 38.5 may be used to correct the permanent seismic displacements predicted by B-T and J simplified procedures in preliminary screening studies addressing the earthquake-induced shear displacements of deepwater slopes.

38.5 Conclusions

The present study demonstrates that simplified displacement procedures may be overly conservative when compared to finite-element based predictions of earthquake-induced permanent shear displacements along the sliding surface of a deepwater slope. Relationships are presented that may be used to improve the predictive capability of the selected simplified procedures in order to make them compatible with the seismic slope performance revealed by more developed numerical models. Simplified procedures are inappropriate for the characterization of seismic slope displacements at the seafloor since the associated sliding block models are unable to capture the yielding behavior of deepwater sediments above and below the potential sliding surface which may result in a significant permanent displacement component at the seafloor in addition to the permanent shear displacement experienced along the sliding surface. The mechanism of the simplified sliding block models allows for plastic yielding to occur only along the sliding surface of the model whereas the overlying sliding mass is treated as a rigid or elastic system. Therefore the application of simplified procedures discussed in this paper should in general be restricted to situations involving submarine slopes likely to experience localized permanent deformations within a relatively narrow band of significantly weaker sediments.

Acknowledgments The authors are grateful to the reviewers Dr. Vernon Schaefer and Dr. Steven Bartlett for improving the paper by providing valuable comments and suggestions.

References

Bray JD, Travasarou T (2007) Simplified procedure for estimating earthquake-induced deviatoric slope displacements. J Geotech Geoenviron Eng ASCE 133(4):381–392

Dobry R, Idriss IM, Ng E (1978) Duration characteristics of horizontal components of strong-motion earthquake records. Bull Seismol Soc Am 68:1487–1520

Forum8 Inc. (2006) UWLC User Manual

Hardin BO, Drnevich VP (1972) Shear modulus and damping in soils: design equations and curves. J Soil Mech Found Div ASCE 98(7):667–692

Jennings PC (1964) Periodic response of a general yielding structure. J Eng Mech Div ASCE 90(2):131–166

Jibson RW (2007) Regression models for estimating coseismic landslide displacement. Eng Geol 91:209–218

Jibson RW, Rathje EM, Jibson MW, Lee YW (2013) SLAMMER—Seismic landslide movement modeled using earthquake records. U.S. Geological Survey Techniques and Methods, Book 12, Chapter B1

Kondner RL, Zelasko JS (1963) A hyperbolic stress-strain formulation of sands. In: Proc. 2nd Pan Am. Conf. on Soil Mechanics and Foundation Engineering, Brazilian Association of Soil Mechanics, Sao Paulo 289–324

Meehan CL, Vahedifard F (2013) Evaluation of simplified methods for predicting earthquake-induced slope displacements in earth dams and embankments. Eng Geol 152:180–193

Rathje EM, Faraj F, Russell S, Bray JD (2004) Empirical relationships for frequency content parameters of earthquake ground motions. Earthq Spectra 20:119–144
Wakai A, Ugai K (2004) A simple constitutive model for the seismic analysis of slopes and its applications. Soils Found 44(4):83–97

Part VI
Fluid Flow and Gas Hydrates

Chapter 39
Deriving the Rate of Salt Rise at the Cape Fear Slide Using New Seismic Data

Levent Akinci and Derek Sawyer

Abstract Recently acquired 2-D seismic data from offshore North Carolina provides images of salt diapirs and landslides in the Carolina Trough that give insight into the interaction between slope sediments and intruding salt from below. The best example of this is the Cape Fear Slide Complex in which at least two salt diapirs are surrounded by the lower headwall of the slide. Here, we present seismic images that were collected for the Eastern North American Margin Community Seismic Experiment. We describe the morphology of the slide and diapirs in order to infer rates of salt rise. We have tentatively estimated a post-slide growth rates of 517 m per million years (m/Ma). However, as the analysis continues, it is possible this estimate will change. This research provides significant insight into the interplay of salt and slope failure processes in an ocean basin setting.

39.1 Introduction

The Cape Fear landslide, located southeast of Cape Fear, North Carolina, is one of the largest mass movement features along the Atlantic continental margin of the United States (Fig. 39.1) (Dillon et al. 1983; Hornbach et al. 2007). The main headwall of this landslide surrounds a salt diapir that has breached the sea floor, possibly due to release of overburden. Utilising newly acquired data, the aim of this research is to investigate the vertical rise of the Cape Fear diapir, and estimate the rate of post-landslide salt rise in order to gain further insight into the interplay of diapirism and slope failure processes. This was done by postulating when in the diapir's history (at what height of the diapir) the main Cape Fear landslide event occurred. Then, correlating this with landslide ages based on previous work (Popenoe 1985; Popenoe et al. 1993; Rodriguez and Paull 2000), we suggest an estimation of the rate of salt rise following the occurrence of the landslide until present day.

L. Akinci (✉) • D. Sawyer
School of Earth Sciences, The Ohio State University, Columbus, OH, USA
e-mail: akinci.2@osu.edu

G. Lamarche et al. (eds.), *Submarine Mass Movements and their Consequences*, Advances in Natural and Technological Hazards Research 41,
DOI 10.1007/978-3-319-20979-1_39

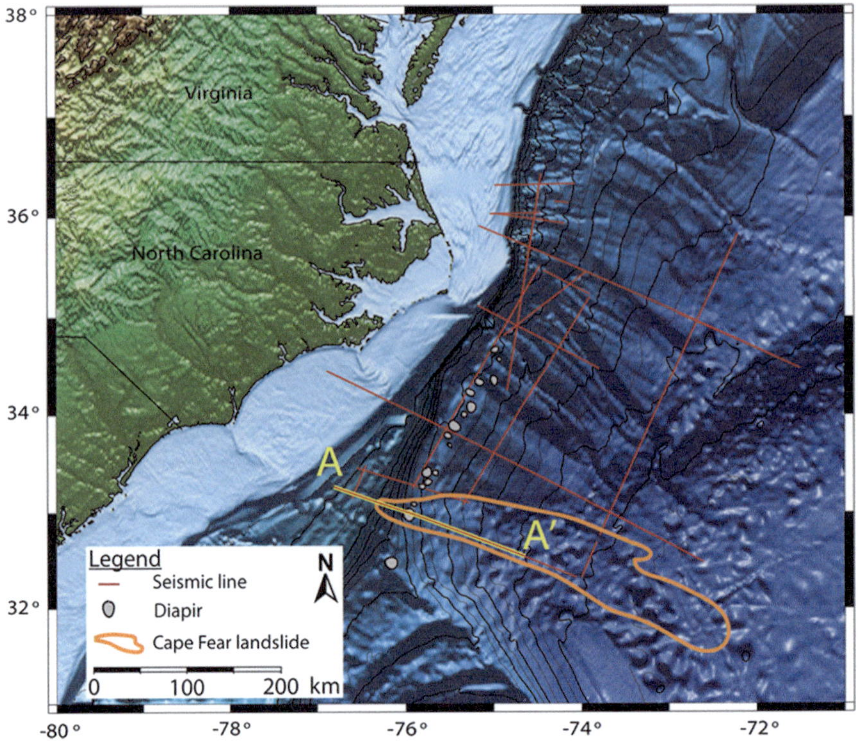

Fig. 39.1 Map of the Eastern North American Margin Community Seismic Experiment, as well as the location of the Cape Fear landslide (From Popenoe et al. 1993) and diapirs (From Dillon et al. 1983) offshore North Carolina, U.S.A. The seismic line A to A′ marked in *yellow* is shown in Fig. 39.3

39.2 Geologic Setting

The main (lower) headwall of the Cape Fear slide is a 50 km long (Popenoe et al. 1993), amphitheatre-shaped scarp that is up to 120 m high at a water depth of approximately 2500 m (Fig. 39.2). From here, mass-movement deposits extend up to 400 km downslope, where water depths exceed 5400 m on the abyssal plain (Popenoe et al. 1993). Furthermore, the main headwall circumscribes two large salt diapirs, one of which has breached the seafloor by as much as 323 m on the downslope side of the diapir (Fig. 39.2).

These diapirs are composed of Jurassic salt that was deposited in the Carolina Trough, which formed during rifting between North America and Africa (Dillon et al. 1983; Hutchinson et al. 1983). Following evaporite deposition, prograding sediments from the continental shelf filled the basin from the west, asymmetrically loading the salt. This caused the salt to flow eastward until shallower basement at the eastern edge of the Carolina Trough forced the salt upwards into a line of diapirs (Cashman and Popenoe 1985; Popenoe et al. 1993).

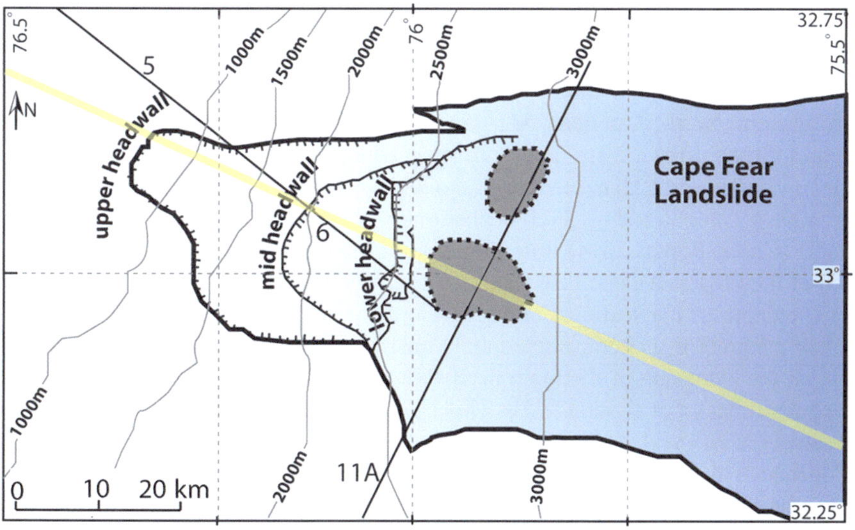

Fig. 39.2 Sketch of the upper part of the Cape Fear landslide, depicting the headwalls and the salt diapirs. The southern diapir is the breached diapir seen in Fig. 39.3. *Yellow line* represents a portion of A – A′ (from Fig. 39.1) (Adapted from Popenoe et al. 1993)

Though several diapirs are associated with the Carolina Trough, the only one thought to be related to a slope failure is the breached diapir in the Cape Fear slide's evacuated zone (Popenoe et al. 1993). As a result, it is no surprise that one of the prevailing theories for the generation of this slope failure implicates the rising salt (Cashman and Popenoe 1985; Popenoe et al. 1993).

Alternatively, Popenoe et al. (1993) and Schmuck and Paull (1993) suggested that slope failures in the Cape Fear slide may have been related to the presence of gas hydrates and free gas, indicated by prominent bottom simulating reflectors (BSR) and bright spots in the seismic data, respectively. According to Popenoe et al. (1993) during lowstand sea levels, decomposition of gas hydrates may have weakened sediments thus making them more prone to slope failure.

39.3 Previous Work

The Cape Fear landslide and nearby diapirs were first discovered and mapped in the late 1970s by the U.S. Geological Survey (USGS) (Popenoe et al. 1993). Further seismic surveys and side-scan sonar imagery were collected by the University of North Carolina in conjunction with the USGS in the late 1980s (Popenoe et al. 1993), while multibeam bathymetry data and seismic Chirp data were acquired during the 2000s (Hornbach et al. 2007). Furthermore, in 1995, the Ocean Drilling Program (ODP), Leg 164, drilled short holes into the flanks of the main Cape Fear diapir to retrieve core samples (Paull and Matsumoto 2000).

39.4 Description and Interpretation of Data

Most recently, new 2-D multi-channel seismic images were collected in 2014 for the Eastern North American Margin Community Seismic Experiment (Cruise Report 2014), using a 2-string air gun array with 3300 in.^{3} volume, 8 km streamer, 6 m tow depth, and 12.5 m bin size. It should be noted that based on multibeam data, the main seismic line (Fig. 39.3) is positioned directly over the highest point of the diapir (Cruise Report 2014). Interpretation of this seismic data enables the evolution of the breached Cape Fear diapir to be investigated.

The depth of the seafloor following the main landslide is approximately 70 m below the present seafloor, immediately upslope of the diapir (Fig. 39.3). We base this on the extension of the main headwall beneath the present day seafloor to an irregular reflector at about 3.395 s. This is interpreted as a detachment at the top of the lower Miocene due to the irregular thickening and thinning of strata seen here (Popenoe et al. 1993). A series of relatively shallow listric faults sole out along this reflector too, further suggesting a detachment at this stratigraphic level. Using this reflector as a guide and assuming the diapir had deformed this weak zone through active diapirism, we approximate a dome across the diapir to represent the palaeo-seafloor.

Then, when the landslide occurred, it would have likely exposed the salt diapir (or perhaps left only a thin veneer of sediment atop the diapir) as the weak layer was

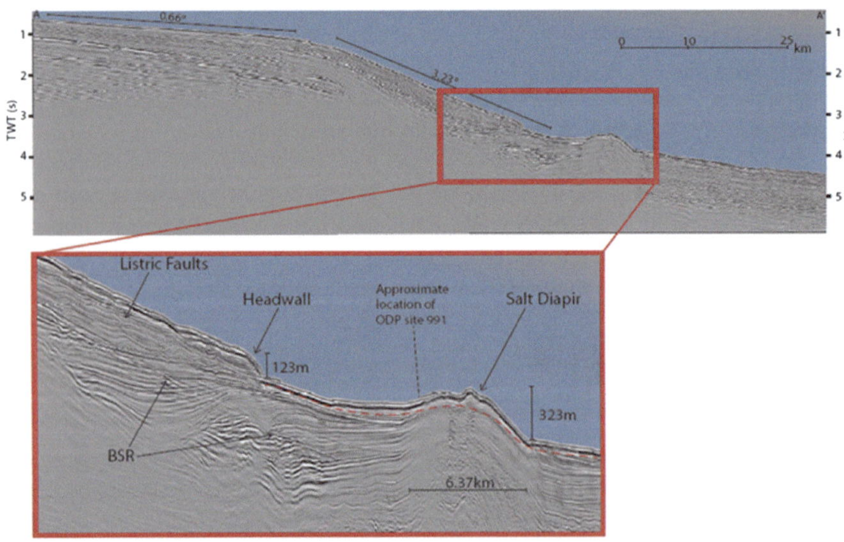

Fig. 39.3 Scaled seismic line through the headwalls of the Cape Fear landslide and the breached Cape Fear diapir. The *inset* shows enlarged view of the breached diapir and the main headwall, and also indicates a clear BSR. Measurements of the diapir and headwall are also indicated. The *dotted red line* represents the assumed palaeo-seafloor immediately following the occurrence of the Cape Fear landslide. Vertical exaggeration is approximately 8:1

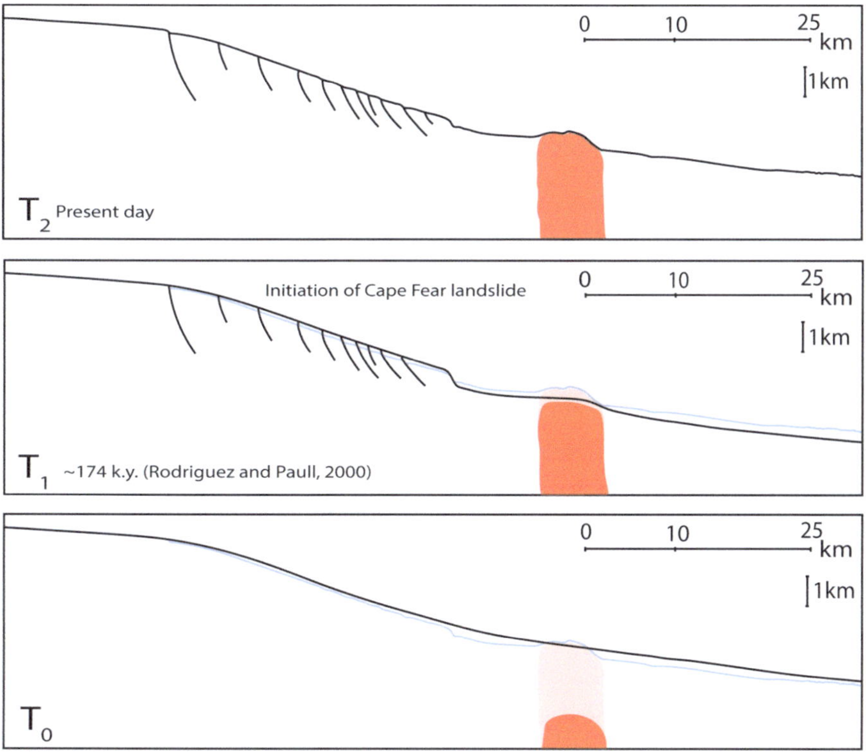

Fig. 39.4 Conceptual sequence of subsurface growth of the Cape Fear diapir and occurrence of the Cape Fear landslide through time from T_0 to T_2. The present day seafloor and diapir morphology is shown in the background for T_0 and T_1. As the diapir grows, it causes deformation above it. Then at T_1 the main landslide occurs, likely exposing the top of the diapir, which then allows for acceleration of the diapir's vertical growth. Subsequent landslides then fill in the space between the diapir and the main headwall and deposit material further downslope as well. This illustration uses the same vertical exaggeration as Fig. 39.3

immediately above it (Fig. 39.4). Therefore, if we assume the estimated dome (Fig. 39.3) represents the approximate level of the salt at the time the main landslide occurred, we can estimate that the salt has risen roughly 90 m since this time to its present-day height.

Moreover, ODP cores indicate several hiatuses and extensive deformation from 2 to 47 m below the seafloor on the flanks of the Cape Fear diapir (Rodriguez and Paull 2000). Thus, this represents the time span that the Cape Fear slide complex was active. Using an average sedimentation rate of 27 cm/k.y. based on ^{14}C ages from core samples (Rodriguez and Paull 2000), we calculate that slide activity has occurred here for a minimum of approximately 174 k.y. Based on this, we can then estimate the nett rate of vertical salt rise post-landslide to be approximately 517 m/Ma. Theoretically this should represent a much faster rate of salt rise than prior to the landslide, as the landslide would have released a significant amount of

Table 39.1 Calculated rates of vertical salt rise based on varying salt heights when the Cape Fear landslide occurred. 174 k.y. used for the elapsed time

Salt position	Vertical growth of salt (m)	Nett rate of vertical salt rise (m/Ma)
At present day height	0	0
5 m below present height	5	29
10 m below present height	10	57
20 m below present height	20	115
50 m below present height	50	287
At estimated palaeo-seafloor	90	517
200 m below present height	200	1149

overlying differential pressure, resulting in an acceleration of salt rise during this time.

This analysis is one estimation that carries significant uncertainty. It is difficult to determine the height of the diapir and roof thickness when the landslide happened, thus we also calculated the rate based on the salt being at various different heights in order to attain the possible range of salt rise rates (Table 39.1).

39.5 Discussion

In the preceding description, we have discounted several different factors that could affect the calculated salt growth rate. For example, given that there is only one 2-D section running through the diapir, and since salt behaves three dimensionally, knowing what is happening out of the plane of this seismic line is impossible. Additionally, increasing salinity in shallow holes drilled on the flanks of the diapir during ODP leg 164 (Paull et al. 1996; Paull and Matsumoto 2000) suggest that the diapir was being dissolved, which could have slowed the nett rate of vertical salt rise. However, over the course of the diapir's history, the amount of dissolution is most likely negligible. Nevertheless, growth rates of diapirs in the Gulf of Mexico range from 75 to 250 m/Ma (Jackson and Talbot 1986), which makes the growth rates of the Cape Fear diapir for a vertical salt rise above 50 m seem high. This could mean that the top of the Cape Fear diapir, at the time of the main landslide, was shallower than initially proposed. Based on the average growth rates from the Gulf of Mexico, this could mean the salt was more likely somewhere in the range of 10–50 m below present height when the landslide occurred.

Another factor that was discounted was the actual timing of the main Cape Fear slide event as the age of the landslide was calculated based on the deepest occurrence of hiatuses, with the assumption that the largest event was the earliest episode of slope failure in the Cape Fear complex. Given the resolution of the seismic data, we are unable to determine which event between the late Miocene and Pleistocene was the main landslide. In addition, the calculated timing of the landslide to

174 k.y. was merely based on constant sedimentation and does not account for removed sediments. However, in reality we realise sedimentation likely fluctuated over the course of that time, and if we could account for the amount of sediment removed, the timing of the landslide would likely be pushed further back.

As such, future work looks at refining these uncertainties. This will include developing a numerical model that couples salt growth with sedimentation rates within a growing sedimentary basin to gain a more precise constraint on the height of the Cape Fear diapir at the time of the landslide. This model will be based on recent models published by Nikolinakou et al. (2014).

39.6 Conclusion

In light of the remaining uncertainties, based on comparisons to published diapir growth rates in other basins, we tentatively estimate post-slide growth rates up to approximately 517 m per million years for the Cape Fear diapir. However as the analysis continues this estimate will be refined. Though this initial study provides only a crude assessment, it still sheds important light on the feedbacks between rising salt diapirs and landslides and may be useful in forming basin evolution models as well as geohazards studies.

Acknowledgments Data for this research was provided by GeoPRISMS Eastern North American Margin Community Seismic Experiment, which was funded by the NSF. We would like to thank our reviewers, including Matt Hornbach, and our editor, Sebastian Krastel, for helping to improve this manuscript.

References

Cashman KV, Popenoe P (1985) Slumping and shallow faulting related to the presence of salt on the continental slope and rise off North Carolina. Mar Petrol Geol 2:260–271

Cruise Report (2014) Eastern North American margin community seismic experiment: cruise MGL1408, R/V Marcus G Langseth

Dillon WP, Popenoe P, Grow JA et al (1983) Growth faulting and salt diapirism: their relationship and control in the Carolina Trough, eastern North America. In: Watkins JS, Drake CL (eds) Studies of continental margin geology. AAPG Memoir, 34, pp 21–46

Hornbach MJ, Lavier LL, Ruppel CD (2007) Triggering mechanism and tsunamogenic potential of the Cape Fear Slide complex, U.S. Atlantic margin. Geochem Geophys Geosyst 8(12):1–16

Hutchinson DR, Grow JA, Klitgord KD, Swift BA (1983) Deep structure and evolution of the Carolina Trough. In: Watkins JS, Drake CL (eds) Studies of continental margin geology. AAPG Memoir, 34. American Association of Petroleum Geologists, Tulsa, pp 129–152

Jackson MPA, Talbot CJ (1986) External shapes, strain rates, and dynamics of salt structures. Geol Soc Am Bull 97(3):305

Nikolinakou MA, Flemings PB, Hudec MR (2014) Modelling stress evolution around a rising salt diapir. Mar Petrol Geol 51:230–238

Paull CK, Matsumoto R (2000) 1. Leg 164 overview. Proceedings of the ocean drilling program, scientific results, 164
Paull CK, Matsumoto R, Wallace PJ et al (1996) Proc. ODP, Init. Repts., 164, College Station, TX (Ocean drilling program)
Popenoe P (1985) Cenozoic depositional and structural history of the North Carolina margin from seismic-stratigraphic analyses. In: Poag W (ed) Geological evolution of the United States Atlantic margin, vol 1. Van Nostrand Reinhold Company, New York, pp 125–187
Popenoe P, Schmuck EA, Dillon WP (1993) The Cape Fear landslide: slope failure associated with salt diapirism and gas hydrate decomposition. In: Schwab WC, Lee HJ, Twichell DC (eds) Submarine landslides: selected studies in the U.S. exclusive economic zone, U. S. geological survey bulletin 2002, Washington D.C., USA, pp 40–53
Rodriguez NM, Paull CK (2000) Data report: ^{14}C dating of sediments of the uppermost cape fear slide plain: constraints on the timing of this massive submarine slide. Proc Ocean Drill Progr Sci Results 164:325–327
Schmuck EA, Paull CK (1993) Evidence for gas accumulation associated with diapirism and gas hydrates at the head of the Cape Fear Slide. Geo-Mar Lett 13:145–152

Chapter 40
Submarine Slope Instabilities Coincident with Shallow Gas Hydrate Systems: Insights from New Zealand Examples

Gareth J. Crutchley, Joshu J. Mountjoy, Ingo A. Pecher, Andrew R. Gorman, and Stuart A. Henrys

Abstract The potential of gas hydrate systems to play a role in submarine slope failure has been well-documented since the late 1970s. Several conceptual models exist for how the gas hydrate-free gas system might weaken submarine sediments, but there is no definitive evidence for gas hydrate-related processes being the primary cause of a particular submarine slope failure. We present a review of coincident gas hydrates and submarine slope instabilities on New Zealand's active margins. The examples we show represent different failure modes in a range of slope environments, including the upper continental slope and tectonic ridges, with the common factor being that the base of gas hydrate stability approaches the seafloor in these regions. We synthesise several proposed sediment weakening mechanisms and draw comparisons to other global models for gas hydrate-related slope instability. This contribution highlights diverse influences that gas hydrate systems could have on submarine sediment strength, while acknowledging gaps in our understanding of the potential role of gas hydrates, free gas and fluid flow on slope stability.

G.J. Crutchley (✉) • S.A. Henrys
GNS Science, Lower Hutt, New Zealand
e-mail: g.crutchley@gns.cri.nz

J.J. Mountjoy
National Institute of Water and Atmospheric Research (NIWA), Wellington, New Zealand

I.A. Pecher
GNS Science, Lower Hutt, New Zealand

University of Auckland, Auckland, New Zealand

A.R. Gorman
University of Otago, Dunedin, New Zealand

G. Lamarche et al. (eds.), *Submarine Mass Movements and their Consequences*, Advances in Natural and Technological Hazards Research 41,
DOI 10.1007/978-3-319-20979-1_40

40.1 Introduction

Gas hydrates are stable at relatively low temperatures and high pressures encountered over a range of depths in submarine sediments. In seismic data, marine gas hydrates are best identified by a reflection known as a bottom simulating reflection (BSR) that marks the base of gas hydrate stability (BGHS). Gas hydrates are stable above the BGHS, and free gas often occurs beneath the BGHS.

Since the first proposed links between gas hydrates and submarine slope failures (McIver 1977), the potential role of hydrates in slope stability has received much attention from many different parts of the world (Bünz et al. 2003; Field and Barber 1993; Kayen and Lee 1993; Paull et al. 1996; Popenoe et al. 1993). The continental slopes of New Zealand's active margins are no exception (Fig. 40.1a) (e.g. Crutchley et al. 2007; Pecher et al. 2005; Mountjoy et al. 2014).

The conventional model of gas hydrates and slope instability is based on excess fluid pressure generation near the BGHS, due to hydrate dissociation occurring in response to a change in environmental conditions (either warming or depressurisation) (Kvenvolden 1993) (Fig. 40.1b). Other conceptual models of gas hydrates and sediment weakening also exist, such as hydrate dissolution at the top of gas hydrate occurrence (Sultan et al. 2004) and repeated sediment fracturing driven by cyclic gas hydrate stability (Pecher et al. 2005), to cite just two further examples. A key criticism of gas hydrates as a submarine landslide trigger has been the lack of direct observational data linking the two. A recent analysis of landslide inventories did not show any clear link to the climate cycle that might implicate large scale triggering of slope instability due to gas hydrate dissociation (Urlaub et al. 2013). Thus, rather than acting as final trigger, the role of gas hydrates may be best considered as a preconditioning mechanism bringing slopes closer to failure.

This paper has three aims: (1) review all known submarine slope instabilities in New Zealand that coincide with gas hydrates, (2) outline the range of gas hydrate-related destabilisation mechanisms proposed for these slope failures, and (3) bring

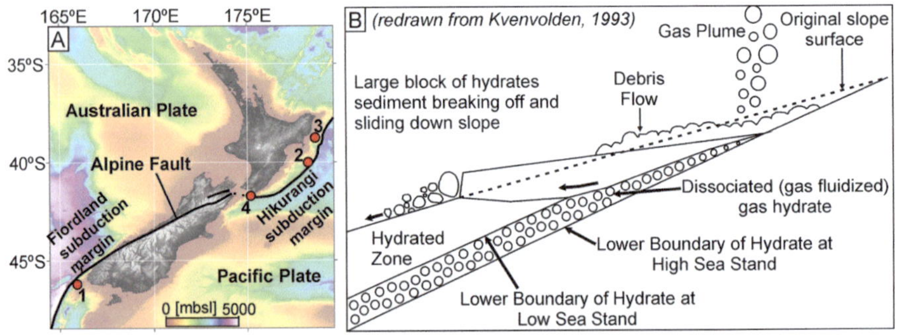

Fig. 40.1 (**a**) Tectonic setting of New Zealand. *Red dots* mark Sites 1–4 of gas hydrate-related slope instabilities discussed in the text. (**b**) Conceptual model of submarine landslides influenced by gas hydrate dissociation in response to a pressure decrease (Redrawn after Kvenvolden 1993)

the New Zealand examples into global context by discussing some other mechanisms from elsewhere around the world.

40.2 Slope Failures Coincident with Gas Hydrates in New Zealand

Three areas of slope instability coincident with gas hydrates have been documented on New Zealand's active margins. Here, we introduce each of these and present a fourth example which, although already documented (Mountjoy et al. 2009a), has not been considered with respect to the gas hydrate system (Fig. 40.1).

40.2.1 Site 1: Puysegur Bank (Fig. 40.2a, b)

Seismic data at Puysegur Bank show a BSR merging towards a seafloor depression (Fig. 40.2b), which Crutchley et al. (2007) interpreted as the headwall of a submarine landslide. Further downslope is hummocky topography typical of landslide deposits. Crutchley et al. (2007) postulated that high fluid pressure beneath the BSR could have contributed to failure, but that seismic loading from the active plate boundary was a likely trigger.

40.2.2 Site 2: Rock Garden (Fig 40.2c, d)

Rock Garden ridge lies in water depths of around 600–700 mbsl and is eroded to a relatively flat morphology (Fig. 40.2c, d). The coincidence of the flattened ridge top with BSRs that merge towards the seafloor (Fig. 40.2d) led Pecher et al. (2005) to hypothesise that gas hydrate dynamics might drive regional erosion here.

40.2.3 Site 3: Tuaheni Landslide Complex (TLC) (Fig. 40.2e, f)

The TLC displays a variety of deformation fabrics, including distinct zones of compression and extension (Mountjoy et al. 2009b). Mountjoy et al. (2009b) recognised kinematic features in the extensional domain that are similar to features of terrestrial earthflows and interpreted the TLC as a slow-moving submarine landslide complex. The base of gas hydrate stability is imaged beneath the TLC's extensional domain (Fig. 40.2f), and Mountjoy et al. (2014) have suggested that gas hydrates play a key role in active, slow-moving deformation.

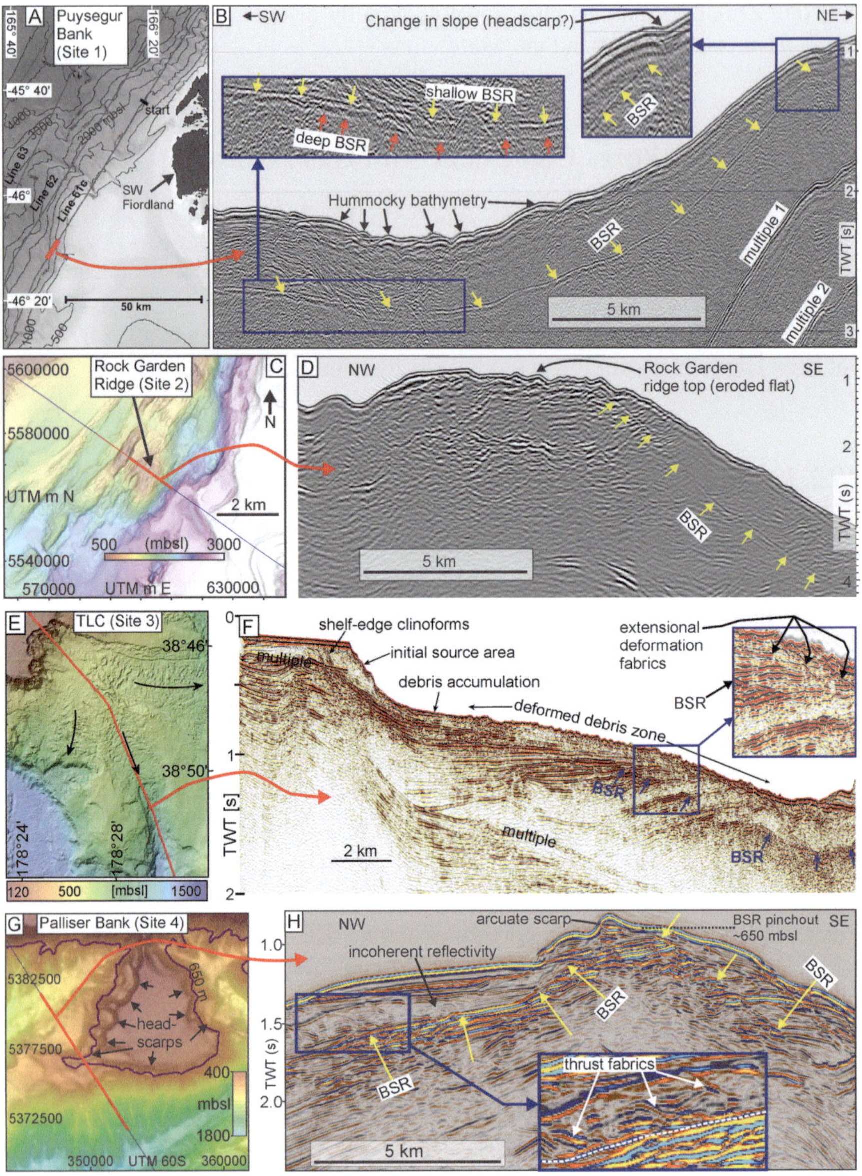

Fig. 40.2 (**a**) Puysegur Bank (Site 1) (refer to Fig. 40.1a for the regional location) (**b**) Seismic data from (**a**) showing the Puysegur Bank submarine landslide. (**c**) Rock Garden Ridge (Site 2). (**d**) Seismic data showing the eroded Rock Garden ridge top. (**e**) Tuaheni Landslide Complex (*TLC*) (Site 3). (**f**) Seismic data through the TLC. (**g**) Palliser Bank (Site 4). (**h**) Seismic data across Palliser Bank showing features of a landslide there

40.2.4 Site 4: Palliser Bank (Fig. 40.2g, h)

Low angle translational landslides occur on the backlimb of the Palliser Bank anticline (Mountjoy et al. 2009a). Seismic data show a BSR pinching out at the seafloor at approximately 650 m water depth and high-amplitude reflections that indicate gas-charged layers beneath the hydrate system (Fig. 40.2h). Bathymetry data show arcuate scarps in the vicinity of the BSR pinchout (Fig. 40.2g). Seismic data also show a unit of relatively disturbed reflectivity, reminiscent of a mass transport deposit, directly downslope of the arcuate scarp (Fig. 40.2h). This unit's internal fabric is characteristic of compressional deformation that presumably formed during deceleration of a moving landslide.

40.3 The Potential Influence of Gas Hydrates on Slope Stability

The conventional model for hydrate-driven slope destabilisation (Fig. 40.1b; Kvenvolden 1993) has relevance for the slope failures at Puysegur Bank, Rock Garden, and the TLC. For example, depressurisation and hydrate dissociation during uplift of Rock Garden may have generated excess fluid pressure and weakened the seafloor (Faure et al. 2006; Pecher et al. 2005). Uplift of Palliser Bank during accretion on the margin means this model could also be relevant to the slope failures there. In the following paragraphs we refer to Fig. 40.3 to describe other mechanisms proposed in New Zealand regarding the role of hydrates in slope destabilisation. We bring these scenarios into a global context by discussing some similar and also distinctly different mechanisms proposed elsewhere.

Pecher et al. (2005) proposed a scenario at Rock Garden (Fig. 40.2d) where sediments are weakened by repeated pore volume expansion and contraction driven by cyclic hydrate dissociation and formation in response to fluctuating ocean bottom temperatures. This scenario (Fig. 40.3b) requires of a zone of cyclic gas hydrate stability where hydrate is stable during cooler periods and unstable during warmer periods. The sediment weakening mechanism is likened to terrestrial frost heave.

1D hydrate formation modelling of Rock Garden shows that ridge uplift can result in concentrated hydrate deposits forming close to the seafloor (Ellis et al. 2010) (Fig. 40.3c). The interplay between upward fluid flow driven by dewatering of the wedge, and permeability reductions caused by gas hydrate formation, generates excess fluid pressure around the hydrate deposits. Hydrate deposits closest to the seafloor have the greatest potential to weaken sediments since overburden pressure is lower, and thus less fluid pressure is required to enable hydrofracturing.

By modelling the influence of buoyant free gas beneath the BSR at Rock Garden, Crutchley et al. (2010) highlighted the potential influence of free gas on pore fluid

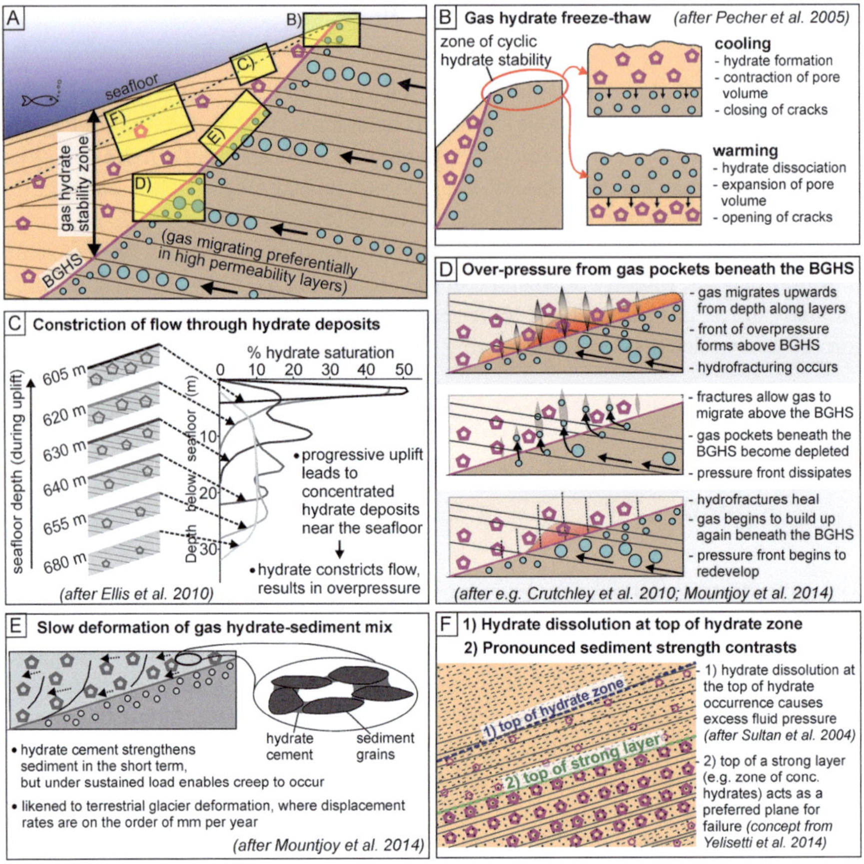

Fig. 40.3 (**a**) Conceptual sketch of a submarine gas hydrate system (*magenta pentagons* = hydrates; *cyan circles* = free gas). *Yellow boxes* labelled "*B*" to "*F*" show approximate positions of gas hydrate-related processes sketched in the corresponding sub-panels of this figure. (**b**) Gas hydrate "freeze-thaw" mechanism (Pecher et al. (2005). (**c**) Constriction of flow through concentrated hydrate deposits (Ellis et al. 2010). Seafloor uplift (in this case from 680 to 605 m) results in an evolving gas hydrate concentration profile beneath the seafloor. Higher gas hydrate saturations associated with later stages of uplift have the potential to significantly perturb fluid flow and cause excess fluid pressure. (**d**) Cyclic overpressure development in response to free gas migrating towards the BGHS. Pressure is relieved by hydrofracturing (e.g. Crutchley et al. 2010 and others in text). (**e**) Slow deformation of hydrate-bearing sediments facilitated by the rheology of the gas hydrate-sediment mix (Mountjoy et al. 2014). (**f**) (1) Hydrate dissolution at the top of gas hydrate occurrence results in excess fluid pressure and sediment weakening (Sultan et al. 2004). (2) Sediment strength contrasts encourage failure at particular stratigraphic levels within the hydrate stability zone (Yelisetti et al. 2014)

pressure—and therefore sediment strength—in the gas hydrate stability zone (Fig. 40.3d). This scenario differs from the conventional model (Fig. 40.1b) in that a distinct phase of hydrate dissociation is not required to generate the free gas and excess pore pressure. Rather, free gas migrating upwards from greater depths accumulates beneath the hydrate layer. The influence of free gas on the structural deformation of hydrate-bearing sediments has been discussed in many different parts of the world in terms of faulting and fracturing of the hydrate layer (Flemings et al. 2003; Hornbach et al. 2004; Hustoft et al. 2009). There is also evidence for free gas beneath the hydrate system causing seafloor erosion at very large scales, e.g. >30 km^3 of eroded sediment offshore Japan (Bangs et al. 2010).

In addition to the interplay between fluid flow and gas hydrates potentially weakening sediments, the existence of gas hydrates themselves may result in a rheology that is prone to slow deformation. At the TLC, Mountjoy et al. (2014) proposed that gas hydrate-bearing sediment could exhibit time-dependent deformation that enables slow, glacial style creeping movement (Fig. 40.3e). More work is required to test this model, which challenges conventional understanding of hydrates acting solely as a strengthening agent in sediments.

A distinctly different mechanism to those outlined above is that of hydrate dissolution within the gas hydrate stability zone (Sultan et al. 2004) (Fig. 40.3f). In contrast to the conventional model (Fig. 40.1b), Sultan et al. (2004) consider the generation of excess fluid pressure at the top of the gas hydrate occurrence zone—i.e. the upper limit of gas hydrates as controlled by the availability of methane in excess of its solubility in seawater. Aside from the possibility of fluid pressure generation at the top of gas hydrate occurrence, it is likely that significant sediment strength contrasts exist within the hydrate stability zone due to variations in lithology and hydrate concentration (Fig. 40.3f). For example, Yelisetti et al. (2014) show evidence at a slide on the Cascadia margin that the top of a zone of sediments strengthened by gas hydrates acted as a glide plane beneath a succession of weaker sediments with little or no gas hydrate.

40.4 Conclusions

By reviewing slope instabilities in New Zealand and providing perspective through some global examples, we draw attention to a range of mechanisms by which hydrates might destabilise sediments. Although hydrate systems can influence hydraulic properties (e.g. permeability and pore fluid pressure) and rheology (e.g. static and dynamic sediment strength), many key questions persist in our quest to understand their role in slope stability. How effectively do gas hydrates clog permeability in nature? How concentrated and interconnected are free gas accumulations beneath shallow hydrate systems—can they significantly influence pore fluid pressure? How does pore scale gas hydrate distribution and concentration influence sediment strength? Future work targeted at addressing such questions and other research gaps will advance our understanding in this important field of research.

Acknowledgements This work was principally supported by GNS Science core funding (Project 530EEZ40-00). We gratefully acknowledge George Spence and Stefan Bünz for their insightful reviews. We also thank Editor Susanne Woelz for handling the manuscript.

References

Bangs NL, Hornbach MJ, Moore GF, Park J-O (2010) Massive methane release triggered by seafloor erosion offshore southwestern Japan. Geology 38(11):1019–1022

Bünz S, Mienert J, Berndt C (2003) Geological controls on the Storegga gas-hydrate system of the mid-Norwegian continental margin. Earth Planet Sci Lett 209(3–4):291–307. http://dx.doi.org/10.1016/S0012-821X(03)00097-9

Crutchley GJ, Gorman AR, Fohrmann M (2007) Investigation of the role of gas hydrates in continental slope stability west of Fiordland, New Zealand. N Z J Geol Geophys 50:357–364

Crutchley GJ, Geiger S, Pecher IA, Gorman AR, Zhu H, Henrys SA (2010) The potential influence of shallow gas and gas hydrates on sea floor erosion of Rock Garden, an uplifted ridge offshore of New Zealand. Geo-Mar Lett 30(3–4):283–303

Ellis S, Pecher IA, Kukowski N, Xu W, Henrys S, Greinert J (2010) Testing proposed mechanisms for seafloor weakening at the top of gas hydrate stability on an uplifted ridge (Rock Garden), New Zealand. Mar Geol 272(1–4):127–140. doi:10.1016/j.margeo.2009.10.008

Faure K, Greinert J, Pecher IA, Graham IJ, Massoth GJ, Ronde CEJD, Wright IC, Baker ET, Olson EJ (2006) Methane seepage and its relation to slumping and gas hydrate at the Hikurangi margin, New Zealand. N Z J Geol Geophys 49:503–516

Field ME, Barber JHJ (1993) A submarine landslide associated with shallow seafloor gas and gas hydrates off Northern California. In: Schwab WC, Lee HJ, Twichell DC (eds) Submarine landslides: selected studies in the U.S. exclusive economic zone. U.S. Geological Survey Bulletin 2002, pp 151–157

Flemings PB, Liu X, Winters WJ (2003) Critical pressure and multiphase flow in Blake Ridge gas hydrates. Geology 31(12):1057–1060. doi:10.1130/g19863.1

Hornbach MJ, Saffer DM, Holbrook WS (2004) Critically pressured free-gas reservoirs below gas-hydrate provinces. Nature 427:142–144

Hustoft S, Bünz S, Mienert J, Chand S (2009) Gas hydrate reservoir and active methane-venting province in sediments on <20 Ma young oceanic crust in the Fram Strait, offshore NW-Svalbard. Earth Planet Sci Lett 284:12–24

Kayen RE, Lee HJ (1993) Slope stability in regions of sea-floor gas hydrate: Beaufort Sea continental slope. In: Schwab, W.C., Lee, H.J. and Twichell, D.C. (eds) Submarine landslides: selected studies in the U.S. exclusive economic zone. US Geological Survey Bulletin 2002: 97–103

Kvenvolden KA (1993) Gas hydrates—geological perspective and global change. Rev Geophys 31(2):173–187

McIver RD (1977) Hydrates of natural gas–important agent in geological processes. Geol Soc Am Abstr Programs 9:1089–1090

Mountjoy JJ, Barnes PM, Pettinga JR (2009a) Morphostructure and evolution of submarine canyons across an active margin: Cook Strait sector of the Hikurangi margin, New Zealand. Mar Geol 260(1–4):45–68. http://dx.doi.org/10.1016/j.margeo.2009.01.006

Mountjoy JJ, McKean J, Barnes PM, Pettinga JR (2009b) Terrestrial-style slow-moving earthflow kinematics in a submarine landslide complex. Mar Geol 267(3–4):114–127. http://dx.doi.org/10.1016/j.margeo.2009.09.007

Mountjoy JJ, Pecher I, Henrys S, Crutchley G, Barnes PM, Plaza-Faverola A (2014) Shallow methane hydrate system controls ongoing, downslope sediment transport in a low-velocity active submarine landslide complex, Hikurangi margin, New Zealand. Geochem Geophys Geosyst 15:4137–4156. doi:10.1002/2014GC005379

Paull CK, Buelow WJ WU III, Borowski WS (1996) Increased continental-margin slumping frequency during sea-level lowstands above gas hydrate–bearing sediments. Geology 24(2):143–146

Pecher IA, Henrys SA, Ellis S, Chiswell SM, Kukowski N (2005) Erosion of the seafloor at the top of the gas hydrate stability zone on the Hikurangi margin, New Zealand. Geophys Res Lett 32: L24603

Popenoe P, Schmuck EA, Dillon WP (1993) The Cape Fear landslide: slope failure associated with salt diapirism and gas hydrate decomposition. In: Schwab WC, Lee HJ, Twichell DC (eds) Submarine landslides: selected studies in the U.S. exclusive economic zone. U.S. Geological Survey Bulletin 2002, pp 40–53

Sultan N, Cochonat P, Foucher JP, Mienert J (2004) Effect of gas hydrates melting on seafloor slope instability. Mar Geol 213(1–4):379–401

Urlaub M, Talling PJ, Masson DG (2013) Timing and frequency of large submarine landslides: implications for understanding triggers and future geohazard. Quat Sci Rev 72(0):63–82. http://dx.doi.org/10.1016/j.quascirev.2013.04.020

Yelisetti S, Spence GD, Riedel M (2014) Role of gas hydrates in slope failure on frontal ridge of northern Cascadia margin. Geophys J Int 199(1):441–458. doi:10.1093/gji/ggu254

Chapter 41
Eel Canyon Slump Scar and Associated Fluid Venting

Roberto Gwiazda, Charles K. Paull, David W. Caress, Tom Lorenson, Peter G. Brewer, Edward T. Peltzer, Peter M. Walz, Krystle Anderson, and Eve Lundsten

Abstract Autonomous underwater vehicles have been used to characterize Eel Slump, a slide scar located south of Eel Canyon, California. The presence of a well developed dendritic network on the headwall with gullies tens of meters deep, thick sediment drape cover on the slide scar sole, and the absence of fresh surfaces on the scarp suggest that the mass failure(s) that produced this feature did not take place in the recent past. Thermogenic oil and gas emanating from a large mound in the sole of the slide scar were sampled with a remotely operated vehicle. Other distinctive morphologies observed from the seafloor of the slide scar indicate fluid seep has occurred at multiple sites within the slide scar sole.

41.1 Introduction

During a surface ship multibeam mapping cruise associated with the US National Oceanic and Atmospheric Administration Ocean Exploration program, large water column acoustic anomalies, extending up to 1,400 m from the seafloor and inferred to be produced by ascending gas bubbles, were encountered above both sides of the Eel Submarine Canyon off the coast of northern California (Gardner et al. 2009).

In order to explore the seafloor sources of these plumes, high-resolution bathymetric mapping and shallow subsurface chirp profiling were conducted with an autonomous underwater vehicle (AUV) on the southern source, which is located on the sole of a slide scar, here called Eel Slump (Fig. 41.1). In addition, fluids emanating from the seafloor were sampled with a remotely operated vehicle (ROV).

R. Gwiazda (✉) • C.K. Paull • D.W. Caress • P.G. Brewer • E.T. Peltzer • P.M. Walz • K. Anderson • E. Lundsten
Monterey Bay Aquarium Research Institute, Moss Landing, CA, USA
e-mail: rgwiazda@mbari.org

T. Lorenson
United States Geological Survey, Santa Cruz, CA, USA

G. Lamarche et al. (eds.), *Submarine Mass Movements and their Consequences*, Advances in Natural and Technological Hazards Research 41,
DOI 10.1007/978-3-319-20979-1_41

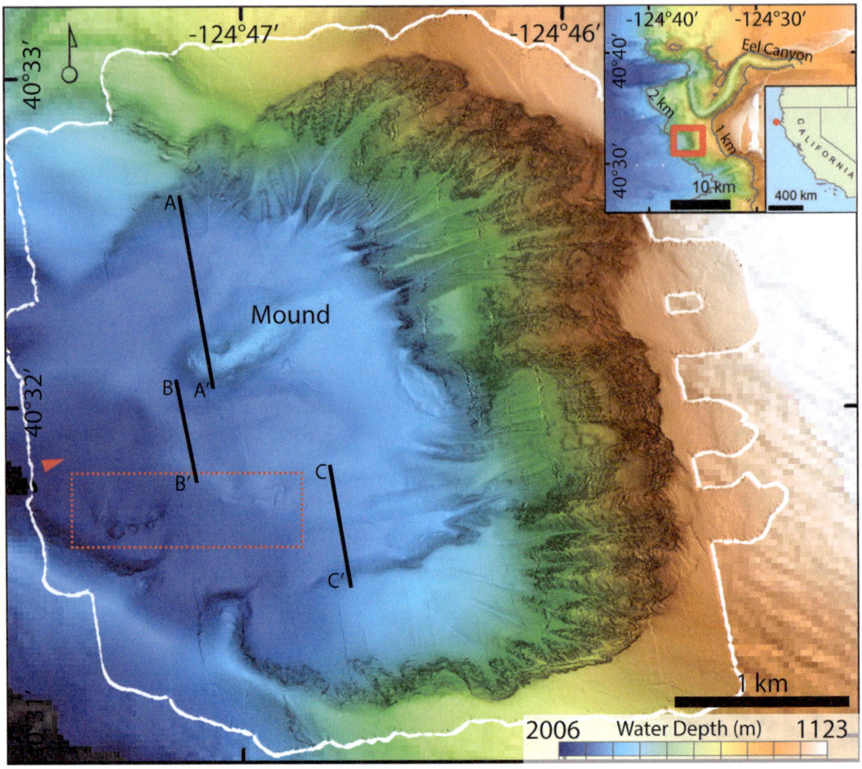

Fig. 41.1 Map of Eel Slump. The *white line* is the limit of the bathymetry collected with the AUV. *Inset* shows the regional location of Eel Slump. Lines A-A′, B-B′, C-C′ are the locations of the chirp profiles shown in Fig. 41.3. The *red triangle* indicates the perspective angle of Fig. 41.2. The *red rectangle* indicates the location of the image shown in Fig. 41.4.

41.2 Methods

A mapping AUV (Caress et al. 2008) was deployed in two missions in July 2011 to survey Eel Slump. The AUV followed a pre-programmed route 50 m above the seafloor at a speed of 3 knots, reaching more than 200 waypoints during the course of two 18-h long dives. The vehicle followed roughly N-S tracklines 150 m apart that provided partially overlapping swaths of the seafloor. A 200-kHz Reson 7125 sonar and an Edgetech 2–16 kHz chirp sub-bottom profiler mounted on the AUV allowed the generation of high resolution maps of the seafloor (vertical precision of 0.15 m and horizontal resolution of 0.7 m) and sub-bottom profiles which imaged up to 70 m below the seafloor with a vertical resolution of 0.11 m. Multibeam data were processed with the MB-system software (Caress and Chayes 1996). Gas and sediment sampling was conducted with the ROV *Doc Ricketts* on a large mound close to the center of the studied area. Gas was collected with an inverted cone into an evacuated cylinder deployed with the robotic arm of the ROV. Gas samples were analyzed for elemental and isotope composition by gas chromatography and dual inlet mass spectrometry, respectively, at Isotech Laboratories, Inc., IL, USA. Oil was extracted from sediments collected in push cores and vibracores. Oil biomarkers and oil carbon isotopic composition were analyzed by gas chromatography-mass spectrometry and by an elemental analyzer at the USGS in Menlo Park, California.

41.3 Results

The landslide scar Eel Slump is located 2.5 km SW of the sharp bend of Eel Canyon in water depths between 1,230 and 1,800 m (Fig. 41.1). This feature is an amphitheater shaped scar of 3.5 km length in a north-south direction. The headwall scarp on the eastern side has 550–600 m of relief with slopes of 31°–36° that decrease to ~25° on the sidewalls. In the center of the scar, there is a ~2 km diameter and 50–100 m deep circular depression, defined by continuous benches on its north and south sides and by a discontinuous ledge jutting out of the headwall. The surface of this depression dips westward 3–6°, and is breached on the western edge (Fig. 41.2). The lower half of the headwall scarp is covered by sediment drape, whereas host rock strata are exposed on the upper half of the headwall. The apparent dip of bedding planes on the exposed wall change along the headwall. Exposed strata on the southern section of the wall dip to the south; those on the north side dip northwards. This indicates that the slump scar cuts through the crest of an anticline.

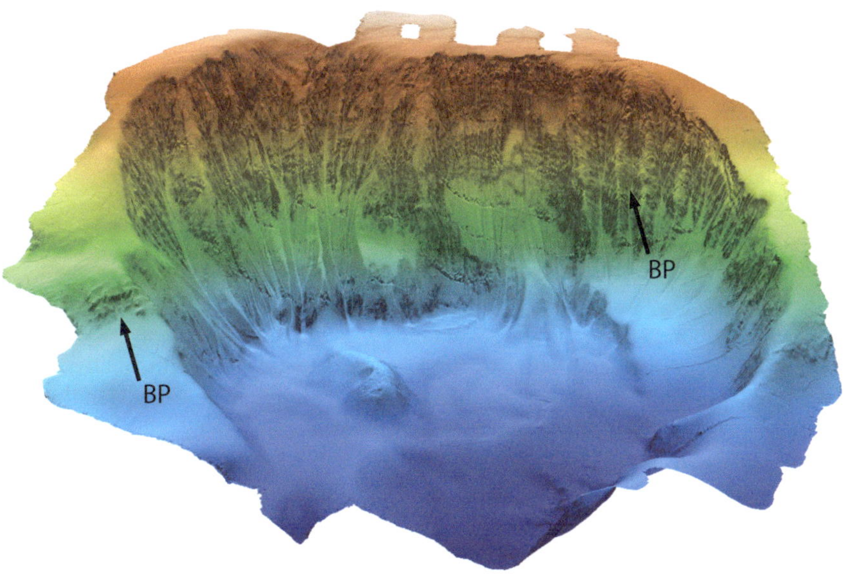

Fig. 41.2 Perspective of Eel slump from WNW. Vertical exaggeration ~2.5. *BP* bedding planes

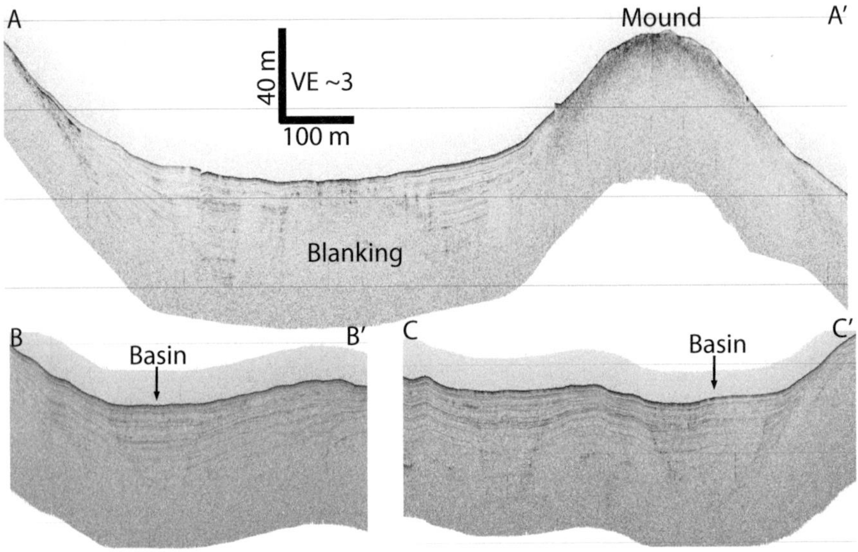

Fig. 41.3 Selected chirp profiles. Locations of profiles are in Fig. 41.1. A-A′ Chirp profile across the mound on the sole of the slide scar. Note the columnar structure of some of the blanking, possibly suggesting ascending gas. No reflectors are observed under the mound. B-B′ Sediment drape accumulates in basins within the sole of the slide scar. C-C′ Sediment filled troughs are seen within the sole of the slide scar

The headwall is incised by drainage channels with 20–40 m of relief between the floor of the gullies and the crest of the intervening straight ridges. The gullies branch out in the upper reaches of the headwall into a dendritic network (Fig. 41.1).

The morphology of the sole of the slide scar is generally smooth with a slightly (~10 m amplitude) undulating seafloor, which forms E-W oriented ridges and troughs within the central depression (Fig. 41.1). Sub-bottom chirp profiles reveal that the base of the slide scar is covered with sediment drape that preferentially accumulated in the basins, in places obtaining thicknesses greater than the imaging limit of the chirp profiles (i.e. >70 m). Chirp profiles show that a similar morphology with relative highs and small basins of ~10 m of relief, occurs in the subsurface (Fig. 41.3). Some of the basins connect to gullies on the headwall scar.

The gas plumes (Gardner et al. 2009) originate from a prominent ~650 m long, 350 m wide mound within the slide scar sole (Figs. 41.1 and 41.2) that stands nearly 60 m higher than the surrounding seafloor. The mound displays a distinctive rough topography known to develop at methane seeps where methane-derived authigenic carbonates overlay and intercalate with gas-hydrate deposits (Paull et al. 2015a). Chirp profiles over the mound do not resolve sub-bottom reflectors (Fig. 41.3). Visual inspection with the ROV detected columns of gas bubbles emanating from the mound and gas-hydrate layers embedded within the sediment (Walz et al. 2014).

Other areas on the slide scar sole to the south of the mound display inflated seafloor morphologies characteristic of gas flow and authigenic carbonate precipitation. Interruption in the continuity of chirp reflectors or no reflectivity is observed

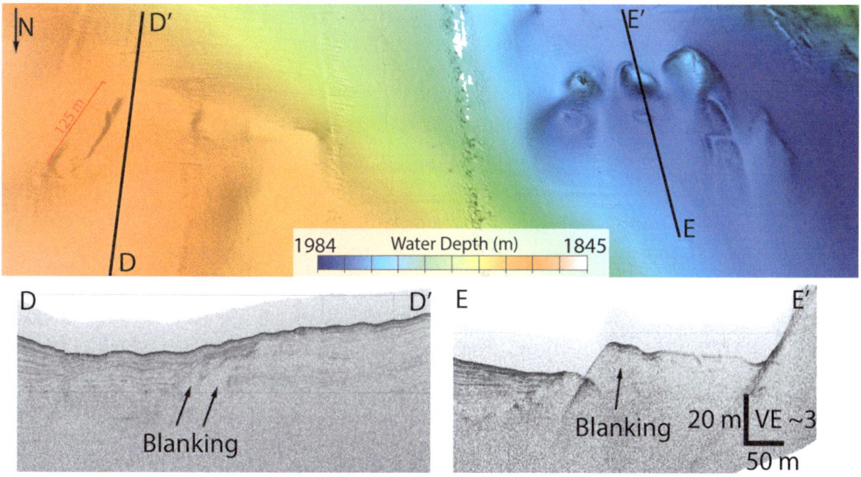

Fig. 41.4 Perspective view from the north of areas with seafloor morphology characteristic of authigenic carbonate precipitation and gas-hydrate presence, and chirp profiles from below these features. Location of this area in the slump scar is noted in Fig. 41.1

Table 41.1 Hydrocarbon abundance and isotopic composition of gas collected over the mound

C_1	C_2	C_3	C_4	C_5	C_{6+}	$\delta^{13}C_1$	δDC_1	$\delta^{13}C_2$	δDC_2	^{14}C
%	%	%	%	%	%	‰	‰	‰	‰	% Modern carbon
87.9	3	1.5	0.7	0.2	0.07	−50	−174.5	−25	−120.5	<0.2

under these features (Fig. 41.4). Chirp profiles also show blanking areas are widespread within the sediment drape covering the slide scar (Fig. 41.3A-A′).

Gas consisted of methane mixed with up to 5 % higher hydrocarbon gases (ethane to hexane), with no apparent recent biogenic contribution since ^{14}C was not detected in the methane (Table 41.1). In addition to gas, oil droplets were observed rising from the seafloor when the gas vents were disturbed with the robotic arm of the ROV. The oil has a distinctly light API gravity. Preliminary biomarker (steranes and hopanes in the saturate fraction of the oil extract) and isotope results indicate that the oil is likely from source rocks of Late Cretaceous or Eocene age as suggested by the whole oil carbon isotopic composition of −27.44 ‰ (Lillis et al. 1999). Thermal maturity biomarkers indicate the oil source rock is currently within the peak of oil generation, or roughly at a temperature of about 100 °C. The gas associated with this oil is also of thermogenic origin and likely co-produced with the oil at depth. Prior work on the adjacent Eel River Basin also found similar minor oil and gas occurrences scattered along the shoreline and inland (Lorenson et al. 1998; Lillis et al. 1999) suggesting these hydrocarbon occurrences are linked and widespread.

41.4 Discussion and Conclusions

The detailed seafloor surveys show that the present seafloor morphology within Eel Slump is the result of extensive post slope-failure alteration. The entire headwall of the scar displays an erosional morphology of a well developed dendritic network of ridges and gullies. There is no evidence of fresher surfaces on the headwall that might suggest secondary late-stage mass wasting events. While unquantified, this post-failure erosional topography took appreciable time to develop.

Because of the circular morphology of the slide scar, the drainage network on the headwall scarp converges downslope into a few channels on the sole of the slide scar (Figs. 41.1 and 41.2), which further merge downslope into two broad channels breaching the western lip of the semi-enclosed depression at the bottom of the scar. The presence of tens-of-meters-deep incised gullies on the headwall with down-slope sediment accumulation on the slide scar floor channels suggests that the post-failure development of the slump landscape has been a sustained process of considerable duration.

Chirp profiles across the sole of the slide scar show that at least 70 m of sediment drape have accumulated in the deepest depressions since the failure or failure events that generated the scar. No evidence was found to indicate secondary failures occurred while this drape was accumulating. Sedimentation rates are not known, however it is expected that they are high due to spillover from sediment laden currents flowing along Eel Canyon, 2.5 km to the northeast. As a first approximation, the time to deposit the thickest imaged sediment drape (70 m) at the average sedimentation rate observed on the Eel River shelf, 0.4 cm $year^{-1}$, (Sommerfield and Nittrouer 1999), and possibly an overestimation of the actual rate on the slump, is on the order of ~17.5 Kyr.

There are few direct indications to determine whether the main landslide scar of Eel Slump was created by a single ancient failure or multiple failure events. However, the geometry of a semi-enclosed depression with only a narrow outlet on the western edge, would argue against a single sudden event. Chirp profiles do not reveal a single sliding plane, which if it were to exist, would be deeper than the imaging reach of the chirp system. The most likely scenario for the formation of the slide scar is that small to medium size slide events took place in the past until the structure reached its present quasi-stable state.

The absence of indicators of recent recurring mass failures, such as sliding failure planes, fresh scarps, and unconformities within the sediment infill, suggests that the processes that have been modifying the slump landscape in the recent past are not gravity driven catastrophic mass wasting events. Instead, the slide scar morphology has been reshaped by erosive process on the headwall, sediment accumulation on the slide scar floor and secondary sediment alteration via gas flow and diagenetic processes.

Extensive seepage of oil and thermogenic gas was observed on the mound within the slide scar. Other areas display distinctive rough morphologies that are indicative of fluid seepage (Fig. 41.4). Furthermore, the frequent blanking observed in chirp profiles underlying these areas of distinctive morphologies as well as within the sediment drape elsewhere is consistent with the occurrence of gassy sediment (Fig. 41.3). Apparently, hydrocarbon seepage is widespread through the floor of this slide scar. This site is the second known occurrence of seafloor oil seepage in the northern California margin. In the other documented case (Kvendolven and Field 1981), the location of the seep was atop a shale diapir in 500 m water depth in the Eel River basin. Similarly, oil and gas seepage at the Eel Slump site is from within an anticlinal-type structure.

The causal relationship between the slide scar, the large mound and other seepage indicators at its base needs to be considered. One possibility is that the active gas and oil flow compromised the sediment strength and facilitated the sediment failure(s) that generated the large-scale morphology of Eel Slump. Once the overburden was removed the mound was created by gas-hydrate formation and authigenic carbonate precipitation. Alternatively, the sediment slide opened pathways for fluid flow, which subsequently derived into gas-hydrate formation, authigenic carbonate precipitation and the blistering that produced the mound (Paull et al. 2015a). The role of gas flow in promoting the sediment destabilization cannot be resolved here. However, it is important to note that under both scenarios the blistering that lead to the mound formation via authigenic carbonate precipitation and gas-hydrate accumulation is a late development that took place after the slide event.

Tubeworm Slump (Paull et al. 2015b), is a slide scar of size similar to Eel Slump, which was surveyed with the same AUV technology utilized in this study. The processes shaping the morphology of these two slides are quite different. Regardless of whether Eel Slump was created in one or more mass failure events, mass wasting has ceased to occur. In contrast, Tubeworm Slump has experienced mass wasting events in the recent past and has not reached a stable state. As such, these two slides may represent an earlier and a later stage in the evolution of slide scars.

Acknowledgements Support for this project from The David and Lucile Packard Foundation is gratefully recognized. Thoughtful reviews by Drs. D. Brothers and K. Coble-Maiers helped improve this manuscript.

References

Caress D, Chayes D (1996) Improved processing of Hydrosweep DS multibeam data on the R/V Maurice Ewing. Mar Geophys Res 18:631–650

Caress D, Thomas H, Kirkwood W et al (2008) High-resolution multibeam, sidescan, and subbottom surveys using the MBARI AUV D. Allan. B. In: Reynolds J, Greene HG (eds) Marine habitat mapping technology for Alaska: Alaska Sea Grant College Program. University of Alaska, Fairbanks. doi:10.4027/mhmta.2008.04

Gardner J, Malik M, Walker S (2009) Plume 2000 m high discovered at the sea floor off Northern California margin. Eos Trans AGU 90(32):275

Kvendolven K, Field M (1981) Thermogenic hydrocarbons in unconsolidated sediment of Eel River basin, offshore Northern California. AAPG Bulletin 65(9):1642–1646

Lillis PG et al (1999) Characterization of Northern California Petroleum by stable carbon isotopes. USGS OFR 99-164. http://pubs.usgs.gov/of/1999/ofr-99-0164/

Lorenson TD et al (1998) Comparison of offshore and onshore gas occurrences, Eel River basin, Northern California. USGS OFR 98-781. http://pubs.er.usgs.gov/publication/ofr98781

Paull C, Caress D, Thomas H, et al (2015a) Seafloor geomorphic manifestations of gas venting and shallow sub-bottom gas-hydrate occurrences. Geosphere 11(2):491–513. doi:10.1130/GES01012.1

Paull C, Anderson K, Caress D et al (2015b) Chapter 15: fine-scale morphology of tubeworm slump, monterey canyon. In: Lamarche G, Mountjoy J, Bull S, Hubble T, Krastel S, Lane E, Micallef A, Moscardelli L, Mueller C, Pecher I, Woelz S (eds) Submarine mass movements and their consequences, vol 41, Advances in Natural and Technological Hazards research. Springer, Dordrecht

Sommerfield C, Nittrouer C (1999) Modern accumulation rates and a sediment budget for the Eel shelf: a flood-dominated depositional environment. Mar Geol 154:227–241

Walz P et al (2014) In situ and laboratory Raman observations of an oil associated gas-hydrate in the Eel River basin. In: Proceedings of the 8th International Conference on gas-hydrates (ICGH8-2014), Beijing, China, 2014

Chapter 42
Shallow Gas and the Development of a Weak Layer in Submarine Spreading, Hikurangi Margin (New Zealand)

Aaron Micallef, Joshu J. Mountjoy, Sebastian Krastel, Gareth Crutchley, and Stephanie Koch

Abstract Submarine spreading is a type of mass movement that involves the extension and fracturing of a thin surficial layer of sediment into coherent blocks and their finite displacement on a gently sloping slip surface. Its characteristic seafloor signature is a repetitive pattern of parallel ridges and troughs oriented perpendicular to the direction of mass movement. We map ~30 km^2 of submarine spreads on the upper slope of the Hikurangi margin, east of Poverty Bay, North Island, New Zealand, using multibeam echosounder and 2D multichannel seismic data. These data show that spreading occurs in thin, gently-dipping, parallel-bedded clay, silt and sandy sedimentary units deposited as lowstand clinoforms. More importantly, high-amplitude and reverse polarity seismic reflectors, which we interpret as evidence of shallow gas accumulations, occur extensively in the fine sediments of the upper continental slope, but are either significantly weaker or entirely absent where the spreads are located. We use this evidence to propose that shallow gas, through the generation of pore pressure, has played a key role in establishing the failure surface above which submarine spreading occurred. Additional dynamic changes in pore pressure could have been triggered by a drop in sea level during the Last Glacial Maximum and seismic loading.

A. Micallef (✉)
University of Malta, Msida, Malta
e-mail: aaron.micallef@um.edu.mt; micallefaaron@gmail.com

J.J. Mountjoy
National Institute of Water and Atmospheric Research (NIWA), Wellington, New Zealand

S. Krastel
Institute of Geosciences, Christian-Albrechts-University at Kiel, Kiel, Germany

G. Crutchley
GNS Science, Lower Hutt, New Zealand

S. Koch
GEOMAR, Kiel, Germany

G. Lamarche et al. (eds.), *Submarine Mass Movements and their Consequences*,
Advances in Natural and Technological Hazards Research 41,
DOI 10.1007/978-3-319-20979-1_42

42.1 Introduction

Spreading entails the finite and downslope surficial displacement of rock/sediment on gently sloping ground, and the fracturing of the displaced mass into coherent blocks (Varnes 1978). Displacement mostly occurs along a shear zone (Rohn et al. 2004), and the deformation may involve subsidence, translation, rotation and disintegration of the upper coherent units (Dikau et al. 1996; Varnes 1978). The ground deformation associated with spreading comprises the extensional fissuring of the surface units in the form of alternating ridges and troughs (Dikau et al. 1996). The literature on spreading is not as extensive and exhaustive as for other types of mass movement, and little is known about the mechanics of the failure process. Deformation in a spread is known to be driven by a combination of transient and static shear stresses, attributed to a loss of shear strength of the underlying sediment, which allows the overlying material to slide downslope as intact blocks. The geological conditions conducive to spreading are usually those where consolidated rocks or sediments overlie a ductile substratum (Dikau et al. 1996; Rohn et al. 2004). In terrestrial environments, spreading is inextricably linked to the build up of pore pressure and associated liquefaction, which may occur in shallow underlying deposits either during an earthquake or due to changes in the height of the water table (Kanibir et al. 2006).

In submarine settings, numerical and mechanical models have indicated that, similarly to terrestrial environments, an increase in pore pressure may be a key preconditioning factor and trigger of spreading (Kvalstad et al. 2005; Micallef et al. 2007). In this paper we address the hypothesis that, by influencing pore pressure in sub-seafloor sediment, shallow gas can promote the development of a weak layer above which submarine spreading can occur. We do this by analysing geophysical data acquired from offshore the east coast of North Island, New Zealand (Fig. 42.1).

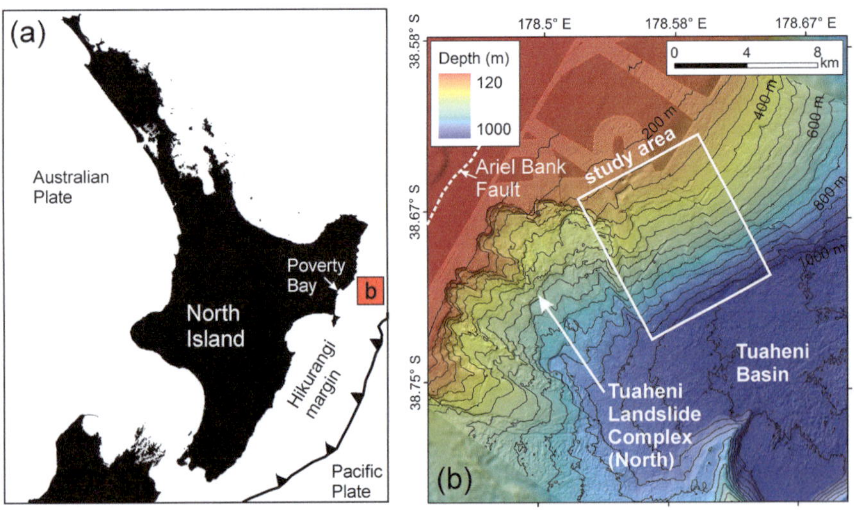

Fig. 42.1 (**a**) Location map. (**b**) Bathymetric map of the continental slope offshore Poverty Bay, showing location of study area. Isobaths at 50 m intervals

In comparison to terrestrial spreading, submarine spreading has received very little attention. First reported from offshore California (Field et al. 1982), most of what we know about submarine spreading comes from studies of the Norwegian passive continental margin (Baeten et al. 2013; Gauer et al. 2005; Kvalstad et al. 2005; Micallef et al. 2007, 2009). Nevertheless, the characteristic submarine spreading morphology, in the form of a recurring pattern of ridges and troughs, can be observed in numerous submarine landslides around the world (Lastras et al. 2003, 2006; Micallef et al. 2013; Piper et al. 1999; Vanneste et al. 2006). This means that submarine spreading is a widespread type of mass movement that has played an important role in the development of submarine landslides in different settings, and which therefore merits more detailed investigation.

42.2 Study Area

Our study area is located on the upper slope of the Hikurangi margin, 45 km east of Poverty Bay, North Island, New Zealand (Fig. 42.1). The east coast of the North Island straddles the boundary between the Pacific and Australian tectonic plates. This margin is characterised by the westward subduction of the Pacific Plate beneath the North Island, at a rate of about 4.5–5.5 cm year^{-1} (Beavan et al. 2002). Across the continental shelf in the region of our study area, active eastward verging splay faults from the plate boundary mega-thrust are known to project to the seafloor (Mountjoy and Barnes 2011). On the mid- to upper-slope, however, there is a lack of active tectonic deformation, which results in a relatively simple facies geometry. The upper continental slope of the Hikurangi margin is underlain by Miocene to recent slope basin sequences with possible Cretaceous and Paleogene sedimentary rocks at depth (Barnes et al. 2002; Mountjoy and Barnes 2011). Overlying these sequences at the shelf break are lowstand clinoforms deposited during the Quaternary glacial cycles (Barnes et al. 2002; Pedley et al. 2010). These deposits are formed of gently dipping, parallel-bedded clay, silt and possibly sandy sedimentary units (Alexander et al. 2010). Modest size (0.01 km^3) to very large (3,000 km^3) submarine landslides have occurred on the Hikurangi margin (Barnes et al. 2010; Kukowski et al. 2010). Some of the best preserved examples of these occur on the upper continental slope directly off Poverty Bay and to the south-west of the study area (Fig. 42.1b). These include the ~30 km^3 Poverty Debris Avalanche, and the ~10 km^3 Tuaheni Landslide Complex (Mountjoy et al. 2009).

42.3 Data and Methods

Our study is based on two types of data. The first is a multibeam echosounder dataset covering 700 km^2 of seafloor (Fig. 42.1b). These data were acquired using a hull-mounted Kongsberg EM300 multibeam system during two cruises (TAN1114 in 2011 and TAN0810 in 2008). The bathymetry data were processed with CARIS Hydrographic Information Processing System (HIPS) by accounting for sound velocity variations, tides and basic quality control. A bathymetry grid with 25×25 m bin size was derived. The second dataset comprises high resolution 2D multichannel seismic reflection data acquired during the TAN1404 cruise in 2014 (Fig. 42.2). The acquisition system entailed a 0.7 l GI Gun and a 150 m long streamer with 96 channels. Processing included crooked line common midpoint (CMP) binning (CMP spacing of 1.5 m), frequency filtering (Butterworth filter with low-cut corner frequencies of 25 and 55 Hz), normal move-out correction, stacking and 2D Stolt migration. All cruises were carried out on board the *R/V Tangaroa*.

42.4 Results

42.4.1 Morphology

The continental slope within the study area has an average slope gradient of 5.5° towards SSE. The morphology is dominated by an elongated scar with a length of 8 km, width of 4 km, and 60 m depth (Fig. 42.2a). The downslope limit of the scar coincides with the regional base of the continental slope where it is contiguous with the Tuaheni sedimentary basin, at 975 m depth. Its headwall is located in the upper continental slope, at a depth of 250 m. Smaller scars, sharing a similar morphology and distal limit, are located 1 km to the north-east of the elongated scar. The seafloor morphology across the upper section of the scar predominantly consists of a subdued, repetitive pattern of ridges and troughs oriented parallel to the isobaths. In the downslope section of the scar, the morphology is smoother and intersected by lineations that are up to 3 km long, 5 m deep, and oriented perpendicular to the isobaths. These lineations and the western boundary of the scar are intersected by a 4.5 km long and 20 m high SW-NE oriented escarpment. Circular depressions that are up to 200 m wide and 30 m deep are located at the headwall of the scar.

42.4.2 Sub-seafloor Architecture

The seismic expression of the sub-seafloor in the study area comprises a sequence of continuous, parallel, gently-dipping seismic reflectors that is at least 150 m thick in places (assuming a seismic P wave velocity of 1600 m s^{-1} for depth conversion). Two reflectors within this sequence are characterised by high amplitude and reverse polarity, and are recorded at an average depth of 55 m below the seafloor (Fig. 42.2b). These high amplitude reflectors occur across the upper continental slope, but are either significantly weaker or entirely absent where the elongated scar with the ridge and trough morphology is present. Here, the upper part of the seismic sequence is generally characterised by a unit of irregular, chaotic, low amplitude

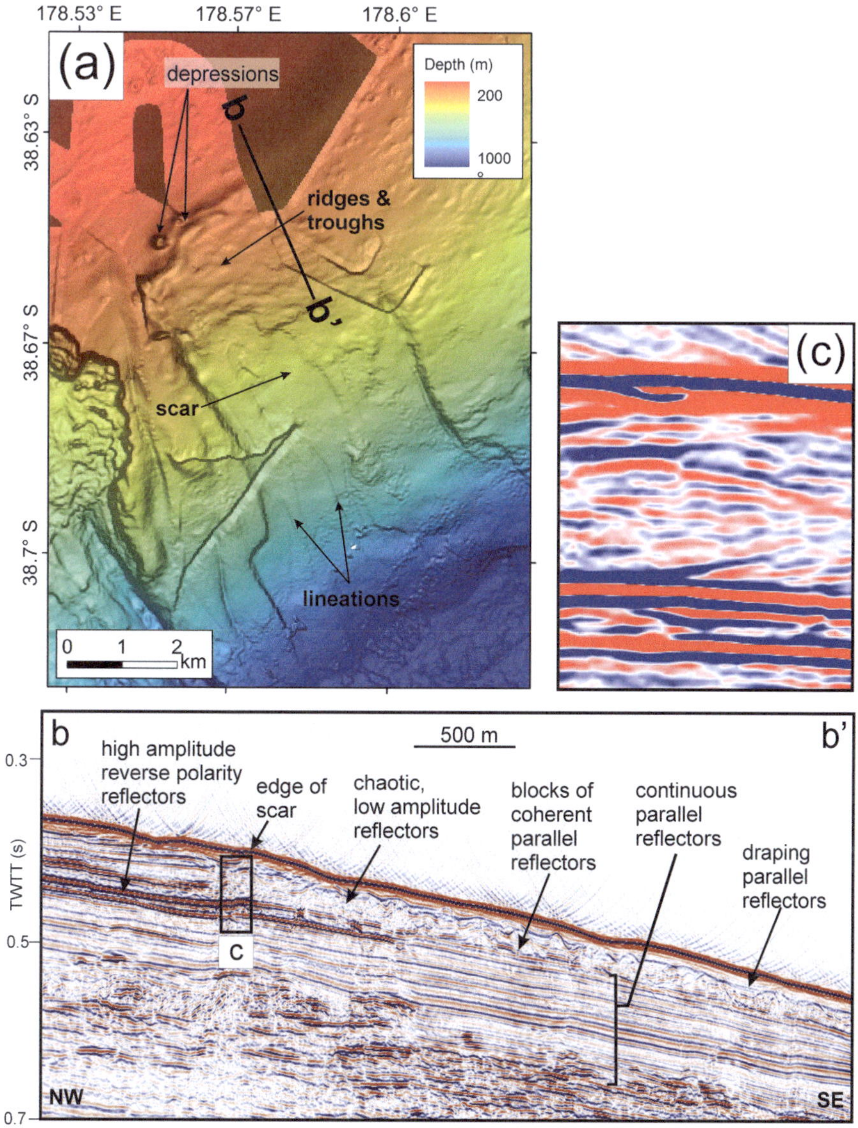

Fig. 42.2 (**a**) Bathymetric map of study area draped on a slope gradient map and showing principal morphologic elements of the scar. (**b**) Seismic reflection profile P3106 across the scar. (**c**) Enlarged section of profile P3106 showing reverse polarity of high amplitude reflector

reflectors, although triangular blocks of coherent, parallel, downslope-dipping reflectors are visible. This unit has a variable thickness, with a maximum of 40 m. The base of this chaotic unit is a planar reflector that connects with the high amplitude reflector further upslope. The chaotic unit is also covered by a draping unit of parallel reflectors, which is characterised by irregular thickness and reaches a maximum thickness of 35 m.

42.5 Discussion and Conclusions

42.5.1 Nature of Mass Movement

The downslope-dipping, parallel seismic reflectors across the study area are interpreted as layers in a stratified sediment package (Mountjoy et al. 2009). We infer that the material in this package is similar to that which failed in the adjacent Tuaheni Landslide Complex. This consisted of muddy sedimentary deposits, which accumulated during periods of eustatic sea level lowering, overlain by a Holocene hemipelagic succession (Carter and Manighetti 2006; Paquet et al. 2009). The ridge and trough morphology, and the chaotic seismic sequence with isolated blocks of coherent reflectors, are signature characteristics of submarine spreading (Micallef et al. 2007). We therefore interpret the upper section of the elongated scar documented across the study area as evidence of a submarine spreading event that comprised thin, extensional deformation of the lowstand units, occurring along stratigraphic surfaces, and which was later draped by Holocene sedimentation. This mode of failure corresponds to model 2 proposed by Micallef et al. (2007) for the Storegga Slide, where a thin slab ruptures under tension into a series of coherent blocks that translate and tilt downslope along a quasi-planar failure plane. The downslope section of the scar has undergone a higher degree of sediment evacuation, likely a result of translational sliding or more plastic deformation. The lineations may correspond to furrows eroded by debris flows into the failure surface.

42.5.2 Role of Shallow Gas

Limit equilibrium modelling by Micallef et al. (2007) showed how spreading can be pre-conditioned or triggered by three processes – loss of support, increase in total weight upslope, and an increase in pore pressure. Loss of support is a potential trigger of spreading in the region because of sediment evacuation in the downslope section of the scar. We exclude increase in total weight upslope as a potential cause because there are no indications of loading of sediment from a slope failure in the seismic data. An increase in pore pressure is also a likely cause of spreading in the study area. We interpret the high amplitude and reverse polarity reflectors in Fig. 42.2b as the top of an accumulation of gas within the sediments. The circular depressions, which we interpret as pockmarks, provide additional evidence of sub-seafloor overpressure. We are not able to determine whether the gas has biogenic or thermogenic origin. In bubble phase, gas is known to markedly increase the pore pressure, which decreases the effective stress of the seafloor sediment, creating weak layers that are prone to failure (Crutchley et al. 2010; Field 1990). In our study area this effect is enhanced by the low permeability of the fine-grained material that failed. The absence or low quantities of shallow gas in the elongated scar, and the correspondence of the depth of failure with that of the shallow gas, indicate that the latter has played a key role in establishing the failure surface. Additionally, dynamic changes in pore pressure may have been triggered by two factors. The first is a drop in sea level during the Last Glacial Maximum, with an associated reduction in effective stress as gas came out of solution due to lower hydrostatic pressures. This could explain why the absence of gas is more noticeable in the deeper part of the slope, where the reduction of the hydrostatic pressure would have been more pronounced. The second factor is seismic loading. The active fault most proximal to the study area is the Ariel Bank Fault (Fig. 42.1b), which is inferred to have a late Quaternary displacement rate in the range of 3.0–6.5 mm year^{-1} (Barnes et al. 2002). Peak ground acceleration, estimated from probabilistic seismic hazard modelling of regional earthquake sources, is in the order of 0.3–0.4 g with a return time of 475 years (Stirling et al. 2002). Some moderately large magnitude historic earthquakes have also occurred in the vicinity (e.g. the 1931 M7.8 Napier earthquake). The escarpment crossing the western boundary of the scar and the lineations may also be interpreted as the location of a recently-active fault.

To evaluate the potential failure mechanisms and perform stability analyses that take into account the role of gas charging and seismic loading, there is the need to acquire long sediment cores and carry out in situ geotechnical measurements from the study area.

Acknowledgments This research was supported by funding from Marie Curie Career Integration Grant PCIG13-GA-2013-618149 within the 7th European Community Framework Programme, New Zealand Ministry for Business Innovation and Employment, NIWA Core Funding under Coasts and Oceans Research Programme 1 (2013/14 SCI), DFG (Deutsche Forschungsgemeinschaft), and the Royal Society of New Zealand International Mobility Fund contract ISATB09-37. We are indebted to the TAN1404 shipboard party, and the captain, crew and technicians of *RV Tangaroa*. We thank Nicole Baeten and David Amblas for their insightful reviews.

References

Alexander CR, Walsh JP, Orpin AR (2010) Modern sediment dispersal and accumulation on the outer poverty continental margin. Mar Geol 270:213–226

Baeten NJ, Laberg JS, Forwick M et al (2013) Morphology and origin of smaller-scale mass movements on the continental slope off northern Norway. Geomorphology 187:122–134

Barnes PM, Nicol A, Harrison T (2002) Late Cenozoic evolution and earthquake potential of an active listric thrust complex above the Hikurangi subduction zone, New Zealand. Geol Soc Am Bull 114:1379–1405

Barnes PM, Lamarche G, Bialas J et al (2010) Tectonic and geological framework for gas hydrates and cold seeps on the Hikurangi subduction margin, New Zealand. Mar Geol 272:26–48

Beavan J, Tregoning P, Bevis M et al (2002) Motion and rigidity of the Pacific Plate and implications for plate boundary deformation. J Geophys Res 107:2261

Carter L, Manighetti B (2006) Glacial/interglacial control of terrigenous and biogenic fluxes in the deep ocean off a high input, collisional margin: a 139kyr-record from New Zealand. Mar Geol 226:307–322

Crutchley GJ, Geiger S, Pecher I et al (2010) The potential influence of shallow gas and gas hydrates on sea floor erosion of Rock Garden, an uplifted ridge offshore of New Zealand. Geo-Mar Lett 30:283–303

Dikau R, Brunsden D, Schrott L et al (1996) Landslide recognition: identification, movement and causes. Wiley, Chichester

Field ME (1990) Submarine landslides associated with shallow seafloor gas and gas hydrates off Northern California, In: AAPG (ed) Fifth circum-pacific energy and mineral resources conference, Honolulu

Field ME, Gardner JV, Jennings AE et al (1982) Earthquake-induced sediment failures on a 0.25° slope, Klamath River delta, California. Geology 10:542–546

Gauer P, Kvalstad TJ, Forsberg CF et al (2005) The last phase of the Storegga Slide: simulation of retrogressive slide dynamics and comparison with slide-scar morphology. Mar Petrol Geol 22:171–178

Kanibir A, Ulusay R, Aydan O (2006) Assessment of liquefaction and lateral spreading on the shore of Lake Sapanca during the Kocaeli (Turkey) earthquake. Eng Geol 83:307–331

Kukowski N, Greinert J, Henrys S (2010) Morphometric and critical taper analysis of the Rock Garden region, Hikurangi Margin, New Zealand: implications for slope stability and potential tsunami generatin. Mar Geol 272:141–153

Kvalstad TJ, Andersen L, Forsberg CF et al (2005) The Storegga slide: evaluation of triggering sources and slide mechanisms. Mar Pet Geol 22:245–256

Lastras G, Canals M, Urgeles R (2003) Lessons from sea-floor and subsea-floor imagery of the BIG'95 debris flow scar and deposit. In: Locat J, Mienert J (eds) Submarine mass movements and their consequences. Kluwer Academic Publishers, Dordrecht, pp 425–431

Lastras G, Canals M, Amblas D et al (2006) Eivissa slides, western Mediterranean sea: morphology and processes. Geo-Mar Lett 26:225–233

Micallef A, Masson DG, Berndt C et al (2007) Morphology and mechanics of submarine spreading: a case study from the Storegga Slide. J Geophys Res 112:F03023

Micallef A, Masson DG, Berndt C et al (2009) Development and mass movement processes of the north-eastern Storegga Slide. Quat Sci Rev 28:433–448

Micallef A, Georgiopoulou A, Le Bas T et al (2013) The Malta-Sicily escarpment: mass movement dynamics in a sediment-undersupplied margin. In: Krastel S et al (eds) Submarine mass movements and their consequences. Springer International Publishing, Switzerland, pp 317–328

Mountjoy JJ, Barnes PM (2011) Active upper-plate thrust faulting in regions of low plate-interface coupling, repeated slow slip events, and coastal uplift: example from the Hikurangi Margin, New Zealand. Geochem Geophys Geosyst 12:Q01005

Mountjoy JJ, McKean J, Barnes PM et al (2009) Terrestrial-style slow-moving earthflow kinematics in a submarine landslide complex. Mar Geol 267:114–127

Paquet F, Proust JN, Barnes PM et al (2009) Inner-forearc sequence architecture in response to climatic and tectonic forcing since 150 Ka: Hawke's Bay, New Zealand. J Sediment Res 79:97–124

Pedley KL, Barnes PM, Pettinga JR et al (2010) Seafloor structural geomorphic evolution of the accretionary frontal wedge in response to seamount subduction, poverty indentation, New Zealand. Mar Geol 270:119–138

Piper DJW, Cochonat P, Morrison ML (1999) The sequence of events around the epicentre of the 1929 Grand Banks earthquake: initiation of debris flows and turbidity currents inferred from sidescan sonar. Sedimentology 46:79–97

Rohn J, Resch M, Schneider H et al (2004) Large-scale lateral spreading and related mass movements in the Northern Calcareous Alps. Bull Eng Geol Environ 63:71–75

Stirling MW, McVerry GH, Berryman KR (2002) A new seismic hazard model for New Zealand. Bull Seismol Soc Am 92:1878–1903

Vanneste M, Mienert J, Bünz S (2006) The Hinlopen Slide: a giant, submarine slope failure on the northern Svalbard margin, Arctic Ocean. Earth Planet Sci Lett 245:373–388

Varnes DJ (1978) Slope movement types and processes. In: Schuster RL, Krisek RJ (eds) Landslides, analysis and control. National Academy of Sciences, Transportation Research Board, Special Report 176, pp 11–33

Chapter 43
Stability of Fine-Grained Sediments Subject to Gas Hydrate Dissociation in the Arctic Continental Margin

Jeffrey A. Priest and Jocelyn L.H. Grozic

Abstract Significant venting of methane gas has been observed on the upper continental slopes of the Arctic Ocean, coinciding with the landward limit of the gas hydrate stability zone. It has been inferred that the methane gas venting is related to the dissociation of methane gas hydrate induced by the unprecedented Arctic warming that has occurred over the last 30 years. Historically, the influence of hydrate dissociation on sediment stability was considered in terms of hydrate dissociation increasing pore pressures, reducing effective stress and therefore sediment strength, leading to slope failures along these over pressurized layers. Recent evidence has shown that gas hydrate readily forms in clay-rich sediments as fracture-filled near-vertical veins, which upon dissociation gives rise to sediments exhibiting high water content, high sediment compressibility and very low shear strength. Thus gas hydrate dissociation in fine-grained sediments may lead to significant slope instabilities, which at present is poorly understood. Slope stability analyses carried out considering the potential influence of hydrate dissociation induced by warming of the Arctic slope suggest that instabilities are likely, with failures ranging from surficial sloughing to deep-seated failures. This paper highlights the importance of understanding the sediment processes in clay soils, and how dissociation of gas hydrate can induce instabilities through sediment softening and generation of excess pore pressure.

J.A. Priest (✉)
Department of Civil Engineering, University of Calgary, ENF 262, 2500 University Drive NW, Calgary, AB T2N 1N4, Canada
e-mail: japriest@ucalgary.ca

J.L.H. Grozic
Department of Civil Engineering, University of Calgary, ENF 216, 2500 University Drive NW, Calgary, AB T2N 1N4, Canada
e-mail: jgrozic@ucalgary.ca

G. Lamarche et al. (eds.), *Submarine Mass Movements and their Consequences*, Advances in Natural and Technological Hazards Research 41,
DOI 10.1007/978-3-319-20979-1_43

43.1 Introduction

It is now globally recognized that anthropogenic induced climate change is occurring, with the greatest change observed in the Arctic region as reported by the Intergovernmental Panel on Climate Change (IPCC 2007). A possible consequence of this warming, especially in the Arctic Ocean, is the dissociation of methane gas hydrate, an ice-like crystalline compound that is stable under certain conditions of low temperature and high pressure that are typically found within deep marine sediments and in and beneath permafrost (Kvenvolden and Lorenson 2001). Methane gas hydrate sequesters large volumes of methane, a greenhouse gas; therefore dissociation of gas hydrate may release significant quantities of methane into the ocean and the atmosphere, exacerbating climate change (Archer et al. 2009). In addition, gas hydrate dissociation can lead to significant reduction in sediment strength leading to slope instability and submarine landslides (Mienert et al. 2005; Grozic 2009).

In this paper we investigate the influence of gas hydrate dissociation, resulting from Arctic Ocean warming, on sediment shear strength and the resulting impact that changes in sediment properties have on slope instabilities.

43.1.1 Climate Change and Arctic Hydrates

Gas hydrates are extensive within the Arctic region (Council of Canadian Academies 2008) located below onshore permafrost, sub-sea permafrost and on slopes of the continental margin. Given predicted warming scenarios hydrate dissociation below onshore permafrost (>300 m below the ground surface) and sub-sea permafrost on the Arctic shelf (~200 m below the current sea level) is unlikely to occur over the next millennia (Ruppel 2011). The cold waters of the Arctic Ocean allow gas hydrate to form within the sediments on the slopes of the Arctic continental margin at shallower water depths than observed on other margins with the upper limit of hydrate stability zone (HSZ) tapering out near the seafloor ~300 m water depth (Fig. 43.1). Therefore, hydrate-bearing sediments in this area are the most sensitive to warming in the near future (Ruppel 2011).

Multiple plumes of gas bubbles, predominantly methane, emanating from the seabed on the West Svalbard margin of the Arctic Ocean have been observed that coincide with the calculated present day landward limit of the HSZ (Westbrook et al. 2009). It is suggested that the 1 °C rise the Arctic Ocean has experienced at this water depth over the last 30 years has led to the dissociation of gas hydrate within the sediment leading to the observed methane gas expulsions from the seabed (Thatcher et al. 2013). In addition, similar observations of active methane gas venting have also been observed within the Beaufort Sea area of the Arctic Ocean at water depths where the HSZ thins out (Paull et al. 2011).

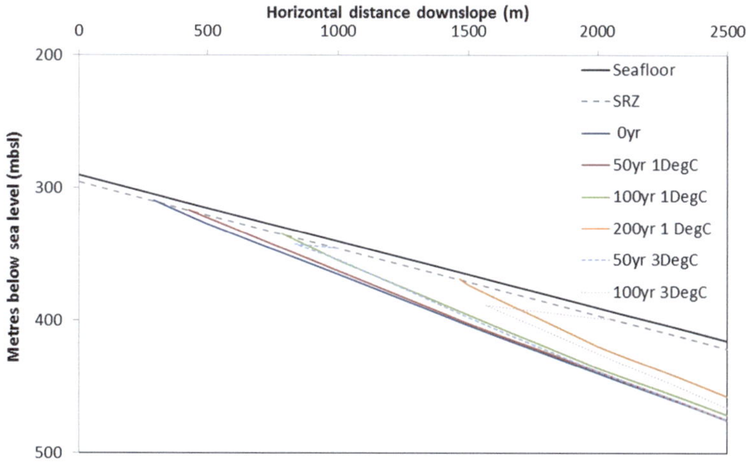

Fig. 43.1 Modeled slope geometry highlighting water depth to seafloor, sulfate reduction zone (*SRZ*), initial base of hydrate stability (0 year) and subsequent changes in the base of hydrate stability for two ocean bottom temperature increases (1 °C and 3 °C) with time

43.1.2 Hydrate in Fine-Grained Sediments

Recent insights from pressure cores where hydrate stability conditions within the sediment during core recovery are maintained have shown that in fine-grained sediments (typical for shallow marine sediments in the Arctic) hydrate readily forms as fracture-filled near-vertical veins with average hydrate saturations of ~20–30 % (Rees et al. 2011). These veins can vary considerably in thickness (<0.5 mm up to 6 mm) and be fibrous in nature with veins showing terminal forking and branching. Results from the limited testing of the pressure core samples suggest that these hydrate veins increase in apparent sediment strength (Yun et al. 2010) inhibiting normal consolidation of the sediment; stiffer hydrate veins support the overburden stress acting on the sediment.

Upon dissociation of the hydrate the host sediment exhibits high void ratios and water contents leading to high compressibility (Kim et al. 2013) and low undrained shear strength, s_u (Priest et al. 2014), compared to normal sediments at corresponding depths. In addition to the low strength of the host sediment, hydrate dissociation can lead to the development of excess pore pressure within the sediment (Sultan et al. 2004; Grozic 2009). Both of these factors may lead to slope instabilities in regions where gas hydrate exists.

43.2 Slope Stability Modeling

Hydrate dissociation through warming of the Arctic Ocean over the next 100 years (IPCC 2007) may lead to significant changes in sediment behavior over this time. To investigate the influences of these changes on slope stability 2-D limit equilibrium slope stability analyses were carried out using Slope/W (a commercially available slope stability analysis program) that considers total shear resistance and mobilized shear stresses along a slip surface to compute factors of safety. In our analyses both undrained and drained soil behavior were considered.

43.2.1 Slope Profile

In our analyses a slope width of 2500 m starting at a water depth to the seafloor of 290 m with an initial slope angle of 3°, corresponding to average slope geometry in the Arctic Ocean, was modeled (Fig. 43.1). To account for typical sulfate reduction processes that inhibit gas hydrate formation in shallow sediments a 5 m thick hydrate-free zone was located directly below the seafloor (Biastoch et al. 2011). The initial base (0 year) of the HSZ was determined using CSMGem code (Sloan and Koh 2008) for an ocean bottom temperature of 0 °C at the seafloor, pore water salinity of 3.5 wt.% and a geothermal gradient of 0.87 °C/m. The geothermal gradient is at the high end of reported values (Phrampus et al. 2014) leading to a reduction in the thickness of the HSZ, which results in a larger volume of sediment subject to hydrate dissociation with time during a warming event, and thus represents a worse case scenario.

Changes in the base of the HSZ with time (Fig. 43.1) due to an increase in ocean bottom temperatures were taken from the results of Reagan et al. (2011) who carried out detailed 2-D simulations of hydrate dissociation. Their simulations took into account the endothermic nature of hydrate dissociation and increase in hydrate stability due to increases in pore pressure for two distinct warming scenarios of 1 °C/100 year and 3 °C/100 year using similar initial conditions (slope geometry, ocean bottom temperature, salinity, geothermal gradient) along with a uniform hydrate saturation of 3 % and sediment permeability of 10^{-15} m^2.

43.2.2 Soil Properties

Limited data exists for the mechanical properties of soils within the Arctic margin, especially for hydrate-bearing sediments. In our analyses soil properties (highlighted in Table 43.1) were based on results from core samples obtained during the National Gas Hydrate Program expedition offshore India (NGHP1, Collett et al. 2008), where hydrate veins were observed in fine-grained marine sediments.

For the undrained analyses different values of s_u were assigned to the three soil layers highlighted in Table 43.1. For the 5 m hydrate-free soil layer the assigned s_u value was based on the results of shear vane tests on near surface sediments with no hydrate present (Winters 2011). The value of s_u for the soil subject to hydrate dissociation was based on the results of Priest et al. (2014) who undertook

Table 43.1 Material properties for the different soils applied during drained and undrained analyses

Soil layer	Undrained s_u	Drained ϕ' and c
Hydrate-free	6 kPa + 0.8 kPa/m[a]	21°, 7.9 kPa[a]
Hydrate-bearing	50 kpa	30°, 50 kPa
Hydrate dissociation	6 kpa[b]	21°, 7.9 kPa[a]

[a]Winters (2011)
[b]Priest et al. (2014)

undrained triaxial tests on samples obtained from hydrate-bearing cores recovered during HGHP1. In these tests the hydrate veins were removed prior to testing such that the host sediment strength was tested at its original in-situ water contents. The value of s_u assigned to the hydrate-bearing layer (this layer has minor influence on slope stability) was notionally chosen to be greater than those for the hydrate-free layer. For the drained analyses a Mohr-Coulomb failure criterion was adopted and so relevant soil properties (friction angle, ϕ' and cohesion, c) were based on the results of drained triaxial tests at different effective stresses conducted on recovered samples from NGHP1 expedition (Winters 2011). As for the undrained analyses, values for the hydrate-bearing layer were notionally chosen to be greater than those for the hydrate-free layer.

A contrast in soil strength will occur at the base of the HSZ during dissociation and so the failure surface was prescribed along this interface, similar to that proposed by McIver (1982). All soils were assigned a saturated unit weight of 16 kN/m^3 relating to a porosity of 60 %, typical values for the near surface sediments recovered during the NGHP1 expedition (Collett et al. 2008).

43.3 Results and Discussion

43.3.1 Slope Stability Using Drained Soil Properties

Initial slope modeling conducted using a drained analysis with all soils parameters set to $s_u = 6$ kPa + 0.8 kPa/m (no hydrate throughout the slope) showed that for a 3° slope the factor of safety (FOS) for the critical slip surface was 3.4. In order for the soil to experience instability without hydrates (i.e. a FOS <1) a slope angle of at least 11° would be required for the applied soil properties. Figure 43.2 shows the calculated FOS for the different warming scenarios (see Fig. 43.1) using the undrained parameters highlighted in Table 43.1. For a 1 °C/100 year warming event the slope fails after 200 years, while for the 3 °C/100 year warming event the slope fails after 100 years. Figure 43.3 highlights the extent of the slope failure and the calculated FOS for the two separate analyses. The critical slip surface for event occurs at the base of the hydrate stability zone, where the weakened soil (due to hydrate dissociation) is in contrast with the stiffer underlying soils. The extent of the critical slip surface for the 3 °C/100 year case extends over 1500 m downslope, resulting in a sliding volume per meter across the slope ~35 × 10^3 m^3.

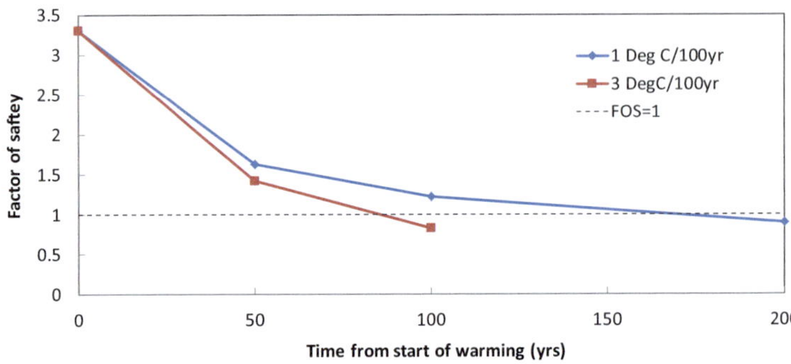

Fig. 43.2 Factor of safety resulting from hydrate dissociation with time due to increases in ocean bottom temperatures

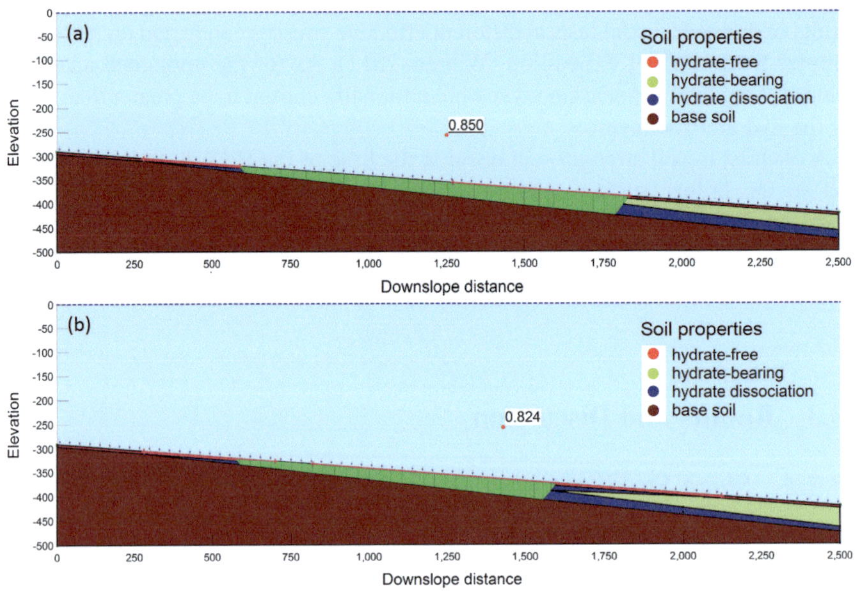

Fig. 43.3 Factor of safety and extent of failure surface resulting from hydrate dissociation due to (**a**) 1 °C/100 year temperature rise in ocean bottom temperatures after 200 year, and (**b**) 3 °C/100 year temperature rise in ocean bottom temperatures after 100 year

In the above analyses a constant s_u is assigned to the soil subject to hydrate dissociation. In reality, if the hydrate veins support the overburden stress then as the dissociation front advances downslope with time (predominantly at the base of the HSZ) the soil will gradually experience an increase in apparent overburden stress leading to consolidation of the soil, thereby increasing s_u. The degree of consolidation, and therefore the potential increase in s_u, will be highly dependent on the resultant soil properties such as; soil permeability, compressibility, drainage path,

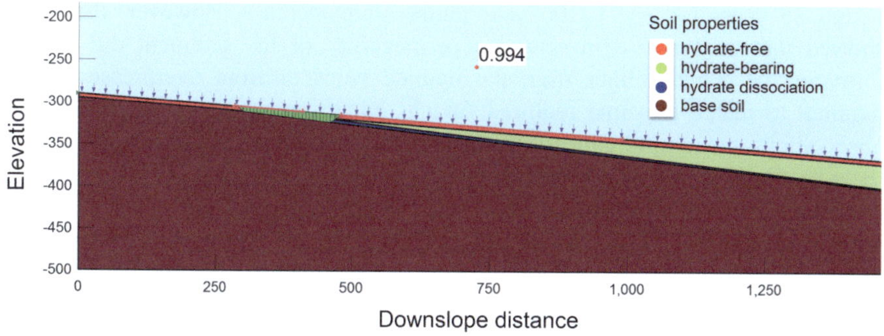

Fig. 43.4 Critical slip surface from drained slope stability analysis for 1 °C/100 year warming over 50 years due to 50 kPa pore pressure increase in hydrate dissociation zone (*colored blue*)

etc. Hydrate dissociation is greatest at the base of the HSZ and so the drainage path for fluid flow increases as the dissociation front moves downslope with time. For the 3 °C/100 year warming event and considering an average coefficient of consolidation, c_v, of 2.2×10^{-8} m^2/s (Winters 2011), which is a function of soil permeability and compressibility, the soil at the base of the HSZ at ~1500 m downslope would experience only ~26 % consolidation after 100 years from the onset of hydrate dissociation. This would suggest that the soil at this depth is likely to remain substantially undrained and exhibit a low s_u.

43.3.2 Slope Stability Using Drained Soil Properties

In addition to the reduction in s_u due to hydrate dissociation, an increase in pore pressure within the sediment can occur due to the release of methane gas. This will lead to a reduction in effective stress, leading to shear failure of the soil. To consider the influence of excess pore pressure on slope stability drained analyses were conducted with increasing pore pressure applied to the region subjected to hydrate dissociation using the drained soil parameters summarized in Table 43.1. Results show (Fig. 43.4) that shallow, near surface, localized failures (extending 100–200 m downslope) with a FOS <1 occur when the excess pore pressure within the soil was >50 kPa.

It has been shown experimentally that during dissociation (for a 5 % hydrate saturation) the rise in pore pressure for an undrained sand sample can be as great as ~170 kPa (Wu and Grozic 2008). The actual increase in pore pressure within sediments on the continental margins during hydrate dissociation over time will depend on the fully-coupled relationship between fluid flow, heat flow, hydrate dissociation and sediment strength/stiffness. Considering similar slope geometry to that modeled, Thatcher et al. (2013) showed for a sediment with a permeability of 10^{-16} m^2 (typical for Arctic sediments), 5 % hydrate saturation and a geothermal gradient of 0.65 °C/m that pore pressure is likely to exceed lithostatic pressures within 25 years for ~3 °C/100 year temperature increase. However, they also showed that an increase in effective permeability of the sediment due to the formation of cracks, either from the hydrate veins or from overpressure, was required to match the time required for gas expulsion at the seafloor to match observations, thus limiting the magnitude of any pore pressure increase. Therefore, localized slope failures due to the generation of excess pore pressures may occur in the short term, depending on actual permeability of the sediment. These localized failures may cause increased loading on soils downslope leading to progressive failure of the slope as a whole, involving both undrained and drained behavior.

43.4 Conclusions

Warming of the Arctic Ocean, through ongoing climate change, is likely to lead to the dissociation of gas hydrate that resides within the sediment on its upper continental slopes. In fine-grained sediments, where the hydrate exists as fracture-filled veins, this will lead to a reduction in soil strength as well as an increase in pore pressure through release of methane gas. Modeling dissociation using a reduction in undrained shear strength indicates large-volume slope failures could occur after some time, when a sufficient mass of sediment has undergone strength reduction. In contrast, drained analyses, where increases in pore pressures as a result of hydrate dissociation were considered, gave rise to shallow localized slope failures at much earlier times than the undrained cases. These shallow failures may progress downslope due to increased loading (from the failing mass above) and ongoing hydrate dissociation.

The 'factors of safety' for slope failures were computed at discrete time steps with a number of simplifying assumptions with regard to soil behavior during dissociation (drained and undrained) and the assumption of an idealized slip surface at the base of HSZ. In reality, during hydrate dissociation, the soil is likely to be neither fully drained nor undrained, with the relationship between the two idealized behaviors being highly dependent on the effective permeability of the soil. This will lead to changes in variations in soil properties with time throughout the slope that was not accounted for in our analyses. In addition, it is implicitly implied that the lateral distribution of hydrate throughout the slope was uniform, which for fine-grained sediments may not be valid. Given that observed submarine slope failures on continental margins do not seem to exhibit large-scale slide plane type failures suggest that the modes of failure due to hydrate dissociation are likely to be more complex than we have modeled.

To correctly assess the influence of hydrate dissociation on slope stability, resulting from climate change in the Arctic Ocean, will requires more detailed characterization of soil properties relevant to the Arctic and the likely changes that may occur during hydrate dissociation, including effective permeability. In addition, more complex analyses, taking account of discrete changes in soil properties, are required to capture realistic failure modes.

Acknowledgements This research was undertaken, in part, thanks to funding from the Canada Research Chairs program. We would also like to thank the reviewers (Prof. Kenichi Soga and Dr William Waite) for their constructive comments and suggestions to help improve this manuscript.

References

Archer D, Buffet B, Brodkin V (2009) Ocean methane hydrates as a slow tipping point in the global carbon cycle. Proc Natl Acad Sci U S A 106(49):20596–20601

Biastoch A, Treude T, Rüpke LH et al (2011) Rising Arctic Ocean temperatures cause gas hydrate destabilization and ocean acidification. Geophys Res Lett 38:L08602. doi:10.1029/2011GL047222

Collett TS et al (2008) National gas hydrate program expedition 01 initial report, Dir. Gen. of Hydrocarbons, Minist. of Pet. and Nat. Gas, New Delhi

Council of Canadian Academies (CCA) (2008) Energy from gas hydrates: assessing the opportunities & challenges for Canada. Report of the expert panel on gas hydrates. J Grace (chair), Ottawa, Canada

Grozic JLH (2009) Interplay between gas hydrates and submarine slope failure. In: Mosher DC, Shipp RC, Moscardilli L et al (eds) Submarine mass movements and their consequences IV. Springer, Dordrecht, pp 11–30

Intergovernmental Panel on Climate Change (IPCC) (2007) Climate change 2007: the physical science basis. Cambridge University Press, New York

Kim HS, Cho GC, Lee JY, Kim SJ (2013) Geotechnical and geophysical properties of deep marine fine-grained sediments recovered during the second Ulleung Basin Gas Hydrate expedition, East Sea, Korea. Mar Pet Geol 47:55–65

Kvenvolden KA, Lorenson TD (2001) A global inventory of natural gas hydrate occurrence. In: Paull CK, Dillon WP (eds) Natural gas hydrates: occurrence, distribution, and detection, Geophys. monogr. ser, vol 124. AGU, Washington, DC, pp 3–18

McIver RD (1982) Role of naturally occurring gas hydrates in sediment transport. AAPG Bull 66:789–792

Mienert J, Vanneste M, Bunz S et al (2005) Ocean warming and gas hydrate stability on the mid-Norwegian margin at the Storegga Slide. Mar Pet Geol 22:233–244

Paull C, Dallimore SR, Hughe-Clarke J et al (2011) Tracking the decomposition of submarine permafrost and gas hydrate under the shelf and slope of the Beaufort Sea. Paper presented at 7th International conference on gas hydrates, Edinburgh

Phrampus BJ, Hornbach MJ, Ruppel CD, Hart PE (2014) Widespread gas hydrate instability on the upper U.S. Beaufort margin. J Geophys Res Solid Earth 119:8594–8609. doi:10.1002/2014JB011290

Priest JA, Clayton CRI, Rees EVL (2014) Potential impact of gas hydrate and its dissociation on the strength of host sediment in the Krishna-Godavari Basin. Mar Pet Geol 58:187–198

Reagan MT, Moridis GJ, Elliott SM, Maltrud M (2011) Contribution of oceanic gas hydrate dissociation to the formation of Arctic Ocean methane plumes. J Geophys Res 116, C09014. doi:10.1029/2011JC007189

Rees EVL, Priest JA, Clayton CRI (2011) The structure of methane gas hydrate bearing sediments from the Krishna–Godavari Basin as seen from Micro-CT scanning. Mar Pet Geol 28(7):1283–1293

Ruppel C (2011) Methane hydrates and contemporary climate change. Nat Educ Knowl 3(10):29

Sloan ED, Koh CA (2008) Clathrate hydrates of natural gases, 3rd edn. Taylor & Francis/CRC Press, Boca Raton

Sultan N, Cochonat P, Foucher JP, Mienert J (2004) Effect of gas hydrate melting on seafloor slope stability. Mar Geol 231:379–401

Thatcher KE, Westbrook GK, Sarkar S et al (2013) Methane release from warming-induced hydrate dissociation in the West Svalbard continental margin: timing, rates, and geological controls. J Geophys Res Solid Earth 118:22–38

Westbrook GK, Thatcher KE, Rohling EJ et al (2009) Escape of methane gas from the seabed along the West Spitsbergen continental margin. Geophys Res Lett 36(15):L15608

Winters WJ (2011) Physical and geotechnical properties of gas-hydrate-bearing sediment from offshore India and the northern Cascadia margin compared to other hydrate reservoirs. Paper presented at 7th International conference on gas hydrates, Edinburgh

Wu L, Grozic JLH (2008) Laboratory analysis of carbon dioxide hydrate-bearing sands. J Geotech Geoenviron Eng 134(4):547–550

Yun TS, Fratta D, Santamarina JC (2010) Hydrate-bearing sediments from the Krishna-Godavari Basin: physical characterization, pressure core testing, and scaled production monitoring. Energy Fuels 24:5972–5983. doi:10.1021/ef100821t

Part VII
Mass Transport Deposits in Modern and Outcrop Sedimentology

Chapter 44
Soft-Sediment Deformation Associated with Mass Transport Deposits of the Ainsa Basin (Spanish Pyrenees)

Christelle Butault, Jakub Fedorik, Francis Odonne, and Patrice Imbert

Abstract Mass-transport deposits (MTDs) are well developed on the slope of the Eocene Sobrarbe delta (Ainsa Basin, Spanish Pyrenees) and are studied in order to improve the understanding of soft sediment deformation in MTDs. The five examples illustrate the distinction between the extensional zone and the compressional domain. The upslope domain is illustrated by a set of three stacked nummulite-rich slid layers. They are deformed by load structures suggesting density inversion and sinking of nummulite gravel down into the underlying silty-shaly material. Post-sliding deformation is evidenced from small-scale roll-overs, neptunian dykes and syn-sedimentary normal faults. Slow extensive deformation has continued in the substratum of the scar after the main sliding episode. Soft-sediment deformation is less obvious in large displaced blocks where deformation is limited at the periphery of the blocks. Imbricate thrusts, formed on the side of a sliding layer, illustrate structures related to displacement. Striations on the thrust planes indicate that sliding occurred along the strike of the imbricate slices. The km scale Castellazo outcrop shows slump folds resting over a basal debris flow.

44.1 Introduction

Mass-Transport Deposits (MTDs) are commonly recognized in the field from the presence of deformed packages of sediment sandwiched between undeformed strata (Posamentier and Martinsen 2011). MTDs typically comprise an extensional headwall, upslope part, a translation domain and a compressive downslope toe domain (Martinsen and Bakken 1990), with or without distal overspill onto the seabed (Frey-Martinez et al. 2006).

C. Butault • J. Fedorik • F. Odonne (✉)
Université de Toulouse, GET-OMP, 14 av E. Belin, F31400 Toulouse, France
e-mail: francis.odonne@get.obs-mip.fr

P. Imbert
TOTAL, CSTJF, av Larribau, F64018 Pau Cedex, France

G. Lamarche et al. (eds.), *Submarine Mass Movements and their Consequences*, Advances in Natural and Technological Hazards Research 41,
DOI 10.1007/978-3-319-20979-1_44

Soft sediment deformation structures (Allen 1982) have been observed in numerous places on delta slopes, they show the variability of deformation styles from the most proximal to the most distal domains of MTDs. In that spirit, five outcrops from the San Vicente Formation (Ainsa basin) illustrate soft sediment deformation structures in MTDs.

44.2 Geological Setting of Ainsa Basin

The Eocene Ainsa Basin (Spanish Pyrenees) has been formed by flexural subsidence due to the southward propagation of thrusts in the South Pyrenees (Puigdefàbregas et al. 1991) (Fig. 44.1). The Basin has progressively developed towards the north-northwest from late Ypresian to Bartonian, guided by the growth of the Boltaña and Mediano lateral-thrust ramp anticlines. It is structured into a large NNW-SSE open syncline, the upper part of which is filled by the infilling of the Sobrarbe delta (Dreyer et al. 1999).

The Ainsa Basin fill is composed of marls and turbidites (Fig. 44.1). The main source area was located in the rising Pyrenean orogen on the NE margin. A minor source is the carbonate platform located on the SW margin. Numerous MTDs are observed (Pickering and Corregidor 2005), in the deeper part of the basin as well as on the coeval slope where numerous submarine gravitational scars of m- to km-scale are observed in each sequence.

44.3 Deformation in the Proximal Part of MTDs

The first example is near an upslope scar located west of Santa Maria de Buil. Soft sediment deformation structures can be observed in the infill of the S4 scar surface of Callot et al. (2009). The background sedimentation consists of a 5-m-thick siltstone succession with two interbedded nummulite-rich layers. This succession is replaced in the study area by three units showing a complex mix of silt and nummulites (Mateu-Vicens et al. 2012). It is exposed on both flanks of a small E-W ridge but all three units are about 1.5 times thicker on the northern flank.

The top of each unit is flat and it is made of poorly-sorted nummulite-rich gravel (Fig. 44.2). Unit 1, at the base, shows a complex mix of contorted silt blocks and nummulite-rich "blades" or wedges. Unit 2 is less deformed: the two types of sediments are organized as large scale load structures, nummulite-rich pockets hanging down, some of them attached to the layer by a thin oblique alignment of small nummulites. Mixing is less developed in Unit 3 but deformation is pervasive. The proportion of mixed sediment decreases both upward and laterally to the west. Abundant evidence for shearing in the most deformed part of the outcrop (Fig. 44.2b, c) suggests that intense shearing may also be responsible for loss of

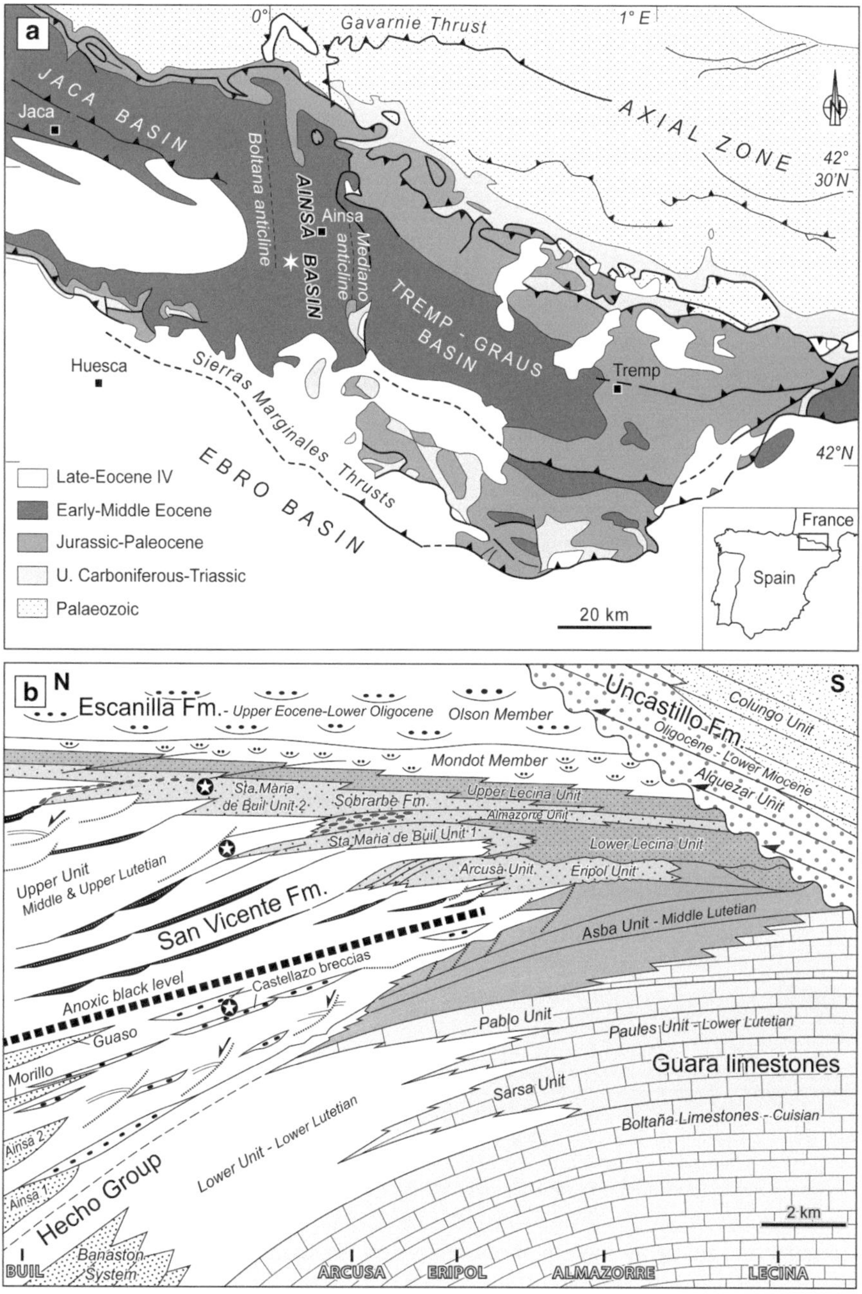

Fig. 44.1 (**a**) General location of the Ainsa Basin, Spanish Pyrenees (Redrawn from Dreyer et al. 1999). (**b**) Transition between Guara Limestone and San Vicente formations (Redrawn from Razin et al. 2010). *Stars* refer to the approximate position of: (1) Nummulite outcrop; (2) Extensional structures; (3) Castellazo MTD

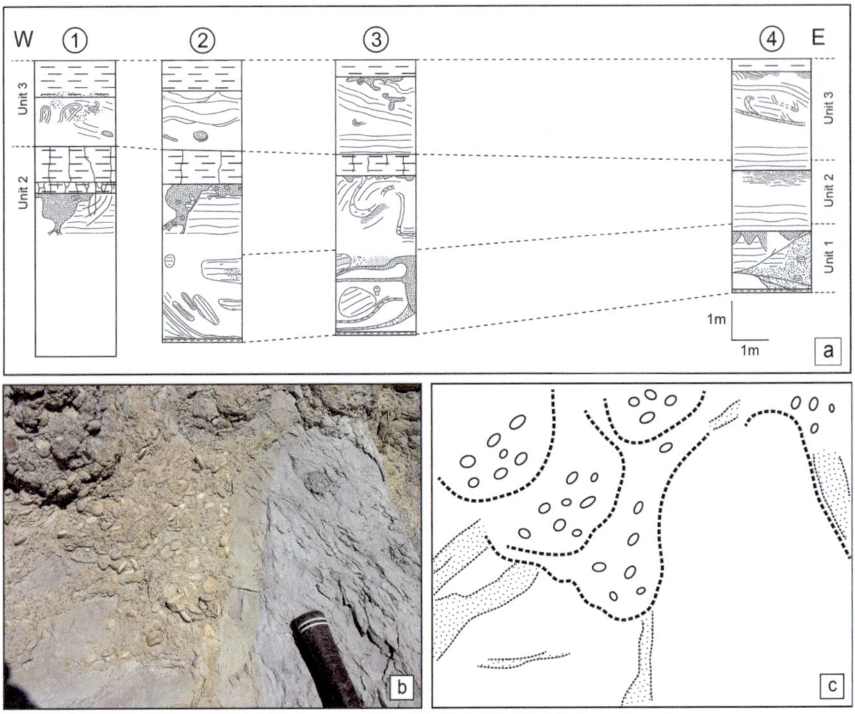

Fig. 44.2 Soft sediment deformation structures in nummulite-rich layers. (**a**) Line drawing of the south side of the outcrop showing the stack of the three slide units. (**b**, **c**) Picture and line drawing of load figures nummulite-rich sediment into silts and injection of small nummulite-rich sediments

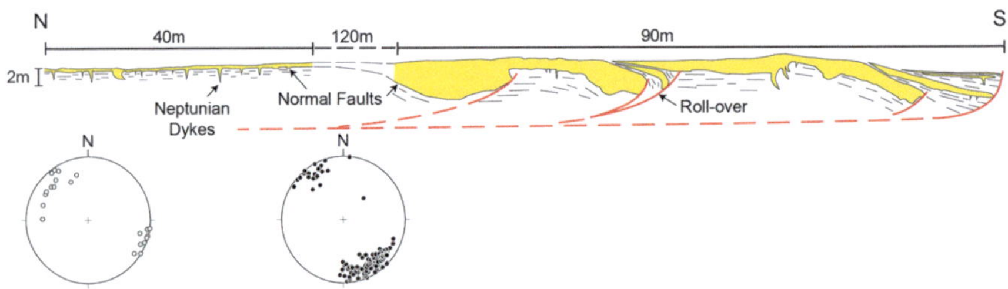

Fig. 44.3 Extensional structures along a scar surface. Neptunian dykes form first, then, they are tilted by m-scale rollover structures the base of which is cut by small scale normal faults. *Open circles* on stereo plots are poles of dykes and *black circles* are poles of small-scale normal faults

strength of sediment, thereby promoting soft sediment deformation fueled by density and viscosity contrasts (Owen 2003).

The second of our examples is a proximal MTD that consists of a small-scale extensional structure located in the sole of the S2 scar surface of Callot et al. (2009). The structures are exposed along 130 m of an almost continuous outcrop (Fig. 44.3). Twenty Neptunian dykes, numerous syn-sedimentary normal faults and some small-scale roll-overs are present in the upper part of the sole of the scar. Neptunian dykes are known as resulting of extensional tectonics in cohesive sediments (Montenat et al. 2007). In the present case they are preserved by coarse-grained sandstones filling the dykes. The same coarse-grained sandstones have filled small-scale half-graben-type structures developed on listric faults. The bases of the sandstone layers are affected by small-scale syn-sedimentary normal faults formed in non-cohesive sediments. When corrected for both the rotation due to the Mediano anticline and that due to the roll-over structures, all the structures show the same NW-SE extension direction (Fig. 44.3). Neptunian dykes form as vertical (Moretti and Sabato 2007) but here, the dykes are curved with their tip bent upslope, suggesting continued downslope shearing.

We propose the following to account for the observations: (1) Neptunian dykes are formed first by stretching, shortly after the major slide above evacuated the overburden; (2) next, small-scale rollover structures formed along m-scale normal faults and both the dykes and the half grabens are filled by sand; (3) finally, sliding of both the substratum and the first infill continues, as evidenced by the bending of the dykes and the small normal faults that affect the sandstone.

44.4 Deformation in the Median Part of a MTD

The third of our examples is a set of large blocks resedimented from the distal shelf domain downwards (Fig. 44.4). The thickness of individual blocks ranges from several meters to several tens of meters. All are made up of alternating marl and turbidite layers that provide depositional polarity. The larger blocks comprise up to 38 m of sedimentary series that have been displaced and tilted. The size of the blocks is compatible with that of scars that have mobilized up to 150 m of sedimentary series on the front of the Sobrarbe Delta (Dreyer et al. 1999). The deformation is concentrated at the periphery of every displaced block. Only some of the smaller blocks record a high degree of deformation (Fig. 44.4).

44.5 Deformation in the Distal Part of MTDs

The Castellazo MTD crops out along a 1 km section and it is about 10 m thick with an increase in thickness in the direction of sliding (Fig. 44.5). The MTD is composed of a stack of three units. Unit 1 at the base is a lenticular debris flow composed of 70 % of bioclastic grainstone debris embedded in 30 % of clayey matrix. Unit 2 is constituted of slumped alternations of silt/sand and claystone (Fig. 44.5a, b). Unit 3 is a homogeneous bioclastic grainstone showing occasional faint stratification. The direction of sliding was deduced from fold axes directions. Many anticlinal hinges are broken by clay injection (Fig. 44.5).

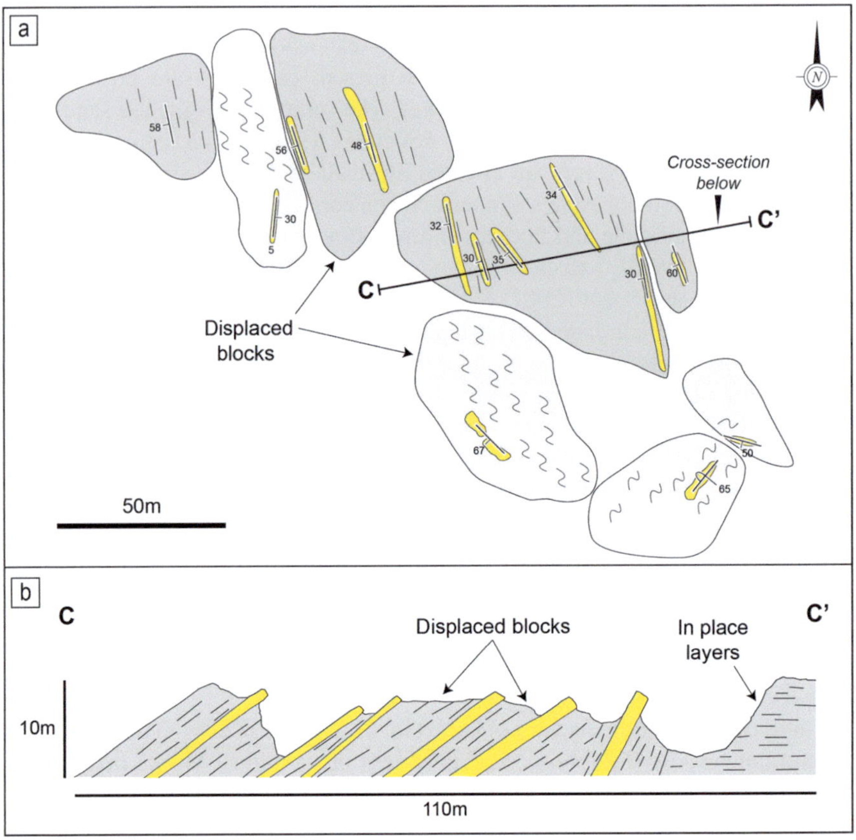

Fig. 44.4 Large displaced blocks. Blocks are rotated and tilted but deformation is concentrated at the border of each one

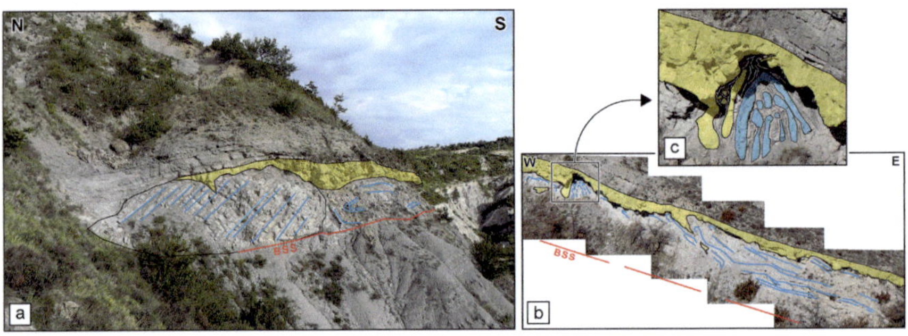

Fig. 44.5 Middle part of the Castellazo MTD. (**a**) Displaced block embedded in slumps, both capped by a grainstone unit; (**b**) Rear part showing slumps and grainstone unit; (**c**) Grainstone unit and broken hinge of slump are folded together. *BSS* stands for "bottom slide surface"

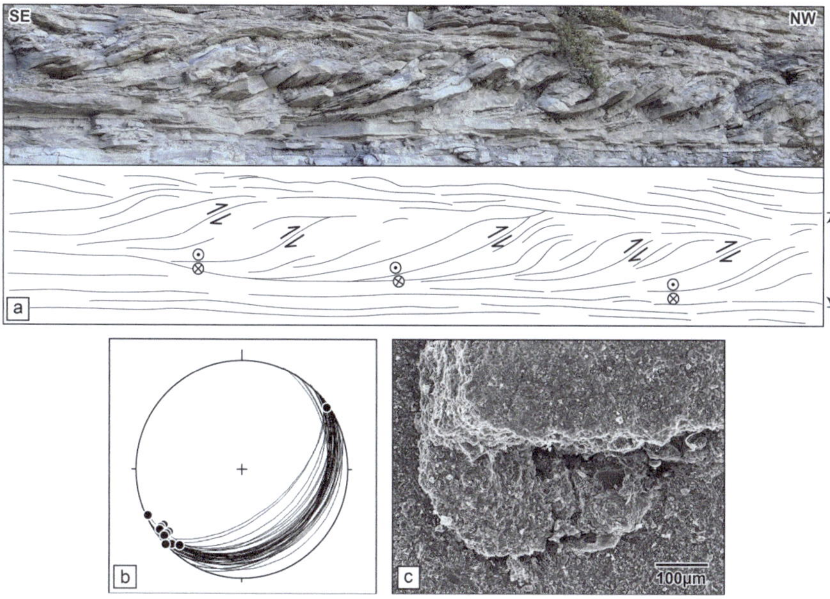

Fig. 44.6 Imbricate thrusts. (a) Picture and line drawing of the sheets; (b) Stereo of thrust sheets and striae; (c) Scanning electron microscope view of a stria, side part on the left and front part below

At the base of the grainstone unit, the bedding is parallel to folded beds of Unit 2, indicating that grainstone deposition started before slumping was completed (Fig. 44.5c); in addition, flat bedding at top indicates that grainstone deposition was still active when folding ceased. The basal shear surface ramps up in distal position, and the MTD reaches its maximum thickness just at the back of this ramp. The shortening of Unit 2 is maximum there and grainstone Unit 3 is missing. Unit 3 is interpreted to disappear by onlap on the peripheral bulge of the MTD, where the front of the transported mass emerged onto the seafloor. All these observations show that the MTD was likely emplaced over a short span of time.

Our fifth example consists of a set of imbricate thrusts which are located 1.5 km NW of Santa Maria de Buil in the distal part of the Sobrarbe delta.

The sedimentary succession is composed of alternating silt and marl layers. Two meter-thick levels are affected by imbricate thrusts. Thrust sheets are oriented N052 with a dip of 23° towards the SE (Fig. 44.6). The lower part of every sheet is stretched and thinned while numerous striae may be observed at the base of the sheets. Striae always strike parallel to the thrust sheets, about N050, and examination under scanning electron microscope shows that striae are formed by displacement without formation of any fiber or new mineral. All these observations indicate early sliding of soft sediment parallel to the sheets towards the NE.

These imbricate thrusts are reminiscent of the compressive thrusts described in the distal part of MTDs (Prior et al. 1984) but all indicate, here, the sliding of a single sedimentary layer that deforms when striking a resting boundary.

44.6 Discussion and Conclusions

The variability of deformation styles observed on these five examples illustrates first the classical distinction between extensional zone and compressional domain, with the first two examples (nummulite mix and rollover) belonging to the former and the last two (Castellazo and imbricate thrusts) to the latter. The third example with juxtaposed blocks may be considered to represent the translation domain.

The first two examples both come from the most upslope part of two MTDs, with limited interpreted displacement. Deformation in the nummulite mix example affected sediments that had never been buried by more than 1 m, resulting in a very soft style of deformation. On the contrary, the glide of rollover structures affected upon failure of the overburden sediments that had been buried by several tens of meters, resulting in a much more brittle style of deformation with faults and mode 1 fractures i.e. neptunian dykes formed in material having reached a higher degree of compaction and lithification.

The distal end examples show contrasting styles of deformation resulting from the amount of displacement undergone by the resedimented package: the frontally-confined MTD in Santa Maria de Buil shows a very regular pattern of brittle deformation corresponding to a very limited amount of displacement. On the contrary, the frontally-emergent MTD in Castellazo exhibits a much more complex pattern with folding and rotational deformation, as well as some disaggregation of the slumped mass. It would be classified as "frontally confined" (Frey-Martinez et al. 2006) and it is supposed to have undergone longer displacement.

Trigger mechanisms are difficult to assess. Nevertheless, bowl-shaped load figures of the nummulite mix outcrop may be interpreted as resulting from seismically triggered soft-sediment deformation structures (Moretti and Sabato 2007). Slumps are driven by gravity (Mastrogiacomo et al. 2012) and soft-sediment deformation structures are favoured by undercompaction of sediments. The high sedimentation rate of the Sobrarbe delta supports these two conditions.

Acknowledgements This work has been supported by TOTAL. This paper benefitted from constructive reviews by Dr. Massimo Moretti and Dr. Samantha Clarke.

References

Allen JRL (1982) Sedimentary structures, their character and physical basis. In: Developments in sedimentology, 30B, vol II, Chapter 9, Soft-sediment deformation structures. Elsevier, Amsterdam, pp 343–393, 663 p

Callot P, Odonne F, Debroas EJ, Maillard A, Dhont D, Basile C, Hoareau G (2009) Three-dimensional architecture of submarine slide surfaces and associated soft-sediment deformation in the Lutetian Sobrarbe deltaic complex (Ainsa, Spanish Pyrenees). Sedimentology 56:1226–1249

Dreyer T, Corregidor J, Arbues P, Puigdefàbregas C (1999) Architecture of the tectonically influenced Sobrarbe deltaic complex in the Ainsa Basin, northern Spain. Sediment Geol 127:127–169

Frey-Martínez J, Cartwright J, James D (2006) Frontally emergent vs. frontally confined submarine landslides: a 3D seismic characterization. Mar Pet Geol 23:585–604

Martinsen OJ, Bakken B (1990) Extensional and compressional zones in slumps and slides in the Namurian of County Clare, Ireland. J Geol Soc Lond 147:153–164

Mastrogiacomo G, Moretti M, Owen G, Spalluto L (2012) Tectonic triggering of slump sheets in the upper cretaceous carbonate succession of the Porto Selvaggio area (Salento peninsula, southern Italy): sedimentary tectonics in the Apulian carbonate platform. Sediment Geol 269–270:15–27

Mateu-Vicens G, Pomar L, Ferràndez-Cañadel C (2012) Nummulitic banks in the upper Lutetian 'Buil level', Ainsa basin, South central Pyrenean zone: the impact of internal waves. Sedimentology 59:527–552

Montenat C, Barrier P, Ott d'Estevenou P, Hibsch C (2007) Seismites: an attempt at critical analysis and classification. Sediment Geol 196:5–30

Moretti M, Sabato L (2007) Recognition of trigger mechanisms for soft-sediment deformation in the Pleistocene lacustrine deposits of the Sant'Arcangelo basin (Southern Italy): seismic shock vs. overloading. Sediment Geol 196:31–45

Owen G (2003) Load structures: gravity-driven sediment mobilization in the shallow subsurface. In: Van Rensbergen P, Hillis RR, Malman AJ, Morley CK (eds) Subsurface sediment mobilization, Special Publication 216. The Geological Society, London, pp 21–34

Pickering KT, Corregidor J (2005) Mass-transport complexes (MTCs) and tectonic control on basin-floor submarine fans, Middle Eocene, south Spanish Pyrenees. J Sed Res 75:761–783

Posamentier HW, Martinsen OJ (2011) The character and genesis of submarine mass-transport deposits: insights from outcrop and 3D seismic data. In: Shipp RC, Weimer P, Posamentier HW (eds) Mass-transport deposits in deepwater settings, Special Publication 96. SEPM Society for Sedimentary Geology, Tulsa, pp 7–38

Prior DB, Bornhold BD, Johns MW (1984) Depositional characteristics of a submarine debris flow. J Geol 92:707–727

Puigdefàbregas C, Muñoz JA, Verges J (1991) Thrusting and foreland basin evolution in the southern Pyrenees. In: McClay K (ed) Thrust tectonics. Chapman & Hall, London, pp 247–254

Razin P, Grélaud C, Odonne F, Debroas EJ (2010) Les systèmes de dépôt tertiaires de la bordure méridionale du bassin sud-pyrénéen dans la Sierra de Guara. RST Bordeaux 2010, Livret d'excursion ASF-AGSO. ISBN 2-907205-66-8

Chapter 45
Synsedimentary Tectonics and Mass Wasting Along the Alpine Margin in Liassic Time

Rüdiger Henrich

Abstract The transition from Triassic to Jurassic strata along the passive northern margin of the Neotethys elucidates typical drowning successions of huge Triassic carbonate platforms. Drowning goes parallel with synsedimentary block tectonics that was initiated by rifting at the southern margin border. An intriguing case study is exposed along the saw-cut wall sections of quarries around the village of Adnet close to Salzburg. The deeply submerged, inherited relief of a drowned reef mound gave rise to pronounced Liassic facies differentiation, i.e. (1) deposition of grey spiculitic cherty limestone and marl beds in the former shelf basin, and (2) red nodular limestones, and red condensed limestones rich in ammonites and Fe-Mn crusts over the slope and top of the former reef mound. Faulting, tilting and submarine erosion of Hettangian drift deposits at the lower slope was followed by repeated down-slope gliding, shearing and multiple opening of fissures with different generations of sediment infill. Renewed tectonics from Late Pliensbachian to Middle Toarcian created deep reaching vertical fissures and triggered multiple mass flow events. At the upper and middle slope the so-called Adnet Scheck breccia, which is a special debrite deeply eroding and incising into well-bedded condensed hemipelagic limestone strata, was deposited. Further down-slope the Scheck breccia evolves into more matrix-rich nodular breccias. Basin sections reveal intercalations of mudflow deposits and were affected by various magnitudes of sliding and mass flow events forming complex mass transport deposits.

45.1 Introduction and Methods

45.1.1 State of the Art and Open Questions

The transition from Triassic to Jurassic strata along the passive northern Alpine margin of the Neotethys elucidates typical drowning successions of Triassic carbonate platforms (1000s km^2 in area). Parts of the platforms were affected by

R. Henrich (✉)
MARUM and Faculty of Geosciences, University of Bremen, Bremen, Germany
e-mail: henrich@uni-bremen.de

G. Lamarche et al. (eds.), *Submarine Mass Movements and their Consequences*,
Advances in Natural and Technological Hazards Research 41,
DOI 10.1007/978-3-319-20979-1_45

karstification during the Rhaetian. During the early Liassic drowning event the platforms were covered by hemipelagic limestone and marlstone. At many localities drowning goes parallel with synsedimentary extensional tectonics that was initiated by rifting at the southern margin of the Vindelician continent and which finally led to the opening of the Penninic Ocean. This is manifested by diverse breccias and a spectrum of neptunian dykes (Henrich et al. 2014). However, there is still a lack of criteria to decipher the exact triggers, flow processes, erosion potential and geometries of different kinds of mass movements. These aspects will be investigated by highlighting the dynamic evolution of the passive Alpine margin at a selected case study area along saw-cut walls at the drowned Rhaetian reef mound at Adnet near Salzburg.

45.1.2 Stratigraphy and Sedimentology of Rhaetian and Liassic Strata

The study area is located in the eastern sector of the Tirolic nappe near Salzburg, where tectonic overprinting is less pronounced and the original paleogeographic setting can be better derived from outcrop descriptions. A southern plateau transits northward into two basins (Glasenbach and Trattberg), which are separated by deep swells (Fig. 45.1a).

Facies patterns developed during the initial phase of drowning and rapid sea level rise (Hettangian/Sinemurian) are closely related to the Rhaetian paleo-relief. Excellent outcrops of Triassic and Jurassic strata (Fig. 45.1b) exist around the village of Adnet. Below the Liassic strata alternating marl-limestone beds of

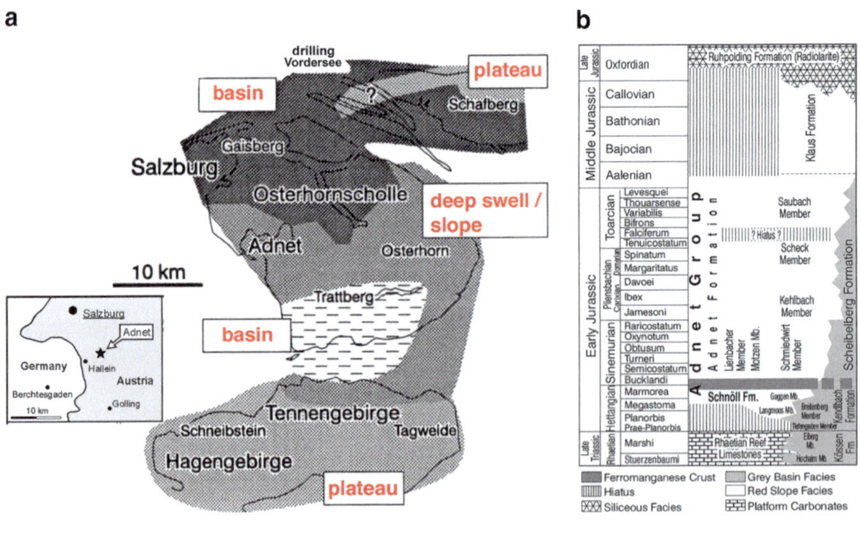

Fig. 45.1 (**a**) Liassic paleogeography, (**b**) stratigraphy (Modified from Böhm et al. 1995)

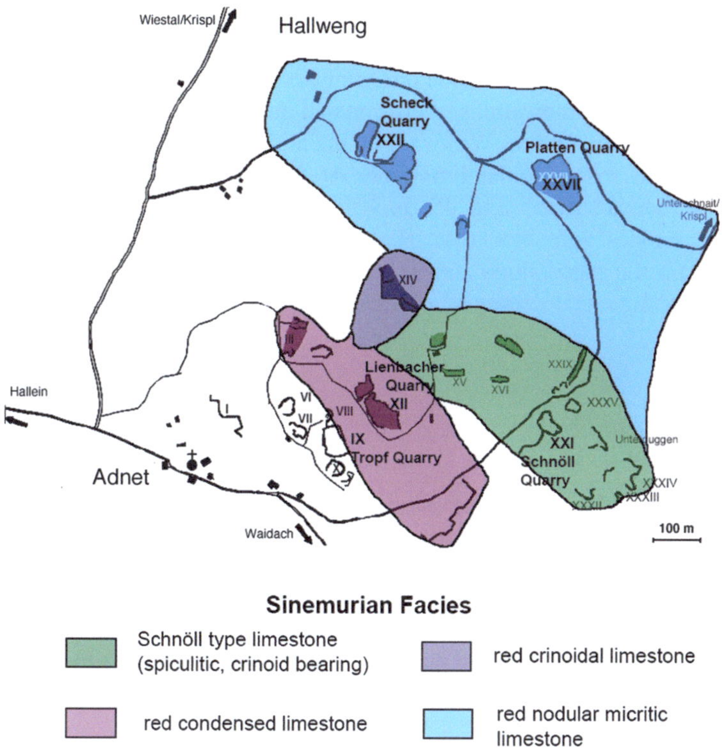

Fig. 45.2 Sinemurian lithologies around the drowned Adnet reef mound

Rhaetian age were deposited in a shelf basin setting (Kössen Formation), which interfingers locally with massive limestone of reef mounds and associated small platforms. The Adnet reef mound reveals a flat-toped to gently northeastwards sloping platform on which several patch reef cores were developed (Schäfer 1979). The reef mound transited northeastwards into deposits of the Kössen basin. At Tropf quarry karst surfaces and cavities filled with lagoonal sediments during the next transgression dissect the reef core (Bernecker et al. 1999). Total elevation of the mound was 50–80 m above the gently inclined seafloor (Böhm et al. 1999). During the end of the Triassic reef-drowning event the deeply submerged, inherited relief gave rise to pronounced Liassic facies differentiation.

Grey spiculitic cherty limestone and marl beds were deposited in the former shelf basin areas (Kendlbach and Scheibelberg formations Fig. 45.1), whereas red nodular limestones, and red condensed limestones rich in ammonites and often intersected with Fe-Mn crusts as well as crinoidal limestones covered the slope and top areas of the former reef mound (Schnöll and Adnet formations Fig. 45.2). Cross-bedded mixed carbonate sand deposits of the Kendlbach Formation presumably represent contourites deposited at the slope-basin transition. Red condensed limestones with numerous Fe-Mn crusts, some of them rich in ammonites, were deposited on top of the reef and in the upper slope region (Fig. 45.2). The

"Marmorea crust" is a thick Fe-Mn crust that marks the top of the Hettangian (Böhm et al. 1995, 1999). Middle slope sections were covered by sponge/crinoidal limestone of the Schnöll Formation including several well-preserved in situ sponge layers and abundant indications of erosion and hard ground formation are present (Delecat and Reitner 2005). Westward on the middle slope the Schnöll Formation transitions into red crinoidal limestone. At the lower slope the typical Adnet Limestone "sensu strictu" (Fig. 45.1b) was deposited (nodular-bedded intraclastic wacke/packstones with bioclasts). Higher up in the section this subfacies was deposited on the paleo-slope and its transition into the adjacent basins. This is documenting a general deepening of the entire area in the Upper Liassic period.

45.1.3 Methods

Jurassic strata exposed in outcrops along road cuts in Glasenbach gorge and saw-cut wall sections of the "marmor" quarries around the village of Adnet were studied. Structural and stratigraphic sections of these outcrops have been mapped and measured in detail. Line drawings of the major geometrical relationships between different sedimentary units were documented. New concepts regarding the nature of the stratigraphic successions and the role of mass wasting events in the evolution of the basin are proposed in this work.

45.2 Results

Debrites and breccias (few cm to a maximum of 40 m thick) are intercalated at various stratigraphic levels with the Liassic red nodular limestone of the Adnet Formation as well as with the grey spiculitic limestones of the Scheiblberg Formation (Fig. 45.1). The oldest reported debrite units are of early Carxian age but erosion and breccia formation continued throughout the early Toarcian (Böhm et al. 1999). However, shedding of breccias quickly decreased during the late Toarcian. Depending on matrix content and stage of consolidation and cementation three basic types of breccias can be distinguished as described in the following sections.

45.2.1 Description Scheck Breccia/Matrix-Poor Debrite: Upper Slope

This unit was deposited during the Late Pliensbachian to Early Toarcian (Böhm et al. 1995). At the type locality a layered Scheck Breccia sits un-conformably on top of Sinemurian nodular Adnet limestone. Three layers can be distinguished from base to top (Fig. 45.3): (1) A 50 cm thick clast-rich muddy debrite, (2) a 4 m thick

Fig. 45.3 Scheck breccia at the type locality (see text for further explanation)

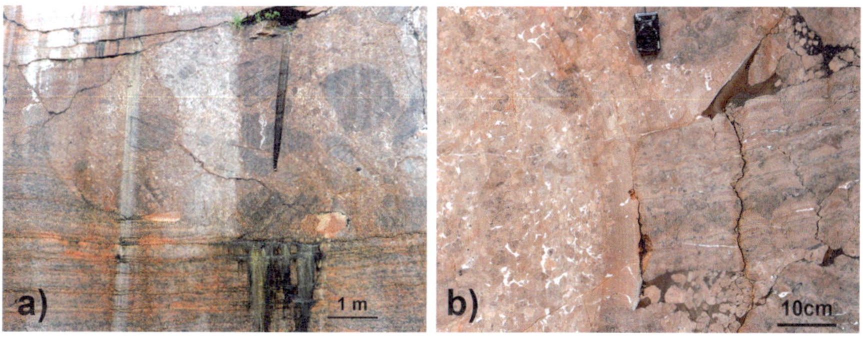

Fig. 45.4 Scoured incision of Scheck breccia (Lienbacher quarry – see text explanation)

main breccia body consisting of a cemented carpet of rounded to sub-rounded clasts, and (3) a 2 m thick unit with blocks of nodular and condensed Adnet limestone. In Lienbacher quarry a scour deeply eroded the underlying nodular Adnet limestone (Fig. 45.5a). The infilling sediments of the scour reveal the presence of metre-scale blocks that are chaotically arranged and embedded within mud-poor matrix. Some of the blocks are angular and contain well-developed deep-water stromatolite structures (Böhm and Brachert 1993), thick Fe-Mn crusts and sediment impregnations (Fig. 45.5b). Previous work in this area (Böhm et al. 1995, 1999) concluded that the Scheck breccia was formed under the influence of strong bottom currents. However we suggest that since the thickness of the Scheck breccia exceeds 3–4 m the explanation claiming outwashing of fines appears rather unlikely. Lack of a muddy matrix indicates that sliding, and shearing of nodular and condensed Adnet limestone, and subsequent disintegration into a chaotic clast carpet formed the Scheck breccia. The blocks at the top were incorporated into the flow either by collapse of overhanging strata at the scour sites (Fig. 45.4) or by upward movement during down-slope movement. Since all layers are transitional the entire succession represents a single flow event, in which the basal muddy debrite (Fig. 45.4) served as glide plane.

Whitish early submarine cements and sparse internal sediments indicate early lithification presumably induced by rapid percolation of seawater driven by strong bottom currents.

45.2.2 Description Nodular Breccia/Clast-Rich Debrites: Middle to Lower Slope

The presence of "nodular breccias" comprising a spectrum of clast-rich debrites is common on the lower slope of the Adnet seamount. At Platten quarry the breccia overlies nodular Adnet limestone. The contact between these units is erosional and has a step-like geometry (Fig. 45.5a). At the base of the breccia body a mud-rich layer with small clasts is observed. In the central core of the breccia disintegration and rotational movement of metre-scale limestone blocks can be observed. These blocks have a thick Fe-Mn crust at the top making it possible to infer that the down-slope flow of the central mass occurred through clockwise rotations of clasts (Fig. 45.5b).

45.2.3 Description Matrix-Rich Pebbly Mudstones: Lower Slope to Basin

Several metre thick red pebbly mudstone beds are intercalated into lower slope and basin sections. Some of them show flat scour-like incision into underlying strata. Abundant limestone nodules and cemented micro-nodules are embedded in a red

Fig. 45.5 Scheck breccia on the middle to lower slope erosively overlies nodular limestone of the Adnet formation (Platten quarry – see text for further explanation)

Fig. 45.6 (**a**) "Knollenbreccie" (Glasenbach gorge). (**b**) Tilted block of drift deposits showing onlap of sponge layers (Schnöll quarry – see text for further explanation)

strongly remobilized mud matrix. There are some intervals that can be 10 m thick ("Knollenbreccie") that contain chunks of grey marls and marly limestone (Fig. 45.6a). Blocks of red nodular limestone bearing Fe-Mn coated ammonites and crinoidal limestone suggest that these blocks were transported for quite some distance and that they were presumably derived from slope settings on the Adnet seamount. In addition, blocks and clasts of greyish basin spiculitic limestone have been incorporated as rip-up clast during emplacement of the mass flow in the basin.

45.2.4 Sediment Infill into Neptunian Dykes

First evidence for syn-sedimentary tectonics around the Adnet seamount is documented in Schnöll quarry, where cross-bedded drift deposits (Kendlbach Formation) were faulted, tilted, eroded at the top, and subsequently onlaped by sponge layers (Fig. 45.6b). Above the sponge-rich section red condensed limestones intercalated with Fe-Mn crusts show truncation by fissures that extend sub-parallel to the bedding planes and that these units are rimed by cement crusts and filled with internal sediments. These features indicate multiple renewed brittle fracturing and down-slope gliding of slope deposits (Fig. 45.7a, b).

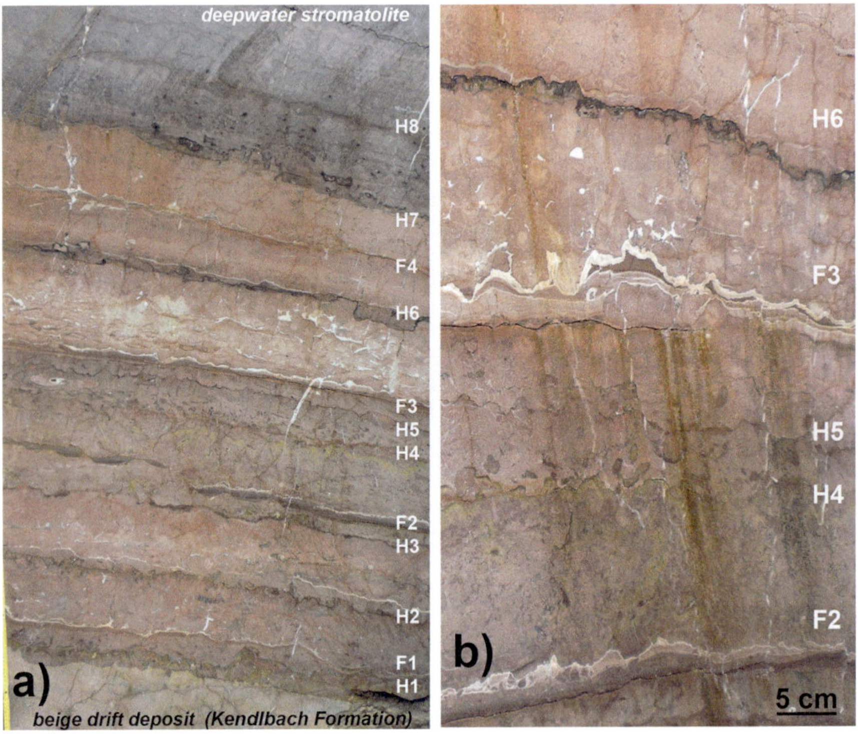

Fig. 45.7 Set of horizontal fissures (*F*) in consolidated slope deposits with Fe-Mn hardgrounds (*H*) (Schnöll quarry)

45.3 Discussion: Triggers, Flow Patterns, and Mass Flows Geometry

The studied mass flow deposits comprise mudflows and debris flows. Their structural and textural features provide important new insights on flow processes and properties. Typical mudflows in lower slope and basin settings are the red pebbly mudstones, which can be classified as a coarse end member of mudflow in transition to debris flow. Because of their high matrix contents they reveal turbulent flow properties. Hydro-planning at the flow-nose accelerates flow speed resulting in rather long run-out distances. In contrast, the studied debrites reveal very low mud contents and high proportions of semilithified clast and lithified rock fragments. This results in extraordinary stiff plastic flow properties with ubiquitous internal friction and shearing of clasts decelerating flow speed. Such features are related to rapid and efficient early cementation processes and thus are typical for carbonate debrites. In contrast, siliciclastic deposits are generally not affected by early lithification and thus can attain higher flow speed and runout distances. “Freezing” in the central part is a common feature of many debris flows. During

Fig. 45.8 Y-shaped tectonic fissure dissecting Rhaetian reef limestone (Tropf quarry – see text for further explanation)

freezing internal shear stress overcomes matrix strength and thus the core of the flow moves like a rigid plug as observed in the Scheck quarry (Fig. 45.3). Other important features of debrites are scouring as instructively illustrated by the stepwise incision at Platten quarry (Fig. 45.5b). Blocks in the main breccia body indicate rotational movement, whereas imbrication of nodular clasts at the base refers to linear motion in the lubrication layer. Furthermore, scour-like incisions of Scheck type flows reveal a high erosion capacity (Lienbacher quarry -Fig. 45.4). Here, the breccia contains deepwater stromatolite blocks. In-situ layers of these stromatolites are found 3 m below the scour indicating erosion of at least 10 m. At Scheck quarry more than 15 m were eroded based on biostratigraphic dating of the in-situ deposits below the breccia (Böhm et al. 1999).

Mass transport deposits (MTDs) at the Jurassic Alpine margin include a spectacular unit at Rote Wand near Waidring as well as several MTDs intercalated in basin deposits (Glasenbach gorge – Henrich et al. 2014). Identification of triggers for mass wasting events (MWE) is often complex. The studied MWE, however, clearly reveal an intimate association to synsedimentary tectonics. Several generations of tectonic fissures (some reaching 50–60 m downward into the reef mound) indicate renewed tectonism from Late Hettangian to Late Toarcian. Interesting aspects about the interaction of synsedimentary tectonics and MWEs can be deduced from sediment infill into the fissures. Sediment infill into a y-shaped fissure truncating the Rhaetian reef mound core testifies that mobilisation and sucking-in from overlying in situ sediments occurred (Fig. 45.8). In contrast, younger vertical fissures truncating nodular Adnet limestone show infill of retextured muds that were dispersed showcasing sharp edged angular rock clasts and Fe-Mn hard ground fragments (Fig. 45.9). This type of deformation indicates that brittle fracturing of rocks followed by downward injection of material under high pressure. This could happen if debris flow tongues overrode open tectonic fissures. At Platten quarry, dying-out flexures and a reverse fault are observed directly below the nodular breccia body indicating that tectonically triggered down-slope dislocation of sediments occurred contemporaneous with breccia emplacement (Fig. 45.10).

Fig. 45.9 Vertical fissures dissecting nodular Adnet limestone in the main wall of Lienbach quarry. Sediment infill of these fissures comprises greyish marls containing fragments of Fe-Mn crusts

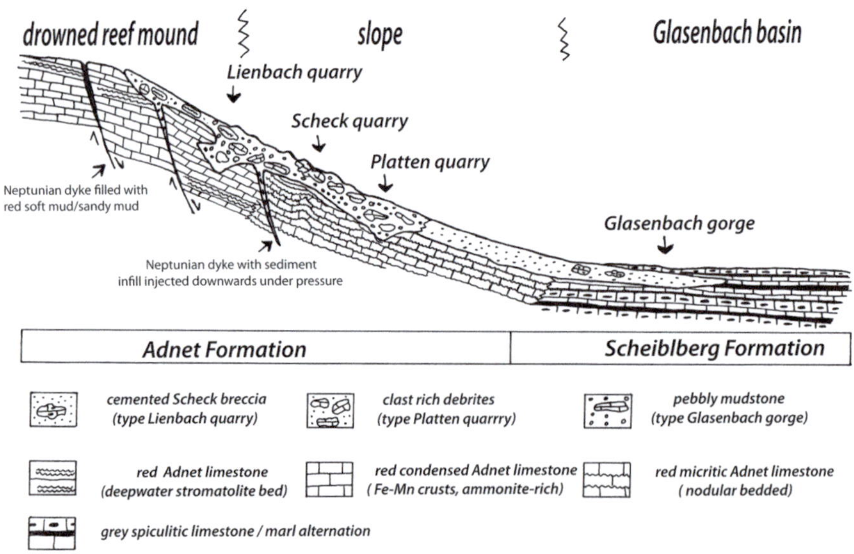

Fig. 45.10 Distribution of breccias and tectonic features along the slope of the drowned Adnet reef mound towards the Glasenbach basin

45.4 Conclusions

The down-slope transition of Scheck breccias into nodular breccias and finally into pebbly mudstones indicates a drastic change in the flow properties. Cohesive Scheck flows are limited in lateral and down-slope extent, whereas the turbulent pebbly mudstone flows reveal long run-out distances due to hydro-planning. Scoured debris tongues of Scheck type testify deep erosion. Our observations show that collapse of scour sidewalls is a new mechanism to explain enrichment of semi-consolidated sediment and lithified rock blocks at top of the flow. The main driving force in Scheck type flows is a thin muddy layer at the base, whereas the main body tends to freeze and pluck the flow.

Acknowledgements Comments by K. H. Baumann, T. Bickert and J. Kuss as well as final drawings by Claudia Henrich are gratefully acknowledged. Very instructive reviews by Lorena Moscardelli and Priska Schäfer significantly improved the manuscript.

References

Bernecker M, Weidlich O, Flügel E (1999) Response of Triassic reef coral communities to sea-level fluctuations, storms and sedimentation: evidence from a spectacular outcrop (Adnet, Austria). Facies 40:229–280

Böhm F, Brachert TC (1993) Deep-water stromatolites and Frutexites MASLOV from the Early and Middle Jurassic of S-Germany and Austria. Facies 28:145–168

Böhm F, Dommergues JL, Meister C (1995) Breccias of the Adnet formation: indicators of a Mid-Liassic tectonic event in the Northern Calcareous Alps (Salzburg/Austria). Geol Rundsch 84:272–286

Böhm F, Ebli O, Krystyn L et al (1999) Fauna, stratigraphy and deposition environment of the Hettangian-Sinemurian (Early Jurassic) of Adnet (Salzburg, Austria). Abh Geol BA 56:143–271

Delecat S, Reitner J (2005) Sponge communities from the Lower Liassic of Adnet (Northern Calcareous Alps, Austria). Facies 51:399–418

Henrich R, Baumann KH, Bickert T (2014) New concepts on mass wasting phenomena at passive and active margins of the Alpine Tethys: famous classical outcrops in the Berchtesgaden – Salzburg Alps revisited. Part A: Jurassic slide/debrite complexes triggered by syn-sedimentary block faulting. In: Krastel S et al (eds) Submarine mass movements and their consequences, Adv Nat Tech Haz Res 37. Springer, Dordrecht, pp 661–674

Schäfer P (1979) Fazielle Entwicklung und palökologische Zonierung zweier obertriadischer Riffstrukturen in den nördlichen Kalkalpen (Oberrhät-Riff-Kalke, Salzburg). Facies 1:3–245

Chapter 46
Meso-Scale Kinematic Indicators in Exhumed Mass Transport Deposits: Definitions and Implications

Kei Ogata, Gian Andrea Pini, Andrea Festa, Željko Pogačnik, and Claudio Corrado Lucente

Abstract In this study we combine observations and analytical data from large-scale (10–100s of m-thick and 100 m^2-extensive), siliciclastic and carbonate MTD/MTCs belonging to the Oligocene – Miocene foredeep and wedge-top successions of the Northern Apennines and the Paleocene – Eocene Friuli basin of the northwestern Dinarides (Italy and Slovenia), to discuss the deformation processes critical to the emplacement of submarine landslides. We focus on the identification of meso-scale structures, used as diagnostic kinematic indicators of local paleo-transport directions. These structures, represented by linear-planar and complex-shaped elements such as tabular shear zones and detached slump-type folds, are the product of ductile-plastic deformation developed at relatively low-confining pressure that involves water-saturated, un- to poorly-lithified sediments, along with liquefaction/fluidization processes. Their final appearance is thus mainly controlled by the mechanical-rheological behavior of deformed sediments, and eventually by tectonic fabrics inherited from deeper structural levels of deformation. Due to this parallelism these structures have been termed and classified accordingly. They reflect strain partitioning due to differential movements within the slide mass, which is in turn controlled by the overall landslide typology. Due to the parallelism with classified tectonic structures and structural associations, we have thus redefined and classified accordingly meso-scale kinematic indicators in ancient MTD/MTCs.

K. Ogata (✉)
Dipartimento di Fisica e Scienze della Terra, Università di Parma, Parma, Italy
e-mail: kei.ogata@gmail.com; kei.ogata@unipr.it

G.A. Pini
Dipartimento di Matematica e Geoscienze, Università di Trieste, Trieste, Italy

A. Festa
Dipartimento di Scienze della Terra, Università di Torino, Torino, Italy

Ž. Pogačnik
Salonit Anhovo, Building Materials, Joint-Stock Co., Deskle, Slovenia

C.C. Lucente
Servizio Tecnico di Bacino Romagna, Regione Emilia-Romagna, Italy

G. Lamarche et al. (eds.), *Submarine Mass Movements and their Consequences*,
Advances in Natural and Technological Hazards Research 41,
DOI 10.1007/978-3-319-20979-1_46

46.1 Introduction

Geophysical imaging of modern seafloor surface and subsurface provide the gross morphology, areal extent and overall internal character of single mass transport deposits (MTDs) and composite mass transport complex (MTCs), outlining their importance in practical terms, such as offshore hydrocarbon exploration/production and coastal geohazard assessment/evaluation (see e.g. Kawamura et al. 2012). Conversely outcrop studies on exhumed "fossil" analogues, allow analyses on the internal deformation mechanisms from micro- to map- (i.e. seismic) scale (see e.g. Lucente and Pini 2003; Alonso et al. 2006; Yamamoto et al. 2009; Ogata et al. 2012). Although scientific drilling provides crucial insights on the anatomy of modern MTD (see e.g. Strasser et al. 2011), the available data are still too punctual, and costly to provide a comprehensive source of information, especially on meso-scale structures, which record the main trace of mass transport-related deformation processes.

We here focus on the meso-scale, syn-sedimentary structures, which spatial-geometric relationships are considered to record the strain direction achieved during and/or immediately after the slide body deposition. Such structures represent important kinematic indicators that can be integrated to unravel the local paleo-transport direction and the internal deformation processes of the slide mass. In this paper we analyze some ancient, seismic-scale MTDs/MTCs analogues, exhumed in the Oligocene – Miocene foredeep and wedge-top sedimentary successions of the Northern Apennines, and the Paleocene – Eocene Friuli basin of the northwestern Dinarides (Fig. 46.1a–c; Italy and Slovenia), expanding punctual observations to a broader and more general picture. The systematic occurrence of the described features in basically all the investigated siliciclastic and carbonate examples justifies this effort.

46.2 Case Studies

The Northern Apennines case studies are the Epiligurian Specchio unit (Ogata et al. 2012) and the Marnoso-Arenacea Formation MTDs (Lucente and Pini 2003; Lucente 2004). The Epiligurian Specchio unit, Rupelian in age, is a MTC composed of at least two different stacked MTDs (i.e., intra- and extra-basinal lower and upper ones, respectively), deposited in an intra-slope, wedge-top basinal system developed atop the evolving proto-Apenninic orogenic wedge. The Miocene Marnoso-Arenacea Formation MTDs are emplaced at different stratigraphic levels within the migrating Northern Apennines foredeep system. These intra-basinal units are sometimes characterized by the occurrence of extra-basinal lithologies (Fig. 46.1d, e).

The northwestern Dinarides case studies are the Paleocene Friuli Basin MTDs (Ogata et al. 2014), which represent carbonate platform collapses from the basin margins (either from the outer foreland and the inner deformation front) into a narrow foredeep basin plain characterized by mixed siliciclastic-carbonate background sedimentation (Fig. 46.1f, g).

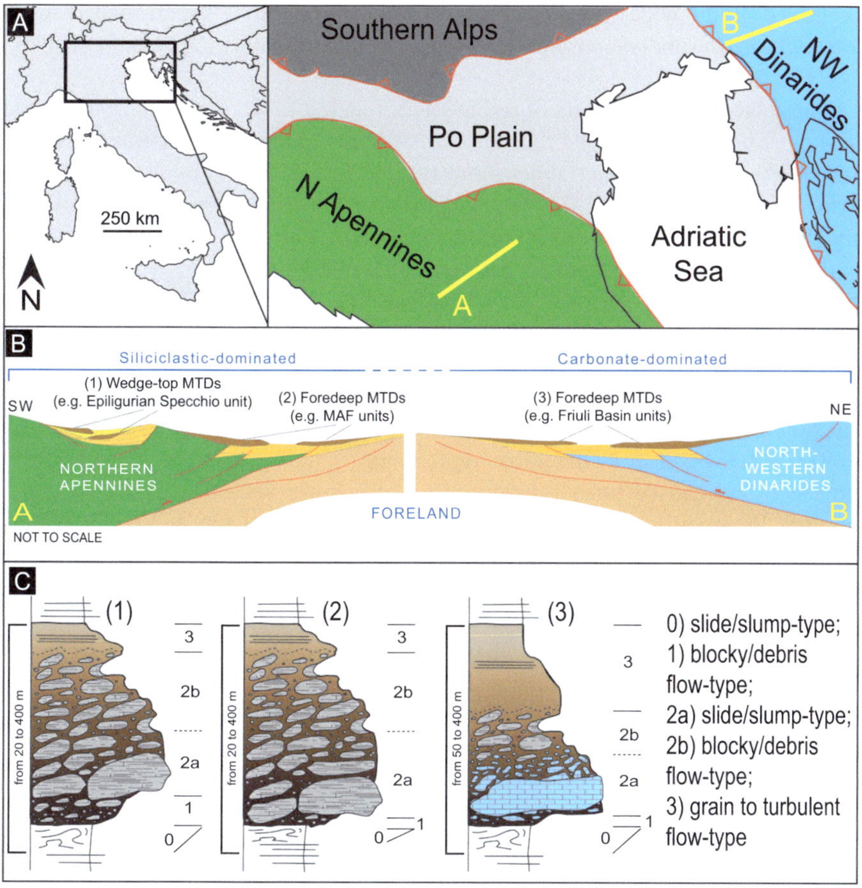

Fig. 46.1 (**a**) Simplified map showing the location of schematic cross sections presented in **c**, reconstructed for the Northern Apennines and northwestern Dinarides orogenic systems. (**b**) Conceptual stratigraphic logs (not to scale) summarizing the internal subdivisions and characterizing facies associations of the investigated MTDs (see text for discussion). (**c**) Schematic cross sections representing the inferred general physiographic setting of the investigated MTDs during the Cenozoic

46.3 Baseline MTD Architecture and Facies Associations

The investigated units share comparable stratigraphy, with laterally/vertically alternating subdivisions characterized by two main specific facies associations: blocky/debris flow- and slide/slump-types (Pini et al. 2012; Ogata et al. 2012). The first implies a viscous-cohesive flow-like deformation accommodated by the yield strength of an hyper-concentrated, fine-grained matrix, whereas the second testifies a coherent, *en-masse* translation of preserved slide blocks driven by localized shearing and folding. In both cases fluid overpressure plays a crucial role as suggested by the widespread occurrence of fluidization/liquefaction-related structures. The main subdivisions are briefly described in the following, from base to top (see Fig. 46.1b).

The subdivision 0 consists of sediments belonging to the substratum, deformed in situ or short-travelled, and thus not formally part of the main, far-travelled slide mass. The characterizing slide/slump-type deformation is likely due to the dynamic/static loading of the overlying slide mass during the syn-depositional and/or early post-depositional phases. Its thickness is highly variable being related to the size, type and down-slope position of the overlying MTD.

The subdivision 1 varies in thickness from few centimeters to tens of meters, characterized by a pervasive highly-sheared, matrix-dominated, blocky/debris flow-type facies association with the dominance of a fluidization/liquefaction-related fabric. This interval, which achieves most of the deformation, represents the "lubricating" layer lowering the basal friction of the major slide mass.

The subdivision 2 is the most represented in terms of vertical and lateral extension and can be further subdivided into a lower sub-unit 2a and an upper sub-unit 2b, sometimes laterally inter-fingering. The subdivision 2a is characterized by block-dominated, slide/slump-type facies associations, represented by meters- to hundreds of meters-sized bedded rafts (i.e. slide blocks), mainly of proximal/marginal origin (i.e. originally located close or within the main evacuation area), in close contact. The subdivision 2b is characterized by matrix-dominated, blocky/debris flow-type facies associations, interpreted as the product of a cohesive-viscous flow enclosing isolated slide block (1–10s meters wide) and comprising mainly distal/basinal origin (i.e. originally located in the transfer zone and incorporated into the slide mass). The relative thickness of both these sub-units generally varies from tens to hundreds of meters.

The subdivision 3 comprises a meter- to tens of meters-thick, grain- to turbulent flow-type facies with evidence of ponding/reflections, representing diluted sediment flow developed atop the moving slide mass. This subdivision is significantly thicker in the carbonate-dominated MTDs.

46.4 Meso-Scale Structures and Kinematic Indicators

Eight types of meso-scale structures (Fig. 46.2a) characterize different parts of the investigated MTDs (Fig. 46.2b), appearing unrelated to syn-kinematic fracturing and mineralization. The absence of metamorphic signature and brittle deformation supports ductile/plastic, soft-sediment deformation mechanisms rather than tectonic processes, as also confirmed by microscopic analyses highlighting independent particulate flow with minor or no grain breakage.

The striking similarities with ductile structures documented in deeper metamorphic rocks allow us to adopt the same descriptive, non-genetic terminology used in structural geology (e.g. Passchier and Trouw 2005). In geometrical terms, these structures can be treated as lines and planes, and their spatial arrangement can be

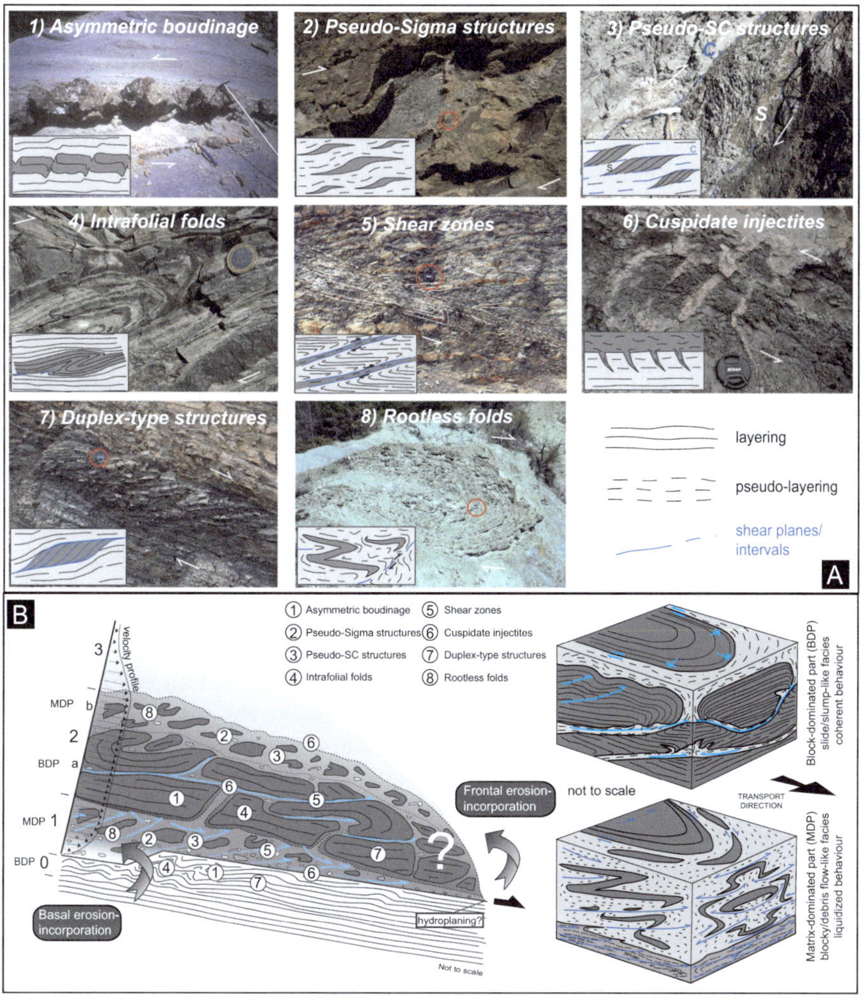

Fig. 46.2 (**a**) Summary of the structures discussed in this work. For a detailed description see Sect. 46.4 (Modified from Ogata et al. 2014). (**b**) Conceptual representation of an evolving slide body with identification and labelling of the main internal subdivisions, facies associations and mesoscale structures, and their typical distribution in relation to the main transport direction. Inferred processes are also highlighted (After Ogata et al. 2012, 2014).

used to infer the local strain, as for the classical plotting and interpretation of structural tectonic data.

Asymmetric boudinage records coupled pure/simple shear-related, layer-parallel extension of poorly-consolidated sandstone-mudstone interbeds with high rheological contrast. Along with a progressive amount of deformation, generally it consists of “pinching-and-swelling” layers to detached and aligned phacoidal/lozenge-shaped elements, whose asymmetry defines the associated stress direction.

These structures may also rework sheet-shaped structures such as sedimentary dykes and matrix injectites (see below) if their elongation direction matches with the local shear stress field.

Pseudo-Sigma structures resemble the typical Sigma structures of the ductile tectonic deformation, consisting of millimeters- to meters-sized, asymmetric, sigma-shaped plastic objects deformed by shearing-related viscous flow, and thus recording directional information. Such elements comprise cohesive lithologies, usually different from the surrounding matrix (e.g. mudstones), coming from the disaggregation of the transported and eroded bedded sediments.

As for the preceding ones, **pseudo-SC structures**, show strong morphological correspondence with the typical tectonic counterparts, comprising discrete zones (mm-to cm-thick shear bands) of high shear strain (i.e. C-surfaces) that bound systematic, pervasive, sigmoidal foliation planes (i.e. S-surfaces). These surfaces are represented by disaggregation-, deformation- and compaction-bands.

Intrafolial folds, centimeters- to meters-sized, are observed to internally deform the layered intervals (i.e. laminae, beds), with the main reference surfaces being traceable laterally into unfolded parts. Typical soft sediment deformation-related thickening of hinge zones is usual observed, showing a sheath-type geometry in three-dimensions. Since these overturned folds commonly form to adsorb the slip related to the upward propagation of a blind and flat-ramp shear fault, an overall fault-propagation fold mechanism is inferred.

Ductile shear zones, low-angled, ranging in thickness from millimeters to meters, with lengths from centimeters to tens of meters are observed throughout the entire slide body. Due to the general absence of geometric references, displacements are usually difficult to resolve, although they seem to increase according to their size. As for the classic tectonic fault architecture, damage zones roughly symmetrical to a core zone accommodating the largest displacement characterize these structures, recording the transition from relatively undeformed to completely disrupted sediments (i.e. matrix) is achieved. The rate and amount of accommodated deformation appear to be proportional to the relative matrix abundance, which reflects the degree of mechanically-induced lithological mixing. No clear slip surfaces are observed in the core zones, and deformation seems distributed through its entire thickness, with the local development of disaggregation-deformation-compaction bands, sometimes arranged in swarms. On the other hand, damage zones show plastically deformed lithologies where the original sedimentary features (e.g. layering) are still partly recognizable. The relative movement is due to the evolutionary phase, type of mass transport process, and lateral/vertical position within the slide mass, with generalized normal faulting in the earlier evolutionary phases and in the upper/up-slope parts, and thrusting in the later phases and the lower/down-slope parts. Mutual reworking and deformation (e.g. re-folding, bending) of these shear zones are commonly observed and interpreted as related to syn- and post-emplacement phases.

Pipe- (i.e. linear) and sheet-, and wedge-shaped (i.e. planar) **cuspidate injections** of matrix are observed in different positions within the slide mass, ranging in thickness from millimeters to meters, and in length from centimeters to tens of meters. They are diagnostic of an original over-pressurized, liquefied/fluidized state of the sedimentary matrix, and thus able to actively penetrate slide blocks via hydrofracturing-type processes and/or passively flow into the low-pressure zone created by their progressive deformation. Such cuspidate injections are developed in randomly-distributed, pervasive, web-like systems or systematically arranged along the bounding surfaces between matrix-rich and coherent lithologies (e.g. blocks' margins, basal contact). In this latter case they provide directional information on the shear sense and the relative movements of the overpressured matrix flow.

Duplex-type structures, centimeter to tens of meters-sized, are made up by the localized imbrication of isolated blocks along flat-ramp to sigmoidal shear surfaces which all tie together, and are bounded above and below by major shear surfaces merging laterally to a single one. Isolated blocks constituting the duplex-type structure belong from a multilayer, progressively stacked and imbricated to accommodate the shortening of the hanging wall, causing consequent formation of roughly symmetrical folds.

Rootless folds, in contrast to intrafolial ones described above, appear as completely detached, isolated slump-type fold hinges dispersed within a fine-grained matrix. The three-dimensional shape of these structures usually resemble strongly asymmetric sheath folds, typically isoclinal, with marked thickening of the hinge zone, and characterized by sedimentary matrix injections along the outer-arc zones, limbs and into the cores. Axial planes, usually low-angled with dominant up-slope dips, become locally steeper toward the inferred margins of the slide mass, reflecting structural confinement effects. These structures, along with the intrafolial counterparts, are important kinematic indicators for the interpretation of paleo-transport directions and several methods have been proposed in the inherent literature to analyze the related structural data (e.g. Lucente and Pini 2003; Ogata et al. 2014).

46.5 Discussion and Conclusions

Mixed pure- and simple-shear deformation mechanisms due to the coupled, cyclic action of the dynamic/static loading and the differential movements of the slide mass and its internal components produce a variety of asymmetrical structures ranging from the hand specimen- to the outcrop-scale (Fig. 46.2a). Due to their shape, spatial arrangement and geometric relationships such structures can be used as standalone kinematic indicators to record the local differential movements between the internal slide parts, or in combination for a robust interpretation of the general paleo-transport directions (Fig. 46.2b). These structures, specifically distributed within the slide body, are interpreted as products of high-rated strain, soft sediment deformation developed at low confining pressure (i.e. superficial conditions) involving undrained, water-saturated, poorly- to un-consolidated sediment, both failed and eroded from the overridden seafloor. These factors, combined with the high strain partitioning, favor localized fluid over-pressure conditions, with consequent fluidization/liquefaction phenomena. The identification of these structures, and integrated outcrop analogue studies in general, allow the correct interpretation of the general and local slide kinematics, bridging the gap between the small-scale, punctual datasets provided drill cores/borehole logging and large-scale, 2D/3D geophysical imaging.

Acknowledgements We are grateful to the associate editor Christof Mueller, and the reviewers Yuzuru Yamamoto and Juan Luis Alonso for their helpfulness and their constructive feedbacks. This work is funded by the FRA2013 research grants of the Università di Trieste and MIUR PRIN grant 2010AZR98L_002.

References

Alonso JL, Marcos A, Suàrez A (2006) Structure and organization of the Porma mélange: progressive denudation of a submarine nappe toe by gravitational collapse. Am J Sci 306:32–65

Kawamura K, Sasaki T, Kanamatsu Y, Sakaguchi A, Ogawa Y (2012) Large submarine landslides in the Japan trench: a new scenario for additional tsunami generation. Geophys Res Lett 39: L05308. doi:10.1029/2011GL050661

Lucente CC (2004) Topography and palaeogeographic evolution of a middle Miocene foredeep basin plain (Northern Apennines, Italy). Sediment Geol 170:107–134

Lucente CC, Pini GA (2003) Anatomy and emplacement mechanism of a large submarine slide within the Miocene foredeep in the Northern Apennines, Italy: a field perspective. Am J Sci 303:565–602

Ogata K, Mutti E, Tinterri R, Pini GA (2012) Mass transport-related stratal disruption within sedimentary mélanges. Tectonophysics 568–569:185–199

Ogata K, Pogačnik Ž, Pini GA, Tunis G, Festa A, Camerlenghi A, Rebesco M (2014) The carbonate mass transport deposits of the Paleogene Julian-Slovenian Basin (Italy/Slovenia): internal anatomy and inferred genetic processes. Mar Geol 356:88–110

Passchier CW, Trouw RA (2005) Microtectonics. Springer, Berlin

Pini GA, Ogata K, Camerlenghi A, Festa A, Lucente CC, Codegone G (2012) Sedimentary mélanges and fossil mass-transport complexes: a key for better understanding submarine mass movements? In: Yamada Y et al (eds) Submarine mass movements and their consequences: advances in natural and technological hazards research, vol 31. Springer, Berlin

Strasser M, Moore GF, Kimura G, Kopf AJ, Underwood MB, Guo J, Screaton EJ (2011) Slumping and mass transport deposition in the Nankai fore arc: evidence from IODP drilling and 3-D reflection seismic data. Geochem Geophys Geosyst 12:1–24. doi:10.1029/2010GC003431

Yamamoto Y, Nidaira M, Ohta Y, Ogawa Y (2009) Formation of chaotic rock units during primary accretion processes: examples from the Miura-Boso accretionary complex, central Japan. Island Arc 18:496–512

Chapter 47
Morphodynamics of Supercritical Turbidity Currents in the Channel-Lobe Transition Zone

George Postma, David C. Hoyal, Vitor Abreu, Matthieu J.B. Cartigny, Timothy Demko, Juan J. Fedele, Kick Kleverlaan, and Keriann H. Pederson

Abstract This study aims to resolve process-facies links at both bed and environmental scales for the channel lobe transition zone (CLTZ). Data comes from existing experimental and modern CLTZ studies and from new outcrop studies. The experiments show that the CLTZ architecture of supercritical turbidity currents is complex and different from their counterparts where flows are subcritical throughout. Supercritical CLTZ's are characterised by erosive channels formed by supercritical turbidity currents, by offset stacked lobes deposited from subcritical turbidity currents and by hydraulic jump related mouth bar deposits and upslope onlapping backfill deposits at the down slope end of the transition zone. Erosive channels and backfill features can be resolved by high resolution seismic data, yet evidence for supercritical flow must come from facies analysis of core data. Outcrop examples of the CLTZ from the Tabernas submarine fan (SE Spain) and the Llorenç del Munt deep-water delta slope (N. Spain) are used to establish such links between seismic scale architecture and facies recognised in cores. The outcrops described here were mapped as transition zone, and show 100 m sized, spoon-shaped scours filled with sediment containing sandy to gravelly backsets up to 4 m in height. Their facies and architecture is indicative of deposition by hydraulic jumps, can be recognized from cores, and is a good proxy for further predicting CLTZ architecture constructed by supercritical turbidity currents.

G. Postma (✉)
Faculty of Geosciences, Utrecht University, PO Box 80.021, 3508TA Utrecht, The Netherlands
e-mail: g.postma@uu.nl

D.C. Hoyal • V. Abreu • T. Demko • J.J. Fedele • K.H. Pederson
ExxonMobil Upstream Research Company, Houston, TX, USA

M.J.B. Cartigny
National Oceanography Centre, University of Southampton Waterfront Campus, Southampton SO14 3ZH, UK

K. Kleverlaan
"Hoop op Welvaart", Zeeburgerdijk 585, 1095 AE Amsterdam, The Netherlands

G. Lamarche et al. (eds.), *Submarine Mass Movements and their Consequences*, Advances in Natural and Technological Hazards Research 41,
DOI 10.1007/978-3-319-20979-1_47

47.1 Introduction

The morphodynamics of turbidity currents at the transition from confined channels to unconfined lobes, commonly referred to as the channel-lobe transition zone (CLTZ), is still a poorly explored part of the submarine turbidite fan environment. Consequently, making inferences about sand body architecture and facies in cored CLTZ successions remains uncertain.

Features of the CLTZ recognized from the modern seafloor vary. Turbidite deposits are commonly patchily distributed and extensively reworked (Wynn et al. 2002). Lobes can have complex architecture with numerous onlap, toplap and downlap surfaces built by numerous events (Gervais et al. 2006). Both seismic and sea floor data reveal that lobes may contain many small channels indicating that gravity flows remain confined at least for their basal part across most of the lobe for some of the turbidite events (Gervais et al. 2006). Under sediment bypass conditions, the CLTZ is characterized by abundant erosional features, including isolated spoon- and chevron-shaped scours up to 20 m deep, 2 km wide, and 2.5 km long, partially filled with coarse-grained material that is trapped in the scour (e.g. MacDonald et al. 2011a, their Table 1). The CLTZ of the Monterey Canyon exhibits elongated sand fingers and shallow channels filled with sand (Klaucke et al. 2004). In contrast, CLTZ's on the slope east of Corsica (Gervais et al. 2004) are characterised by (1) small proximal isolated lobes (PILs) connected with local slope gullies and deposited at the slope break ($3^\circ \rightarrow 1.5^\circ$), and (2) somewhat larger composite midfan lobes (CMLs) on slopes of 1.5–1° that are connected with erosive and leveed channels issuing from large canyons. The CMLs were probably river fed during sea-level low stands and have more complex architecture than the PILs (see further Deptuck et al. 2008).

Sediment bypass features of inferred CLTZ similar to those described by Wynn et al. (2002) and Deptuck et al. (2008) have occasionally been inferred from ancient turbidite successions. In particular the bypass zones in the turbidite series of the Hecho Basin (Pyrenees, see Mutti and Normark 1987) and the Ross Formation (Ireland, see Chapin et al. 1994; Macdonald et al. 2011b), the latter displaying chevron shaped scours (Elliott 2000) are well known.

It is hypothesized in this paper that some of the CLTZ depositional complexities can be significantly clarified by considering the flow domain of turbidity currents. Recent experimental fan and field studies suggest that sub- and supercritical turbidity currents each have a characteristic architecture (Cantelli et al. 2011; Fernandez et al. 2014; Hoyal and Sheets 2009; Hamilton et al. 2015) and facies associations (Postma and Cartigny 2014). The objectives of our CLTZ studies include highlighting the relevance of the critical densimetric Froude number for facies and architectural development in the CLTZ.

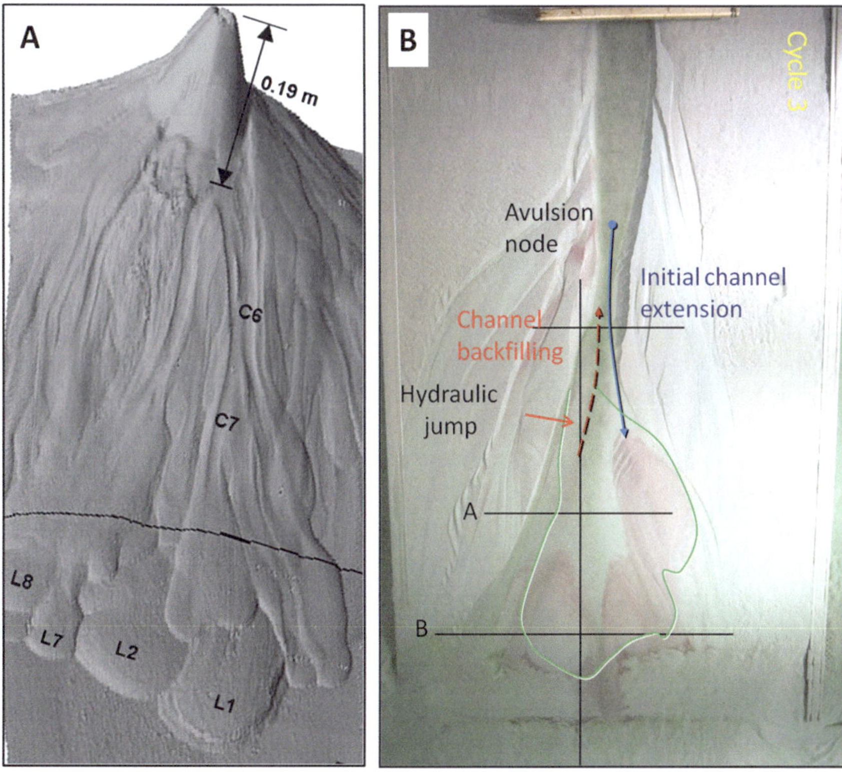

Fig. 47.1 Experimental studies of channel lobe systems. Table length in both **a** and **b** is 3 m: (**a**) Subcritical channel lobe system of Fernandez et al. (2014) with 5–45 μm silica particle slurries of 5–10 % volume concentration over 4° slope. The 0.19 measure is the height drop onto the slope; (**b**) supercritical channel lobe system of Hoyal and Sheets (2009) and Hamilton et al. (2015) with saline underflow over 10° slope and transporting 180–250 μm crushed plastic particles over a similar substrate. Cross-sections **a** and **b** are given in Fig. 47.5a, b

47.2 Experimental Analogues of Channelized Submarine Fans

Experiments with *subcritical* kaolinite suspensions showed numerous channels formed by levee building, which avulsed near the steep cone as the fan evolved (Fig. 47.1a, Cantelli et al. 2011; Fernandez et al. 2014). Lobes formed once the channels crossed the break of slope, and channel-lobe tracts were typically maintained over several flows. Compensational deposition occurred at a time scale that was longer than that of the individual runs (Fernandez et al. 2014). The resulting architecture was mainly aggradational with channel levee elements above and lobes below the slope break.

In experiments with *supercritical* saline density flows, channels were not formed by levee building, but through erosion, thus forming a ‘bypass channel’ (Hoyal and Sheets 2009; Hamilton et al. 2015). Eventually, an extending channel stagnates and aggrades a mouth bar ahead of itself thus creating an obstacle, which progressively slows down the supercritical flow to subcritical through a hydraulic jump. Backsets develop in the scour just upstream of the mouth bar, backfilling the channel. It is interesting to note that channel stagnation occurred in the experiments on the sloping portion of the domain, i.e., a slope break did not play a role in restraining the system as it did in the subcritical case. Eventually, the backfill of the channel leads to avulsion and the creation of a new channel (Fig. 47.1b).

The conditions for forcing a hydraulic jump on a slope depends on the flow's densimetric Froude number and thus on its density and velocity, which is a function of slope. If the slope is mild then a downstream subcritical normal flow boundary condition is likely to force the jump. If the slope is steep, then it is likely that the jump is forced by a stepped obstruction (see Hamilton et al. 2015). Flow density plays a role in the required height of the obstruction: dilute flows with high velocities require a very large stepped obstruction, many times the flow thickness, to trigger a jump, while dense flows with relatively low velocities require very small steps (Hamilton et al. 2015).

47.3 Facies Associations

The difference in morphodynamics of the sub- and supercritical CLTZ experiments is striking. The subcritical CLTZ is mainly progradational and aggradational, forming tabular, compensationally stacked channels and lobes, while the supercritical CLTZ shows bypass zones characterized by erosion and hydraulic jump related deposits. The stacking of channel and lobes in the supercritical case is complex showing channel erosion, offset stacked lobes and backfilling patterns including hydraulic jump deposits. Each cycle (see Fig. 47.1b) starts with avulsion and channel elongation, followed by onlap (mouth bar), flow stagnation (hydraulic jump deposits) and bypass features and ends with channel abandonment (mud cover). Hydraulic jump related deposits can be easily recognised by their backset architecture filling an asymmetrical scour. Facies associations related to a hydraulic jump is typically Ta, Tb4 – Tb3a (see Fig. 47.2, Postma and Cartigny 2014; Russell and Arnott 2003) indicating high concentration and little shear of the flow during deposition. Ta facies is often in combination with soft sediment deformation structures in particular flames at the base of the bed (Postma et al. 2009) and angular mud chips ripped up locally from the substrate (Postma et al. 2014).

47.4 Field Observations

Superb examples of supercritical CLTZ turbidites have been found in the late Miocene submarine fan deposits of the Tabernas basin, mapped by Kleverlaan (1989a). The inferred transition zone stretches over several kilometres from the

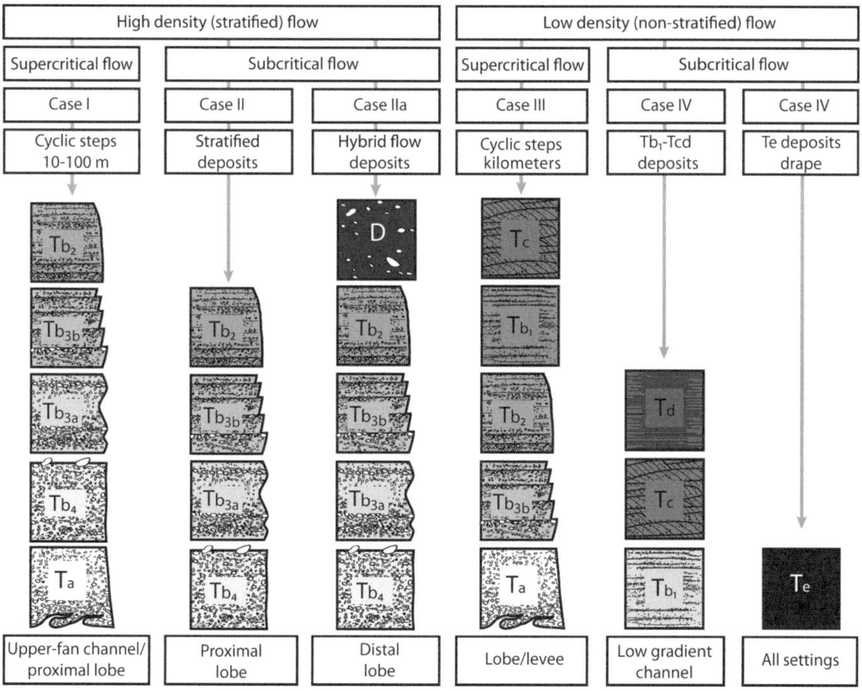

Fig. 47.2 Facies associations for the channel lobe transition zone, which can be read from cores (From Postma and Cartigny 2014). T unit legend: Ta = coarse-tail normally graded with flame structures at the base; Tb4 = mainly massive and inversely graded; Tb3a = crude stratification; Tb3b = spaced stratification; Tb2 = <0.5 cm planar stratification; Tb1 = plane bed lamination with parting lineation; Tc = ripples; Td = plane bed of the lower flow regime and Te = hemipelagic fall out; D = debrite

canyon ('Buho Canyon') to the lobe deposits (Kleverlaan 1989b). Its facies is characterized by large, asymmetric gravel lenses enveloped in bioturbated mud (Fig. 47.3). The gravel units are characterized by upcurrent dipping crude stratification (Tb4 and Tb3a) and structureless gravel nests (Ta). This up-flow dipping crude stratification forms backset beds. They are seen locally to grade upslope into a structureless gravel, with vertical clast fabrics. Imbricated clasts dip slightly steeper than the upslope dipping cross stratifications pointing to flow traction, hence distinguishing the deposit from a debris flow origin. The top of the gravel units is sharp (grain size change from gravel to mud) and wavy possibly pointing to a bypassing supercritical flow. Clusters of outsized clasts drape the wavy surface that is overlain by bioturbated, thin sand and mud layers.

Sedimentation units in the transition zone comprise stacks of up to several meters thickness of the above described gravel lenses, and alternating dm's thick sandy (Tb4-2) and homogenised muddy turbidite beds. The stacking has an aggradational (little vertical change in bedding and grain size) to progradational (overall coarsening) character. The lobe consist of turbidite sheets with rare Ta facies, some Tb4-1 and abundant Tcde (see Cartigny 2012).

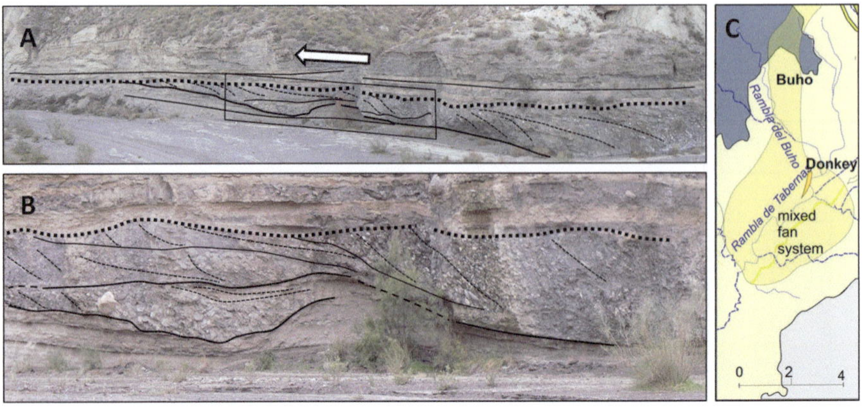

Fig. 47.3 Supercritical CLTZ bypass zone of the mixed sand-mud turbidite fan (Tabernas Basin); (**a**) Amalgamated spoon-shaped scours filled with cross stratified gravel (backsets); Paleo-flow is from right to the left and obliquely into the outcrop. Outcrop width on picture is about 200 m; (**b**) Detail of **a** showing large, trough-shaped sets of gravelly backsets with well defined wavy erosive surface at the top of the bed; Note the multiple event infill of the scour by variously stacked backsets, and chaotic massive fill at the base. Outcrop height is about 5.5 m, (**c**) Map of the area North of the village Tabernas showing main features of mapped CLTZ time slice in *yellow* (see Kleverlaan 1989a, b for details). South is to the left

Another example comes from the deep-water fan deltas of the Vilomara section of Llorenç del Munt fan delta complex, mapped and studied for its sequence stratigraphy by López Blanco et al. (2000a, b, Fig. 47.4). At the base of the Vilomara section, sandy turbidite beds are seen onlapping an erosive surface on the muddy slope of the delta. The outcrop has no obvious break in slope (Fig. 47.4). Facies that includes Ta and Tb4-1 are believed to indicate supercritical flow (see Fig. 47.2). These turbidite beds are truncated by a thick sand deposit up to 3 m thick at its centre, which has a large backset and a distinct bypass surface at its top (dotted white line in Fig. 47.4). Note the change in relief (c. 0,80 m) between the upstream section and the backset.

47.5 Discussion and Conclusion

The diagnostic features for supercritical flow controlled CLTZ that emerged from the experimental studies of Hoyal and Sheets (2009) and Hamilton et al. (2015) are (1) extending erosive channels that feed and prograde the lobe and (2) a hydraulic jump that heralds the backfilling of the channel and the development of a mouth bar. The latter process is likely marked by scours filled with steep backset bedding, but also by stacks of more gently upslope dipping and onlapping beds. The proximal isolated lobes (PILs) and the complex midfan lobes (CMLs) of the Golo fan on the lower slope east of Corsica (Gervais et al. 2004; Deptuck et al. 2008) might be a

Fig. 47.4 Example of scour filling sediment body on a continuous sloping surface (*white solid line*) of the Llorenç del Munt fan delta (Vilomara section, N. Spain). The internal architecture of the body consists of variously oriented backsets (not clear from photograph), which are interpreted to have formed by hydraulic jumps in successive flow events. Beds within the backsets show abundant soft sediment deformation structures. The *dotted white line* marks a bypass surface with characteristic grain size jump. Detailed logs illustrate the grain size variation which is up to granules (G) in the thicker units and down to very fine lower sand (vF) in the tabular turbidite beds. South is to the *right*

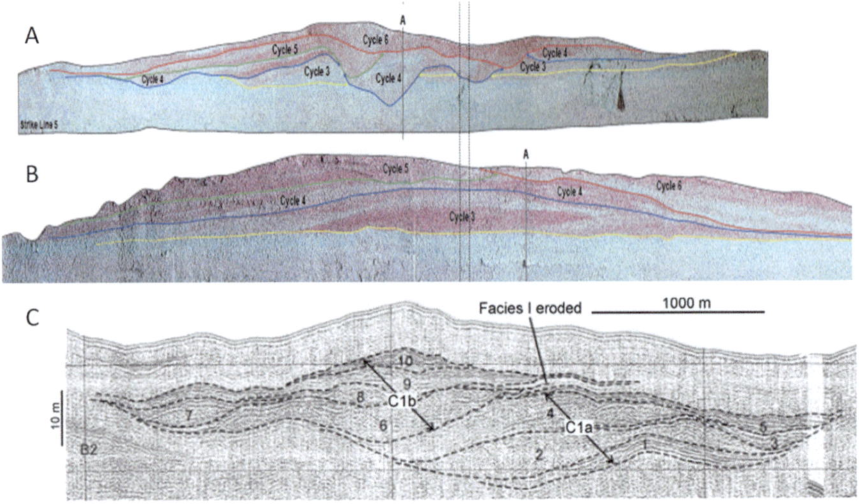

Fig. 47.5 Cross-sections lines *A* and *B* (see Fig. 47.1b) of the supercritical CLTZ experiments (Hoyal and Sheets 2009, with thanks to Roger Bloch). Vertical exaggeration is about 1.8; *C*. Cross-section from the C1 complex midfan lobe (Deptuck et al. 2008)

nice analogue for supercritical CLTZ systems. Both lobe systems are deposited on relatively steep slopes in excess of 1° favouring development of supercritical turbidity currents (Sequeiros 2012). In case of the PILs most abrupt deposition is taking place within a few kilometres of the gully mouth near the slope break. Seismic facies of the proximal parts of the lobe is typical for thick massive and probably amalgamated sand beds onlapping up slope, while cores through the proximal front of the bar show nearly 5 m of continuous massive coarse-grained sands with clay chips probably consisting of a series of amalgamated beds. These lobes may have formed when the outer shelf was inundated during the last transgression and, hence, may have been supplied from inefficient surge like flows triggered when sandy shoals or bars near the heads of gullies failed (see Deptuck et al. 2008). The architecture of CMLs in Golo fan is much more complex and is analogous to the complexity observed in the fans formed in the supercritical flow experiments. Cross sections through these fan lobes have combinations of channel erosion and offset stacking of lobes similar to those observed in experimental studies (Fig. 47.5). Cores and seismic data reveal predominant massive and amalgamated beds in the proximal lobe regions. Sustained mixed-load hyperpycnal flows were most likely to reach the fan during periods of low sea-level stands, but surge-like flows were probably also triggered after a temporary period of storage in the canyon head (Deptuck et al. 2008).

Facies and architecture of the outcrop examples shown by Figs. 47.3 and 47.4 support an interpretation that the sediment bodies were formed in a hydraulic jump zone (see also Russell and Arnott 2003; Postma and Cartigny 2014). Their size, however, is an order of magnitude smaller than the features described from the PILs and CMLs of the east Corsican slope. Hence, these scour fills are not the mouth bar themselves, but rather represent features of the erosive channel floor upslope of the mouth bar in the transition zone similar to the scours found by Wynn et al. (2002), Chapin et al. (1994), and Macdonald et al. (2011b).

In conclusion, by combining experimental studies with studies on small modern fans predictive models emerge for facies and process studies on outcrops in the CLTZ. We infer that the Froude number of turbidity currents is an important parameter to consider when dealing with the CLTZ environments, and perhaps most promising is that this parameter has characteristic facies that can be identified from core.

References

Cantelli A, Pirmez C, Johnson S, Parker G (2011) Morphodynamic and stratigraphic evolution of self-channelized subaqueous fans emplaced by turbidity currents. J Sediment Res 81:233–247

Cartigny MJB (2012) Morphodynamics of supercritical high-density turbidity currents. Utrecht Studies in Earth Sciences, PhD thesis, 153pp

Chapin MA, Davies P, Gibson JL, Pettingill HS (1994) Reservoir architecture of turbidite sheet sandstones in laterally extensive outcrops, Ross formation, western Ireland. In: Weimer P et al (eds) Submarine fans and turbidite systems: GCSSEPM Foundation 15th annual research conference, London, pp 53–68

Deptuck ME, Piper DJW, Savoye B, Gervais A (2008) Dimensions and architecture of late Pleistocene submarine lobes off the northern margin of East Corsica. Sedimentology 55:869–898

Elliott T (2000) Mega flute erosion surfaces and the initiation of turbidite channels. Geology 28:119–122

Fernandez RL, Cantelli A, Pirmez C, Sequeiros O, Parker G (2014) Growth patterns of subaqueous depositional channel lobe systems developed over a basement with a downdip break in slope: laboratory experiments. J Sediment Res 84:168–182

Gervais A, Savoye B, Piper DJW, Mulder T, Cremer M, Pichevin L (2004) Present morphology and depositional architecture of a sandy confined submarine system: the Golo turbidite system (eastern margin of Corsica). In: Lomas SA, Joseph P (eds) Confined turbidite systems, Geological Society special publication, 222. Geological Society, London, pp 59–89

Gervais A, Savoye B, Mulder T, Gonthier E (2006) Sandy modern turbidite lobes: a new insight from high resolution seismic data. Mar Petrol Geol 23(2006):485–502

Hamilton PB, Strom KB, Hoyal DCJD (2015) Hydraulic and sediment transport properties of autogenic avulsion cycles on submarine fans with supercritical distributaries. J Geophys Res doi:10.1002/2014JF003414, in press

Hoyal DCJD, Sheets BA (2009) Hydraulic jumps as controls on the evolution of distributary channel networks on experimental submarine fans. In: The 33rd international association of Hydraulic Research Congress, London

Klaucke I, Masson DG, Kenyon NH, Gardner JV (2004) Sedimentary processes of the lower Monterey Fan channel and channel-mouth lobe. Mar Geol 206:181–198

Kleverlaan K (1989a) Neogene history of the Tabernas Basin (SE Spain) and its Tortonian submarine fan development. Geol Mijnbouw 68(4):421–432

Kleverlaan K (1989b) 3 distinctive feeder-lobe systems within one time slice of the Tortonian Tabernas Fan, SE Spain. Sedimentology 36:25–45

López-Blanco M, Marzo M, Burbank DW, Vergés J, Roca E, Anadon P, Piña J (2000a) Tectonic and climatic controls on the development of foreland fan deltas: Montserrat and Sant Llorenç del Munt systems (Middle Eocene, Ebro Basin, NE Spain). Sed Geol 138:17–39

López-Blanco M, Marzo M, Piña J (2000b) Transgressive-regressive sequence hierarchy of foreland, fan-delta clastic wedges (Montserrat and Sant Llorenç del Munt, Middle Eocene, Ebro Basin, NE Spain). Sed Geol 138:41–69

Macdonald HA, Wynn RB, Huvenne VAI, Peakall J, Masson DG, Weaver PPE, McPhail SD (2011a) New insights into the morphology, fill and remarkable longevity (>0.2 m.y.) of modern deep-water erosional scours along the northeast Atlantic margin. Geosphere 7:845–867. doi:10.1130/GES00611.1

Macdonald HA, Peakall J, Wignall PB, Best J (2011b) Sedimentation in deep-sea lobe-elements: implications for the origin of thickening-upward sequences. J Geol Soc 168:319–331. doi:10.1144/0016-76492010-036

Mutti E, Normark WR (1987) Comparing examples of modern and ancient turbidite systems: problems and concepts. In: Leggett JK, Zuffa GG (eds) Marine clastic sedimentology: concepts and case studies. Graham and Trotman, London, pp 1–38

Postma G, Cartigny M (2014) Supercritical and subcritical turbidity currents and their deposits- a synthesis. Geology 42:987–990

Postma G, Cartigny M, Kleverlaan K (2009) Structureless, coarse-tail graded Bouma Ta formed by internal hydraulic jump of the turbidity current? Sed Geol 219:1–6

Postma G, Kleverlaan K, Cartigny M (2014) Recognition of cyclic steps in sandy and gravelly turbidite sequences and consequences for the Bouma facies model. Sedimentology 61:2268–2290. doi:10.1111/sed.12135

Russell HAJ, Arnott RWC (2003) Hydraulic-jump and hyperconcentrated-flow deposits of a glacigenic subaqueous fan: Oak ridge's moraine, southern Ontario. Can J Sediment Res 73:887–905. doi:10.1306/041103730887

Sequeiros OA (2012) Estimating turbidity current conditions from channel morphology: a Froude number approach. J Geophys Res 117, C04003. doi:10.1029/2011JC007201

Wynn RB, Kenyon NH, Masson DG, Stow DAV, Weaver PPE (2002) Characterization and recognition of deep-water channel-lobe transition zones. AAPG Bull 86:1441–1462

Chapter 48
Tiny Fossils, Big Impact: The Role of Foraminifera-Enriched Condensed Section in Arresting the Movement of a Large Retrogressive Submarine Landslide in the Gulf of Mexico

Derek E. Sawyer and Bailee Hodelka

Abstract A 3.4-m thick condensed section enriched in foraminifera formed the final detachment horizon of a retrogressive submarine landslide, in the Ursa Basin, northern Gulf of Mexico. The high concentration of foraminifera produces a high porosity (up to five porosity units) layer distinct from the background clay. Integrated Ocean Drilling Program Expedition 308 Site U1324 cored and logged this layer. We conducted a sedimentological analysis on 31 samples across this zone and the overlying and underlying background clay. CT images show that foraminifera are individuals dispersed within the clay, unbroken, and have retained a significant amount of intraskeletal void space. The assemblage is expected for this time interval in the Late Pleistocene (~24 kya). We interpret the layer is a result of a pause in terrigenous sedimentation. The condensed section was a preferred detachment horizon but only minimal sliding occurred before further movement ceased. One possible mechanism to explain this is the presence of foraminifera results in a dilational shear strengthening behavior, which arrested movement. Further work will be required to test this, however. On a broader scale, condensed sections with abundant microfossils, may play a key role in landslide mechanics because they can alter the shearing properties of the background material.

D.E. Sawyer (✉)
School of Earth Sciences, The Ohio State University, Columbus, OH, USA
e-mail: sawyer.144@osu.edu

B. Hodelka
Department of Earth and Environmental Sciences, University of Kentucky, Lexington, KY, USA

G. Lamarche et al. (eds.), *Submarine Mass Movements and their Consequences*, Advances in Natural and Technological Hazards Research 41,
DOI 10.1007/978-3-319-20979-1_48

48.1 Introduction

Submarine mass movements transport large volumes of sediment across the ocean floor and are capable of generating deadly tsunami. They occur on all margin types and many of the largest slides are retrogressive (Gauer et al. 2005; Masson et al. 2006; Prior 1978; Sawyer et al. 2009). These repetitive, back-stepping failures record long-term instability. Failure surfaces observed in seismic and outcrop are often complex that cross-cut stratigraphy, which makes interpretation of the sequence of events very difficult. Furthermore, once a retrogressive failure can proceed for long distances, but an important question is what arrests motion once a landslide has started retrogressively unraveling? Does the slip plane jump to a weaker layer as it unravels? Weak layers are understood to be an important aspect of slope failures but are elusive to find (Locat et al. 2013; Masson et al. 2010).

In the Ursa Basin of the upper Mississippi Fan, a large retrogressive slide occurred approximately 24,000 years ago in the Gulf of Mexico. Integrated Ocean Drilling Program (IODP) Expedition 308 collected cores and petrophysical logs through this landslide. Cores were collected both in the down-dip deposit and, importantly, through the in-situ stratigraphy above the main headwall where the landslide stopped its retrogression. Much of the original material was deformed and moved far downslope by the landslide. However, at Site U1324 the original stratigraphy is fortuitously preserved, including the final detachment horizon. The final detachment here corresponds to a significant decrease in p-wave velocity, bulk density, resistivity, and is a prominent reflection in seismic.

We conduct a detailed sedimentological characterization of this final detachment layer. We show that this layer is enriched in foraminifera relative to background, which results in a high-porosity layer. We suggest that this high-porosity reduced shear strength and formed a weak layer.

48.2 Geologic Setting

The Ursa Basin lies 210 km southeast of New Orleans, Louisiana (USA), on the Mississippi Fan, in water depths from 800 to 1500 m (Fig. 48.1). IODP Expedition 308 cored and logged sediments at three locations (Fig. 48.2). A large proportion of the succession sampled by Expedition 308 is composed of clay-rich mass transport deposits (MTDs) (Fig. 48.1). Landsliding in this area is associated with high pore pressure (Flemings et al. 2008) driven primarily by high sedimentation rates. The largest MTD, MTD-2, has been previously characterized as a large, retrogressive landslide with a complex basal surface that cross-cuts stratigraphy at multiple detachment levels (Sawyer et al. 2009). The final (shallowest) detachment layer and original stratigraphy are fortuitously preserved and corresponds with the same depth of a 3-m thick low-velocity, high porosity unit and a large amplitude horizon (Fig. 48.3). Gamma ray and resistivity also vary across this unique layer.

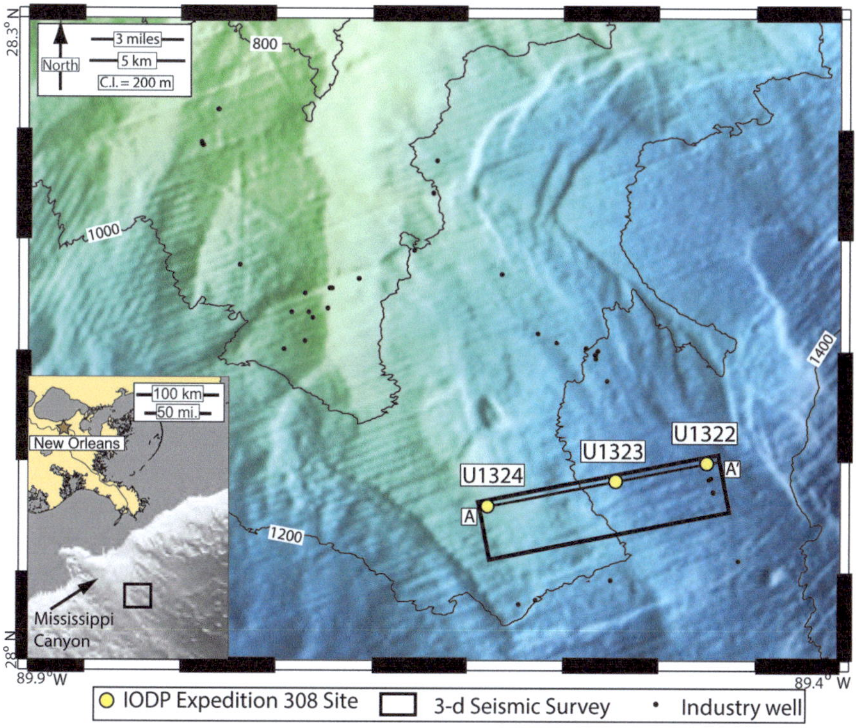

Fig. 48.1 The Ursa Basin in the northern Gulf of Mexico. IODP Expedition 308 cored and logged Sites U1324, U1323, and U1322 on the eastern flank of the Mississippi Canyon through numerous mass transport deposits

48.3 Methods

We selected a total of 31 samples across the 3-m thick zone of interest for detailed investigation of why this layer was a preferred detachment layer: 22 samples within the zone, and 9 samples in the surrounding 'background' clay. Samples were obtained from the Gulf Coast Core Repository (GCCR) as 20 cm^3 plugs. Prior to shipping, GCCR scanned two samples (a background sample and a sample from within the zone of interest) with a micro-CT unit at the University of Texas at Austin.

Each plug sample was first weighed to determine wet mass and then placed into an oven at 100 °C for 8–12 h to obtain the total dry mass. The dried sample was then submerged in deionized water and deflocculant to soak for 8–12 h, The sample was then sonicated for 1 h and gently wet-sieved through a 62.5 m sieve. Foraminifera and any siliciclastic sand were retained on the sieve while the mud fraction (silt and clay size) drained into a 1000-mL graduated cylinder for particle size analysis. The sieve was then dried to determine the total dry mass of the >62.5 m fraction. Mass weight percentages of foraminifera were calculated by the ratio of the dry mass of the >62.5 m fraction to the dry total sample mass. The >62.5 m fraction contained very few siliciclastic grains, which is consistent with previous particle size analyses from this core interval (Sawyer et al. 2008).

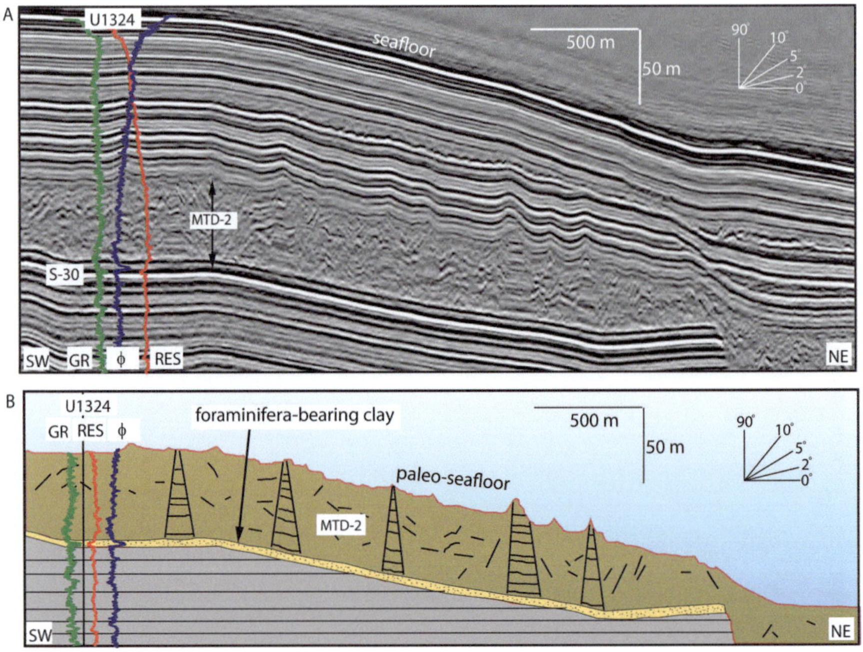

Fig. 48.2 (**a**) Cross-section A-A′ within three-dimensional seismic data showing approximately 200 m of sedimentary formation and a 60-m thick landslide deposit. Well logs show that there is a layer at the landslide base with unique properties: high porosity (*f*), high gamma ray (*GR*), and low resistivity (*RES*). This unique layer is 3.5 m thick and is enriched with foraminifera. The material in this region of the landslide is slightly deformed but is original in-situ material that began to move but was arrested before complete failure and mobilization could occur. (**b**) Interpreted seismic line showing the paleo-seafloor at the time of the landslide. The foraminifera-bearing clay is preserved and mappable along the entire base of the landslide

We created randomly mixed splits of the >62.5 m fraction from which foraminifera were picked and arranged on a 50-grid (grid size = 2 mm × 2 mm) slide and covered with a glass plate. Photographs were taken of selected samples using a microscope with 2× lens (Fig. 48.4).

We performed particle size analysis using the hydrometer method of the mud fraction. These grain size data were merged with previous hydrometer data from the same site (Sawyer et al. 2008).

48.4 Results

Foraminifera concentration by weight increases by an average of eight times over background values (Fig. 48.3). The mean weight percent of foraminifera in the background clay is 0.06 %. The mean weight percent of the foraminifera-enriched zone is 0.47 %, with a range of 0.08–0.97 % and standard deviation of 0.27 %.

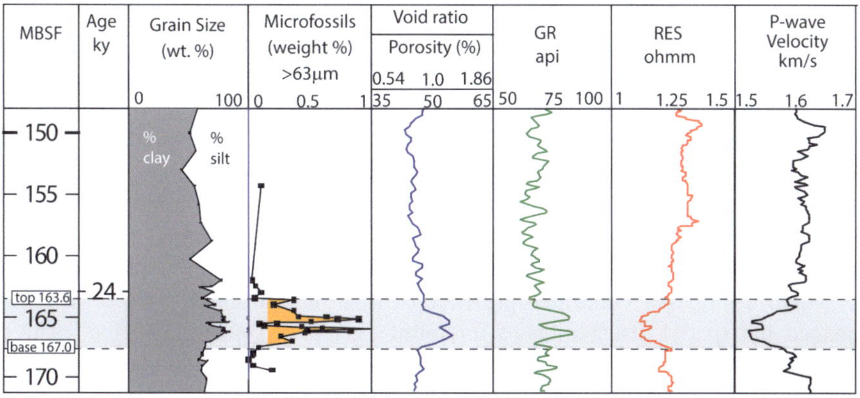

Fig. 48.3 Abundance of foraminifera from 163.6 to 167 mbsf at Site U1324 controls physical properties by creating high porosity. The high porosity controls the decreases in resistivity and compressional velocity. The foraminifera-enriched layer has two peaks in concentration separated by a return to near-background values. This two-peak character is recorded by the logs and grain size. Log data was acquired during Exp. 308 (Gamma Ray (*GR*), Resistivity (*RES*), Porosity (from bulk density log), and compressional P-wave velocity). Grain size track is a merged dataset of hydrometer results (From Sawyer et al. 2008 with new data generated for the current study)

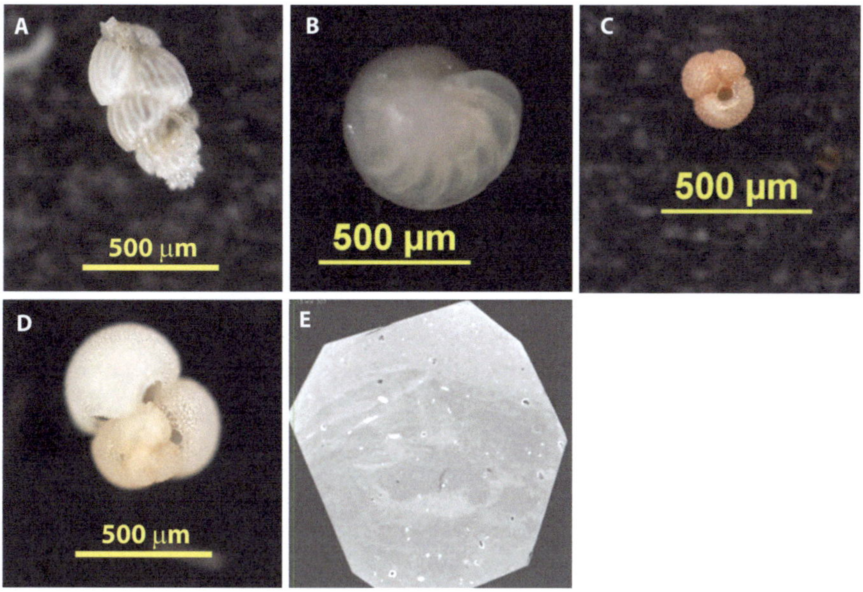

Fig. 48.4 Selected images of foraminifera from enriched zone. (**a**) Benthic Foram Uvigerina peregrina. (**b**) Benthic Foram Cibicidoides pseudoungerianus. (**c**) Planktonic Foram Globigerinoides ruber (*pink*). (**d**) Planktonic Foram Globigerinoi- des ruber (*white*). (**e**) Micro-CT scan of sample from 165 mbsf. Sample is 1 cm × 1 cm. High density is lighter color and lower density is dark. Foraminifera are individuals dispersed in clay and interiors are low-density (*dark*)

Foraminifera concentrations vary within the enriched zone: two prominent peaks are separated by an intervening drop back to background values (Fig. 48.3). This two-peak character is also observed in the percent clay, porosity, resistivity, gamma ray, and compressional wave velocity logs.

The assemblage contains both planktonic and benthic species and are consistent with the assemblage expected based on biostratigraphy established at this site during IODP Expedition 308 (Flemings et al. 2006). Most common planktonic species are *Globigerinoides sacculifer*, and *Globigerinoides ruber* (Fig. 48.4). Most common benthic species are *Bolivina-Bulimina* assemblage.

Micro-CT images confirmed presence of foraminifera in the zone of interest, imaged the spatial distribution of foraminifera, and (3) determined the relative density of foraminifera compared to the surrounding clay. We find that, in general, foraminifera are unbroken and dispersed as individuals throughout the surrounding clay (Fig. 48.4), which was also observed after wet-sieving. Importantly, they also document that each foraminifera and the space it encloses is low-density relative to the surrounding clay (Fig. 48.4).

48.5 Discussion

Our results demonstrate that the clay at the base of the submarine landslide contains elevated concentrations of foraminifera relative to the background values. The increased foraminifera creates a high-porosity unit that distinct from the background Ursa clay (Fig. 48.3). The high porosity is a result of preserved intraskeletal porosity. The elevated porosity in turn controls the low p-wave velocity and low resistivity. The product of low velocity and low bulk density results in a strongly negative impedance boundary marked by the high-amplitude seismic reflector with polarity that is opposite of the seafloor reflector (Fig. 48.2).

The relative percentage of foraminifera by weight significantly increases in this zone relative to background foraminifera concentrations, however it should be noted that the absolute values of weight percent of foraminifera are still less than 1 % of the entire sample. Even though by mass the foraminifera enrichment is not particularly dramatic, the volume of the additional foraminifera in this zone accounts for the significant deviation of 5 porosity units from the background value (Fig. 48.3).

The increased clay percent within the foraminifera-enriched zone suggests that the layer is a condensed section and records a time of reduced terrigenous clastic input, leaving only the finest-grained hemipelagic clays and microfossils to fall out of the water column. Further work is ongoing to age-date and conduct isotope stratigraphy to further place this layer in stratigraphic context.

All movement of the large retrogressive landslide that created MTD-2 ceased at the stratigraphic level coincident with the foraminifera-enriched layer. It is not clear if the layer played a role in stopping the landslide but it is conspicuous that only minimal sliding occurred along this layer (Sawyer et al. 2009). The foraminifera-enriched layer was not the original shear zone/slip plane, but, as retrogressive failure backstepped upward, it eventually chose this layer to detach upon and turned out to be the final detachment layer. If it was a weak layer it is unclear why landslide motion ceased once it started. One potential mechanism by which this layer could have aided in arresting landslide movement is due to the abundance of coarse-grained, spherical particles (foraminifera). This addition of large particles could alter the bulk shearing properties of the overall layer such that it would volumetrically dilate upon shear (Iverson et al. 2000). Dilation during shear would decrease pore pressure and increase effective stress, which would inhibit further deformation. This is exactly opposite of what occurs in clay that is free of large grains: clay will contract upon shear, which increases pore pressure and decreases effective stress, which promotes further shear deformation to occur. The Ursa Clay (background clay) is a contractional clay. It is possible that the peak shear strength of the foraminifera-enriched layer is low because it is a high-porosity unit, which might explain why the detachment chose that surface to fail along. However, as soon as the peak strength was exceeded, it is possible that dilational shear behavior acted to suppress further and resulted in a global arrest of the entire landslide process. Further work is ongoing to determine if the foraminifera enrichment was enough to transform the otherwise contractional clay to a dilational clay.

48.6 Conclusions

A 3.4-m thick condensed section enriched in foraminifera occurs at the base of a retrogressive submarine landslide, in the Ursa Basin, northern Gulf of Mexico. The high concentration of foraminifera produces a high porosity (up to five porosity units), low velocity, low resistivity, and high gamma ray zone relative to the background clay properties. This layer was a preferred detachment horizon but only minimal sliding occurred before further movement ceased. One potential mechanism to explain this is the presence of foraminifera causes a dilational shear strengthening behavior, which acts to arrest movement. On a broader scale, condensed sections with abundant microfossils, may play a key role in landslide mechanics by altering the shearing properties of the parent material.

Acknowledgements We thank John Firth (Gulf Coast Core Repository) for conducting CT scans of samples. This study relied on core samples from Integrated Ocean Drilling Program Expedition 308. We thank David Harris (Kentucky Geological Survey) for access to microscope. We thank Peter Talling and Morelia Urlaub for constructive and thoughtful reviews that improved this paper.

References

Flemings PB, Behrmann JH, John CM, Scientists E (2006) Gulf of Mexico hydrogeology. In: Flemings PB, Behrmann JH, John CM (eds) Proceedings of the integrated ocean drilling program, vol 308. Integrated Ocean Drilling Program Management International, Inc., College Station

Flemings PB, Long H, Dugan B, Germaine J, John C, Behrmann JH, Sawyer D, Scientists I. E. (2008) Pore pressure penetrometers document high overpressure near the seafloor where multiple submarine landslides have occurred on the continental slope, offshore Louisiana, Gulf of Mexico. Earth Planet Sci Lett 269(3–4):309–324

Gauer P, Kvalstad TJ, Forsberg CF, Bryn P, Berg K (2005) The last phase of the Storegga Slide: simulation of retrogressive slide dynamics and comparison with slide-scar morphology. Mar Pet Geol 22(1–2):171–178. doi:10.1016/j.marpetgeo.2004.1010.1004

Iverson RM, Reid ME, Iverson NR, LaHusen RG, Logan M, Mann JE, Brien DL (2000) Acute sensitivity of landslide rates to initial soil porosity. Science 290(5491):513–516

Locat J, Leroueil S, Locat, A, Lee, H (2013) Weak layers: their definitions and classification from a geotechnical perspective. In Krastel S, Behrmann JH, Volker D, Stipp M, Berndt C, Urgeles R, Chaytor J, Huhn K, Strasser M, Harbitz CB (eds) Submarine mass movements and their consequences, 6th International symposium, vol 37. Springer International Publishing, Cham

Masson DG, Harbitz CB, Wynn RB, Pedersen G, Lovholt F (2006) Submarine landslides: processes, triggers and hazard prediction. Philos Trans R Soc Math Phys Eng Sci 364(1845):2009–2039

Masson DG, Wynn RB, Talling PJ (2010) Large landslides on passive continental margins: processes, hypotheses and outstanding questions. In Mosher DC, Shipp RC, Moscardelli L, Chaytor JD, Baxter CDP, Lee HJ, Urgeles R (eds) Submarine mass movements and their consequences, vol 28. Springer, Dordrecht, pp 153–165

Prior DB (1978) Disintegrating retrogressive landslides on very-low-angle subaqueous slopes, Mississippi Delta. In: Coleman JM (ed) Marine geotechnology, vol 3: International. Crane, Russak & Co., New York, pp 37–60

Sawyer DE, Jacoby R, Flemings PB, Germaine JT (2008) Data report: particle size analysis of sediments in the Ursa Basin, IODP Expedition 308 Sites U1324 and U1322, Northern Gulf of Mexico. In: Flemings PB, Behrmann JH, John CM, Scientists a. t. E (eds) Proc. IODP, 308: College Station, Integrated Ocean Drilling Program Management International, Inc. doi:10.2204/iodp.proc.308.205.2008

Sawyer DE, Flemings PB, Dugan B, Germaine JT (2009) Retrogressive failures recorded in mass transport deposits in the Ursa Basin, Northern Gulf of Mexico. J Geophys Res 114:1–20

Chapter 49
Inclusion of Substrate Blocks Within a Mass Transport Deposit: A Case Study from Cerro Bola, Argentina

Matheus S. Sobiesiak, Ben Kneller, G. Ian Alsop, and Juan Pablo Milana

Abstract The preservation of large, relatively undeformed blocks is a characteristic feature of mass transport deposits (MTD). We examine a well-exposed succession at Cerro Bola in La Rioja Province, western Argentina, which comprises mid to late Carboniferous fluvio-deltaic sediments, turbidites and MTD's. The main MTD, which is up to 180 m thick and crops out over 8 km, is characterized by allochthonous sandstone blocks that range in size from metres to 100s of metres in length, and are up to tens of metres in thickness. Blocks are preserved throughout the entire MTD, but are typically larger and much more abundant towards its base where they comprise up to ~30 % of the unit, and become progressively smaller and less frequent upward. Blocks were eroded from the underlying unlithified deltaic sands, and incorporated into the MTD during its transport and emplacement, resulting in local gouges and grooves in the substrate along the basal contact of the MTD. Sandstone blocks are interpreted to have undergone progressive abrasion and fragmentation as they rose through the MTD, thereby creating smaller blocks in the upper parts of the unit. We suggest that buoyancy-driven rise combined with the synchronous fragmentation of sandstone blocks that are entrained within a finer matrix, provides a mechanism for the observed distribution of blocks during overall downslope transport of the MTD.

49.1 Introduction

Submarine gravity-driven failures are widely described from modern and ancient deep water sedimentary basin, with their deposits commonly representing 50 % or more of the deep-water stratigraphic succession (e.g. Moscardelli and Wood 2008). They are considered to be created by a highly complex process comprising creep, slide, slump, debris flow and multiphase granular flow, and the resulting deposits are termed MTD's

M.S. Sobiesiak (✉) • B. Kneller • G.I. Alsop
School of Geosciences, University of Aberdeen, Aberdeen AB24 3UE, UK
e-mail: sobiesiak.ms@abdn.ac.uk

J.P. Milana
Universidad Nacional de San Juan, Mitre Este, San Juan, Argentina

G. Lamarche et al. (eds.), *Submarine Mass Movements and their Consequences*,
Advances in Natural and Technological Hazards Research 41,
DOI 10.1007/978-3-319-20979-1_49

(mass transport deposits) or MTC's (mass transport complexes). Such processes may be initiated in shallow water and transport sediments long distances downslope into the deep water environment, with their size ranging from a few hundred square metres to tens of thousands square kilometres (e.g. Hampton et al. 1996).

During translation, the mass movement can incorporate sediment through erosion, or bypass via a process of hydroplaning. Based on scaled laboratory experiments, several authors have advocated extensive hydroplaning as a means to explain the exceptional mobility and long-run-out of some subaerial mass movements, (e.g. Mohrig et al. 1998). Conversely, erosional features at the base of MTDs are commonly described both from seismic data (e.g. Moscardelli and Wood 2008) and from outcrop (e.g. Dykstra et al. 2011). The whole erosional process is not well understood due to the lack of data on the mechanical and physical properties, which can only be deduced by theory, limited physical experiments or by circumstantial evidence, and therefore involves substantial uncertainty.

49.2 Geological Setting

The study area is located in the southwestern part of La Rioja Province in northwestern Argentina, and consists of a large N-S trending west-vergent doubly-plunging anticline located in the hanging wall to a thrust that dips eastward at about 24° related to the Late Tertiary to Quaternary Pampean Range orogenic deformation (Fig. 49.1) (Zapata and Allmendinger 1996). The sedimentary succession of Cerro Bola exposes a thickness of c.1 km of Carboniferous rocks of the Guandacol Fm. (Gulbranson et al. 2010) that are organized into three individual 200–400 m thick shallowing-upwards packages (Milana et al. 2010) (Fig. 49.1).

The first of these packages has an overall thickness of 200 m and comprises an MTD (MTDI) overlain by a 90 m thick sequence of fluviodeltaic sediments. The second package has an overall thickness of 375 m and is composed of an MTD (MTDII) with sandstone blocks (the focus of this study), overlain by ponded turbiditic sandstones, a black siltstone (maximum flooding surface), a turbidite sandstone unit, and ending in a fluviodeltaic unit. The third package has an overall thickness of 400 m and is quite similar to the second but it is capped by Permian red Beds (Milana et al. 2010). The succession is controlled by high-amplitude glacio-eustatic sea level changes; the upper surfaces of the fluviodeltaic units were thus flooded to substantial depths before the emplacement of overlying MTDs (Dykstra et al. 2011).

49.3 MTDII

MTDII is exposed in a seismic-scale outcrop (Dykstra et al. 2011); the "protolith" is ice-rafted debris (IRD) consisting of fine-grained (silt-size) matrix, with granitoid and metamorphic clasts interpreted as drop-stones. Laminated siltstone rafts are

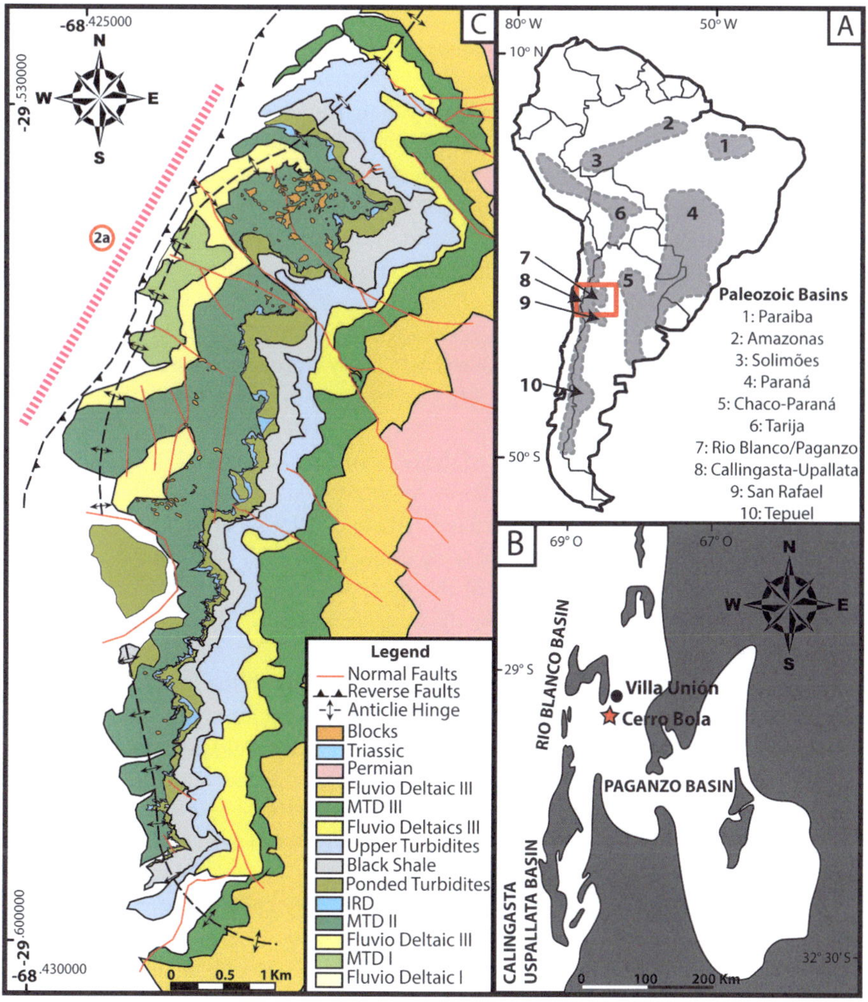

Fig. 49.1 (**a**) Outline map of South America showing the late Paleozoic sedimentary basins. The study area is marked by the *red rectangle* (Modified from Gulbranson et al. 2010). (**b**) Paleogeographic map of the late Paleozoic sedimentary basins of western Argentina (Modified from Gulbranson et al. 2010). (**c**) Geological map of Cerro Bola showing the location of Fig. 49.2a (Modified from Dykstra et al. 2011)

interpreted to be the least-deformed remnants of the "protolith", and are up to several metres thick and 200 m in length. MTDII possesses an erosive contact with the underlying fluviodeltaic rocks (Dykstra et al. 2011). MTDII at Cerro Bola has been stratigraphically divided into three distinct zones (lower, middle and upper) according to its internal variations in textures and structures (Dykstra et al. 2011). The lower part is characterized by numerous blocks of sandstone, that petrographically resembles the underlying fluviodeltaic sandstone deposits (Garyfalou 2015).

The blocks range in size from a few to tens (occasionally hundreds) of metres wide, and up to ~10 m thick, and decrease in abundance upward. Such textures have been described by Mutti et al. (2006) (cited in Ogata et al. 2012a) as blocky flows.

These blocks are interpreted as being ripped up from the underlying fluviodeltaic sediments during mass movement and incorporated into the moving mass (cf. Ogata 2010). The blocks would have been unlithified upon incorporation into the MTD hence it is possible to observe original bedding in some blocks where it has not being destroyed by shearing during transport. The boundary between the MTD and the underlying deposits is markedly irregular (Fig. 49.2a). Individual blocks exhibit fragmentation by stretching and boudinage (Fig. 49.2b) in the inferred transport direction. Material sheared from the blocks forms sand streaks or blebs within the matrix (Fig. 49.2c) that decrease in abundance away from the blocks. In areas of the lower part that are poorer in sandstone blocks there is also a lack of sand in the matrix.

Stratigraphically, the middle part of the deposit consists largely of an apparently structureless siltstone, containing granule- to boulder- size crystalline clasts of IRD, sandstone blocks and rafts of laminated siltstone. Blocks of sandstone are also present, but show evidence of less shearing at their margins, with consequently less sand in the matrix. The upper part of the deposit consists of folded and sheared siltstones, that are imbricated along discrete thrust-sense shear planes, dipping approximately 20° east (Dykstra et al. 2011).

The textural differences between the sand-rich matrix of the lower zone and the rest of the deposit may be explained in part by the dominance of simple shear in the lower part of the flow. This would result in the rotation and fragmentation of the blocks, and the shear-stripping of their outer surfaces to introduce sand into the matrix. Conversely, the upper part of the deposit may have less simple shear and therefore have been more dominated by pure shear, or acted as a plug for some of its movement.

The structures within the MTD are very complex, with compressional and extensional features occurring alone or overprinting one another (Alsop and Marco 2014). They are more evident in the lower zone due to sand streaks in the matrix which act as markers. Structural analysis reported in Dykstra et al. (2011) indicates movement towards the WNW.

49.4 Blocks

The distribution of sandstone blocks is shown in Fig. 49.4, in which the block's midpoint height is measured upwards from the base of the MTD (normalized to the local thickness of the MTD), and is plotted against block size (taken as the square root of the block's outcrop area). The outcrop essentially represents a strike section of the deposit, so this plot represents the vertical distribution of blocks at one point along the transport pathway. It is evident that: (a) blocks, though present throughout the MTD, diminish in frequency upwards (in contrast, for example, to the deposits

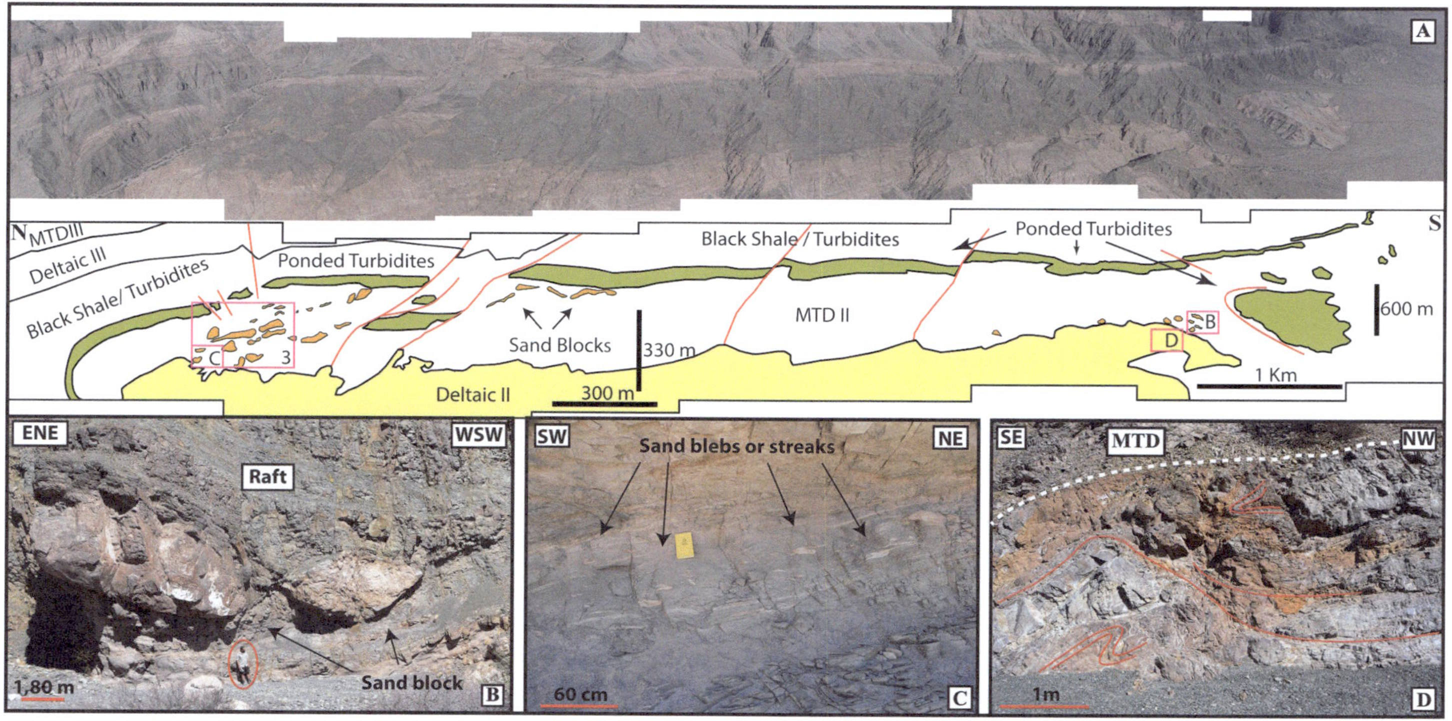

Fig. 49.2 (**a**) Oblique photo-mosaic looking east at Cerro Bola and showing the erosive boundary between the deltaic rocks and the MTD, as well as the distribution of sandstone in the MTD. Note the variation in thickness of the ponded turbidites overlying the MTD. Location shown in Fig. 49.1, while the units legend is the same as Fig. 49.1, showing the location of (**b–d**) and 49.3. (**b**) Two large scale sandstone boudins overlain by a siltstone raft. (**c**) Sand streaks sheared from sand blocks. (**d**) Deformation in the upper most zone of the fluvio-deltaic unit immediately underlying the MTD

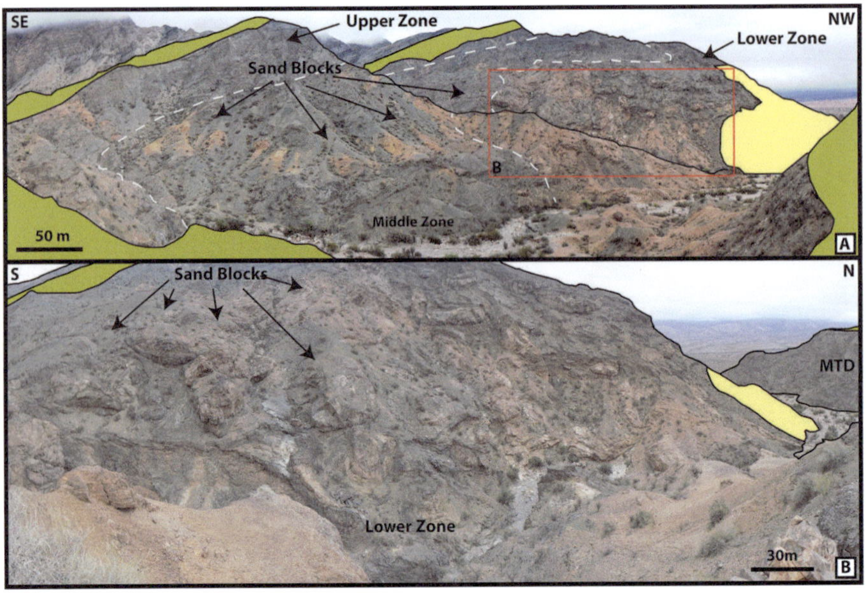

Fig. 49.3 (**a**) Photo-mosaic taken parallel to the inferred transport direction and showing the distribution of the sand blocks throughout the stratigraphy of the MTD. Note how the block size diminishes towards the left-hand side. *Dotted lines* mark the zones boundaries (**b**) Photo-mosaic taken normal to the transport direction and showing the accumulation of blocks in the lower zone of the MTD. The location is shown on (**a**) and unit legends are the same as Fig. 49.1

described by Ogata et al. 2012b, 2014), especially in the upper half of the MTD; and (b) maximum block size diminishes upwards, though blocks near the base have a wide range of sizes.

Two questions arise concerning the distribution of the blocks:

- Given that they were apparently derived by erosion at the lower boundary, how were they transported to the upper part of the MTD? and
- What process explains the block size distribution within the deposit?

Addressing the first of these issues; since the MTD is stratigraphically zoned, it seems unlikely that wholesale imbrication could be invoked to transport sandstone blocks from the base towards the top. An alternative mechanism is buoyancy, requiring that the blocks have a lower density than the matrix. The mineral density of the arkosic sand that makes up the blocks is likely to be in the range of 2620–2640 kg m^{-3}. The porosity of poorly compacted, well sorted sand may be as much as 42–45 % (Rogers and Head 1961), leading to a possible bulk density in the range of 1905–1964 kg m^{-3} for the sandstone blocks.

The matrix of the MTD consists of quartz-rich siltstone with varying proportions of crystalline lithic fragments (IRD), largely of granodioritic or schistose composition. Estimation of the original density of the matrix is subject to substantial uncertainties concerning the initial porosity and the proportion of crystalline lithic

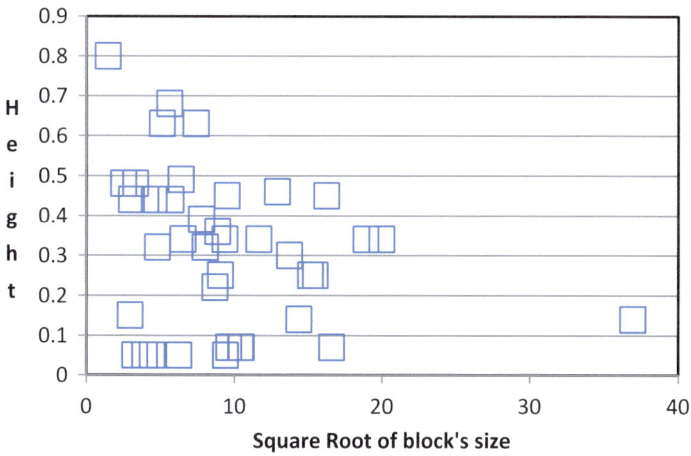

Fig. 49.4 Graph showing the height of sand blocks normalized from the base of the MTD (*vertical axis*), versus the square root area of blocks (*horizontal axis*). The plot illustrates how the blocks diminish in size and number upwards through the MTD (n = 32)

clasts, so must be based upon analogues. MTDs in the Amazon Fan within about 50 m of the sea floor have mean porosities of c.48 % (Busch and Brister 1997), and several authors (e.g. Elverhoi et al. 2000) suggest bulk densities of as little as 1500–1600 kg m^{-3} for subaqueous debris flows, yielding possible bulk densities lower than those of the sandstone blocks. Conversely Pratson et al. 2000 cite a range of 1200–2000 kg m^{-3} for subaqueous debris flows, and Iverson (1997) estimates saturated densities of subaerial debris flows to be in the range 2100–2400 kg m^{-3}. These figures suggest that if buoyancy is the mechanism by which the blocks ascended, the bulk density of the MTD matrix must have been at the upper end of the published range, with porosities perhaps as low as 40 %, though probably varying with depth within the flow. In reality, the majority of the sandstone blocks occur in the lower part of the MTD, perhaps suggesting close to neutral buoyancy for many of them, and only small positive buoyancy for those that did rise.

Positive buoyancy should exert a greater force on larger blocks, yet the blocks higher in the deposit tend to be smaller (Figs. 49.3 and 49.4). We suggest that this is due to progressive fragmentation of the blocks as they rose through the shearing flow. That block fragmentation did occur is indicated by the presence of numerous blocks frozen in the process of boudinage (Fig. 49.2b). The presence of smaller blocks within the lower zone of the MTD may either be due to fragmentation of more-or-less neutrally buoyant blocks that were not rising, or indicates that variably sized blocks were originally plucked from the substrate.

For the relatively low shear rates in natural debris flows, such materials behave as Bingham fluids (Govier and Aziz 1982) rather than as Herschel-Bulkley (shear-thinning) materials. Experimental data from O'Brien and Julien (1988) and Major and Pierson (1992) give some indications of the ranges of plastic viscosity and yield strength in muddy subaqueous debris flows, and their relationship with solids content. Even with small differences in density between the blocks and the matrix, the buoyancy force on the blocks would be great enough to overcome the yield strength of the matrix. The sheared margins of blocks clearly indicate that there was no stagnant fluid around them (cf Valentik and Whitmore 1965), suggesting rather low matrix yield strength. Given the linear relationship between shear stress and strain rate for Bingham fluids once their yield strength is exceeded, one can assume Stokes law for the rate of ascent of large blocks at low particle Reynolds numbers. Order of magnitude calculations, assuming viscosities of a few hundreds of Pa S (appropriate to solids concentrations of 0.5–0.6), block dimensions of the order of only 10 m and density difference of only a few kg m^{-3} suggest vertical velocities of the order of a metre per second, sufficient to allow some blocks to ascend through most of the flow even during a relatively short-lived event.

49.5 Basal Interaction

The presence of sandstone blocks, combined with the irregular gouged upper surface of the underlying fluvio-deltaic sediments, provides clear evidence that the MTD was interacting vigorously with the substrate. The incorporation of substrate as blocks requires the initiation of failure planes below the basal surface of the mass movement, within the underlying sand unit (unlithified at the time). This indicates that shear stress must have been transmitted through the interface into the upper part of the substrate. Evidence for this is demonstrated by strain within the uppermost few tens of metres of the fluvio-deltaic unit below the contact with the MTD, including abundant recumbent soft-sediment folds (Fig. 49.2c), and local large scale fragmentation and block rotation of the uppermost few tens of metres of the sandstone unit. It is interesting to note that a similar debate regarding the interaction and deformation of substrate below ice sheets and glaciers is currently taking place, and involves models of glacier-bed decoupling resulting in both recumbent soft-sediment folds and blocks (e.g. see discussion in Lesemann et al. 2010).

49.6 Conclusions

In common with many MTDs, this deposit shows clear evidence that the flow interacted with the substrate, removing material from the underlying unit and incorporating it into the mass movement. The strain front related to the mass movement therefore does not coincide with the base of the MTD, but occurs a considerable distance into the substrate. We propose a model whereby large sandstone blocks have risen through the MTD matrix via a density-driven mechanism. This is likely to have consequences for the interpretation of such deposits in more poorly exposed areas of outcrop, or in the subsurface. It also has considerable implications for the disruption of potential hydrocarbon reservoirs that occur below MTDs.

Acknowledgments This work was carried out with support from CNPq (Conselho Nacional de Desenvolvimento Científico e Tecnológico) – Brazil and BG-Brazil. We would like to thank the reviewers Claudio Corrado Lucente and Kei Ogata for their suggestions and criticism that helped to improve the manuscript.

References

Alsop GI, Marco S (2014) Fold and fabric relationships in temporally and spatially evolving slump systems: a multi-cell flow model. J Struct Geol 63:27–49. doi:10.1016/j.jsg.2014.02.007

Busch WH, Brister MR (1997) Fabric of fine-grained Amazon Fan sediments: influence of depositional processes and burial transformations. Proc Ocean Drill Program Sci Results 155:447–464

Dykstra M, Garyfalou K, Kertznus V, Kneller B, Milana JP, Milinaro M, Szuman M, Thompson P (2011) Mass-transport deposits: combining outcrop studies and seismic forward modeling to understand lithofacies distributions, deformation, and their seismic expression. SEPM Spec Publ 95:1–25

Elverhoi A, Harbitz CB, Dimakis P, Mohrig D, Marr J, Parker G (2000) On the dynamics of subaqueous debris flows. Oceanography 13:109–125

Garyfalou K (2015) Integrated analysis of mass-transport deposits: outcrop, 3D seismic interpretation and fast fourier transform. Unpublished PhD thesis, University of Aberdeen

Govier GW, Aziz K (1982) The flow of complex mixtures in pipes. Krieger Publ. Co., Melbourne

Gulbranson EL, Montanez IP, Schmitz MD, Limarino CO, Isbell JL, Marenssi SA, Crowley JL (2010) High-precision U-Pb calibration of Carboniferous glaciation and climate history, Paganzo Group, NW Argentina. Geol Soc Am Bull 122:1480–1498. doi:10.1130/B30025.1

Hampton MA, Lee HJ, Locat J (1996) Submarine landslides. Rev Geophys 34:33. doi:10.1029/95RG03287

Iverson RM (1997) The physics of debris flows. Rev Geophys 35:245. doi:10.1029/97RG00426

Lesemann J-E, Alsop GI, Piotrowski J (2010) Incremental subglacial meltwater sediment deposition and deformation associated with repeated ice-bed decoupling: a case study from the Island of Funen, Denmark. Quat Sci Rev 29:3212–3229

Major JJ, Pierson TC (1992) Debris flow rheology: experimental analysis of fine-grained slurries. Water Resour Res 28:841. doi:10.1029/91WR02834

Milana JP, Kneller B, Dykstra M (2010) Mass-transport deposits and turbidites, syn-to-post-glacial carboniferous basins of Western Argentina. ISC 2010F Guid 01–88

Mohrig D, Ellis C, Parker G, Whipple KX, Hondzo M (1998) Hydroplaning of subaqueous debris flows. Geol Soc Am Bull 110:387–394. doi:10.1130/0016-7606(1998)110<0387:HOSDF>2.3.CO;2

Moscardelli L, Wood L (2008) New classification system for mass transport complexes in offshore Trinidad. Basin Res 20:73–98. doi:10.1111/j.1365-2117.2007.00340.x

Mutti, E, Carminatti M, Moreira JLP, Grassi A (2006) Chaotic deposits: examples from the Brazilian offshore and from outcrop studies in the Spanish Pyrenees and Northern Apennines, Italy. A.A.P.G. annual meeting, 9–12 Apr, Houston

O'Brien JS, Julien PY (1988) Laboratory analysis of mudflow properties. J Hydrol Eng 114:877–887

Ogata K (2010) Mass transport complexes in structurally-controlled basins: the Epiligurian Specchio Unit (Northern Apennines, Italy). PhD thesis, University of Parma

Ogata K, Mutti E, Pini GA, Tinterri R (2012a) Mass transport-related stratal disruption within sedimentary mélanges: examples from the northern Apennines (Italy) and south-central Pyrenees (Spain). Tectonophysics 568–569:185–199. doi:10.1016/j.tecto.2011.08.021

Ogata K, Tinterri R, Pini GA, Mutti E (2012b) The Specchio Unit (Northern Apennines, Italy): an ancient mass transport complex originated from near-coastal areas in an intra-slope setting. 5th Submar Mass Mov Conseq 31:595–605. doi:10.1007/978-3-319-00972-8

Ogata K, Mountjoy JJ, Pini GA, Festa A, Tinterri R (2014) Shear zone liquefaction in mass transport deposit emplacement: a multi-scale integration of seismic reflection and outcrop data. Mar Geol 1–15. doi:10.1016/j.margeo.2014.05.001

Pratson L, Imran J, Parker G, Syvitski JP, Hutton E (2000) Debris flows vs. turbidity currents: a modeling comparison of their dynamics and deposits. AAPG Mem 72/SEPM Spec Publ 68:57–72

Rogers JJ, Head W (1961) Relationships between porosity, median size, and sorting coefficients of synthetic sands. J Sediment Petrol 31:467–470

Valentik L, Whitmore RL (1965) The terminal velocity of spheres in Bingham plastics. Br J Appl Phys 16:1197–1203. doi:10.1088/0508-3443/16/8/320

Zapata TR, Allmendinger RW (1996) Thrust-front zone of the Precordillera, Argentina: a thick-skinned triangle zone. Am Assoc Pet Geol 80:359–381. doi:10.1306/64ED87E6-1724-11D7-8645000102C1865D

Part VIII
Numerical and Statistical Analysis

Chapter 50
GIS Catalogue of Submarine Landslides in the Spanish Continental Shelf: Potential and Difficulties for Susceptibility Assessment

Natalia Borrell, Luis Somoza, Ricardo León, Teresa Medialdea, Francisco J. Gonzalez, and Carmen J. Gimenez-Moreno

Abstract This study presents the first version of a GIS catalogue of submarine landslides affecting the Spanish Continental Shelf pursuant UNCLOS (United Convention for the Law of the Sea) that comprise continental-type margins (Atlantic and Mediterranean) as well as hot-spot type volcanic islands and seamounts (Canary Archipelago). This first version compiles a total of 223 submarine landslides within an area of 1.5 million square kilometres. This catalogue, developed in a geographic information system, compiles information of each submarine landslide such as the name, location, typology, age, volume, source, dimensions (width, elongation), minimum and maximum height, minimum and maximum seafloor slopes. As a direct application and advantage of this digital catalogue, a first approach to a susceptibility map has been elaborated using GIS analysis tools taking into account the following available factors sourced from the geological maps of the Spanish continental margins: (i) faults map, (ii) active faults map, (iii) earthquakes density map, (iv) seafloor slope map and (v) seabed composition map. The potential and difficulties of this GIS catalogue of landslide as a first step in the submarine risk analysis are discussed. This first version of the GIS catalogue is conceived as the origin of a national submarine landslides database within the European network, which is being built as part of the EMODNET-Geology project.

50.1 Introduction

In order to assess the geological risks associated to submarine landslide, one of the main issues concerns what types of mechanisms may trigger submarine landslides (Locat et al. 2003). Submarine landslides are some of the most frequent processes affecting the evolution and morphology of the continental margins, although they also occur in other geological environments such as active volcanic islands and

N. Borrell • L. Somoza (✉) • R. León • T. Medialdea • F.J. Gonzalez • C.J. Gimenez-Moreno
Marine Geology Dv., Geological Survey of Spain, IGME, Rios Rosas 23, 28003 Madrid, Spain
e-mail: l.somoza@igme.es

G. Lamarche et al. (eds.), *Submarine Mass Movements and their Consequences*, Advances in Natural and Technological Hazards Research 41,
DOI 10.1007/978-3-319-20979-1_50

oceanic seamounts (Masson et al. 2006). In continental margins, these instabilities play a major role in the transport and distribution of sediments from the coastal zones with a high rate of sedimentation, such as deltas, through the continental slope to deep ocean floor. Instabilities of oceanic islands may mobilise hundreds of thousands of km^3 of sediment and rock more than 500 km downslope into the deep water abyssal plains (Masson et al. 2006). Several mechanisms have been proposed as causes of submarine landslides, such as rapid rates of sediment accumulation on slopes, seismicity, gas hydrates dissociation, occurrence of sedimentary layers with low resistance, dikes intrusion and volcanic expansion (Canals et al. 2004). Mapping and understanding how these submarine landslides have been triggered is important in order to assess future associated geologic hazards (Masson et al. 2006; Løvholt et al. 2008). This paper presents the first version of the Geographic Information System (GIS) catalogue of submarine landslides affecting the Spanish Continental Shelf (according the legal definition of Continental Shelf of UNCLOS article 76, United Convention for the Law of the Sea) that comprise continental-type margins (Atlantic and Mediterranean) as well as volcanic islands and seamounts (Canary Archipelago). Furthermore, a submarine landslide susceptibility map has been generated from this catalogue. The catalogue is conceived as the first step of a future national submarine landslides database within the European framework and a baseline for geohazard assessment for marine infrastructure.

50.2 Regional Geology

The study area covers the seafloor of the whole Spanish Continental Shelf, including the Iberian continental margins and the entire Canary Islands (Fig. 50.1). At least four geological regions can be distinguished within the Iberian continental margins (e.g. Medialdea and Terrinha 2015). The Atlantic margins can be subdivided into two: (1) The Galician-Cantabrian margin and (2) The Gulf of Cadiz margin. Likewise, the Mediterranean margins can be subdivided into two: (3) The Alborán margin and (4) The Catalonian/Balearic margin within the Western Mediterranean.

The Galician-Cantabrian margin formed after the oceanic spreading of the northern Atlantic and the opening of the Bay of Biscay. Later, the northern Iberian margin (Cantabrian margin) overthrusted the oceanic crust of the Biscay Bay crust ceasing at mid Miocene times. Thus, this margin is characterised by a narrow shelf incised by deep submarine canyons and associated deep-sea turbiditic systems. Towards the west, the northeast Iberian margin (Galicia margin) is a hyperextended margin characterised morphologically by a wide shelf, with submarine canyons that feed deep-water valleys intercalated between shallow banks as the Galicia Bank. The Gulf of Cádiz and the Alborán margin are located along the African–Eurasian plate boundary. The Gulf of Cádiz continental margin is characterised by the emplacement of the huge allochthonous unit (the so-called "Olistostrome" of the Gulf of Cadiz) that generated a wedge-like highly unstable body on the continental

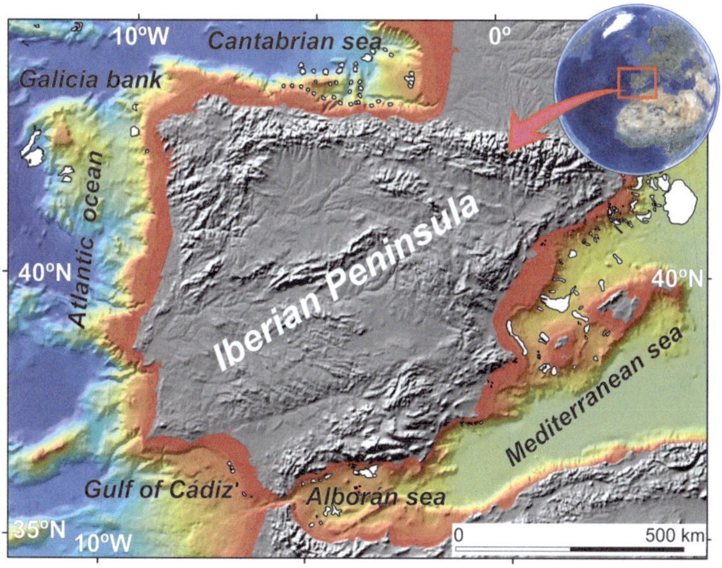

Fig. 50.1 GIS catalogue (inventory map) of the submarine landslides (denoted by *white polygons*) on the Iberian continental margin

slope. On the contrary, the Alborán margin shows a complex morphotectonic structure with sedimentary basins crossed by volcanic ridges. In these two margins, active strike-slip faults visibly affect the seafloor. In contrast, the Catalonian/ Balearic continental margins are passive margins generated in the Oligocene–Early Miocene rifting phase. This extensional phase led to the formation of sub-basins that control the morphostructural configuration of the western Mediterranean Basin.

Finally, the Canary Islands sector encompasses a 1350 km long NE-SW oceanic volcanic ridge, volcanic islands and seamounts located 100 km off the northwestern African coast (Fig. 50.1), where more than 20 mega-landslides have been documented (Masson et al. 2006).

50.3 Data and Methods

50.3.1 GIS Catalogue

The GIS catalogue has been implemented with ArcGIS ESRI® software. The landslides have been digitised from geo referenced paper and digital maps using a Geographic coordinate system and Datum ETRS89. In the majority of cases, the digitised area corresponds to scars, bypass and depositional areas (Table 50.1). The main source of cartographic information of the catalogue is the Geological Map of

Table 50.1 Description of items related to each submarine landslide stored in the data base

Item	Description
FID	Numeric internal identifier
Latitude	Y Location. Geographical coordinates
Longitude	X Location. Geographical coordinates
Name	Name published of the submarine landslide
Setting	Geographical setting (general toponymy)
Location	Local toponymy, margin name
Failure_Ty	Failure type published
Period	Geologic period
Age	Age (Ka)
Water_Dep	Water depth (m)
Thickness	Thickness of the deposits
Volume	Volume of the deposits
Slope	Slope gradient of the seafloor (over deposit)
Shape_Area	Area of the gravitational instability (m^2)
Length	Distance from the headwall scarp to the distal toe
Lateral_Extension	Maximum lateral extent of a landslide
Source	Source of cartographic boundaries
References	Bibliographic references cited

Spain at 1:1,000,000 scale (Medialdea and Terrinha 2015). This information has been complemented with submarine landslides mapped under the National Marine Mapping FOMAR by Programme at 1: 200,000 scale, published papers and PhD theses. Attributes of Table 50.1 for each landslide are sourced from published data.

50.3.2 Susceptibility Assessment

In this work, we define the susceptibility as the likelihood of seafloor to be affected by submarine landslides. Following this definition, we have considered that the main factors that might affect the occurrence of submarine landslides in a determined sea-floor area are the following: proximity to faults, proximity to active faults, earthquake density, sea-floor slope gradient and sea-floor composition. The quality and quantity of information for each of these factors depends on the available information:

- Fault maps: Faults older than Quaternary have been taken from the Geological Map 1:1,000,000 of the continental margin (Medialdea and Terrinha 2015). In addition, fluids generally migrate along faults, weakening its geotechnical properties.

- Map of Active Faults: Information of active faults has been taken from the Quaternary Active Faults Database of Iberia (QAFI) (IGME 2013). This map provides information of faults with geological evidence of activity in the Quaternary period (i.e. during the last 2.6 Ma).
- Earthquake density map: This map has been taken from Borrell (2013) and show density of earthquakes with depths ranging from 0 to 30 km and magnitudes >3.5 Richter scale.
- Seafloor Slope gradient: Theoretically, slopes become gravitationally unstable when the gradient exceeds the internal angle of friction and cohesion of the deposit. Slope gradient has been calculated from the GEBCO DTM (GEBCO 2012).
- Seafloor composition: The seafloor composition maps define mainly the granulometry of sediment. Based on the textural map (IGME 1989) six classes have been defined: gravel, sand, mud, sand and mud, silt and mud, and rock.

The susceptibility assessment has been calculated in the Iberian Peninsula continental margins only (where all factor maps were available), using the Wi-index bivariate method (van Westen 1997). The Wi index is calculated as the natural logarithm of the landslide density in each factor maps class, divided by the landslide density in the whole area under study (Eq. 50.1).

$$W_i = L_n\left[\frac{DensityClass}{DensityMap}\right] = L_n\left[\frac{{}^{Area}(Si)\big/_{Area(Ni)}}{\sum_{Area(Si)}\Big/\sum_{Area(Ni)}}\right] \quad (50.1)$$

Where "Wi" is the weight assigned to a specific parameter, "DensityClass" is density of the parameter class, "DensityMap" is the landslide density for the whole map, "Area (Si)" is the Area affected by landslides in the factor map and "Area (Ni)" is the surface of a class in the factor map.

The final susceptibility map is the result of the sum of each of the weights (Wi) for each weighted factor map (Eq. 50.2).

$$W_t = \sum W_i = W_{if} + W_{iA} + W_{iDT} + W_{iPF} + W_{iL} \quad (50.2)$$

Where "Wt" is ocean floor susceptibility, "WiF" is Wi index value for each class in the Faults factor map, "WiAF" is Wi index value for each class in the Active Faults factor map, "WiED" is "Wi" index for each class in the Earthquake Density factor map and "WiL" is Wi index value for each class in the Lithology factor map.

The weight calculation for each class in the factor maps (Wi) was performed using Eq. (50.1). Results are summarised in Table 50.2. A final susceptibility map (Fig. 50.3) is obtained using map algebra adding the value of the index (Wi) of each weighted factor map (Eq. 50.2). This has been done with "Union" tool of ArcGIS. Legend has been segmented by quantiles.

Table 50.2 Table of weights. Wi index calculated with Eq. (50.1) for each class in the factor maps

	Factor map: seafloor slope gradient		Landslides vs Factor map		Weight (Wi)
	Class (°)	Area (Ha)	Class	Area (Ha)	
Seafloor slope gradient	0 – 0.8	18,816,560	0 – 0.8	913,774	−0.19
	0.8 – 2.2	11,329,705	0.8 – 2.2	833,150	0.23
	2.2 – 4.0	5,901,788	2.2 – 4.0	449,512	0.26
	4.0 – 6.3	3,381,527	4.0 – 6.3	178,939	−0.10
	6.3 – 9.3	2,153,800	6.3 – 9.3	101,381	−0.22
	9.3 – 13.3	1,073,673	9.3 – 13.3	41,669	−0.41
	13.3 – 22.9	258,037	13.3 – 22.9	795	−2.95
	Total	42,915,089	Total	2,519,221	

	Factor map: texture of seafloor		Landslides vs Factor map		Weight (Wi)
	Class	Area (Ha)	Class	Area (Ha)	
Seafloor composition	Sand	5,971,239	Sand	66,722	−1.66
	Mud and sand	8,910,501	Mud and sand	772,389	0.39
	Mud	19,587,112	Mud	644,064	−0.58
	Gravel	878,300	Gravel	2,754	−2.93
	Volcanic rocks	67,707	Volcanic rocks	16,427	1.42
	Lime and mud	7,500,229	Lime and mud	1,016,864	0.84
	Total	42,915,089	Total	2,519,220	

	Factor map: faults		Landslides vs Factor map		Weight (Wi)
	Class	Area (Ha)	Class	Area (Ha)	
Faults	Fault	6,086,862	Fault	370,604	0.04
	No fault	36,828,227	No fault	2,148,617	−0.01
	Total	42,915,089	Total	2,519,221	

	Factor map: density of earthquakes		Landslides vs Factor map		Weight (Wi)
	Class (10^{-4}Earthquake/Km^2)	Area (Ha)	Class (10^{-4}Earthquake/Km^2)	Area (Ha)	
Earthquake density	0	1,765,620	0	0	−241.81
	0 – 2.14	24,448,744	0 – 2.14	1,686,346	0.16
	2.14 – 6.41	7,374,005	2.14 – 6.41	504,088	0.15
	6.41 – 2.13	4,456,856	6.41 – 2.13	154,984	−0.52
	21.3 – 546.7	4,869,863	21.3 – 546.7	173,803	−0.50
	TOTAL	42,815,089	TOTAL	2,519,221	

	Factor map: active faults		Landslides vs Factor map		Weight (Wi)
	Class	Area (Ha)	Class	Area (Ha)	
Active faults	Fault	358,945	Fault	15,944	−0.28
	No fault	42,556,144	No fault	2,503,276	0.00
	Total	42,915,089	Total	2,519,221	

50.4 Results

A total of 223 submarine landslides have been compiled within the GIS Catalogue developed, 195 of them around the Iberian Peninsula continental margins (Fig. 50.1) and 28 from the Canary Islands (Fig. 50.2). In the Galicia margin, 13 instabilities have been recognised, which include landslides in the broad sense (*s.l)* and creeps. The average area affected by the seafloor instabilities is 513 km^2 with average slopes of 5 %. In the Cantabrian continental margin, 38 instabilities were identified that are mainly composed of landslides (*s.l.)* with similar sizes. The average surface area affected by seafloor instabilities is 70 km^2 with average slopes of 3 %. In the Gulf of Cadiz, only eight instabilities have been mapped, classified as landslides (*s.l.)* according to the authors. They are small movements with an

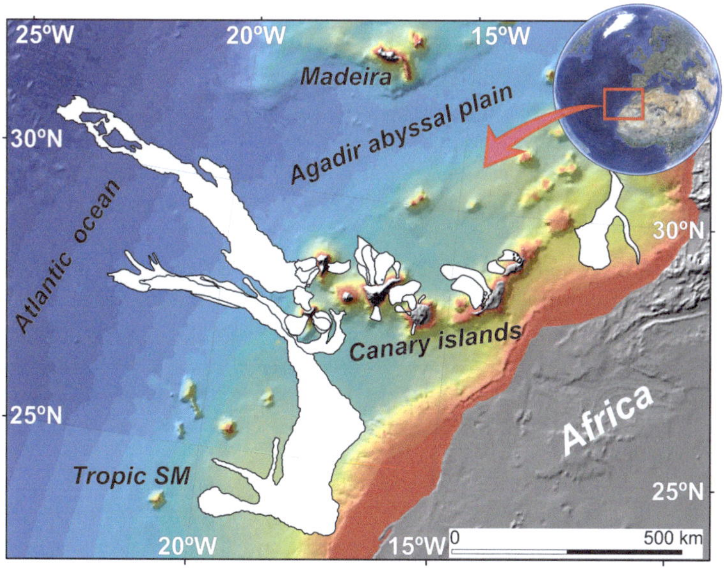

Fig. 50.2 GIS catalogue (inventory map) of the submarine landslides (denoted by *white polygons*) on the Canary Islands area

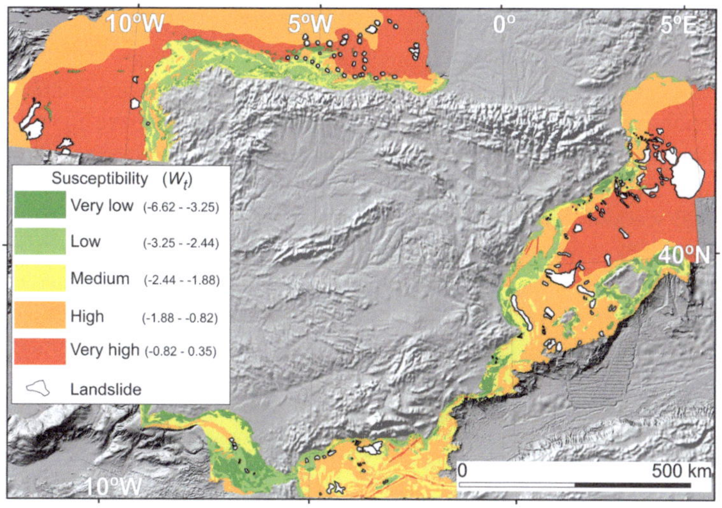

Fig. 50.3 Susceptibility map of the Spanish continental margin

average surface area of 3 km^2, located on the upper continental slope with an average slope of 1.6 %.

In the Catalonian margin, at least 81 seafloor instabilities are composed of buried landslides, debris flow, creep and landslides. The most are classified as creep and translational slides covering an average area of 123 km^2 and an average slope of

3.6 %. This region contains the largest cohesive submarine landslides of the Iberian continental margin, named as the Western Debris Flow and BIG'95 (Canals et al. 2004; Lastras et al. 2007) covering an area of 5000 km^2 and 1540 km^2 respectively. The Balearic margin (40 instabilities) is mainly affected by creeps and translational slides with an average surface area of 115 km^2 and an average slope of 2.4 %. In the Alborán margin, at least 34 instabilities have been identified being constituted by translational slides, creep and debris flow affecting to a total area of 74 km^2 with average slopes of 2.7 %.

In the Canary archipelago area, 28 seafloor instabilities have been identified and mainly consist of cohesive debris flows. The largest gravitational movements are found in this region, extending 650 km westwards from the volcanic islands of the El Hierro and La Palma to 5200 m water depths (Masson et al. 2006). They are complex movements (debris-avalanche-debris flow) with an average surface area of 5300 km^2 developed on average slopes of 3.5 %.

50.5 Discussion: Potential and Difficulties in the Susceptibility Assessment

The direct application and one advantage of a GIS digital catalogue is its statistical analysis, in this case, the susceptibility assessment. We have to take into account that submarine landslide include a great variety of processes affecting sea-floor such as failure, erosion, high-energy transport and high-rate sedimentation.

Susceptibility shows the correlation between elements of the inventory map (i.e. mapped seafloor areas affected by submarine landslide process) and the control factors (i.e. factors that might affect the occurrence of submarine landslide in a determined seafloor area). In this sense, two main difficulties have been found for the susceptibility assessment. Firstly, the representation of the inventory map: if the catalogue is not fully completed or when only depositional areas are mapped but not the full affected area (e.g. slide scars, erosional, bypass and deposits). Second, the limited available control factor maps: in our case, these factors were constrained to (i) proximity to pre-Quaternary faults, (ii) proximity to active faults, (iii) earthquake density, (iv) sea-floor slope gradient, and (v) sea-floor composition.

Along narrow continental margins, the resulting susceptibility map (Fig. 50.3) depict clearly how the areas labelled as "very high" susceptibility coincide with those displaying the highest development of landslides, particularly on the continental slope of the Galician, Cantabrian, Catalonian and the northern Balearic margins. Therefore, the continental slope shows the "higher" values, whereas the areas labelled as "medium", "low" and "very low" are located on the shelf (Fig. 50.3).

Besides physiographic domains, susceptibility is highly sensible to the proximity to basement faults more than to active faults. Thus, the positive value of the Wi index in the "fault" class in the factor map of "faults" indicates a clear geographical correlation between the presence of landslides and proximity to faults. We suggest that anisotropies generated by faults in the basement and the sedimentary column could condition the future instability of the seafloor either because the sub-seabed geotechnical conditions are more unstable (e.g. partially fractured rock) or because they constitute preferred conduits for migration of fluids. Otherwise, the absence of earthquakes is related to absence of landslides (earthquake density and Wi value of zero and -241, respectively; Table 50.2).

With respect to the slope gradient and seafloor composition, Wi index values point out that the relationship between seabed sediment classes and landslide occurrence is closely linked and dependent on the sedimentary nature of each physiographic province. Landslides are more frequent in the continental slope between 0.8 and 4 % of slope gradient. However, maximum values of slope gradient are not related to major density of landslide, as Canals et al. (2004) and Hühnerbach and Masson (2004) have observed in North Atlantic ocean and Mediterranean sea.

It is clear there is a need to gather more data and extend the study area to have a more conclusive statistical analysis. This version of the GIS catalogue of submarine landslide is the first step in developing a database of submarine landslides at national scale and a baseline for future research on marine infrastructure hazards. In the next future, we will include further factors influencing sea-floor mass transport such as seafloor gas hydrate occurrence, seabed fluid flow, submarine (mud and lava) volcanoes and tsunamigenic structures.

Acknowledgments This work has been supported by IGCP-640 S4SLIDE, European marine observation and data network-geology (EMODNET) and the Spanish projects SUBVENT (CGL2012-39524-C02) and "Project for Extension of the Spanish Continental Shelf" (CTM2010-09496-E). The authors would like express their appreciation to Aaron Micallef and Francesco Chiocci and for their useful reviews and suggestions.

References

Borrell N (2013) Catálogo de deslizamientos y análisis de la susceptibilidad en el Margen Continental Español. MSc Thesis, Universidad Complutense de Madrid, Madrid, Spain

Canals M, Lastras G, Urgeles R, Casamor JL, Mienert J, Cattaneo A, De Batist M et al (2004) Slope failure dynamics and impacts from seafloor and shallow sub-seafloor geophysical data: case studies from the COSTA project. Mar Geol 213:9–72

GEBCO (2012) General bathymetric chart of the oceans. Query: 2 Aug 2012. http://www.gebco.net/

Hühnerbach V, Masson DG (2004) Landslides in the North Atlantic and its adjacent seas: an analysis of their morphology, setting and behaviour. Mar Geol 213:343–362

IGME (1989) Mapa del Cuaternario de España a escala 1:1000000. Instituto Tecnológico y Geominero de España. Instituto Tecnológico y Geominero de España, Madrid

IGME (2013) Quaternary active faults database of Iberia. Query: 27 Jan 2013. http://www.igme.es/infoigme/aplicaciones/qafi/

Lastras G, Canals M, Amblas D, Frigola J, Urgeles R, Calafat AM, Acosta J (2007) Slope instability along the northeastern Iberian and Balearic continental margins. Geol Acta 5:35–47

Locat J, Mienert J, Boisvert L (2003) Submarine mass movements and their consequences I. Kluwer Academic Publishers, Dordrecht

Løvholt F, Pedersen G, Gisler G (2008) Oceanic propagation of a potential tsunami from the La Palma island. J Geophys Res 113:C09026

Masson DG, Harbitz CB, Wynn RB, Pedersen G, Løvholt F (2006) Submarine landslides: processes, triggers and hazard prediction. Philos Trans Royal Soc Lond A Math Phys Eng Sci 364:2009–2039. doi:10.1098/rsta.2006.1810

Medialdea T, Terrinha P (2015) Mapa Geológico del Margen Continental de España y Portugal. In: Rodríguez LR, Oliveira JT (eds) Mapa Geológico de España y Portugal, escala 1.000.000. Instituto Geológico y Minero de España, Madrid

Van Westen CJ (1997) Statistical landslide hazard analysis. ILWIS 2.1 for windows application guide. ITC Publication, Enschede, pp 73–84

Chapter 51
Tempo and Triggering of Large Submarine Landslides: Statistical Analysis for Hazard Assessment

Michael A. Clare, Peter J. Talling, Peter G. Challenor, and James E. Hunt

Abstract Due to their potential volume and speed, large submarine landslides can generate destructive tsunamis or damage expensive seafloor structures. Understanding their timing is therefore important for hazard assessments; however, dating large numbers of landslides close to their origin is logistically difficult. Previous landslide studies are typically limited to fewer than ten observations of ages. To address this we analyse extensive, continuous and long-term turbidite records from four deep-sea basins which are interpreted to be the distal deposits of large, disintegrative landslides. Our records include sufficient numbers of turbidites ($N = 151–1571$) for robust statistical analysis of long-term controls on event timing and testing for relationships with triggering mechanisms such as earthquakes, sea level, climate change and volcanic activity. We explore statistical methods developed by medical, economic and biological disciplines and show how they can be applied to analysis of submarine landslide frequency and triggering. Frequency analysis of field data reveals two different distribution forms for landslide recurrence – exponential and log-normal. We discuss possible individual and combined effects of controlling factors that result in these distributions. Rescaled range and Gaussian finite mixture models determine whether and how landslides are clustered in time. Parametric Generalised Linear Models and non-parametric Proportional Hazards Models are used to test for the significance and influence of different variables and their rate of change. We demonstrate the value of unusually detailed long-term landslide records, and show how statistical analysis provides quantitative inputs for future hazard assessments, landslide-climate studies and understanding the tempo of deep-sea sediment flux.

M.A. Clare (✉) • P.J. Talling • J.E. Hunt
National Oceanography Centre Southampton, Waterfront Campus,
Southampton SO17 3ZH, UK
e-mail: michael.clare@noc.ac.uk

P.G. Challenor
College of Engineering, Mathematics and Physical Sciences, University of Exeter,
North Park Road, Exeter EX4 4QF, UK

G. Lamarche et al. (eds.), *Submarine Mass Movements and their Consequences*,
Advances in Natural and Technological Hazards Research 41,
DOI 10.1007/978-3-319-20979-1_51

51.1 Introduction

Submarine landslides on continental margins include the largest mass flows on Earth, which can potentially generate damaging tsunamis and produce long run-out turbidity flows (Talling et al. 2014). Determining the past frequency distribution of large slides is therefore important and can inform our estimates of future hazard rate (i.e. the rate of landslide recurrence during any point in time). Despite this, few studies have attempted to analyse submarine slide frequency and probability statistically. This is largely due to uncertainties in dating individual landslides (Urlaub et al. 2013), lack of sediment for dating (Geist et al. 2013) and insufficiently large sample sizes for robust analysis (Urlaub et al. 2013; Harbitz et al. 2014). Published landslide catalogues include less than 68 dated landslides and in most cases many fewer (Urlaub et al. 2013). This is not enough to establish a confident view on past and future recurrence.

Notable developments in quantifying slide probability include techniques used in palaeoseismology (Geist and Parsons 2009; Geist et al. 2013). These novel studies present robust approaches accounting for different frequency distributions and address uncertainties using Bayesian and Monte Carlo methods for sub-optimal sample sizes (N < 14). Geist and Parsons (2009) outlined the need for more age dates of global and local landslide sequences to determine longer-term variations. We address this by studying slide deposits from multiple sites (Fig. 51.1), but unlike most previous studies that sample close to the slide initiation point, we use a method for documenting the time periods between large slides around a basin margin. We analyse turbidites inferred to be generated by large slides as detailed in Clare et al. (2014). The recurrence time of slides is inferred from intervals of hemipelagic fallout between turbidity currents, and the average accumulation rate of hemipelagic mud between dated horizons. Therefore, our method provides information on timing of many different slides from a small number of cores or outcrops, with large numbers of slides needed for robust statistical analysis ($N > 100$).

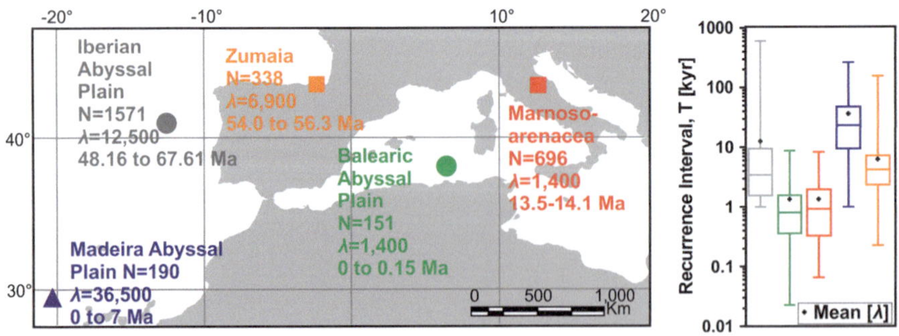

Fig. 51.1 Location of sites and time windows referenced in this paper. λ refers to the mean recurrence interval of turbidites in years, N is the number of landslide-turbidites sampled at each site. Box (25–75 %) and whisker (5–95 %) plots on right are annotated with a black cross for λ

It avoids the need to date prohibitively large numbers of slides, each in a different location on the margin.

51.2 Data

The landslide-turbidite records analysed here come from distal basin plain locations that include recent cored sediments, ancient cored records, and ancient rock outcrops. The sampled time windows range from 0.15 to 19.45 Ma and feature up to 1571 recurrence intervals. We specify a minimum threshold of $N = 100$ observations for each dataset to ensure that we have confidence in the results. This is a conservative approach based on conclusions from various statistical studies (Hunt et al. 2014). Geist et al. (2013) provide excellent methodologies for how errors and uncertainties in dating can further be addressed by statistical analysis.

51.3 Statistical Analysis of Slide Recurrence

First, we introduce statistical methodologies to characterise slide recurrence that are routinely used in other disciplines, including medicine, economics, and manufacturing. These techniques are used to determine the time before fatality, major depreciation or failure, so in the same manner they can be applied to assessment of slides. We present a summary of some results from recent studies to show how the techniques can be applied. As we focus on the method it is not our intention to provide detailed results here. Second, we test whether slides occur randomly in time, or if they show time-dependence. Recent studies have indicated that large slides may be random (or at least indiscernible from a Poisson distribution) in time (Geist et al. 2013; Urlaub et al. 2013; Clare et al. 2014). If landslide recurrence shows a Poisson distribution, hazard rate does not change with time. For other distributions, hazard rate varies with time (i.e. time-dependence). Finally, we test whether explanatory variables, such as sea level and global temperature exert control on recurrence intensity (i.e. hazard rate). This provides new insights as to whether future climate change may increase, or even decrease, the likelihood of slide recurrence.

51.3.1 What Is the Characteristic Frequency of Slide Recurrence?

We identify the frequency distribution of recurrence intervals by determining goodness of fit on exceedance plots (e.g. Van Daele et al. 2014). We consider

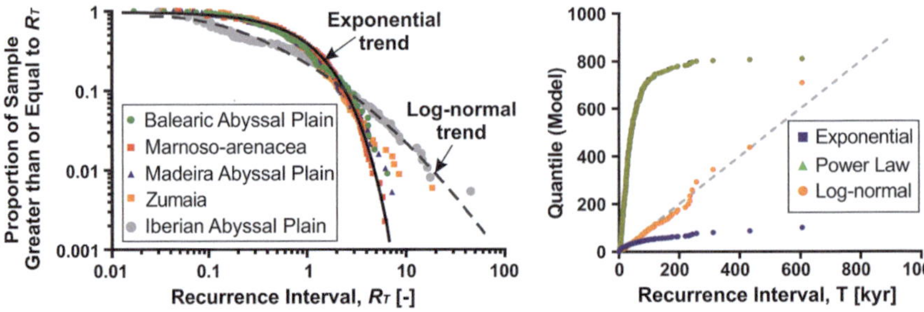

Fig. 51.2 Comparison of log-normal and exponentially distributed recurrence intervals normalised to the mean recurrence value (λ) to determine R_T (*left*). Quantile-quantile plot for log-normal Iberian Abyssal Plain recurrence intervals in kyr (*right*)

three main forms as this is appropriate for assessing variables conforming to the majority of common statistical distributions (Van Rooij et al. 2013). This includes Gaussian forms relating to addition of independent variables, log-normal distributions arising from multiplication of normally or log-normally distributed independent variables, and a Power Law where interdependent processes cause an amplification of effects (Van Rooij et al. 2013). A straight line trend indicates an exponential distribution on a log-linear plot, log-normal distribution on a log-probability plot, and power law distribution on a log-log plot. Trendlines are applied by least squares method to provide a coefficient of determination (R^2) to test their fit. This method has limitations for Power Law distributions, however, hence we also use Quantile-quantile (Q-Q) plots to compare the shape of theoretical frequency distributions against the observed data. In this paper we do not consider segmented-Power Law, nor bimodal or polymodal overlapping log-normal distributions but focus instead on simple end-member distribution scenarios (Fig. 51.2).

We observe two distinct trends of recurrence distribution. The dominant form, seen in four of the five data sets is a Poisson (or exponential) distribution; indicative of random, time-independent behaviour. This may indicate a random trigger, or may relate to the additive effect of different normally-distributed effects (e.g. different input sources around the basin margin; Clare et al. 2015). The Iberian Abyssal Plain data conform closely to a log-normal distribution, however, which indicates non-random recurrence; and therefore some degree of time-dependence. This may arise from a log-normally distributed trigger or due to multiplicative effects (e.g. probability of a trigger multiplied by the probability of failure and/or probability of the flow reaching the basin plain). To a first order it may be appropriate to use the mean recurrence value for the exponentially distributed data to characterise future hazard rate on a regional basis (i.e. hazard rate is constant with time). This is not the case for the log-normally distributed data which requires a greater understanding of the forcing variables in order to determine how recurrence may change with time.

51.3.2 Are Slides Clustered in Time?

The rescaled range exponent (Hurst 1951) provides a method of determining the degree of clustering of low and high values within a sequence and has previously been applied to studies of turbidite bed thickness (Chen and Hiscott 1999). It is used in economics to assess potential for stock market crashes and does not require any prior knowledge on the form of frequency distribution (Kristoufek 2012). Hurst (1951) presented the following relationship, using K as an estimator for a modified Hurst exponent, h:

$$K = \frac{log_{10}(R/S)}{log_{10}(N/2)} \tag{51.1}$$

where R is the maximum range in cumulative departure from the mean, N is the number of observations, and S is the standard deviation. Values that are closer to $K = 1$ show persistent or trend reinforcement (Mandelbrot and van Ness 1968), and those that are closer to $K = 0$ are mean-reverting or anti-persistent (Barkoulas et al. 2000). Values that approximate $K = 0.5$ are deemed to be randomly distributed (Hurst 1951).

For the four exponentially distributed sites, Hurst exponent values range from $K = 0.5$ to 0.6, which indicates no significant trend-reinforcement. Therefore the datasets can be viewed as randomly distributed or non-clustered with respect to the timing of preceding or succeeding events. For the Iberian Abyssal Plain site (log-normal distribution) $K = 0.96$, which indicates a high degree of trend-reinforcement (i.e. periods of clustering).

To define how the data are clustered in time we further apply a Gaussian finite mixture model (Fraley et al. 2012). The model uses an Expectation-Maximisation (EM) algorithm (Dempster et al. 1977) to iteratively search for the maximum likelihood estimate of parameters including number of clusters, shape, volume and orientation. The EM algorithm alternates between performing an expectation (E) step, which creates a function for the expectation of the log-likelihood evaluated using the current estimate for the parameters, and a maximisation (M) step, which computes parameters maximising the expected log-likelihood found on the E step (Dempster et al. 1977). The number of components (i.e. clusters) and their parameterisation are assigned according to the optimal Bayesian Information Criterion (BIC; Konishi and Kitagawa 2008) output by the model (Fig. 51.3).

Applying the finite mixture model to the Iberian Abyssal Plain data we detect significant periods of clustering, with the optimal parameterisation being a diagonal equal shape (VEI) model (Fraley et al. 2012) with in excess of nine clusters. The best model lies outside the available range of possible clusters (i.e. >9); however, the Bayesian Information Criterion starts to plateau when five to six clusters are considered so this is considered reasonable. This analysis can thus be used to identify general temporal trends. We identify several clustered periods of relatively long (~0.02 Myr) to short (~0.002 Myr) characteristic recurrence intervals. These

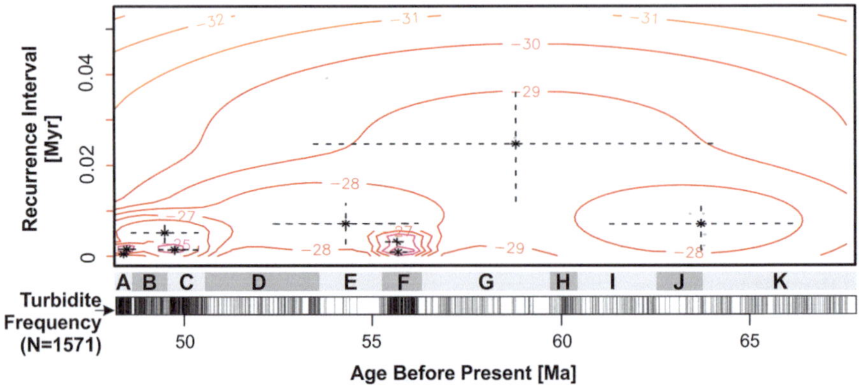

Fig. 51.3 Time series plot of the Iberian Abyssal Plain record showing individual turbidites (*below*), visually identified periods of characteristic recurrence (*middle*), and log-likelihood density contour plots of Gaussian finite mixture model (*above*). Cluster centres from the finite mixture model are shown as *stars*. *Dashed lines* correspond to the 95 % confidence intervals

clusters appear to be closely related to variations in climate and global temperatures (Talling et al. 2014).

51.3.3 Is Sea Level a Dominant Control on Slide Recurrence?

If events happen at random, the distribution of the numbers of events in a period is Poissonian and the recurrence intervals have an exponential distribution. This is given by Eq. (51.2):

$$P = (X < x) = 1 - e^{\frac{-x}{\lambda}} \tag{51.2}$$

where P is the Probability Function that a discrete random recurrence interval, X, is less than a specific value for the data series, x. The solution is related to an exponential function, e^{-x}, and the rate parameter, λ. λ is the rate at which the events on average occur (MacDiarmid et al. 1995). We are not restricted to a constant rate and it is trivial to make λ a function of time, $\lambda(t)$. We can further generalise the model and make λ a function of covariates. In principle this function can be anything, but for this analysis we consider only linear models. Linear models include polynomials and Fourier models, among others. There is a lot of latitude in our choice of model. We therefore have:

$$\lambda(t) = \mu + \alpha_1 s_1(t) + \ldots \tag{51.3}$$

where μ is the mean of the distribution, α is the dispersion parameter, s is one of any possible explanatory variables, and t is time. The framework we use is known as the

Generalised Linear Model (GLM). These models include standard linear models with Gaussian errors as well as exponential models. Here we fit a more general model with Gamma (rather than exponential errors, as the exponential is a special case of Gamma). We run an exponential regression initially to test the significance of an explanatory variable such as sea-level (and its rate of change) against event timing. The Generalised Linear Model derives an output dispersion parameter, α, for a Gamma distribution comparing event timing and sea-level. In addition to testing the timing of events in relation to sea-level, we consider hazard rate (also termed hazard function), which is the probability that an event will occur at time t given that one occurred at time $t = 0$.

$$h(t) = \frac{f(t)}{S(t)} \tag{51.4}$$

where $f(t)$ is the failure density function and $S(t)$ is the survival function (Kleinbaum and Klein 2005). The term survival derives from biological studies of mortality. These are in turn derived by the following:

$$S(t) = P(T > t) \tag{51.5}$$

$$f(t) = F'(t) \tag{51.6}$$

where T is a random variable that denotes the time of an event, P is probability, and t is a specified time (Kleinbaum and Klein 2005). The survival function $S(t)$ is therefore the probability that the time of an event occurring is later than some specified time. $f(t)$ is the rate of events, F', per unit time expressed as t. If we assume that the hazard rate, $h(t)$, is proportional to an explanatory variable x (e.g. sea-level or its rate of change) we can estimate the effect it has regardless of the form of $h(t)$. We do this by applying the Cox Proportional Hazards Model (PHM; Cox 1972) which considers the following generalisation.

$$h(t, x) = h_0(t, \alpha)\exp\left(\beta^T x\right) \tag{51.7}$$

The hazard is decomposed into a product of two items: $h_o(t, a)$ is a term that depends on time but not the covariates – α are some parameters affecting the baseline hazard function; and, $exp(\beta^T x)$, is a term that depends on the covariates but not time – β refers to baseline hazard function, x represents an explanatory variable. The null hypothesis tested is that the hazard function does not vary in response to changes in the explanatory variable. Using a model of this form tests the effects of covariates on survival and provides an estimate of β, without needing to specify the form of $h_o(t, a)$. Therefore we do not need to make any assumption about the distribution of the time between events for this model. In this model, hazards are proportional to each other and do not depend on time (Smith et al. 2003).

Studies in the Balearic and Madeira Abyssal Plains have indicated that neither sea level, nor its rate of change are shown to be statistically significant even at the

90 % level using both Generalised Linear and Proportional Hazards Models (Clare et al. 2014; Hunt et al. 2014). This is reasonable given their exponential distribution and apparent lack of clustering. In contrast, the log-normal Iberian Abyssal Plain data shows a strong degree of significance ($p \ll 0.05$) when considering sea level as a dominant control on recurrence assuming a 1 Myr lag effect, as identified by autocorrelation. The hazard ratio, E^b, which is output from this model, indicates that turbidity current activity becomes 1 % (±0.3 % based on 95 % confidence intervals) more likely with each 1 m drop in sea level. Thus, where correlation is identified, the Proportional Hazards Model may provide a useful tool for future hazard assessments.

51.4 Conclusions

We have outlined a variety of statistical analyses that can be used to characterise recurrence of submarine slides if sufficient number of observations ($N > 100$) exist. This is achieved using extensive records of large volume distal turbidite deposits. Our approach allows for quantification of links between recurrence and variables such as sea level or global climate, and can therefore be used to guide the inference of future hazard rate and how it may be affected by future sea level change.

Acknowledgements This paper significantly benefited from the reviews of Eric Geist and Richard Hiscott. We thank both for their time and insightful comments. This research used samples and data provided by the Ocean Drilling Program (ODP). ODP is sponsored by the U.S. National Science Foundation (NSF) and participating countries under management of Joint Oceanographic Institutions (JOI), Inc.

References

Barkoulas J, Baum C, Travlos N (2000) Long memory in the Greek stock market. Appl Financ Econ 177–184. doi:10.1080/096031000331815

Chen C, Hiscott RN (1999) Statistical analysis of facies clustering in submarine-fan turbidite successions. J Sediment Res 69:505–517

Clare MA et al (2014) Distal turbidites reveal a common distribution for large (>0.1 km^3) submarine landslide recurrence. Geology 42:263–266. doi:10.1130/G35160.1

Clare MA, Talling PJ, Hunt JE (2015) Implications of reduced turbidity current and landslide activity for the initial eocene thermal maximum–evidence from two distal, deep-water sites. Earth Planet Sci Lett 420:102–115. doi:10.1016/j.epsl.2015.03.022

Cox DR (1972) Regression models and life-tables. J R Stat Soc Ser B Methodol 34(2):187–220

Dempster AP, Laird NM, Rubin B (1977) Maximum likelihood from incomplete data via the EM algorithm. J R Stat Soc Ser B Methodol 39(1):1–38

Fraley C, Raftery AE, Murphy TB, Scrugga L (2012) Technical report no. 597, Dep. Stat. Uni. Washington. https://www.stat.washington.edu/research/reports/2012/tr597.pdf

Geist EL, Parsons T (2009) Estimating the empirical probability of submarine landslide occurrence. In: Mosher DC et al (eds) Submarine mass movements and their consequences. Adv Nat Technol Hazards Res 377–386. doi:10.1007/978-90-481-3071-9_31

Geist EL et al (2013) Estimation of submarine mass failure probability from a sequence of deposits with age dates. Geosphere 9:287–298. doi:10.1130/GES00829.1

Harbitz CB, Løvholt F, Bungum H (2014) Submarine landslide tsunamis: how extreme and how likely? Nat Hazards 72:1341–1374. doi:10.1007/s11069-013-0681-3

Hunt JE, Talling PJ, Clare MA, Jarvis I, Wynn RB (2014) Long-term 17 Ma turbidite record of the timing and frequency of large flank collapses of the Canary Islands. Geochem Geophys Geosyst 15:3322–3345. doi:10.1002/2014GC005232

Hurst HE (1951) Long term storage capacity of reservoirs. Am Soc Civ Eng 116:770–808

Kleinbaum DG, Klein M (2005) Survival analysis: a self-learning text. Statistics in the health sciences. Springer, New York

Konishi S, Kitagawa G (2008) Bayesian information criteria. Inf Criteria Stat Model 211–237. doi:10.1007/978-0-387-71887-3_9

Kristoufek L (2012) How are rescaled range analyses affected by different memory and distributional properties? A Monte Carlo study. Physica A 391:4252–4260. doi:10.1016/j.physa.2012.04.005

MacDiarmid PR et al (1995) Reliability toolkit: commercial practices edition. Rome Laboratory, Rome

Mandelbrot BB, Van Ness JW (1968) Fractional Brownian motions, fractional noises and applications. SIAM Rev 10:422–437. doi:10.1137/1010093

Smith T, Smith B, Ryan MA (2003) Survival analysis using cox proportional hazards modelling for single and multiple event time data. In: Proceedings of the twenty-eighth annual SAS users group international conference, SAS Institute, Inc, Cary, paper, vol 2003, pp 254–228. http://www2.sas.com/proceedings/sugi28/254-28.pdf

Talling PJ et al (2014) Large submarine landslides on continental slopes: geohazards and role in methane release and climate change. Oceanography 27:32–45. doi:10.5670/oceanog.2014.38

Urlaub M, Talling PJ, Masson DG (2013) Timing and frequency of large submarine landslides: implications for understanding triggers and future geohazards. Quat Sci Rev 72:63–82. doi:10.1016/j.quascirev.2013.04.020

Van Daele M et al (2014) The 600 yr eruptive history of Villarica Volcano (Chile) revealed by annually laminated lake sediments. Geol Soc Am Bull 126(3):481–498. doi:10.1130/B30798.1

Van Rooij MMJW et al (2013) A fractal approach to dynamic inference and distribution analysis. Front Physiol 4:1–16. doi:10.3389/fphys.2013.00001

Chapter 52
Morphological Controls on Submarine Slab Failures

Oliver J.N. Dabson, John Barlow, and Roger Moore

Abstract An understanding of submarine mass movements is of great importance to the hydrocarbon industry due to the risk they pose to sea floor infrastructure. Technological developments in deepwater surveying methods have produced datasets of the sea floor that rival the best terrestrial ones; however, the study of submarine mass movements remains poorly-developed. Multivariate statistical analysis has a well-established track record for producing quantitative estimates of associated risk for terrestrial landslides and given the often homogenous nature of sea floor sediments, a morphological control on mass movements seems viable. In this study, we perform a statistical analysis on an inventory of shallow slab slides in the West Nile Delta to identify morphometric controls on failure. We find that slopes with planar plan curvature and slope angles $<6^{\circ}$ account for approximately 95 % of observed landslides and that, beyond this, increasing plan concavity stabilises submarine slopes. This presents a foundation to ultimately reconcile geomorphological observation with geotechnical modelling, and provide additional insight on the controls on submarine instability.

52.1 Introduction

Recent technological developments have produced bathymetric datasets with resolutions that parallel those of their terrestrial counterparts. Despite this, there is still a reliance on terrestrial analogues to explain the sea floor movements, and submarine process mechanics are deemed complex and poorly understood (Masson et al. 2006; Urgeles 2012). Although the effects of submarine instabilities have been well-documented (e.g. Hampton et al. 1996; Locat and Lee 2002;

O.J.N. Dabson (✉)
CH2M, Elms House, 43 Brook Green, London W6 7EF, UK
e-mail: oliver.dabson@ch2m.com

J. Barlow
Department of Geography, University of Sussex, Brighton BN1 9RH, UK

R. Moore
CH2M, Lyndon House, 62 Hagley Rd, Birmingham B16 8PE, UK

G. Lamarche et al. (eds.), *Submarine Mass Movements and their Consequences*,
Advances in Natural and Technological Hazards Research 41,
DOI 10.1007/978-3-319-20979-1_52

Greene et al. 2006), as exploration for hydrocarbon reserves extends further offshore, an understanding of sea floor processes is essential to mitigate against substantial risk.

Methods of assessing submarine slope stability typically employ both geomorphological and geotechnical facets. Geomorphological mapping provides a qualitative appreciation of landscape form and process, which is a fundamental tool in terrestrial risk management (Hough et al. 2011; Otto and Smith 2013). Despite this, geomorphological mapping often takes a supporting role in seabed projects, possibly as it yields few quantitative results. Factor of Safety (FoS) analysis from geotechnical testing provides this quantitative element (Li et al. 2014), but outputs commonly disregard visual indicators of instability such as previously-failed slabs and crown scarps. This results in contradictions in the delineation of 'stable areas' when frameworks are applied together, but the inherent limitations of each mean that they are also not entirely reliable in isolation. This paradox reflects the state of understanding of submarine processes and suggests the need for fresh consideration of sea floor dynamics.

In this study, we investigate morphometric controls on shallow translational slab failures in the West Nile Delta (WND). Surface characteristics from an inventory are analysed to identify combinations pertinent to slab sliding. The use of inventory maps to quantify our understanding of mass movements is common in terrestrial studies to dynamically link form and process (Barlow et al. 2009; Mazzanti and De Blasio 2010); consequently, the methodology presented here has been adapted from standard procedure. Here, we largely consider morphological factors given the observed lack of geotechnical variability over the site. It is hoped that this study will shed some light on key morphometric controls on submarine slope stability to reduce uncertainty in submarine risk assessment.

52.2 Site Description

The WND lies on the Nile Deep Sea Fan (NDSF), a sedimentary wedge of *c*.100,000 km^2 and thickness of over 9 km in places (Fig. 52.1) (Mascle et al. 2003). The present-day state of the fan originates from the late Miocene (Sharp and Samuel 2004). Evaporite deposits from the Mediterranean desiccation have been covered by clastic sedimentation from the Nile, interrupted by episodes of erosion and landsliding over a least the last 250,000 years (Moore et al. 2007). Deposited material flows down two fluvial/sedimentary pathways: the Damietta (east) and Rosetta (west), to a maximum depth of approximately 3500 m BSL (Lonke et al. 2006).

The western province of the NDSF is the best-documented and has been comprehensively surveyed with focus on geohazard mitigation (e.g. Hough et al. 2011; Dimmock et al. 2012). Soils are soft marine clays of terragenous origin and hemipelagic clays whose mechanical properties are influenced by sedimentation and post-depositional history (Moore et al. 2007; Lonke et al. 2009). These are

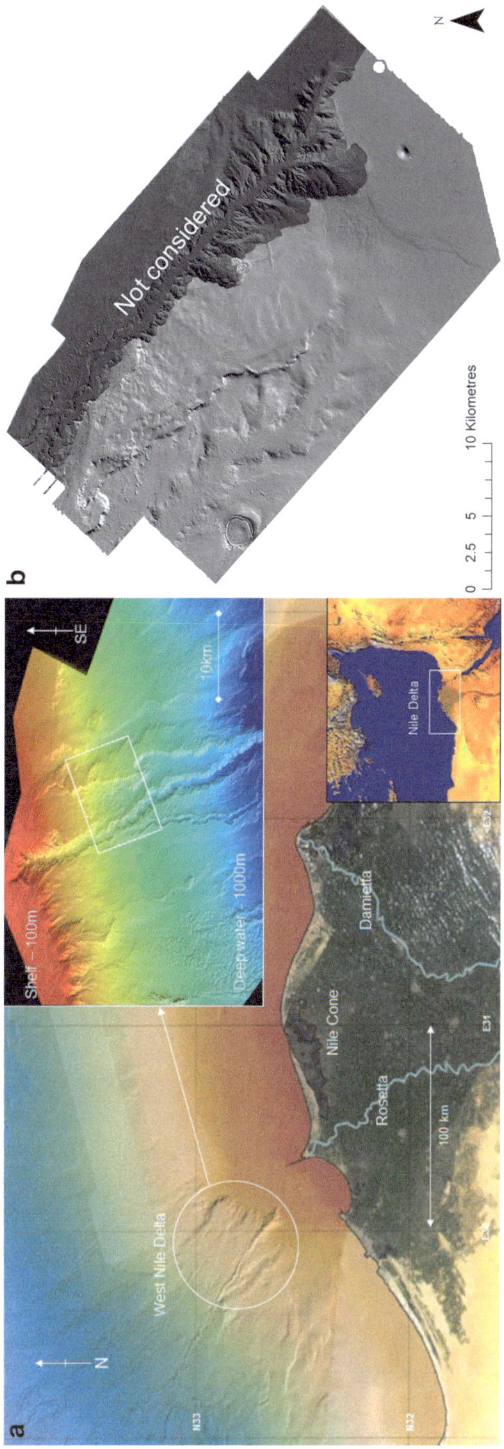

Fig. 52.1 (**a**) Location of the WND (After Moore et al. 2007); (**b**) Multi-beam Echosounder (MBES) hillshade image of the study site

underlain by two evaporite facies marking the dessication, and lago-marine sediments from a lacustrine event preceding the refilling of the Mediterranean (Lonke et al. 2006). The Rosetta canyon is also a conduit for sediment transfer processes, which explains the frequency of turbidite deposits in this area of the fan (Sharp and Samuel 2004).

The study site is to the west of the Rosetta canyon between 300 and 900 m BSL. The area is characterised by first-time slab failures with a shallow shear surface of <20 m, triggered either seismically or from the release of gas hydrates (Dimmock et al. 2012; Urgeles 2012). These shallow geomorphological processes have had a greater role in shaping the area than larger processes, such as local fault displacement, which are more influential in western areas of the WND (Evans et al. 2007).

52.3 Methods

52.3.1 Data Sources

3D seismic and MBES datasets were used to determine sea floor topography (further information on data quality is provided in Table 52.1). Polygons delimiting the locations of all shallow translational landslides were drawn to produce an inventory of 267 events.

52.3.2 Determining Pre-failure Conditions

Perhaps the most commonly-used method to determine pre-failure conditions is through back analysis (Hu et al. 2009), which has mainly been used in study of large-scale sea floor events (e.g. De Blasio et al. 2005; Vardy et al. 2012). This approach is unfeasible here given the number of events and scale of the area of interest. Hence, we use data of different spatial resolutions as a proxy for pre- and post-failure states of the environment. This is based on observations made during mapping rather than from literature, which would be difficult to obtain given the scant publication of high-resolution datasets. Coarser data partially mask post-failure morphologies such as hummocky terrain and lateral scarps; consequently, 3D seismic acts as the pre-failure proxy, and MBES represents the current state.

Table 52.1 Typical data resolution and penetration (After Hough et al. 2011)

	Resolution (m)		Penetration below sea bed (m)
Dataset	Horizontal	Vertical	
3D seismic[a]	12.5	10	5000
MBES[b]	3	None	None

[a]Limited resolution for top 15 m
[b]Multi-beam echosounder

During mapping, it was noted that large-scale deformations found at the slope toe were not hidden by coarser resolution data. To avoid incorporating these structures into analysis, data was sampled from the initiation zone, defined as material upslope of the crown, where the least distortion occurs following failure (Fleming et al. 1999). These areas were outlined using MBES, and 25 random sample points per slide were extracted from 3D seismic to obtain 'pre-failure' metrics.

Surface morphometry was ascertained from the DEM. Various first- (slope, aspect) and second- (plan, profile and gross curvature) order slope derivatives were calculated and extracted in GIS. These were examined alongside elevation and attribute data from landslide polygons which included depth and aerial extent.

52.3.3 *Statistical Analysis*

A magnitude-frequency analysis was carried out using linear axes and appropriate linear bins. Frequency density was calculated after Malamud et al. (2004):

$$f(E_s) = \delta N_s / \delta V_s \tag{52.1}$$

where $f(E_s)$ is the frequency density of an event with parameter value E_s, δN_s is the number of events that fall within bin E_s and δV_s is bin width. Frequency density was then normalised over area to calculate frequency density per square kilometre.

52.4 Results and Analysis

The quality of the frequency distributions was variable and only those producing significant trends are discussed here. Results for aspect appeared to show a strong correlation between slope direction and frequency density of failure, but this reflected failures down the delta slope (NW) or canyon levees (NE-E). Consequently, aspect was also omitted from further analysis.

The distribution for the slope metric showed the clearest trend. Plotting normalised frequency density against 1° bins indicated a three-stage relationship with slab sliding (Fig. 52.2). Figure 52.2a shows failure frequency increase until 6°. The distribution of this may be fitted with an exponential trendline ($r^2 = 0.9978$) (Fig. 52.2b). The second section shows a downward trend, followed by rising limb as frequency density begins to increase. This rise may also be fitted with an exponential curve ($r^2 = 0.9997$).

It is plausible that, at 6°, an internal system threshold is exceeded and that the limiting factor controlling failure changes. Landslide frequency directly increases with slope angle for gradients of 0–6°; hence these failures are slope-limited. It

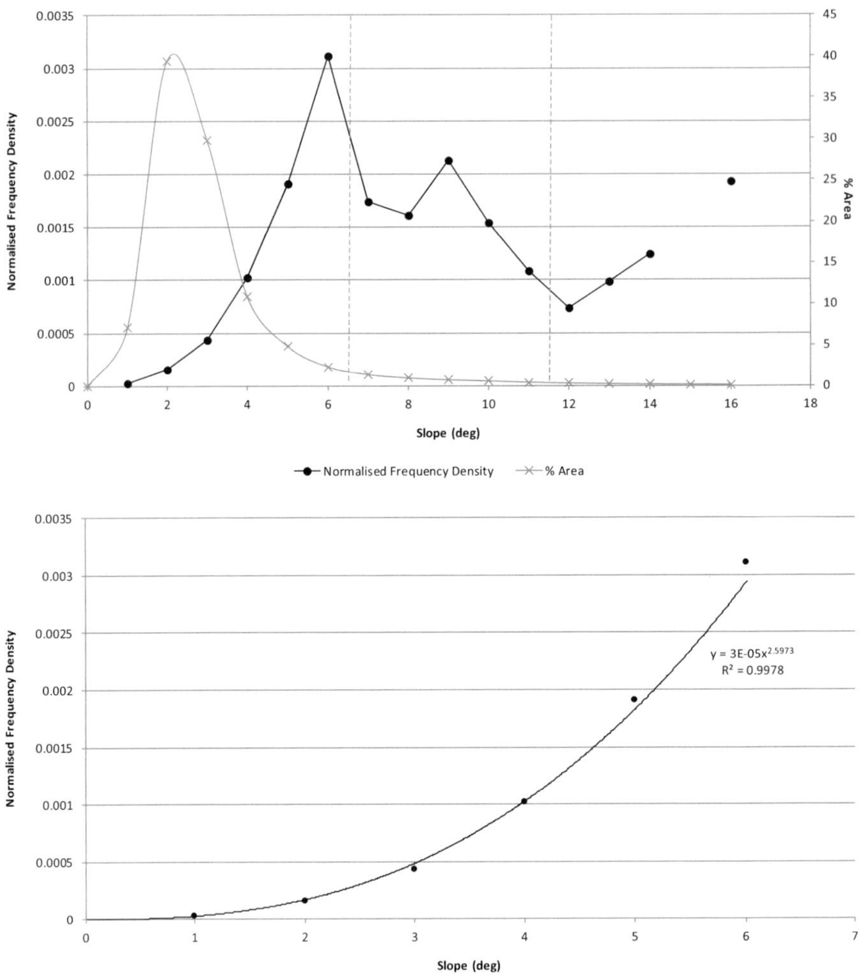

Fig. 52.2 (**a**) Normalised frequency density against slope; (**b**) emphasis on the exponential trend

would seem that the relative importance of another limiting factor increases beyond this point, stabilising the slope. The boundary between the second and third sections of the graph indicate that the destabilising effects of slope overcome this new factor, continuing the original trend. This third threshold is tentative as each bin is populated by a singular landslide as slopes of this angle become increasingly rare.

To assess what may have become the dominant control after 6°, data was sorted by slope and the arithmetic mean of each variable was calculated per degree. Of particular interest is the plan curvature output (Fig. 52.3). There is a difference between 6° and 7° and a negative trend indicating increased plan concavity. The secondary threshold at 11° is also apparent as this trend becomes less distinct; however, this may be due to the dataset size.

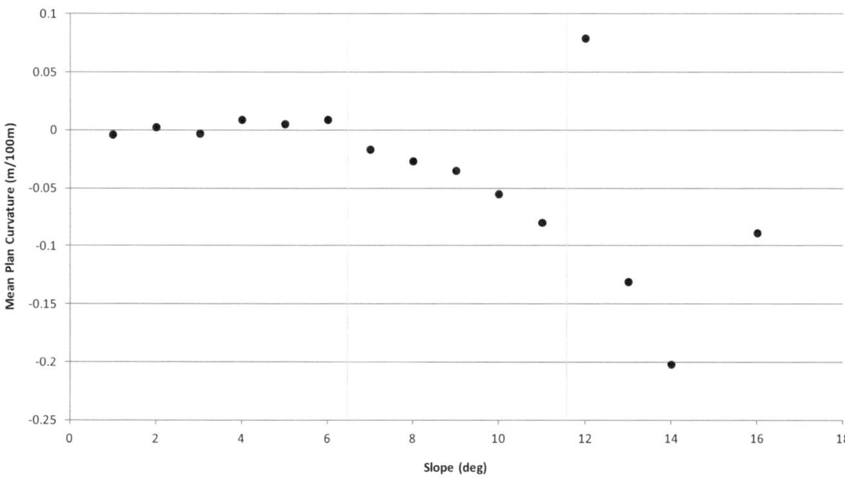

Fig. 52.3 Distribution of mean plan curvature against slope bins

52.5 Discussion

The distribution in Fig. 52.2b may be an important finding as it suggests that slope angle may be a primary control on slab sliding until a threshold at 6°. Existing mapping approaches use a working value of 5° to denote the boundary between 'shallow' and 'moderate' slopes, and current infinite slope models converge ~7.5° suggesting that the threshold is here for geotechnical classifications. It has been documented that a difference in slope of $<1°$ can be significant in modelling factor of safety (e.g. Locat and Lee 2002; Owen et al. 2007), so it can be expected that these findings will be of use in fine-tuning modelling frameworks to better represent deepwater slope stability.

The proposed influence of plan curvature on stability is also notable. Terrestrial landslide susceptibility analyses indicate that failures are partly controlled by hillslope geometry, and by extension hillslope curvature should play a key role (Rowbotham and Dudycha 1998). However, in many cases only the gross curvature metric is used (e.g. Lee et al. 2004). Where curvature is subdivided, effects on stability differ, which seems to be a function of topography and the mode of slope failure. For shallow slides in complex topography, plan curvature has a strong effect on subsurface groundwater flow, which differentially alters soil strength (Talebi et al. 2008). As plan shape changes from convex to concave, soils become more saturated as groundwater is focused into hollows (Gao 1993). This increases pore water pressure and decreases shear strength, reducing stability. By contrast, Pike and Sobieszczyk (2008) observe no plan shape preference for shallow landslides on soil-mantled hillslopes following heavy rainfall, but this is likely due to rapid soil saturation and discontinuous sheet flow which may prolong elevated positive pore water pressures to a stage where failure occurred without control from plan geometry.

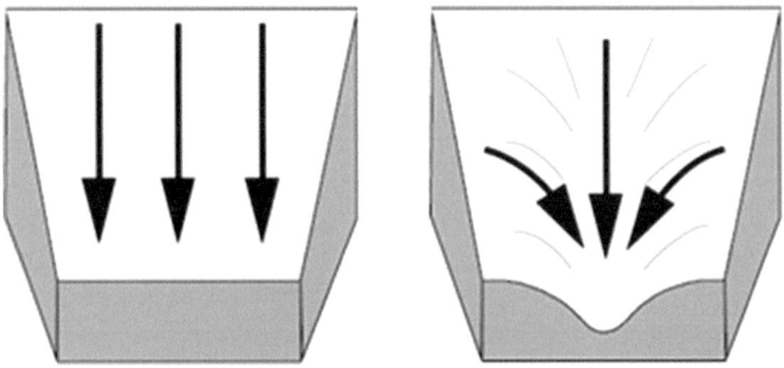

Fig. 52.4 Cohesive sediment convergence on concave plan slopes, increasing upslope stability

Hence, these curvature results are a complete reversal from what might be expected for terrestrial locations. This may be due to the relative homogeneity of submarine pore water pressures. In the absence of significant subsurface water movement, there may be a convergence or divergence of matter causing differences in stability. As particles move in the direction of steepest descent, they congregate in concave plan curvatures, resulting in lateral forces out of the plane of landslide motion. The effects of lateral forces on stability have been documented; Hutchinson (1961) includes friction on the lateral boundaries of landslides as a factor which increases shear strength, and slope stability analyses incorporating lateral forces indicate a greater factor of safety against those that do not (Hungr et al. 1989). Cohesive sediments allow for further lateral friction (Gens et al. 1988). To explain the mechanics, Ohlmacher (2007) uses the analogy of a bridge composed of trapezoidal blocks which represent sections of cohesive sediment. The plan curvature causes sediment to converge in the same way that blocks are driven together under gravity. This increases friction between sedimentary sections, increasing stability upslope (Fig. 52.4). However, this is a purely speculative suggestion, as there is little literature on the effects of curvature on slope stability that does not explicitly cite groundwater hydrology. This is seen as a viable direction for future research to better characterise this process domain.

As well as outlining new directions for academic research, the results of this study have some commercial applications. The statistical results may inform a numerical modelling exercise to map the observed frequency of failure across a site on a pixel-by-pixel basis, in a manner similar to Barlow et al. (2009). This would add a quantitative element to geomorphological sea floor studies, and also highlight unstable areas that are more difficult to detect 'by eye'.

Some limitations have been identified. The omission of geotechnical parameters may result in an incomplete picture of the system. Given the lack of observable subpopulations within these data it was thought that the influence of geotechnics on these modes of mass movement is relatively slight, but this has yet to be investigated. Similarly, landslides of different spatial magnitudes were characterised using the same methodology, which may inadequately represent larger failures. This may be rectified in future by using a number of sampling points proportional to the magnitude of the slide.

52.6 Conclusions

This paper provides the basis for quantitative geomorphological landslide susceptibility approach through statistical analysis of an inventory. It outlines a novel technique to determine sea bed stability and possible controls within the system. The threshold at 6° for slope-limited failures, which explains some 95 % of inventory landslides, is a key finding. The proposed influence of plan curvature on slope stability has also not been widely recognised in previous publications.

Roll-out of the methods in this study to other sites and other forms of submarine mass movement may result in a more complete understanding of the controls on submarine mass movements based on their morphometry. The statistical analysis of additional sites may also identify patterns in the dominant controls on failure. This will provide the foundations upon which to reconcile geotechnical and geomorphological methods for determining submarine hazard and risk.

Acknowledgments The authors wish to acknowledge BP for providing the data, and Dr Andy Mills (AM Geomorphology Ltd) for further correspondence. We also appreciate the helpful manuscript reviews by Dr Stephen Howell and Dr Matthew Brain, and the additional comments of Dr Joshu Mountjoy and Dr Geoffroy Lamarche during the editorial process.

References

Barlow J, Martin Y, Franklin S (2009) Evaluating debris slide occurrence using digital data: paraglacial activity in Chilliwack Valley, British Columbia. Can J Earth Sci 46(3):181–191

De Blasio FV, Elverhøi A, Issler D, Harbitz CB, Pryn P, Lien R (2005) On the dynamics of subaqueous clay rich gravity mass flows – the giant Storegga slide, Norway. Mar Petrol Geol 22:179–186

Dimmock P, Mackenzie B, Mills A (2012) Probabilistic slope stability analysis in the West Nile Delta, offshore Egypt. In: Proceedings of the 7th International conference on offshore site investigation and geotechnics: integrated technologies - present and future, pp 535–542

Evans T, Usher N, Moore R (2007) Management of geotechnical and geohazard risks in the West Nile Delta. In: Proceedings of the 6th offshore site investigation and geotechnics conference: confronting new challenges and sharing knowledge, London, 8pp

Fleming RW, Baum RL, Giardino M (1999) Map and description of the active part of the Slumgullion landslides, Hinsdale County, Colorado, Geologic investigations series map I-2672. U.S. Geological Survey, Denver

Gao J (1993) Identification of topographic settings conducive to landsliding from DEM in Nelson County, Virginia, USA. Earth Surf Proc Land 18:579–591

Gens A, Hutchinson JN, Cavounidis S (1988) Three-dimensional analysis of slides in cohesive soils. Géotechnique 38:1–23

Greene HG, Murai LY, Watts P, Maher NA, Fisher MA, Paull CE, Eichhubl P (2006) Submarine landslides in the Santa Barbara Channel as potential tsunami sources. Nat Hazards Earth Syst 6:63–88

Hampton MA, Locat J, Lee H (1996) Submarine landslides. Rev Geophys 34(1):33–59

Hough G, Green J, Fish P, Mills A, Moore R (2011) A geomorphological mapping approach for the assessment of seabed geohazard and risk. Mar Geophys Res 32(1–2):151–162

Hu H, Fernandez-Steeger TM, Azzam R, Arnhardt C (2009) 3D modelling of landslide in open-pit mining on basis of LIDAR data. Geotech Res Abstr 11:EGU2009–EGU5378

Hungr O, Salgado FM, Byrne PM (1989) Evaluation of a three-dimensional method of slope stability analysis. Can Geotech J 26:679–686

Hutchinson JN (1961) A landslides on a thin layer of quick clay at Furre, central Norway. Géotechnique 40:139–143

Lee S, Ryu JH, Won JS, Park HJ (2004) Determination and application of the weights for landslide susceptibility mapping using an artificial neural network. Eng Geol 71:289–302

Li W, Wu S, Völker D, Zhao F, Mi L, Kopf A (2014) Morphology, seismic characterization and sediment dynamics of the Baiyun Slide Complex on the northern South China Sea margin. J Geol Soc Lond 171:865–877

Locat J, Lee HJ (2002) Submarine landslides: advances and challenges. Can Geotech J 39:193–212

Lonke L, Gaullier V, Mascle J, Vendeville B, Camera L (2006) The Nile deep-sea fan: an example of interacting sedimentation, salt tectonic, and inherited subsalt paleotopographic features. Mar Pet Geol 23:297–315

Lonke L, Gaullier V, Droz L, Ducassou E, Migeon S, Mascle J (2009) Multi-scale slope instabilities along the Nile deep-sea fan, Egyptian margin: a general overview. Mar Pet Geol 26:633–646

Malamud B, Turcotte D, Guzzetti F, Reichenbach P (2004) Landslide inventories and their statistical properties. Earth Surf Proc Land 29:687–711

Mascle J, Camera L, Chamot-Rooke N, Costis C, Gaullier V, Loncke L, Nielsen C, Operto S, Ribondetti A, Sage F, Sallares V, Schenini L (2003) New constraints on the deep structure of the eastern Mediterranean sea from new MSC seismic reflection data, EGS-AGU-EIG Joint Assembly. ISSN: 1029-7006, Abstract Number: EAE03-A-09172

Masson DG, Harbitz CB, Wynn RB, Pedersen G, Løvholt F (2006) Submarine landslides: processes, triggers and hazard prediction. Phil Trans R Soc A 364(1845):2009–2039

Mazzanti P, De Blasio FV (2010) Peculiar morphologies of subaqueous landslide deposits and their relationship to flow dynamics. In: Mosher DC, Shipp RC, Moscardelli L, Chaytor JD, Baxter CDP, Lee HJ, Urgeles R (eds) Submarine mass movements and their consequences, vol 28, Advances in natural and technological hazards research., pp 141–151

Moore R, Usher N, Evans T (2007): Integrated multidisciplinary assessment and mitigation of West Nile Delta geohazards. In: Proceedings of the 6th Offshore site investigation and geotechnics conference: confronting new challenges and sharing knowledge, London, 10pp

Ohlmacher GC (2007) Plan curvature and landslide probability in regions dominated by earth flows and earth slides. Eng Geol 91:117–134

Otto JC, Smith MJ (2013) Geomorphological mapping. In: Clarke LE, Nield JM (eds) Geomorphological techniques (online edition). British Society for Geomorphology, London, ISSN: 2047-0371

Owen M, Day S, Maslin M (2007) Late Pleistocene submarine mass movements: occurrence and causes. Quat Sci Rev 26:958–978

Pike RJ, Sobieszczyk S (2008) Soil slip/debris flow localized by site attributes and wind-driven rain in the San Francisco Bay region storm of January 1982. Geomorphology 94:290–313

Rowbotham DN, Dudycha D (1998) GIS modelling of slope stability in Phewa Tal watershed, Nepal. Geomorphology 26:151–170

Sharp A, Samuel A (2004) An example study using conventional 3D seismic data to delineate shallow gas drilling hazards from the West Delta Deep Marine Concession, offshore Nile Delta, Egypt. Pet Geosci 10:121–129

Talebi A, Troch PA, Uijlenhoet R (2008) A steady-state analytical slope stability model for complex hillslopes. Hydrol Process 22:546–553

Urgeles R (2012) Fluid flow focusing in passive continental margins: significance to submarine slope instability. In: Eberhardt E, Froese C, Turner K, Leroueil S (eds) Landslides and engineered slopes: protecting the public through improved understanding. Taylor & Francis, London, pp 73–82

Vardy ME, L'Heureux JS, Vanneste M (2012) Multidisciplinary investigation of a shallow near-shore landslide, Finneidfjord, Norway. Near Surf Geophys 10(4):267–277

Chapter 53
Incorporating Correlated Variables into GIS-Based Probabilistic Submarine Slope Stability Assessments

William C. Haneberg

Abstract First-order, second-moment (FOSM) formulations are useful tools for assessing uncertainty in GIS based submarine slope stability models. In the simplest applications, variables are assumed to be uncorrelated. In some cases, however, correlation among variables may be significant enough to require inclusion. Correlations among variables can be incorporated by creating an empirical covariance matrix and combining it with analytically derived expressions for partial derivatives of the factor of safety equation. Example calculations show that ignoring correlated variables over-predicts the probability of sliding for gentle slopes and under-predicts the probability of sliding for steep slopes, with small differences for moderate slopes. GIS-based application is illustrated using a hypothetical example motivated by an actual deepwater geohazard assessment, showing areas in which the use of uncorrelated rather than correlated variables over-predicts the likelihood of instability.

53.1 Introduction

Evaluation of submarine geohazards often requires recognition of previous slope instability as well as the likelihood of future instability associated with triggers such as major earthquakes. This can be done qualitatively based on the presence or absence of past landslide deposits, semi-quantitatively using correlations with empirical measures such as slope angle, or quantitatively using limit equilibrium slope stability methods. Limit equilibrium methods are attractive because they integrate, albeit in a simplified way, the essential physics of slide initiation and allow evaluation of rare or unprecedented conditions that may not be represented by the geologic history of the site. They also require, however, knowledge of geotechnical variables such as sediment shear strength, thickness, and unit weight in addition some description of slope geometry (minimally the slope angle). This

W.C. Haneberg (✉)
Fugro GeoConsulting, Inc., 6100 Hillcroft, Houston, TX 77081, USA
e-mail: whaneberg@fugro.com

G. Lamarche et al. (eds.), *Submarine Mass Movements and their Consequences*, Advances in Natural and Technological Hazards Research 41,
DOI 10.1007/978-3-319-20979-1_53

can hinder their use over wide areas and during early stages of investigations, when geotechnical data may be sparse, uncertain, or both.

Using first-order, second-moment (FOSM) approximations to estimate uncertainty in calculated factors of safety is an attractive alternative to computationally intensive Monte Carlo simulations, especially for reconnaissance studies. FOSM approximations have long been accepted in both geotechnical and GIS applications (e.g. Harr 1987; Wu et al. 1996; Burrough and McDonnell 1998). Van Westen and Terlien (1996), Haneberg (2004a, 2012), and Haneberg et al. (2009) discuss their application to spatially distributed slope stability assessment.

In the simplest FOSM applications, variables are assumed to be uncorrelated and the variance for an appropriate factor of safety equation is found through straightforward but tedious derivations (e.g. Haneberg 2012). This can be an acceptable in some situations. In other cases, however, correlation among variables may be significant enough to require inclusion. This paper shows how correlated geotechnical variables can be incorporated into a GIS-based FOSM infinite slope stability assessment.

53.2 Theory

The factor of safety against sliding for an infinite slope under pseudostatic seismic conditions used for this analysis is (e.g., Morgenstern 1967; Haneberg 2012)

$$F = \frac{S_u}{H\ \gamma_b \sin\theta \cos\theta + H\ k(\gamma_b + \gamma_w)\ \cos^2\theta} \tag{53.1}$$

In Eq. (53.1), S_u is the undrained shear strength of the sediment comprising the slope (kPa), γ_b is the buoyant or submerged unit weight of the sediment (kN/m^3), γ_w is the unit weight of water (kN/m^3), H is sediment thickness above a potential slip surface (m), θ is the slope angle (degrees), and k is the pseudostatic seismic coefficient (as a fraction of gravitational acceleration g). Although one was not added here, some authors add an additional bias term to the numerator to account for effects, such as shear strength along lateral boundaries, not included in the infinite slope approximation (e.g., Dimmock et al. 2012). The nature of k has also been subject to debate among practitioners; however, it is generally conceded that it must be reduced to a fraction of the peak horizontal acceleration in order to produce credible results. Discussion of the approaches to estimating k advocated by different authors is beyond the scope of this paper.

To calculate the probability of sliding, Prob $[F < 1]$, Eq. (53.1) is used to calculate a mean factor of safety using mean values for all of the variables. The variance of the factor of safety is approximated using a first-order, second-moment (FOSM) expansion that, allowing for correlation among variables and written in compact form, is (e.g., Harr 1987; Wu et al. 1996; Haneberg 2012)

$$\mathrm{Var}(F) = \mathbf{b}^{\mathrm{T}} \cdot \mathbf{C} \cdot \mathbf{b} \tag{53.2}$$

The variance is the square of the standard deviation. In Eq. (53.2), **C** is the covariance matrix for the independent variables in Eq. (53.1) and the T indicates transposition. For i observations or measurements, the sample covariance between the two variables X_j and X_k, with mean values denoted by overbars, is

$$C_{j,k} = \frac{1}{N-1}\sum_{i=1}^{N}(X_{i,j} - \bar{X}_j)(X_{i,k} - \bar{X}_k) \tag{53.3}$$

Equation (53.3) gives the variance of a variable when $j = k$. The **b** vector consists of derivatives of the factor of safety with respect to the independent variables evaluated at the mean values of the variables, or

$$\mathbf{b}^{\mathrm{T}} = \left\{ \frac{\partial F}{\partial H}, \frac{\partial F}{\partial S_u}, \frac{\partial F}{\partial \gamma_b}, \frac{\partial F}{\partial k}, \frac{\partial F}{\partial \theta} \right\} \tag{53.4}$$

Equation (53.2) assumes a considerably simpler form if there is no correlation among variables or certain ratios of variables, and all of the off-diagonal elements of the covariance matrix vanish (e.g., Haneberg 2004b, 2012; Haneberg et al. 2013). In such a case, only the variances of the uncorrelated variables are considered. Once the factor of safety mean and variance are calculated using Eqs. (53.1) and (53.2), the probability of sliding can be calculated as the cumulative distribution function (CDF) of F at the limiting value of unity. In this paper, F is assumed to be log-normally distributed on the basis of Monte Carlo simulations (e.g., Hammond et al. 1992; Haneberg 2004a, b; Griffiths et al. 2011).

53.3 Example Model Parameterization

Figures 53.1 and 53.2 illustrate an example data set used to evaluate the importance of including correlation among variables in the analysis. The example data are based upon typical deepwater marine data but were simulated with the aid of a random number generator so as not to represent any actual project location. As is typical, both undrained shear strength (S_u) and buoyant unit weight (γ_b) increase noticeably with depth (z). Likewise, the pseudostatic coefficient k is simulated with strong depth dependence to reflect the fact that the low frequency (i.e., long wavelength) seismic waves necessary to trigger thick slides generally have lower spectral accelerations than the higher frequency waves capable of triggering thin slides.

There is also uncertainty inherent in slope angles calculated from seafloor DEMs. Therefore, a slope angle variance of $(1°)^2$, equivalent to 0.0003046 rad^2, was assumed based on previous experience with terrestrial digital elevation model errors (Haneberg 2006; also see Haneberg (2012) and Haneberg et al. (2013)).

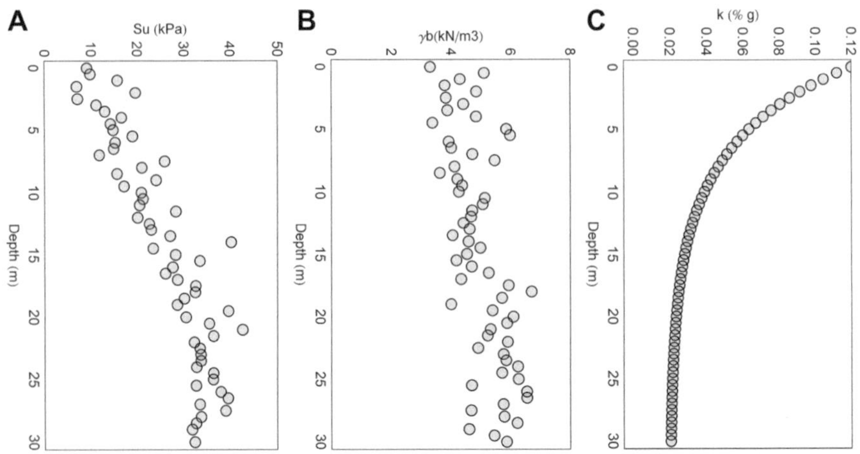

Fig. 53.1 Simulated geotechnical variable distributions used for the example calculations described in this paper. (**a**) Undrained shear strength. (**b**) Saturated unit weight (kN/m^3). (**c**) Pseuodstatic coefficient k (g)

The covariance matrix used for the example in this paper, calculated from the data shown in Fig. 53.1 using the computer program Mathematica, is:

$$\mathbf{C} = \begin{bmatrix} 78.79 & 86.14 & 4.875 & -0.1955 & 0 \\ 86.14 & 111.0 & 5.869 & -0.2310 & 0 \\ 4.875 & 5.869 & 0.6388 & -0.0131 & 0 \\ -0.1955 & -0.2310 & -0.0131 & 0.0006635 & 0 \\ 0 & 0 & 0 & 0 & 0.0003046 \end{bmatrix} \quad (53.5)$$

The zero values in the covariance matrix indicate an assumed lack of correlation of any of the other geotechnical properties with slope angle.

53.4 Results

53.4.1 Comparison of Correlated and Uncorrelated Results

Results for uncorrelated variables can be calculated by setting to zero all of the off-diagonal elements of the covariance matrix. Figure 53.3 shows calculated factor of safety distributions for mean slope angles of 5°, 10°, 15°, and 20° using both correlated and uncorrelated variables. In both cases, the factor of safety distributions shift to the left—becoming narrower and taller—as the mean slope angle increases, but at different rates. The shaded areas beneath the curves are the probability of sliding, Prob [$F < 1$], for each of the calculations. Figure 53.4 summarizes the probability of sliding, Prob [$F < 1$], calculated for both correlated

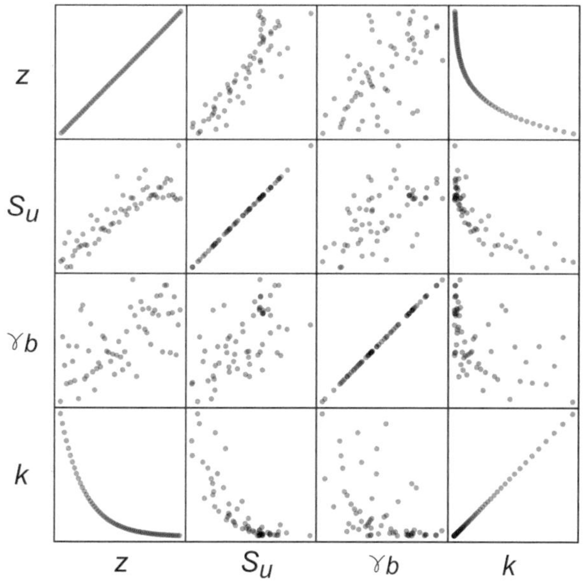

Fig. 53.2 Scatterplots illustrating correlations between geotechnical variables and depth (z)

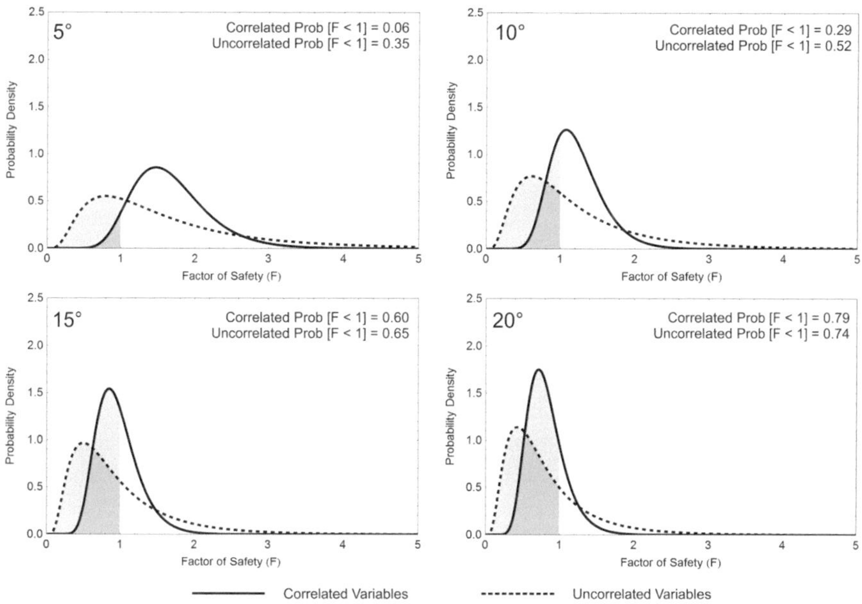

Fig. 53.3 Lognormal factor of safety (F) distributions for mean slope angles of 5°, 10°, 15°, and 20° and the correlated and uncorrelated geotechnical variable distributions shown in Fig. 53.1

and uncorrelated variables as a function of mean slope angle. One would expect an infinitely high factor of safety and therefore zero probability of sliding for a 0° slope; however, inclusion of slope angle uncertainty allows for the possibility of a non-zero slope and non-zero probability of sliding. For intermediate slopes, the

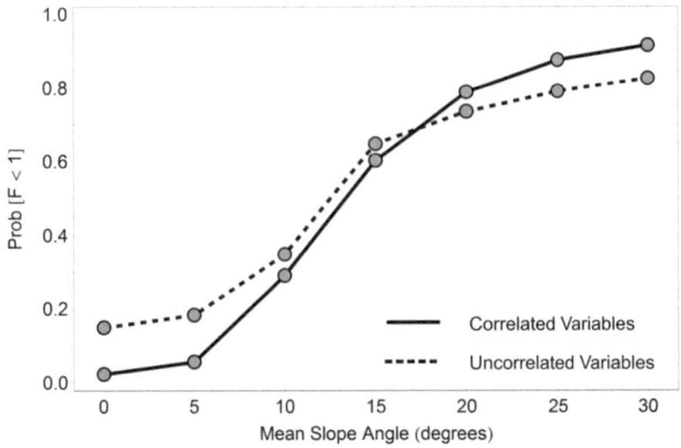

Fig. 53.4 Probability of sliding, defined as Prob [$F < 1$]>0.5 calculated using correlated and uncorrelated geotechnical variables as a function of mean slope angle. The calculations incorporated a slope angle standard deviation of ±1°

differences between correlated and uncorrelated variables in this example are 5 % or less. Slope angles for which the differences increase are either very low (for which potential instability would typically not be a concern) or very high (for which potential instability would typically be suspected).

53.4.2 GIS Application

The FOSM method can be extended to map-based analyses by using the slope angle calculated for each raster in a digital elevation model (DEM) as the mean slope for that raster and then performing a calculation as described above (e.g., Haneberg 2004b, 2012; Haneberg et al. 2009). Experience has shown that the calculations may be more conveniently performed using software with more powerful mathematical capabilities than are typical in GIS programs. A workflow might be to create a slope angle map from a DEM using GIS, export the grid of slopes, import the slope grid into computational software such as Mathematica to perform the slope stability calculations, then export and re-import the results back into the GIS software for visualization and integration with other information.

One GIS-specific consideration is the need to smooth either the slope angle map or the DEM prior to calculation of slope angles if the terrain being evaluated is rough, in order to calculate slope angles that reflect a relatively smooth underlying potential slip surface rather than a highly variable seafloor. If the slope is relatively constant, however, smoothing may not be necessary.

Figure 53.5 illustrates a comparison of GIS-based FOSM slope stability calculations using both correlated and uncorrelated variables. The left image (Fig. 53.5a) shows a portion of the seafloor from an undisclosed deepwater location and with geomorphological evidence of past slope instability. The example area covers about 25 km^2, the DEM raster size is 10 m, and the contour interval (CI) is 25 m. The probability of sliding, Prob [$F < 1$] was calculated using the geotechnical information in Fig. 53.1, both with and without including correlation among variables, and

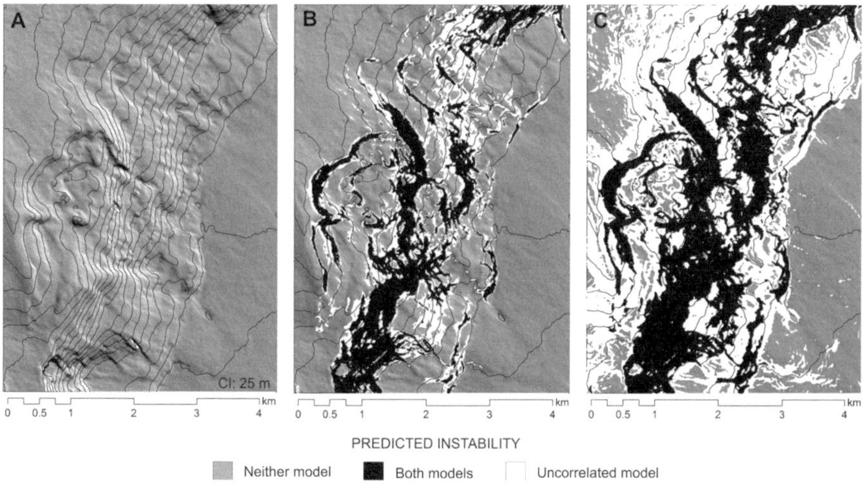

Fig. 53.5 Results from a GIS implementation of the FOSM infinite slope stability model using both correlated and uncorrelated variables. (**a**) Hillshade image of the 25 km^2 example area with 25 m water depth contours. (**b**) Predicted stable and unstable areas using the Prob[$F < 1$]>0.5 criterion. (**c**) Predicted stable and unstable areas using unsupervised iso-classification of Prob [$F < 1$] results for with the assumption of two classes

the results draped over the hillshade image. The objective of the modeling is not to account for the distribution of previous landslides (which would have required a pre-slide seafloor DEM) but rather to predict the distribution of future slides if the example area were subjected to an earthquake represented by the k values in Fig. 53.1c.

Figure 53.5b shows the results if instability is defined by the criterion Prob [$F < 1$] $\geq$0.5. Areas for which both the correlated and uncorrelated variable models predicted sliding, only the uncorrelated variable model predicted sliding, and neither model predicted sliding are indicated. There were no rasters for which only the correlated variable model predicted sliding. Using this criterion, the correlated model predicts 15.5 % of the example area to be unstable whereas the uncorrelated model predicts about twice as much, 30.5 %, of the example area to be unstable.

The classification process was repeated using iso-clustering unsupervised classification in ArcGIS 10.2 (Fig. 53.5c). Two classes were specified under the assumption that stable and unstable portions of the seafloor would form two clusters. As above, areas for which both the correlated and uncorrelated variable models predicted sliding, only the uncorrelated variable model predicted sliding, and neither model predicted sliding are indicated. Iso-clustering produced a more conservative result, with the correlated model predicting 28.9 % of the example area to be unstable and the uncorrelated model predicting 70.6 % of the example area to be unstable. As with the Prob [$F < 1$] $\geq$0.5 criterion, iso-clustering classification predicted about twice the amount of unstable seafloor for uncorrelated as for correlated variables.

53.5 Summary

First-order, second-moment (FOSM) approximations are useful tools that facilitate GIS-based probabilistic slope stability analyses of large areas. As illustrated in this paper, assuming that geotechnical variables are uncorrelated can either over- or under-predict the probability of sliding relative to that calculated using correlated variables. Differences between results for correlated and uncorrelated variables in the example described here were small but significant enough to double the proportion of the study area classified as unstable by two different methods. Thus, a lack of correlation must be demonstrated, not simply assumed in order to simplify calculations, if unnecessarily conservative assessments are to be avoided.

References

Burrough PA, McDonnell RA (1998) Principles of geographical information systems. Oxford University Press, Oxford, 333 pp

Dimmock P, Mackenzie B, Mills AJ (2012) Probabilistic slope stability analysis in the West Nile Delta, offshore Egypt. In: Allan P, 9 others (eds) Offshore site investigation and geotechnics. Proceedings of the 7th International Offshore Site Investigation and Geotechnics Conference, London, pp 535–542

Griffiths DV, Huang J, Fenton GA (2011) Probabilistic infinite slope analysis. Comput Geotech 38(4):577–584

Hammond C, Hall D, Miller S, Swetik P (1992) Level I Stability Analysis (LISA) documentation for version 2.0. US Forest Service Intermountain Research Station General Technical Report INT-285, Moscow, 190 pp

Haneberg WC (2004a) Computational geosciences with Mathematica. Springer, Berlin, 381 pp

Haneberg WC (2004b) A rational probabilistic method for spatially distributed landslide hazard assessment. Environ Eng Geosci 10:23–47

Haneberg WC (2006) Effects of digital elevation model errors on spatially distributed seismic slope stability calculations: an example from Seattle, Washington. Environ Eng Geosci 12:247–260

Haneberg WC (2012) Spatially distributed probabilistic assessment of submarine slope stability. In Allan P, 9 others (eds) Offshore site investigation and geotechnics. Proceedings of the 7th International Offshore Site Investigation and Geotechnics Conference, London, pp 551–556

Haneberg WC, Cole WF, Kasali G (2009) High-resolution lidar-based landslide hazard mapping and modeling. UCSF Parnassus Campus, San Francisco, USA. Bull Eng Geol Environ 68:263–276

Haneberg WC, Bruce B, Drazba MC (2013) Using qualitative slope hazard maps and quantitative probabilistic slope stability models to constrain least-cost pipeline route optimization. In: Proceedings of the 2013 Offshore Technology Conference, Paper OTC-23980 (in press)

Harr ME (1987) Reliability based design in civil engineering. McGraw-Hill, Englewood Cliffs, 291 pp

Morgenstern NR (1967) Submarine slumping and the initiation of turbidity currents. In: Richards A (ed) Marine geotechnique. University of Illinois Press, Urbana, pp 189–220

van Westen CH, Terlien MTJ (1996) Deterministic land-slide hazard analysis in GIS: a case study from Manizales (Columbia). Earth Surf Proc Land 21:853–868

Wu TH, Tang WH, Einstein HH (1996) Landslide hazard and risk assessment. In: Turner AK, Schuster RL (eds) Landslides, investigation and mitigation. Transportation Research Board special report 247. Transportation Research Board, Washington, DC, pp 106–118

Chapter 54
Quantifying the Key Role of Slope Material Peak Strength – Using Discrete Element Simulations

Katrin Huhn, Frank Strozyk, and Ingo Kock

Abstract This study investigates how progressive oversteepening and fault kinematics impact on slope failure initiation and subsequent landsliding along subsiding basin flanks using 2D DEM simulations. We use large assemblages of granular particles to simulate the deformation behaviour of slope sediments with varying peak strength. Sediments with high peak strength deform preferentially on major faults and produce a stepped topography and a stable slope in the long-term. Mass failures in these sediments occur as large, compact slides of short run out. In contrast, slopes with lower peak strength deform diffusely and present large numbers of faults that fail frequently and maintain the slope at its critical angle of inclination. The resulting slope topography is smoother and laterally more elongated. These differences in mass movements are governed by (i) characteristic fault patterns, and (ii) repeated oversteepening during ongoing basin subsidence, which is an important prerequisite for failure initiation. Our experiments indicate quantitatively that the failure distribution, dimension, and transport mechanism, as well as the recurrence rate of landslides are essentially controlled by the peak strength of the failed material.

54.1 Introduction

The evolution of sedimentary basins is often associated with slope failure events occurring along steepening basin flanks due to ongoing basin subsidence. Hence, tectonic movements and oversteepening exert an important control on the

K. Huhn (✉)
MARUM – Center for Marine Environmental Sciences, Universität of Bremen, Leobener Str., 28359 Bremen, Germany
e-mail: khuhn@marum.de

F. Strozyk
RWTH Aachen, Geologisches Institut, Wüllnerstr. 2, 52062 Aachen, Germany

I. Kock
GRSmbH, Schwertnergasse 1, 50667 Köln, Germany

G. Lamarche et al. (eds.), *Submarine Mass Movements and their Consequences*, Advances in Natural and Technological Hazards Research 41,
DOI 10.1007/978-3-319-20979-1_54

occurrence of gravitational mass movements (Hampton and Lee 1996). Although other mechanisms, e.g. transients in excess pore pressure (Stegmann et al. 2007) or seismic loading (Biscontin and Pestana 2006), are generally proposed as major trigger of failure events, at least a small inclination is required to cause slope destabilisation. The critical inclination depends on sediment physical properties (Hampton and Lee 1996). Hence, in general terms tectonic oversteepening plays a major role in failure pre-condition whereas sediment mechanical attributes exert an important constraint on slope failure development (Lee et al. 2007). However, the interplay between sediment physical properties and localisation and transport mechanism of landslides is still not fully quantitatively understood.

The major aim of this study is to highlight the key role of sediment physical properties, particularly the peak strength of the failed material, on slope deformation, e.g. fault activity, failure events and topography evolution. Therefore, we test two end-member materials with contrasting peak strength: a sandy-like slope sediment with high strength (material-A), versus a muddy slope with lower strength (material-B). Tectonic movement is linked to landslide patterns along different sedimentary slopes by utilizing a numerical particle based simulation technique – the Discrete Element Method (DEM). We simulate the evolution of basin flanks resulting from ongoing subsidence along a single normal fault in a 2D experiment.

54.2 Theory and Model Configuration

54.2.1 The Discrete Element Method

The Discrete Element Method (DEM) is a numerical simulation technique based on the description of granular materials. In this study, we use the commercial Particle Flow Code *PFC2D* in 2-dimensions (Itasca, Inc. ® (Itasca 2004)) which utilizes the theory by Cundall and Strack (1979) (see also e.g. Morgan and Boettcher 1999).

The numerical 'slope sediments' are simulated by an assemblages of discrete, rigid, frictionally coupled, spherical particles which interact as common contact points in accordance to simple physical contact laws (Fig. 54.1a). As we are aiming to simulate a Mohr-Coulomb material, particle micro-properties, e.g. particles' density (ρ), coefficient of friction (μ_P) and shear and normal stiffnesses (k_S, k_N) have to be defined (Table 54.1). In addition, variations in peak strength are simulated by bonding of particles (via normal (B_N) and shear bond stiffnesses (B_S)).

Particles can be displaced in any direction forced by the given boundary conditions, while interacting and overlapping (ΔD) with their neighbours (Fig. 54.1a). Resulting forces (normal – F_N; shear – F_S) are calculated using simple physical laws. Subsequently, all forces acting on a single particle are summed up; and Newton's 2nd law is used to calculate the resulting particle displacement and the new particle position. Furthermore, shear forces are used to define if particles are gliding away. Hence, maximum shear forces are simulating the peak strength of the

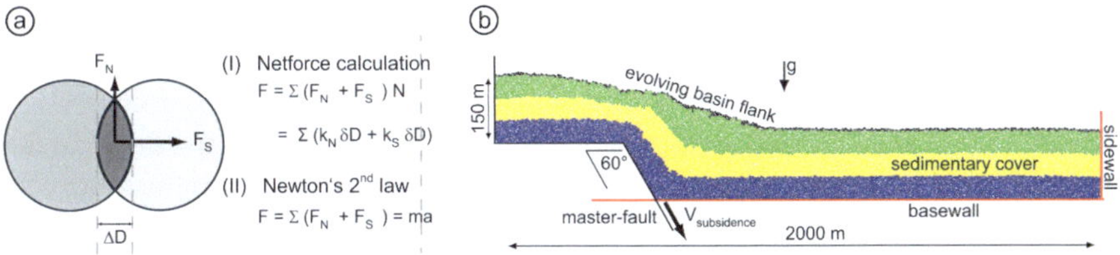

Fig. 54.1 (**a**) DEM principles of contact force calculation and displacement calculation, (**b**) basic model setup used to generate a 2D sedimentary slope simulating basin subsidence along a 60° master-normal fault. The models are performed with 0.1 [m/s] base-wall off-set until the total vertical off-set equals the 150-m-thickness of the particle ('sediment') assemblage. 'Sediment' cover is marked by coloured particles. Moving walls are marked in *red*

Table 54.1 Defined micro-(particle)-properties and calibrated macro-(material)-properties to simulate two slope sediment end-members

Micro-properties	**Material-A** ('**sand**')	**Material-B** ('**mud**')
Density ρ [kg/m^3]	2000.0	2000.0
Coefficient of particle friction μ_p	0.6	0.1
Particle stiffnesses k_N k_S [N/m^2]	1.0×10^{10}	1.0×10^{10}
Bond stiffnesses $B_{N;}$ B_S [N/m^2]	1.25×10^5	1×10^1
Wall stiffness k_N; k_S [N/m^2]	1×10^9	1×10^9
Coefficient of wall friction μ_{wall}	0.6	0.1
Macro-properties		
Coefficient of friction μ_{bulk}	0.37	0.18
Peak strength τ [kPa]	49.8	4.64

material. So a progressive breaking of contacts to discrete planes reproduces fractures and faults in nature. Besides, the micro-particle-properties of a single particle do not correspond to the bulk macro-properties of the entire particle assemblage (e.g. Kock and Huhn 2007). Hence, the values of the bulk material properties have to be defined to simulate natural material behaviour.

54.2.2 Numerical 'Slope Sediments'

As the micro-properties in the DEM do not correspond with the macro-properties of the particle assemblage, e.g., we utilize numerical direct shear tests (e.g. Kock and Huhn 2007), and standard 2-D biaxial shear tests (Itasca 2004) to calibrate micro-properties of the particle assemblage to reproduce the deformation behaviour and mechanical macro-properties in accordance to natural slope materials.

In order to test the key role of material strength on failure events, we designed two numerical sediment end-members: (A) sandy 'sediments' and (B) mud-rich 'sediments' (Table 54.1). The physical and mechanical behaviour of material-A simulates typical siliciclastic sediment with higher strength as observed e.g. in the northern Gulf of Mexico (Balsam and Beeson 2003) and the slope off Norway (Jackson et al. 2008). In contrast, material-B resembles muddy slope sediments with lower strength, e.g. from the northern Cretan margin (Strozyk et al. 2010).

54.2.3 Model Configuration

We generate a rectangular fixed box model of 2000 m length and 200 m height to simulate a flat basin. All walls are defined as impermeable and stiff boundaries (Fig. 54.1b). Approximately 22,000 particles with a Gaussian particle size distribution and diameters between 1 and 3 m are generated in this fixed box. These particles settle down under gravity generating the 150-m-thick undeformed basin 'sediment' layer. 'Sediment' thickness and particle resolution enable modelling of slide events of meter to several tens of meters thickness in a reasonable calculation time.

Basin subsidence is performed by a continuous downward motion of two thirds of the box bottom wall along a 60° inclined master-fault (Fig. 54.1b). Movement along this master-fault simulates basin flank evolution. The angle of master-fault inclination represents the average orientation of brittle material failure in nature. The basin wall subsides until the total vertical off-set equals the initial material thickness of 150 m (Fig. 54.1b). Subsidence velocity was dynamically chosen to ensure a force and momentum equilibrium at each calculation time (Cundall and Strack 1979); and low strain rates are ensured.

During each experimental run, particle configuration as well as forces are saved every 0.7–1 m subsidence. These data are processed to gain geological interpretable data, e.g. fault zones, fault kinematics and mass transfer patterns (Figs. 54.2 and 54.3).

54.3 Results and Interpretation

In all experiments, numerous fault zones evolve, basin slope flanks oversteep and gravitational mass movements occur in consequence of basin subsidence, outlining a highly deformed slope segment (Fig. 54.2). In particular, a tectonically inactive footwall and hanging wall could be identified which are separated by major faults embedding a highly deformed slope segment. This segment is always characterized by near-surface mass movements and shows further segmentation into an upper, extension-dominated, and a lower, compression-dominated slope. However, slope evolution differ significantly according to slope type, especially material strength.

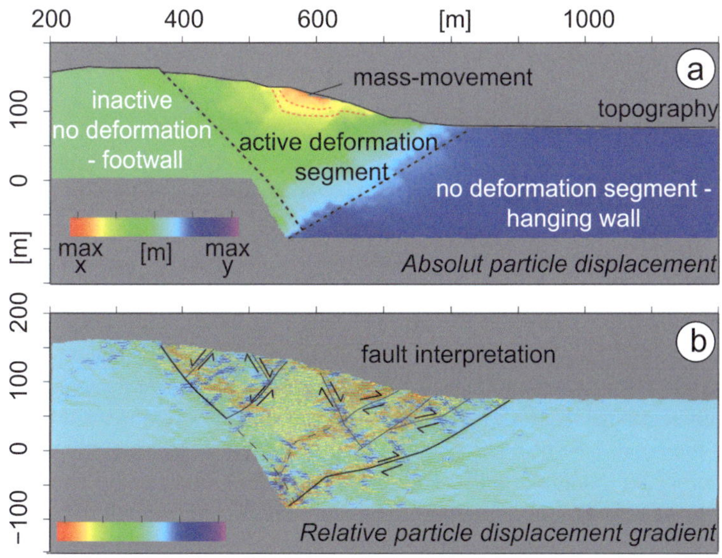

Fig. 54.2 Example of data visualization styles within a calculation step of 0.7 m subsidence: (**a**) absolute particle displacement plot depicts three deformation segments and gravitational mass transport along distinct slid masses, (**b**) relative displacement field calculated from offset between single neighbouring particles enables to identify and track fault evolution. *Black* indicate active faults. *Dashed lines* mark older faults only active at initial subsidence

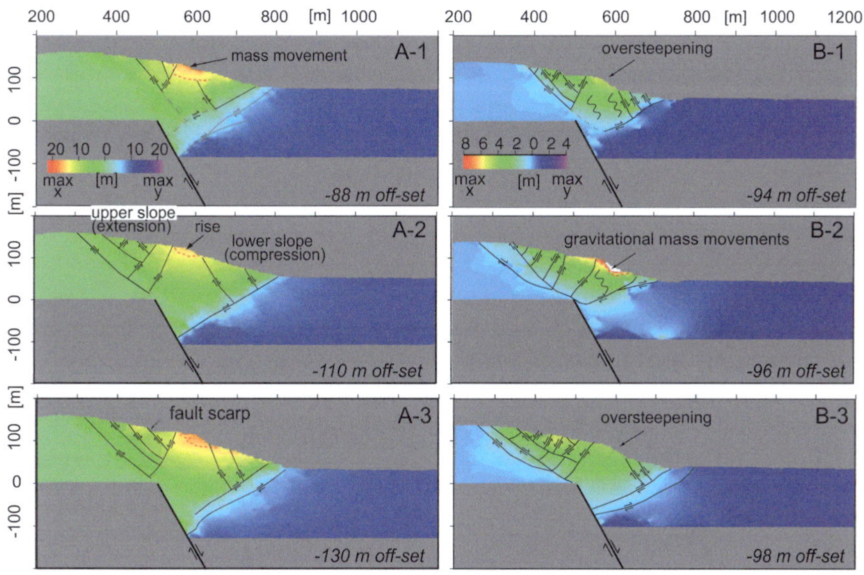

Fig. 54.3 Time series of incremental particle displacement plots for 20-m basin subsidence for material-A (A1-3) and material-B (B1-3). *Black lines* indicate potential faults; which were at least for a short term active, but accumulating not necessarily significant offsets. *Yellow* to *orange* colours indicate the occurrence of near-surface mass movements; also marked by *red dotted line*

54.3.1 High Peak Strength, 'Sandy' Slope Covers (Material-A)

(*i*) Evolution of deformation segments and fault system: The active deformation segment evolves immediately after basin subsidence. After 150 m subsidence, this segment exhibits a width of 600 m. This lateral extension is primarily caused by a basin-ward migration of the large-scaled, antithetic thrusts which establish towards the foot of the slope (Fig. 54.3a). Furthermore, the internal deformation of the active slope segment is mainly controlled by numerous normal faults, which often produce steep fault scarps at the surface (Fig. 54.3A-2). Sub-ordinated, mostly antithetic, minor faults occur only over short time-periods. Hence, extensional behaviour is the most prominent deformation mechanism, represented by extensional sub-segments of 10–50 m width. Compression is only observed in a narrow range at the basin-side of the active segment where slope material is overthrusting the undeformed hanging wall (Fig. 54.3A-3). This indicates that material-A deforms preferentially along major faults, while internal deformation of the embedded sub-segments is low. (*ii*) Slope topography: A typical tripartite evolves: the nearly flat hanging and footwall inactive segments as well as the embedded steeper, even deforming slope (Fig. 54.3A). Slope angles increase continuously until a typical gently dipping upper ($\alpha_{upper} = 4\text{–}8°$) and steeper lower slope ($\alpha_{lower} = 15\text{–}28°$) could be identified after 90 m subsidence. This change in slope taper marked by a rise correlates always with the lower edge of the master fault. Hence, the rise moves continuously basin-ward simultaneously to basin subsidence. Furthermore, based on long-term activity of major normal faults and high stability of the small-scaled extensional sub-segments, the upper slope is characterized by an alternation of smoother topography of flat sub-segments and areas of highest slope gradients correlating with fault scarps (~35°; Fig. 54.3-A3). Thus, the outcropping scarps of major faults lead to a steep, stepped topography and irregular morphology of the upper, extension-dominated slope (Fig. 54.3A).

(*iii*) Gravitational mass movement pattern: Sliding always occurs at the transition between upper and lower slope (Fig. 54.3). In general, headwalls of slides generate steep normal fault scarps outcropping at the upper slope. The destabilized masses occur as deep-incised, wedge-shaped, compact bodies with partial downward concave shear planes. Hence, mass transport is dominated by low internal deformation and short runout. In addition, sliding events show a recurrence with at max. one event per 20-m subsidence. In addition, failure alternates with periods of slope stability. The dimension of the described mass movements is linked to the stage of basin subsidence. So, as slope failure headwalls migrate in footwall direction (re-)activating younger normal faults, failure events mobilize larger 'sediment' portions. The largest slope failure event of >200 m width and about 50 m in thickness could be observed in the final model stages.

54.3.2 Low Peak Strength, 'Muddy' Slope Covers (Material-B)

(*i*) *Evolution of deformation segments and fault system*: Deformation segments establish immediately and the active slope reaches its largest lateral extension of ~650 m after more than 100 m basin subsidence (Fig. 54.3B-3). Again, major, long-term active normal faults and thrust faults continuously define the outer limits of the active segment. However, while the thrust faults are more or less static in position, they vary in number and activity. Internal deformation of these slope segment tends to be dominated by dense sets of short-term (i.e. <2 m subsidence intervals), active normal and antithetic thrust faults (Fig. 54.3B-2). This overall more diffuse and widespread deformation strongly contrasts with the establishment of less-deformed sub-segments in model-A. (*ii*) *Slope topography*: Surface topography is similar to model-A, but with slightly lower-inclination ($\alpha_{upper} = 4–6°$ and $\alpha_{lower} = 14–18°$), and a well-defined rise is not observable. Furthermore, slope angle transitions at both active-inactive-segment boundaries are smoother. Hence, the entire slope maintains a smoother topography. Nevertheless, the upper slope periodically develops faults scarps due to ongoing normal faulting. (*iii*) *Gravitational mass movement pattern*: After a basin subsidence of 100 m, mass movements occur almost continuously along the different slope segments. However, the size of single events decrease, while their number and the transport distance increase. Slides with a maximum thickness of 10–20 m and 20–50 m in length with transport distances of >200 m occur (Fig. 54.3B-2). Their mean recurrence rate is about 1–2 events per 4-m subsidence interval. Further, some of the detected mass-movement events correlate to a temporary activation of minor low-angle normal faults at the continental rise similar to model-A. However, we observe that oversteepening is immediately balanced by a high number of mass movements. This contrasts strongly with the long-term stability of sub-segments in model-A.

54.4 Summary and Conclusions

In this study, we have shown that the long-term evolution of extensional basin flanks (particularly the associated fault pattern and mass movements) is strongly predisposed by the mechanical behavior of the slope sediments. This modelling approach enables investigation and testing of common hypotheses and conceptual models (e.g., Hampton and Lee 1996; Locat and Lee 2002). Nevertheless, we are aware that there are numerous differences between natural slopes and models. However, our very simplified model already demonstrates the key role of the peak strength of the slope sediments and enables a deeper insight into the controlling factors of failure kinematics. Comparing our observations and interpretations, we conclude that:

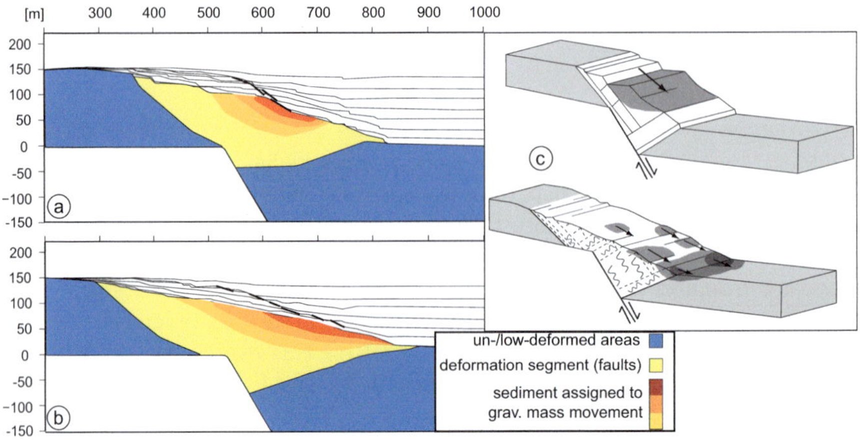

Fig. 54.4 The displacement that occurs from 150 m basin subsidence ((**a**) material-A; (**b**) material-B). Slope topography evolution is shown in 20-m subsidence intervals. *Thicker black lines* mark over steepened slopes. (**c**) Conceptual sketches for extensional basin slopes bearing in a high strength (*top*) vs. a low strength sedimentary cover (*bottom*)

- The activation and location of larger normal, slope-parallel faults are a limiting factor for the occurrence as well as the dimension, location, and the failure mechanism of slope instability in high strength sedimentary slopes. In this case, normal faults position and control the development of slide headwalls due to ongoing basin subsidence. Hence, stresses from the tectonic movement are accommodated by single major faults, while the slope is stable about long periods of time, and only occasional collapses occur (Fig. 54.4). In summary, these kind of slope flanks will exhibit less frequent mass wasting events, whereas sediment destabilization migrates upslope towards younger normal faults resulting in larger thick slides. Consequently, due to their short runout, the overall slope shape remains compact with a steep, stepped topography.
- In contrast, gravitational mass movements on low strength slopes occur along the lower-inclined slope parts, where deformation is accommodated by a diffuse faulting. In this case, material tends to fail more frequently along the entire portion of the slope but involves less material over larger run-outs. The high number of near-surface mass movements provides taper adjustment on one hand, and a high sediment supply to the slope foot region on the other hand which is associated to compressional deformation at the lower slope.

Acknowledgments This work has been funded by the DFG – MARUM. We gratefully acknowledge the constructive reviews by Dr Gian Andrea Pini and Dr Roger Urgeles. We also wish to thank Lina Podszun for her help with the figures.

References

Balsam WL, Beeson JP (2003) Sea-floor sediment distribution in the Gulf of Mexico. Deep-Sea Res 50:1421–1444

Biscontin G, Pestana JM (2006) Factors affecting seismic response of submarine slopes. Nat Hazards Earth Syst Sci 6:97–107

Cundall PA, Strack ODL (1979) A discrete numerical method for granular assemblies. Geotechnique 29:47–65

Hampton MA, Lee HJ (1996) Submarine landslides. doi:10.1029/96RG03287

Itasca Consulting Group (2004) Particle flow code in 2-D (PFC-2D) user's manual. Version 3.0

Jackson CAL, Barber GP, Martinsen OJ (2008) Submarine slope morphology as a control on the development of sand-rich turbidite depositional systems. Mar Petrol Geol 25:663–680

Kock I, Huhn K (2007) Numerical investigation of localization and micromechanics in a stratified soil specimen. J Struct Geo 29:1679–1694. doi:10.1016/j_jsg.2007.07.013

Lee HJ, Locat J, Desgagnes P et al (2007) Submarine mass movements on continental margins. In: Nittrouer CA et al (eds) Continental margin sedimentation. Blackwell, Oxford, pp 213–274

Locat J, Lee HJ (2002) Submarine landslides: advances and challenges. Can Geo J 39:193–212

Morgan JK, Boettcher MS (1999) Numerical simulations of granular shear zones using the distinct element method. J Geophys Res 104:2,703–2,719

Stegmann S, Strasser M, Anselmetti F, Kopf A (2007) Geotechnical in situ characterization of subaquatic slopes. Geophys Res Lett 34:L07607. doi:10.1029/2006GL029122

Strozyk F, Strasser M, Krastel S, Meyer M, Huhn K (2010) Reconstruction of retreating mass wasting in response to progressive slope steepening of the northeastern Cretan margin, eastern Mediterranean. Mar Geol 271(1–2):44–54

Chapter 55
Correction Factors for 1-D Runout Analyses of Selected Submarine Slides

Rafael Rodríguez-Ochoa, Farrokh Nadim, and José M. Cepeda

Abstract Numerical modelling of submarine mass movements is often used to estimate gravity mass flow runout distances, velocities, and the final shape of the sediments in offshore geohazards studies. This paper proposes an approach for the use of numerical models in a practical way, based on calibration against back analyses of known events, to obtain meaningful and reliable results. The proposed approach consists of estimating correction factors that quantify the approximations incurred by the numerical modelling, assuming that the input anchoring facts (i.e. geometrical, geotechnical and rheological information) are based on reliable information, therefore the difference between field evidence and simulations is merely due to the limitations of the numerical model. The approach is exemplified in the paper by simulating submarine debris flows, focused on runout distances, using the numerical model BING with the bilinear rheological model. The anchoring facts are obtained from a database of seafloor slope stability and empirical correlations. The results show the need to assess the level of uncertainty of the assumed anchoring facts in order to narrow the range of the proposed correction factors and use them for predicting runout distances in practical applications. At present the estimated correction factors range from 0.60 to 1.10.

55.1 Introduction

Submarine mass gravity flows are a major concern that affects several economic sectors, threatening lives and properties as well as the environment. Regarding numerical simulation of submarine debris flows, since the pioneering work by Edgers and Karlsrud (1981), there have been important developments.

R. Rodríguez-Ochoa (✉)
Department of Geosciences, University of Oslo (UiO), P.O. Box 1047, Blindern 0316, Oslo, Norway
e-mail: rafaelr@ulrik.uio.no; rafaro33@hotmail.com

F. Nadim • J.M. Cepeda
Norwegian Geotechnical Institute (NGI), Postboks 3930, Ullevål Stadion 0806, Oslo, Norway
e-mail: Farrokh.Nadim@ngi.no; Jose.Cepeda@ngi.no

G. Lamarche et al. (eds.), *Submarine Mass Movements and their Consequences*, Advances in Natural and Technological Hazards Research 41,
DOI 10.1007/978-3-319-20979-1_55

Imran et al. (2001a) proposed a 1-D numerical model called BING to simulate the downslope spreading of a finite-source subaqueous debris flow. The model considers three types of fluid rheology: Bingham, Herschel-Bulkley and bilinear; and is able to compute runout distance, down slope velocity and thickness of the deposit. BING is used as a typical numerical analysis tool for the analyses presented in this paper, but the approach described herein can be adopted with other numerical tools as well.

One of the main challenges of numerical modelling of debris flows is finding the adequate input parameters to obtain meaningful results. A traditional way of using numerical models in debris flow analyses is to perform back analyses, and doing parametric studies to obtain the input parameters that match the field observations (Locat and Lee 2005; Locat et al. 2004). One pitfall of this approach is that, in order to match the field observations, sometimes the input parameters that have physical meaning need to be assigned unrealistic values for the sake of matching the field evidence (De Blasio et al. 2005; Hance 2003; Marr et al. 2002; Niedoroda et al. 2006). This approach also implicitly hides the limitations of the model.

This paper suggest a new approach to perform numerical modelling simulations called the "anchoring facts" approach. It consists of setting realistic, meaningful values to all the input parameters that have a physical sense, based on field and laboratory evidence. The focus of this approach is on estimation of correction factors to match field observations, and quantifying the limitations of the numerical model. The proposed anchoring facts approach is explained in this study using the numerical model BING (Imran et al. 2001a) and the rheological bilinear model (Locat 1997). The computed runout distance from the simulations are compared with field observations obtained from a submarine slide database produced by Hance (2003). The intention is to estimate the correction factors that quantify the limitations of the numerical model BING and use those to make predictions.

The results of this paper suggest that the main assumption of the anchoring facts approach may not always be satisfied, leading to a large scatter of correction factor values. Therefore, it is important to assess the degree of uncertainty of the assumed anchoring facts (i.e. geometrical, geotechnical and rheological information of slide events) and weight the correction factors accordingly.

55.2 Methodology

55.2.1 Submarine Slide Selection

To apply the anchoring facts approach, the submarine slope failure database developed by Hance (2003) was used. This database compiled 534 submarine slide events based on published literature.

In order to carry out the back analyses, the minimum basic information needed for each slide event are seven fields or categories from the database (Locat and Lee

2005): soil type, soil properties, volume, thickness, length, width and slope angle. From the 534 submarine slide events in the database, only 11 events met the basic fields to perform the back analyses using BING.

55.2.2 Geometrical Characterization of Slides

The required geometrical characterization of the submarine slides to run BING are as follows: (1) position of the tail of mud deposit; (2) initial length of mud deposit; and (3) maximum thickness of mud deposit with a parabola shape.

In order to set the essential geometrical inputs to perform numerical experiments in BING, the procedure described below was developed to obtain the geometrical input data based on information available from the database.

The procedure to estimate the maximum thickness of mud deposit is described in the following five steps:

1. Cross section area of slices with 1 km width are obtained by dividing the volume (i.e. *volume* is a number field in the database that defines the approximate volume of displaced soil in km^3) over the width (i.e. *width* is the average width of the landslide from an aerial view in km).
2. The obtained cross section area is set equal to the parabola area (Parabola area = 2/3 bh; where: b = base, h = height).
3. A spreading coefficient $S = H_f/H_i$, which is the ratio of the final height H_f to the initial height H_i of the failed mass, is assumed. This value is usually not easy to estimate unless there are enough field information and laboratory results to run numerical simulations capable of matching the field evidence. Therefore, to overcome this lack of information, the results obtained from the parametric analysis shown in Locat and Lee (2005) were used (Fig. 55.1). The curve in Fig. 55.1 is based on modelling debris flows in the Mississippi fan done by Locat et al. (1996). Based on this parametric analysis, a mean value of 0.15 was assigned to the spreading coefficient S in all simulations.
4. The initial height of the failed mass H_i was estimated based on the spreading coefficient S and the thickness (i.e. *thickness* is a number field in the database and represents the average thickness of the landslide based on geophysical data and geologic core dating).
5. It is assumed that the initial height of the failed mass H_i is equal to the maximum thickness of mud deposit (i.e. height of the parabola). It is believed that as long as this assumption is applied systematically in all cases it can give an estimation of the model uncertainty in the BING model. Additionally, according to Imran et al. (2001b) the initial shape has negligible bearing on the final runout shape and distance.

And the procedure to estimate the initial length of mud deposit is described below:

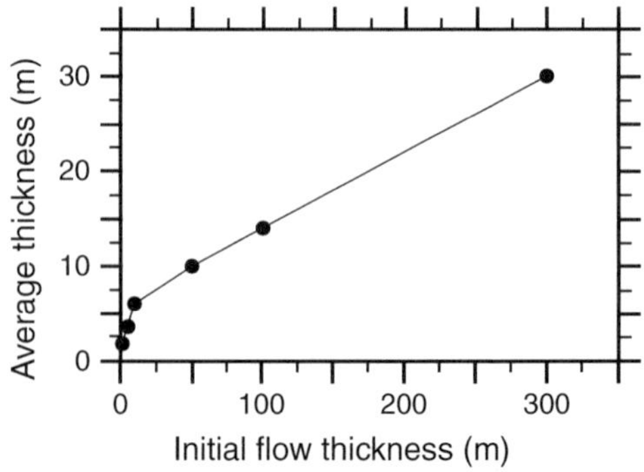

Fig. 55.1 Relationship between the initial flow thickness and the average flow thickness based on modelling debris flows in the Mississippi fan (After Locat et al. 1996) (Reproduced from Locat and Lee (2005), with permission from Springer Science + Business Media and Dr. Jacques Locat)

After equating the estimated cross section area of the slice with the parabola area, the base value of the parabola *b* is calculated. The *b* value of the parabola is assumed to be the initial length of the mud deposit.

While setting the geometrical characterization of the submarine slides, three additional events were removed from the study. Slide events #5, #27 and #72 were removed because they showed inconsistency between the features of volume and width, resulting in anomalous initial length of mud deposits of 0.32 m, 675,000 m and 46,957 m respectively. The remaining eight slides were used to perform the back analyses.

Regarding the position of the tail of mud deposit, it was set equal to zero in all cases.

55.2.3 Geotechnical and Rheological Characterization of Slides

In general, there is not enough geotechnical and rheological information in the database to perform mass gravity flow back analyses. Therefore, field observations and empirical correlations were used to estimate the geotechnical and rheological parameters.

The quantity and quality of geotechnical information vary from case to case. In this study, the required geotechnical data for debris flow back analyses are the water content w and the specific gravity G_s of the failed sediments. These parameters are used to assess the mud density ρ by means of soil gravimetric and volumetric relationships. The ambient fluid density ρ_w was set as 1025 kg/m^3 in all cases.

In this study, the bilinear rheological model developed by (Locat 1997) was preferred for the numerical simulations in BING, since it has shown to give acceptable results in previous studies (Locat 1997; Imran et al. 2001b; Jeong et al. 2010) and it is relatively more stable than its predecessors since the numerical point of view (Imran et al. 2001a).

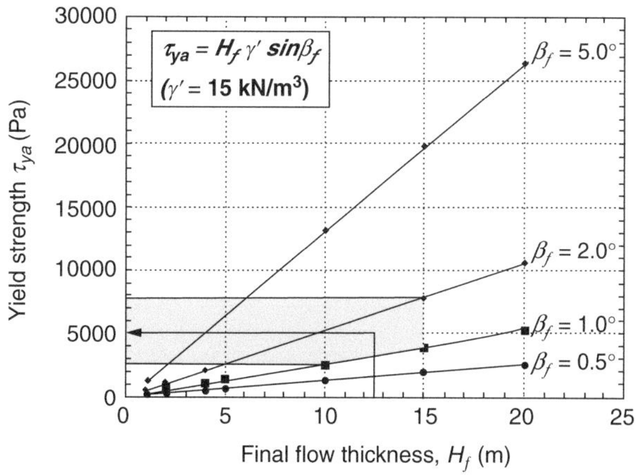

Fig. 55.2 Parametric analysis to estimate the apparent yield strength τ_{ya} of the mud flow, based on the relationship proposed by Johnson (1984) for the Palos Verdes debris avalanche (After Locat et al. 2004) (Reproduced from Locat and Lee (2005), with permission from Springer Science + Business Media, Elsevier and Dr. Jacques Locat)

The rheological input parameters for the bilinear model in BING were estimated from field observations available in the database and empirical correlations. The bilinear model is defined by three rheological parameters: (1) the apparent yield strength τ_{ya}, which controls the runout distance; (2) the plastic viscosity μ_{dh}, which controls the flow velocity; and (3) the shear strain rate at the transition from a Newtonian to a Bingham behaviour γ_0, which also controls the runout distance.

As recommended by Locat and Lee (2005), the apparent yield strength τ_{ya} was estimated based on the thickness of the failed sediments in the runout zone and by performing parametric analyses (Fig. 55.2) with the relationship given in Eq. (55.1). This equation was first proposed by Johnson (1984):

$$\tau_{ya} = H_f \gamma' \sin \beta_f \tag{55.1}$$

where:

τ_{ya} = apparent yield strength (kPa)
H_f = Flow deposit thickness (m)
γ' = submerged unit weight of flow deposits (kN/m^3)
β_f = slope angle at deposition zone (degrees)

The plastic viscosity μ_{dh} was estimated from empirical correlations proposed by Locat (1997) for clayey mixtures, and Jeong et al. (2010) for silt-rich soils and sands, Eqs. (55.2), (55.3), and (55.4) respectively,

$$\eta_h = 0.52\tau_c^{1.12} \tag{55.2}$$

$$\eta_h = 5.77\tau_c^{1.1} \tag{55.3}$$

$$\frac{\tau_c}{\eta_h} = 1 \text{ for clays, } 0.1 \text{ for silts, and } 0.01 \text{ for sands} \tag{55.4}$$

where η_h = plastic viscosity (mPa.s) and τ_c = yield strength (Pa).

The shear strain rate at the transition from a Newtonian to a Bingham behaviour γ_0 was set with two values for each back analysis case, in order to overcome the significant uncertainty in this value. Therefore, each slide case was analysed using two values of γ_0, resulting two runout distances, where 0.1 was considered the lower bound and 1.0 was considered the upper bound of the runout distance.

55.3 Numerical Simulations Using BING

In total, 16 numerical simulations were carried out, 2 runs for each submarine slide event under investigation, providing lower and upper bound estimates of the runout distance.

55.3.1 Runout Distances

The lower bound (LB) and upper bound (UB) runout distances calculated with BING are shown in Fig. 55.3 together with the database runout distances for each slide event to allow comparisons.

Figure 55.3 shows a large dispersion of data between the numerical and field observation results. In general the numerical results present four submarine slides with longer runout distances than the field observations (slides No. 1, No. 2, No. 4 and No. 8), one submarine slide with shorter runout distances than the field observations (slide No. 6), and three slides with somewhat similar runout distances compared to the field observations (slides No. 3, No. 5 and No. 7).

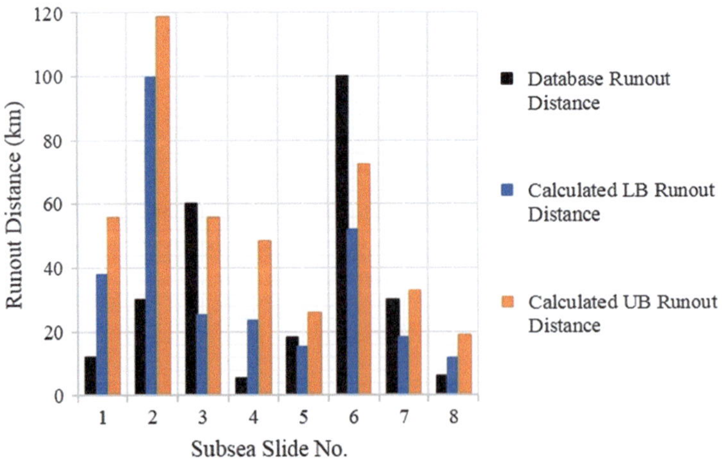

Fig. 55.3 Comparison of runout distances from numerical simulations and field observations documented in the subsea slope failure database

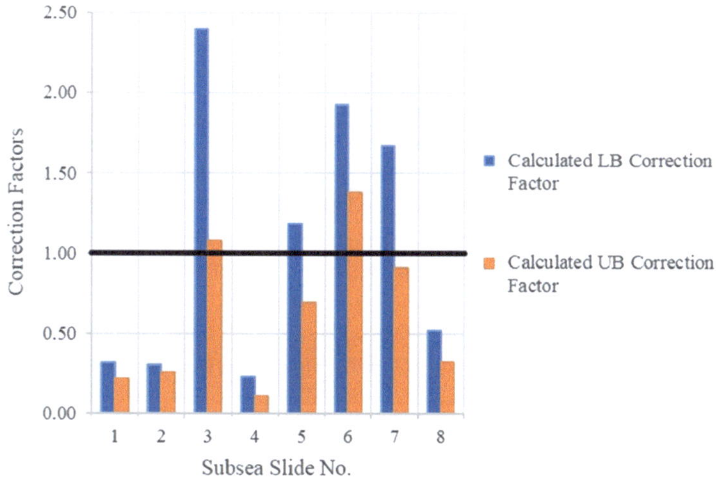

Fig. 55.4 Estimation of BING correction factors based on the anchoring facts approach, for lower bound (*LB*) and upper bound (*UB*) runout distances

55.3.2 Correction Factors

The correction factor is defined as the ratio of observed to calculated runout distance. The estimation of BING correction factors by the anchoring facts approach is shown in Fig. 55.4 for the eight submarine slide events modelled.

Figure 55.4 shows a large scatter of data. This makes it hard to draw conclusions regarding BING model and answering the question: Is BING conservative or nonconservative?

The results of the study are summarized as follows:

Overall arithmetic mean (μ_T) = 0.84
Overall standard deviation (σ_T) = 0.68
Overall coefficient of variation (CV_T) = 0.81

Based on these results, it is suggested to apply a correction factor in the range of 0.60–1.10 with an average value of 0.85 to numerically estimated runout distances using BING, as a first approximation.

55.4 Discussion and Conclusions

The proposed anchoring facts approach, to estimate correction factors for numerical models, assumes that the information used to establish the input parameters for the BING model and the applied empirical correlations and assumptions made in the modelling are feasible. Thus the discrepancy between the numerically calculated runout distances and field observations may be attributed to the limitations in the numerical model. However, during the study, some apparently inconsistent data from the database did not allow the analysis of three subsea slide cases (slides #5, #27 and #72). In these three cases, the discrepancy between the documented average width and mass volume resulted in anomalous initial lengths of the mud deposits. The level of precision for each number field in the subsea slide database is provided by Hance (2003). For example, the level of precision of volume (km^3) varies from whole numbers to 10^{-6}, and the same applies for the geotechnical characterisation of the sediments. This indicates that the level of precision available in the database does not fulfil the accuracy requirements of the anchoring facts approach.

Regarding the rheological empirical correlations used to assess the input parameters for the bilinear model, it is believed that these empirical correlations are based on high quality laboratory experiments carried out by their proposers to characterize the dynamics of the mass gravity flows. However, the rheological characterization of mud mixtures containing large size clasts is still a challenge. So is the measurement of rheological mud properties in the lower range of the shear strain rate (Jeong et al. 2010).

In conclusion, this study suggests that the accuracy of correction factors using the anchoring facts approach is a function of the precision of the input parameters that are used to feed the model. The results obtained using Hance's database, rheological empirical correlations, and proposed assumptions suggest a correction factor in the range of 0.60–1.10, with an average value of 0.85, for the estimated runout distance. However, it is expected that this range of correction factors could be narrowed when the precision of the characterization of the submarine mass movement is improved.

Acknowledgments The authors want to thank James J. Hance for sharing the database of seafloor slope stability developed during his master's studies at The University of Texas at Austin, as well as supervising committee member Robert B. Gilbert. The authors also thank Jasim Imran for making the numerical model BING available to the general public and providing useful comments to improve the manuscript during the review process as well as Byron Quan Luna. This work was partially supported by the Mexican National Council for Science and Technology (CONACYT) by means of PhD scholarship No. 308676 to the first author.

References

De Blasio FV, Elverhøi A, Issler D, Harbitz CB, Bryn P, Lien R (2005) On the dynamics of subaqueous clay rich gravity mass flows – the giant Storegga slide, Norway. Mar Pet Geol 22(1–2):179–186, doi:http://dx.doi.org/10.1016/j.marpetgeo.2004.10.014

Edgers L, Karlsrud K (1981) Viscous analysis of submarine flows. Norwegian Geotechnical Institute, Norway

Hance JJ (2003) Development of a database and assessment of seafloor slope stability based on published literature. University of Texas at Austin

Imran J, Harff P, Parker G (2001a) A numerical model of submarine debris flow with graphical user interface. Comput Geosci 27(6):717–729, doi:http://dx.doi.org/10.1016/S0098-3004(00)00124-2

Imran J, Parker G, Locat J, Lee H (2001b) 1D numerical model of muddy subaqueous and subaerial debris flows. J Hydraul Eng 127(11):959–968

Jeong SW, Locat J, Leroueil S, Malet J-P (2010) Rheological properties of fine-grained sediment: the roles of texture and mineralogy. Can Geotech J 47(10):1085–1100. doi:10.1139/T10-012

Johnson AM (1984) Debris flows. In: Brunsden D, Prior DB (eds) Slope instability. Wiley, New York, pp 257–361

Locat J (1997) Normalized rheological behaviour of fine muds and their flow properties in a pseudoplastic regime. In: Chen C-l (ed) Debris-flow hazard mitigation: mechanics, prediction, and assessment. American Society of Civil Engineers, San Francisco, pp 260–269

Locat J, Lee H (2005) Subaqueous debris flows. In: Jakob M, Hungr O (eds) Debris-flow hazards and related phenomena. Springer Praxis Books, Springer, Berlin, pp 203–245. doi:10.1007/3-540-27129-5_9

Locat J, Lee HJ, Nelson HC, Schwab WC, Twichell DC (1996) Analysis of the mobility of far reaching debris flows on the Mississippi Fan, Gulf of Mexico. In: Senneset K (ed) 7th International symposium on landslides. A.A. Balkema, Rotterdam, pp 555–560

Locat J, Lee HJ, Locat P, Imran J (2004) Numerical analysis of the mobility of the Palos Verdes debris avalanche, California, and its implication for the generation of tsunamis. Mar Geol 203 (3–4):269–280, doi:http://dx.doi.org/10.1016/S0025-3227(03)00310-4

Marr JG, Elverhøi A, Harbitz C, Imran J, Harff P (2002) Numerical simulation of mud-rich subaqueous debris flows on the glacially active margins of the Svalbard–Barents Sea. Mar Geol 188(3–4):351–364, doi:http://dx.doi.org/10.1016/S0025-3227(02)00310-9

Niedoroda A, Reed C, Das H, Hatchett L, Perlet A (2006) Controls of the behavior of marine debris flows. Nor J Geol 86(3):265–274

Part IX
Tsunami Generation from Slope Failure

Chapter 56
Volcanic Generation of Tsunamis: Two New Zealand Palaeo-Events

Willem de Lange and Vicki Moon

Abstract Rapid emplacement of a mass via pyroclastic flows, or edifice failure, generates volcanic tsunamis. Physical modelling demonstrates that the efficiency of tsunami generation is influenced by the angle the mass enters the ocean. Efficiency decreases with increasing slope angle from 20° to 60°, before increasing to a maximum at 90°, which corresponds to a mass falling directly into the ocean without interacting with the slope (impact tsunami). Further, in the case of surging pyroclastic flows or regressive failures, successive closely spaced events may generate larger tsunami waves than a single event of comparable volume.

It is difficult to assess if physical model results are meaningful for real world tsunami events due to limited observational data. Two New Zealand palaeo-events – pyroclastic flows from Mt Tarawera and edifice failure at Whakaari (White Island) – can be linked to tsunami deposits, which constrains numerical simulations of the source mechanisms. The Mt Tarawera event involved multiple pyroclastic flows entering a lake during the AD 1314 ± 12 Kaharoa Eruption. The interaction of multiple closely spaced pyroclastic flows is necessary to generate the 6–7 m maximum wave height inferred from near source tsunami deposits. Tsunami deposits in the Bay of Plenty, dated to 2962 ± 52 BP, are consistent with edifice failure at Whakaari. In this case a single event with a volume of 0.23 km^3 is sufficient to account for the tsunami deposits. Hence, if the failure was regressive, the successive stages were sufficiently close together to be indistinguishable from a large single event.

56.1 Introduction

Tsunami generation during volcanic eruptions has been observed within the historical record, but the mechanisms involved are not always obvious and there has been significant debate over the primary processes generating volcanic tsunami (Latter 1981; Watts and Waythomas 2003). Most studies indicate that rapid displacement

W. de Lange (✉) • V. Moon
School of Science, The University of Waikato, Hamilton, New Zealand
e-mail: w.delange@waikato.ac.nz; v.moon@waikato.ac.nz

G. Lamarche et al. (eds.), *Submarine Mass Movements and their Consequences*, Advances in Natural and Technological Hazards Research 41,
DOI 10.1007/978-3-319-20979-1_56

of water by mass flows is the most effective generation mechanism, and involves either pyroclastic flows (*viz*. de Lange et al. 2001; Watts and Waythomas 2003; Maeno and Imamura 2011), or debris flows and avalanches associated with edifice failure (*viz*. Keating and McGuire 2000; Tinti et al. 2001). However, there are limited observations of the processes involved resulting in uncertainty about the appropriate modelling approach (*viz*. Maeno and Imamura 2007, 2011). Attribution of palaeotsunami deposits to specific volcanic sources (*viz*. de Lange et al. 2008; Goff et al. 2011) is difficult without numerical modelling.

56.2 Modelling Tsunami Generation by Pyroclastic Flows

Prasetya (1998) undertook a series of physical simulations of different generation mechanisms, including 2-dimensional flume experiments of subaerial flows entering water based on simulations undertaken by Monaghan et al. (1997). The simulations indicated that a primary solitary wave was generated by the head of the advancing flow, with a much smaller secondary set of waves formed by turbulence behind the head. Sustained flow did not generate significant additional waves. Hence, a sliding block appeared to be a suitable approximation for the generating process.

The magnitude of the generated wave, behaviour of the flow, and the characteristics of deposits produced varied with the entry angle of the flow (Fig. 56.1). At low entry angles ($<50°$), the flow separated into a low-density component that travelled near the fluid surface, and a high-density component that travelled along the bed. The coarsest sediment was deposited first, followed by progressively finer sizes settling from suspension, producing a graded deposit that fined upwards. The wave height generated initially decreased with increasing entry angle. This matches the observations of McLeod et al. (1999) who used a fixed 20° entry angle.

The minimum wave height was generated for entry angles around 60°. At this angle, the flow retained more of the fine components resulting in a weaker surface plume. The near-bed flow scoured sediment from the base of the slope, producing a sediment free moat within the flume, with the bulk of the sediment deposited as a mound adjacent to the moat. The finer sediment in the surface flow predominantly was deposited on the bed on the far side of the mound (Fig. 56.1).

With increasingly steep entry angles $>60°$, the generated wave height increased, reaching a maximum at 90° (a vertical drop analogous to an impact tsunami). The coarsest sediment within the flow was deposited directly at the base of the slope, with a secondary area of finer sediment deposition occurring further along the flume (Fig. 56.1).

Prasetya (1998) was unable to include hot materials in his model. However, Freundt (2003) was able to undertake physical modelling using hot granular sediment at temperatures ranging from 55 to 403 °C, and two-stage ramp with an initial slope of 36° and an entry angle of 26°. The mechanism for discharging the hot sediment produced an unsteady flow. The simulations for temperatures

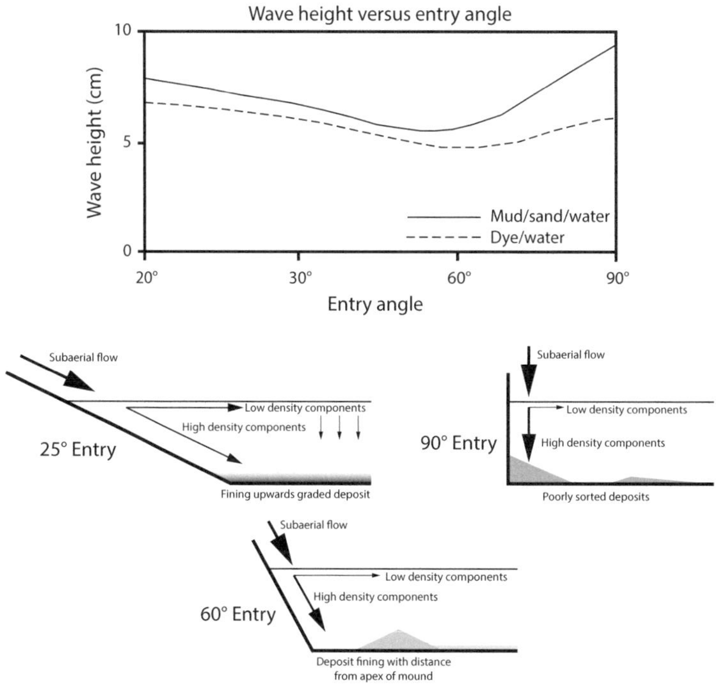

Fig. 56.1 Summary of the results of the physical modelling of granular and dyed water flows entering a flume at different angles based on the results of Prasetya (1998)

<250 °C are very similar to those of Prasetya (1998). Hotter flows tended to produce much more steam that supported the flow above the water surface, and additional wave formation was associated with hydrothermal explosions. Freundt (2003) concluded that the temperature of the flow was more important that the flow density for determining if the flow predominantly moved offshore above or below the water surface.

The observed patterns of behaviour are consistent with field observations of pyroclastic flows entering the sea and deposits attributed to this mechanism. Pyroclastic flows with relatively low entry angles tend to mix with seawater, sometimes with secondary hydrothermal explosions, and produce near-bed turbidity currents, and subaerial surges that flow above the water (*viz.* Trofimovs et al. 2006, 2008). The patterns of behaviour are also consistent with the generation of tsunami by edifice failure (*viz.* Freundt et al. 2007).

Observations of pyroclastic flows at steeper angles of entry (>45°) are based on interpretation of deposits (*viz.* Kano 2003), mostly associated with shallow submarine volcanoes. The deposits produced by steeper entry angles correspond to the Type 1 deposits of Kano (2003), and the main effect of increasing entry angle appears to be movement of the location of the main centre of deposition for the coarsest sediment. The centre of the deposit initially moves offshore, before moving onshore to the base of the incline.

Prasetya (1998) identified that pyroclastic flows involve multiple flows or surges, which has not been replicated in physical models; although the mechanism used by Freundt (2003) did produce some pulsing in a single flow. Prasetya (1998) undertook some numerical simulations for sequences of separate pyroclastic flows and found that it was possible to generate larger waves than for a single flow of an equivalent volume. However, he was unable to calibrate his model against observations. Freundt et al. (2007) also suggested that pyroclastic flows and debris avalanches associated with edifice failure entering a shallow water body over a wide front could produce waves that exceed the normal breaking-wave limit, and hence be larger than expected. They note that this may account for landslide-induced waves of 15 m and 60 m height in the Kasu Tephra Cone Crater Lake, Papua New Guinea, and Vaiont Reservoir, Italy, respectively.

The following two examples compare numerical models developed from those of Prasetya (1998) with palaeotsunami deposits in New Zealand.

56.3 Kaharoa Eruption Pyroclastic Tsunamis

Lake Tarawera (Fig. 56.2) is a volcanogenic lava-dammed marginal moat lake formed by the development of volcanic complexes within the Haroharo caldera, Okataina Volcanic Zone (Manville et al. 2007). The lake has varied significantly in extent and depth over the last 300 ka, with the present-day extent forming after the 5.6 ka Whakatane Eruption. The deposits from the Whakatane Eruption appear to overlie a palaeosol with trees in growth position >30 m (~270 m asl) below present lake level at 298 m asl (Hodgson and Nairn 2005), and produce a distinctive horizon in seismic reflection images of the lake sediments (Davy and Bibby 2005).

After the Whakatane Eruption, Lake Tarawera rapidly rose to an elevation >330 m asl, forming extensive terraces around the lake, before falling slowly to an elevation of ~315 m asl at the time of the Kaharoa Eruption (Hodgson and Nairn 2005). The 1314 ± 12 AD Kaharoa Eruption is a key stratigraphic marker for New Zealand as it coincides with the earliest evidence of human occupation (Hogg et al. 2003). The eruption was a complex event, with multiple episodes occurring over ~4 years from at least two vents along an 8-km long fissure (Sahetapy-Engel et al. 2014). The main pyroclastic flows were produced in the first ~7.5 days of an initial 2 week long explosive phase, with a series of flows entering Lake Tarawera over approximately 1.6 km of shoreline (Nairn et al. 2001). The flow deposits show no evidence of secondary hydrothermal explosions at the 315 m asl shoreline, and multi-beam bathymetric data suggest the flows continue 100–200 m offshore from the present-day 298 m asl shoreline. Davy and Bibby (2005) did not identify Kaharoa pyroclastic flows in their seismic reflection surveys, but they did not approach closer than 100 m to the shore and did not have a shallow seismic survey in the area affected by the pyroclastic flows.

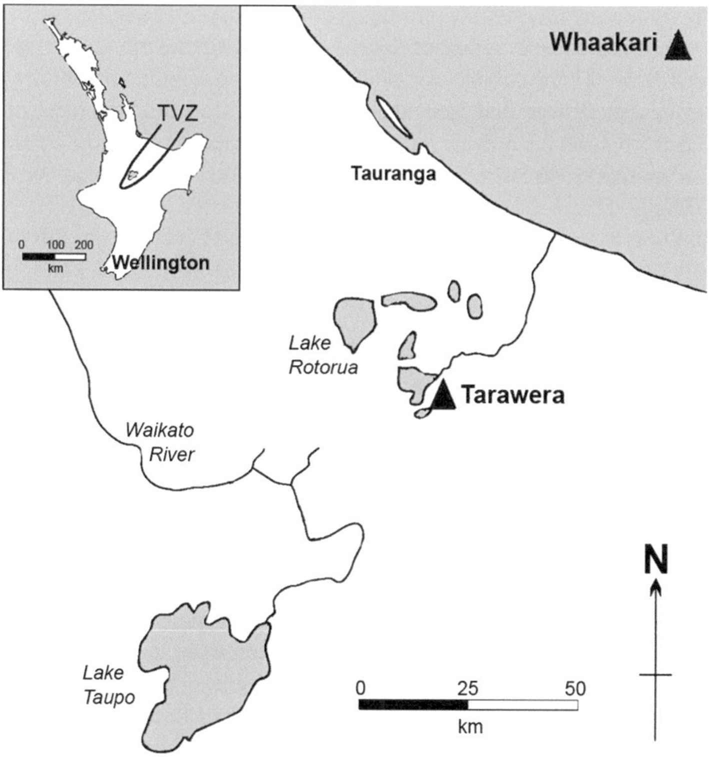

Fig. 56.2 Locations of the two volcanic sources within the Taupo Volcanic Zone (TVZ) discussed in the text

Hodgson and Nairn (2005) state that the lake level fell to <303.5 m asl during the initial explosive phase of the eruption, before rising ~30 m after the eruption due to damming of the lake outlet by the eruptive deposits. They suggest that this abrupt fall was due to tsunami waves overtopping the natural pre-eruption dam, and scouring out a deeper outlet channel.

Magill (2001) examined deposits near the 315 m asl terrace and described a palaeo-tsunami deposit consisting of a sharp erosional contact on lacustrine silts underneath two fining upwards units, overtopped by Kaharoa Tephra. The lowest unit contains cobbles that were identified as being derived from a cobble beach deposit at a lower elevation downslope. The upper unit contained gravels, which were consistent with the older beach deposits on the 315 m asl terrace. Based on the threshold velocities for entraining the largest clasts in each unit, Magill proposed that two tsunami waves of ~7 m and ~1 m height were responsible for the separate units.

Magill (2001) simulated the generation of tsunami waves by a range of different pyroclastic flows within the volume constraints provided by the known subaerial deposits between 315 and 298 m asl, and an assumed submarine terminal deposit below 298 m asl constrained by bathymetric contours. She calibrated the numerical model by simulating measured wind-induced seiches within Lake Tarawera. Magill (2001) then undertook a series of sensitivity tests to determine the effects of pyroclastic flow velocity, thickness, width, submerged length, and duration of the flow. Her results showed that the wave characteristics were largely determined by the initial impact at the water's edge, which depended on the flow velocity and width of shoreline affected. This is consistent with the findings of Freundt et al. (2007).

Simulations based on single pyroclastic flows with characteristics inferred from the flow deposits (10–30 $m.s^{-1}$ flow velocity, 0.5–2 km shoreline width), produced wave heights too small to produce the observed deposits. However, the sensitivity tests suggested that multiple flow impacts on the lake would be a more effective wave generation mechanism than single pyroclastic flows. Therefore, a series of simulations were undertaken using multiple occurrences of a 20 $m.s^{-1}$ velocity flow affecting 1 km of shoreline with a range of time intervals between 50 and 725 s at 5 s steps.

Considering only two pyroclastic flows, the changing time interval between flows created conditions dominated by constructive or destructive interference. There was no obvious relationship between the interference effects and the seiche periods identified for the lake, so it was assumed that this was due solely to the generating process. Overall, constructive interference occurred more often, so that most time intervals between flows modelled resulted in an increased wave height.

Using a fixed time interval of 75 s between flows, between two and eight successive flows were simulated. There was an overall trend of increasing wave height with increasing number of flows from two to six, and then there was no further change. The largest increase in height occurred between a single flow and two flows. It was found that two successive flows of 30 $m.s^{-1}$ velocity and 2 km shoreline width were sufficient to produce the 7 m tsunami required to form the lower palaeo-tsunami deposit. Additional flows would be required for slower and/or narrower flow fronts.

56.4 Whakaari Edifice Failure Tsunami

Whakaari (White Island) is an active stratovolcano located on the continental margin, ~50 km offshore in the Bay of Plenty, New Zealand. Moon et al. (2009) identified that the present geomorphology is consistent with at least two major episodes of regressive edifice failure. Smaller failures are ongoing, with the last significant failure occurring in 1914, resulting in the deaths of all 11 inhabitants of the island (Moon et al. 2005).

The age of prehistoric eruptions and edifice failures at Whakaari is not known. However, Shane et al. (2006) collected 20 piston cores from the continental slope and rise offshore of the Bay of Plenty and identified 6 tephra units and a sequence of turbidity current deposits at the terminus of the White Island Canyon that matched the composition of Whakaari eruptives. Based on known stratigraphic markers within the cores, they dated the oldest tephra unit at ~21 ka, and the turbidity current deposits as <9.5 ka.

Palaeotsunami deposits attributed to local or regional sources have been identified at various sites around the Bay of Plenty coastline (Bell et al. 2004). One regional event, associated with up to 7 km of inundation, is identified at multiple sites with ages ranging from 2.2 to 3.0 ka. The best-constrained age is 2962 ± 52 BP, although Goff et al. (2011) prefer an age of 2800 BP, which correlates with palaeotsunami deposits in Vanuatu. The association of sea-rafted pumice with the deposits in Vanuatu suggests a volcanic source (*viz.* de Lange and Moon 2007). There is also evidence of a significant earthquake that produced up to 2 m of coastal subsidence in the central Bay of Plenty (Bell et al. 2004). Such an event is capable of generating the 0.17–0.45 g accelerations required to induce edifice failure at Whakaari (Moon et al. 2009).

Hansford (2006) simulated tsunamis generated by edifice failure of Whakaari. Unpublished side-scan and multi-beam echo sounder data from the University of Hawaii identified a possible debris avalanche deposit downslope of Whakaari, and linking to the White Island Canyon. Hansford (2006) estimated a total submarine volume of 0.23 km^3 within the mapped deposit boundaries, which was in good agreement with the 0.24 km^3 that Bradshaw (2004) estimated as the subaerial volume lost from Whakaari due to edifice failure.

Hansford (2006) undertook numerical simulations of tsunami generation by one or more debris avalanches resulting from edifice failure, by treating them as translational landslides and applying the methodology of Watts (1997) to produce an initial waveform. He undertook sensitivity analysis of the input parameters, since the prehistoric debris avalanche characteristics were not known, and found that entry angle and flow volume were the most important parameters. Entry angles from 5 to 25° at 2.5° increments were tested, with flow volumes of 0.14, 0.28 and 0.56 km^3 (assuming that the flow volume was slightly greater than the final deposit). The width of the debris avalanche was controlled by the shape of the edifice failure as evident in the present-day geomorphology (Moon et al. 2009), and hence was kept constant at 860 m.

The initial tsunami waveforms were then propagated onshore into the Bay of Plenty, and the resulting wave heights were compared to inferred wave heights based on the palaeo-tsunami deposits assuming that all the events dated between 2.2 and 3.0 ka were a single event. The best match was obtained for entry angles of 20–25° and a flow volume of 0.23 km^3, suggesting that even though there is evidence for regressive failure, the successive failures behaved as a single generating event. Although, a significant component of the tsunami energy was reflected offshore by the continental slope, radial dispersion limits the potential magnitude of tsunami waves travelling northwards from New Zealand. If the palaeotsunami deposits in Vanuatu and New Zealand represent a single event around 2800 BP as suggested by Goff et al. (2011), edifice failure at Whakaari is unlikely to have been the source.

56.5 Conclusions

Physical modelling and numerical simulations linked to palaeo-tsunami deposits provided useful insights into tsunami generation by mass flows. Key parameters affecting the generation process are the entry angle of the flow and flow width on entry. These determine the nature of the initial impact, and appear to be the dominant control on the characteristics of the wave leaving the generating zone, as suggested by Freundt et al. (2007). Therefore, a simple moving block model for generation, as discussed by de Lange et al. (2001), is adequate for simulating tsunami generation.

An additional consideration for modeling is the likelihood of multiple impacts due to a series of pyroclastic flows, or regressive edifice failure. Applying a sequence of simple block impacts provided a sensible prediction of tsunami wave height for Lake Tarawera. However, the sinusoidal forcing approach of Maeno and Imamura (2011) may be a useful alternative.

References

Bell RG, Goff J, Downes GL, Berryman KR, Walters RA, Chagué-Goff C, Barnes PM, Wright IC (2004) Tsunami hazard for the bay of plenty and Eastern Coromandel Peninsula: stage 2. Environment Waikato Technical Report 2004/32, Hamilton

Bradshaw JD (2004) Aspects of slope stability for White Island Volcano, New Zealand. MSc thesis, University of Waikato, 194 pp

Davy B, Bibby H (2005) Seismic reflection imaging of the Haraharo Caldera boundary beneath Lake Tarawera, Okataina Volcanic Centre, New Zealand. N Z J Geol Geop 48:153–166

de Lange WP, Moon VG (2007) Tsunami washover deposits, Tawharanui, New Zealand. Sediment Geol 200:232–247

de Lange WP, Prasetya GS, Healy TR (2001) Modelling of tsunamis generated by pyroclastic flows (ignimbrites). Nat Hazards 24:251–266

de Lange, WP, Prasetya, GS, Spiers, KC, Moon, CG (2008). Utilising palaeotsunami data for hazard assessment: numerical modelling to identify credible sources. Solutions to Coastal Disasters 2008 – Tsunamis, Turtle Bay, Oahu, Hawaii. ASCE, pp 248–255

Freundt A (2003) Entrance of hot pyroclastic flows into the sea: experimental observations. Bull Volcanol 65:144–164

Freundt A, Strauch W, Kutterolf S, Schmincke H-U (2007) Volcanogenic tsunamis in lakes: examples from Nicaragua and general implications. Pure Appl Geophys 164:527–545

Goff J, Chagué-Goff C, Dominey-Howes D, McAdoo B, Cronin S, Bonté-Grapetin M, Nichol S, Horrocks M, Cisternas M, Lamarche G, Pelletier B, Jaffe B, Dudley W (2011) Palaeotsunamis in the pacific islands. Earth-Sci Rev 107:141–148

Hansford AJ (2006) Tsunami generation by island edifice failure, for White Island and Motuhora, New Zealand. MSc thesis, University of Waikato, 92 p

Hodgson KA, Nairn IA (2005) The *c.* AD 1315 syn-eruption and AD 1904 post-eruption breakout floods from Lake Tarawera, Haroharo caldera, North Island, New Zealand. N Z J Geol Geophys 48:491–506

Hogg AG, Higham TFG, Lowe DJ, Palmer JG, Reimer PJ, Newnham RM (2003) Wiggle-match date for Polynesian settlement of New Zealand. Antiquity 77:116–125

Kano K (2003) Subaqueous pumice eruptions and their products: a review. In: White JDL, Smellie JL, Clague DA (eds) Explosive subaqueous volcanism. Geophysical monograph 140. American Geophysical Union, Washington, DC, USA, pp 213–229

Keating BH, McGuire WJ (2000) Island edifice failures and associated tsunami hazards. Pure Appl Geophys 157(6–8):899–955

Latter JH (1981) Tsunamis of volcanic origin: summary of causes, with particular reference to Krakatoa, 1883. Bull Volcanol 44:468–490

Maeno F, Imamura I (2007) Numerical investigation of tsunamis generated by pyroclastic flows from the Kikai caldera. Jpn Geophys Res Lett 34(L23303):1–5

Maeno F, Imamura I (2011) Tsunami generation by a rapid entrance of pyroclastic flow into the sea during the 1883 Krakatau eruption, Indonesia. J Geophys Res 116(B09205):1–24

Magill CR (2001) Numerical modelling of tsunami generated by mass movement. MSc thesis, University of Waikato, 198 pp

Manville V, Hodgson KA, Nairn IA (2007) A review of break-out floods from volcanogenic lakes in New Zealand. N Z J Geol Geophys 50:131–150

McLeod P, Carey S, Sparks RSJ (1999) Behaviour of particle-laden flows into the ocean: experimental simulation and geological implications. Sedimentology 46:523–536

Monaghan JD, Cas RAF, Kos AM, Hallworth M (1997) Tsunami generation by pyroclastic flows, applied mathematics report and preprints. Monash University, Melbourne, 47 p

Moon VG, Bradshaw JD, Smith RT, de Lange WP (2005) Geotechnical characterisation of stratocone crater wall sequences, White Island Volcano, New Zealand. Eng Geol 81:146–178

Moon V, Bradshaw J, de Lange W (2009) Geomorphic development of White Island Volcano based on slope stability modelling. Eng Geol 104:16–30

Nairn IA, Self S, Cole JW, Leonard GS, Scutter C (2001) Distribution, stratigraphy, and history of proximal deposits from the c. AD 1305 Kaharoa eruptive episode at Tarawera Volcano, New Zealand. N Z J Geol Geop 44:467–484

Prasetya GS (1998) Modelling volcanic tsunamis. MSc thesis, University of Waikato, 299 p

Sahetapy-Engel S, Self S, Carey RJ, Nairn IA (2014) Deposition and generation of multiple widespread fall units from the c. AD 1314 Kaharoa rhyolitic eruption, Tarawera. N Z Bull Volcanol 76(836):1–28

Shane P, Sikes EL, Guilderson TP (2006) Tephra beds in deep-sea cores off northern New Zealand: implications for the history of Taupo Volcanic Zone, Mayor Island and White Island volcanoes. J Volcanol Geoth Res 154:276–290

Tinti S, Bortolucci E, Chiavettieri C (2001) Tsunami excitation by submarine slides in shallow-water approximation. Pure Appl Geophys 158:759–797

Trofimovs J, Amy L, Boudon G, Deplus C, Doyle E, Fournier N, Hart MB, Komorowski JC, Le Friant A, Lock EJ, Pudsey C, Ryan G, Sparks RSJ, Talling PJ (2006) Submarine pyroclastic deposits formed at the Soufrère Hills volcano, Montserrat (1995–2003): what happens when pyroclastic flows enter the ocean? Geology 34(7):549–552

Trofimovs J, Sparks RSJ, Talling PJ (2008) Anatomy of a submarine pyroclastic flow and associated turbidity current: July 2003 dome collapse, Soufrière Hills volcano, Montserrat, West Indies. Sedimentology 55(3):617–634

Watts P (1997) Water waves generated by underwater landslides. PhD thesis, California Institute of Technology, 319 pp

Watts P, Waythomas CF (2003) Theoretical analysis of tsunami generation by pyroclastic flows. J Geophys Res 108(B12):2156–2202

Chapter 57
Tsunami-Genesis Due to Retrogressive Landslides on an Inclined Seabed

Finn Løvholt, Geir Pedersen, and Carl B. Harbitz

Abstract Clay-rich landslides commonly involve retrogressive mass and momentum release mechanisms. Motivated by the retrogressive behaviour of major landslides offshore Norway, previous studies have demonstrated substantial effects of the release rate on the generation of the tsunami. However, the few existing models are limited to overly idealized conditions. In the present study, we explore further the wave generation due to a continuous retrogressive landslide model, quantifying the effects of the wave model, landslide configuration, and the continental slope. In the present examples, we find that the landslides involve large accelerations that may be crucial for tsunami-genesis. Tsunami footprints due to individual short blocks comprising the landslide are smeared out by dispersion. Keeping landslide material properties constant, we investigate mobilised landslide mass and maximum tsunami crest elevations for three different slopes: 1°, 1.5°, and 2° respectively. In the present examples, the smaller volume landslides are stronger tsunami generators than the larger ones because they are situated in shallower water, thereby clearly demonstrating the importance of the water depth on the tsunami generation.

57.1 Introduction

Tsunamis induced by submarine landslides and volcanic collapses constitute the second most frequent cause of tsunamis worldwide (e.g. Harbitz et al. 2014). The generation and propagation of tsunamis due to fully submerged landslides have been subject to a range of studies, including block slides (Løvholt et al. 2005), slumps (e.g. Lynett et al. 2003; Tappin et al. 2014), and deformable landslides with

F. Løvholt (✉) • C.B. Harbitz
Norwegian Geotechnical Institute, P.O. Box 3930, Ullevål Stadion, 0806 Oslo, Norway

Department of Mathematics, University of Oslo, P.O. Box 1053, Blindern, 0316 Oslo, Norway
e-mail: finn.lovholt@ngi.no

G. Pedersen
Department of Mathematics, University of Oslo, P.O. Box 1053, Blindern, 0316 Oslo, Norway

G. Lamarche et al. (eds.), *Submarine Mass Movements and their Consequences*,
Advances in Natural and Technological Hazards Research 41,
DOI 10.1007/978-3-319-20979-1_57

various rheologies such as viscous, visco-plastic, or granular landslides (e.g. Jiang and LeBlond 1992; Ma et al. 2013; Løvholt et al. 2014; Kim 2014). While the landslide volume is of importance for the tsunami-genic power, a rapid emplacement process is equally important for the generation of destructive tsunamis in terms of the wave amplitude in the proximal field (for a review on landslide generation mechanisms, see Løvholt et al. in press).

Retrogression is a phenomenon involving gradual mobilisation of the landslide mass starting from the slide toe. As the landslide progresses the head of the landslide moves backwards until it reaches its final position. Many large mass-failures, such as the clay-rich landslides offshore Norway involve retrogression. The most well-known examples comprise the Storegga (Solheim et al. 2005; Bryn et al. 2005) and Trænadjupet (Laberg and Vorren 2000) landslides. Storegga slide observations comprise clear evidences for retrogressive failure, particularly striking due to the terrace shaped bathymetry close to the headwall. Geophysical observations revealing submerged blocks inside remoulded clay have been reproduced by numerical simulations (Gauer et al. 2005). Retrogressive effects are interpreted as important also for volcanic flank collapses (Watt et al. 2014), and may have a massive influence on the tsunami genic potential. Most landslide tsunami studies assume that landslides fail in a single event, neglecting initiation processes such as retrogression. Using a series of block slide elements as tsunami genic sources, Haugen et al. (2005) quantified how the time lag between the individual elements with prescribed motion on a horizontal seafloor influenced the tsunami-genesis. In that study, the landslide was discontinuous, in the sense that all individual blocks followed the same time history, separated by a constant time lag. Giachetti et al. (2011), simulating a volcanic flank collapse on Tenerife using a series of landslide elements having a time lag. Here, we investigate the run-out and coupled tsunami generation using a retrogressive landslide model for clay-rich submarine landslides (Kvalstad et al. 2005), on an inclined seabed in one horizontal dimension (1HD). This is the first attempt to date linking a dynamic retrogressive landslide model to the tsunami generation, in contrast to the first retrogressive landslide tsunami analysis by Haugen et al. (2005) where block motion and time lag was prescribed, and a horizontal bathymetry was applied.

57.2 Methodology

57.2.1 Retrogressive Landslide Modelling

We use the energy-balance approach from Kvalstad et al. (2005) for simulating the retrogressive landslide motion in one horizontal dimension. The principle of the procedure is sketched in Fig. 57.1, and is briefly reviewed below (for more details we refer to Kvalstad et al. 2005). First, we assume that the landslide consists of equally shaped blocks having two triangular parts that are not deforming during

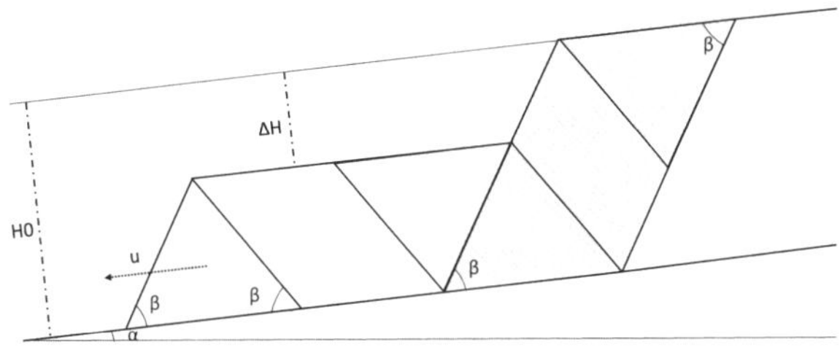

Fig. 57.1 Wedge shaped geometry for the retrogressive landslide. The *triangular shaped* blocks are not deformed during motion, while the central block is deforming. The *white* area of the landslide depicts the first mobilized block, while the shaded depicts the next block to be mobilized. Friction along the slide and failure surface contributes to the energy dissipation

motion, and a central parallelogram shaped body that is subject to simple shear in the initial deformation. Such geometries are revealed in field observations of the Storegga landslide, as well as numerical simulations of its last phase by a more sophisticated CFD analysis including a strain-softening material (Gauer et al. 2005). The potential energy E_P due to the released mass drop is computed for each block, and likewise the energy loss E_{int} due to internal deformation. In addition, we compute the dissipative energy loss E_{base} over the sliding surface. As long as the released potential energy exceeds the internal energy loss, the retrogressive process is allowed to continue with a new block being released. In the opposite case, the process is terminated. For each step, we calculate the energy by the end of each block deformation, and set the kinetic energy equal to the difference between the potential and dissipative energy according to the expression:

$$E_K = \frac{1}{2}(m + m_w)u^2 = E_P - E_{int} - E_{base}$$

Here, m and m_w are the landslide and added mass per unit width respectively (for m_w, an added mass coefficient of 1 is assumed). We obtain the block velocity u at the end of each block release as a function of the run-out distance r. For the tsunami source representation, we are dependent on a densely sampled time history of the landslide surface. To obtain this, we refine $u(r)$ using quadratic interpolation, and calculate the time step between the spatial step j and $j+1$ as $\Delta t_{j+1/2} = \Delta r_{j+1/2}/\overline{u}_{j+1/2}$. For the initial acceleration of the first block, it turned out that keeping the time-step constant over a one-tenth fraction of the run-out of the first block provided a stable measure for the acceleration for arbitrary mesh size.

Material parameters and geometric assumptions are largely motivated from the Ormen Lange landslide and tsunami hazard study (see Kvalstad et al. 2005 and Solheim et al. 2005 including references therein). As geometric input parameters, we choose the total height of the wedge $H_0 = 400$ m, drop height ΔH, the wedge slope relative to the seabed $\beta = 35°$, and the seabed slope angle α (see Fig. 57.1). The drop height and the slope angle are both subject to parametric variation. Joint material input parameters for both the solid blocks and the base comprise the submerged unit weight $\gamma' = (\rho_s - \rho_w)g = 10\,\mathrm{kN/m^3}$, the normalized consolidated shear strength $\alpha_C = 0.39$, excess pore-pressure ratio 0.1, the undrained shear strength at the seabed $s_{u,0} = 25$ kPa, and the shear-to-compression-ratio of undrained shear strength $s_{u,d}/s_{u,c} = 0.8$. The deformable block sensitivity relative to s_u is set to $S_{block} = 2$, while the sensitivity for the base is set to $S_{base} = 6$. The latter sensitivity is set somewhat higher than in general practice based on laboratory experiments (Kvalstad et al. 2005 employed $S_{base} = 5$). We use this high value of S_{base} for the following reasons: the block model is simplified, and does not take into account the transition to flow that is expected during the latter phase of motion; the frictional forces are uncertain and subject to change with time due to remoulding and changes in material properties, and are difficult to measure under these conditions; and finally, a high value of S_{base} will allow larger run-out values more similar to those observed in the field, such as Storegga and Trænadjupet, among others.

Based on the above material parameters and geometry, the individual energy terms E_P, E_{int}, and E_{base} read:

$$E_P = H_0 \Delta H (H_0 - \Delta H) \cdot n \cdot \cot(\beta) \gamma' \cdot [(2n+1) \cdot \cot(\beta) \cdot \sin(\alpha) + \cos(\alpha)]$$

$$E_{int} = n \cdot \left[\beta \cdot s_{u,av,B} \cdot 2\Delta H \cdot \cot(\beta)(H_0 - \Delta H) + s_{u,rem,C} \frac{1}{(\sin(\beta))^2} \Delta H (H_0 - \Delta H) \right]$$

$$E_{base} = \frac{s_{u,base}}{S_{base}} \Delta H \left[2n \cdot \cot(\beta) \frac{\Delta H}{\sin(\beta)} + \frac{(n+1)n}{2} (2\cot(\beta))^2 (H_0 - \Delta H) + \frac{1}{2} \Delta, H (2\cot(\beta))^2 \right]$$

Here, the term n denotes the number of released blocks. $s_{u,av,B}$ and $s_{u,rem,C}$ are average and remoulded shear strength of the blocks, and are functions of the material undrained shear strength, the sensitivity, the excess pore pressure ratio, and the normalized consolidated shear strength (see Kvalstad et al. 2005 for a discussion of the interpretation of the undrained strength factors).

57.2.2 Tsunami Modelling

The tsunami generation takes place through the volume displacement due to the landslide. Landslide-induced volumetric fluxes are computed by taking the difference in landslide surface height at discrete variable time steps. Short scales are smoothed using the full potential method of Kajiura (1963), as described in Løvholt et al. (2012) and Løvholt et al. (in press). We use the linear version of the Boussinesq model Globouss (Løvholt et al. 2008; Pedersen and Løvholt 2008) for simulating the wave propagation. Model assumptions include hydrostatic (LSW), standard dispersion, and higher order dispersion. We refer to Løvholt et al. (in press) for validation, limitations, and further details related to the modelling procedure.

57.2.3 *Model Setup*

The bathymetry consists of two levels of constant depth being $h_{min} = 200$ m and $h_{max} = 5000$ m, respectively. The two levels are connected by a slope that may take an arbitrary angle. Herein, we investigate landslide run-out using three different angles, namely $\alpha = 1°$, $1.5°$, and $2°$. We add the landslide volume on top of the slope, and adjust its location, putting the initial headwall at the horizontal coordinate $x = 0$ and the final headwall (after run-out) at the edge of the shallowest plane. On each slope, we simulate the landslide evolution and the tsunami generation for two different drop height configurations, namely $\Delta H = 200$ m and $\Delta H = 300$ m. In the tsunami simulations, we use a horizontal grid resolution of $\Delta x = 160$ m and a Courant number of $Cr = (gh_{max})^{1/2}\Delta t/\Delta x = 0.6$ ($\Delta t = 0.4$ s). Furthermore, we calculate volume fluxes as a non-uniform function of time, depending of the duration of each block release. Convergence to a few percent accuracy for both leading and trailing wave amplitudes were ensured through grid refinement tests, both for the spatial grid spacing and the time resolution of the fluxes.

57.3 Results

Simulated frontal landslide velocities as a function of time are depicted in Fig. 57.2. In the present model, the run-out is very sensitive to the slope angle (note that the bed friction is kept constant through equal undrained shear strength and sensitivity). A larger amount of blocks are released for the larger slope angles; for $\Delta H = 200$ m, we get 13 blocks for $\alpha = 1°$ and 95 blocks for $\alpha = 2°$, i.e. more than 7 times larger volume for the steepest slope. Figure 57.2 further shows that the maximum landslide velocity is less sensitive to the slope angle, and that it is higher for $\Delta H = 300$ m than for $\Delta H = 200$ m. Finally, we obtain an almost twice as high mean acceleration for $\Delta H = 200$ m compared to $\Delta H = 300$ m. However, it is remarked that the accelerations obtained here are rough estimates, and that a model resolving the failure process in more detail would be needed to more accurately quantify the acceleration.

The principles of the wave generation are demonstrated for the case of $\alpha = 1.5°$ and $\Delta H = 200$ m, Fig. 57.3. The initiation is shown in the upper panel, with a single depression wave forming over the rear graben formation and an elevated wave over the forward moving front. The mid panel shows the surface elevation after 120 s, after the release of the third block. The wave simulations reveal signs of two distinct waves in the left direction, which are due to the first two blocks, respectively. At

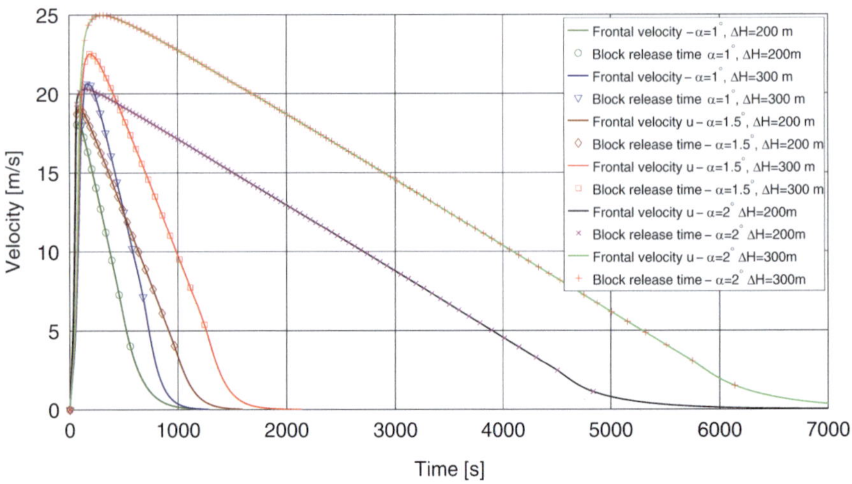

Fig. 57.2 Simulated landslide front velocity as a function of time. The markers show the release time of each individual block

this stage, the different wave models behave fairly similarly, although frequency dispersion is noticeable already. At the landslide termination, the difference between the models are much more pronounced. In the LSW model, the footprints of the individual blocks are visible as distinct waves, but these footprints are partly smeared out in the dispersive models. Clearly, the LSW assumption is not adequate for these retrogressive landslides. However, the standard and higher order dispersive models also differ. The higher frequency trailing waves move faster and are more coherent in the higher order than in the standard dispersive model. Their wavelength-to-depth-ratio is close to unity. As discussed by Pedersen and Løvholt (2008), the higher order dispersive model introduces a finite group velocity for small wavelengths that may typically lead to overestimation for wavelengths-to-depth ratios $\lambda/h < 2$. To remove such spurious oscillations, either selective filtering or model modifications will be needed. We therefore present results using standard dispersion that involve less pronounced trailing oscillations in the analyses below. Finally, we point out that for the waves entering the shallowest bathymetric plane, appreciable non-linearities are expected, but studying non-linear effects are beyond the scope of the present paper.

The upper panel in Fig. 57.4 compares the generated tsunami assuming standard dispersion at $t = 600$ s for the 1.5° slope angle for cases of $\Delta H = 200$ m and $\Delta H = 300$ m, respectively. At this time, the main part of the leading elevation (moving leftwards) and leading depression (moving rightwards) has formed. Although the depth configuration is slightly different, we clearly see that the landslide with $\Delta H = 200$ m is the more efficient wave generator. We relate this to a higher initial acceleration. In the lower panel of Fig. 57.4, the leftward moving waves are compared for different slope angles. We see that higher waves are generated by landslides with smaller slope angles. It seems immediate somehow

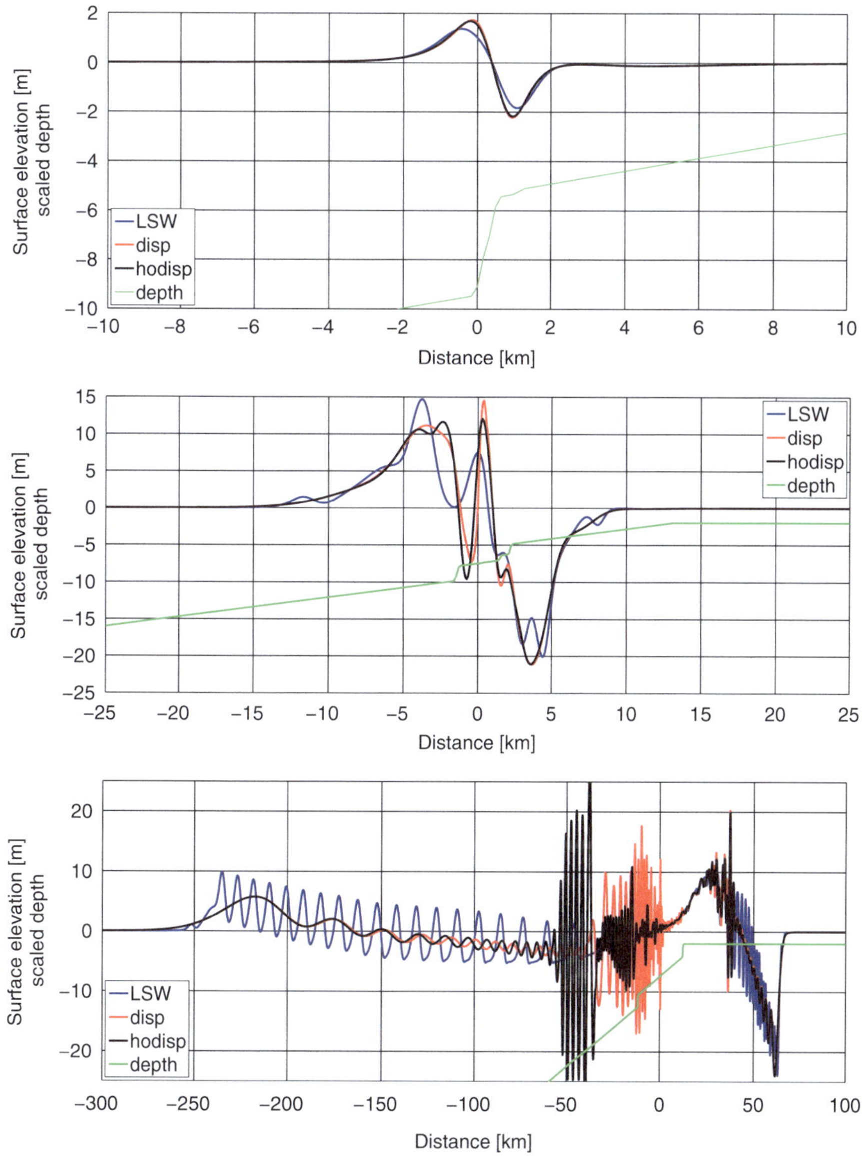

Fig. 57.3 Wave and retrogressive landslide evolution for $\alpha = 1.5°$, $\Delta H = 200$ m. The tsunami is simulated using GloBouss under linear shallow water mode (LSW), standard dispersion (disp), higher order dispersion (hodisp). Surface elevation and seabed profile after 10 s (*upper*), 120 s (*mid*), and at final run-out after 1430 s (*lower*). The depth is scaled by a factor of 100

counterintuitive that the landslides generated on the gentler slopes and involving the smallest volumes generates the larger waves. However, it is noted that in the present examples, the landslides on the gentlest slopes are located in shallower

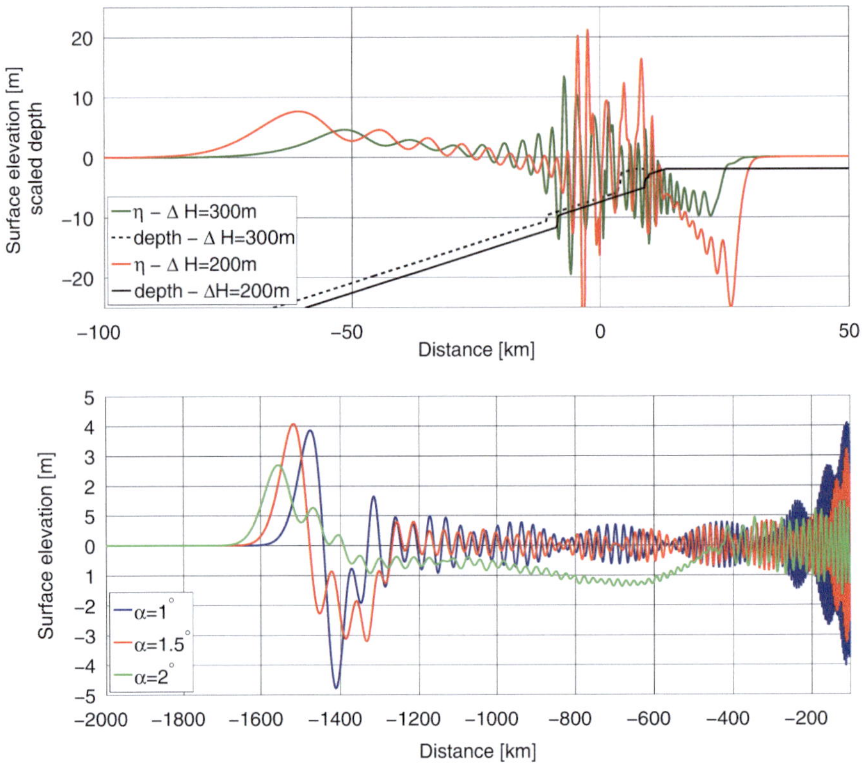

Fig. 57.4 *Upper panel*: Shapshot of the generated wave for two different drop heights after 600 s, using the standard dispersive model and $\alpha = 1.5°$. The depth is scaled by a factor of 100. At this stage, the main part of the leading wave generation have taken place. *Lower panel*: Shapshots of the generated wave at the time of landslide termination for $\alpha = 2°$ (2 h 3 min) for three different slope angles using the standard dispersive model and $\Delta H = 200$ m

waters, meaning that their Froude number $Fr = u/(gh)^{1/2}$ is increased, with resulting increased relative tsunami-genic power (in terms of induced crest heights). To this end, we add that the main cause of the large initial acceleration is the steep slope of the wedge, generating similar accelerations for different base slopes. The large initial acceleration combined with relatively large Froude numbers, renders the shallow and smaller volume scenarios more efficient tsunami generators than a larger landslide volume example being situated on larger water depth in the present example.

57.4 Concluding Remarks

Clay-rich submarine landslides and volcanic flank collapses often involve retrogression, distributing the landslide mass mobilisation over a significant time span. Such effects are in most cases neglected when modelling landslide-induced tsunamis. The present analysis is a first attempt to link a continuous retrogressive block model for the landslide run-out to the tsunami generation. Due to the short horizontal scales of the block source, the hydrodynamic response is distributed over a distance comparable to the water depth or larger. Frequency dispersion is further crucial for an adequate description of the wave generation and propagation. The retrogressive model gives a large initial landslide acceleration of importance for the wave generation, followed by a slow retardation. As the initial acceleration and Froude number dominates the tsunami generation, the tsunami genic strength of the landslide involving largest volumes is in the present example a weaker tsunami generator than the smaller ones in the present examples. Although the 2° slope landslide involves more than 7 times the landslide volume compared to 1° slope landslide, the 1° slope landslide still gives about 1.5 times higher tsunami crest height. This demonstrates that shallow landslides are more effective tsunami generators than deep ones. In a real three dimensional case however, radial spread is expected to contribute to relatively stronger distance attenuation for the smaller landslides.

The initial acceleration is strongly linked to geometry and material properties of the block and to a lesser extent the (gentle) slope angle. The retrogressive landslides share mechanisms with both slumps (involving high initial acceleration) and translational landslides (involving long run-out), and appear as more complex due to this combination as well as the gradual mass mobilisation. However, this study provides just some first few examples, and a more systematic work is needed to link both geometric and material properties to the tsunami-genic potential.

Acknowledgments The work has been supported by the Norwegian Research Council under project no. 231252 and no. 205184, as well as by the project ASTARTE (Assessment, STrategy, And Risk reduction for Tsunamis in Europe), Grant 603839, 7th FP (ENV.2013.6.4-3, ENV.2013.6.4-3). We also thank Tore Kvalstad for his assistance in setting up the retrogressive landslide model, and for his comments to the section describing the landslide model. Finally, Mohammad Heidarzadeh and Raphaël Paris are both thanked for their constructive review comments.

References

Bryn P, Berg K, Forsberg CF, Solheim A, Kvalstad TJ (2005) Explaining the Storegga Slide. Mar Pet Geol 22(1–2):11–19

Gauer P, Kvalstad TJ, Forsberg CF, Bryn P, Berg K (2005) The last phase of the Storegga Slide: simulation of retrogressive slide dynamics and comparison with slide-scar morphology. Mar Pet Geol 22(1–2):171–178

Giachetti T, Paris R, Kelfoun K, Pérez-Torrado FJ (2011) Numerical modelling of the tsunami triggered by the Güímar debris avalanche, Tenerife (Canary Islands): comparison with fieldbased data. Mar Geol 284(1–4):189–202

Harbitz CB, Løvholt F, Bungum H (2014) Submarine landslide tsunamis: how extreme and how likely? Nat Hazards 72(3):1341–1374

Haugen KB, Løvholt F, Harbitz CB (2005) Fundamental mechanisms for tsunami generation by submarine mass flows in idealised geometries. Mar Pet Geol 22(1–2):209–219

Jiang L, LeBlond PH (1992) The coupling of a submarine slide and the surface waves which it generates. J Geophys Res 12:731–744

Kajiura K (1963) The leading wave of a tsunami. Bull Earthq Res Inst 41:535–571

Kim J (2014) Finite volume methods for tsunamis generated by submarine landslides. PhD thesis, Univ Washington

Kvalstad TJ, Andresen L, Forsberg CF et al (2005) The Storegga slide: evaluation of triggering sources and slide mechanics. Mar Pet Geol 22(1–2):245–256

Laberg JS, Vorren TO (2000) The Trænadjupet Slide, offshore Norway—morphology, evacuation and triggering mechanisms. Mar Geol 171(1–4):95–114

Løvholt F, Harbitz CB, Haugen KB (2005) A parametric study of tsunamis generated by submarine slides in the Ormen Lange/Storegga area off western Norway. Mar Pet Geol 22(1–2):219–233

Løvholt F, Pedersen G, Gisler G (2008) Oceanic propagation of a potential tsunami from the La Palma Island. J Geophys Res 113:C09026. doi:10.1029/2007JC004603

Løvholt F, Pedersen G, Bazin S, Bredesen R, Harbitz CB (2012) Stochastic analysis of tsunami runup due to heterogeneous coseismic slip and dispersion. J Geophys Res 117(C1):2156–2202

Løvholt F, Harbitz CB, Vanneste M et al (2014) Modeling potential tsunami generation by the BIG'95 landslide. Submarine Mass Move Conseq 37:507–515

Løvholt F, Pedersen G, Harbitz CB, Glimsdal S, Kim J (in press) On the characteristics of landslide tsunamis, Phil Trans R Soc A

Lynett PJ, Borrero JC, Liu PL-F, Synolakis CE (2003) Field survey and numerical simulations: a review of the 1998 Papua New Guinea tsunami. Pure Appl Geophys 160:2119–2146

Ma G, Kirby JT, Shi F (2013) Numerical simulation of tsunami waves generated by deformable submarine landslides. Ocean Model 69:146–165

Pedersen G, Løvholt F (2008) Documentation of a global Boussinesq solver. Technical report. http://urn.nb.no/URN:NBN:no-27775

Solheim A, Bryn P, Sejrup HP, Mienert J, Berg K (2005) Ormen Lange—an integrated study for the safe development of a deep-water gas field within the Storegga Slide Complex, NE Atlantic continental margin; executive summary. Mar Pet Geol 22(1–2):1–9

Tappin D, Grilli ST, Harris JC, Geller RJ, Masterlark T, Kirby JT, Shi F, Ma G, Thingbaijam KKS, Mai PM (2014) Did a submarine landslide contribute to the 2011 Tohoku tsunami? Mar Geol 357:344–361

Watt SFL, Talling PJ, Hunt JE (2014) New insights into the emplacement dynamics of volcanic island landslides. Oceanography 27(2):46–57, http://dx.doi.org/10.5670/oceanog.2014.39

Chapter 58
Geothermal System as the Cause of the 1979 Landslide Tsunami in Lembata Island, Indonesia

Yudhicara, Phillipson Bani, and Alwin Darmawan

Abstract A tsunami landslide which caused hundreds casualties and lots of damage took place on Lembata Island in 1979. In order to understand the characteristics of the landslide mechanism, a field survey was conducted in 2013 which sampled both the origin soil and landslide material, and the water from hotspring around the landslide site. The physical properties of the soil were obtained and show that the origin soil has dominantly coarser grain than the landslide material (80.5 % coarser grain compared to 11.8 % coarse grain respectively) which indicates that the soil has been altered to be finer and softer. Hot spring analysis determined that the mineral content of the water was 99.48 % SO_4. This shows that magmatism process are involved which caused the soil to become acidic and may have fragilised the system. Results of X-ray Diffraction Mineralogy Analysis (XRD) show that the origin soil is composed of minerals of cristobalite, quartz, and albite, while the landslide material consists of clay minerals such as quartz, saponite, chabazite, silicon oxide and coesite which are typical minerals in a hydrothermal environment. Based on these results, it can be concluded that the area is influenced by an active geothermal system that could be the main source mechanism behind this disastrous event.

Yudhicara (✉)
Geological Agency, Jl. Diponegoro No. 57, Bandung, West Java, Indonesia

Padjadjaran University, Jl. Dipati Ukur No. 35, Bandung, West Java, Indonesia
e-mail: yudhicara@yahoo.com

P. Bani
Geological Agency, Jl. Diponegoro No. 57, Bandung, West Java, Indonesia

Laboratory of Magmas and Volcanoes, Univ. Blaise Pascal-CNRS-IRD-OPGC, 63000 Clermont-Ferrand, France

A. Darmawan
Geological Agency, Jl. Diponegoro No. 57, Bandung, West Java, Indonesia

G. Lamarche et al. (eds.), *Submarine Mass Movements and their Consequences*,
Advances in Natural and Technological Hazards Research 41,
DOI 10.1007/978-3-319-20979-1_58

58.1 Introduction

On July 18, 1979, Lembata Island, East Nusa Tenggara province, Indonesia, was struck by a sudden tsunami, induced by a massive landslide. Fifty million cubic meters of material was displaced and one third of it tumbled into the sea and generated a 7–9 m height tsunami which spread along the Waiteba Bay (Lassa 2009; Hadian et al. 1979). Four villages were buried by the landslide material and 539 of inhabitant were reported killed by both landslide and tsunami. This event is not well known by most Indonesian people, even though newspapers "KOMPAS" of 21 July 1979 and "Herald Tribune" 24 July 1979 reported this event (Lassa 2009; Arif 2010).

Lembata is an 80 × 30 km island that hosts numerous massive volcanic edifices, three of which remain active: Ili Lewotolo in the north and Ili Labalekan and Ili Werung in the south (see Fig. 58.1). Most of these edifices extend to the sea, exposing steep slopes and unconsolidated material. Adding the regular heavy rain falls, and the intense tectonic activity in that region, Lembata Island thus possesses many factors capable triggering land displacement with a significant potential for landslide-tsunami. The 1979 event being the unfortunate example.

To gain further insights into the origin of the disastrous 1979 event, a field investigation was carried out in April 2013, looking at landslide material and the surrounding edifice morphology, evaluating potential links to volcanic activities and determining tectonic structures in the study area. Soil and water samples were taken for laboratory analysis to constrain the physical properties of the landslide materials (soil density, specific gravity, cohesion, shear friction and the nature of expanded soil) and other factor influencing the system.

The landslide site is located at 08°28′44″S – 123°33′34.3″E, a hill ca. 200 m above sea level, while the landslide material reach the sea of 08°28′30.8″S – 123°33′24.7″E (Fig. 58.1).

According to the reports published in 1979 and compiled in Lassa (2009), the Volcanological team from Bandung reported that there was no volcano activity during the 2 days preceding the landslide (16–17 July 1979), while meteorological and geophysical teams noted no earthquakes during that time either. According to Lamanepa (2013), in the month of July 1979 the Lembata Island and its surroundings was in a rain season and had heavy rainfall in some areas of East Nusa Tenggara. The study is intended to understand what caused this landslide given that lack of an obvious triggering mechanism.

58.2 General Review

Geologically, the region where the landslide site is located consists of a Tertiary lithological unit called Kiro Formation, lying beneath the Quaternary old volcanic rocks and Quaternary young volcanic products (Koesoemadinata and Nova 1990). Old volcanic rock units consist of lava, breccia, agglomerate, volcanic sandy tuff,

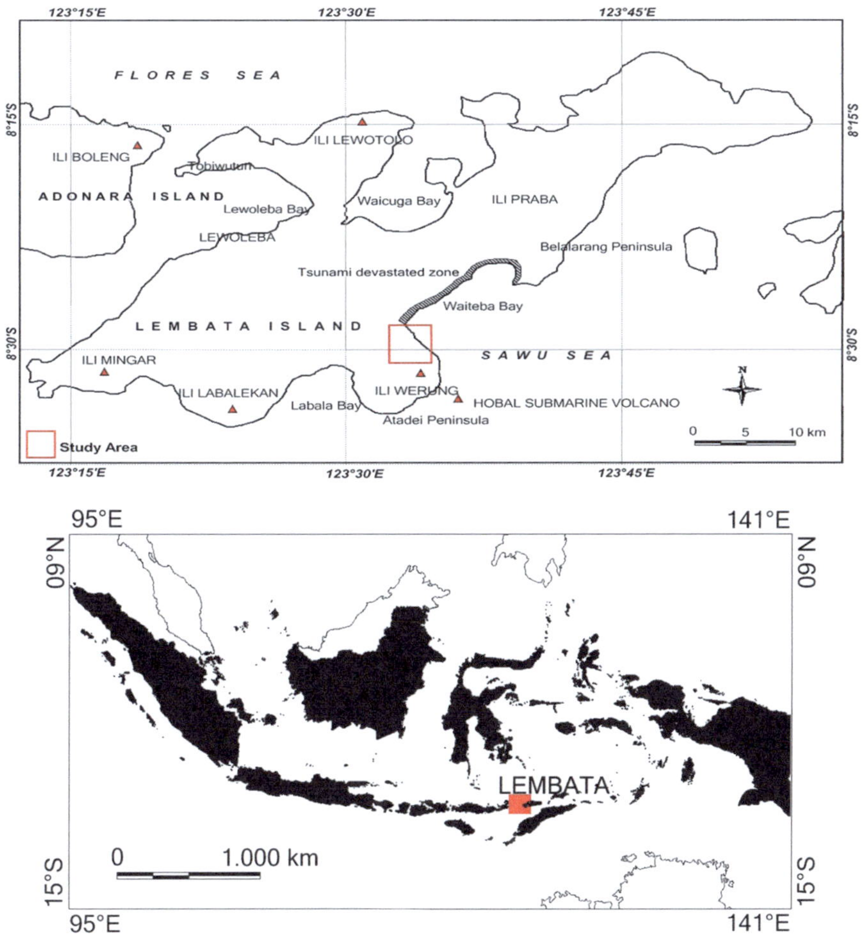

Fig. 58.1 *Top*: Lembata Island showing active volcanos, study area and tsunami affected area. *Bottom*: location of Lembata Island within Indonesia

and pumiceous sandy tuff as products of volcanoes which are no longer active, such as Wikiriwak, Ili Lewung, Ili Minggar, Ili Ujolewung, Pura, Ternate and Treweg. These form high mountains along the southern of Lembata Island. There are also locally younger volcanic rocks, consisting of lava, agglomerates, bombs, sands and volcanic ashes which are products of younger active volcanoes, such as Watuomi, Ili Boleng, Ili Lewotolo, Ili Werung, Ili Topaki/Sirung, Batutara and Komba (Fig. 58.2).

There are two major geological structure in Lembata Island, which have trends of southwest-northeast and northwest-southeast. One of them is normal faulting in a northwest-southeast direction from the Plio-Pleistocene which has produced a scattering of hotsprings near to the landslide site that may influence to the soil condition (Fig. 58.2).

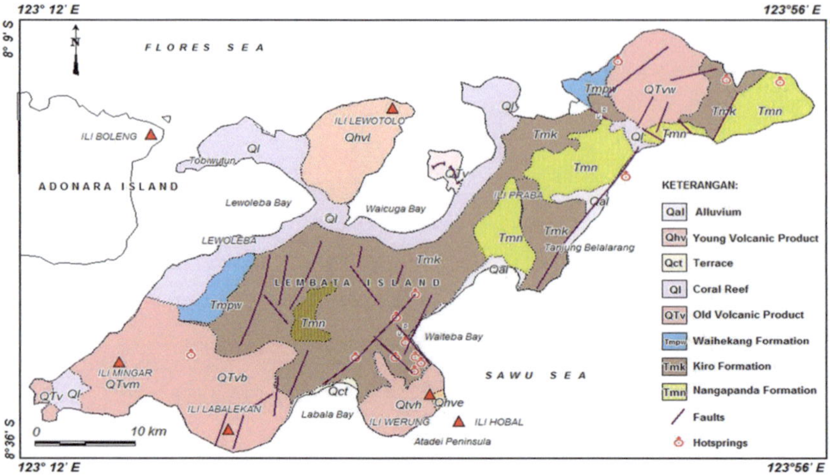

Fig. 58.2 Geological map of Lembata Island and location of hotsprings (vents) (Koesoemadinata and Nova 1990)

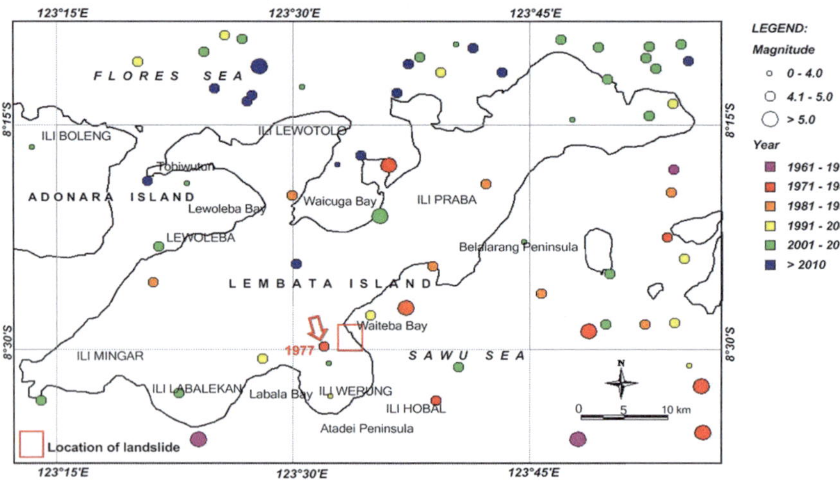

Fig. 58.3 Seismicity in Lembata Island and its surroundings (USGS 2013). *Red arrow* indicates the epicentre of the earthquake that occurred in 1977 close to the landslide site

There is also a lineament showing a volcanic chain which has moved to the southwestward, and a submarine volcano named Ili Hobal at the end of faulting which has been observed since 1972.

A dense distribution of epicentres around Lembata Island indicates that the area is seismically very active. One earthquake which occurred in 1977 (USGS 2013) located near to the landslide site and resulted in surface rupture. This may have fragilised soil at the site, although the earthquake magnitude was less than five (Fig. 58.3).

According to the eruptive history around the Lembata Island, the last eruption occurred on Ili Hobal which started in 1972 and ended in 1974. Ili Hobal is a volcano which located at the southern end below the sea (CVGHM 2013 in http://www.volcano.si.edu/).

58.3 Landslide Induced Tsunami

Based on the scientific report related to the landslide and tsunami incident of 18 July 1979, the event occurred in the middle of the night and the landslide swept four villages that were located at the coast of the Waiteba Bay (Hadian et al. 1979). Geological condition shows that the site is composed of alluvium with loose sand and gravel. Both land and the seafloor near the shore are morphologically very steep and easy to be eroded. The eyewitnesses said that the landslides were observed at the site since 1976 and occurred progressively until 1979 when the biggest event occurred (Hadian et al. 1979). The landslide body has a length of 3000 m along the slide direction starting at elevation of 500 m and width of 300 m to the east-southeast. Part of landslide material had tumbled down into the sea, generated the tsunami wave that came about three times on land and devastated about 50 km along the Waiteba Bay (Fig. 58.4). Remnants of these landslides can still be traced from natural features observed in the field (Figs. 58.5 and 58.6) and satellite images.

Fig. 58.4 Location of landslide and tsunami impacted area (Taken from Iryanto 1980)

Fig. 58.5 Evidences of the 1979 landslide taken on April 2013, the crown (*left*) and the landslide body (*right*) (Yudhicara et al. 2013)

Fig. 58.6 The landslide product protrudes into the sea (*red circle*), location of water sample is shown by *red arrow* (Yudhicara et al. 2013)

58.4 Methods

In order to understand the nature of the landslide and what caused this event, a field survey was carried out in April 2013. Visual observations were made and samples taken of both disturbed (i.e. landslide) and undisturbed (origin) soil material. This was followed by soil mechanics tests to obtain the physical properties of both the origin soil and the landslide material.

XRD Analysis was undertaken of the origin soil and landslide material to obtain the mineral content, which can prove the involvement of hydrothermal system and an ion chromatograph analysis was conducted on water samples taken from geothermal sites scattered around the study area to confirm that evidence.

58.5 Results

The landslide area is located at an old crater of Iliwerung complex. The crown of landslide is found at Atakore Village, named Bauraja Hill (Fig. 58.5). Bauraja Hill form a cliff that is unstable and prone to landslide, particularly after a heavy rain. The study area is about 1.2 km^2. It has many volcanic cones around and close to the landslide site, active faults cutting across it and is often subjected to heavy rain fall. The study area has a hilly morphology with a very steep slope and is covered by dense forest and trees, however the landslide scar is still barren and not been revegetated.

Some in situ samples were taken around the hill in areas assumed to be undisturbed. Megascopic appearance shows that the soil has coarse sand to gravel grain size, yellow to reddish colour, loose, bad sorting, and open fabric. The landslide material was taken near to the sea (Fig. 58.6), which has finer grain volcanic ashes, yellow to white colour, very soft, clayey, loose, slippery and easy to move.

Results of the soil mechanics analysis shows that the grain size curve of origin soil consists of 6.9 % gravel, 80.5 % coarser grain and 12.60 % the fine grain (Fig. 58.7). This means that the soil is dominantly composed of coarser sandy grains. In contrast, the landslide material has fine grains which compose 88.2 % of the soil, while the coarse grains represent 11.8 % (Fig. 58.7). This indicates that soil has been altered to be softer, which can be differentiated from surrounding soil.

According to the physical properties of the origin soil, it has cohesion value of 0.04 kg/cm^2 and deep shear angle of 33.72°, so for the safety factor of 1.0, by this soil, it has potential to be slide even without ground shaking on slope of 15–24°.

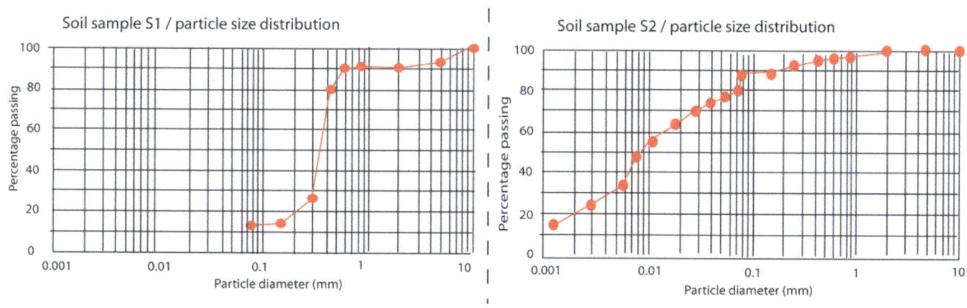

Fig. 58.7 Grain size analyses of the origin soil (*left*) and the landslide material (*right*)

Table 58.1 Chemical content of water sample from hotspring close to the landslide site

No.	Ret. Time (min)	Peak Name	Height µS	Area µSxT	Rel. Area %	Amount ppm
1.	3.09	F	0.097	0.012	0.09	1.54
2.	4.57	Cl	0.402	0.059	0.43	10.36
3.	11.91	SO4	41.835	13.597	99.48	3458.61

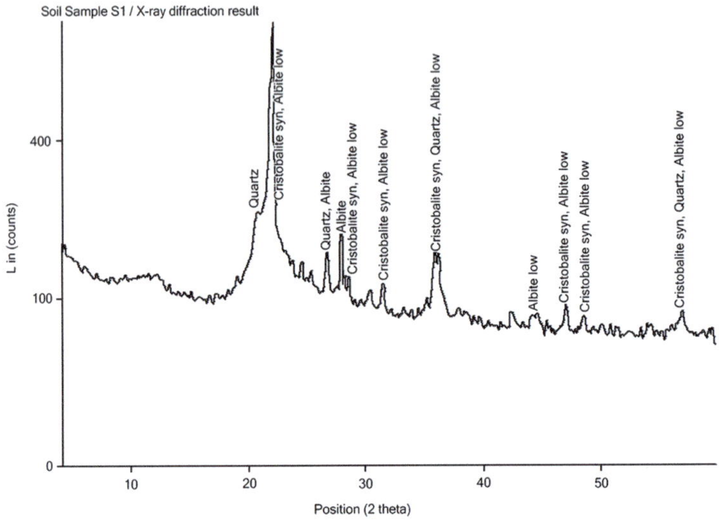

Fig. 58.8 Spectral record of the origin soil (Geological Agency 2014)

Some hot springs were found around the landslide site indicating a fumarole system. A water sample was taken from a hot spring at the foot hill of the landslide material at position of 08°28′44″S – 123°33′34.3″E (Fig. 58.6), which coincides with a fault line. The mixing of water and hot material from inside can alter the rocks, and some of the material could be changed to be clay minerals that may fragilised the system and be responsible for the slide occurrence.

Results of ion chromatograph analysis on the hot spring water sample show that the SO_4 content is 99.48 % of the total chemical content (Table 58.1). This indicates that this water is associated with the magmatism beneath the area and take the role in the hydrothermal system. It is also the reason why soil are fragile, loose, unconsolidated and easy to move. The sulfate content of 3458.61 ppm contained in this hot water may also explain why vegetation cannot grow in the landslide material, even after 35 years. This is due to the high acidity properties of the soil in that location.

According to the result of the XRD analysis, the origin soil contains minerals such as cristobalite, quartz, and albite (Fig. 58.8), while the landslide material contain minerals of quartz, saponite, chabazite, silicon oxide and coesite (Fig. 58.9). It seems that minerals in the origin soil have altered and some of the

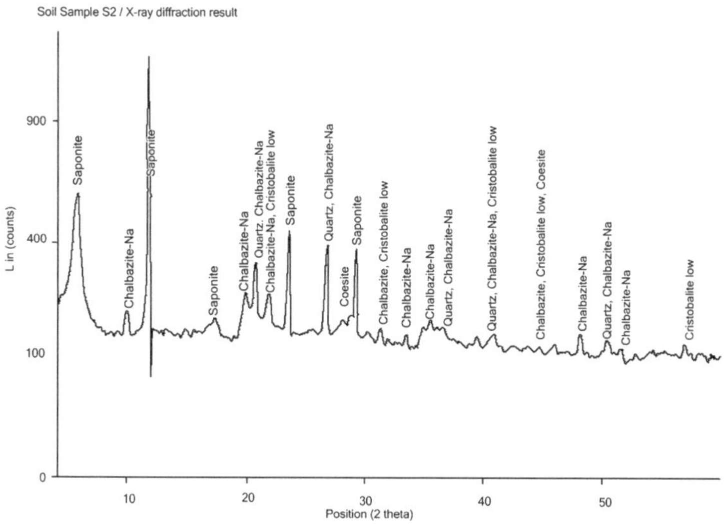

Fig. 58.9 Spectral record of the landslide material (Geological Agency 2014)

altered minerals contain clay minerals that could influenced the soil physical properties thus becoming loose.

Saponite belongs to the smectite group, which is mineralized by hydrothermal processes. Smectite is a clay mineral that can swell when immersed in water or some organic liquids, while chabazite belongs to zeolite group, a clay mineral which is formed in a volcanic complex. Coesite is an alteration mineral formed from quartz under high temperature and pressure (Inoue 1995).

58.6 Discussion and Conclusion

In general, factors that may cause landslides include steep slope angle, earthquakes shaking, prolonged rainfall and volcanic eruptions. These factors can all be found in the study area. The region is hilly with steep slope with relatively high rainfall. Despite the fact that earthquakes are common, before the landslide event, there had been only earthquake which occurred in 1977 with magnitude less than five. The last eruption which occurred before the landslide was Ili Hobal in 1976. Both earthquake and volcanic activities do not directly relate to the 1979 landslide events but they may have fragilised the system.

The Waiteba landslide area is located at a volcano complex where there are scattered geothermal fields, indicated by numerous hot springs. Rocks and soils which are altered by the geothermal activity could lose their underlying structure. The altered rocks are loose, slightly light and consist of clay minerals. Clay minerals expand when they absorb water, so in the season of heavy rainfall, the area would susceptible to landslide occurrence. Clay minerals content found in the landslide material, such as smectite and zeolite are typical minerals in hydrothermal environments. Hence, there is a high likelihood that the 1979 Lembata landslide and resulting tsunami was primed by hydrothermal alteration of the rocks and soil in the geothermal environment. This correlation should be kept in mind when assessing the hazard posed by landslides in other geothermally active areas.

Acknowledgements Our deep gratitude to the Geological Agency which has funded this survey. We thank you to Wilson, for his assistance during the survey. Our appreciation To Iman K. Sinulingga, Sri Hidayati, Rana Suparan, Djadja, Prettina Sitinjak, Godwin and all the people who helped and give their support during the research. We are greatly appreciative of the support provided by sub editor Emily Lane and in her substantial work in improving the English of this manuscript.

References

Arif A, (2010) Jurnalisme Bencana – Bencana Jurnalisme: Kesaksian dari tanah bencana (Disaster Journalism – Journalism Disaster: Testimony from the disaster area). Gramedia Popular, Indonesia, 238 p

Center for Volcanological and Geological Hazard Mitigation (CVGHM) (2013) Eruptive volcano history. http://www.volcano.si.edu/

Geological Agency (2014) X-ray diffraction mineral analysis. Internal report

Hadian R, Sudradjat A, Elifas JJ (1979) Laporan Singkat Pemeriksaan Ili Werung Bulan Juli 1979 (Brief examination of Ili Werung Month July 1979), Internal publication, Volcanological Survey of Indonesia

Inoue A (1995) Formation of clay minerals in hydrothermal environments in origin and mineralogy of clays. Springer, Berlin, pp 268–329

Iryanto (1980) Pemeriksaan G. Sirung, G. Ili Werung dan G. Hobal di Nusa Tenggara Timur, Bulan Oktober 1980 (Examination of volcanoes of Sirung, Ili Werung and Hobal in East Nusa Tenggara, on October 1980). Volcanological Survey of Indonesia, Indonesia

Koesoemadinata S, Nova Y (1990) Geology of the Lomblen Quadrangle, East Nusa Tenggara. Geological Agency, Indonesia

Lamanepa F (2013) Larantuka, hujan deras, dan banjir 1979 (Larantuka, heavy rain fall and flood 1979). http://franklamanepa.blogspot.com/2013/12/larantuka-deras-dan-banjir-1979.html

Lassa J (2009) Bencana yang terlupakan? Mengingat kembali bencana Larantuka dan Lembata 1979–2009 (The forgotten disaster? Remembering the Larantuka and Lembata disaster 1979–2009). J NTT Stud

United State of Geological Survey (USGS) (2013) The Preliminary Determination of Epicenters (PDE) Bulletin 1976–2013. Online source of http://www.earthquake.usgs.gov/data/pde.php

Yudhicara, Hidayati S, Suparan R (2013) Laporan Penyelidikan Tsunamigenik di Pulau Lembata, Nusa Tenggara Timur (The investigation report of tsunamogenic in Lembata Island, East Nusa Tenggara). Internal report of Center for Volcanology and Geological Hazard Mitigation, Geological Agency

Chapter 59
Towards a Spatial Probabilistic Submarine Landslide Hazard Model for Submarine Canyons

Christof Mueller, Joshu Mountjoy, William Power, Emily Lane, and Xiaoming Wang

Abstract The Cook Strait Canyon of central New Zealand was identified as a priority area to quantify landslide-generated tsunami hazard in a national study in 2005. Therefore the canyon system has seen increasing research interest over the last decade. Landslide scars have been mapped throughout the whole of the Cook Strait Canyon area and analysis of landslide morphology demonstrates that the majority of landslides have some dependence on the topography of the canyon system. Axial downcutting destabilising lower canyon walls is proposed as the principal factor preconditioning slopes for failure. The canyons occur in an active tectonic environment and earthquakes are inferred to be the overriding failure triggering mechanism.

To develop an underpinning magnitude frequency model for probabilistic landslide-tsunami hazard assessment we have created a Monte Carlo based framework to assess the spatial distribution earthquake triggered slope stability within the canyon. The framework is object-oriented and the individual components required to calculate slope stability are implemented in a modular fashion. We drive the model using a long term synthetic earthquake catalogue based on known earthquake parameters for upper-plate and subduction zone fault sources. It is using an empirically derived landslide volume distribution for the Cook Strait Canyon. The model predicts about 1.35 potential slope failures in the Cook Strait Canyon over a period of 500 years with a volume exceeding 0.1 km^3.

C. Mueller (✉) • W. Power • X. Wang
GNS Science, Lower Hutt, New Zealand
e-mail: c.mueller@gns.cri.nz

J. Mountjoy
National Institute of Water and Atmospheric Research (NIWA), Wellington, New Zealand

E. Lane
National Institute of Water and Atmospheric Research (NIWA), Christchurch, New Zealand

G. Lamarche et al. (eds.), *Submarine Mass Movements and their Consequences*, Advances in Natural and Technological Hazards Research 41,
DOI 10.1007/978-3-319-20979-1_59

59.1 Introduction

Landslides have been implicated as contributing to approximately only 10 % of historical tsunami. In certain situations however the impact of landslide tsunami can be very dramatic. The 1958 Lituya Bay landslide generated tsunami in Alaska remains the largest tsunami wave ever documented, and in 1998 a deep-water landslide in Papua New Guinea triggered a 10–15 m high wave that killed more than 2000 people. In New Zealand coastal areas near the Cook Strait seaway (i.e. Wellington and Marlborough) have a number of factors that make them vulnerable to landslide-generated tsunami hazard: (1) the deep Cook Strait canyons result in steep seafloor slopes in close proximity to the coast which lead to decreased attenuation of tsunami, (2) numerous high slip active faults exist near the canyon, which can be a direct tsunami source or can trigger submarine landslides that cause tsunami, (3) Wellington's south coast includes areas of high coastal population density and critical infrastructure.

An ultimate goal for the region is to establish the tsunami hazard that stems from submarine landslide activity.

In a previous study (Mountjoy et al. 2014) the tsunami hazard for the coastal regions near Cook Strait has been assessed by estimating the volume-frequency relationships and probabilities of landslide occurrence within the canyon system. The probabilities were then directly combined with a synthetic tsunami catalogue, with slope failures of preselected volumes and a set of source points on the canyon slopes, to produce the hazard curves. Source points were assumed on slope transect midpoints as shown in Fig. 59.1.

Here we report on the first stages of a complementary approach in which a synthetic earthquake catalogue is used to drive a probabilistic model of submarine slope failure, which then in turn can be used to estimate the tsunami hazard curves. Such an approach will allow us to assess the spatial distribution of potential mass failures as well.

59.2 Empirical Cook Strait Canyons Landslide Scar and Volume Frequency Distribution

Previously a population of 85 landslides was mapped in the canyons (Micallef et al. 2012, Fig. 59.2a). A power law regression provides a representative probability distribution for the landslide volumes. To estimate the annual probability of exceeding a given failure volume as shown in Fig. 59.2b we infer a time period over which the landslides have occurred. The physiography and sedimentary dynamics of the Cook Strait seaway have changed significantly in response to glacio-eustatic sea level rise.

During the lowstand period a land bridge between Taranaki and NW South Island reduced the tidal energy transfer between the Tasman Sea and Pacific Ocean

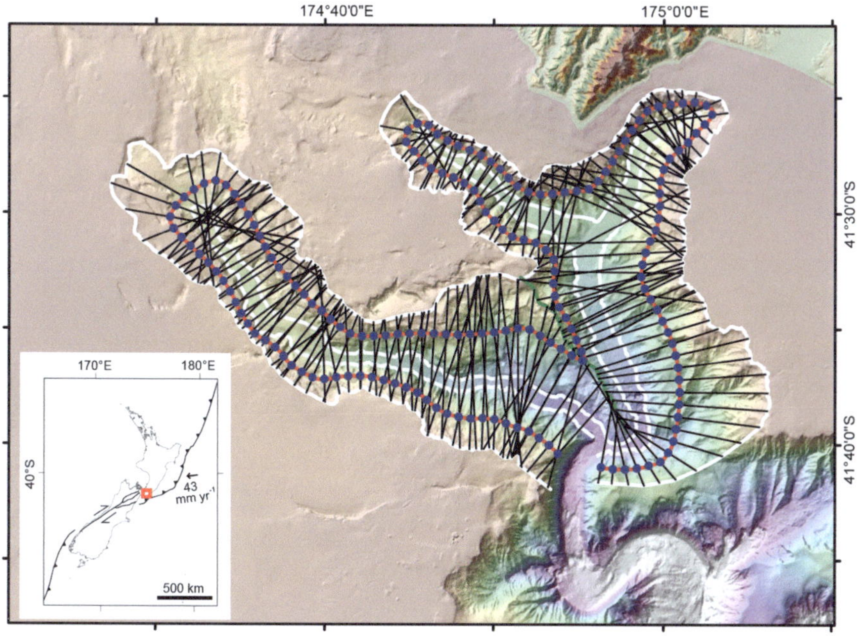

Fig. 59.1 Set of transects spanning the Cook Strait canyons system (*black lines*); *blue dots* indicate the slope midpoints of the transects, which are assumed to be the sources of potential slope failures in this study

that occurs through the strait under present conditions, but maintained strong currents and bed shear stress. This was in place at 20 kyr BP. By 15 kyr BP the land bridge was breached and a period of high sediment flux through the canyon is expected to have resulted in a period of peak canyon modification. By 10 kyr BP sea level was close to the present day shoreline and canyon dynamics would be similar to that seen today (Proctor and Carter 1988). Dating of three landslides indicates ages of approximately 14 kyr BP, 2.5 kyr BP and 1855 AD (Power et al. 2011). Based on this information we assume the most probable age for the oldest mapped landslide to be 15 kyr with and estimation error margin of plus or minus 5000 years. From this estimate and the observed distribution of failure volumes the annual probability of exceedance is derived (Fig. 59.2b).

59.3 Probabilistic Landslide Failure Assessment Methodology and First Results

Complimentary to the deterministic landslide magnitude frequency relationships developed earlier we propose here a model that is driven by known earthquake sources from the New Zealand National Seismic Hazard Model (Stirling

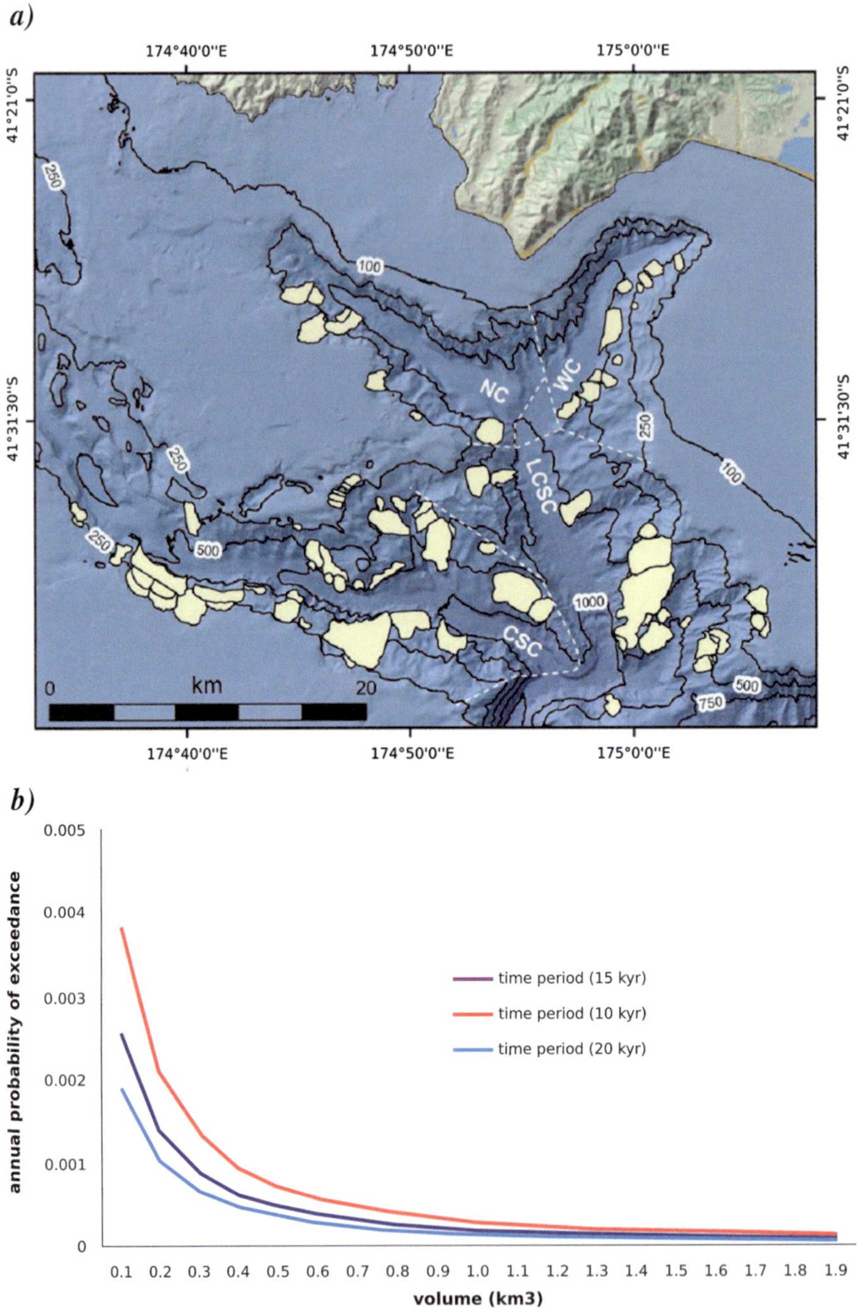

Fig. 59.2 (**a**) Landslide distribution in the upper Cook Strait canyon system (After Micallef et al. 2012). *Yellow polygons* are mapped canyon scar areas. *Dashed white lines* are canyon boundaries used for spatial analysis. *NC* Nicholson Canyon, *WC* Wairarapa Canyon, *CSC* Cook Strait canyon, *LCSC* Lower Cook Strait Canyon. (**b**) Empirical annual probability of exceeding a given landslide volume for Cook Strait canyons

et al. 2012). The advantage of this model is that it provides information on how the rate of landslide occurrence varies according to location. This builds on work completed in an earlier New Zealand Natural Hazards Research Platform project (Power et al. 2011).

The model is implemented in a modular python framework that allows us to exchange components of the model as required. We generate synthetic earthquake catalogues with the program SYNCAT, which was encapsulated by the python framework and builds on the source catalogue of the National Seismic Hazard Model for New Zealand (Stirling et al. 2012). The catalogue holds information about source mechanisms, locations, magnitudes and conditional recurrence intervals. For each source scenario peak ground acceleration (PGA) is predicted on a grid which covers our region of interest. We use the McVerry ground motion prediction equations (McVerry et al. 2006) as implemented in OPENQUAKE (Pagani et al. 2014) assuming a site class 'C' (rock). We calculate the probability of exceeding a given PGA for each mid-slope point on a set of transects (shown in Fig. 59.1). The probability of failure is taken as the probability of exceedance for PGAs that are equal to the yield accelerations (k_y) estimated for the given slope mid-points.

The stability of a slope can be determined numerically in a number of ways. The most widely used method is the limit equilibrium model that considers the *factor of safety* (F_s) for a slope in terms of transient driving forces (Eq. 59.1).

$$F_s = \textit{forces resisting failure/forces driving failure} \qquad (59.1)$$

$F_s = 1.0$ indicates equilibrium, $F_s > 1$ indicates the slope is stable and $F_s < 1$ indicates the slope is unstable (failure has occurred). Three primary geomechanical domains are identified within the canyon system, (1) canyon floor infill; (2) canyon walls eroded into sedimentary basin sequences and (3) canyon walls inferred to be eroded into Mesozoic greywacke (after Mountjoy et al. 2009). To model the stability of a simple translational failure we determine the stability of a block failing along a planar surface, taking account of gravitational forces as well as horizontal seismic loading as an earthquake load (Fig. 59.3).

Sediment strength responds primarily to effective stress parameters and this negates the need to consider the submerged depth of the slope in the equation as the derived effective stress accounts for pore pressure effects (Hampton et al. 1996).

Using the Mohr Coulomb failure criterion, both the frictional and cohesive strength of the soil can be derived from laboratory testing and incorporated into the infinite slope stability equation. Equation (59.2) calculates the yield acceleration (k_y) for the slope, i.e. the level of seismic shaking required so that $F_s = 1$ (after Bray et al. (1998)).

$$k_y = \tan\left(\phi' - \beta\right) + c'/\left(\gamma H \cos 2\beta\left(1 + \tan\phi' \tan\beta\right)\right) \qquad (59.2)$$

The input parameters required to define k_y are the geometric parameters H (block height or failure depth) and β (failure angle), and the effective strength parameters

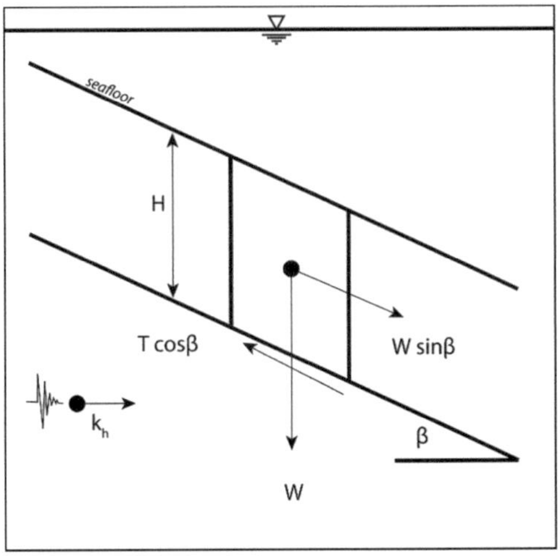

Fig. 59.3 Diagrammatic representation of the infinite slope model (Modified from Hampton et al. (1996)). *W* weight force, *H* block height, β failure angle, *T* shear stress and k_h seismic coefficient

ϕ' (internal angle of friction), *c*' (cohesion) and γ (soil density). Given that slope stability needs to be calculated across the area of the canyon, values for these parameters are required to be defined continuously across the upper Cook Strait canyon system.

For each of the 85 landslides mapped across the upper Cook Strait canyon system, the geometric parameters *H* and β have been defined by the headwall height and scar slope, respectively. Interpolation between mapped landslides enables definition of *H* and β values for the canyon slopes by assuming that the previous landslides are characteristic of their canyon-slope region. It was necessary to extrapolate values out to some parts of the canyon slope where no landslide scars are preserved in the landscape. A smoothing algorithm was applied to the data to remove outlier values and the results gridded at 1 km, resulting in a quasi-normal distribution of values across the canyon system. The 1 km grid is determined to be appropriate for the stability calculations to avoid erroneous values in a too detailed grid, while keeping a scale that resolves the majority of landslides. Parameters for calculating k_y are sampled from the parameter grids using spline interpolation.

The rate of exceedance is then calculated using the conditional recurrence interval given in the earthquake catalogue for each event. A sum over all events is taken to give the total annual rate of failure R_i for any potential volume and for each transect midpoint *i*.

Because the annual probabilities P_v of exceeding a given volume *V* are relatively small we can assume that the rates R_v will be almost equal to the probabilities. For example the annual rate of exceeding a volume of 0.1 km^3 (for the 15 kyr time period) as given by Fig. 59.2b will be $R_{0.1} \sim 0.0027$. We can calculate a scaling factor *f* for the annual rates of failure R_i at each transect midpoint *i* by relating the annual rate of exceeding a certain volume to the sum over all failure rates in the canyon:

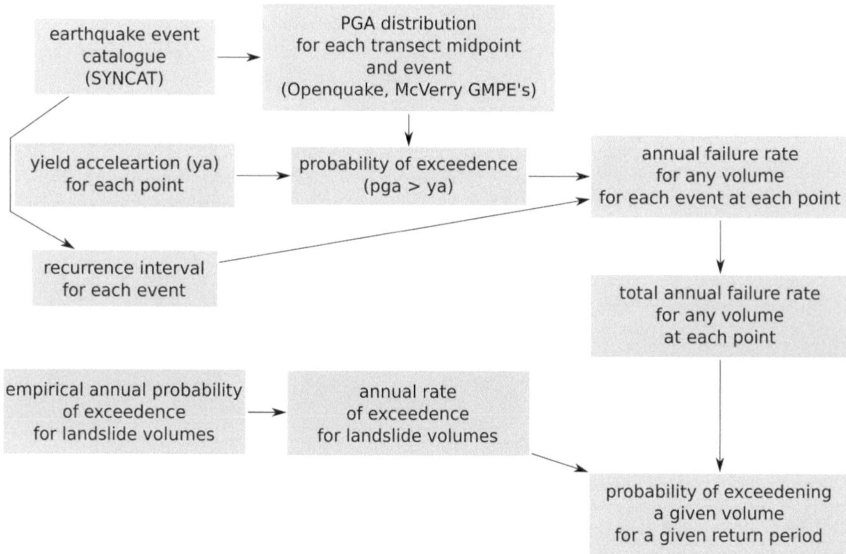

Fig. 59.4 Workflow diagram, further description in text

$$f = R_v / \sum_i R_i$$

When scaling the individual rates R_i with this factor we will obtain an estimate for the annual rate of exceeding the given volume at each midpoint. This approach assumes that the volume distribution given in Fig. 59.2b is the same at each transect midpoint. Figure 59.4 shows a diagram of the complete work flow.

Figure 59.5 shows the Poissonian probability of failures larger than 0.1 km^3 for the 500 year interval. The expected number of slope failures in this time span is ~1.35. This is in accordance with the annual rate of exceeding a volume of 0.1 km^3 ($R_{0.1}$ ~ 0.0027) as given in the empirical volume distribution in Fig. 59.2b.

The south west part of the Canyon exhibits up to a factor of two higher probabilities of exceedance compared to the northern and eastern parts of the Canyon. It should be noted that there is no strong visual correlation between the distribution of landslide scars determined from field study (Fig. 59.2a) and the spatial distribution of failure rates and probabilities in our model.

59.4 Discussion

The results presented here are only based on one realisation of the earthquake catalogue. In a true Monte Carlo simulation approach we should consider many realisations of this catalogue to assess the variability of the probabilities for failure.

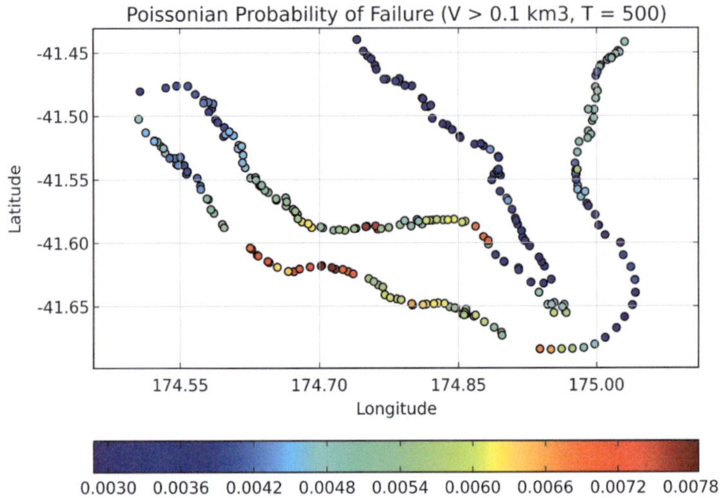

Fig. 59.5 Poissonian probability of exceeding a volume of 0.1 km^3 for a period of 500 years. Please note that the northern part of the canyon has not enough data coverage to allow for a calculation of slope stability. Therefore the transect midpoints from that area are not shown

Another source of uncertainty in the above assessment is the unknown distribution of parameters that enter into the slope stability equation. To what extent such variation will influence the result still needs to be investigated. The same holds for the influence of topographic amplification on the locally encountered PGA. This effect is not directly considered in the ground motion prediction equations.

Most of the other parameters entering into the equation for k_y are currently only known on a fairly coarse grid and further sampling might be required if their local variability should prove to be significant.

59.5 Conclusion

The results presented here have to be treated as preliminary. Our model helps to assess the probability of failures occurring above a certain landslide volume and the spatial distribution of these failures. We predict about 1.35 potential slope failures in the Cook Strait Canyon over a period of 500 years with a volume exceeding 0.1 km^3. This assessment is consistent with previous studies. Based on slope failure magnitude frequency distributions from field observation in the Cook Strait Canyons it has to be concluded that failures will predominantly be small. However the chances of some of these failures producing a landslide volume that will generate a significant tsunami with in this time frame are not negligible. Further investigation is required to confine the error bounds of this assessment and to study the influence of variability in the parameters entering into the slope stability equation and the effects of topographic amplification.

Acknowledgement The authors would like to thank the reviewers Nick Horspool and David Rhoades for valuable comments that helped to improve this paper greatly. We would also like to thank the editor Geoffroy Lamarche for his time and valuable insights.

References

Bray JD, Rathje EM, Augello AJ, Merry SM (1998) Simplified seismic design procedure for geosynthetic-lined, solid-waste landfills. Geosynth Int 5(1–2):203–235

Hampton MA, Lee HJ, Locat J (1996) Submarine landslides. Rev Geophys 34:33–59

McVerry G, Zhao J, Abrahamson NA, Somerville P (2006) New Zealand acceleration response spectrum attenuation relations for crustal and subduction zone earthquakes. Bull N Z Soc Earthq Eng 39(1):1–58

Micallef A, Mountjoy JJ, Canals M, Lastras G (2012) Deep-seated bedrock landslides and submarine canyon evolution in an active tectonic margin: Cook Strait, New Zealand. In: Yamada Y, Kawamura K, Ikehara K et al (eds) Submarine mass movements and their consequences, vol 31. Springer, Dordrecht, pp 201–212

Mountjoy JJ, Barnes PM, Pettinga JR (2009) Morphostructure and evolution of submarine canyons across an active margin: cook strait sector of the Hikurangi Margin, New Zealand. Mar Geol 260(1–4):45–68

Mountjoy JJ et al (2014) Quantifying the landslide-generated tsunami hazard in central New Zealand: a workflow for probabilistic landslide tsunami hazard assessment. 2012-NIW-03-NHRP project completion report

Pagani M, Monelli D, Weatherill G et al (2014) OpenQuake engine: an open hazard (and risk) software for the global earthquake model. Seismol Res Lett 85:692–702. doi:10.1785/0220130087

Power W, Mountjoy JJ, Delaux S et al (2011) Towards a probabilistic landslide tsunami model for New Zealand. Unpublished project completion report to the Natural Hazards Research Platform, contract CS GNS 25

Proctor R, Carter L (1988) Tidal and sedimentary response to the late quaternary closure and opening of cook strait, New Zealand: results from numerical modelling. Paleoceanography 4:167–180

Stirling MW, McVerry GH, Gerstenberger MC et al (2012) National seismic hazard model for New Zealand: 2010 update. Bull Seismol Soc Am 102(4):1514–1542. doi:10.1785/0120110170

Chapter 60
Coupled Modelling of the Failure and Tsunami of a Submarine Debris Avalanche Offshore Central New Zealand

Xiaoming Wang, Joshu Mountjoy, William L. Power, Emily M. Lane, and Christof Mueller

Abstract Evidence of previous submarine mass failures in the form of excavation scars has been widely documented in the Cook Strait Canyons of New Zealand. Recent bathymetry surveying has identified a well-defined submarine landslide scar and its associated debris deposit on the northern slope of southern Hikurangi Trough. The newly acquired multi-beam data allowed determination of the location and extent of the deposit, estimation of its volume, as well as reconstruction of both the pre-failure bathymetry and the initial state of the mass failure. A dynamically coupled two-layer model was used to numerically investigate this submarine debris avalanche and its resulting tsunami impact on the coasts of central New Zealand. The modeling results show a fairly good overall agreement with the observed debris deposition and also suggest that tsunami associated with the debris avalanche quite possibly inundated the coasts of central New Zealand, with maximum run-up elevations of between 3 and 5 m in several nearby locations.

60.1 Introduction

The Southern Hikurangi Margin is where the Pacific plate is obliquely subducting beneath the Australian plate at ~20 mm/year, transforming and colliding with the intra-continental Alpine Fault of South Island, New Zealand (Wallace et al. 2004). The abundant supply of sediment from rapidly rising mountains in association with strong ocean currents, active faults and several submarine canyon systems, provides an ideal set of circumstances favouring frequent occurrences of submarine mass

X. Wang (✉) • W.L. Power • C. Mueller
GNS Science, Lower Hutt, New Zealand
e-mail: x.wang@gns.cri.nz

J. Mountjoy
National Institute of Water and Atmospheric Research (NIWA), Wellington, New Zealand

E.M. Lane
NIWA, Wellington, New Zealand

G. Lamarche et al. (eds.), *Submarine Mass Movements and their Consequences*, Advances in Natural and Technological Hazards Research 41,
DOI 10.1007/978-3-319-20979-1_60

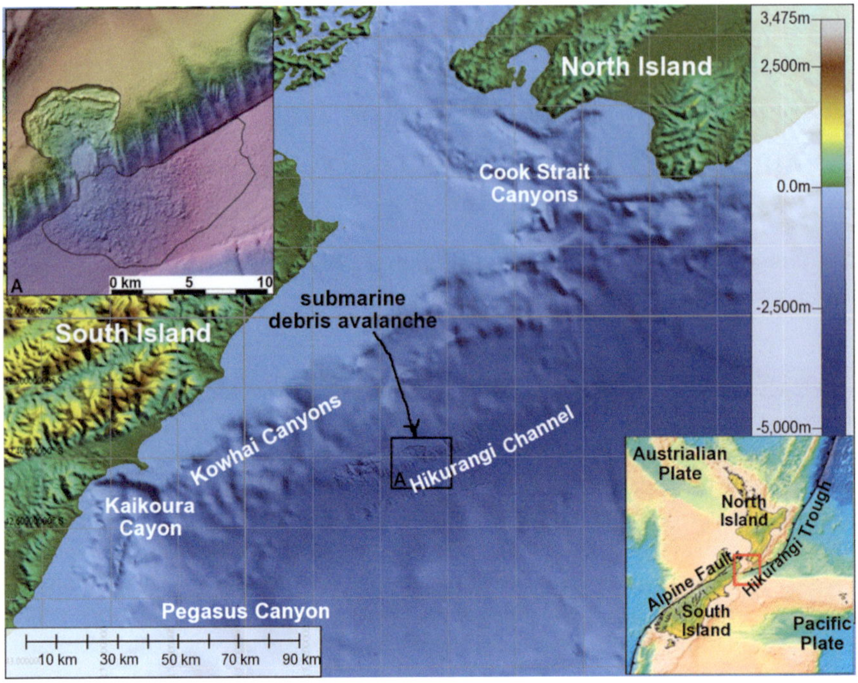

Fig. 60.1 Submarine canyons in central New Zealand. The submarine debris avalanche is shown in *A*. The *black contour* in *A* outlines the rough range of the debris deposit

failure (Fig. 60.1). Numerous past submarine mass failures have been identified in this region, especially in Cook Strait Canyons, Kaikoura Canyon and Pegasus Canyon (Lewis et al. 1998; Lewis and Pantin 2002; Mountjoy et al. 2009; Mitchelle et al. 2012), and these canyon systems are recognised to present potential landslide tsunami threat to the coast of central New Zealand.

Recently, a very well defined scar and its matching deposit were revealed in unprecedented detail on the southern side of the Hikurangi channel (Fig. 60.1a). Both the scar and the debris deposit remain well preserved, which provides a rare opportunity to study the failure and the tsunami impact. In this study, we investigated this submarine debris avalanche and the resulting tsunami with a dynamically coupled two-layer model. The model simultaneously calculates the debris avalanche and its resulting tsunami with a fully two-way coupled approach.

60.2 The Submarine Debris Avalanche

The submarine debris avalanche was identified off the northern apex of the southern Hikurangi Channel, lying about 2500 m under water (Fig. 60.1), among other submarine mass failures discovered during a recent Southern Hikurangi mapping

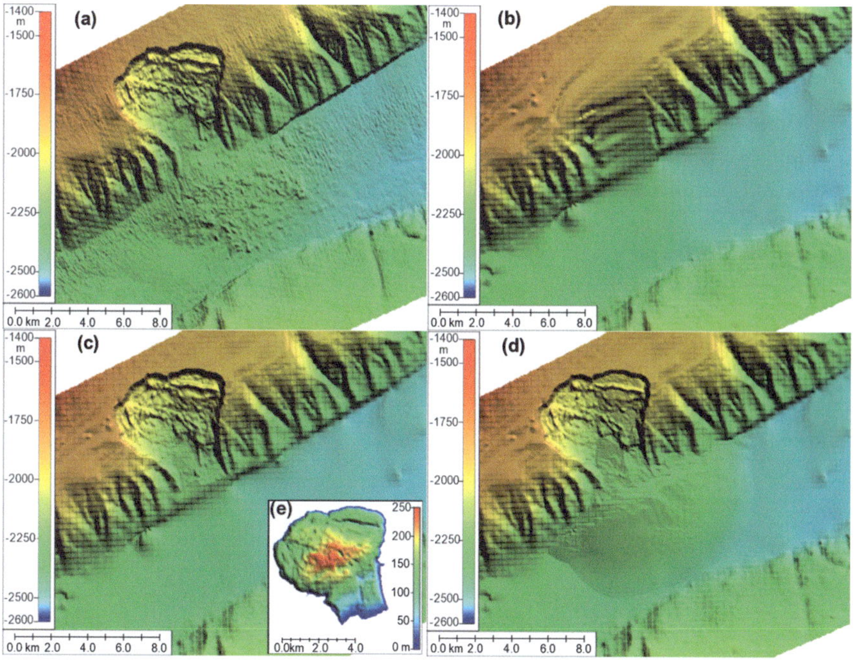

Fig. 60.2 (**a**) Post-failure bathymetry; (**b**) reconstructed pre-failure bathymetry; (**c**) bed surface of the failure; (**d**) computed post-failure bathymetry; and (**e**) initial debris thickness before failure

cruise (Mitchell et al. 2012). Onboard the RV a Kongsberg EM302 swath mapping instrument, operating at 30 kHz with 288 beams, provided a resolution better than in the previous imaging (Lewis and Pantin 2002). The high-resolution data indicates that the displaced debris remains well preserved on the channel floor, suggesting that it must have occurred within the timeframe of present geomorphology. The pristine status of the amphitheater-like scar and the matching deposit on the relatively flat floor allows outlining the range of debris deposition, restore the pre-failure bathymetry, and estimate the volume of the initial mass (Fig. 60.2). The initial mass was estimated as approximately 2.5 km^3, with a maximum thickness about 250 m and over 6.0 km wide (Fig. 60.2e).

Past studies suggest that the sediment in this region ranges from fine-grained sand to coarser granule sand and gravels (Lewis and Pantin 2002). The low cohesion of such material and the widespread debris deposition on the channel floor imply that the failure disintegrated and transformed into a debris avalanche after its initiation. But it is not clear where the failure started and what caused it. Due to its proximity to active faults, it is suspected that earthquake-induced strong ground shaking was likely the trigger.

60.3 Numerical Model and Modeling Results

Debris avalanche has often been modelled with depth-integrated equations, owing to the fact that the thickness of debris flow is usually over one order-of-magnitude smaller than its spatial scales (Jiang and LeBlond 1992). In particular, Heinrich et al. (2001) and Kelfoun and Druitt (2005) solved these types of equations in a topography-linked local coordinate system. This approach reduces difficulties in solving vertically integrated equations over relatively steep terrains.

In this study, a fully-coupled two-layer model was proposed to study this submarine debris avalanche and its consequent tsunami. The debris avalanche in the lower layer serves as forcing boundaries for tsunami modeling in the upper layer. For simplicity while remaining effectiveness, the thickness-averaged Eq. (60.1), similar to those of Heinrich et al. (2001), were used to describe the debris avalanche with granular flow assumption.

$$\begin{aligned} &\frac{\partial h}{\partial t}+\frac{\partial(hu_s)}{\partial\tilde{x}}+\frac{\partial(hv_s)}{\partial\tilde{y}}=0 \\ &\frac{\partial(hu_s)}{\partial t}+\alpha\left[\frac{\partial(hu_su_s)}{\partial\tilde{x}}+\frac{\partial(hu_sv_s)}{\partial\tilde{y}}\right]=khgsin\theta_x-khgcos\theta\frac{\partial h}{\partial\tilde{x}}+f_{\tilde{x}} \\ &\quad+f\tilde{x}'\ \frac{\partial(hv_s)}{\partial t}+\alpha\left[\frac{\partial(hv_su_s)}{\partial\tilde{x}}+\frac{\partial(hv_sv_s)}{\partial\tilde{y}}\right] \\ &\quad=khgsin\theta_y-khgcos\theta\frac{\partial h}{\partial\tilde{y}}+f_{\tilde{y}}+f\tilde{y}' \end{aligned} \tag{60.1}$$

in which h, u_s and v_s denote the slope-normal thickness and the velocity components in slope-parallel directions $\tilde{x}$ and $\tilde{y}$, respectively; θ is hydrologic slope angle, and $\theta_{\tilde{x}}, \theta_{\tilde{y}}$, are slope angles in $\tilde{x}$ and $\tilde{y}$; α relates to the over-thickness velocity profile. In this study, $\alpha = 1.0$, assuming a uniform over-thickness velocity profile. Parameter k reflects the effect of water. $k = 1.0$ for terrestrial avalanches and $k = 1-1/\rho_s$ for submarine avalanches in which ρ_s is the density ratio of mass to water. $\rho_s = 2.15$ was used here, which is an average of the values 1.85~2.45 found in other studies. g = 9.801 m^2/s.

In Eq. (60.1), $f'_{\tilde{x}}$ and $f'_{\tilde{y}}$ are counter-components of top interface friction determined from Eq. (60.4) which also introduce two-way momentum exchange between (60.1) and (60.3). $f_{\tilde{x}}$ and $f_{\tilde{y}}$ are frictional force components, determined from a Coulomb-type friction law commonly used for granular flow simulations as

$$f_{\tilde{x}}=khg\cos\theta\tan\varphi\left(u_s/\left|\vec{u}_s\right|\right), f_{\tilde{y}}=khgcos\theta tan\varphi\left(v_s/\left|\vec{u}_s\right|\right) \tag{60.2}$$

where $\vec{u}_s=(u_s,v_s)$; φ is Coulomb friction angle, approximately encapsulating the effects of basal and internal frictions. Without additional momentum input, debris does not move over a slope less than this angle. $\varphi = 4.0°$was adopted in this study, estimated from the repose angle of the debris deposit.

This debris avalanche model was further coupled with the depth-averaged Nonlinear Shallow Water Equation (NSWE) model in the well-established COMCOT simulation package (Wang and Power 2011) to simulate the resulting tsunami in the upper layer. In the Cartesian system (χ,y,z) with χ, y and z pointing eastward, northward and upward respectively, the NSWEs read

$$\begin{aligned}
&\frac{\partial H}{\partial t} + \frac{\partial P}{\partial x} + \frac{\partial Q}{\partial y} = 0 \\
&\frac{\partial P}{\partial t} + \frac{\partial P^2}{\partial x} + \frac{\partial (PQ)}{\partial y} + gH\frac{\partial z_w}{\partial x} + f_x = 0 \\
&\frac{\partial Q}{\partial t} + \frac{\partial (PQ)}{\partial x} + \frac{\partial Q^2}{\partial y} + gH\frac{\partial z_w}{\partial y} + f_y = 0
\end{aligned} \qquad (60.3)$$

where H stands for the vertical thickness of water column, $H = z_w - z_s$, in which Z_w denotes water surface elevation and Z_s is the top surface of debris. $z_s = z_b + h/cos\theta$ where Z_b represents the bed surface where the debris moves (Fig. 60.2c), and h and θ are obtained from the debris avalanche model. $P = Hu_w$ and $Q = Hv_w$ represent volume flux components, and u_w and v_w are velocity components of tsunami in x and y, respectively.

In Eq. (60.3), f_x and f_y are friction components at the water bottom determined from Manning's formula in Eq. (60.4), based on the relative velocity between debris avalanche and tsunami. Manning's roughness n = 0.015 was chosen for the sandy/muddy floor in this study. The friction along the interface of avalanche and tsunami motions, updated at every time step, serves as a feedback mechanism of momentum exchange between the two processes.

$$\begin{aligned}
f_x &= gn^2H^{-1/3}(u_w - u_s cos\,\theta_x)\sqrt{(u_w - u_s cos\,\theta_x)^2 + (v_w - v_s cos\,\theta_y)^2} \\
f_y &= gn^2H^{-1/3}(v_w - v_s cos\,\theta_y)\sqrt{(u_w - u_s cos\,\theta_x)^2 + (v_w - v_s cos\,\theta_y)^2}
\end{aligned} \qquad (60.4)$$

The modified explicit staggered leap-frog Finite Difference (FD) scheme in Wang and Liu (2011) together with Arakawa C-grids was implemented to solve the NSWEs for tsunami evolutions. One advantage of this scheme is that numerical dispersion, part of truncation errors in the discretized NSWEs, is manipulated to resemble the weak physical dispersion property of classic Boussinesq equations, with a high computational efficiency. This was used to account for the potential dispersion effect of the tsunami generated from the deep channel. Based on Watts et al. (2003), the characteristic wavelength of the tsunami with solid-block motion assumption was estimated about 10–15 times larger than the channel depth, i.e., weakly dispersive waves. The characteristic wavelength will be even longer for the fully deformable landslide in this case study.

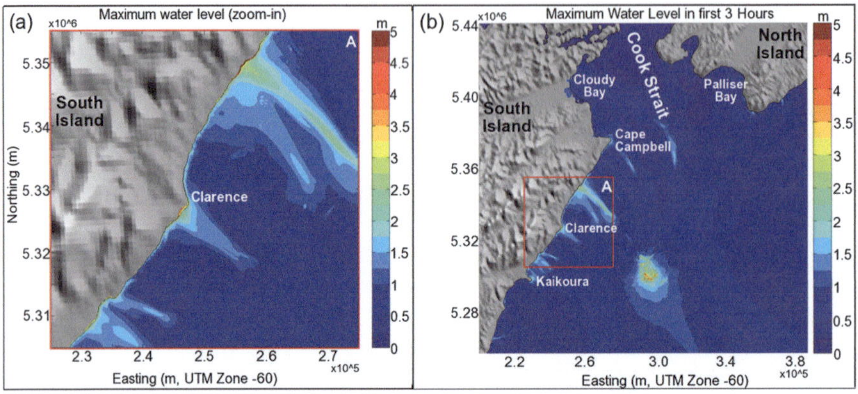

Fig. 60.3 (**a**) Computed max tsunami elevation near the coast (zoom-in view of box A); (**b**) Computed max tsunami elevation in central New Zealand

Equation (60.1) for avalanche-type motions were solved with the explicit upwind FD schemes proposed by Kelfoun and Druitt (2005), however with more general quadrilateral grids to discretize the terrain surface. The rectilinear grids in Heinrich et al. (2001) and in Kelfoun and Druitt (2005) are hardly achievable for complex terrains. The topography-linked quadrilateral grids were constructed by vertically projecting the horizontal Arakawa C-grids onto the topographical surface. Our tests indicated that this type of grid implementation allows easy coupling with the existing tsunami model, and the model also produces satisfactory agreements with analytical studies, e.g., those in Mangeney et al. (2000). One hundred-meter horizontal resolution was used in the simulation, enough to capture leading-order features of the debris avalanche and tsunami evolutions. In total, a 3-h duration of avalanche and tsunami was simulated to investigate the resulting tsunami impact on the coast of central New Zealand.

The computed final debris deposition and the maximum tsunami elevation are shown in Figs. 60.2d and 60.3, respectively. The computed post-failure bathymetry (Fig. 60.2d) reproduced the observation (Fig. 60.2a) fairly well, except for some details. Close comparisons reveal that the largest discrepancies over 25 m occur near the toe of the excavation scar and between the upper crown and bar-like scarp, probably due to inaccurate reconstruction of the bed surface (Fig. 60.2c). Without additional coring data, such reconstruction proves to be particularly challenging. Furthermore, in this simulation the entire initial mass was assumed to fail simultaneously at the beginning. In reality, the failure might start from one part then spread to other parts, e.g., in the type of a progressive or retrogressive failure.

Figure 60.4 shows the computed sequential snapshots of the debris avalanche with the assumption of simultaneous failure of the entire initial mass. The results suggest that in this failure scenario the most energetic process likely completed in the first 5–6 min, however the whole process might last over 10 min to reach the new equilibrium (Fig. 60.4). Tsunami might have been generated and affected most of the coasts of central New Zealand, with maximum elevations up to 3–5 m above

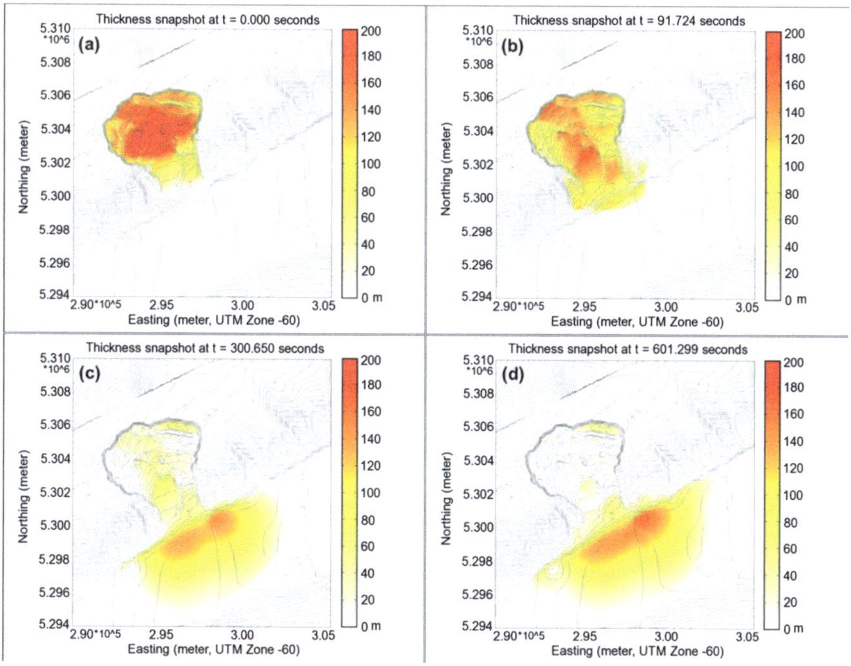

Fig. 60.4 Computed sequential snapshots of vertical thickness of debris at about (**a**) 0.0, (**b**) 1.5, (**c**) 5.0 and (**d**) 10.0 min into the simulation. *Grey lines* illustrate the bathymetry contours of the bed floor at a 20-m depth interval

ambient sea level in a few places from Kaikoura to Cape Campbell (Fig. 60.3). This estimate of the tsunami size is conservative, i.e., a worst case, because of the assumption that the failure started as a single large mass. Further investigation and modeling are required to improve our understanding of the pre-failure bathymetry, probable failure sequence and the characteristics of the resulting tsunami.

60.4 Conclusion

A well-defined submarine debris avalanche in the southern Hikurangi Channel was investigated to study its failure and the resulting tsunami impact on the coast of central New Zealand, using a fully coupled two-layer numerical model. Newly acquired high-resolution multi-beam data was used to reconstruct the pre-failure bathymetry and restore the initial condition of the mass failure. The numerical model generates a reasonable correspondence with the observed debris deposit caused by the failure, although it is recognized that this initial model requires further work and refinement. The modeling results also suggest that a related tsunami of moderate size may have been generated. If this was the case, it would have affected the coasts of central New Zealand, with elevations up to 3–5 m above ambient water level in a few areas from Kaikoura to Cape Campbell.

The proposed model calculates debris avalanche and tsunami evolutions in a fully coupled two-layer system. The avalanche-type mass failure in the lower layer is computed with thickness-averaged equations in topography-linked coordinate system, serving as forcing boundaries for tsunami modeling in the upper water layer. Friction along the interface acts as a mechanism of momentum exchange between the two processes. This case study demonstrates that the coupled model can successfully model tsunamis by this type of mass failure.

Acknowledgments The authors gratefully acknowledge the invaluable comments and suggestions from the reviewers Dr Nabil Sultan and Dr David Tapppin and the Editor Dr Thomas Hubble during the preparation of this manuscript. Their contributions are highly appreciated.

References

Heinrich P, Piatensi A, Hebert H (2001) Numerical modelling of tsunami generation and propagation from submarine slumps: 1998 Papua New Guinea event. Geophys J Int 145:97–111

Jiang L, LeBlond PH (1992) The coupling of a submarine slide and the surface waves which it generates. J Geophys Res 97((C8) 12):731–744

Kelfoun K, Druitt TH (2005) Numerical modeling of the emplacement of Socompa rock avalanche. Chile J Geophys Res 110:B12202. doi:10.1029/2005JB003758

Lewis KB, Pantin HM (2002) Channel-axis, overbank and drift sediment waves in the southern Hikurangi Trough, New Zealand. Mar Geol 192:123–151

Lewis KB, Collott JY, Lallemand SE (1998) The dammed Hikurangi Trough: a channel-fed trench blocked b subducting seamounts and their wake avalanches (New Zealand-France GeodyNZ Project). Basin Res 10:441–468

Mangeney A, Heinrich P, Roche R (2000) Analytical solution for testing debris avalanche numerical models. Pure Appl Geophys 157:1081–1096

Michelle J, Mountjoy J, Barnes P (2012) OS20/20 southern Hikurangi voyage report prepared for land information NZ. NIWA client report no: WLG2012-21, May 2012

Mountjoy JJ, Barnes PM, Pettinga JR (2009) Morphostructure and evolution of submarine canyons across an active margin: cook strait sector of the Hikurangi Margin, New Zealand. Mar Geol 260(1–4):45–68

Wallace LM, Beavan J, McCaffrey R, Darby D (2004) Subduction zone coupling and tectonic block rotations in the North Island, New Zealand. J Geophys Res 109:B12406. doi:10.1029/2004JB003241

Wang X, Liu P (2011) An explicit finite difference model for simulating weakly nonlinear and weakly dispersive waves over slowly varying water depth. Coast Eng 58:173–183

Wang X, Power W (2011) COMCOT: a tsunami generation propagation and run-up model, GNS science report 2011/43, 129p

Watts P, Grilli ST, Kirby JT, Fryer GJ, Tappin DR (2003) Landslide tsunami case studies using a Boussinesq model and a fully nonlinear tsunami generation model. Nat Hazards Earth Syst Sci 3:391–402

Chapter 61
Observations of Coastal Landslide-Generated Tsunami Under an Ice Cover: The Case of Lac-des-Seize-Îles, Québec, Canada

Jonathan Leblanc, Dominique Turmel, Julie Therrien, and Jacques Locat

Abstract On April 15 2014, a landslide occurred on the east shore of a lake in the municipality of Lac-des-Seize-Îles, about 100 km north of Montréal, Québec, Canada. The coastal landslide, with a length of 94 m and a width of 55 m, had a volume of about 30,500 m^3 of sandy to gravelly till. It was likely triggered by a significant amount of water infiltration caused by a heavy rain coincident with an accelerated snow melt. The displaced mass spread on the lake floor and triggered a tsunami that broke and partly lifted the ice cover. Water and ice damaged several seasonal residences and boathouses on the periphery of the lake, in an area extending 450 m north and 500 m south of the landslide debris location. Observations on aerial photographs taken shortly after the event revealed the existence of radial structures on the ice cover in the affected area resulting from the evolution of the tsunami wave. Investigations carried out on east and west shores showed evidences of net marks on tree trunks explained by a maximum inundation height which can be as much as 1.8 m directly in front of the landslide position on the west shore.

61.1 Introduction

According to the Ministère des Transports du Québec database, landslide-generated tsunamis are not common in Québec (e.g. 2004 St-Étienne-des-Grès, 2005 Longue-Rive, 2006 Nicolet) and except for few cases, their impacts are generally limited. Still, historical cases did have significant effects, like the La Grande River at the James Bay in 1987 where the landslide-generated tsunami devastated a large portion

J. Leblanc (✉) • D. Turmel • J. Locat
Laboratoire d'études sur les risques naturels, Département de géologie et de génie géologique, Université Laval, Québec, QC, Canada
e-mail: Jonathan.leblanc.4@ulaval.ca

J. Therrien
Section des mouvements de terrain, Service de la géotechnique et de la géologie, Ministère des Transports, Québec, QC, Canada

G. Lamarche et al. (eds.), *Submarine Mass Movements and their Consequences*, Advances in Natural and Technological Hazards Research 41,
DOI 10.1007/978-3-319-20979-1_61

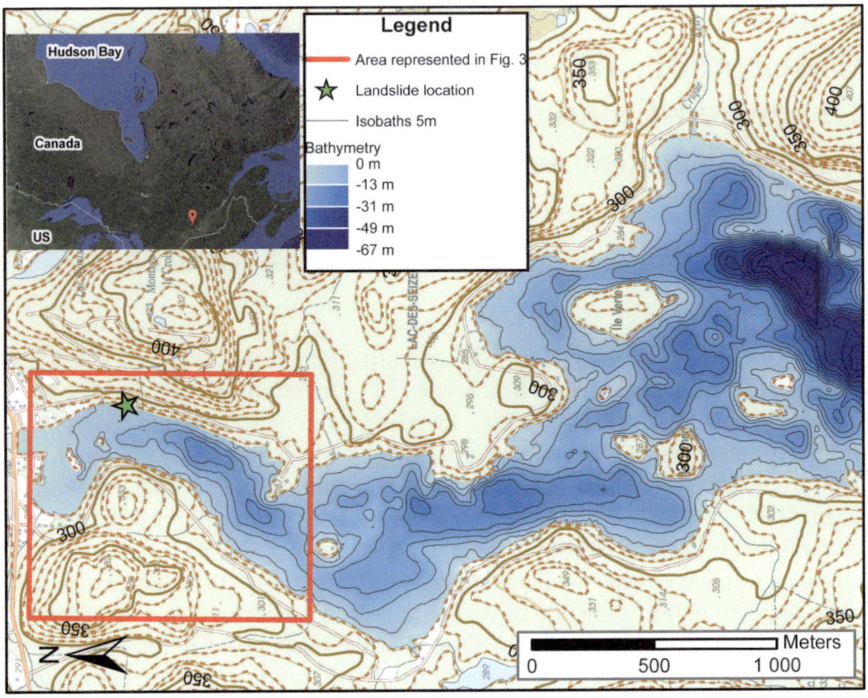

Fig. 61.1 Location and bathymetric map of the lake in the municipality of Lac-des-Seize-Îles, Québec, Canada

of the opposite bank to an elevation of 14 m above the river level (Lefebvre et al. 1991). Another notable case, the Notre-Dame-de-la-Salette in 1908, is unique because water and ice displaced by the landslide in quick clays destroyed a portion of the village and caused 33 deaths (Ells 1908). More recently, on April 15th 2014, a coastal landslide with an estimated volume of 30,500 m^3 generated a tsunami in Lac-des-Seize-Îles (Fig. 61.1). A unique aspect of this event is that it shattered and lifted the ice cover. More details are presented below but one of the interesting aspects of this event is the interaction between the wave propagation and the ice cover. There is little literature on the effect of ice cover on tsunami wave propagation, particularly for shallow depths (Jørstad 1968; Murty and Polavarapu 1979; Vanneste et al. 2010; Wang et al. 2015). The objectives of the paper are to illustrate and analyse the signature left on the ice cover by the tsunami wave and underline the role of the ice cover as a potential aggravating factor for coastal infrastructures.

61.2 Methodology

Lake floor bathymetric data, aerial photographs, field surveys and witness interviews are used in the analysis of the April 15th, 2014 Lac-des-Seize-Îles landslide and tsunami. Bathymetric data are from a June 2013 survey conducted by researchers at the Université de Montréal. The survey was carried out using a BioSonics single beam echo sounder connected to a GPS with a precision of 10 cm. A series of digital aerial photographs were taken by the Ministère des Transports du Québec immediately after the event, largely covering the area of interest. They have a resolution of 7 cm per pixel. A field survey was carried out on April 21st and 22nd to characterize the landslide scar and the mobilised material. Field evidence was also compiled along with witness interviews.

61.3 Morphology of the Coastal Area and Lake

The sediment type in the area is associated with glacial deposit of various thicknesses overlaying gneissic bedrock. The landslide mobilised a sandy to gravely till containing angular to sub-angular boulders with a diameter between 1 and 2 m. The inclination of the east shore subaerial slope in the affected zone is about 25°. The Lac des Seize-Îles is located at a mean elevation of 278 m above sea level and covers an area of 3.5 km^2. The bathymetric survey carried out in June 2013 (Fig. 61.1) shows the presence of a 67 m deep trench, located far away from the slide area. Focusing on the affected area, the maximum water depth is 23 m and is located at the southern limit. Pre-slide water depth in the slide deposition zone is nearly constant at 6 m with a small 10 m deep depression 130 m away from the eastern shore. The ice blocks observed on the shore during the field investigation indicate that the ice cover had an ice thickness of about 0.66 m before the event.

Islands can be observed in the lake and two of them, located north of the slide, are particularly interesting as they will influence the propagation pattern of the wave that will be described below. Witnesses also claimed that the lake level was 0.5 m higher than normal, which is common in early spring. Coastal infrastructures at the north end of the lake consist of seasonal residences located on the shoreline as well as a small harbour. These residences are all equipped with docks or boathouses. Apart from narrow trails on both side of the lake, the only access between the main part of the municipality and the infrastructures on the shoreline is by boat on the lake in summer and by an ice road during the winter.

61.4 The Slide Event

The sliding material consisted of a sandy to gravely till containing 1–2 m diameter boulders (Fig. 61.2). The length of the failure surface is 94 m and the width at the shoreline is 55 m. A mean thickness of 6.5 m of material was mobilised for an approximate volume of 30,500 m^3. The displaced material emerges from the water surface and can be seen on the aerial photographs to a distance of 68 m from the shore with a runout angle of 13°. This event can be considered as a translational

Fig. 61.2 Landslide scar, with a length of 94 m and a width of 55 m, as seen from the west shore of the lake

slide of granular material over a relatively planar and polished underlying bedrock surface.

Before the event, the region experienced heavy rainfall associated with warmer temperatures resulting in rapid melting of the snow cover. This water intake was observed during the field trip with a run-off water accumulation behind the crown of the landslide. Flow coming from groundwater seepage was also present in the scar during the investigation. This sudden modification in pore water pressures could be a reasonable triggering factor leading to the slide (Turner and Schuster 1996).

61.5 The Tsunami Event

Evidence of the tsunami wave propagation can be observed on the aerial photographs with the fractured ice cover (Fig. 61.3). Indeed, a radial pattern of fissures was generated by the wave starting from the landslide deposition zone. This wave propagation pattern on the ice is seen 500 m south and 450 m north of the slide impact zone. In a 75 m radius area near the landslide, the propagation pattern is not as easy to highlight since the ice was shattered by the impact. Farther away, we can see a regular pattern with a fairly constant spacing of 10–15 m between fractures. The spacing, reaching 20 m, is observed in the south limit and coincides with the deep trench of 23 m.

The first island located at 170 m northwest of the landslide does not seem to affect the propagation pattern. Indeed, the cracks are relatively continuous beyond

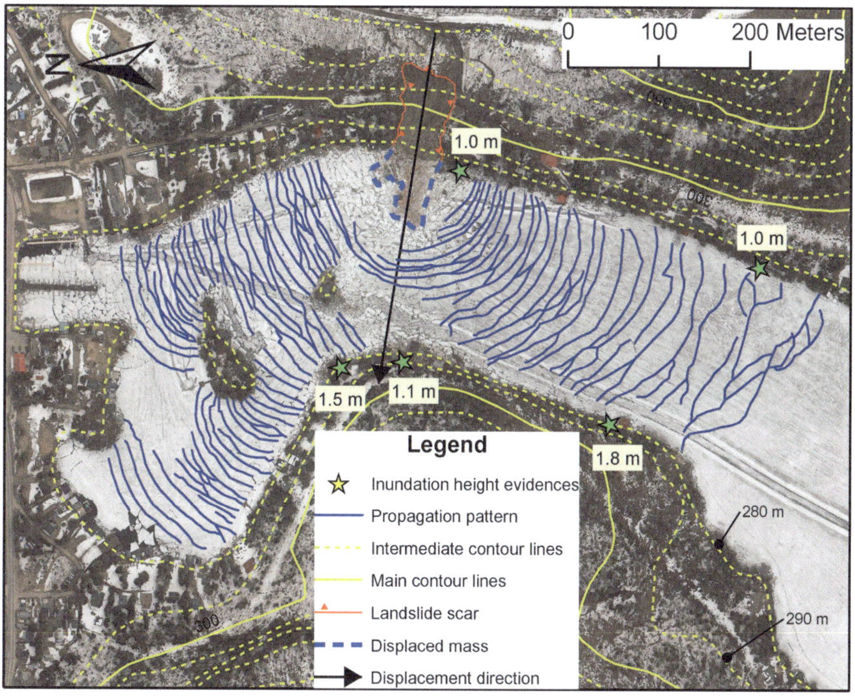

Fig. 61.3 Aerial photographs taken just after the event and showing the affected zone by the landslide-generated tsunami that broke and partly lifted the ice cover on a 450 m radius area

it and there is no apparent change in the propagation direction. This is quite different for the second island located at 300 m northwest of the landslide. We can observe diffraction in the shadow of the island, where there is no fracture on the ice cover. Thomson (1981) pretends that this can be explained because only a portion of the wave energy makes its way behind the island who acts as a barrier. Movements similar to shear bands, perpendicular to the shoreline, are observed mainly on the access roads visible as darker lines on the ice cover of Fig. 61.3. They are believed to be associated with the water mass oscillation following the inundation process. The mean shear on both side of the lake is about 2.6 m.

Inundation heights are estimated by marks left on tree trunks made by ice impact during the inundation and is observed on both shores. On the western shore, the inundation height is 1.1 m above the lake level at a distance of 11 m from the shoreline opposite to the slide. Fifty metres further north (Fig. 61.3), such marks are 1.5 m high above lake level (Fig. 61.4a). Inundation also seems to reach a height of 1.8 m, 260 m south of the landslide position on the west shore (Fig. 61.3). There is much less evidence of ice impact on trees on the east shore. Twenty metres south of the landslide scar, the inundation seems to have reached a height of 1 m (Fig. 61.3). There are also indications of a 1 m high inundation on the east shore, 370 m south of the landslide scar (Fig. 61.3). Inundation distance and run-up elevation above sea

Fig. 61.4 (**a**) Marks on tree trunks located on the west shore. Those marks are explained by the ice impact during the inundation process. (**b**) Seasonal residence badly damaged by the water and the ice during the inundation process

level were measured from aerial photographs using observations of wave damage and from digital elevation model of the affected area. The run-up elevation corresponds to the maximal elevation reached by the wave during the inundation of the coast at the limit of penetration. In front of the landslide position on the west shore, the inundation distance is about 10–15 m and the run-up elevation corresponds to the 290 m contour line. Beyond this area, the inundation distance decreases to around 5–6 m with a run-up elevation of 280 m. It is about the same on the east shore where the inundation distance is 5 m north of the landslide, and 10 m to the south. The run-up elevation above sea level is fairly constant at about 280 m.

In terms of structural damage, one seasonal residence was smashed and carried into the lake by the sliding debris. Four other houses, located in front of the landslide on the west shore, were severely damaged by ice and water impact. One house was pulled off its foundation (Fig. 61.4b). The majority of the docks and boathouses in the affected area were also damaged or destroyed. The affected area as indicated by damage goes further south than the propagation pattern where ice fractures are still observed along the shorelines. The few residents were preventively evacuated after the event. However, two residents were temporarily isolated since their unique road access was destroyed.

Two witnesses saw the evolution of the wave lifting the ice cover at the time of the event. The first one claimed that the wave was about 2 m high during its propagation. Moreover, he saw the reflection of the wave from the western to the eastern shore. The second witness also identified two waves, which seems to also correspond to the wave reflection. He claimed that the wave were about 2.5–3.0 m high during propagation. He also noted that the wave height had diminished when the latter reached the marina at the north part of the affected area.

61.6 Discussions

The landslide debris impact through the ice cover initiated a wave that propagated under the ice cover towards the entire north portion of the lake and at least 500 m to the south. While propagating, the wavefront broke the ice cover creating a radial fracture pattern starting at the impact location. The wavefront caused the ice to bend, inducing enough strain to initiate structural failures (Liu and Mollo-Christensen 1988; Kohout and Meylan 2008). Squire (1993) explained that the maximum strain applied on the ice sheet is located at constant distance from the previous fracture created by the wavefront. The ice cover should then break into floes of approximately the same size as the wavefront is propagating, which is what has been observed at the Lac des Seize-Îles. The wave also diffracted around an island located 300 m northwest of the generation zone. Fracture patterns indicative of shear were observed along the access roads and witness interpretations of the event indicate at least one episode of wave reflection. As was the case of Notre-Dame-de-la-Salette in 1908 (Ells 1908), the ice cover was an aggravating factor during the event and also a major cause of infrastructure damage along the coast. However, the ice may also have reduced the areal impact by dampening the wave energy.

Regarding future work, it will be interesting to conduct numerical modelling of the landslide-generated tsunami to recreate the observed propagation pattern, wave height, run-up and inundation height. Moreover, the physical experiments proposed by Mohammed and Fritz (2012) may help provide more insight on the process of impact of granular slide on tsunami generation. The major challenge will be to consider the unknown effect of the resistance and the behaviour of the ice cover on the wave propagation in pre-existing numerical models. To perform numerical modelling, kinematic analysis of the landslide will be necessary to estimate the initial acceleration and the maximum velocity of the displaced mass.

61.7 Conclusions

This paper presents the case of Lac-des-Seize-Îles, Québec, Canada in which a landslide of 30,500 m^3 occurred on the east shore of the northern portion of the lake. The displaced mass impacted the ice covered lake surface and generated a wave that propagated over an area extending 450 m north and 500 m south of the impact zone. As it was propagating, the water mass broke and lifted the ice which caused serious damage on boathouses, docks and seasonal residences located on the shoreline of the affected zone.

This event showed that even small landslide-generated tsunamis can have consequences on the coastal infrastructures both in and around a water basin. The ice cover creates serious consequences while it acts as an aggravating factor in the inundation process. These kinds of events should be considered in landslide susceptibility planning in inhabited areas. There is a need for a more comprehensive analysis of the impact and the behaviour of river or lake ice cover in landslide consequences, especially in the case of wave generation, propagation and inundation processes. The role of ice cover should also be considered in future numerical modelling of tsunami events in cold regions.

Acknowledgments This work has been possible with the support of the Section des mouvements de terrain of the Ministère des Transports du Québec who rapidly realised and shared the aerial photographs of the affected zone. Thanks also to Richard Carignan of the Université de Montréal who has given us the access to the bathymetric data of the lake. We also thank the two reviewers of the manuscript, David Mosher and Carl B. Harbitz, who provided constructive comments.

References

Ells RW (1908) Report on the landslide at Notre-Dame de la Salette, Lièvre River, Quebec. Department of Mines, Geological Survey Branch, Ottawa

Jørstad FA (1968) Waves generated by landslides in Norwegian fjords and lakes. Norw Geotech Inst 79:13–32

Kohout AL, Meylan MH (2008) An elastic plate model for wave attenuation and ice floe breaking in the marginal ice zone. J Geophys Res. doi:10.1029/2007JC004434

Lefebvre G, Rosenberg P, Paquette J, Lavallée JG (1991) The 5 september, 1987, landslide on the La Grande River, James Bay, Quebec, Canada. Can Geotech J 28:263–275

Liu AK, Mollo-Christensen E (1988) Wave propagation in a solid ice pack. J Phys Oceanogr 18:1702–1712

Mohammed F, Fritz HM (2012) Physical modeling of tsunamis generated by three-dimensional deformable granular landslides. J Geophys Res. doi:10.1029/2011JC007850

Murty TS, Polavarapu RJ (1979) Influence of an ice layer on the propagation of long waves. Mar Geod. doi:10.1080/15210607909379342

Squire VA (1993) The breakup of shore fast sea ice. Cold Reg Sci Technol 21:211–218

Thomson RE (1981) Oceanography of the British Columbia coast, vol 56. Canadian Special Publication of Fisheries and Aquatic Sciences, Ottawa

Turner AK, Schuster RL (1996) Landslides: investigation and mitigation. National Academy Press, Washington, DC

Vanneste M, Harbitz CB, DeBlasio FV, Glimsdal S, Mienert J, Elverhoi A (2010) Hinlopen-Yermak landslide, Arctic Ocean geomorphology, landslide mechanics, and tsunami simulations. Mass-Transport Deposits in Deepwater Settings. SEPM Special Publication No. 95, pp 1–19

Wang Z, Parau EI, Milewski PA, Vanden-Broeck JM (2015) Numerical study of interfacial solitary waves propagating under an elastic sheet. Proc R Soc A 470:1–17

Index

G. Lamarche et al. (eds.), *Submarine Mass Movements and their Consequences*,
Advances in Natural and Technological Hazards Research 41,
DOI 10.1007/978-3-319-20979-1

C

D

E

MIX
Papier aus verantwortungsvollen Quellen
Paper from responsible sources
FSC® C105338

If you have any concerns about our products,
you can contact us on
ProductSafety@springernature.com

In case Publisher is established outside the EU,
the EU authorized representative is:
Springer Nature Customer Service Center GmbH
Europaplatz 3, 69115 Heidelberg, Germany

Printed by Libri Plureos GmbH
in Hamburg, Germany